T0180076

Lecture Notes in Computer Science 12112

More information about this series at http://www.springer.com/series/7409

Yunmook Nah · Bin Cui ·
Sang-Won Lee · Jeffrey Xu Yu ·
Yang-Sae Moon · Steven Euijong Whang (Eds.)

Database Systems for Advanced Applications

25th International Conference, DASFAA 2020
Jeju, South Korea, September 24–27, 2020
Proceedings, Part I

Springer

Editors
Yunmook Nah
Dankook University
Yongin, Korea (Republic of)

Sang-Won Lee
Sungkyunkwan University
Suwon, Korea (Republic of)

Yang-Sae Moon
Kangwon National University
Chunchon, Korea (Republic of)

Bin Cui
Peking University
Haidian, China

Jeffrey Xu Yu
Department of System Engineering
and Engineering Management
The Chinese University of Hong Kong
Hong Kong, Hong Kong

Steven Euijong Whang
Korea Advanced Institute of Science
and Technology
Daejeon, Korea (Republic of)

ISSN 0302-9743 ISSN 1611-3349 (electronic)
Lecture Notes in Computer Science
ISBN 978-3-030-59409-1 ISBN 978-3-030-59410-7 (eBook)
https://doi.org/10.1007/978-3-030-59410-7

LNCS Sublibrary: SL3 – Information Systems and Applications, incl. Internet/Web, and HCI

This Springer imprint is published by the registered company Springer Nature Switzerland AG
The registered company address is: Gewerbestrasse 11, 6330 Cham, Switzerland

Preface

It is our great pleasure to introduce the proceedings of the 25th International Conference on Database Systems for Advanced Applications (DASFAA 2020), held during September 24–27, 2020, in Jeju, Korea. The conference was originally scheduled for May 21–24, 2020, but inevitably postponed due to the outbreak of COVID-19 and its continual spreading all over the world. DASFAA provides a leading international forum for discussing the latest research on database systems and advanced applications. The conference's long history has established the event as the premier research conference in the database area.

To rigorously review the 487 research paper submissions, we conducted a double-blind review following the tradition of DASFAA and constructed the large committee consisting of 16 Senior Program Committee (SPC) members and 212 Program Committee (PC) members. Each valid submission was reviewed by three PC members and meta-reviewed by one SPC member who also led the discussion with the PC members. We, the PC co-chairs, considered the recommendations from the SPC members and looked into each submission as well as its reviews to make the final decisions. As a result, 119 full papers (acceptance ratio of 24.4%) and 23 short papers were accepted. The review process was supported by the EasyChair system. During the three main conference days, these 142 papers were presented in 27 research sessions. The dominant keywords for the accepted papers included neural network, knowledge graph, time series, social networks, and attention mechanism. In addition, we included 4 industrial papers, 15 demo papers, and 3 tutorials in the program. Last but not least, to shed the light on the direction where the database field is headed to, the conference program included four invited keynote presentations by Amr El Abbadi (University of California, Santa Barbara, USA), Kian-Lee Tan (National University of Singapore, Singapore), Wolfgang Lehner (TU Dresden, Germany), and Sang Kyun Cha (Seoul National University, South Korea).

Five workshops were selected by the workshop co-chairs to be held in conjunction with DASFAA 2020: the 7th Big Data Management and Service (BDMS 2020); the 6th International Symposium on Semantic Computing and Personalization (SeCoP 2020); the 5th Big Data Quality Management (BDQM 2020); the 4th International Workshop on Graph Data Management and Analysis (GDMA 2020); and the First International Workshop on Artificial Intelligence for Data Engineering (AIDE 2020). The workshop papers are included in a separate volume of the proceedings also published by Springer in its *Lecture Notes in Computer Science* series.

We would like to thank all SPC members, PC members, and external reviewers for their hard work to provide us with thoughtful and comprehensive reviews and recommendations. Many thanks to the authors who submitted their papers to the conference. In addition, we are grateful to all the members of the Organizing Committee, and many volunteers, for their great support in the conference organization. Also, we would like to express our sincere thanks to Yang-Sae Moon for compiling all accepted papers and for working with the Springer team to produce the proceedings. Lastly, we acknowledge the generous financial support from IITP[1], Dankook University SW Centric University Project Office, DKU RICT, OKESTRO, SUNJESOFT, KISTI, LG CNS, INZENT, Begas, SK Broadband, MTDATA, WAVUS, SELIMTSG, and Springer.

We hope that the readers of the proceedings find the content interesting, rewarding, and beneficial to their research.

September 2020

Bin Cui
Sang-Won Lee
Jeffrey Xu Yu

[1] Institute of Information & communications Technology Planning & Evaluation (IITP) grant funded by the Korea government (MSIT) (No. 2020-0-01356, 25th International Conference on Database Systems for Advanced Applications (DASFAA)).

Organization

Organizing Committee

General Chair

Yunmook Nah	Dankook University, South Korea

Program Co-chairs

Bin Cui	Peking University, China
Sang-Won Lee	Sungkyunkwan University, South Korea
Jeffrey Xu Yu	The Chinese University of Hong Kong, Hong Kong

Industry Program Co-chairs

Jinyang Gao	Alibaba Group, China
Sangjun Lee	Soongsil University, South Korea
Eenjun Hwang	Korea University, South Korea

Demo Co-chairs

Makoto P. Kato	Kyoto University, Japan
Hwanjo Yu	POSTECH, South Korea

Tutorial Chair

U. Kang	Seoul National University, South Korea

Workshop Co-chairs

Chulyun Kim	Sookmyung Women's University, South Korea
Seon Ho Kim	USC, USA

Panel Chair

Wook-Shin Han	POSTECH, South Korea

Organizing Committee Chair

Jinseok Chae	Incheon National University, South Korea

Local Arrangement Co-chairs

Jun-Ki Min	Koreatec, South Korea
Haejin Chung	Dankook University, South Korea

Registration Chair

Min-Soo Kim — DGIST, South Korea

Publication Co-chairs

Yang-Sae Moon — Kangwon National University, South Korea
Steven Euijong Whang — KAIST, South Korea

Publicity Co-chairs

Yingxia Shao — Beijing University of Posts and Telecommunications, China
Taehyung Wang — California State University Northridge, USA
Jonghoon Chun — Myongji University, South Korea

Web Chair

Ha-Joo Song — Pukyong National University, South Korea

Finance Chair

Dongseop Kwon — Myongji University, South Korea

Sponsor Chair

Junho Choi — Sunjesoft Inc., South Korea

DASFAA Steering Committee Liaison

Kyuseok Shim — Seoul National University, South Korea

Program Committee

Senior Program Committee Members

K. Selcuk Candan — Arizona State University, USA
Lei Chen — The Hong Kong University of Science and Technology, Hong Kong
Wook-Shin Han — POSTECH, South Korea
Christian S. Jensen — Aalborg University, Denmark
Feifei Li — University of Utah, USA
Chengfei Liu — Swinburne University of Technology, Australia
Werner Nutt — Free University of Bozen-Bolzano, Italy
Makoto Onizuka — Osaka University, Japan
Kyuseok Shim — Seoul National University, South Korea
Yongxin Tong — Beihang University, China
Xiaokui Xiao — National University of Singapore, Singapore
Junjie Yao — East China Normal University, China
Hongzhi Yin — The University of Queensland, Australia
Ce Zhang — ETH Zurich, Switzerland

Qiang Zhu	University of Michigan, USA
Eenjun Hwang	Korea University, South Korea

Program Committee Members

Alberto Abello	Universitat Politècnica de Catalunya, Spain
Marco Aldinucci	University of Turin, Italy
Akhil Arora	Ecole Polytechnique Fédérale de Lausanne, Switzerland
Jie Bao	JD Finance, China
Zhifeng Bao	RMIT University, Australia
Ladjel Bellatreche	LIAS, ENSMA, France
Andrea Calì	University of London, Birkbeck College, UK
Xin Cao	The University of New South Wales, Australia
Yang Cao	Kyoto University, Japan
Yang Cao	The University of Edinburgh, UK
Barbara Catania	DIBRIS, University of Genoa, Italy
Chengliang Chai	Tsinghua University, China
Lijun Chang	The University of Sydney, Australia
Chen Chen	Arizona State University, USA
Cindy Chen	University of Massachusetts Lowell, USA
Huiyuan Chen	Case Western Reserve University, USA
Shimin Chen	ICT CAS, China
Wei Chen	Soochow University, China
Yang Chen	Fudan University, China
Peng Cheng	East China Normal University, China
Reynold Cheng	The University of Hong Kong, Hong Kong
Theodoros Chondrogiannis	University of Konstanz, Germany
Jaegul Choo	Korea University, South Korea
Lingyang Chu	Simon Fraser University, Canada
Gao Cong	Nanyang Technological University, Singapore
Antonio Corral	University of Almeria, Spain
Lizhen Cui	Shandong University, China
Lars Dannecker	SAP SE, Germany
Ernesto Damiani	University of Milan, Italy
Sabrina De Capitani	University of Milan, Italy
Dong Den	Rutgers University, USA
Anton Dignös	Free University of Bozen-Bolzano, Italy
Lei Duan	Sichuan University, China
Amr Ebaid	Google, USA
Ju Fan Renmin	University of China, China
Yanjie Fu	University of Central Florida, USA
Hong Gao	Harbin Institute of Technology, China
Xiaofeng Gao	Shanghai Jiao Tong University, China
Yunjun Gao	Zhejiang University, China
Tingjian Ge	University of Massachusetts Lowell, USA

Boris Glavic	Illinois Institute of Technology, USA
Neil Gong	Iowa State University, USA
Zhiguo Gong	University of Macau, Macau
Yu Gu	Northeastern University, China
Lei Guo	Shandong Normal University, China
Long Guo	Alibaba Group, China
Yuxing Han	Alibaba Group, China
Peng Hao	Beihang University, China
Huiqi Hu	East China Normal University, China
Juhua Hu	University of Washington Tacoma, USA
Zhiting Hu	Carnegie Mellon University, USA
Wen Hua	The University of Queensland, Australia
Chao Huang	University of Notre Dame, USA
Zi Huang	The University of Queensland, Australia
Seung-Won Hwang	Yonsei University, South Korea
Matteo Interlandi	Microsoft, USA
Md. Saiful Islam	Griffith University, Australia
Di Jiang	WeBank, China
Jiawei Jiang	ETH Zurich, Switzerland
Lilong Jiang	Twitter, USA
Cheqing Jin	East China Normal University, China
Peiquan Jin	University of Science and Technology of China, China
Woon-Hak Kang	e-Bay Inc., USA
Jongik Kim	JeonBuk National University, South Korea
Min-Soo Kim	KAIST, South Korea
Sang-Wook Kim	Hanyang University, South Korea
Younghoon Kim	Hanyang University, South Korea
Peer Kröger	Ludwig Maximilian University of Munich, Germany
Anne Laurent	University of Montpellier, France
Julien Leblay	National Institute of Advanced Industrial Science and Technology (AIST), Japan
Dong-Ho Lee	Hanyang University, South Korea
Jae-Gil Lee	KAIST, South Korea
Jongwuk Lee	Sungkyunkwan University, South Korea
Young-Koo Lee	Kyung Hee University, South Korea
Bohan Li	Nanjing University of Aeronautics and Astronautics, China
Cuiping Li	Renmin University of China, China
Guoliang Li	Tsinghua University, China
Jianxin Li	Deakin University, Australia
Yawen Li	Beijing University of Posts and Telecommunications, China
Zhixu Li	Soochow University, China
Xiang Lian	Kent State University, USA
Qing Liao	Harbin Institute of Technology, China

Zheng Liu	Nanjing University of Posts and Telecommunications, China
Chunbin Lin	Amazon Web Services, USA
Guanfeng Liu	Macquarie University, Australia
Hailong Liu	Northwestern Polytechnical University, China
Qing Liu	CSIRO, Australia
Qingyun Liu	Facebook, USA
Eric Lo	The Chinese University of Hong Kong, Hong Kong
Cheng Long	Nanyang Technological University, Singapore
Guodong Long	University of Technology Sydney, Australia
Hua Lu	Aalborg University, Denmark
Wei Lu	Renmin University of China, China
Shuai Ma	Beihang University, China
Yannis Manolopoulos	Open University of Cyprus, Cyprus
Jun-Ki Min	Korea University of Technology and Education, South Korea
Yang-Sae Moon	Kangwon National University, South Korea
Mikolaj Morzy	Poznan University of Technology, Poland
Parth Nagarkar	New Mexico State University, USA
Liqiang Nie	Shandong University, China
Baoning Niu	Taiyuan University of Technology, China
Kjetil Nørvåg	Norwegian University of Science and Technology, Norway
Vincent Oria	New Jersey Institute of Technology, USA
Noseong Park	George Mason University, USA
Dhaval Patel	IBM, USA
Wen-Chih Peng	National Chiao Tung University, Taiwan
Ruggero G. Pensa	University of Turin, Italy
Dieter Pfoser	George Mason University, USA
Silvestro R. Poccia	Polytechnic of Turin, Italy
Shaojie Qiao	Chengdu University of Information Technology, China
Lu Qin	University of Technology Sydney, Australia
Weixiong Rao	Tongji University, China
Oscar Romero	Universitat Politènica de Catalunya, Spain
Olivier Ruas	Peking University, China
Babak Salimi	University of Washington, USA
Maria Luisa Sapino	University of Turin, Italy
Claudio Schifanella	University of Turin, Italy
Shuo Shang	Inception Institute of Artificial Intelligence, UAE
Xuequn Shang	Northwestern Polytechnical University, China
Zechao Shang	The University of Chicago, USA
Jie Shao	University of Electronic Science and Technology of China, China
Yingxia Shao	Beijing University of Posts and Telecommunications, China
Wei Shen	Nankai University, China

Yanyan Shen	Shanghai Jiao Tong University, China
Xiaogang Shi	Tencent, China
Kijung Shin	KAIST, South Korea
Alkis Simitsis	HP Labs, USA
Chunyao Song	Nankai University, China
Guojie Song	Peking University, China
Shaoxu Song	Tsinghua University, China
Fei Sun	Huawei, USA
Hailong Sun	Beihang University, China
Han Sun	University of Electronic Science and Technology of China, China
Weiwei Sun	Fudan University, China
Yahui Sun	Nanyang Technological University, Singapore
Jing Tang	National University of Singapore, Singapore
Nan Tang	Hamad Bin Khalifa University, Qatar
Ismail Toroslu	Middle East Technical University, Turkey
Vincent Tseng	National Chiao Tung University, Taiwan
Leong Hou	University of Macau, Macau
Bin Wang	Northeastern University, China
Chang-Dong Wang	Sun Yat-sen University, China
Chaokun Wang	Tsinghua University, China
Chenguang Wang	IBM, USA
Hongzhi Wang	Harbin Institute of Technology, China
Jianmin Wang	Tsinghua University, China
Jin Wang	University of California, Los Angeles, USA
Ning Wang	Beijing Jiaotong University, China
Pinghui Wang	Xi'an Jiaotong University, China
Senzhang Wang	Nanjing University of Aeronautics and Astronautics, China
Sibo Wang	The Chinese University of Hong Kong, Hong Kong
Wei Wang	National University of Singapore, Singapore
Wei Wang	The University of New South Wales, Australia
Weiqing Wang	Monash University, Australia
Xiaoling Wang	East China Normal University, China
Xin Wang	Tianjin University, China
Zeke Wang	ETH Zurich, Switzerland
Joyce Whang	Sungkyunkwan University, South Korea
Steven Whang	KAIST, South Korea
Kesheng Wu	Lawrence Berkeley Laboratory, USA
Sai Wu	Zhejiang University, China
Yingjie Wu	Fuzhou University, China
Mingjun Xiao	University of Science and Technology of China, China
Xike Xie	University of Science and Technology of China, China
Guandong Xu	University of Technology Sydney, Australia
Jianliang Xu	Hong Kong Baptist University, Hong Kong

Jianqiu Xu	Nanjing University of Aeronautics and Astronautics, China
Quanqing Xu	A*STAR, Singapore
Tong Yang	Peking University, China
Yu Yang	City University of Hong Kong, Hong Kong
Zhi Yang	Peking University, China
Bin Yao	Shanghai Jiao Tong University, China
Lina Yao	The University of New South Wales, Australia
Man Lung Yiu	The Hong Kong Polytechnic University, Hong Kong
Ge Yu	Northeastern University, China
Lele Yu	Tencent, China
Minghe Yu	Northeastern University, China
Ye Yuan	Northeastern University, China
Dongxiang Zhang	Zhejiang University, China
Jilian Zhang	Jinan University, China
Rui Zhang	The University of Melbourne, Australia
Tieying Zhang	Alibaba Group, USA
Wei Zhang	East China Normal University, China
Xiaofei Zhang	The University of Memphis, USA
Xiaowang Zhang	Tianjin University, China
Ying Zhang	University of Technology Sydney, Australia
Yong Zhang	Tsinghua University, China
Zhenjie Zhang	Yitu Technology, Singapore
Zhipeng Zhang	Peking University, China
Jun Zhao	Nanyang Technological University, Singapore
Kangfei Zhao	The Chinese University of Hong Kong, Hong Kong
Pengpeng Zhao	Soochow University, China
Xiang Zhao	National University of Defense Technology, China
Bolong Zheng	Huazhong University of Science and Technology, China
Kai Zheng	University of Electronic Science and Technology of China, China
Weiguo Zheng	Fudan University, China
Yudian Zheng	Twitter, USA
Chang Zhou	Alibaba Group, China
Rui Zhou	Swinburne University of Technology, Australia
Xiangmin Zhou	RMIT University, Australia
Xuan Zhou	East China Normal University, China
Yongluan Zhou	University of Copenhagen, Denmark
Zimu Zhou	Singapore Management University, Singapore
Yuanyuan Zhu	Wuhan University, China
Lei Zou	Peking University, China
Zhaonian Zou	Harbin Institute of Technology, China
Andreas Züfle	George Mason University, USA

External Reviewers

Ahmed Al-Baghdadi
Alberto R. Martinelli
Anastasios Gounaris
Antonio Corral
Antonio Jesus
Baozhu Liu
Barbara Cantalupo
Bayu Distiawan
Besim Bilalli
Bing Tian
Caihua Shan
Chen Li
Chengkun He
Chenhao Ma
Chris Liu
Chuanwen Feng
Conghui Tan
Davide Colla
Deyu Kong
Dimitrios Rafailidis
Dingyuan Shi
Dominique Laurent
Dong Wen
Eleftherios Tiakas
Elena Battaglia
Feng Yuan
Francisco Garcia-Garcia
Fuxiang Zhang
Gang Qian
Gianluca Mittone
Hans Behrens
Hanyuan Zhang
Huajun He
Huan Li
Huaqiang Xu
Huasha Zhao
Iacopo Colonnelli
Jiaojiao Jiang
Jiejie Zhao
Jiliang Tang
Jing Nathan Yan
Jinglin Peng
Jithin Vachery
Joon-Seok Kim
Junhua Zhang
Kostas Tsichlas
Liang Li
Lin Sun
Livio Bioglio
Lu Liu
Luigi Di Caro
Mahmoud Mohammadi
Massimo Torquati
Mengmeng Yang
Michael Vassilakopoulos
Moditha Hewasinghage
Mushfiq Islam
Nhi N.Y. Vo
Niccolo Meneghetti
Niranjan Rai
Panayiotis Bozanis
Peilun Yang
Pengfei Li
Petar Jovanovic
Pietro Galliani
Qian Li
Qian Tao
Qiang Fu
Qianhao Cong
Qianren Mao
Qinyong Wang
Qize Jiang
Ran Gao
Rongzhong Lian
Rosni Lumbantoruan
Ruixuan Liu
Ruiyuan Li
Saket Gurukar
San Kim
Seokki Lee
Sergi Nadal
Shaowu Liu
Shiquan Yang
Shuyuan Li
Sicong Dong
Sicong Liu
Sijie Ruan
Sizhuo Li
Tao Shen
Teng Wang
Tianfu He
Tiantian Liu
Tianyu Zhao
Tong Chen
Waqar Ali
Weilong Ren
Weiwei Zhao
Weixue Chen
Wentao Li
Wenya Sun
Xia Hu
Xiang Li
Xiang Yu
Xiang Zhang
Xiangguo Sun
Xianzhe Wu
Xiao He
Xiaocong Chen
Xiaocui Li
Xiaodong Li
Xiaojie Wang
Xiaolin Han
Xiaoqi Li
Xiaoshuang Chen
Xing Niu
Xinting Huang
Xinyi Zhang
Xinyu Zhang
Yang He
Yang Zhao
Yao Wan
Yaohua Tang
Yash Garg
Yasir Arfat
Yijian Liu
Yilun Huang
Yingjun Wu
Yixin Su
Yu Yang

Yuan Liang
Yuanfeng Song
Yuanhang Yu
Yukun Cao
Yuming Huang
Yuwei Wang
Yuxing Han
Yuxuan Qiu
Yuyu Luo
Zelei Cheng
Zhangqing Shan
Zhuo Ma
Zicun Cong
Zili Zhou
Zisheng Yu
Zizhe Wang
Zonghan Wu

Financial Sponsors

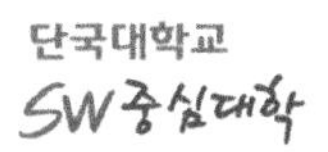

Academic Sponsors

Contents – Part I

Big Data

Machine Learning

Contents – Part II

Data Mining

Graph Data

Contents – Part III

Social Network

Query Processing

Embedding Analysis

Recommendation

Industrial Papers

Demo Papers

Big Data

A Data-Driven Approach for GPS Trajectory Data Cleaning

Lun Li[1], Xiaohang Chen[1], Qizhi Liu[1](✉), and Zhifeng Bao[2]

[1] State Key Laboratory for Novel Software Technology, Nanjing University, Nanjing, China
{mg1733027,141210002}@smail.nju.edu.cn, lqz@nju.edu.cn

[2] RMIT University, Melbourne, Australia
zhifeng.bao@rmit.edu.au

Abstract. In this paper, we study the problem of GPS trajectory data cleaning, aiming to clean the noises in trajectory data. The noises can be generated due to many factors, such as GPS devices failure, sensor error, transmission error and storage error. Existing cleaning algorithms usually focus on certain types of noises and have many limitations in applications. In this paper, we propose a data-driven approach to clean the noises by exploiting historical trajectory point cloud. We extract road information from the historical trajectories and use such information to detect and correct the noises. As compared to map matching techniques, our method does not have many requirements such as high sampling rates, and it is robust to noises and nonuniform distribution in the historical trajectory point cloud. Extensive experiments are conducted on real datasets to demonstrate that the proposed approach can effectively clean the noises while not utilizing any map information.

1 Introduction

With the spread of smart phones and other GPS-enabled devices, large-scale collections of GPS trajectory data are widely available. These data have plentiful valuable information which has given rise to location-based services (LBS) and many other areas such as urban planning [3,20], intelligent transportation [5,23], and athlete behavior analysis [22]. However, due to various factors such as GPS devices failure, sensor error, transmission error and storage error, most raw trajectory data contain a lot of noises which may lead to different or even opposite results in downstream mining and search tasks over trajectories. Hence, it is necessary and important to clean the noises before propagating to those downstream tasks.

Among existing approaches for GPS trajectory data cleaning, map matching [10,21] is one of the most widely used approaches. Map matching utilizes maps and road networks as reference, by mapping each GPS point to the nearest road in the map, noises can be detected and corrected. However, map matching methods usually use a fixed predefined distance threshold to detect the noisy

Y. Nah et al. (Eds.): DASFAA 2020, LNCS 12112, pp. 3–19, 2020.
https://doi.org/10.1007/978-3-030-59410-7_1

points while different roads usually have different widths in reality. In addition, map matching techniques may suffer from incomplete or unmatched map information, which may cause inaccuracy or inconsistency in applications, as also confirmed by our experimental study later. Some approaches such as [16] first use historical trajectories to generate or fix maps, then use the generated maps for the cleaning task. Our approach differs in that we not only extract detailed route skeletons from the historical trajectories, but also use these route skeletons as anchors and compute safe areas according to the distribution of the historical trajectory points for noise cleaning.

Another class of widely-used approaches for trajectory data cleaning is to utilize the features of the trajectory itself, such as distance, velocity, direction, time and density [4,12,13]. These approaches usually work under the premise that the trajectory points are uniformly sampled with high sampling rates. However, according to our research, most available trajectory data may have the problems of sparsity and nonuniform sampling rates. For example, in the T-Drive [25] dataset, it shows that 55% of the neighboring points have the time gap of more than 2 min, and the number for more than 3 min is 40%. Besides, the trajectory shapes of the vehicles are largely decided by the drivers who control the vehicles. Hence it is hard to design a general algorithm based on trajectory features to decide whether a GPS point is a noise or not. In some conditions, correct GPS points may be cleaned by mistake, which may negatively impact further downstream query and mining tasks on these data [1,26].

The aforementioned approaches focus only on certain types of trajectory data, and only have good performance under certain conditions which they are designed for. However, most available trajectory data do not have the required qualities due to device errors or limitations. As a result, these approaches are not generally applicable.

To tackle these problems, we propose a data-driven approach to clean the noises in GPS trajectory data using a concept called historical trajectory point cloud. Our method is purely data-driven and does not have many requirements for the raw trajectory data, such as high sampling rates, and it is robust to noises and nonuniform distribution in the historical trajectory point cloud.

In our approach, we consider both global and local factors. Globally, we extract the main route from a large amount of historical trajectory point clouds, and then from the main route, we extract the route skeleton which is smooth, continuous and uniformly distributed. Locally, we use the extracted route skeleton points as anchors, and consider the local distribution of the raw point cloud to construct a local safe area for each skeleton point, which can separate noisy points from correct points. With the extracted route skeleton and the safe areas, we can detect and correct the noises in the trajectory data.

Our method has three main advantages – (1) Our method is purely data-driven, without utilizing any additional information such as maps. (2) Our method does not have many requirements on the input data which makes our method more general and widely applicable. (3) Our method considers both global and local factors to detect and clean the noises.

In summary, the main contributions of this paper lie in the following aspects:

- **Data-driven and generally applicable model**. We propose a data-driven trajectory cleaning approach that does not rely on any additional information, and can handle trajectories with different qualities.
- **Route skeleton extraction**. We propose a method to extract the main route from the historical trajectory point cloud (Sect. 4). Then we smooth and contract the main route to get the continuous and uniformly-distributed route skeleton (Sect. 5).
- **Safe area construction**. We construct local safe areas for each skeleton points according to the distribution of historical trajectory points. With the skeleton and the safe areas as reference, we can detect and clean the noises in the raw trajectory data (Sect. 6).
- **Extensive experiments**. We conduct extensive experiments on two real datasets to demonstrate that the proposed approach can effectively clean the noises in GPS trajectory data while not utilizing any map information (Sect. 7).

2 Related Work

Existing data cleaning methods can be broadly categorized into three categories: map matching which aims to remove noises, pattern recognition which aims to identify outliers, and trajectory fusion which aims to remove inconsistent points. The first two target for a single trajectory while the last targets for fusing multiple trajectories of the same object. Map matching is the most closely related one to this work.

Map Matching. The major challenge here is how to handle noises, and existing studies usually address it by utilizing additional information as reference, such as maps, road networks and points of interests (POI). Newson et al. [10] described a new map matching algorithm based on Hidden Markov Model (HMM). Instead of naively matching each noisy point to the nearest road, it used HMM to find the most likely route. Yang et al. [21] improved this work by presenting fast map matching (FMM) that integrated HMM with precomputation. An origin-destination table is precomputed to store all pairs of shortest paths within a certain length in the road network. As a benefit, repeated routing queries known as the bottleneck of map matching are replaced with hash table search. Su et al. [17] defined several reference systems for calibrating heterogeneous trajectory data. A reference system can be built on stable anchor points such as POIs and turning points, and then it aligned raw trajectory points to the defined anchor points to do the calibration.

However, map matching methods usually use a fixed predefined distance threshold to detect the noises while different roads usually have different widths in reality. In addition, map matching techniques may suffer from incomplete or unmatched map information, which may cause inaccuracy or inconsistency in applications. In contrast, our work extracts road information from the historical trajectories and uses the information to detect the noises directly.

Some approaches such as [16] first use historical trajectories to generate or fix maps, then use the generated maps for the cleaning task. Our approach differs in that we not only extract detailed route skeletons from the historical trajectories, but also use these route skeletons as anchors and compute safe areas according to the distribution of the historical trajectory points for noise cleaning.

Pattern Recognition. This line of work tries to formalize the patterns of noises in the trajectories by utilizing trajectory features, such as distance, velocity, direction, time and density. Patil et al. [12] proposed a Z-test based secure point anomaly detection method using the combination of distance, velocity, and acceleration for GPS trajectory data preprocessing. Lee et al. [7] considered both distance and density to detect outliers in GPS trajectory data. Potamias et al. [13] used the previous two points to predict the position of the next point according to the velocity, direction and time. If one point can be predicted by its previous two points, then this point is considered as redundant. Chen et al. [4] used the heading change degree of a GPS point and the distance between this point and its adjacent neighbors to weigh the importance of the point.

However, they work based on a general assumption that the trajectory data is dense and sampled in a uniform manner; and they can only have good performance under certain conditions. Unfortunately, most real-world trajectory data are sparse and nonuniformly sampled [15,19,24], which is also evident from our observations in experiments. In contrast, our work does not have many requirements on the raw trajectory data, making it generally applicable.

Trajectory Fusion. If multiple sensors record the same trajectory, by merging these trajectories we can reduce noises and uncertainties in the raw trajectories to get a new trajectory with better quality. Pumpichet et al. [14] deployed a belief parameter to select the helpful neighboring sensors to clean the target trajectory data. The belief parameter is based on sensor trajectories and the consistency of their streaming data correctly received at the base station. After a group of sensors is selected, a cleaning process based on time and distance will compute the cleansed value to replace the value of the dirty sample. Wang et al. [18] studied large-scale trajectory clustering, by combining map matching with an efficient intermediate representation of trajectories and a novel edge-based distance measure, they presented a scalable clustering method to solve k-paths which can cluster millions of taxi trajectories in less than one minute.

3 Problem Formulation and Approach Overview

In this section, we first present preliminaries and the problem formulation, followed by an overview of our approach. Frequently used notations are summarized in Table 1.

3.1 Definitions

Definition 1 (GPS trajectory). *A GPS trajectory T is a sequence of GPS points, i.e.,* $T = \{p_1, p_2, ..., p_n\}$. *Each GPS point p_i is in form of a triple* $\{Longitude, Latitude, ts\}$ *indicating the position and the timestamp of the point.*

Definition 2 (Trajectory point cloud). *Trajectory point cloud (PC) consists of historical GPS trajectory points within a certain area.*

Table 1. Summary of notations

Symbol	Description
T	A raw GPS trajectory
PC	A trajectory point cloud
MR	The main route extracted from the point cloud
SK	The route skeleton sampled from the main route
SA	A safe area
p_i	A GPS point in the trajectory
$c_{i,j}$	A grid cell

In trajectories as defined in Definition 1, there can be *noises* due to various factors (See Sect. 1). *Noises* are those points which are inconsistent with other points in the trajectories. Their positions usually deviate from the normal routes and do not have many neighboring points around. For example, Fig. 1a shows a trajectory point cloud with noises, some sample noises are highlighted in red color. This definition is also in line with that of [8,9]. To our best knowledge, there is no one-size-fits-all formal definition for noise, and the thresholds to decide a noise vary w.r.t. different contexts where trajectories are generated, not to mention the GPS device has an error of 7.8 m in average [2].

Problem Formulation. Given a raw GPS trajectory T with noises, and collections of historical trajectory point cloud PC which belong to the same areas as T, the goal is to use PC to detect noises in T and correct these noises with proper values.

3.2 Approach Overview

Our approach, namely *DTC* (Data-driven Trajectory Cleaning), consists of three main steps:

Main Route Extraction. We first extract the main route from the historical GPS trajectory point cloud by constructing a gird system on the trajectory area. We select representative cells in the gird system to indicate the main route. In the resulted main route, there still exist several outlying cells that do not belong to the main route and may have the problems of disconnectivity and nonuniform distribution (Sect. 4).

Route Skeleton Extraction. We use neighbors to smooth the main route and propose a graph-based method to clean those outlying cells and hence get a more smoothed, contracted and connective main route. Then, we sample the contracted main route with uniform distance intervals to get the route skeleton (Sect. 5).

Algorithm 1: DTC: Data-driven Trajectory Cleaning

Input: historical trajectory point cloud $PC = \{T_1, T_2, ..., T_N\}$, a trajectory with noises $T = \{p_1, p_2, ..., p_n\}$, parameters: $\alpha, d, r, r', d_{sk}, p$
Output: cleaned trajectory $\overline{T}$

```
1  Initialization: main route set MR = ∅, skeleton set SK = ∅
2  /* Main route extraction (Sec. 4)*/
3  Construct the grid system Grid = {c_{i,j}} on the area of PC;
4  foreach c_{i,j} ∈ MR do
5      Built a neighborhood for c_{i,j}: α × α cells;
6      Compute the density center Ω_{i,j} of the neighborhood (See Formula 1);
7      if ||Ω_{i,j} − (i, j)|| ≤ d then
8          Add c_{i,j} into MR;
9  /* Route skeleton extraction (Sec. 5) */
10 Smooth cells in MR using neighbors (with parameter r);
11 Filter outliers in MR by applying a graph-based method (with parameter r');
12 Sample MR with uniform distance interval d_sk to get SK;
13 /* Safe area construction (Sec. 6) */
14 foreach A_i ∈ SK do
15     Build a cover set Ψ_i for A_i (See Definition 3);
16     Refine the radius r_i of Ψ_i (with parameter p);
17 /* Noise detection and correction (Sec. 6.2) */
18 foreach each p_i ∈ T do
19     Match p_i to the nearest A_j ∈ SK;
20     if dist(p_i, A_j) ≥ r_j then
21         Mark p_i as noisy point;
22         Correct p_i;
```

Safe Area Construction. We use skeleton points as anchors and consider the local distribution of the historical points around each anchor to construct local safe areas. The safe areas can separate noisy points from correct points in historical trajectory point cloud. With the extracted route skeleton and the safe areas as reference, we can detect and correct noises in the trajectory data (Sect. 6)

To summarize, we present a complete workflow of the proposed trajectory cleaning approach in Algorithm 1.

4 Main Route Extraction

The main route of the historical GPS trajectory point cloud should cross the regions that lie on the road centerlines as many as possible and can summarize the raw point cloud. Since there exist many noises in the raw historical trajectory point cloud which do not belong to the main route, we propose an approach to extract the main route from the raw point cloud by constructing a grid system on the point cloud area, and then select representative cells in the grid system to indicate the main route (Lines 2–8 in Algorithm 1).

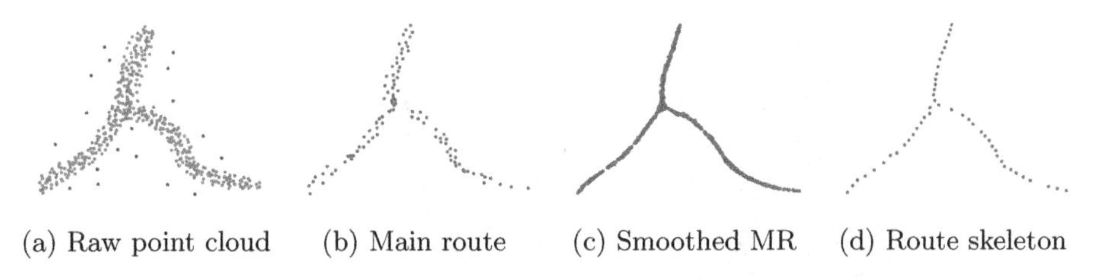

(a) Raw point cloud (b) Main route (c) Smoothed MR (d) Route skeleton

Fig. 1. The process of route skeleton extraction (Best viewed in color).

To construct the grid system, we partition the bounding rectangle that covers the whole point cloud into equal-sized rectangular cells, and all the points in the point cloud lie in these cells. The size of each cell should satisfy that one cell would not cover more than one road. In our paper, each cell is set as $5\,\text{m} \times 5\,\text{m}$ because the minimum width for the main road is about $5\,\text{m}$ according to our research. It is important to note that the bounding rectangle is not the minimum bounding rectangle for the whole point cloud, but a rectangle with some margins around, which are helpful for neighborhood finding (to be illustrated in next paragraph). We denote a cell by $c_{i,j}$ and use $|c_{i,j}|$ to denote the number of points in $c_{i,j}$, where i and j are indexes in the grid system. Since the cells are well fine-grained, we can choose important cells instead of the raw points to represent the main route.

We observe that regions with higher density are more likely to be the main route. To decide if a cell $c_{i,j}$ should be chosen as the main route, we consider both $|c_{i,j}|$ and the number of points in the neighboring cells around $c_{i,j}$ to weigh the importance of the cell $c_{i,j}$.

For each cell $c_{i,j}$, its neighborhood is defined as a region of $\alpha \times \alpha$ cells, where α is a positive odd number, and $c_{i,j}$ is the center of the neighborhood. We set $\alpha = 9$ in our paper. The margins aforementioned are used to build neighborhoods for cells near the edges. For each cell $c_{i,j}$, we formulate the density center $\Omega_{i,j}$ of its $\alpha \times \alpha$ neighboorhood as:

$$\Omega_{i,j} = \frac{\sum_{i'=i-l}^{i+l} \sum_{j'=j-l}^{j+l} (|c_{i,j}| \times (i', j'))}{\sum_{i'=i-l}^{i+l} \sum_{j'=j-l}^{j+l} |c_{i,j}|}, \tag{1}$$

where $l = \lfloor \frac{\alpha}{2} \rfloor$, and (i, j) is the index of cell $c_{i,j}$ in the grid system. If the density center $\Omega_{i,j}$ is close enough to the cell $c_{i,j}$, that is,$||(i, j) - \Omega_{i,j}|| \leq d$, where $d < \frac{\alpha}{2}$, then $c_{i,j}$ is chosen into the main route MR. A small d may cause some main route points to be left out, while a large d may make some noises chosen into MR. In our paper, we set $d = 0.2\alpha$ to strike a balance.

5 Route Skeleton Extraction

In this step, we aim to smooth the extracted main route MR to get a more smoothed, contracted and connective route skeleton SK. Route skeleton is a

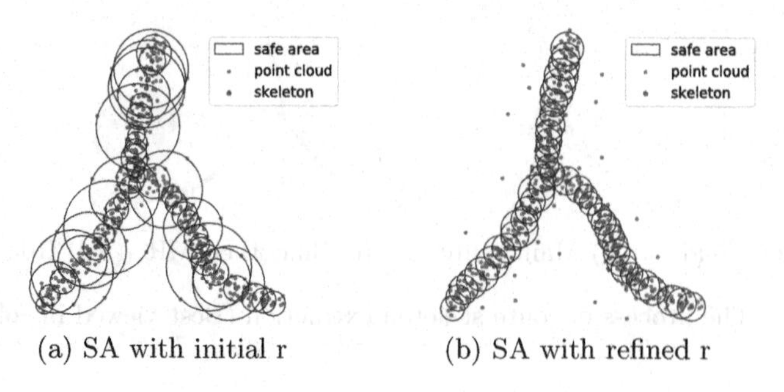

(a) SA with initial r (b) SA with refined r

Fig. 2. The process of safe area construction (Best viewed in color).

uniform sample of the road centerlines, and it indicates the shape and the connection of the underlying routes. Figure 1d shows the route skeleton of the original point cloud in Fig. 1a (Lines 9–12 in Algorithm 1).

We use neighborhood information to smooth the cells in the main route MR. For each cell $c_{i,j} \in MR$, we build a neighbor set, denoted by $NS_{i,j}$, which consists of cells in MR whose distances to $c_{i,j}$ are less than a radius r, i.e.,

$$NS_{i,j} = \{c_{i',j'} \in MR | dist(c_{i',j'}, c_{i,j}) \leq r\},$$

where $dist(c_{i,j}, c_{i',j'})$ means the Euclidean distance between the centers of the two cells. We use the coordinate of a cell's center as the cell's position, then we compute the average position of all the cells in $NS_{i,j}$ as $c_{i,j}$'s smoothed position.

Among the smoothing results, there may still exist some outlying cells that are not connected to other cells. We solve this problem by applying a graph-based method. We consider each cell as a vertex in the graph, and there is an edge between two vertexes when $dist(c_{i,j}, c_{i',j'}) \leq r'$, where r' is a predefined value. In this way, we can find all the connected components. A connected component is outlying when the number of cells in this connected component is less than $0.01 \times |MR|$, and these outlying components need to be dropped.

To get the route skeleton SK, we can simply sample the contracted main route MR with a uniform distance interval d_{sk}. The sampled cells form the route skeleton set SK and the distance between each pair of the SK points is no less than d_{sk}, as shown in Fig. 1d.

6 Safe Area Construction

With the route skeleton SK and the raw historical trajectory point cloud PC, we can consider the cells in SK as anchors, denoted by $\mathcal{A}_i$, and construct a safe area SA_i for each $\mathcal{A}_i$. Every SA_i is a circle with different radius r_i. The radius r_i is refined according to the distribution of the historical points around $\mathcal{A}_i$, then SA_i can separate the noises from correct points in PC, as shown in Fig. 2b (Lines 13–16 in Algorithm 1).

6.1 Radius Refinement

Before constructing the safe areas, we first provide the definition of cover set.

Definition 3 (*CoverSet*). *For each $\mathcal{A}_i \in SK$, it has a cover set, denoted by Ψ_i, which consists of points $p_j \in PC$ satisfying that the distance from p_j to $\mathcal{A}_i$ is the minimum among distances from p_j to all the anchors, i.e.,*

$$\Psi_i = \{p_j \in PC | \forall \mathcal{A}_l \in SK, dist(p_j, \mathcal{A}_i) \leq dist(p_j, \mathcal{A}_l)\}$$

To build the cover sets, we match each $p_j \in PC$ to its nearest anchor in SK. If p_j's nearest anchor is $\mathcal{A}_i$, then p_j is added into Ψ_i. After matching all the points in PC, each $\mathcal{A}_i$ has its own cover set Ψ_i. It is straightforward to infer that every Ψ_i is a subset of PC and all the cover sets are disjoint to each other.

Then we construct the safe area SA_i for every $\mathcal{A}_i$ based on the cover set Ψ_i. We initialize the radius r_i of SA_i with r_i^0, which is the maximum distance from $\mathcal{A}_i$ to the points in Ψ_i, so that the initial safe areas of all the anchors can cover the whole PC, see Fig. 2a.

To this end, the next task is to adjust the radius r_i of SA_i according to the distribution of the historical points around $\mathcal{A}_i$ which is actually $\mathcal{A}_i$'s cover set Ψ_i to separate noises from correct points. We have an observation that, in a cover set, noises are those points which do not have many neighbors and are scattered in the areas away from the main region, as illustrated in Sect. 3.1. Based on this observation, we can adjust the radius r_i by gradually decreasing r_i by a small value. In our paper, every time we decrease the radius r_i by $p \cdot r_i$, where $p = 0.01$, that is, $r_i = r_i^0, (1-p) \cdot r_i^0, (1-p)^2 \cdot r_i^0, (1-p)^3 \cdot r_i^0, ...$ We use Δ_i to denote the number of points removed from the safe area SA_i due to the decreased radius, the decrease stops when $\Delta_i \geq p|\Psi_i|$.

6.2 Noise Detection and Correction

Given a raw GPS trajectory $T = \{p_1, p_2, p_3, ..., p_n\}$ with noises, we use the extracted route skeleton $SK = \{\mathcal{A}_i\}$ and the corresponding safe area SA_i for each $\mathcal{A}_i$ as reference to detect and clean the noises (See Sect. 3.1) in T (Lines 17–22 in Algorithm 1).

For each $p_i \in T$, we match it to the nearest $\mathcal{A}_j \in SK$. To decide if p_i is a noisy point is to see if p_i is within the safe area SA_j, that is, if $dist(p_i, \mathcal{A}_j) \geq r_j$, then p_i is noise and need to be corrected, else reserve p_i.

When a point $p_i \in T$ is detected as a noise, we need to correct it with a proper value. We utilize the interdependent relationship between consecutive points in the trajectory T to find the proper value. For a noisy point p_i, we take its preceding point p_{i-1} and its following point p_{i+1} which are within their own safe areas, and get the average point p_m whose position is the average of p_{i-1} and p_{i+1}. Then, we match p_m to its nearest anchor point in SK, which is $\mathcal{A}_m$, and we use the position of $\mathcal{A}_m$ to correct the noisy point p_i.

7 Experiments

7.1 Settings

Datasets. We use two real-world GPS trajectory datasets: T-Drive [25] data set and S-Taxi data set, which are trajectory data of taxis in two main cities in China. Basic information of the two datasets are summarized in Table 2.

In the T-Drive dataset, the sampling rate varies from 1 s per point to 10+ minutes per point, and 55% of the neighboring points have the time gap of more than 2 min, and the number for more than 3 min is 40%. In the S-Taxi dataset, the sampling rate varies from 1 s to 15 s per point. We can see that T-Drive is very sparse while S-Taxi has high sampling rates.

Table 2. Summary of the datasets

	# of taxis	Duration	# of points	# of trajectories	Sampling rate
T-Drive	10357	One week	17 million	164 thousand	1 s to over 10 min
S-Taxi	110	One day	784 thousand	4000	1 s to 15 s

Table 3. Parameter selection.

	MR extraction (Sect. 4)	*SK* extraction (Sect. 5)			*SA* construction (Sect. 6)
	α	r	r'	d_{sk}	p
Alternatives	7	20	**20**	10	**0.01**
	9	**25**	45	**20**	0.05
	11	30	70	30	0.1
T-Drive	9	25	20	20	0.01
S-Taxi	9	25	70	20	0.01

Implementations. All experiments were performed on a computer with Intel Core i5-8500 CPU with 16 GB RAM running windows 10, implemented in Python 3.6.

Parameters. Table 3 summarizes the parameter selection in the experiments on the two datasets with default values in bold.

7.2 Evaluation Approaches

The evaluation has two main parts: *extracted skeleton evaluation* and *cleaning effectiveness evaluation*. For the extracted skeleton evaluation, we compare the

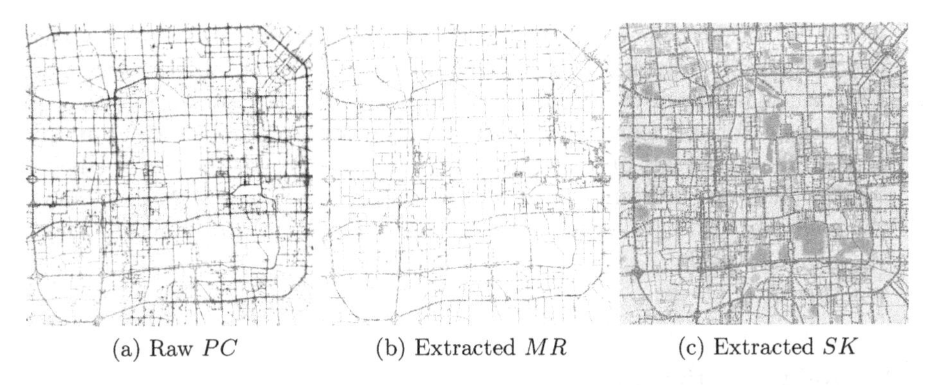

(a) Raw PC (b) Extracted MR (c) Extracted SK

Fig. 3. Skeleton extraction for T-Drive.

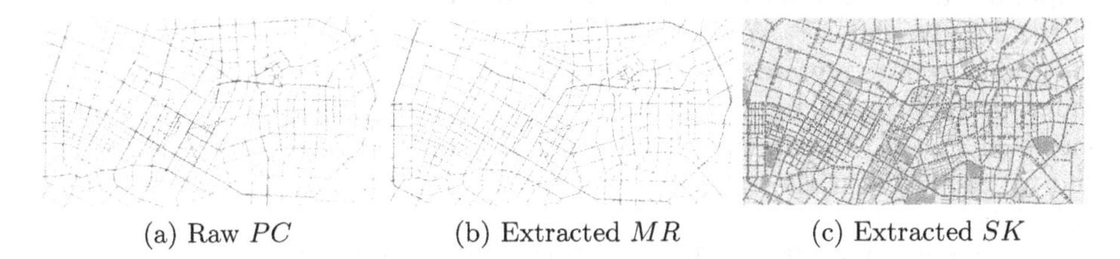

(a) Raw PC (b) Extracted MR (c) Extracted SK

Fig. 4. Skeleton extraction for S-Taxi.

extracted skeleton with the real map obtained from OpenStreetMap [11] both visually and statistically to verify that the extracted route skeleton is consistent with the real map. For the cleaning effectiveness evaluation, we compare our approach DTC (Data-driven Trajectory Cleaning) with map matching [6,10] to verify that our approach can effectively detect the noise in the trajectory data while not utilizing any map information. We also perform our approach on trajectories with different qualities to show that our approach's robustness to different kinds of trajectory data.

Ground Truth Construction. For the extracted skeleton evaluation, we map trajectories to the real map and take the mapping result as the ground truth. For the cleaning effectiveness evaluation, we perform both our approach and map matching on the same dataset serially, then we take the result dataset as the ground truth and label each point in the dataset with $\mathcal{T}$, which means the point is normal. After that, we randomly shift some points in this dataset and label these shifted points with $\mathcal{F}$, which means the point is a noise. We use N_1 to count all the $\mathcal{F}$ points in this dataset, and use N_2 to count all the $\mathcal{T}$ points.

Evaluation Metrics. We propose three metrics. The first metric perpendicular distance is for the extracted skeleton evaluation; the other two metrics, accuracy and recall, are for the cleaning effectiveness evaluation.

- **Perpendicular distance.** For the extracted skeleton evaluation, we compare the extracted skeleton with the ground truth by calculating the perpendicular distances between the extracted skeleton points and the real road network which is obtained by using map matching [6,10] on the original trajectories. The results represent the deviation of the extracted skeleton.

- **Accuracy.** For the cleaning effectiveness evaluation, we perform our approach on the labeled dataset obtained before, we use n_1 to count the points labeled by $\mathcal{F}$ which are successfully detected in our approach, and use n_2 to count the points labeled by $\mathcal{T}$ which are successfully judged in our approach. Then, $Accuracy = \frac{n_1+n_2}{N_1+N_2}$.
- **Recall.** $Recall = \frac{n_1}{N_1}$.

7.3 Evaluation Results

Experiment 1: Extracted Skeleton Evaluation
In the T-Drive dataset, we select 8168 cells from the total 4440×4440 cells in the grid system to represent the route skeleton of the raw point cloud. Figure 3b shows partially extracted main route and its corresponding raw point cloud is shown in Fig. 3a. We draw the extracted skeleton on the real map in Fig. 3c, and we can see that the extracted skeleton is visually consistent with the real underlying route.

The statistical results are shown in Table 4. From Table 4, we can see that, in the T-Drive dataset, 67.7% of the total 8168 extracted skeleton points have the perpendicular distances of less than 5m compared with the ground truth, while 44.1% less than 3 m. The average perpendicular distance between the extracted skeleton and the ground truth is 6.086 m, and the median perpendicular distance is 3.418 m. The average value is larger because our approach can extract some detailed roads which are not specified on the map and the perpendicular

Table 4. Extracted skeleton evaluation on two datasets.

Perpendicular distance	<5 m	<3 m	Average	Median
T-Drive	67.7%	44.1%	6.085 m	3.418 m
S-Taxi	70.0%	42.9%	7.07 m	3.55 m

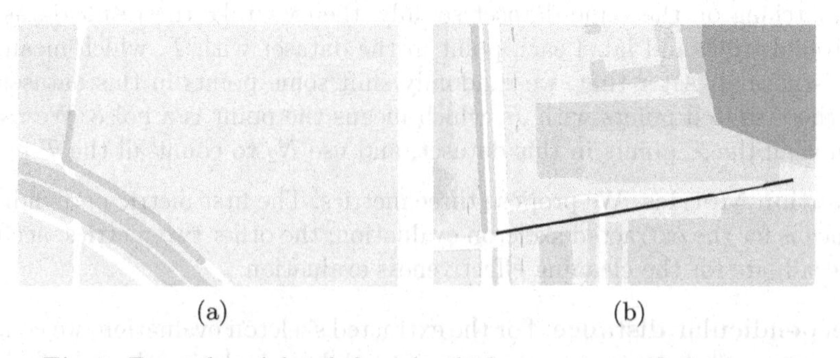

(a) (b)

Fig. 5. Examples of detailed roads which are not specified on map.

distances of such skeleton points can be about 180 m, which pull the average value up. Figure 5 shows examples of the detailed roads which are not specified on OpenStreetMap [11]. The black curves are the detailed roads extracted by our approach and the green curves are the map matching results of the detailed roads. Since the map does not have the records of these detailed roads, it maps the roads to the nearest specified roads on the map. As a result, the median distance is more referable than the average distance. Since the real road width varies from 10 m to 70 m in real life, the deviation of the extracted skeleton is totally acceptable compared with the ground truth.

The results of the S-Taxi dataset are shown in Fig. 4 and Table 4. We select 3122 cells from the total 2040 × 3240 cells in the grid system. Compared with the ground truth, about 70.0% of the total extracted skeleton points have the perpendicular distance of less than 5.0 m, and the number for less than 3 m is 42.9%. The average perpendicular distance is 7.07 m and the median distance is 3.55 m, which also show consistency.

This experiment verifies that our approach can successfully extract the route skeleton from the historical trajectory point cloud, and the results are consistent with the real underlying routes. It demonstrates that the extracted skeletons are reliable to be the reference while detecting the noise in the raw trajectory data.

Table 5. Cleaning effectiveness evaluation on two datasets.

		Accuracy(%)	Recall(%)
T-Drive	DTC	97.4	46.4
	Map-matching	96.2	26.4
S-Taxi	DTC	98.5	46.9
	Map-matching	98.3	38.9

Table 6. Recall on trajectories with different sampling rates.

Average sampling interval (s)	50	100	150	200	250	300	350
Recall(%)	45.6	49.4	49.1	50.9	43.3	45.2	47.5

Experiment 2: Cleaning Effectiveness Evaluation

We select 1467 GPS trajectories in the T-Drive dataset to construct the ground truth for the cleaning effectiveness evaluation. The total GPS point number of the ground truth is 85939. Then we randomly shift 4379 points of them to 1–200 m and label these points as noise points, that is, for the T-Drive dataset, $N_1 = 4379, N_2 = 85939 - N_1 = 81560$. While in the S-Taxi dataset, we select 2592 trajectories to construct the ground truth and the total point number is 277162. We randomly shift 7786 points, so we have $N_1 = 7786, N_2 = 277162 - N_1 = 269376$ for the S-Taxi dataset.

We apply both our approach and map matching [10] to detect the noise points in the datasets. Table 5 shows the statistical results of the experiments with regard to the two metrics proposed before, and the parameter selection in the experiment can refer to Table 3.

In the T-Drive dataset, 46.4% of all the 4379 noise points are detected by our approach, while only 26.4% are detected by map matching. In the S-Taxi dataset, our approach detects 46.9% noise points while map matching detects 38.9%. By comparing the results in Table 5, we can verify that our approach can effectively clean the noise in the trajectory data and the results of both accuracy and recall are better than map matching, even though we do not utilize any extra map information. This is because map matching uses a fixed predefined distance threshold to judge whether a point is noise or not, while road widths are different in real world. In contrast, our approach solves this by constructing local safe areas with different radiuses to detect noises.

To demonstrate our approach's robustness to trajectories with different qualities, we randomly remove some points in the trajectory datasets and get their average sampling rates, then we perform our approach on these data. Table 6 shows the results of recall on trajectories with different average sampling intervals. We can see that the our approach is not sensitive to the sampling rates of the input trajectory data.

Experiment 3: Efficiency Evaluation

We process 17 million points in the T-Drive dataset and 784 thousand points in the S-Taxi dataset. The time costs of the three steps of our approach DTC are shown in Fig. 6. We use MRE to represent the main route extraction part (Sect. 4), SKE to represent the skeleton extraction part (Sect. 5) and SAC for safe area construction part (Sect. 6).

We can see from Fig. 6 that, in our approach, the most time-consuming step is the SAC part. That is because, when building the cover set Ψ_i for each anchor $\mathcal{A}_i \in SK$, if $|PC| = N, |SK| = M$, then the time complexity is $O(M \times N)$.

The time costs in Fig. 6 are for extracting information from the historical trajectory point cloud. With the extracted information, we can detect and correct noises in target trajectories. As for the efficiency of the cleaning step, our approach can process about 53 thousand points per second.

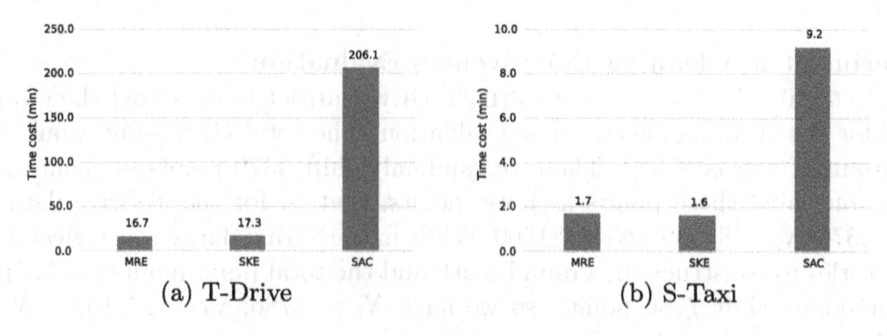

(a) T-Drive (b) S-Taxi

Fig. 6. Time cost of DTC on two datasets.

Experiment 4: Parameter Tuning

We discuss the selection of some important parameters in this part. Figure 7 compares the number of main route cells for T-Drive after smoothing when using different values of the parameter r. Before smoothing, that is when $r = 0$, the number of the main route cells is 1456890 and the number decreases as the value of r increases. When $r = 25$, the number of main route cells is minimal, which means that the main route is more contracted and smoothed.

Figure 8 shows partially the main route of the S-Taxi dataset using different values of the parameter r' which is used to find those outlying cells in MR. Comparing Fig. 8a with Fig. 8b, we can see that when r' is too small, some main route cells may be judged as outlying cells and are dropped, making the main route incomplete. In contrast, when r' is too large, some outlying cells may not be filtered.

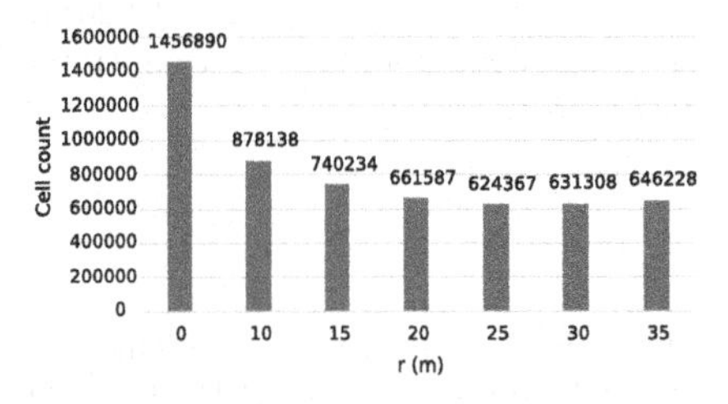

Fig. 7. The number of MR cells after smoothing with parameter r.

Table 7. Cleaning effectiveness with parameter p

p	Average radius (m)	Recall(%)
0.005	82.6	42.3
0.01	77.0	46.9
0.05	66.4	55.9
0.1	58.3	62.9

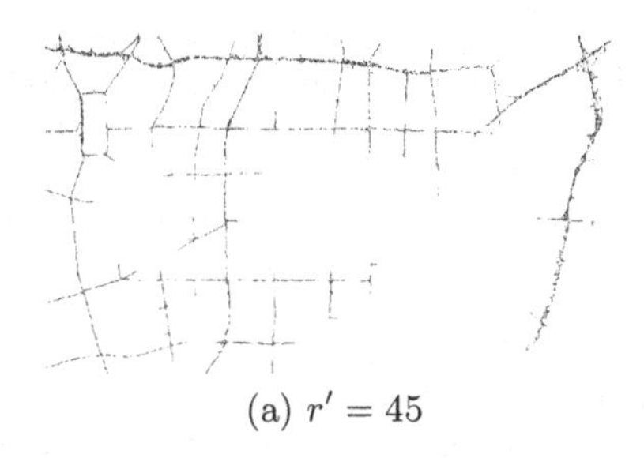

(a) $r' = 45$

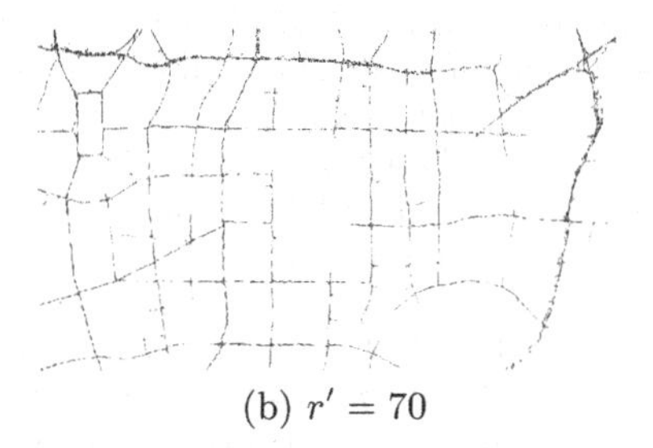

(b) $r' = 70$

Fig. 8. Main route with different values of parameter r'.

Table 7 shows the average radius of the safe areas and the results of recall when using different values of the parameter p. We can find that the average radius of the safe areas decreases as p increases, and as a result, more noises can be detected. However, as the average radius decreases, there may be gaps between adjacent safe areas which may cause some points on the roads to be detected as noise by mistake.

8 Conclusion

In this paper, we proposed a data-driven approach for cleaning GPS trajectory data. We considered both global and local factors to clean the noise by utilizing historical trajectory point cloud. We extracted route skeletons from historical point cloud and exploited the local distribution of the point cloud to construct local safe areas. Then, using skeleton and safe areas as reference, we can detect and clean the noises in the raw trajectories. It is worth noting that our approach does not require high and regular sampling rate of the raw trajectory data, and it is robust to noise and nonuniform distribution in history trajectory point cloud; moreover, our approach does not utilize any additional information such as map or road network. Experiments on real datasets verified the effectiveness of our approach.

Acknowledgements. This research is supported in part by ARC DP200102611, DP180102050, NSFC 91646204, and we appreciate the advice from Mr. Yu Yang (Pancar Technology Ltd.).

References

1. Ali, M.E., Eusuf, S.S., Abdullah, K., Choudhury, F.M., Culpepper, J.S., Sellis, T.: The maximum trajectory coverage query in spatial databases. PVLDB **12**, 197–209 (2018)
2. de Almeida, V.T., Güting, R.H.: Indexing the trajectories of moving objects in networks. GeoInformatica **9**, 33–60 (2005). https://doi.org/10.1007/s10707-004-5621-7
3. Castro, P.S., Zhang, D., Chen, C., Li, S., Pan, G.: From taxi GPS traces to social and community dynamics: a survey. ACM Comput. Surv. **46**, 17:1–17:34 (2013)
4. Chen, Y., Jiang, K., Zheng, Y., Li, C., Yu, N.: Trajectory simplification method for location-based social networking services. In: LBSN, pp. 33–40. ACM (2009)
5. de Graaff, V., de By, R.A., van Keulen, M., Flokstra, J.: Point of interest to region of interest conversion. In: SIGSPATIAL, pp. 378–381. ACM (2013)
6. GraphHopper. www.graphhopper.com
7. Lee, J., Han, J., Li, X.: Trajectory outlier detection: a partition-and-detect framework. In: ICDE, pp. 140–149. IEEE Computer Society (2008)
8. Lee, W., Krumm, J.: Trajectory preprocessing. In: Zheng, Y., Zhou, X. (eds.) Computing with Spatial Trajectories, pp. 3–33. Springer, New York (2011). https://doi.org/10.1007/978-1-4614-1629-6_1
9. Lin, K., Xu, Z., Qiu, M., Wang, X., Han, T.: Noise filtering, trajectory compression and trajectory segmentation on GPS data. In: ICCSE, pp. 490–495. IEEE (2016)
10. Newson, P., Krumm, J.: Hidden Markov map matching through noise and sparseness. In: ACM-GIS, pp. 336–343. ACM (2009)
11. OpenStreetMap. https://www.openstreetmap.org
12. Patil, V., Singh, P., Parikh, S., Atrey, P.K.: GeoSClean: secure cleaning of GPS trajectory data using anomaly detection. In: MIPR, pp. 166–169. IEEE (2018)
13. Potamias, M., Patroumpas, K., Sellis, T.K.: Sampling trajectory streams with spatiotemporal criteria. In: SSDBM, pp. 275–284. IEEE Computer Society (2006)

14. Pumpichet, S., Pissinou, N., Jin, X., Pan, D.: Belief-based cleaning in trajectory sensor streams. In: ICC, pp. 208–212. IEEE (2012)
15. Shang, Z., Li, G., Bao, Z.: DITA: distributed in-memory trajectory analytics. In: SIGMOD, pp. 725–740. ACM (2018)
16. Stanojevic, R., Abbar, S., Thirumuruganathan, S., Chawla, S., Filali, F., Aleimat, A.: Robust road map inference through network alignment of trajectories. In: SDM, pp. 135–143. SIAM (2018)
17. Su, H., Zheng, K., Wang, H., Huang, J., Zhou, X.: Calibrating trajectory data for similarity-based analysis. In: SIGMOD, pp. 833–844. ACM (2013)
18. Wang, S., Bao, Z., Culpepper, J.S., Sellis, T., Qin, X.: Fast large-scale trajectory clustering. PVLDB **13**, 29–42 (2019)
19. Wang, S., Bao, Z., Culpepper, J.S., Xie, Z., Liu, Q., Qin, X.: Torch: a search engine for trajectory data. In: SIGIR, pp. 535–544. ACM (2018)
20. Widder, L., Ko, J., Braden, J., Steinfeld, K.: Spatial behaviors of individuals in cities: case studies in data tracking and scaling. In: Urb-IoT, pp. 98–101. ACM (2016)
21. Yang, C., Gidófalvi, G.: Fast map matching, an algorithm integrating Hidden Markov model with precomputation. Int. J. Geogr. Inf. Sci. **32**, 547–570 (2018)
22. Yang, Y., Wang, X., Zhao, D.: Attack trajectory extraction based on focus player detection in broadcast sports videos. In: ICIMCS, pp. 57:1–57:6. ACM (2015)
23. Yoon, J., Noble, B., Liu, M.: Surface street traffic estimation. In: MobiSys, pp. 220–232. ACM (2007)
24. Yuan, H., Li, G.: Distributed in-memory trajectory similarity search and join on road network. In: ICDE, pp. 1262–1273. IEEE (2019)
25. Yuan, J., et al.: T-drive: driving directions based on taxi trajectories. In: ACM-GIS, pp. 99–108. ACM (2010)
26. Zhang, Z., Qi, X., Wang, Y., Jin, C., Mao, J., Zhou, A.: Distributed top-k similarity query on big trajectory streams. Front. Comput. Sci. **13**, 647–664 (2019). https://doi.org/10.1007/s11704-018-7234-6

Heterogeneous Replicas for Multi-dimensional Data Management

Jialin Qiao[1,2], Yuyuan Kang[1,2], Xiangdong Huang[1,2(✉)], Lei Rui[1,2], Tian Jiang[1,2], Jianmin Wang[1,2], and Philip S. Yu[3]

[1] KLiss, MOE; BNRist; School of Software, Tsinghua University, Beijing, China
{qjl16,kyy19,rl18,jiangtia18}@mails.tsinghua.edu.cn,
{huangxdong,jimwang}@tsinghua.edu.cn

[2] Research Center for Big Data, Tsinghua University, Beijing, China

[3] University of Illinois, Champaign, IL, USA
psyu@uic.edu

Abstract. Multi-dimensional data is widely used in different scenarios, such as cluster monitoring and user behavior analysis for web services. The data is usually managed by distributed databases with a replication strategy, which enhances the availability, fault-tolerance, and I/O throughput. Normally, these replicas share the same physical layout on the disk, which is designed by database administrators according to the target workload. However, it is critical to derive an optimal layout that benefits as many queries as possible, because a layout that accommodates only some queries can negatively impact the others. To tackle this limitation, we propose heterogeneous replicas for multi-dimensional data that provide a higher query throughput without additional disk occupation and without slowing down the writing speed, while still ensuring high availability and load balance. The proposed replication method allows different replicas to be logically identical while having different physical data layouts on the disk. We verified the efficiency of our method in a NoSQL system, Cassandra, with the TPC-H dataset and with a synthetically generated dataset. The results show that our method outperforms state-of-the-art solutions.

Keywords: Multi-dimensional · Heterogeneous · Replica · Layout

1 Introduction

With the development of big data technologies, multi-dimensional data is becoming increasingly popular. For meteorological monitoring, for example, a meteorological station collects metrics with various dimensions, such as temperature, rainfall, wind direction, and timestamps for weather forecasts [21]. The data in all of these has multiple dimensions that are sortable and allow for filtering.

To support fast queries, the physical layout of data on a disk plays an important role, especially for multi-dimensional data management. This is because the

Y. Nah et al. (Eds.): DASFAA 2020, LNCS 12112, pp. 20–36, 2020.
https://doi.org/10.1007/978-3-030-59410-7_2

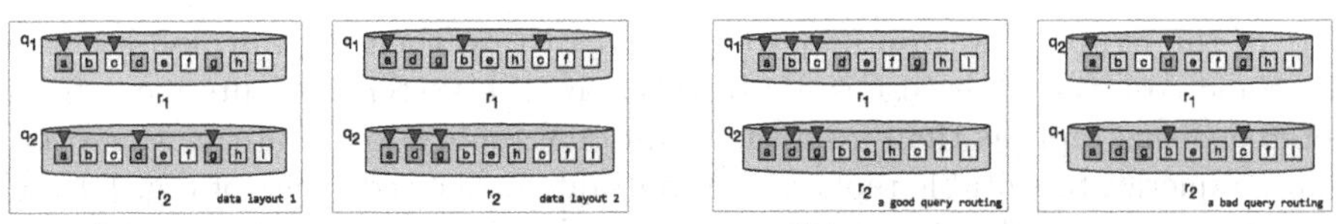

(a) Layouts of uniform replicas

(b) Two routing strategies of heterogeneous replicas

Fig. 1. Uniform replicas and heterogeneous replicas (Color figure online)

data can be sorted in multiple dimensions, and the order of these dimensions will ultimately impact the number of I/O operations, which can be time consuming. Therefore, one duty of database administrators (DBAs) is to find the best schema (and therefore the best physical data layout on the disk) for the data according to the query workload. Indeed, it is critical to find an optimal layout that benefits all queries, because a layout that accommodates only some queries can negatively impact the others.

Although distributed storage systems that use a replication strategy are widely used, DBAs benefit little from replication strategies when designing the schema: replicas behave as slaves, and a read operation can be routed to any one of them such that the query load is spread across the nodes. However, if a query runs slowly on one node because of an unsuitable physical layout, rather than the overhead of the node, routing the query to other nodes is of no use. This is because the physical layout of the data on the disk on all nodes is the same; that is, the replicas are uniform.

For example, Fig. 1(a) shows two kinds of data layouts. In the figure, r_1 and r_2 are two replicas of a dataset, and each data item has two attributes: the character (*a*–*i*) and the color (blue, red and yellow). There are two queries on the dataset: q_1 selects the data whose character precedes "*d*" in alphabetical order, and q_2 selects the data whose color is "blue". The results are marked by the red triangle in the figure. In data layout 1, the data from replicas r_1 and r_2 is serialized on the disk by the order of the characters. The cost of q_1 is 4 (by scanning from "a" until "d") and the cost of q_2 is 9 (by scanning from "a" until "i"), regardless of which replica serves the two queries. In data layout 2, by contrast, the data from the two replicas is serialized by the order of color. Inversely, however, the cost of q_2 is 4 and the cost of q_1 is 9. That is, uniform replicas cannot benefit all queries, regardless of the data layout on the disk.

Because the "*one size fits all*" model fails to take full advantage of replicas, we propose a new replication mechanism for multi-dimensional data, called **heterogeneous replicas**, to maximize query throughput (i.e., the number of queries completed in a period of time). With our method, replicas are logically identical, despite having different physical layouts. As such, different replicas can be customized for different queries to make full use of the replicas for queries. Figure 1(b) shows an example of heterogeneous replicas: r_1 serializes data in alphabetical order, and r_2 serializes the same data clustered by color. In the left part of the figure, if we route q_1 to r_1 and q_2 to r_2, then both q_1 and q_2 have the same cost (i.e., the cost is 4). In this way, the system benefits more from heterogeneous replicas than from the traditional replication strategy. Moreover, routing queries to suitable replicas is important. In the right part of Fig. 1(b), if we route q_1 to r_2 and q_2 to r_1, then both q_1 and q_2 have a cost of 9, degenerating to uniform replicas.

The above example shows how heterogeneous replicas can maximize the query throughput. To use the mechanism, two questions must be answered: "how do we design the data layout for each replica?", and "how do we route queries to suitable replicas?".

Current methods of accelerating queries inevitably bring additional costs. They either optimize limited kinds of queries by adjusting the layout of data, or they optimize more kinds of queries by reduplicating data or building indices at the cost of disk space and write performance. For example, NoSE [11] designs the best schema of column families in NoSQL systems according to a given conceptual model and query workload. NoSE can be seen as offering an optimal uniform replica layout. Considering the diversity of replicas, DivgDesign [4] was proposed to generate different replica configurations. It treats the underlying database as a black box and can be used in our scenario. We compared these two methods to our proposed method in the experiments. Trojan [7] and diversified cache [22] focus on replica design, respectively adopting a Hadoop distributed file system (HDFS) and a distributed cache system. However, these methods cannot be applied to multi-dimensional data management. Another work [18] generates different secondary indices for different replicas such that they are specialized for a specific subset of the workload. However, this incurs overhead on the write speed and disk occupation.

Unlike these approaches, our method of constructing heterogeneous replicas is designed to maximize the query throughput of multi-dimensional data management systems. It can be applied to databases that support multi-dimensional data management, and it avoids additional disk occupation more effectively than the traditional replication strategy. We summarize our contributions in this paper as follows:

- We define and formulate the heterogeneous replica construction problem for multi-dimensional data to achieve the best query throughput for a given workload.
- We propose a new heterogeneous replica construction algorithm and an efficient routing strategy to derive a near-optimal layout that contains different replicas.
- We verified our method on a NoSQL system and conducted extensive experiments that show that our method outperforms state-of-the-art solutions.

The remainder of this paper is organized as follows. We introduce related work in Sect. 2. The workload and problem definition are introduced in Sect. 3. In Section 4, we present a cost model and an efficient routing strategy. The experimental evaluation is reported in Sect. 5. Finally, we conclude the paper in Sect. 6.

2 Related Work

Partition Attributes Across (PAX) is a typical method of adjusting the data layout to improve queries [1]. It stores different attributes in a columnar format inside each page on the disk to utilize the CPU cache fully. A column ordering strategy [2] was proposed for large-scale log data stored in Parquet [12] on HDFS [3]. By adjusting the column order stored on the disk based on the query access pattern, it reduces the overall seek cost when accessing multiple columns to accelerate queries. In the work of Rabl et al. [13], data is partitioned by different granularities according to the known workload. An optimal replication factor (i.e., the number of replicas) is selected for each partition, and a partition allocation strategy maximizes system throughput. HYRISE DBMS automatically partitions tables into vertical partitions of varying widths depending on how the columns of a table are accessed by queries [6]. NoSE [11] was proposed to guide and support the schema design of Cassandra. Given a conceptual model of the data required by the application and the workload, NoSE recommends the best schema and query plans based on it. These approaches optimize data layout for queries without considering replicas. Unlike these approaches, we focus on constructing a heterogeneous replica inside each partition. Therefore, data partitioning methods are orthogonal to our work, and can work together with heterogeneous replicas.

DB2 Advisor [19] is an index recommender for IBM's DB2 universal database. Given a query workload and the statistics of the database, it recommends indices by modeling the index selection problem as a variant of the Knapsack Problem. RITA is an index advisor for fully replicated databases [18]. Multiple indices are selected in different replicas. Improvements made by generating materialized views or selecting indices come at the cost of a large extra space budget in addition to the basic data size. Furthermore, maintaining indices and the materialized view can slow down the insertion speed. For example, Cassandra provides a limited form of secondary indexing, and many applications do not use this option for performance reasons [11]. Heterogeneous replicas can be thought of as a restricted form of materialized views [16] in that they are the only data layout, rather than auxiliary structures [10]. Classical materialized views also contain aggregation, joins, and other query constructs that do not exist in replicas.

Fractured mirrors [14] is a method that generalizes RAID 1 to use a hybrid DBMS architecture, where it keeps both the N-ary Storage Model and the Decomposition Storage Model [5] inside each mirror. To the best of our knowledge, this is the first work that applies a layout with different replicas, although it is limited to scenarios that contain two replicas. Trojan generates a layout for each replica in HDFS to optimize a subset of the workload [7]. It benefits from grouping the frequently accessed columns together in each replica, a process known as vertical partitioning [15]. Distorted replicas [8] is a method of restructuring replicas for document-stores. However, these works are specialized to specific data models, which lose applicability in the multi-dimensional data. C-Store [17] and its commercial database, Vertica [10], leverage projections (column groups) to avoid overhead from record reconstruction. However, it is unclear how the number of projections is determined and how projections are generated—the main topics of our study. Divergent design [4] for leveraging replication was proposed to tune databases more effectively. It works much like the k-means clustering algorithm. We compare our proposal with this approach in Sect. 5.

3 Problem Statement

We first introduce the data model and query workload. Then, we define the problem and prove its hardness.

3.1 Data Model

Multi-dimensional Data: Multi-dimensional data refers to a dataset with multiple sortable dimensions representing the attributes of the data. Each record in the dataset al.so contains columns for metrics. As multi-dimensional data is usually organized on the disk according to the order of the dimensions, rather than metrics, we omit the metrics. We use $P = \{d_1, d_2, ..., d_n\}$ to represent the multi-dimensional data model, in which d_i is a dimension. Data is sorted by d_1, followed by d_2, etc. Given a **data item**, $p_j \in P$, p_j can be formalized as

$$p_j = (p_j.d_1, p_j.d_2, ..., p_j.d_n)$$

where $p_j.d_i$ is the value of p_j in dimension d_i. Given two data items p_1 and p_2, the values in each dimension are comparable. Given a data model with n dimensions, $\{d_1, d_2, ..., d_n\}$, $A = \{k_1, k_2, ..., k_n\}$ is a permutation of the set of dimensions, i.e., $\forall i \in [1, n], \exists j, d_i = k_j$.

Workload: The target workload is described as a set of query and insert operations. We focus here on queries, and we defer discussion of insertions to Sect. 5.5. We use Q to represent the known query workload, which is defined as a sequence of query instances. Each query instance q consists of arbitrary predicates connected with and/or. We transform the predicate into a disjunctive normal form that contains x conjunction normal forms (CNFs). The predicate $\sigma(d_i)$ on each dimension is either a range predicate ($d_i \in [s_i, e_i)$) or an equality predicate ($d_i = v_i$). Dimensions in a CNF that do not have a predicate are seen having a range predicate with 100% selectivity. Therefore, a query q is

$$q = \{CNF_1 \vee CNF_2, ..., \vee CNF_x\}, CNF = \{\sigma(d_1) \wedge \sigma(d_2), ..., \wedge\sigma(d_n)\}$$

This query model captures the functionality that is commonly present in multi-dimensional stores.

3.2 Problem Definition

Following the intuition that a query has different latencies when executed on replicas with different layouts, our goal is to find an optimal layout for all replicas to maximize query throughput, called the **M**ulti-dimensional data **R**eplica **C**onstruction **P**roblem (MRCP). We introduce the procedure in the following section and provide a formal definition of the MRCP.

Let $R = \{r_1, r_2, ..., r_N\}$ be the N replicas, each with a specific data layout on the disk. Given a workload Q, the query router sends each query instance q to a specific replica. Then, each replica (e.g., r_i) is assigned with a subset workload (e.g., Q_i). We use $Cost(q, r)$ to represent the cost of query instance q on replica r. The total overhead for Q_i on r_i is defined as

$$Cost(Q_i, r_i) = \sum_{q \in Q_i} Cost(q, r_i) \tag{1}$$

The total processing time depends on the slowest replica:

$$Cost(Q, R) = max(Cost(Q_i, r_i)) \ , \ i \in [1, N] \tag{2}$$

The layout of the replica and the routing strategy considerably impact query performance. Therefore, finding an optimal layout of heterogeneous replicas and an appropriate routing strategy is essential. The MRCP is defined as follows:

Definition 1. *MRCP: Given a multi-dimensional dataset P with n dimensions, a query workload Q, and the replication factor N, we find a layout of heterogeneous replicas R^* (the permutation of dimensions) and an adaptive routing strategy, such that the $Cost(Q, R)$ is minimized:*

$$R^* = arg \min_{R} \{Cost(Q, R)\} \tag{3}$$

Hardness: The MRCP is an NP-hard problem. Therefore, trying all possible permutations of the dimension columns is infeasible. For example, 3 replicas with 7 dimensions contain 128 billion permutations.

Proof. To analyze the hardness of MRCP, we first introduce the column ordering problem (COP) [2]: given a group of queries Q and a set of columns in a column store, we find an optimal column ordering that has minimum query costs. This problem is NP-hard. Given a COP instance, we construct an MRCP instance with one replica. Each column in the column store is a dimension in our multi-dimensional data model. For each query in Q, we construct a query instance on this replica. Thus, finding an optimal layout such that the $Cost(Q, R)$ is minimized is equivalent to a COP instance. We reduce the COP to the MRCP with one replica in polynomial time. Therefore, the MRCP is NP-Hard.

4 Heterogeneous Replicas

In this section, we first model the query cost on multi-dimensional data. Then, we define the routing strategy and describe the construction of a near-optimal heterogeneous replicas layout.

4.1 Cost Model

First, we establish the query cost model on a replica with a specific layout. For a query, a range of data needs to be scanned, which is denoted as the **candidate set** *Row* and calculated as follows:

Given a layout of replica, $A = \{k_1, k_2, ..., k_n\}$, for each subquery CNF, suppose the m-th dimension in A is the first dimension with a range predicate, i.e., (1) $\forall i \in [1, m-1]$, $\sigma(k_i)$ is an equality predicate; and (2) $\sigma(k_m)$ is a range predicate. Because predicates on k_1–k_{m-1} are equality predicates, these predicates can form a prefix $(v_1, v_2, ..., v_{m-1})$. Rather than scanning all of the data items, the database can leverage the prefix to prune the data to be scanned. For predicate $\sigma(k_m) = \{p | p.k_m \in [s, e)\}$, we further derive a range of data items whose values in the dimension m are between s and e. Therefore, the database can quickly locate the two data items

$$p_s^A = (v_1, v_2, ...v_{m-1}, s_m, \_), \; p_e^A = (v_1, v_2, ...v_{m-1}, e_m, \_)$$

in which p_s^A and p_e^A are the first points whose values of the first m dimensions are equal to the given values. It is difficult to use predicates on the latter dimensions $\{k_i | i > m\}$ to prune the scanned data. Therefore, the remaining values of the $n-m$ columns are omitted and denoted as "_". We call the two points **boundary points**. In the query process, all data items between p_s^A and p_e^A need to be scanned to derive the final result, which forms the candidate set.

We can measure the query cost by using the statistics of the dataset. Given a dataset S, $|S|$ is the number of data items in S. For each dimension d_i in P, the **distribution function** of the value in d_i is $F_{d_i}(x)$, and the **probability density function** is $f_{d_i}(x)$, where $f_{d_i}(x) = dF_{d_i}(x)/dx$. The major time cost of a subquery CNF on replica r depends on the candidate set Row_r^{CNF} (i.e., the number of rows that need to searched), which can be estimated as follows:

$$Row_r^{CNF} = |S| \times \prod_{i=1}^{m-1} f_{k_i}(v_i) \times (F_{k_m}(e_m) - F_{k_m}(s_m)) \tag{4}$$

In this equation, we first estimate the number of data items by multiplying the probability density of each value in the equality predicate $f_{k_i}(v_i)$. Then, we multiply the proportion of data in dimension k_m: $F_{k_m}(e_m) - F_{k_m}(s_m)$. The total cost of a query instance q on replica r is

$$Row_r^q = \sum_{p=1}^{x} Row_r^{CNF_p} \tag{5}$$

The data that a query needs to scan may be smaller than what we estimate. It is possible to further reduce the number of data items in the estimate, by redefining p_s^A as $(s_1, s_2, ...s_n)$, where s_i is the value of the equality predicates or the start value of the range predicates on k_i. This complicates the estimate of the candidate set and is not especially beneficial. Therefore, we do not change our model further.

In the query process of a database, the system needs to locate the data, scan the candidate set on the disk, and apply predicates to the data. Finally, it transfers the result to the client through the network. We use function $t()$ to model the total cost of q on replica r:

$$Cost(q, r) = t(Row_r^q) \tag{6}$$

The function $t()$ depends on the actual environment of the system, including the disk throughput, the size of each data item, and other configurations of the system, which should be modeled in a real system. Given specific data layouts of N replicas and a query q, there are many routing strategies that can be applied. Our routing strategy is as follows. We route the query to the replica that has minimal cost. The minimal time cost of a query is defined as

$$Cost_{min}(q, R) = \{Cost(q, r) | \nexists r_j \in R, Cost(q, r_j) < Cost(q, r_i)\} \tag{7}$$

Although it seems that this might cause a load imbalance, our optimization goal helps to avoid this. Furthermore, we found that this strategy performed best with our proposed heterogeneous replicas construction algorithm.

4.2 Replica Construction

We here propose an algorithm, called Simulated annealing-based [2,9] Multi-dimensional data Replica Construction (SMRC), to find an approximation of the optimal heterogeneous replicas. Algorithm 1 provides details, in which a specific layout of all heterogeneous replicas R corresponds to a **state**. The inputs include the query workload Q, an arbitrary state as the initial state R_0, and two parameters: the initial temperature t_0 and cooling_rate. In this algorithm, the temperature shrinks at a rate of $1 - cooling\_rate$. The SMRC returns an approximate optimal state R such that the average query latency is minimized.

Lines 2–9 in Algorithm 1 are the main searching process in simulated annealing. A "good" state will always be accepted, whereas a "bad" state will be accepted probabilistically to avoid the local optimum. The new state generation function $NewState(R)$ is implemented as follows. We randomly choose two columns k_i and k_j, and apply one of the following three operations:

- $swap(k_i, k_j)$: swap the two columns k_i, k_j.
- $inverse(k_i, k_j)$: invert the columns between k_i and k_j.
- $insert(k_i, k_j)$: insert k_i into the position of another column k_j.

Algorithm 1. Simulated annealing-based Multi-dimensional data Replica Construction (SMRC)

Input:
  Q: Query workload
  $R_0 = \{r_1, r_2, ...r_N\}$:Initial structure of replicas
Output:
  R: Optimized heterogeneous replicas
1: $t := t_0, R := R_0, C := Cost(Q, R_0)$
2: **for** $k = 1$ *to* k_{max} **do**
3: $t := t \times (1 - cooling\_rate)$
4: $R' := NewState(R)$
5: $C' := Cost(Q, R')$
6: **if** $C' < C || e^{\frac{C-C'}{t}} > random(0, 1)$ **then**
7: $R := R', C := C'$
8: **end if**
9: **end for**
10: **return** R

Although inverse and insert could be replaced with multiple swap operations, we retain them for fast traversal. When $NewState(R)$ is called, it performs one of the above operations on a random selected replica r_j in R. We illustrate the state generation process using three operations above in Fig. 2. The initial state is R_0. We then use $swap(k_3, k_2)$ on r_2 to generate a new state R_1. Suppose that all new states are better in this example. Then, the current state is R_1, and we use $Inverse(k_1, k_4)$ on r_1 to generate R_2. Finally, we use $Insert(k_2, k_1)$ on r_3 to generate state R_3.

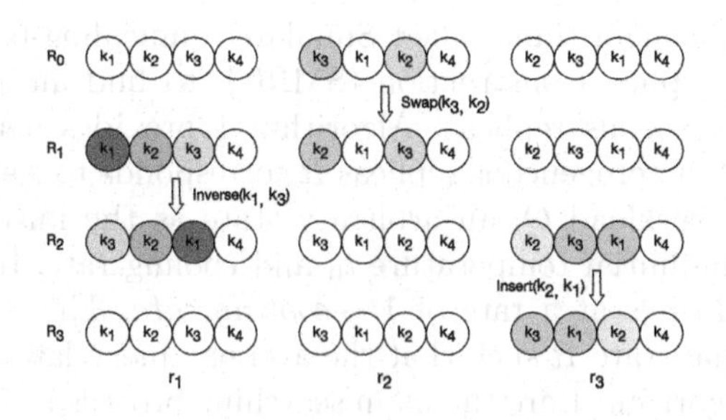

Fig. 2. Example of state generation using three operations

5 Experiments

5.1 Implementation

We implemented our method on Apache Cassandra, which is widely used at Hulu, GitHub, eBay, and over 1500 companies for mission-critical data.

In Cassandra, each tuple contains a partition key and some clustering keys. The partition key is used to partition data across the nodes. Clustering keys are the dimensions of data. Cassandra serializes data items according to the order of the clustering column values. By controlling the order of clustering columns, we can control the data layout on the disk. We did not change the partition key, and we optimized the data layout in each partition. For a table with n replicas, we generated n replica layouts by SMRC and constructed the heterogeneous replicas in Cassandra. All writes and queries were routed by a middle layer that deployed a request router.

5.2 Experimental Methodology

Hardware: The experiments were carried out on a Cassandra (version 3.11.4) cluster with 5 nodes. Each node had 2 Intel Xeon E5-2697 CPUs with 36 cores in total, 8 GB memory, and a 7200-rpm HDD. The operating system was 64-bit Ubuntu Server 16.04.4 LTS with Linux kernel version 4.15.0-36.

Comparison: We compared our method to two state-of-the-art methods:

- NoSE [11]: In this strategy, all replicas had the same layout, which can be seen as the best layout for uniform replicas.
- Divergent Design (DIVG) [4]: In this method, the **balance factor** m (each query cost was evenly shared by m replicas) ranged from 1 to $N-1$, where N is the replication factor. When $m = N$, DIVG was equal to NoSE because each query was routed to N replicas. Thus, we omitted this case. When designing each replica layout, the DBAdv() was our method under one replica.
- Simulated Annealing based Multi-dimensional data Replica Construction (SMRC): With SMRC, k_{max} was set to 1000, and *cooling_rate* was set to 0.1 to ensure that the temperature cooled down to nearly zero for a near-optimal result. 100 budgets were used in the histogram for each dimension.

Dataset: There were two datasets:

- TPC-H: We used the "lineitem" table in TPC-H. The "lineitem" table has 17 columns, in which 7 columns are the dimensions (*l_quantity*, *l_partkey*, *l_orderkey*, *l_linenumber*, *l_extendedprice*, *l_suppkey*, *l_discount*). We used a scale factor of 20–100, resulting in different data sizes, ranging from 120 million to 600 million data items in the table.
- Synthetic dataset: To simulate big data, we generated a synthetic dataset that contained 1000 dimensions. The value scope of each dimension was 1–100. The data was uniformly distributed in the space. The total number of data items was 1 billion.

Workload: Two workloads were generated for each dataset. Each workload contained 1000 query instances. Supposing that the dataset has n dimensions, each query instance contains one CNF with n predicates. The predicate is either a full range predicate or an equality predicate. The selectivity of a range predicate is 10% on its dimension. We generated a skewed workload Q_s where some dimensions had more range predicates than others. We also generated a uniform query workload, called Q_u, where the range predicates were uniformly distributed on each dimension.

Metrics: We used $Cost(Q, R)$ to measure the performance of the different methods. The lower $Cost(Q, R)$ is, the higher the query throughput is. To guarantee accuracy, we ran each algorithm three times and used the average values.

In the following experiments, we answer six questions:

1) How can we get the cost model in a real cluster? Sect. 5.3
2) How does our method perform under different circumstances? (Sect. 5.4).
3) What additional cost will heterogeneous replicas bring? (Sect. 5.5).
4) Can simulated annealing be replaced by other heuristic algorithms? (Sect. 5.6).
5) What is the SMRC's performance when facing node failure? (Sect. 5.7).

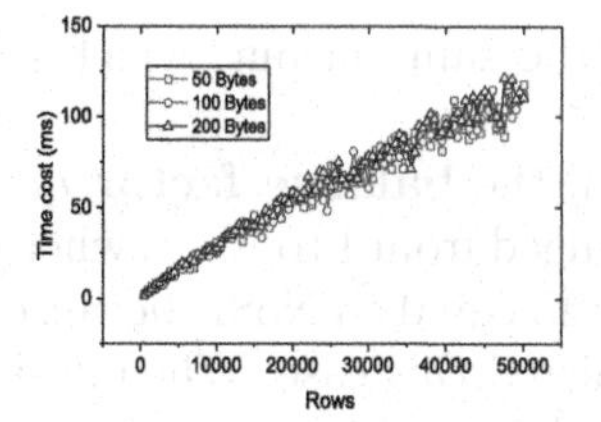

(a) with a different size of value column

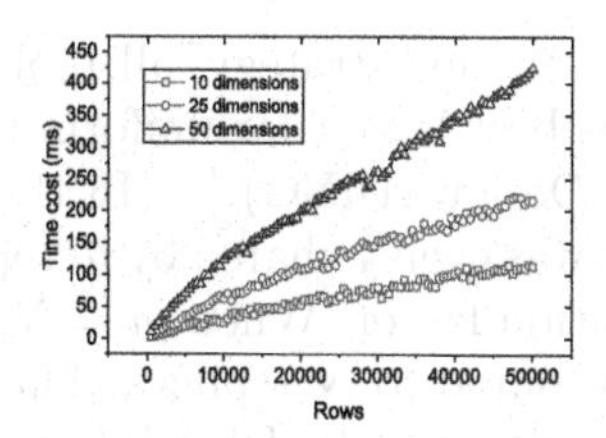

(b) with a different number of clustering columns

Fig. 3. Modeling $t()$ on Cassandra

5.3 Model Robustness

Deriving an exact cost function is challenging, because the cost function not only depends on the configurations of the system but also on hardware performance. Therefore, we treated it as a black box and evaluated the function $t()$ statistically. We found that $t()$ is a linear function under different workloads.

We generated a number of queries where the size of the candidate set Row_r^q differed on a simulation dataset. We enabled **tracing** to profile the time cost in each query using the *"tracing on"* command[1] and recorded the query latency.

We evaluated the time cost function $t()$ with different sizes of data items on a synthetic dataset with ten dimensions. First, we changed the size of the metrics columns to control the size of each data item from 50 bytes to 200 bytes. Figure 3(a) shows the result. The points show the roughly linear relationship

[1] https://docs.datastax.com/en/cql/3.3/cql/cql_reference/cqlshTracing.html.

between candidate set Row_r^q and the time cost. The cost did not change significantly when the size of the data items increased.

Second, we studied the impact of the number of dimensions on the cost function. We changed the number of dimensions from 10 to 50. As can be seen in Fig. 3(b), the cost function remained linear with different numbers of clustering columns. Unlike the increasing size of the value column, the slope of the cost function increased considerably when the number of dimensions increased. Thus, the cost function $C(q, r)$ can be replaced by Row_r^q without impacting the result of the SMRC algorithm.

5.4 Performance of SMRC

The query time cost of the different methods on the TPC-H dataset is shown in Fig. 4. The scale factor was 20, the replication factor was 3, and the workload was Q_s. With DIVG, we used a balance factor m of 1, because we found that a higher m value degraded the query throughput of DIVG. For example, when m was 1, the Cost(Q, R) was 26,800 s, whereas it was 54,800 s when m was 2. The reason for this is that when the parameter m is set higher, more query types need to be served by a replica, resulting in less diversity in each replica. We set balance factor to 1 for DIVG in the following experiments.

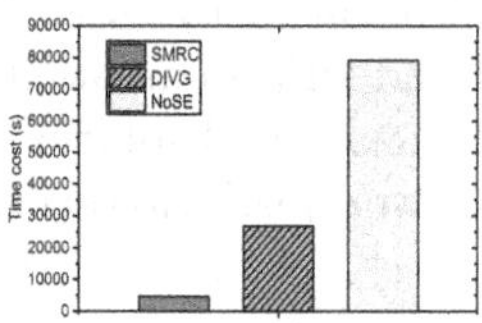

Fig. 4. Time cost with Q_s on the TPC-H dataset

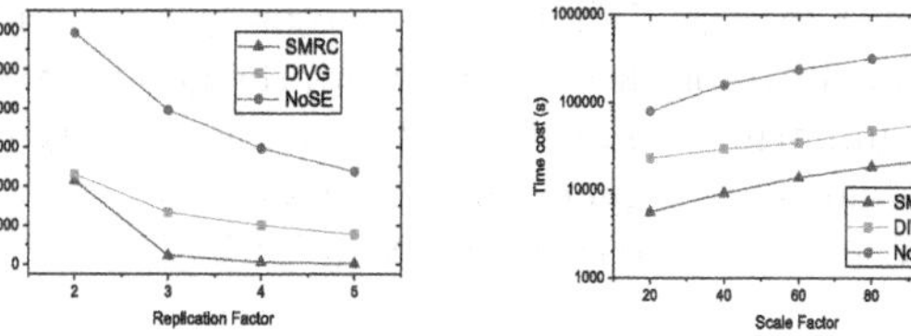

Fig. 5. Replication factor with Q_s on the TPC-H dataset

Fig. 6. Data size with Q_s on the TPC-H dataset

To evaluate the performance and scalability of the different methods further, we performed the following experiments under different replication factors, different dataset scales, and different workloads.

Replication Factor: The benefits of heterogeneous replicas depend on the replication factor. For most applications, the replication factor is between two and five to support high availability when facing node failure. We evaluated the impact of the replication factor on query time cost with the TPC-H dataset and workload Q_s. The scale factor was 20, and the results are shown in Fig. 5. When the replication factor was two, SMRC was similar to DIVG, because the alternative replica layouts were limited and both methods could find a good state. As the replication factor increased, however, the number of potential layouts grew

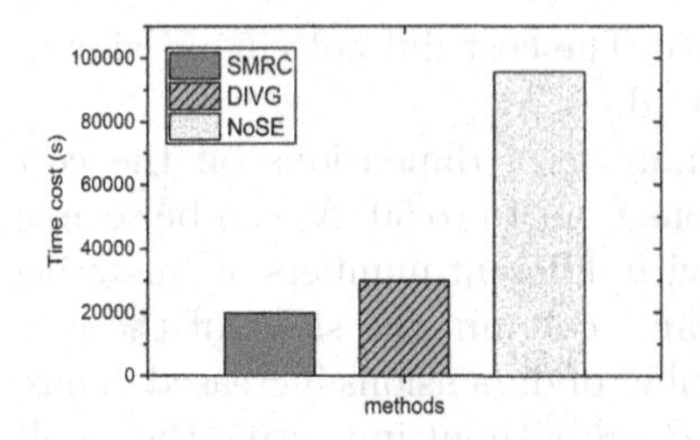

Fig. 7. Workload Q_u on the TPC-H dataset

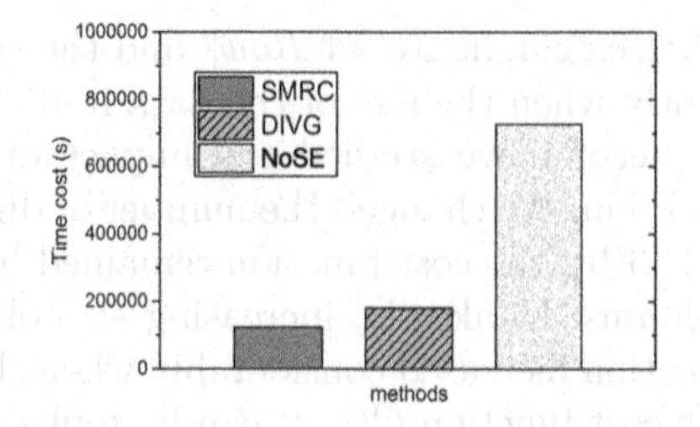

Fig. 8. Large-scale problem with Q_s on the synthetic dataset

exponentially. The advantage of SMRC is more obvious. Furthermore, even two heterogeneous replicas greatly improved query throughput compared to NoSE.

Data Size: We evaluated query time cost on the TPC-H dataset with a scale factor from 20 to 100, and a replication factor of 3. Figure 6 shows the time cost of the different strategies. As the size of the dataset increased, the size of the result set grew proportionally. Under different scale factors, SMRC always required the least time among the methods, demonstrating that SMRC has good scalability.

Uniform Workload: The performance of workload Q_u on the TPC-H dataset is shown in Fig. 7. The scale factor was 20 and the replication factor was 3. Compared to the skewed workload, the gap between SMRC and DIVG narrowed. For a uniform workload, DIVG was more likely to find a near-optimal solution, which means that query clustering was more suitable for that workload. However, SMRC still performed better than DIVG.

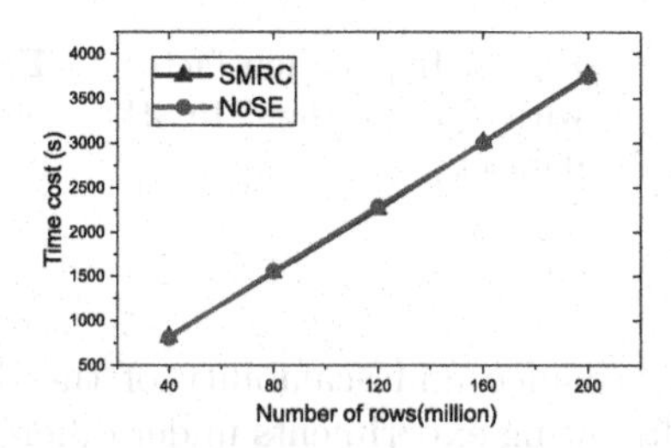

Fig. 9. Insertion time of the TPC-H dataset

Problem Scale: The scale of the problem depends on the searching space, except for the TPC-H dataset with seven dimensions. We also used the synthetic dataset with 1000 dimensions. The replication factor was 3 and the query cost of the different methods is in Fig. 8. Even with hundreds or thousands of dimensions, SMRC performed the best. Thus, SMRC can solve large-scale problems.

5.5 Overhead of Heterogeneous Replicas

Write Speed: We evaluated the writing speed of heterogeneous replicas by loading the TPC-H data into different layouts generated by SMRC and NoSE. In real applications, the data insertion order is difficult to guarantee. We loaded the data in order of origin. The result is shown in Fig. 9. Our evaluation shows that heterogeneous replicas bring no additional cost to the insertion speed, because we do not maintain any extra structure or dataset.

Memory Consumption and Routing Overhead: The memory consumption of our method is low. Unlike [18], we do not store the training workload. Only the statistics of the data are kept in memory. For a dataset with 7 dimensions, if 100 budgets (each budget stores the value count) are used in each histogram, the total memory usage is 5 KB ($7 \times 100 \times 8B$). When routing a query, we only need to execute N calculations to find the most efficient replica, where N is the replication factor. In our experiments, the routing time occupied less than 0.5% of the query processing, because they are in-memory calculations.

5.6 Other Heuristic Algorithms

In addition to simulated annealing, we also attempted to use the genetic algorithm [20], insofar as it is also suitable for multi-dimensional data replica construction. However, the genetic algorithm required more time to converge—approximately five times more than SMRC. Although other heuristic algorithms may work, we focused on SMRC in this paper for efficiency.

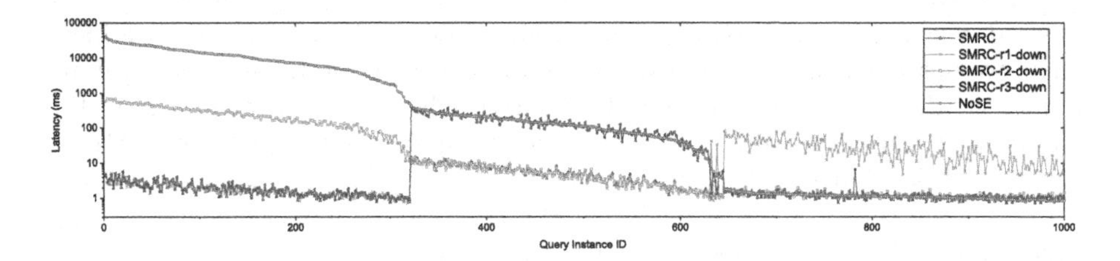

Fig. 10. Latency of each query instance when facing node failure (Color figure online)

5.7 Failure and Recovery

To explore the behavior of our method when facing failure, we conducted the following experiments. The dataset we used was TPC-H, the workload was Q_s, and the replication factor was set to three.

Replication is used to guarantee system availability. Given three replicas, r1, r2, and r3, we can choose only two replicas to route queries when a node is down. Intuitively, many queries may be worse than uniform replicas in this

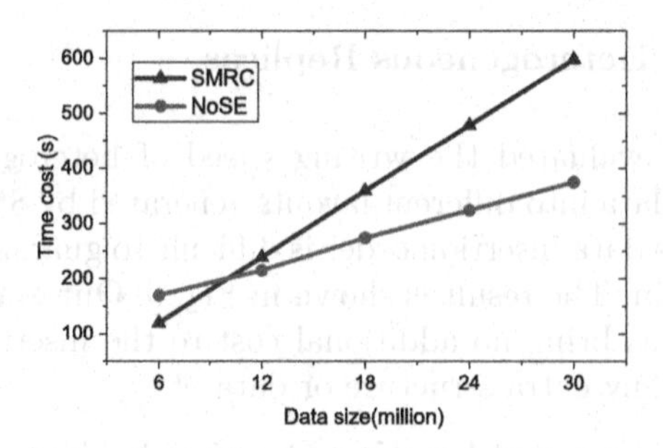

Fig. 11. Data recovery speed

situation. We traced the latency of each query in this scenario, and the results are shown in Fig. 10. The query instances are sorted according to descending order of latency on NoSE (red line). The blue line represents all replicas that were alive in SMRC. In this case, each replica served approximately a third of the query instances. Query instances 0–320 were routed to r2, 321–632 to r3, and 633–1000 to r1. The other three lines represent each replica that is down in SMRC. Obviously, when a replica is down, approximately a third of the queries will be affected. However, the query latency was still acceptable and the query time cost was greater than NoSE. Improvements to query time cost in the case of node failure were discussed in [18], and the solutions offered there can be applied to our method.

We recover a replica by rewriting the data to another replica. We measured the recovery speed of NoSE with the repair command in Cassandra, and we measured the recovery speed of SMRC with rewritten data. For NoSE, we first stopped one node in a Cassandra cluster and removed that node's data folder. Then, we restarted the Cassandra process and called the *nodetool repair -full* to launch the data recovery process in Cassandra. The result is shown in Fig. 11. The recovery speed of NoSE was 60% faster than that of the proposed SMRC. However, disk failures occur infrequently, and recoverability is more important than recovery speed. Thus, considering the tremendous improvement in query performance, the speed of recovery with SMRC is acceptable.

When the query workload changes dramatically, query throughput can decrease with current replica layouts. When the query performance decreases to a certain threshold, the algorithm can generate a new replica layout for restructuring data. There are two ways to restructure the data: (1) restructuring all historical data [18], or (2) restructuring recent data [2]. These approaches are both applicable to our methods.

6 Conclusions

In this paper, we studied and defined the problem of how to leverage heterogeneous replicas to improve query throughput on multi-dimensional data, which is widely used in applications. Existing approaches to accelerate queries either optimize limited kinds of queries by adjusting the layout of the data, or they bring additional overhead by introducing auxiliary structures. The proposed method, however, does not introduce any additional disk cost when optimizing the existing replica layout on the disk.

We modeled the query cost with multi-dimensional data, and introduced a routing strategy and a replica construction algorithm. The proposed method outperformed state-of-the-art solutions. Furthermore, our solutions did not incur additional overhead, such as extra disk occupation or slowed insertion speeds. When node failures occurred, our heterogeneous replicas worked well and could recover in a reasonable time. We believe that our solutions can be easily applied to other multi-dimensional data management systems apart from Cassandra. The future work is finding some rules to accelerate the searching of SMRC algorithm.

Acknowledgments. The work was supported by the Nature Science Foundation of China (No. 61802224, 71690231), and Beijing Key Laboratory of Industrial Bigdata System and Application. We also thank anonymous reviewers for their valuable comments.

References

1. Ailamaki, A., DeWitt, D.J., Hill, M.D., Skounakis, M.: Weaving relations for cache performance. VLDB **1**, 169–180 (2001)
2. Bian, H., Yan: Wide table layout optimization based on column ordering and duplication. In: Proceedings of the 2017 ACM International Conference on Management of Data, pp. 299–314. ACM (2017)
3. Borthakur, D., et al.: HDFS architecture guide. Hadoop Apache Project 53 (2008)
4. Consens, M.P., Ioannidou, K., LeFevre, J., Polyzotis, N.: Divergent physical design tuning for replicated databases. In: Proceedings of the 2012 ACM SIGMOD International Conference on Management of Data, pp. 49–60. ACM (2012)
5. Copeland, G.P., Khoshafian, S.: A decomposition storage model. In: SIGMOD Conference (1985)
6. Grund, M., Krüger, J., Plattner, H., Zeier, A., Cudré-Mauroux, P., Madden, S.: Hyrise - a main memory hybrid storage engine. PVLDB **4**, 105–116 (2010)
7. Jindal, A., Quiané-Ruiz, J.A., Dittrich, J.: Trojan data layouts: right shoes for a running elephant. In: Proceedings of the 2nd ACM Symposium on Cloud Computing, p. 21. ACM (2011)
8. Jouini, K.: Distorted replicas: intelligent replication schemes to boost I/O throughput in document-stores. In: 2017 IEEE/ACS 14th International Conference on Computer Systems and Applications (AICCSA), pp. 25–32 (2017)
9. Kirkpatrick, S., Gelatt, D., Vecchi, M.P.: Optimization by simulated annealing. Science **220**, 671–680 (1983)
10. Lamb, A., et al.: The vertica analytic database: C-store 7 years later. PVLDB **5**, 1790–1801 (2012)
11. Mior, M.J., Salem, K., Aboulnaga, A., Liu, R.: NoSE: schema design for NoSQL applications. IEEE Trans. Knowl. Data Eng. **29**(10), 2275–2289 (2017)
12. Home page P (2018). http://parquet.apache.org/documentation/latest/

13. Rabl, T., Jacobsen, H.A.: Query centric partitioning and allocation for partially replicated database systems. In: Proceedings of the 2017 ACM International Conference on Management of Data. pp. 315–330. ACM (2017)
14. Ramamurthy, R., DeWitt, D.J., Su, Q.: A case for fractured mirrors. VLDB J. **12**, 89–101 (2002)
15. Saccà, D., Wiederhold, G.: Database partitioning in a cluster of processors. ACM Trans. Database Syst. **10**, 29–56 (1983)
16. Staudt, M., Jarke, M.: Incremental maintenance of externally materialized views. In: VLDB (1996)
17. Stonebraker, M., et al.: C-store: a column-oriented DBMs. In: Proceedings of the 31st International Conference on Very Large Data Bases, pp. 553–564. VLDB Endowment (2005)
18. Tran, Q.T., Jimenez, I., Wang, R., Polyzotis, N., Ailamaki, A.: RITA: an index-tuning advisor for replicated databases. In: Proceedings of the 27th International Conference on Scientific and Statistical Database Management, p. 22. ACM (2015)
19. Valentin, G., Zuliani, M., Zilio, D.C., Lohman, G., Skelley, A.: DB2 advisor: an optimizer smart enough to recommend its own indexes. In: Proceedings of 16th International Conference on Data Engineering (Cat. No. 00CB37073), pp. 101–110. IEEE (2000)
20. Whitley, D.: A genetic algorithm tutorial (1994)
21. Xiang-dong, H., Jian-min, W., Si-han, G., et al.: A storage model for large scale multi-dimension data files. Proc NDBC **1**, 48–56 (2014)
22. Xu, C., Tang, B., Yiu, M.L.: Diversified caching for replicated web search engines. 2015 IEEE 31st International Conference on Data Engineering, pp. 207–218 (2015)

Latency-Aware Data Placements for Operational Cost Minimization of Distributed Data Centers

Yuqi Fan[1(✉)], Chen Wang[1], Bei Zhang[1], Donghui Hu[1], Weili Wu[2], and Dingzhu Du[2]

[1] School of Computer Science and Information Engineering, Anhui Province Key Laboratory of Industry Safety and Emergency Technology, Hefei University of Technology, Hefei 230601, Anhui, China
{yuqi.fan,hudh}@hfut.edu.cn, chenw@mail.hfut.edu.cn, bei0lei0zhang@gmail.com

[2] Department of Computer Science, University of Texas at Dallas, Richardson 75080, USA
{weiliwu,dzdu}@utdallas.edu

Abstract. A large amount of data are stored in geographically distributed data centers interconnected by the Internet. The power consumption for running the servers storing the data and inter-data center network transport for data update among the multiple data copies stored in different data centers impose a significant operational cost on cloud service providers. Few research places data replicas in the data centers by jointly taking both electricity consumption and network transport into account under the user access latency requirement constraints. However, there is an intrinsic tradeoff between the electricity consumption of the data centers and the network transport cost for data update. In this paper, we tackle the problem of data placement in data centers with the aim to minimize the operational cost under user access latency requirements, assuming all the data have K copies. We propose an effective algorithm, Latency-aware and operational Cost minimization Data Placement (LCDP), which partitions the data into multiple data groups according to the data access rates and greedily selects K data centers incurring the minimum cost for each data in each data group. We prove that algorithm LCDP is $\frac{1}{2}\ln|\mathrm{U}|$-approximation to the data placement problem, where $|\mathrm{U}|$ is the number of users. Our simulation results demonstrate that the proposed algorithm can effectively reduce the power consumption cost, the network transport cost, and the operational cost of the data centers.

Keywords: Latency · Power consumption · Network transport · Data placement · Operational cost

Supported partly by the National Natural Science Foundation of China under grant U1836102.

Y. Nah et al. (Eds.): DASFAA 2020, LNCS 12112, pp. 37–53, 2020.
https://doi.org/10.1007/978-3-030-59410-7_3

1 Introduction

Fast Internet development encourages the boom of data-intensive applications, which leads to a large amount of data stored in geographically distributed data centers and a high frequency of data access requests from hundreds of millions of end users. For most applications, latency is a critical performance metric, and the end-to-end network latency accounts largely for the user-perceived latency, especially for data-intensive applications [1,2].

On one hand, it is desirable to place the data in as many data centers as possible in order to achieve low data access latency, since the users can access the data from the nearby data centers to reduce the data access delay. On the other hand, the power required to run the servers storing the data and the network transport necessitated for data synchronization among different data copies at different data centers impose a significant operational cost on the service providers [3,4]. The more the data copies, the more the operational cost. Therefore, cloud service providers often simultaneously place multiple copies of the data in a limited number of data centers to satisfy the user latency requirements and achieve data availability and reliability.

A key problem for cloud service providers is data placement, which is to select the proper data centers to place the data replicas. There has been some work on data placements to improve the performances of user access latency, electricity consumption of the data centers, network transport cost, etc., in recent years. However, few research places the data in geo-distributed data centers by jointly considering both electricity consumption and network transport under the user access latency requirement constraints. Note that there is an intrinsic tradeoff between the electricity consumption of the data centers and the network transport cost for data update, when the data are placed under the user access latency requirements as illustrated by Example 1.

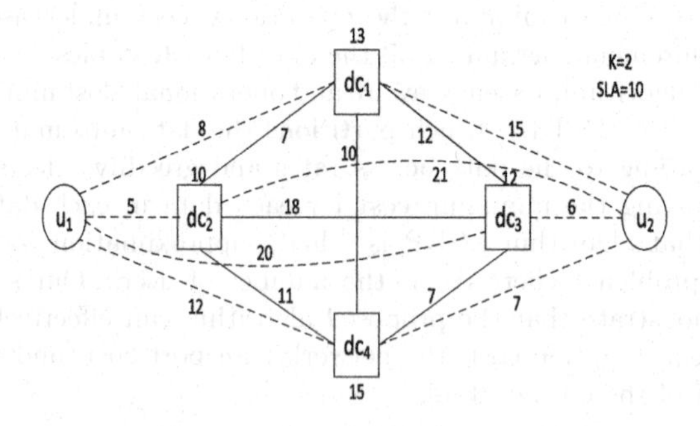

Fig. 1. An example of the tradeoff between power consumption and network communication.

Example 1. Two users of u_1 and u_2 access some data from the four data centers of dc_1, dc_2, dc_3, dc_4 as shown in Fig. 1. The number of data copies of each data is 2, and the latency requirement of the two users is 10 ms. The power cost of the four data centers is 13, 10, 12, and 15, respectively. The network transmission delay between the users and the data centers are shown beside the dashed lines, while the network transmission cost between the data centers is shown beside the solid lines in Fig. 1. There are three scenarios during the data placements under the user access latency requirements. (1) Scenario 1: We only consider the power consumption cost of the data centers, and the data will be placed in data centers dc_2 and dc_3. The power consumption cost is $10 + 12 = 22$, while the network transmission cost and operational cost are 18 and 40, respectively. (2) Scenario 2: We only take into account the network communication cost between the data centers, and the data will be placed in data centers dc_1 and dc_4. The power consumption cost is $13 + 15 = 28$, while the network transmission cost is 10 and the total operational cost is 38. (3) Scenario 3: We consider both power consumption cost and network communication cost, and the data will be placed in data centers dc_2 and dc_4, which leads to the power consumption cost, network transmission cost and operational cost as $10 + 15 = 25$, 11, and 36, respectively. It can be observed from Scenario 1 and Scenario 2 that there is a tradeoff between power consumption and network communication. We can reduce the total operational cost by considering both of the factors during data placements as shown in Scenario 3. □

The main contributions of this paper are as follows. We investigate the data placement problem to minimize the total operational cost of power consumption and network communication under the user access latency constraint, assuming that the number of data replicas is K. We propose an effective algorithm, Latency-aware and operational Cost minimization Data Placement (LCDP), which partitions the data into multiple data groups according to the rates of data read and write requests and greedily selects K data centers incurring the minimum cost for each data in each data group. We prove that algorithm LCDP is $\frac{1}{2}\ln|\mathrm{U}|$-approximation to the data placement problem, where $|\mathrm{U}|$ is the number of users. Our simulations results demonstrate the proposed algorithm LCDP can effectively reduce the power consumption cost, the network transport cost, and the operational cost of the data centers.

2 Related Work

Some studies focus on the impact of data placements on the user access latency. The data assignment to multiple storage devices was studied with an aim to minimize the total latency cost of serving each request for all the data, under the storage device capacity constraints [5]. A system to determine the data placement for geo-distributed storage systems was proposed to minimize the overall cost while satisfying the inter- and intra-data center latency requirement, and mixed integer linear programming (MILP) was used to determine the data placement [6].

Some research improves the system performance by jointly considering the electricity consumption of the data centers and the user access latency. The electricity cost, service level agreement requirement, and emission reduction budget were jointly considered by exploiting the spatial and temporal variabilities of the electricity carbon footprint [7]. A request-routing scheme, FORTE, was proposed to allow operators to strike a tradeoff among electricity costs, access latency, and carbon emissions [8]. Another request-routing scheme was devised to minimize the electricity cost of multi-data center systems [9], and the scheme was improved in [10] on multi-region electricity markets to better capture the fluctuating electricity price to reduce electricity cost.

Some research studies the communication cost optimization of data placements. The problem of placing user data into a distributed cloud was introduced with the aim to minimize network communication cost and power consumption cost of data centers, and a community fitness metric for data placement was proposed [4]. The cost optimization of data storage and inter-cloud communication was studied while providing satisfactory Quality of Service to users[11]. The data placement problem of minimizing the data communication costs was proposed with a reduction of the data placement problem to the well-studied Graph Partitioning problem [12]. A hypergraph-based model was employed to partition the set of data which were placed in distributed nodes to achieve the multiple objectives of inter data center traffic, latency, storage cost, and node span [13]. The data placement in geographically distributed clouds was modeled as a multi-objective optimization problem with the objectives of inter data center traffic, latency, storage cost, and data center span [14].

It can be observed that few research places the data in geo-distributed data centers by jointly taking both electricity consumption and network transport into account, under the user access latency requirement constraints. However, there is an intrinsic tradeoff between the electricity consumption of the data centers and the network transport cost.

3 Problem Formulation

We model a distributed cloud as $\mathbb{C} = (\mathcal{DC}, E)$, where $\mathcal{DC} = \{dc_1, dc_2, dc_3, \ldots, dc_N\}$ is the set of N geographically distributed data centers interconnected by link set E. We assume that each data center has a limited storage capacity, and the servers in a data center are homogeneous. Each user accesses the data from one of the data centers. For network links, we assume that the bandwidth capacities of such inter-data center links are unlimited [4]. The symbols and notations used in the paper are listed in Table 1.

We assume the average read and write rates from each user can be learned in advance [15], and each data has K replicas stored in different data centers. Without loss of generality, if a write request from a user updates one of the data replicas, the other $K-1$ replicas also need to be updated. The electricity is used to run the servers storing the data so that the data can be accessed. A read request incurs only power consumption cost, while a write request incurs communication cost as well as power consumption, because the network resources are used to update all the data copies among the data centers. The data placement for service providers should minimize the operational cost and meet the user access latency SLA requirements.

3.1 Power Consumption Cost Model

The power consumption of a server running at speed μ can be defined by $a \cdot \mu^{\rho} + b$ [2,16], where a is a factor related to peak power consumption, b is the average idle power consumption of the server, and parameter ρ is empirically determined as $\rho \geq 1$. The number of active servers M_j required by data center j is $M_j = \left\lceil \frac{A_j}{S_j} \right\rceil$. The actual service rate of each active server in data center j, μ_j, is calculated as $\mu_j = \frac{A_j}{M_j}$.

Given the power usage efficiency metric PUE_j for each data center j [1], we model the power consumption cost ψ_e of all data centers as

$$\psi_e = \sum_{j \in \mathcal{DC}} p_j \cdot PUE_j \cdot M_j \cdot (a_j \cdot \mu_j^{\rho} + b_j) \tag{1}$$

Table 1. Table of symbols and notations

U	The set of users;
$\mathcal{D}$	The set of data;
$x_{i,m,j}$	Indicate whether user i accessing data m is assigned to data center j $(= 1)$ or not $(= 0)$;
$y_{m,j}$	Indicate whether data m is placed in data center j $(= 1)$ or not $(= 0)$;
$s(m)$	The size of data m;
$r_{i,m}$	The request rate that user i reads data m;
$w_{i,m}$	The request rate that user i writes data m;
K	The number of replicas of each data;
cap_j	The storage capacity of data center j;
p_j	The average unit-power electricity price at the location of data center j;
S_j	The maximum service rate of each server in data center j;
A_j	The number of requests arriving at data center j;
ψ_e	The total power consumption cost of all the data centers;
λ_w^m	The amount of data generated by a single write request for data m;
c_e	The cost of transmitting one unit of data along link $e \in E$;
ψ_w	The total communication cost between data centers holding the data replicas;
$d_{i,j}$	The latency between user i and data center j;

3.2 Network Communication Cost Model

We find the routing paths from a data center with a replica to the remaining $K - 1$ data centers hosting the data replicas, and use a multicast tree for the data replica synchronization, such that the updated data are transmitted only once on each required network link. The total communication cost ψ_w for all the data is

$$\psi_w = \sum_{m \in \mathcal{D}} \left(\sum_{i \in \mathrm{U}} w_{i,m} \sum_{e \in path_m} c_e \cdot \lambda_w^m \right) \tag{2}$$

where $path_m$ is the multicast routing tree to connect all the data centers hosting the K replicas of data m.

3.3 Latency SLA Model

The user may have different SLAs for different data, and different users may have different SLAs for the same data. We assume that there are Q different levels of latency requirements, denoted by L^q, where $L^1 < L^2 \cdots < L^q \cdots < L^Q$. Latency requirement $L_{i,m}^q$ of user i accessing data m is satisfied if the end-to-end access latency $d_{i,j}$ is no greater than L^q. That is, we have the SLA constraint defined by Eq. (3).

$$d_{i,j} \leq L_{i,m}^q, \forall j \in \mathcal{DC} \tag{3}$$

3.4 Data Placement Model

With the SLA and the storage capacity constraints, our objective is to place the K replicas of each data into K data centers, such that the operational cost is minimized, where the cost consists of intra-data center power consumption cost and inter-data center network communication cost. Assuming $K \leq\mid \mathcal{DC} \mid$, the problem is to minimize

$$\psi = \psi_e + \psi_w \tag{4}$$

subject to:

$$\sum_{m \in \mathcal{D}} s(m) \cdot y_{m,j} \leq cap_j, \forall j \in \mathcal{DC} \tag{5}$$

$$d_{i,j} \cdot x_{i,m,j} \leq L_{i,m}^q, \forall i \in \mathrm{U}, m \in \mathcal{D}, j \in \mathcal{DC} \tag{6}$$

$$x_{i,m,j} \leq y_{m,j}, \forall i \in \mathrm{U}, m \in \mathcal{D}, j \in \mathcal{DC} \tag{7}$$

$$\sum_{j \in \mathcal{DC}} y_{m,j} = K, \forall m \in \mathcal{D} \tag{8}$$

$$\sum_{j \in \mathcal{DC}} x_{i,m,j} = 1, \forall i \in \mathrm{U}, m \in \mathcal{D} \tag{9}$$

$$x_{i,m,j} \in \{0,1\}, y_{m,j} \in \{0,1\}, \forall i \in \mathrm{U}, m \in \mathcal{D}, j \in \mathcal{DC} \tag{10}$$

Equation (5) dictates the total size of the data stored in a data center cannot exceed the capacity of the data center. Equation (6) defines that the access latency between user and data center cannot exceed the SLA requirement. Equation (7) ensures that a data center from which a user accesses a data must host one of the data replicas. Equation (8) requires each data to be placed in K data centers. Equation (9) determines that user request for a data is assigned to one and only one data center. Equation (10) dictates that the mapping relationship of the users, the data, and the data centers is represented by binary variables.

4 Latency-Aware and Operational Cost Minimization Data Placement Algorithm

We propose a Latency-aware and operational Cost minimization Data Placement algorithm (LCDP) for the large-scale data placement problem, taking latency, power consumption and network transport into account. Algorithm LCDP consists of two stages:

Stage One: We divide data set $\mathcal{D}$ into multiple data groups according to the data read and write rates. The data with similar read and write rates potentially incurs the similar operational cost and hence are put into the same group.

Stage Two: A data group with high accumulated read and write rates leads to a high cost, which brings more opportunities for reducing the operational cost. Therefore, the data groups are sorted by the non-ascending order according to the accumulated read and write rates, and the data in the data group with high accumulated access rates are given high priority during placement. We select a data center subset with K data centers for each data, such that the operational cost is minimized under the user access latency requirements and data center storage capacity constraints.

4.1 Data Group Division

We introduce a data similarity metric for data grouping, which takes the data read and write rates into account. During the data group division process, we decide whether to put data m into a data group by comparing the read and write rates of the data with the average read and write rates of the data group. The similarity metric $L(m, g)$ for data m and the data in group g is defined as Eq. (11).

$$L(m,g) = \alpha \cdot |r_m - \overline{r_g}| + \beta \cdot |w_m - \overline{w_g}| \tag{11}$$

where $\overline{r_g}$ and $\overline{w_g}$ represent the average read and write rates of the data in group g, respectively, and α and β ($0 \leq \alpha \leq \beta$) are the weights of read and write requests, respectively.

The process of data group division is shown in Algorithm 1. Given the similarity threshold $\Delta > 0$, we initially choose a data randomly from data set $\mathcal{D}$

Algorithm 1. *Data Group Division*

Input: Set of data $\mathcal{D}$, Set of users U, Group similarity threshold Δ,
Read and Write Requests from users Matrix $R(i,m)/W(i,m)$.
Output: Data group set G.
1: Select a random data m' from data set $\mathcal{D}$ as the first group g, $G \leftarrow \{g\}$;
2: **for** each data $m \in \mathcal{D}$ **do**
3: **for** each data group $g \in G$ **do**
4: Calculate $L(m,g)$ via Equation (11);
5: **end for**
6: $g^* = \underset{g \in G}{\operatorname{argmin}} L(m,g)$;
7: **if** $L(m,g^*) \geq \Delta$ **then**
8: Create a new group $g' \leftarrow \{m\}$;
9: $G \leftarrow G \cup g'$;
10: **else**
11: $g^* \leftarrow g^* + m$;
12: **end if**
13: **end for**
14: **return** data group set G;

as the initial group g, and the data group set G includes only one group g. For each remaining data m, we calculate the similarity $L(m,g)$ between data m and the data in each group g, and find group g^* with the minimum $L(m,g)$ from all the groups. That is, $g^* = \underset{g \in G}{\operatorname{argmin}} L(m,g)$. If $L(m,g^*) < \Delta$, we add data m into group g^*. Otherwise, data m does not belong to any groups, and hence it forms a group by itself. Notice that the value of similarity threshold Δ determines the number of data groups. Assuming r_{max} and w_{max} are the upper bounds of read and write rates of all the requests, respectively, we set $\Delta = \delta \cdot |\alpha \cdot r_{max} + \beta \cdot w_{max}|$, where δ is a constant with $\delta \leq 1$ to tune the number of groups.

4.2 Data Center Subset Selection

If the data access latency requirements of user i can be satisfied by placing data m in data center j, we say data center j can serve user i for data m, and data center j is a candidate data center for user i accessing data m. All the users that data center j can serve for the accesses of data m compose user collection $U_{m,j}$.

Each data should be placed in a data center subset which consists of K data centers. If a data center subset $\mathbb{S}$ cannot serve all the users in user set U for accessing data m, data m should not be placed in data center subset $\mathbb{S}$ and $\mathbb{S}$ is not a candidate data center subset for placing data m; otherwise, data m can be accommodated by data center subset $\mathbb{S}$ and $\mathbb{S}$ is a candidate data center subset for placing data m.

For each data m, given user set U, and $\mathbb{U} = \{U_{m,1}, U_{m,2}, \cdots, U_{m,j}, \cdots, U_{m,N}\}$, where element $U_{m,j}$ is the user collection that data center j can serve for the accesses of data m, we need to find a candidate data center subset $\mathbb{S}$ which consists of K data centers for placing data m under the user access latency requirements and all the users in U can be served by the data centers in $\mathbb{S}$, which is called *data center subset selection problem.* If there are multiple such candidate data center subsets, we select the one with the minimum operational cost, which is called *data center subset selection for operational cost minimization problem.*

Theorem 1. *Data center subset selection problem is $\mathcal{NP}$-complete.*

Proof. We can show that the set cover decision problem, a well-known $\mathcal{NP}$-complete problem, can be reduced to the data center subset selection problem. Details are omitted to conserve space, due to page restrictions. □

Obviously, the data center subset selection for operational cost minimization problem and the data placement problem defined in this paper are also $\mathcal{NP}$-complete, since data center subset selection problem is a sub-problem of these two problems.

We denote the cost of placing a single data m in data center subset $\mathbb{S}$ as $\varpi(\mathbb{S})$, the weight of data center subset $\mathbb{S}$, which can be calculated by Eq. (4). We also denote the weight increase by adding data center j in data center subset $\mathbb{S}$ ($j \notin \mathbb{S}$) as $\Delta\varpi(\mathbb{S}, j) = \varpi(\mathbb{S}+j) - \varpi(\mathbb{S})$. We define the increased number of users that data center j can serve per unit weight increase as θ_j by Eq. (12).

$$\theta_j = \frac{|U_{m,j} \cap \mathrm{U}|}{\Delta\varpi(\mathbb{S}, j)} \tag{12}$$

For data m, algorithm DCSS shown in Algorithm 2 depicts the process of selecting K data centers which cause the least operational cost and meet the latency requirements of all users accessing data m. Initially, the selected data center subset $\mathbb{S} = \phi$, and all the users are to be served (line 1). Algorithm 2 then obtains the user collection U_j for each data center $j \in \mathcal{DC}$ (line 2). The algorithm proceeds iteratively by finding a data center with the largest θ_j and enough capacity for data m to be added into data center subset $\mathbb{S}$ in each iteration (lines 3–7). The iterative process continues until the number of data centers in subset $\mathbb{S}$ is increased to K.

4.3 Algorithm LCDP

We introduce a group-ranking metric that integrates the impact on the operational cost from both the read and write requests. Each data group g is assigned a rank $\Omega(g)$ which is defined as $\Omega(g) = \frac{\widehat{w_g}}{\widehat{r_g}}$, where $\widehat{w_g}$ and $\widehat{r_g}$ are the accumulated read and write rates of data group g, respectively.

After dividing the data into multiple data groups via Algorithm 1, algorithm LCDP sorts the data groups in the non-ascending order of the rank values $\Omega(g)$. The data in the data group with high $\Omega(g)$ are given priority during data placement, as the operational cost incurred by the data access in the data group with high read and update rates will potentially be significant. Algorithm LCDP then deals with the data groups in turn. For each data group, the algorithm proceeds

Algorithm 2. *DCSS(m)*

Input: Set of users U, Data centers $\mathcal{DC}$, User latency requirements Matrix $L_{i,m}$.
Output: The data center subset $\mathbb{S}$.
1: $\mathbb{S} \leftarrow \phi$; $\mathrm{U}' \leftarrow \mathrm{U}$; $c \leftarrow 0$ /* The number of data centers in subset $\mathbb{S}$ */;
2: Obtain all user collections U_j $(\forall j \in \mathcal{DC})$;
3: **while** $\mathrm{U}' \neq \phi$ and $c < K$ **do**
4: Calculate θ_j $(\forall j \in \mathcal{DC} - \mathbb{S})$;
5: Select data center j^* with the largest θ_j and enough capacity for accommodating data m;
6: $\mathbb{S} \leftarrow \mathbb{S} + j^*$; $\mathrm{U}' \leftarrow \mathrm{U}' \setminus U_{j^*}$; $c \leftarrow c + 1$;
7: **end while**
8: **return** data center subset $\mathbb{S}$;

iteratively. For a data group g, algorithm LCDP keeps placing the data of data group g in a data center subset in the non-descending order of the data sizes using Algorithm 2, until all the data of data group g are placed in some data center subsets with K data centers. The iterative process continues until all the data are placed in some data centers.

4.4 Algorithm Analysis

We now analyze the performance and time complexity of the proposed algorithm LCDP.

Theorem 2. *Algorithm DCSS shown in Algorithm 2 is a* $\ln|\mathrm{U}|$*-approximation algorithm to the data center subset selection for operational cost minimization problem.*

Proof. In the process of data center subset selection, the weight of data center subset $\mathbb{S}$ can be obtained by Eq. (4). Based on θ_j defined by Eq. (12), Algorithm 2 selects data center j^* with the largest θ_j among all the data centers, i.e., $j^* = \arg\max_{j \in \mathcal{DC}} \theta_j$, until K data centers are found. For each user i, let j' be the first selected data center that can serve user i, and we define the cost of serving user i, $C(i)$, via Eq. (13).

$$C(i) = \frac{\Delta\varpi(\mathbb{S}, j')}{|U_{j'} \cap \mathrm{U}'|} \tag{13}$$

When the l-th user needs to be served, which means $l-1$ users before l-th user have been served, unserved user set U' contains at least $|\mathrm{U}| - l + 1$ users. Assume $\mathbb{S}^*$ is the optimal solution for data center subset selection for operational cost minimization problem, and the same $|\mathrm{U}| - l + 1$ users can be served by a data center subset $\mathbb{S}^*_k \subseteq \mathbb{S}^*$, where $\mathbb{S}^*_k$ consists of k data centers. The number of users that data center subset $\mathbb{S}^*_k$ can serve per unit weight is $\frac{|\mathrm{U}|-l+1}{\varpi(\mathbb{S}^*_k)}$.

During the process of iteratively selecting the data centers, we select a data center which can serve the most number of unserved users per unit weight among all the unselected data centers in each iteration. Assume j is the selected data center by Algorithm 2 in some iteration, and the l-th user, which is unable to be served by the other selected data centers, can be served by the selected data center j. For data center j, we have

$$\frac{|U_j \cap \mathrm{U}'|}{\Delta\varpi(\mathbb{S}, j)} \geq \frac{|\mathrm{U}| - l + 1}{\varpi(\mathbb{S}_k^*)} \geq \frac{|\mathrm{U}| - l + 1}{\varpi(\mathbb{S}^*)} \tag{14}$$

and the cost of serving user l, $C(l)$, is as follows.

$$C(l) = \frac{\Delta\varpi(\mathbb{S}, j)}{|U_j \cap \mathrm{U}'|} \leq \frac{\varpi(\mathbb{S}^*)}{(|\mathrm{U}| - l + 1)} \tag{15}$$

The cost of serving all the users by the data centers chosen via Algorithm 2 satisfies

$$\sum_{l=1}^{|\mathrm{U}|} C(l) \leq \sum_{l=1}^{|\mathrm{U}|} \frac{\varpi(\mathbb{S}^*)}{(|\mathrm{U}| - l + 1)} \leq \varpi(\mathbb{S}^*) \cdot \ln |\mathrm{U}| \tag{16}$$

That is, $\frac{\sum_{l=1}^{|\mathrm{U}|} C(l)}{\varpi(\mathbb{S}^*)} \leq \ln |\mathrm{U}|$. Therefore, the approximation ratio of the solution produced by Algorithm 2 to the data center subset selection for operational cost minimization problem is $\ln |\mathrm{U}|$. □

Theorem 3. *Algorithm LCDP is a* $\frac{1}{2} \cdot \ln |\mathrm{U}|$*-approximation algorithm to the data placement problem.*

Proof. To minimize the operational cost of data centers, for each data group g, we place the data in group g into the data centers in the non-descending order of the data sizes. If any data center in the selected data center subset $\mathbb{S}$ has not enough capacity for a data, the data cannot be put into data center subset $\mathbb{S}$. We assume that $m \in g$ is the first data that cannot be placed in subset $\mathbb{S}$ due to the data center storage constraint, and we also assume data m can be fractionally placed in subset $\mathbb{S}$. Let $\mathbb{Q}$ be the solution obtained by Algorithm LCDP, and $\mathbb{Q}'$ be the solution for fractional placement. Since we try to minimize the operational cost, it is trivial to show that $\psi(\mathbb{Q}') \leq \psi(\mathbb{Q})$, where ψ is the operational cost calculated via Eq. (4).

Assuming all the data in subset $\{m_1, m_2, m_3 \ldots m_{k-1}\} \subseteq g$ are placed in the same data center subset $\mathbb{S}$, and data m_k cannot be placed in subset $\mathbb{S}$ due to the data center storage capacity constraint. In solution $\mathbb{Q}'$, data m_k is fractionally placed in subset $\mathbb{S}$. We construct a solution $\mathbb{Q}''$ by removing data m_k from solution $\mathbb{Q}'$, and the remaining data in solution $\mathbb{Q}'$ are kept in solution $\mathbb{Q}''$.

Assuming that the maximum cost of placing a data in the data placement problem is *maxcost*, the cost reduction of placing any data m in a data center subset is $p(m) = maxcost - \psi(m)$. Note that the data with small sizes are placed first, which makes as many data as possible be placed with the least cost. So, the cost reduction of any data m'' in solution $\mathbb{Q}''$ is no less than that of data m_k. That is, $p(m'') \geq p(m_k)$, and $p(m_1) \geq p(m_2) \geq \cdots \geq p(m'') \ldots p(m_{k-1})$.

Let $p(m^*) = max\{p(m_1), p(m_2), \ldots, p(m''), \ldots, p(m_{k-1})\} = p(m_1)$, data m^* must be placed before data m_k. Therefore, assuming $f()$ is the cost reduction function, the cost reduction by solution $\mathbb{Q}''$ is

$$f(\mathbb{Q}'') = \sum_{t=1}^{k-1} p(m_t) \geq p(m^*) \tag{17}$$

and

$$\begin{aligned} f(\mathbb{Q}'') &= \sum_{t=1}^{k-1} p(m_t) = max\{\sum_{t=1}^{k-1} p(m_t), p(m^*)\} \\ &\geq \frac{1}{2}(\sum_{t=1}^{k-1} p(m_t) + p(m^*)) \geq \frac{1}{2}(\sum_{t=1}^{k-1} p(t) + p(m_k)) \\ &= \frac{1}{2} f(\mathbb{Q}') \geq \frac{1}{2} f(\mathbb{Q}) \end{aligned} \tag{18}$$

According to Theorem 2, the operational cost of placing a data in a data center subset $\mathbb{S}$ obtained by Algorithm 2 is $\ln|\mathrm{U}|$-approximation to the data center subset selection for operational cost minimization problem. Therefore, let $\mathbb{Q}^*$ be the optimal solution to the data placement problem, the cost reduction with solution $\mathbb{Q}''$ is as follows.

$$f(\mathbb{Q}'') \geq \frac{1}{2} f(\mathbb{Q}) \geq \frac{1}{2} \ln|\mathrm{U}| f(\mathbb{Q}^*) \tag{19}$$

For each data subset in each data group, corresponding solutions $\mathbb{Q}'$, $\mathbb{Q}''$ and $\mathbb{Q}^*$ will be constructed. Therefore, algorithm LCDP is a $\frac{1}{2} \cdot \ln|\mathrm{U}|$-approximation algorithm to the data placement problem. □

Theorem 4. *Given distributed cloud* $\mathbb{C} = (\mathcal{DC}, E)$, *data set* $\mathcal{D}$, *and user set* U, *The time complexity of Algorithm LCDP shown in Algorithm LCDP is* $O(|\mathcal{D}|^2 + |\mathcal{D}| \cdot |\mathrm{U}| \cdot log|\mathrm{U}|)$.

Proof. We can construct a graph in which the data are connected with each other by an edge, with the edge weight representing the similarity of the read and write rates between the two data. The number of edges is $E \leq |\mathcal{D}| \cdot (|\mathcal{D}| - 1)/2$. To identify the data groups, all edges will be examined, and the time spent on data group division is $O(E) \leq O(|\mathcal{D}|^2)$. Assuming the number of data groups is $|G|$, the sorting of all the data groups in Algorithm LCDP takes $O(|G| \cdot log|G|)$ time. The algorithm finds a data center subset for each data iteratively, and the number of iterations is the number of data $|\mathcal{D}|$. The time spent on data center subset selection for each data m is $O(\sum_{U_{m,j} \in \mathrm{U}} |U_{m,j}|) = O(|\mathcal{DC}| \cdot |\mathrm{U}|)$. For all the data, it takes $O(|\mathcal{D}| \cdot |\mathrm{U}| \cdot log|\mathrm{U}|)$ time to find data center subsets. Therefore, Algorithm LCDP runs in $O(|\mathcal{D}|^2 + |G| \cdot log|G| + |\mathcal{D}| \cdot |\mathrm{U}| \cdot log|\mathrm{U}|)$ time. Since the number of data groups is less than the amount of data, i.e., $|G| \leq |\mathcal{D}|$, the time complexity of Algorithm LCDP is $O(|\mathcal{D}|^2 + |\mathcal{D}| \cdot |\mathrm{U}| \cdot log|\mathrm{U}|)$. □

5 Performance Evaluation

5.1 Simulation Setup

In our simulation, the data centers in cloud $\mathbb{C}$ are located across 8 regions according to Amazon AWS [6]. We also use GT-ITM tool [4] to generate 30 clouds with different topologies, in which there is an edge between each pair of data centers with probability of 0.4. The cost of transmitting $1GB$ data for each data center is a random value in the range of [\$0.01, \$0.09] [4]. The read rate $r_{i,m}$ and write rate $w_{i,m}$ of each user i accessing the data m are randomly generated in [0, 20] [4]. Each update operation involves $256MB$ data [4]. There are 5 levels of latency requirements ($Q = 5$) from 15 to 19 milliseconds [17]. The end-to-end network latency $d_{i,j}$ between user i and data center j is approximated by the geographical distance between them [17].

The relative importance of write to read is set as 5. We then use Analytic Hierarchy Process (AHP) [18] to decide the weights α and β, and we get the values of α and β as $\alpha = 0.17$ and $\beta = 0.83$. The parameters of the data centers and the data sizes are similar to those in [1,2]. The PUE for each data center is randomly generated in the range of [1.3, 1.7]. We choose a typical setting of parameter $\rho = 2$ in Eq. (1) for all the data centers. The parameter a in Eq. (1) is in [0.18,0.3], and the idle power consumption b of a server in each data center in Eq. (1) varies in [100, 125]. The service rate of each server is represented by the number of requests processed per second, which varies from 20 to 32.5. The storage capacity of each data center is randomly generated from 50 to 200 TB.

The proposed data placement problem is a typical 0–1 nonlinear programming problem. To evaluate the performance of algorithm LCDP, we adopt a widely-adopted benchmark [5]: replacing integer variables with appropriate continuous constraints, which is called NPR in the simulation, since there is no work similar to ours. Algorithm NPR consists of three stages: (1) transform the 0–1 programming problem to a continuous optimization model by relaxing the variable value $x_{i,m,j}$,$y_{m,j}$ to range [0, 1]; (2) after solving the continuous optimization model, iteratively fix the variables closest to 1 as 1, substitute the fixed variable for the continuous variable, and repeat the model solving and variable fixing process, until all variables in the solution are fixed as 0 or 1; (3) if the number of data replicas produced by the continuous optimization model is not K, add or remove data replicas to guarantee that the number of data replicas is K.

We investigate the performance of the algorithms in terms of operational cost, power consumption cost and network communication cost, which are denoted as case 1, case 2 and case 3, respectively.

5.2 Performance Evaluation of the Proposed Algorithm

Impact of the Number of Data. We first evaluate the performance of algorithms LCDP and NPR by varying the number of data, assuming the number of users is 1000.

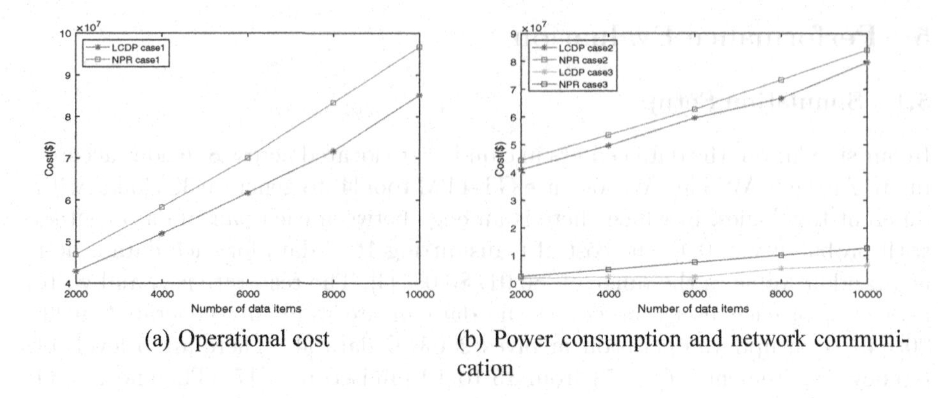

(a) Operational cost

(b) Power consumption and network communication

Fig. 2. The performance with different number of data

Figure 2(a) shows that the operational cost (case 1) increases with the growing number of data, as we need more servers to accommodate the data for user access requests and more network transfer for data synchronization. The operational cost with algorithm LCDP is better than that with algorithm NPR from 9.8% to 14.9%. Figure 2(b) illustrates that both power consumption cost (case 2) and network communication cost (case 3) increase with the growing number of data. The power consumption cost with algorithm LCDP is 8%–23% better than that with algorithm NPR. The network communication cost with algorithm LCDP is about 7.5% to 28% better than that with algorithm NPR. Moreover, the cost reduction in power consumption cost and network communication cost by algorithm LCDP increases obviously when the number of data increases from 2000 to 10000. With algorithm LCDP, the data in higher ranked groups are placed to data center subsets before the data in lower ranked groups, and data placement in the selected subsets can minimize the operational cost while meeting the latency requirements. Algorithm NPR relaxes the integer variable constraints and minimizes the cost through the variable fixing. However, the optimal solution of the relaxed continuous optimization problem is usually no better than that of the integer programming problem, and the number of data replicas resulted from the fixing process may not be K. Therefore, we need to add or remove data centers in order to satisfy the number of data replicas constraint, which potentially increases the operational cost.

Impact of the Number of Users. We study the impact of the number of users on the performance of different algorithms, assuming the number of data is set as 3000.

The simulation results in Fig. 3 show that the cost in all the three cases increases with the increasing number of users. The operational cost (case 1) with algorithm LCDP is 12.9%–24.5% better than that with algorithm NPR. When the number of data is fixed, more users generate more data access requests, and the large amount of request workloads from the users require more servers and

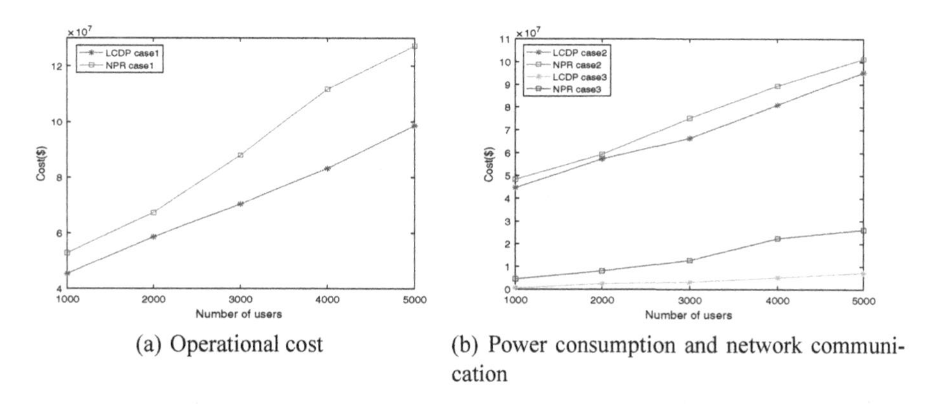

(a) Operational cost

(b) Power consumption and network communication

Fig. 3. The performance with different number of users

more data transfer, resulting in the increase of both power consumption cost (case 2) and network communication cost (case 3). With algorithm LCDP, the data with higher access rates are placed before the data with lower access rates, and the data with higher write rates will be placed in the data centers close to each other with short routing paths and low power consumption. Therefore, algorithm LCDP achieves better performance than algorithm NPR by reducing the cost in all the three cases.

Impact of the Number of Data Replicas on Operational Cost. We investigate the impact of K, the number of data replicas, on the performance of different algorithms. Figure 4 and Fig. 5 illustrate the performance with different number of data and users, respectively. The number of data are set as 6000 and 10000 in Figs. 4(a) and 4(b), respectively, and the number of users in Figs. 5(a),and 5(b) are respectively set as 3000 and 5000.

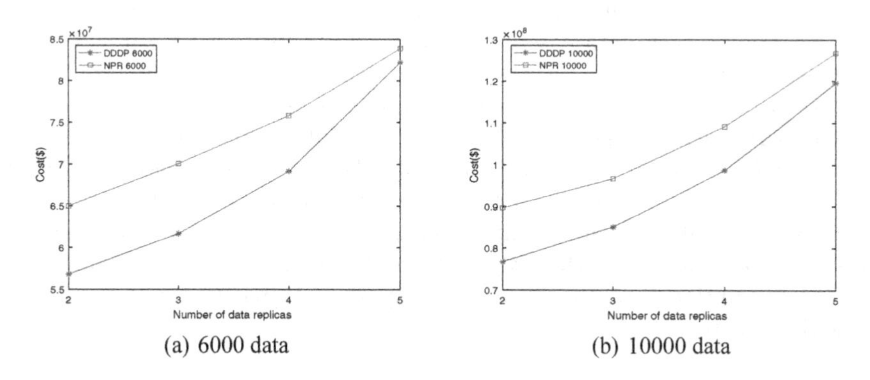

(a) 6000 data

(b) 10000 data

Fig. 4. The impact of the number of data replicas on operational cost under different number of data

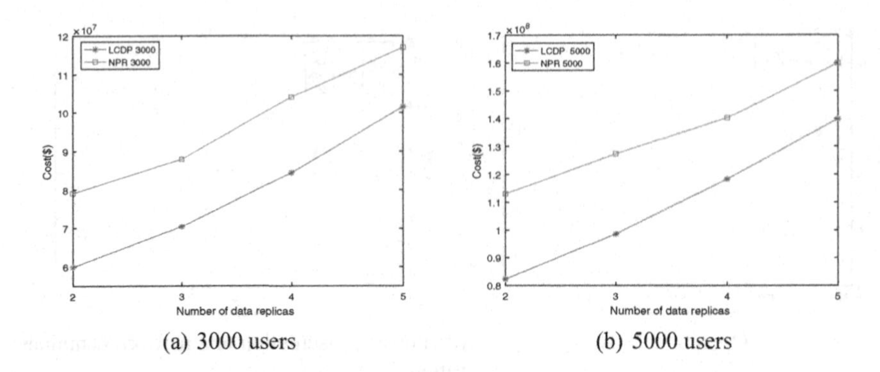

(a) 3000 users (b) 5000 users

Fig. 5. The impact of the number of data replicas on operational cost under different number of users

The operational cost of both algorithms LCDP and NPR increases with the increase of K, as the increase of the number of data replicas is equivalent to the increase of the number of data. Algorithm LCDP achieves less operational cost than algorithm NPR. When $K = 3$, algorithm LCDP outperforms algorithm NPR up to 29% and 26% as depicted in Figs. 4 and 5, respectively. However, the performance of algorithms LCDP and NPR gets close to each other with the increasing number of data replicas. With the increase of K, the total number of data replicas increases, more power is required, and the communication cost for updating data replicas will significantly increase. Therefore, algorithm LCDP achieves better performance with a smaller K than with a larger K.

6 Conclusions

In this paper, we studied the data placement problem with the aim to minimize the operational cost incurred by intra-data center power consumption and inter-data center network communication, while satisfying the user latency requirements, assuming each data has K data replicas. We proposed a Latency aware and Cost minimization Data Placement algorithm (LCDP) for the problem by partitioning the data into multiple data groups according to the rates of data read and write requests and greedily selecting K data centers incurring the minimum cost for each data in each data group. We proved that algorithm LCDP is $\frac{1}{2}\ln|\mathrm{U}|$-approximation to the data placement problem, where $|\mathrm{U}|$ is the number of users. We also conducted experiments through simulations. Experimental results demonstrated that the proposed algorithm could effectively reduce the power consumption cost, the network transport cost, and the operational cost of the data centers.

References

1. Fan, Y., Ding, H., Wang, L., Yuan, X.: Green latency-aware data placement in data centers. Comput. Netw. **110**(C), 46–57 (2016)

2. Zhi, Z., Liu, F., Zou, R., Liu, J., Hong, X., Hai, J.: Carbon-aware online control of geo-distributed cloud services. IEEE Trans. Parallel Distrib. Syst. **27**(9), 2506–2519 (2016)
3. Shao, H., Rao, L., Wang, Z., Liu, X., Ren, K.: Optimal load balancing and energy cost management for internet data centers in deregulated electricity markets. IEEE Trans. Parallel Distrib. Syst. **25**(10), 2659–2669 (2014)
4. Xia, Q., Liang, W., Xu, Z.: The operational cost minimization in distributed clouds via community-aware user data placements of social networks. Comput. Netw. **112**, 263–278 (2017)
5. Ghandeharizadeh, S., Irani, S., Lam, J.: The subset assignment problem for data placement in caches. Algorithmica **80**(7), 2201–2220 (2018)
6. Oh, K., Chandra, A., Weissman, J.: TripS: automated multi-tiered data placement in a geo-distributed cloud environment. In: ACM International Systems and Storage Conference, Haifa, Israel, 22–24 May 2017, pp. 1–11 (2017)
7. Zhou, Z., Liu, F., Zou, R., Liu, J., Xu, H., Jin, H.: Carbon-aware online control of geo-distributed cloud services. IEEE Trans. Parallel Distrib. Syst. **12**(3), 1–14 (2015)
8. Gao, P.X., Curtis, A.R., Wong, B., Keshav, S.: It's not easy being green. In: ACM SIGCOMM 2012, Helsinki, Finland, 13–17 August 2012, pp. 211–222 (2012)
9. Qureshi, A., Weber, R., Balakrishnan, H., Guttag, J., Maggs, B.: Cutting the electric bill for internet-scale systems. In: ACM SIGCOMM 2009, Barcelona, Spain, 17–21 August 2009, pp. 123–134 (2009)
10. Rao, L., Liu, X., Xie, L., Liu, W.: Minimizing electricity cost: Optimization of distributed internet data centers in a multi-electricity-market environment. In: INFOCOM 2010, San Diego, CA, USA, 15–19 March 2010, pp. 1145–1153 (2010)
11. Lei, J., Li, J., Xu, T., Fu, X.: Cost optimization for online social networks on geo-distributed clouds. In: The 20th IEEE International Conference on Network Protocols (ICNP 2012), Austin, TX, 30 October–2 November 2012 (2012)
12. Golab, L., Hadjieleftheriou, M., Karloff, H., Saha, B.: Distributed data placement to minimize communication costs via graph partitioning. In: International Conference on Scientific & Statistical Database Management (2014)
13. Yu, B., Pan, J.: A framework of hypergraph-based data placement among geo-distributed datacenters. IEEE Trans. Serv. Comput. **13**(3), 395–409 (2017)
14. Atrey, A., Seghbroeck, G.V., Mora, H., Turck, F.D., Volckaerts, B.: Full-system poweranalysis and modeling for server environments. In: The 9th International Conference on Cloud Computing and Services Science (CLOSER 2019), Boston, MA, 18 June 2006 (2006)
15. Tang, J., Tang, X., Yuan, J.: Optimizing inter-server communication for online social networks. In: IEEE International Conference on Distributed Computing Systems, Columbus, OH, USA, 29 June–July 2 2015, pp. 215–224 (2015)
16. Lin, M., Wierman, S., Andrew, L.L.H., Thereska, E.: Dynamic right-sizing for power-proportional data centers. In: IEEE INFOCOM, Shanghai, China, 10–15 April 2011, pp. 1098–1106 (2011)
17. Power-demand routing in massive geo-distributed systems. Ph.D. Thesis, Massachusetts Institute of Technology (2010)
18. Saaty, T.L.: Axiomatic foundation of the analytic hierarchy process. Manage. Sci. **32**(7), 841–855 (1986)

Verify a Valid Message in Single Tuple: A Watermarking Technique for Relational Database

Shuguang Yuan[1,2], Jing Yu[1,2(✉)], Peisong Shen[1], and Chi Chen[1,2]

[1] State Key Laboratory of Information Security, Institute of Information Engineering, Chinese Academy of Science, Beijing 100089, China
{yuanshuguang,yujing,shenpeisong,chenchi}@iie.ac.cn

[2] School of Cyber Security, University of Chinese Academy of Sciences, Beijing 100043, China

Abstract. The leakage of sensitive digital assets is a major problem which causes huge legal risk and economic loss. Service providers need to ensure data security for owners. Robust watermarking techniques play a critical role in ownership protection of relational databases. In this paper, we proposed a new Double-layer Ellipse Model called DEM that embeds a valid message in each candidates tuples. Each watermark can independently prove ownership. The main idea of DEM is to use watermarks itself to locate and make the most of contextual information to verify validity of watermarks. Under the framework of DEM, we propose a robust and semi-blind reversible watermarking scheme. Our scheme handles non-significant data for locating and embedding. The scheme generates watermark by exchanging data groups extracted from scattered attributes using Computation and Sort-Exchange step. Key information such as primary key, most significant bit (MSB) become a assistant feature for verifying the validity of embedded watermark. Our scheme can be applied on all type of numerical attributes (e.g. Integer, float, double, boolean). In robust experiments, the scheme is proved to be extremely resilient to insertion/detection/alteration attacks in both normal and hard situations. From a practical point of view, our solution is easy to implement and has good performance in statistics, incremental updates, adaptation.

Keywords: Watermarking · Relational database · Robust

1 Introduction

With the arrival of data era, business interest in database are mined and analyzed. At the same time, attackers also covet the values of database. Thus, copyright protection of relational database is an essential requirement for data owners and service providers. Watermarking is mainly used for ownership protection and tamper proofing. Watermarking techniques are widely used in many fields like

Y. Nah et al. (Eds.): DASFAA 2020, LNCS 12112, pp. 54–71, 2020.
https://doi.org/10.1007/978-3-030-59410-7_4

images [14] , multimedia [6], text [11], databases [1] and applications [20]. In relational database, the robust watermarking [3,16,18,21–23] and fragile watermarking [4,8,10,13] are used for ownership and integrity checking respectively.

The watermarking process mainly comprised of locating phase, embedding phase, and detection phase. In locating phase, positions of watermarks are identified. There are three ways of locating: pseudorandom sequence generator (PSG), hash function (HF), and statistical feature. The [1] relys on fixed order generated by PSG and constant starting/end point. A number of techniques like [2,3,17,21–23] use HF (Example of candidates for HF are the MD5 and SHA) with key information (e.g. Primary Key, Virtual Primary Key, MSBs) for locating. The [7] embeds watermarks on non-significant effect features which locates by mutual information(MI). When it comes to embedding phase, there are two typically approaches. The type of approaches in [2,21–23] is that one embedding validates one bit (OEVOB). In detection phase, OEVOB is a pattern that every iteration on one tuple validates one bit. The other approaches in [3,16,18] are multiple embedding validates one bit (MEVOB). One bit of watermarks is embedded per group of tuples by modifying the values of one or several digits (according to the variables that are bounded by constraints).

In this paper, we propose a new idea that embeds a valid message in every selected single tuple and declares ownership by counting numbers of valid message. The detection can be reduced to counting problem which includes many standalone matching process on embedded tuples. This type model is one embedding validates one message (OEVOM). We propose a innovative Double-layer Ellipse Model (DEM) based on OEVOM. The main idea of DEM is that each watermark will have an independent process which includes actual embedding/detection and verification. DEM has flexible mechanism that can adjust single or combined feature with different weight according to the application scenarios.

Based on DEM, we implement a robust and semi-blind reversible watermarking scheme. The scheme can be applied on all type of numerical attributes (e.g. Integer, float, double, boolean). The embedding phase has two major processes. Firstly, the scheme extract digits from scattered attributes. The selected digits are scattered and unsignificant. Secondly, watermark will be generated in Computation and Sort-Exchange step. Based on the Computation result by key, two data groups are exchanged. The values in the corresponding original tuples are modified. Meantime, the scheme extracts context information for verification. Our watermarking scheme has following characteristics: **1.** Our scheme use watermarks itself to locate. Contextual information including key attributes becomes an enhanced factor of watermark. In our case, the extra locating placeholders (e.g. Primary Key, Virtual Primary Key, MSBs) are optional. Thus, our scheme has weak dependence on key information and improved robustness. **2.** Our scheme can generate a valid message of ownership on each selected tuple. These embedded watermarks are independent of each other. The embedding process supports on-demand incremental updating while has changing previous record. Data can be watermarked separately and stored centrally. Or it is to embed watermark centrally, store separately. In detection phase, our scheme counts the number of valid message. **3.** Our scheme has little impact on statistical result of data. By setting proper positions of watermarking, experiment shows the statistical impact can be reduced to negligible. **4.** Our scheme is a reversible watermarking technique. It can recover watermarked data to original data by inverse embedding progress.

Our contributions are summarized as follows:

1. We propose a model DEM for OEVOM. It includes complete locating, embedding and detection processes. DEM abstracts components of watermarking and evaluation framework.
2. We implement a new watermarking scheme which addresses practical factors such as robust, low false hit rate, incremental updates and low statistical impact.

The paper is organized as follows: In Sect. 2, we describe the related work. In Sect. 3, we present Double-layer Ellipse Model. In Sect. 4, the implementation of scheme including detail of process is shown. In Sect. 5, data-driven experiments are showed. In Sect. 6, the performance of the robustness against subset addition/deletion/alteration attack is presented. In Sect. 7, we conclude our work.

2 Related Work

The watermarking techniques can be classified into three broad categories [9]: Bit-Resetting Techniques (BRT), Data Statistics-Modifying Techniques (DSMT) and Constrained Data Content-Modifying Techniques (CDCMT). The first relational databases watermarking technique was published by Agrawal *et al.* [1,15], which depends on embedding bit string as watermarks. By Hash and secret key, it selects a fraction of tuples, attributes and bit locations for embedding. A number of following techniques like [2,21–23] continue to improve this model method. Cui *et al.* [22] proposed a weighted watermarking algorithm which assigns different weights to attributes. Guo *et al.* [2] proposed a twice-embedding scheme for identifying both the owner and the traitor. Zhou *et al.* [23] presented a scheme that embeds image (BMP file) into the relational databases, and an error correction approach of BCH (BoseChaudhuri-Hocquenhem) coding is used for enhancing the robustness of the algorithms. Wang *et al.* [21] introduced speech signal as watermark. The speech signal is more meaningful and correlative to the data owner. In [17,19], Sion *et al.* provided a new idea for categorical data. It establishes a secret correspondence between category attributes according to a certain rule. Above mentioned techniques belongs to Bit-Resetting Techniques (BRT).

Recently, Data Statistics-Modifying Techniques have evolved gradually. In [18], Sion *et al.* proposed a method that encoding of the watermark bit relies on altering the size of the "positive violators" set. This solution addresses data re-sorting, subset selection, linear data changes attacks. Shehab *et al.* [16] formulated the watermarking techniques of relational database as a constrained optimization problem. They embedded watermarks with constraints on partitioned tuple groups. They presented two techniques to solve the formulated optimization problem based on genetic algorithms and pattern-searching techniques. In terms of [7], Saman *et al.* developed an information techniques that embeds watermarks on non-significant effect features. The attribute selection relys on mutual information(MI) for controlling data distortions. In [3], Javier *et al.* raised a scheme that modulates the relative angular position of the circular histogram center of mass of one numerical attribute for message embedding. Compared with the other two categories of watermarking techniques, fewer Constrained Data Content-Modifying Techniques (CDCMT) are proposed. Li *et al.* [12] proposed a publicly verified watermarking scheme of which detection and ownership proof can be effectively performed publicly by anyone. The idea is that the sets of selected MSBs (selected by pseudorandom sequence generator) of attributes are jointed together to form the watermarks.

3 Model

3.1 Attacker Model

Alice marks database D to generate a watermarked database D_w. The attacker Mallory can operate several types of attacks. The target of attacks is deleting the watermarks. In terms of robust watermarking techniques, survivability is important gist of technological choices of watermarking. We assumes Mallory haven't secret key of watermarking process and original data. Mallory's malicious attack may take various forms: **1. Deletion Attacks.** For the purpose of destroying watermarks, Mallory can delete subset of watermarked tuples from database D_w. Attackers haven't secret key for locating position of watermarks. Deletion will damage tuples which may has watermarks. **2. Alteration Attacks.** In this type of attack, Mallory can change the value of tuples on database D_w. Mallory randomly select 1 to 5 attributes and modify their values. We assume that the attacker dose not know the real positions where the watermarks was embedded. **3. Insertion Attacks.** Mallory adds similar tuples that may disturb detection.

3.2 Double-Layer Ellipse Model

The main components of DEM is presented in Fig. 1. The embedding and detection use the same model but different components. As a model for OEVOM, the main proposal of DEM is to use watermarks itself to locate positions and use contextual information to verify watermarks.

The model is divided into two layer. The watermark should be generated primarily by inner layer. The inner layer defines six components: **Filter**, **Preprocessor**, **Calculator**, **Indexer**, **Modifier**, **Recover**. The components of Filter and Preprocessor select tuple and generate processed data. Calculator computes processed data by parameters and key to generate watermarks. Indexer stores watermarks with affiliated feature extracted from outer layer. Modifier

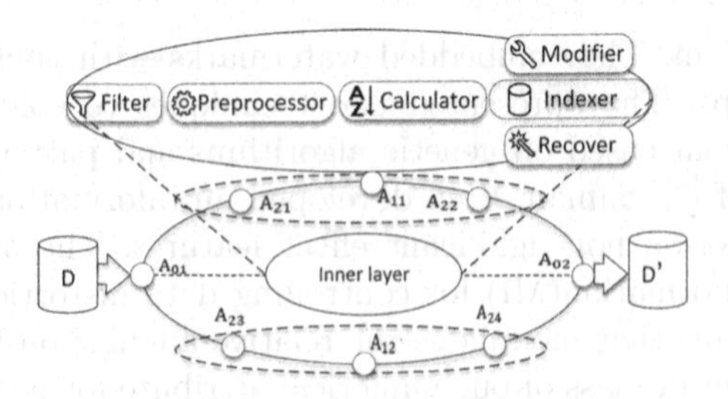

Fig. 1. The Components of DEM.

and Recover both execute SQL to finish modification task on databases. The locating and embedding use the same position of data, which needs a trade-off that makes as few changes as possible for embedding but uniqueness for locating. How to find a suitable bandwidth on tuples for both locating and embedding becomes a priority issue.

The outer layer extracts contextual information for verification. The score functions on outer layer ranks features with different weight. Facing different usage scenario, developers can adapt different strategy of features selection and weight assignment. In outer layer, DEM defines vertex, co-vertex, normal points as affiliated feature according to relative position compared with inner ellipse. To do so, strong dependence on key information become a enhanced verification instead of necessity. In general, unique feature, like primary key, has high credibility for verifying watermark. Correspondingly, fractional or circumjacent information has low credibility with low weight.

The below formula measures probability of watermarking of database D_w by verifying every tuples from 0 to R. When inner layer detects a watermark with $p_i = 1$, the score of every feature is calculated by $w_0 q_{i0}, ..., w_n q_{in}$. The score of watermarked tuple is $w_{pi} + max\{w_0 q_{i0}, ..., w_n q_{in}\}$. However, if current tuple can't be identified as watermarks with $p_i = 0$, the score of current tuple is 0. The q_i is the score function of feature A_j with weight w_i. The result of weighted accumulative detection is compared with the threshold τ. The synthetic evaluation weight of watermarking is given by

$$Score(D_w) = \frac{\sum_{i=0}^{R} p_i \left(w_{pi} + max\{w_0 q_{i0}, ..., w_n q_{in}\}\right)}{S} * 100\%. \tag{1}$$

The $f(A, K)$ should be piecewise function where value is 0/1 to determine whether the condition of the feature is satisfied. The vertex and co-vertex verify current and non-neighbor information of watermarked tuple respectively. The normal points have geometrically symmetric. A couple of normal points can represent adjacent feature in opposite directions of context of watermarked tuple in the vertical. And in the horizontal, a couple of normal points have the relation of "AND" or "OR". For example, the relation of A_{j1} and A_{j2} have relation of "And", so they obtain one weight when both conditions are met. Secret key K

is added to salt the result for protecting leakage of feature information. The feature A_j is computed as follows

$$q_{ij} = \begin{cases} f_{j1}(A_{j1}, A_{j2}, K) & j = 0, 1 \\ \frac{f_{j1}(A_{j1}, A_{j2}, K) + f_{j2}(A_{j3}, A_{j4}, K)}{2} & j \neq 0, 1. \end{cases} \quad (2)$$

The m is total number of feature. Besides features can be combined for obtaining a higher weight value

$$q_{ik} = \sum_{1 \leq n \leq m} \sum_{1 \leq i_1 i_2 ... i_n \leq m} (q_{i_1} + q_{i_2} + ... + q_{i_n}). \quad (3)$$

The notations present in Table 1 for reference.

Table 1. Notations

Sym	Description	Sym	Description
D	Original database	A	The feature of contextual information
D_w	Watermarked database	N_w	Watermark groups
R	Number of tuples	p_i,q_j	Score function
S	Number of watermarked tuples	τ	Threshold of detection
r	A tuple	w_i	Weight of i-th feature
r_w	A watermarked tuple	W_i	Watermark of i-th entity
r_{adj}	Adjacent tuples of r	n	Number of feature of outer layer
K	Secret key	l	Number of attributes in a tuple
C	Configuration of watermarking process	l_g	Number of digits in a group
M	The value of modulo	L_g	Number of groups
I	Interval of watermark	K_g	Key groups
Pos_w	Available length of attributes	N_g	Data groups

4 Proposed Scheme

We proposed a robust and semi-blind reversible watermarking scheme under DEM. The scheme consist consists of three subsystems: Watermark Embedding, Watermark detection and Data Recovery. Scheme simplifies six components for unfolding our algorithm. The scheme embeds private message which stands on exchanging positions of scattered digits by key, and generating index for locating. Our scheme uses the scattered data on tuples as the carrier of watermarks. It includes a configure set $C = \{l_g, I, M, P_1, P_2, Pos_w\}$ and secret key K.

4.1 Watermark Embedding

The embedding process is showed below.

Algorithm 1 Preprocessor

```
Require: Pos_w, l_g, l, r
Ensure: N_g
1: N_g ← Array[], result ← ""
2: for i = 0 → l do
3:     if r[i] is not integer/float or Pos_w[i] <= 0 then
4:         continue
5:     end if
6:     spliteDigit ← extractDigits(r[i])
7:     if length(splitDigit) > Pos_w[i] then
8:         result.concat(spliteDigit.lastSubString(Pos_w[i]))
9:     else
10:        result.concat(spliteDigit)
11:    end if
12: end for
13: N_g ← groupBySize(result, l_g)
14: return N_g
```

Algorithm 2 Modifier

```
Require: W_i, Pos_w, r, l
Ensure: r_w
1: r_w ← r, iter ← 0
2: for i = 0 → l do
3:     if r[i] is not integer/float or Pos_w[i] <= 0 then
4:         continue
5:     end if
6:     spliteDigit ← extractDigits(r[i])
7:     len ← length(spliteDigit)
8:     if len > Pos_w[i] then
9:         subMark ← W_i.subString(iter, iter + Pos_w[i])
10:        value ← value.subString(0, len − Pos_w[i]) + subMark
11:        iter ← iter + Pos_w[i]
12:    else
13:        subMark ← W_i.subString(iter, iter + len)
14:        value ← subMark
15:        iter ← iter + len
16:    end if
17:    r_w[i] ← fillDigit(value)
18: end for
19: replaceTuple(D, r_w)
20: return r_w
```

1. Filter. Scheme generates random number in the range of 0 to I to pick a candidate every I tuples. The selected tuple will be embedded a watermark. The parameter I controls the density of watermark in watermarked database D_w. The number of watermarked tuples S is R divided by the interval I. By comparing the actual quantity with the theoretical quantity, scheme has a measurable baseline by interval I for watermarking detection.

2. Preprocessor. In Fig. 2 A, the process of Preprocessor is described. The embedding uses digits across different attributes. For preserving statistic, the last few place of integer/decimals of attributes are considered. Only one or two digits are extracted in a single attribute. Pos_w is a predetermined list that contains available length of each attribute. Extraction work makes reference to Pos_w. The extracted digits are grouped into the N_g of size l_g. Algorithm 1 describes the steps involved in Preprocessing. l is the number of attributes in tuple r. In line 3, the attribute $r[i]$ is determined whether it is suitable for extracting digits. This ensures that suitable type (e.g. Integer, float, double, boolean) under restrained available length will not be skipped. Line 6 splits the decimal place of float/integer. Finally, integrated digits are divided into groups. Meanwhile, context of selected tuple enters outer layer of DEM.

3. Calculator. The watermark W will be generated in this component. The groups N_g forms watermark W which is resulted by Computation and Sort-Exchange steps. During Calculator on inner layer, scheme has four parameters key K, modulo M, and exchange position P_1, P_2. The key K is transformed to K_g as the same format of N_g.

Firstly, Computation step defines operator $\bigoplus$ that each element in N_g is multiplied by same-position element in K_g respectively, then result of $\bigoplus$ is added up. Because the number of K_g is less than N_g, K is recycled on rounds in $\bigoplus$ operation. The modulo operator with the value M is prevent from order-preserving. Secondly, according Computation result, the Sort-Exchange step swaps P_1-th largest value and P_2-th largest value using Quick-sort algorithm. The spliced result forms watermark W_i. Only if attackers possess C and K, can this process be repeated. Figure 2 B describes the steps involved in Calculator.

After generating W_i on inner layer, affiliated features are extracted from contextual tuples. Four features are selected with score function. Feature A_{01} defines q_1 that relies on hash function of MSB (Most Significant Bit) with key. This feature is assigned a high score because of uniqueness of MSB. A_{21} and A_{23} compare significant information of adjacent tuples and are defined by function q_2, q_3. A_{11} records the distance from the previous same entry during traversal. When two embedded tuples where surrounding tuples have same content, feature A_{11} helps distinguish correct one. A_{21}, A_{23} and A_{11} together form a high score evaluation function q_4. All score functions use K for salting in order to prevent original information disclosure.

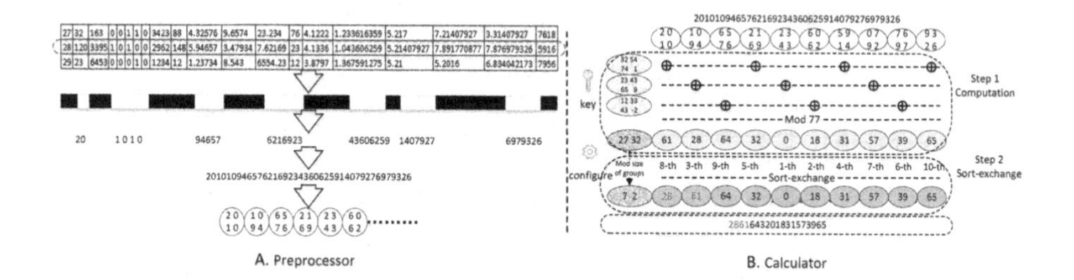

Fig. 2. The schematic diagram of Preprocessor and Calculator.

4. Indexer and Modifier. Receiving the watermark from Calculator, Modifier continues the work to change the database. The Algorithm 2 shows Modifier process. Firstly, algorithm measures the usable length of attributes. Then algorithm takes substrings of watermark fill back into the tuple in turn on the basis of usable length Pos_w and actual length of attributes. It plays a role in producing and executing SQL for embedding on database. Indexer stores/extracts the record of watermarks.

The watermarking process is shown in algorithm 3. In embedding phase, the algorithm traverses the whole database. Because A_{11} requires comparing the distance between the current tuple and the most recent same one. In every interval I, algorithm generates a random number between 0 and I. This random number picks a tuple to embed watermark. After Preprocessor step, in line 8–13, watermark will be generated by Calculator. Finally, modification of tuple is transformed to SQL for executing on D. In Fig. 3 A, the whole process is presented.

4.2 Watermark Detection and Recovery

In the watermark detection process, the first step is to locate the watermarks. The scheme uses adaptable data structure HashMap for matching extracted watermarks. Because the same watermark exists but has different affiliated feature A. Thus, the key of HashMap is watermarks, and the value is a list of different A_i. This process contains only Preprocessor and Calculator components. Because detection is a traverse process on database without modifying the original content. After extracting, based on the result of matching, detection compute the score function by predefined weight using Eq. 1. For example, when watermark W' and affiliated A are both matched, the watermark is 100% confidence. If A can be matched partially, the watermark has the confidence of the corresponding part. If only the watermark is detected and no attached features are fond, confidence is low. Our scheme defines High, Middle and Low scores. The detection process is presented in Fig. 3B.

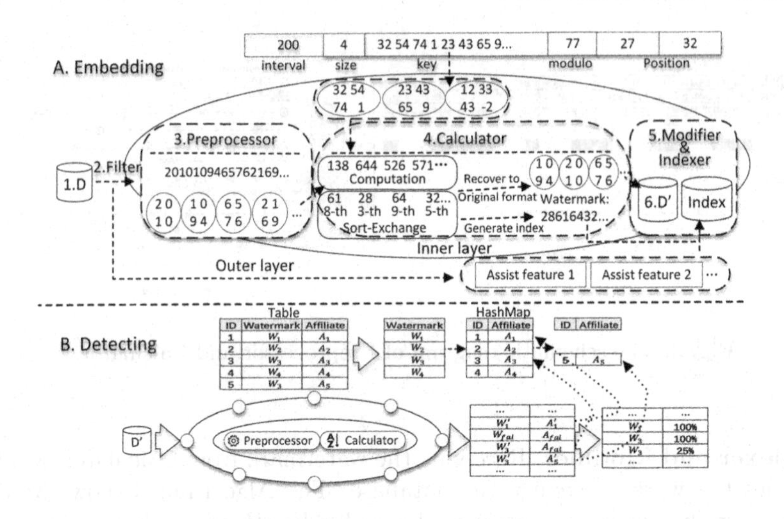

Fig. 3. The process of Detection.

The detection algorithm is presented in Algorithm 4. The detection mainly consists of three steps: 1. Scheme extracts the index of watermarks saved in the database as matching reference. 2. After building hashMap data structure, scheme traverses all tuples to calculate watermark W_i for a preliminary verification. If W_i can't be match with hashMap of matching reference, program will move on to the next one. If W_i can be matched on map_W, the extracted affiliated feature A is compared with every items using Eq. 1 in the list of $HashMap(W)$ to get the highest score. Finally, comparing with τ, the program returns a deterministic result. In our scheme, the assigned value of τ is 50%. Whether to perform a recovery operation is optional. And this action only works on the tuples of which are 100% confidence of detection. If user chooses to runs this operation, based on the detection, one more component (Recover) is added. The function of Recover is similar to Modifier that executes SQL for recovering the embedded tuples.

5 Experiments and Discussion

Experiments are conducted on Intel Core i7 with CPU of 3.60 GHz and RAM of 16 GB. Algorithms were implemented on Postgresql11.0 with JDK 1.7. Experiments were performed using the Forest Cover Type data set, downloaded from Archive.[1] The data set has 581,012 rows with 61 attributes. Each tuple contains 10 integer attributes, 1 categorical attribute, and 44 Boolean attributes. We added an extra attribute ID as the primary key, eight attributes of float type, eight attributes of double type. The data set has four type: boolean, integer, float, double. We chose 7 integer attributes, 16 float/double attributes, and 2 boolean attributes as candidates for watermarking. The experiments were repeated 15 times and the average of result for each trial was calculated.

5.1 Overhead

Equation 4 explains the total computational time of watermarking. The R/I represents the number of embedded tuples, $(|W_i|/l_g)^2$ is the time of single embedding/detection. $|W_i|$ is the length of watermark W_i. l_g is the number of digits in a group. So $|W_i|/l_g$ is equal to the number of groups. Because Sort-Exchange operation needs Quick-Sort algorithm, execution time is $(|W_i|/l_g)^2$. For every embedded tuple are calculated and added together as $Time(D_w)$.

$$Time(D_w) = (R/I) * (|W_i|/l_g)^2 \tag{4}$$

There are three experiments for assessing the computational cost of embedding, detection and recover. In Fig. 4, two aspects (tuple number and interval) of the execution times are shown. In Fig. 4(a), figure shows the computational cost of embedding, detection, recover for different number of tuples with interval $I = 200$. The range of tuple numbers is 100000 to 581012. The time consumption increases along with the number of embedded tuples. As the number of watermarks increases, the trend is not a straight line but a gradually steep curve because of $(|W_i|/l_g)^2$ operation. In Fig. 4(b), figure displays the effect of Interval I for executing time. The larger the interval, the less time it takes because the number of embedded tuples becomes less. As value of interval goes above 500, the gradient of time curve decreases because of watermarks decreases. Thus, the batch size of processing tuples should not be large, which results in a disproportionate increase in time consumption. These results shows that our scheme perform well enough to be used in off-line processing.

5.2 False Hit Rate

Watermarking techniques may detect message of ownership from un-embedded database without correct key. In our scheme, detected watermarks without correct affiliated feature dose not affect score because of score *Low* has weight 0.

[1] kdd.ics.uci.edu/databases/covertype/covertype.html.

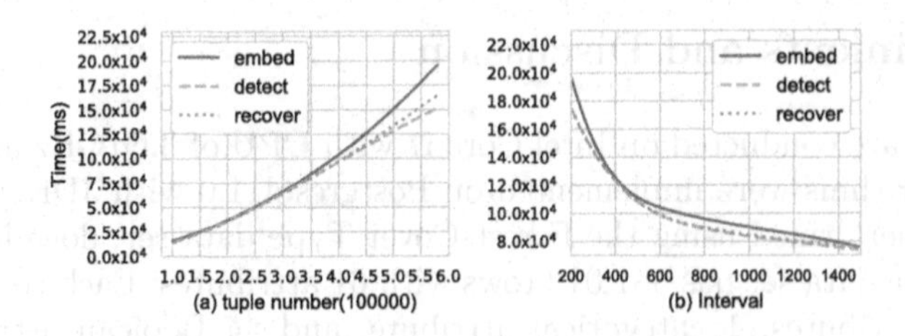

Fig. 4. The computation time of watermarking.

Therefore, only when watermark and affiliated feature are detected correctly can scores be obtained. In Table 2. We represents the experiments in different parameters. D_1, D_2, D_3 are test databases that embedded watermarks with parameter I of 200, 500, 1000 and independent key. SCDK is the abbreviation of Same Configure but Different Key. And DCDK is the abbreviation of Different Configure and Different Key. All the tests scored low which is in line with expectations. The difference between SCDK and DCDK can be neglected. As parameter I increases, the error rate decreases slightly because the total number of watermarks is reduced at the same time. In practice, the False Hit Rate can be completely ignored in our watermarking techniques.

Table 2. False hit rate experiments

Database		D_1	D_2	D_3
I		200	500	1000
SCDK	Suc	N	N	N
	Score	1.093%	1.054%	0.861%
DCDK	Suc	N	N	N
	Score	1.452%	0.813%	0.807%

5.3 Effect on Statistics

We evaluated statistical distortion through the variations of mean and standard deviation on embedded database. Experiments were performed on independent database with parameters $I = 200, 500, 1000$ and $Pos_w = 18, 24, 36$. The 24 attributes are selected as candidates for embedding. We choose 6 attributes of different numeric data type for presenting distortion. The type of $a1$ and $a2$ is a three-digit and four-digit integer respectively. The $a3$ is a boolean. The $a4$,$a5$ are float type, $a6$ is double. In Table 3, table shows the variation rate of Mean and Std deviation between watermarked data and original data. The variation in attributes $a1$, $a2$ was tiny in all cases. But in $a3$, the fluctuation increases significantly because of the length of $a3$ is 1. Thus, boolean attributes should not be an option for embedding watermarks. According to the experiments on three attributes of $a1$, $a2$ and $a3$, the larger the length of integer data is, the more suitable it is to embed watermarks. Attributes like $a4$, $a5$, $a6$ are the best candidates for watermarking. Embedding has almost no statistical effect on $a4$, $a5$, $a6$. In summary, if watermark is suitably embedded in the last few digits, which has little impact on the individual and the overall statistics for almost all type attributes.

The measures of Mean and Std decrease with the number of interval I but the change is not significant. Thus, we adjust the parameter I as needed for practical application rather than reduce statistical perturbations. With increase of Pos_w, perturbations decreased significantly. But because the perturbations are so small, sometimes there's a little bit of anomaly. Perturbation increased with a maximum value in attribute $a4$ at $I = 200$ and Pos_w.

5.4 Incremental Updates and Adaptation

Incremental updates is an important feature for watermarking technique. The data is continuously produced in many environments. The watermark depends only on groups N_g extracted from scattered attributes and secret key. Thus a tuple can be deleted, inserted without examining or altering any other tuples. When updating candidates attributes of embedding, we recompute current tuple using Calculator component and update new watermark index in the database. When updating non-candidates attribute, nothing need be done. Our scheme defines a parameter I that controls the interval of each embedding. When the amount of incremental data is much larger than I, it won't be any problems. However, if the amount of each incremental update is less than I, a batch may not be embedded with a watermark which will lead to the decrease of watermarks R/I. Thus, the basic requirement is that batch size of increments should be greater than I.

The data is usually stored on multiple nodes, which organization and management is not in a single server. Therefore, the watermarking techniques adapted to the application need to meet data cutting or consolidation problems. Our scheme relys on the watermarks to locate. Scheme needs to share the index records generated by the watermark in the storage node to the detection server, which can use a single key to complete the ownership identification for different databases. In Table 4. We use a secret key to detect data that is fused across multiple database sources. First column is the number of database from multiple data sources. The second column is R/I that the right number of total watermarks. Then next 3 columns are detection results of High/Middle/Low score. The column of Suc is Y or N which represents the successful or failed detection. And column of Score is the detection score, Recover is the percentage of recovered watermarks. The result shows that Multiple databases embedded separately which do not affect accuracy and recovery for detection with the same key. In Table 5. We tested the ability of detection for partial data. The first column is the split ratio on a original embedded database. The results shows that the accuracy and recovery rates were impressive in this case. The result represents our watermarking scheme has ability of facing partial and consolidated data for adaptation.

Table 3. Statical distortion experiments

I		200			500			1000		
Pos_w		18	24	36	18	24	36	18	24	36
Mean.	a1	0.001202744%	0.000647662%	0.000122770%	0.000266471%	0.000194822%	0.000054743%	0.000286597%	0.000130015%	0.000011270%
	a2	0.000025436%	0.000011588%	0.000017904%	0.000002723%	0.000005359%	0.000005736%	0.000003215%	0.000001680%	0.000003273%
	a3	3.508439731%	3.512046454%	1.509413546%	2.303794272%	1.280386640%	0.865613503%	1.028817716%	1.082918560%	0.516663059%
	a4	0.000062976%	0.000015910%	0.000018888%	0.000009228%	0.000008995%	0.000009036%	0.000020242%	0.000018056%	0.000002479%
	a5	0.000045576%	0.000004733%	0.000005368%	0.000033804%	0.000010773%	0.000032949%	0.000021699%	0.000005888%	0.000001224%
	a6	0.000000459%	0.000000952%	0.000000063%	0.000000015%	0.000000001%	0.000000049%	0.000000086%	0.000000005%	0.000000012%
Std.	a1	0.000601960%	0.000515174%	0.000537451%	0.000332161%	0.000161452%	0.000115328%	0.000150423%	0.000001484%	0.000098659%
	a2	0.000015376%	0.000002622%	0.000017030%	0.000003017%	0.000003500%	0.000001998%	0.000000007%	0.000004535%	0.000000061%
	a3	11.024910177%	11.285514891%	4.953004830%	7.338794350%	4.280060337%	2.923639995%	3.371868926%	3.638831796%	1.718041449%
	a4	0.000021213%	0.000027466%	0.000001033%	0.000027849%	0.000049301%	0.000018759%	0.000036018%	0.000042024%	0.000005003%
	a5	0.000021884%	0.000021995%	0.000025646%	0.000007663%	0.000001051%	0.000051026%	0.000013529%	0.000002588%	0.000038540%
	a6	0.000000770%	0.000002712%	0.000000057%	0.000000005%	0.000000000%	0.000000305%	0.000000056%	0.000000003%	0.000000039%

Table 4. Detection for consolidated data

databases	R/I	RESULT			Suc	Score	Recover
		High	Mid	Low			
2	5810	5810	0	0	Y	100%	100%
3	8715	8715	0	0	Y	100%	100%
4	11620	11620	0	0	Y	100%	100%
5	14525	14525	0	0	Y	100%	100%

Table 5. Detection for partial data

%	R/I	RESULT			Suc	Score	Recover
		High	Mid	Low			
10%	290	289–291	0	0	Y	100%	100%
20%	581	580–582	0	0	Y	100%	100%
50%	1452	1451–1453	0	0	Y	100%	100%
70%	2033	2032–2034	0	0	Y	100%	100%
90%	2614	2613–2615	0	0	Y	100%	100%

6 Robustness

In this section, we experiment the robustness of our watermarking scheme under three attacks. Mallory will try their best to remove watermarks to disturb detection. The robustness requirement should be higher to adapt to more rigorous environment. For example, after stealing order data of market, Mallory deletes sensitive information such as order id (most likely primary key), name, phone number, and keeps the non-sensitive valuable information. Our scheme embeds watermarks on non-significant data for locating and embedding. Thus, under a more severe condition, our scheme has a stronger ability of robustness. Attacks are conducted in two situations: 1. Normal situation which refers to an attack test under original watermarked database. 2. Hard situation where key information is deleted in experiments. Database D was watermarked using different parameter I and ratios of influence for attacks. Figures presented in this section have score on vertical axis, which represents the scores of detection using Eq. 1.

6.1 Deletion Attacks

Mallory can randomly delete subset of watermarked tuples from database D_w. Figure 5(a)(c) shows the experiment result with parameter $I = 200, 500, 1000$ by randomly deleting different ratios of tuples from D_w in normal situation and hard situation. The detection threshold is $\tau = 50\%$ with red dotted line in (a)(c). Score is greater than the τ which has a successful detection. In Fig. 5(a), even when up to 90% of the tuples are deleted, the detection is successful. The Fig. 5 (b) shows the proportions of the high/Middle/Low scores, with the percentage of high scores being 100%. In Fig. 5(c), suppose that we obtained a file containing suspected part of our business data. But the order id and sensitive attributes (e.g. name, age, address) were deleted. It still has semi-structured or structured formats that can separate data. In this hard case, our watermarking method still works. When the deletion ratio is less than 50% with $I = 200, 500, 1000$, the detection is successful. In Fig. 5(d), the proportion of Middle score increases first and then decreases. And the ratio of high score keeps going down but low score increases. When the ratio of deletion is more than 50%, almost all of mutual relation for tuples are destroyed which converts High/Middle score to Low score.

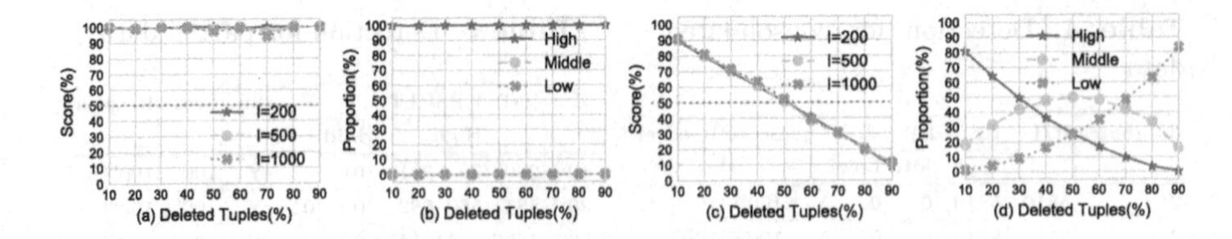

Fig. 5. The experiments of deletion attacks

6.2 Insertion Attacks

The source of new tuples for insertion is original database or similar fake data. The experiments didn't disturb the original tuples of database D_w. The degree of insertion ranges from 10% to 250%. Since the effect is consistent in both normal situation and hard situation, only two diagrams are shown. In Fig. 6(a), when Mallory tries to insert more than 100% tuples, the detection can't sure who owns database because the result is smaller than value of τ. The curves with different I follow a similar trend in figure. In experiments, the value of τ is 50%. But 50% is not a fixed value. If obtaining a large number of watermarks, detection can claim ownership. If detecting 10,000 watermarks in a database with tens of thousands of tuples, owner may still claims ownership even if the ratio of watermarks is less than τ. In Fig. 6(b), the radio of Middle is 0%. The ratio of Low goes up, the ratio of High goes down with increase of insertion radio. Although the new tuples are replica of the original tuples, there is only a watermark W_i with Low score.

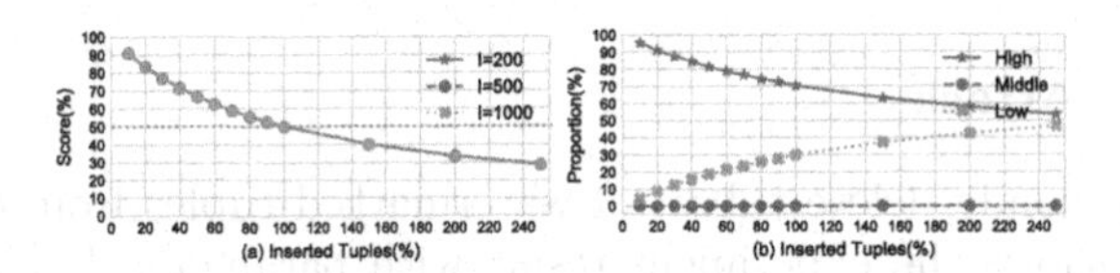

Fig. 6. The experiments of insertion attacks.

6.3 Alteration Attacks

Mallory changes the value of tuples on database D_w. Mallory randomly selects 1 to 5 attributes and modify their values. The experiments assume that the attacker dose not know the real positions where the watermarks was embedded. Figure 7(a) shows the result of attacks under different ratio of alteration and parameter $I = 200, 500, 1000$ in normal situation. All results of experiment are higher than τ which represents successful detections for ownership identification. Figure 7(b) shows that alteration can disrupt the watermark locating by decreasing the amount of watermarks in normal situation. Figure 7(c) shows that the result curve drops more steeply because of lacking of key information for verification of watermarks in hard situation. Even the ratio of alteration is up to 100%, all detection are successful. In Fig. 7(d), the proportion of High score keeps going down over increase of altered tuples. And the ratio of Middle/Low keeps going up.

Table 6 shows a comparison among our scheme and Sion's [18] technique and DEW [5] for robustness in normal situation. Our scheme is highly robust as compared to Sion's and DEW techniques in three types of attacks.

Table 6. Comparison among DEM and other techniques

	Our technique	Sion's technique	DEW
Deletion attack	Resilient to random tuple insertion attacks; 95% watermark score even when 90% of the tuples are deleted	Not resilient to random tuple deletion attack; watermark accuracy deteriorates to 50% when only 10% of the tuples are deleted	Not resilient to random tuple deletion attack; watermark accuracy deteriorates to 50% when only 50% of the tuples are deleted.
Insertion attack	Resilient to random tuple insertion attacks; 50% watermark score even when 200% of the tuples are inserted	Not resilient to random tuple insertion attacks; watermark accuracy deteriorates to 50% when only 10% of the tuples are inserted	Resilient to random tuple insertion attacks; 85% watermark accuracy even when 50% of the tuples are inserted.
Alteration attack	Resilient to random tuple Alteration attacks; 70% watermark score even when 90% of the tuples are altered		Resilient to random tuple Alteration attacks; 90% watermark accuracy even when 50% of the tuples are altered

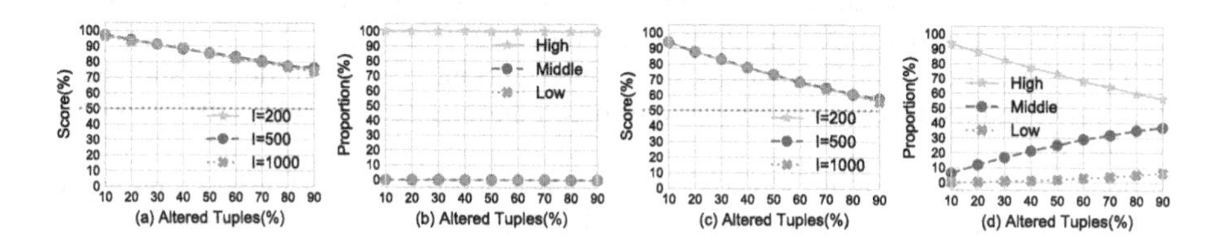

Fig. 7. The experiments of alteration attacks.

7 Conclusion

In this paper, we defined a new method of watermarking and designed a general model DEM. A novel robust and semi-blind reversible watermarking scheme on numerical attributes was proposed. Experiments presented the robustness under three attacks and excellent properties in many aspects such as implementation, false hit rate, incremental updates, statistic distortion and adaptation.

Acknowledgments. This work has been supported by National Key R&D Program of China (No.2017YFC0820700), National Science and Technology Major Project (No.2016ZX05047003), the Beijing Municipal Science & Technology Commission Funds for Cyberspace Security Defense Theory and Key Technology Project (No.Z191100007119003).

References

1. Agrawal, R., Kiernan, J.: Watermarking relational databases. In: Very Large Data Bases, pp. 155–166 (2002)
2. Fei, G., Wang, J., Li, D.: Fingerprinting relational databases. In: ACM Symposium on Applied Computing (2006)
3. Franco-Contreras, J., Coatrieux, G.: Robust watermarking of relational databases with ontology-guided distortion control. IEEE Trans. Inf. Forensics Secur. **10**(9), 1939–1952 (2015)
4. Guo, H., Li, Y., Liu, A., Jajodia, S.: A fragile watermarking scheme for detecting malicious modifications of database relations. Inf. Sci. Int. J. **176**(10), 1350–1378 (2006)
5. Gupta, G., Pieprzyk, J.: Reversible and blind database watermarking using difference expansion. Int. J. Digit. Crime Forensics **1**(2), 42–54 (2009)
6. Hartung, F., Kutter, M.: Multimedia watermarking techniques. Proc. IEEE **87**(7), 1079–1107 (1999)
7. Iftikhar, S., Kamran, M., Anwar, Z.: RRW a robust and reversible watermarking technique for relational data. IEEE Trans. Knowl. Data Eng. **27**(4), 1132–1145 (2015)
8. Kamel, I.: A schema for protecting the integrity of databases. Comput. Secur. **28**(7), 698–709 (2009)
9. Kamran, M., Farooq, M.: A comprehensive survey of watermarking relational databases research. arXiv preprint arXiv:1801.08271 (2018)
10. Khan, A., Husain, S.A.: A fragile zero watermarking scheme to detect and characterize malicious modifications in database relations. Sci. World J. **2013**, 796726 (2013)
11. Kim, Y.W., Moon, K.A., Oh, I.S.: A text watermarking algorithm based on word classification and inter-word space statistics. In: International Conference on Document Analysis & Recognition (2003)
12. Li, Y., Deng, R.H.: Publicly verifiable ownership protection for relational databases. In: ACM Symposium on Information (2006)
13. Li, Y., Guo, H., Jajodia, S.: Tamper detection and localization for categorical data using fragile watermarks. In: ACM Workshop on Digital Rights Management (2004)
14. Podilchuk, C.I., Zeng, W.: Image-adaptive watermarking using visual models. IEEE J. Sel. A. Commun. **16**(4), 525–539 (2006). https://doi.org/10.1109/49.668975
15. Rakesh, A., Peter, H., Jerry, K.: Watermarking relational data: framework, algorithms and analysis. VLDB **12**, 157–169 (2003). https://doi.org/10.1007/s00778-003-0097-x
16. Shehab, M., Bertino, E., Ghafoor, A.: watermarking relational databases using optimization-based techniques. IEEE Trans. Knowl. Data Eng. **20**(1), 116–129 (2007)

17. Sion, R., Atallah, M., Prabhakar, S.: Rights protection for categorical data. IEEE Trans. Knowl. Data Eng. **17**(7), 912–926 (2005)
18. Sion, R., Atallah, M., Prabhakar, S.: Rights protection for relational data. IEEE Trans. Knowl. Data Eng. **16**(12), 1509–1525 (2004)
19. Sion, R.: Proving ownership over categorical data (2004)
20. Stern, J.P., Hachez, G., Koeune, F., Quisquater, J.-J.: Robust object watermarking: application to code. In: Pfitzmann, A. (ed.) IH 1999. LNCS, vol. 1768, pp. 368–378. Springer, Heidelberg (2000). https://doi.org/10.1007/10719724_25
21. Wang, H., Cui, X., Cao, Z.: A speech based algorithm for watermarking relational databases. In: International Symposiums on Information Processing (2008)
22. Cui, X., Qin, X., Sheng, G.: A weighted algorithm for watermarking relational databases. Wuhan Univ. J. Nat. Sci. **12**(1), 79–82 (2007). https://doi.org/10.1007/s11859-006-0204-0
23. Zhou, X., Huang, M., Peng, Z.: An additive-attack-proof watermarking mechanism for databases' copyrights protection using image. In: ACM Symposium on Applied Computing (2007)

WFApprox: Approximate Window Functions Processing

Chunbo Lin[1], Jingdong Li[1], Xiaoling Wang[1,2], Xingjian Lu[1(✉)], and Ji Zhang[3]

[1] Shanghai Key Laboratory of Trustworthy Computing, East China Normal University, Shanghai, China
{cbl,jdl}@stu.ecnu.edu.cn
xlwang@sei.ecnu.edu.cn, xjlu@cs.ecnu.edu.cn

[2] Shanghai Institute of Intelligent Science and Technology, Tongji University, Shanghai, China

[3] Zhejiang Lab, Hangzhou, China
zhangji77@gmail.com

Abstract. Window functions, despite being supported by all major database systems, are unable to keep up with the steeply growing size of data. Recently, some approximate query process (AQP) systems are proposed to deal with large and complex data in relational databases, which offer us a flexible trade-off between accuracy and efficiency. At the same time, Machine Learning has been adopted extensively to optimize databases due to its powerful ability in dealing with data. However, there have been few publications that consider using AQP techniques especially model-based methods to accelerate window functions processing. This work presents WFApprox, an AQP system based on Machine Learning models aims at efficiently providing an approximate answer for window functions. WFApprox uses Machine Learning models instead of massive data for query answering. Our experimental evaluation shows that WFApprox significantly outperforms the mainstream database systems over TPC-H benchmark.

Keywords: Window functions · Approximate query processing · Machine learning

1 Introduction

Window functions, introduced firstly in SQL:1999 and then formally specified in SQL:2013, have become an important part of SQL. Due to its simple semantics and powerful expression ability, window functions were supported by almost all mainstream commercial databases in recent years. Unlike aggregation with *group by*, window functions allow the evaluation of aggregate functions for the window corresponding to every single tuple. Besides, comparing to correlated subqueries and self-joins, window functions improve the readability of the SQL, and at the same time, lead to much more efficient query processing.

Y. Nah et al. (Eds.): DASFAA 2020, LNCS 12112, pp. 72–87, 2020.
https://doi.org/10.1007/978-3-030-59410-7_5

However, in the era of big data, the traditional implementation of window functions is too expensive to keep up with the growing data size. Apart from this, the rapidly growing size of data tend to downplay the improvements proposed by a few pioneering studies [4,10] that concentrate on optimizing the query processing of window functions, such as avoiding unnecessary sorting or partitioning. As a result, these works can't meet the demand for real-time query processing, which is becoming a basic requirement for many services.

At the same time, there are abundant researches focusing on approximate query processing (AQP) [2,8,13,14]. Sampling, which is the dominating approach in this area, speeds up query answering by carefully choosing a part of representative tuples. Since the precise result is not always needed, it is more desirable to quickly offer an approximate result, together with an error bound such as confidence interval. Despite the usefulness and prevalence of AQP in query acceleration, only a few pioneering papers such as [20] adapted AQP to window functions using uniform sampling. However, the algorithms proposed in [20] perform sampling after all tuples are loaded into memory, which means a limited speedup of query processing.

Recently, a lot of literature applies machine learning (ML) to optimize database systems [3,9,11,17]. ML models are used for efficient query processing, workload forecasting, storage space saving and so on. Compared with the data or data structure in the traditional databases, ML models can represent the statistical characteristics and distribution of data more accurately, at the same time occupy less storage space and provide more efficient computing performance.

In this work, we integrate density estimator and regression model into window functions processing. We studied how to use these two models to support various types of window functions and how to optimize them to improve their efficiency. And best to our knowledge, we are the first applying ML to deliver approximate answers for window functions.

The main contributions of this paper are summarized as follows:

- We elaborate on the ways of adopting density estimator and regression model to approximately process window functions. We show how to train models and how they can be used in query answering.
- We present WFApprox, a model-based window functions processing system, which utilizes density estimator and regression model to effectively answer window functions queries. In addition, WFApprox adopts model approximation and incremental computing for accelerating integral computing.
- We compare WFApprox to MySQL, PostgreSQL and Sampling proposed in [20] with TPC-H benchmark, of which the results demonstrate that WFApprox offers substantially improved performance and small errors.

The rest of the paper is organized as follows: Sect. 2 briefly introduces window functions in SQL and related work. Section 3 details our proposed approaches for approximately processing window functions. In Sect. 4 we experimentally evaluate our method and compare it with other implementations. Finally, we summarize the paper in Sect. 5.

2 Background

2.1 Window Functions

Window functions are part of the SQL: 2003 standard, accommodating complex analytical queries. The evaluation of window functions is based on three concepts: partitioning, ordering, and framing. Figure 1 illustrates these concepts and the process of window functions evaluation.

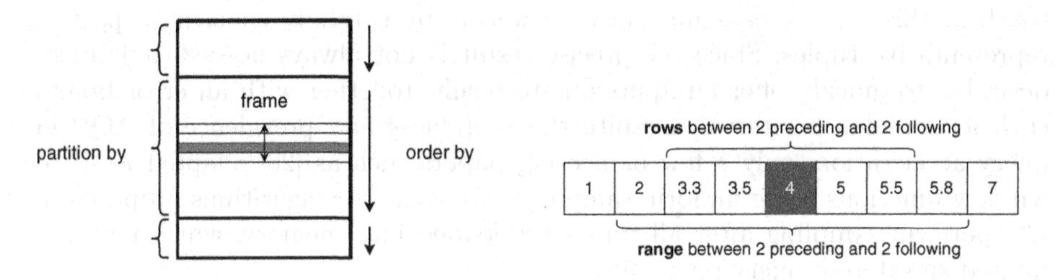

Fig. 1. Window function concepts

Fig. 2. Range and rows modes

The *partition by* clause divides the input table into independent groups and thereby building the boundaries of the window of the current tuple. The *order by* clause defines how tuples are ordered within each partition. Apart from the *partition by* clause, the *framing* clause further restricts the window of tuples. There are two frame modes to identify window boundaries which are *rows* and *range*. Figure 2 illustrates these two modes:

1. In the *rows* mode, the boundaries of the frame is directly expressed as the number of tuples before or after the current tuple belonging to the frame. As Fig. 2 implicates, the two rows before or after the current row belong to the frame, which also includes the current row therefore consisting of values 3.3, 3.5, 4, 5, 5.5.
2. *Range* mode specifies the maximum difference between the value of the current tuple and its value in preceding or following tuples of the frame. As shown in Fig. 2, the window frame boundaries are 2 (4 − 2) and 6 (4 + 2), thus 2, 3.3, 3.5, 4, 5, 5.5, 5.8 are included in current frame.

After three steps mentioned above, the calculation of specified window function is performed using tuples related to the current tuple. Window functions can be divided into two categories: aggregate window functions and nonaggregate window functions. The former includes functions such as COUNT, SUM, AVG, MIN, MAX. The latter contains functions specially designed for window functions such as ROW_NUMBER, RANK, CUME_DIST, NTH_VALUE.

2.2 Related Work

There are a number of optimization papers that relate to the window functions. Cao et al. developed a novel method [4] for the multi-sorting operations which could avoid repeated sorting for multiple window functions in a single query. A general execution framework of window functions was proposed by Leis et al. [10], which maintains data structures named segment tree to store aggregates for sub-ranges of an entire group, contributing to the reduction of redundant calculations. However, their works are only for distributive and algebraic aggregates. Wesley et al. [21] proposed an incremental method for three holistic windowed aggregates. Additionally, Song et al. proposed several sampling-based methods [20] for approximate evaluation of window functions. Since all tuples are required in the final result, the sampler can not be pushed down to the disk access layer, which indicating their sampling algorithms can only be performed after tuples are loaded into memory. Therefore their approach has a minimal improvement in performance due to little reduction of computation and I/O. Compared with sampling-based methods, this work relies on ML models to answer window functions without massive computing and disk access.

With respect to AQP, existing techniques can be broadly categorized into two categories: online aggregation and offline synopses generation. Online aggregation [5,6,12,22] selects samples when queries arrive and use them to answer the query. Offline synopses generation uses prior knowledge on the data statistics or query workload to generate synopses. And sampling[1,2,8,13,14] is the dominating approach in AQP. BlinkDB [2] proposed by Sameer et al. studies the ways of selecting samples based on the previous workload and uses them to answer future queries. VerdictDB [13] acts as a middleware allowing users to select samples from underlying databases for query answering. VerdictDB also proposes fast error approximation techniques, which offer error guarantees with low costs. However, traditional AQP solutions still suffer from the dilemma caused by the growing data size, therefore, some methods based on Machine Learning are proposed [3,11,16–19]. Jayavel et al. [19] builds low-error density estimators using clustering techniques and adopts them for calculating COUNT/SUM/AVG. DBEst [11] adds regression models therefore it is able to handle more query types. This work adopts models similar to DBEst. But unlike DBEst utilizing these models to answer aggregate functions, the processing of window functions is much more complicated due to a lot of repeated computation and different functions.

3 The WFApprox Query Engine

Different from the traditional AQP systems which directly use samples to answer queries, WFApprox obtains samples using uniform sampling from the original table to train the models, specifically density estimator and regression model, and then utilize these models to answer queries. We assume queries do not contain *partition by* clause, except where mentioned, and the handling of *partition by* will be described separately. This section details the calculation processes of supported functions. Table 1 shows the notations we used in this section.

Table 1. Notations

T	Original table
AGG	Supported aggregate functions, include COUNT, AVG and SUM
S	The size of table T
x	Column used to determine the window
y	Column used to compute aggregate result
D(x)	The density estimator training over column x
R(x)	The regression model training over column pair (x , y)
lo	The lower (preceding) offset in frame clause
uo	The upper (preceding) offset in frame clause
x_{min}	The min value of column x
x_{max}	The max value of column x

3.1 System Overview

Figure 3 presents WFApprox's high-level architecture. WFApprox is a middleware which uses density estimator and regression model to quickly provide approximate answers. To illustrate more, the overview of four major components of WFApprox are presented in the following part:

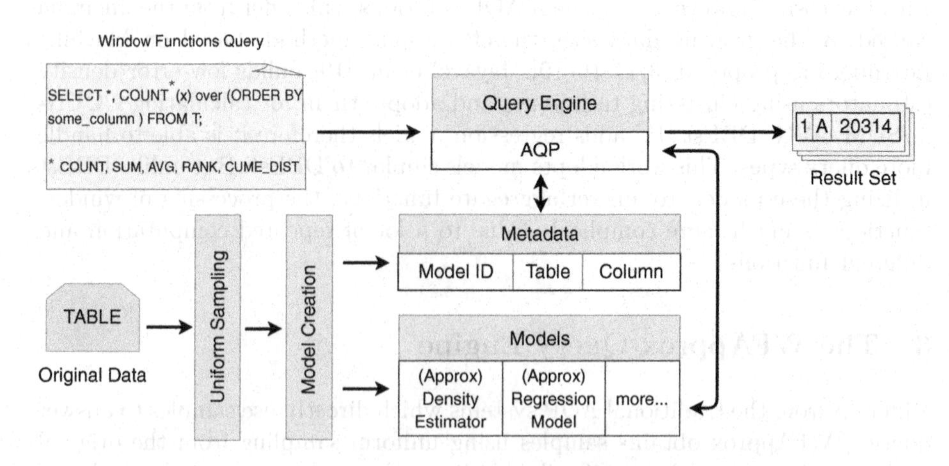

Fig. 3. WFApprox Architecture

- **Samples.** In a big data environment, considering the tables in the database usually contain millions or more tuples, it is not feasible to train the model based on the entire table. WFApprox adopts uniform sampling to generate samples, which are used for model training.
- **Models.** WFApprox relies on users to select which models to build and trains models in advance based on the sampled data. WFApprox uses density estimator based on KDE from *scikit-learn* [15] and regression model selected from Linear Regression, Decision Tree, SVM, and KNN. Cross validation from *scikit-learn* is used for both bandwidth of KDE determination and regression model selection. In addition, we approximate the models to improve computational efficiency. WFApprox allows users to add more models to improve the accuracy of query processing.
- **Metadata.** Metadata stores the information of available models and their corresponding columns in tables. Metadata provides the query engine with information to select the appropriate models.
- **Query Engine.** When a user submits a query, the query engine uses the metadata to determine whether there exists a corresponding model to process the query. If qualified models exist, the query engine calculates approximate results based on these models; if not, the query is directly handed over to the underlying database system for accurate results.

WFApprox supports the *partition by* and the *order by* clause in window functions, as well as both the *rows* and *range* modes of *frame* clause. The functions supported by WFApprox include aggregate functions: COUNT, SUM, AVG, and nonaggregate functions: CUME_DIST, RANK. Supporting for remaining functions is under consideration. In addition, window functions queries are divided into the following four types based on function type and frame type:

1. Aggregate functions without frame. Aggregate functions are evaluated based on the entire partition to which the current tuple belongs.
2. Aggregate functions with range mode frame. Aggregate functions are evaluated based on windows specified by *range frame* clause.
3. Aggregate functions with rows mode frame. Aggregate functions are evaluated based on windows specified by *rows frame* clause.
4. Nonaggregate functions. CUME_DIST and RANK are always evaluated on the entire partition and are independent of *frame* clause.

The following describes how these four types of queries mentioned above are calculated via density estimator and regression model.

3.2 Aggregate Functions Without Frame

WFApprox supports queries of the form:

```
SELECT *, AGG(x) over
(ORDER BY some_column ) FROM T;
```

This SQL seems to be meaningless since it gets almost the same result as ordinary aggregate functions. However, after adding *partition by* clause, which is handled in following, the SQL becomes the most commonly used form of window functions. Since no window range is specified for this type of query, COUNT represents the number of tuples in each partition, which can be recorded directly, therefore it is meaningless to provide the approximate answer for COUNT.

AVG. The average value of x is equal to the expectation of x. Since $\int_{x_{min}}^{x_{max}} D(x)dx$ is equal to 1, we only need to compute $\int_{x_{min}}^{x_{max}} D(x)xdx$.

$$AVG(x) = E[x] \approx \frac{\int_{x_{min}}^{x_{max}} D(x)xdx}{\int_{x_{min}}^{x_{max}} D(x)dx} = \int_{x_{min}}^{x_{max}} D(x)xdx \tag{1}$$

SUM. The sum of x equals the product of the size of the table and the expectation of x. Similar to AVG, the results can be calculated directly from $\int_{x_{min}}^{x_{max}} D(x)xdx$.

$$SUM(x) = S \cdot E[x] \approx S \cdot \frac{\int_{x_{min}}^{x_{max}} D(x)xdx}{\int_{x_{min}}^{x_{max}} D(x)dx} = S \cdot \int_{x_{min}}^{x_{max}} D(x)xdx \tag{2}$$

3.3 Aggregate Functions with Range Frame

The form of this kind of queries is as follows:

```
SELECT *, AGG( x or y ) over
( ORDER BY x range BETWEEN
lo PRECEDING AND uo FOLLOWING ) FROM T;
```

The bounds of the window are determined by x and the aggregate can be calculated based on both x or y. The calculation methods are given below separately.

COUNT. The results based on x or y are exactly the same for COUNT. We calculate the integration of density function in the range between lo and uo, yielding the proportion of tuples that lies within this window. S is used to scale up the integration to get an approximate representation of the total number of tuples in this window.

$$COUNT(x \text{ or } y) \approx S \cdot \int_{lo}^{uo} D(x)dx \tag{3}$$

AVG. In the case of calculating the average value based on x, since the average value of x is equal to the expectation of x, we only need to use the density function to calculate the expectation through integration. The calculation based on y is similar except that the regression model is used to estimate the value of y.

$$AVG(x) = E[x] \approx \frac{\int_{lo}^{uo} D(x)xdx}{\int_{lo}^{uo} D(x)dx} \tag{4}$$

$$AVG(y) = E[y] \approx E[R(x)] = \frac{\int_{lo}^{uo} D(x)R(x)dx}{\int_{lo}^{uo} D(x)dx} \tag{5}$$

SUM. The sum of y equals the product of the count and the average value of x. With the help of Eq. 3, 4 and 5, we can get a simplified formula: $S \cdot \int_{lo}^{uo} D(x)xdx$, which is easy to compute. As mentioned above, the calculation based on y is similar to the calculation based on x.

$$\begin{aligned} SUM(x) &= COUNT(x) \cdot AVG(x) \\ &\approx S \cdot \int_{lo}^{uo} D(x)dx \cdot \frac{\int_{lo}^{uo} D(x)xdx}{\int_{lo}^{uo} D(x)dx} = S \cdot \int_{lo}^{uo} D(x)xdx \end{aligned} \tag{6}$$

$$\begin{aligned} SUM(y) &= COUNT(y) \cdot AVG(y) \\ &\approx S \cdot \int_{lo}^{uo} D(x)dx \cdot \frac{\int_{lo}^{uo} D(x)R(x)dx}{\int_{lo}^{uo} D(x)dx} = S \cdot \int_{lo}^{uo} D(x)R(x)dx \end{aligned} \tag{7}$$

3.4 Aggregate Functions with Rows Frame

Following form of SQL is also supported by WFApprox:

```
SELECT *, AGG( y ) over
(ORDER BY some_column ROWS BETWEEN
lo PRECEDING AND uo FOLLOWING )FROM T;
```

This kind of query is similar to the query with range frame, except that x becomes the index of the current tuple. Since the distribution of the index is already known (which is uniform distribution), regression models themselves are enough and the density estimator is no longer needed. Unfortunately, one regression model can only deal with one sort condition due to the correlation between indexes and sort criteria. COUNT is meaningless in *rows* mode, thus we just ignore it.

AVG. The calculation of the average value only needs the integral of the continuous function R in a given window to get the sum of y, and then divide by the number of tuples in the window to get the final result.

$$AVG(y) \approx \frac{\int_{lo}^{uo} R(index)d(index)}{uo - lo + 1} \tag{8}$$

SUM. The calculation of SUM has been explained AVG.

$$SUM(y) \approx \int_{lo}^{uo} R(index)d(index) \tag{9}$$

3.5 Nonaggregate Functions

WFApprox supports RANK and CUME_DIST.

RANK. The rank function gives the rank of the current tuple within the partition. However, providing an approximate ranking for each tuple is impractical and may lead to great error, therefore, WFApprox mainly supports top-k query based on rank function. The SQL form is as follows:

```
SELECT * FROM
(SELECT *, rank () over (ORDER BY x) AS rank FROM T)
WHERE rank < k;
```

The above SQL is a nested query, which is used to get the top-k tuples sorted by x in table T. In order to quickly obtain the top-k results, we use binary search to approximately find the x value of the tuple ranked k with the help of density estimator. In addition, we establish a hash table in advance which records x as the key and the disk location of the corresponding tuple as the value in order to efficiently fetc.h the corresponding tuple according to the x value obtained previously. If there exists an index on column x, we can use it directly without building the hash table mentioned above. To avoid occupying large disk space, the hash table only needs to maintain the information of a part of tuples, for example, top 100 thousand tuples. We adjust the binary search algorithm to ensure that the number of tuples returned is always greater than k, so that users can get al.l the tuples they need.

CUME_DIST. The function CUME_DIST gives the cumulative distribution value of the current tuple. The result of this function can be obtained directly through the density estimator by calculating the integral from the minimum value of x's column to x:

$$CUME\_DIST(x) \approx \int_{x_{min}}^{x} D(x)dx \tag{10}$$

3.6 Discussion on Supporting "partition By"

WFApprox adopts a straightforward strategy to handle *partition by* clause. Since data blocks segmented by *partition by* clause are no longer associated with each other, WFApprox treats each data block as a separate table. That is to say, WFApprox establishes corresponding models in advance on each data block, which are evaluated separately to get results and integrated to reach the final results.

3.7 Computation Efficiency Optimization

As mentioned above, for each window, we need to calculate the integral to get its results, so the key of efficiency improvement depends on the achievement of faster integral calculation. The straightforward idea is to use the integral evaluation function (WFApprox adopts the function from the *integrate* module in *SciPy* [7]) to call the corresponding model repeatedly to calculate the integral. Although one call to the model is very fast, the processing of window functions involves repeated calls to the model, which is far from efficient. In this regard, the model is approximated to improve the efficiency, we slice the original model, pre-calculated the integration value of the model on each slice, and store them in an array. After that, for each window, WFApprox can simply sum the array values within the window range without invoking the original models. In addition, the array can be dumped into file for reuse.

WFApprox also adopts an incremental calculation strategy. Assuming the range of the previous window is lo_{pre} to uo_{pre} and the current window is lo to up, we only need to calculate the integral between lo_{pre} to lo and uo_{pre} to uo to get the integral of the current window:

$$\int_{lo}^{uo} f(x)dx = \int_{lo_{pre}}^{uo_{pre}} f(x)dx - \int_{lo_{pre}}^{lo} f(x)dx + \int_{uo_{pre}}^{uo} f(x)dx \qquad (11)$$

4 Experiments

In this section, we experimentally evaluate our implementations and compare its performance with baselines including MySQL, PostgreSQL and Sampling from [20].

4.1 Experimental Setup

We ran experiments on an Ubuntu 18.04 Server with 8 Intel i7-9700K cores, 64GB RAM and 1TB SSD. We compare WFApprox against MySQL 8.0.18, PostgreSQL 12.0 and Sampling, which is implemented base on Incremental Sampling Algorithm proposed in [20]. However, Sampling is not applicable to CUME_DIST and RANK, since they require all tuples to participate in the calculation to get the answer, and the missing tuples in Sampling will cause an unacceptable error. Therefore we only use MySQL and PostgreSQL as baselines for these two functions. WFApprox is implemented based on Python 3.6.8. We use TPC-H benchmark to evaluate WFApprox. Our experiments are mainly conducted on table ORDERS with 1.5 million, 15 million and 150 million rows and table LINEITEM with about 6 million, 60 million and 600 million rows.

4.2 Experimental Results

We build all models based on 100 thousand rows sampled from original table. For convenience, in the following charts, WFApprox, Sampling and relative error are abbreviated as WFA, SAM, RE respectively.

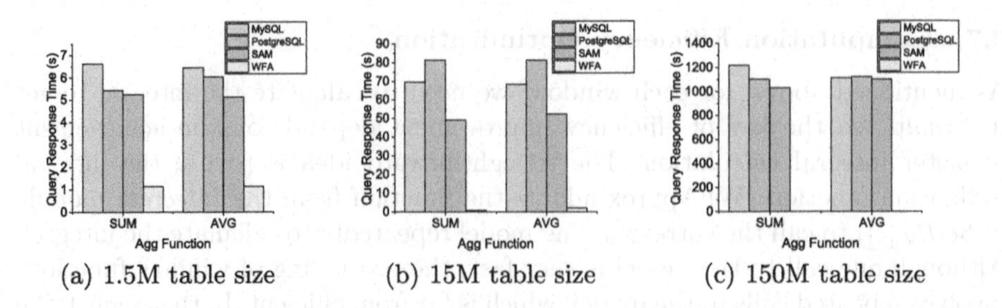

(a) 1.5M table size (b) 15M table size (c) 150M table size

Fig. 4. Response time of aggregate functions without frame

Table 2. Relative error of aggregate functions without frame

System	Function					
	SUM			AVG		
	1.5M	15M	150M	1.5M	15M	150M
SAM	0.49%	0.61%	0.53%	0.21%	0.19%	0.23%
WFA	0.96%	1.11%	1.01%	1.01%	0.56%	0.64%

Aggregate Functions Without Frame. Figure 4 and Table 2 show the response times and relative errors produced by following SQL:

```
SELECT *, AGG( o_totalprice ) over
( PARTITION by o_orderpriority ) FROM orders;
```

WFApprox achieves a significant result of time-saving (5x for 1.5M, 50x for 15M, 1000x for 150M), compared to other baselines. At the same time, The relative errors are only around 2x larger than that of Sampling. Besides, as shown in Fig. 4, since all ML models are trained based on the same sample size, WFApprox is not sensitive to the growth of data scale, which is an ideal feature in handling big data.

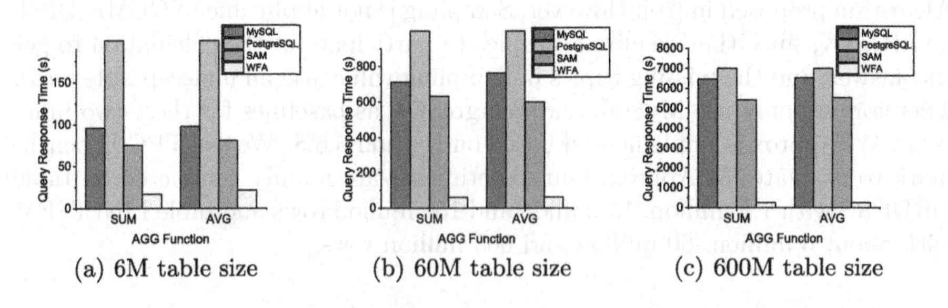

(a) 6M table size (b) 60M table size (c) 600M table size

Fig. 5. Response time of aggregate functions with range frame

Table 3. Relative error of aggregate functions with range frame

System	Function					
	SUM			AVG		
	1.5M	15M	150M	1.5M	15M	150M
SAM	0.54%	0.44%	0.41%	0.24%	0.26%	0.31%
WFA	0.85%	0.65%	0.93%	0.54%	0.23%	0.43%

Aggregate Functions with Range Frame. We adopt the following two SQLs for the experiment which involve one column and two columns respectively:

```
SELECT *, AGG( o_totalprice ) over
( ORDER BY o_totalprice range BETWEEN
10000 PRECEDING AND 10000 FOLLOWING ) FROM orders;

SELECT *, avg( l_quantity ) over
( ORDER BY l_extendedprice
RANGE BETWEEN 10000 PRECEDING AND 10000 FOLLOWING )
FROM lineitem;
```

Here we only show the results of SQLs involve two columns since this type of query is more complex than SQLs involve one column and the experimental results of the two types of queries are basically the same. Figure 5 and Table 3 show the response times and relative errors respectively. PostgreSQL can't complete the query in all three table scales within the time limit. WFApprox is 4x faster than Sampling for 1.5M table size, 10x faster for 15M and 20x faster for 150M, and the errors are slightly larger than that of Sampling within 1%. And again, the response times of WFApprox is less sensitive to the data scale compared with baselines.

Aggregate Functions with Rows Frame. We use the following query for experiment:

```
SELECT *, AGG( o_totalprice ) over
( ORDER BY o_totalprice rows BETWEEN
10000 PRECEDING AND 10000 FOLLOWING ) FROM orders;
```

Same as *range* mode, with the increase of data scale, WFApprox has achieved more and more remarkable query acceleration effect, up to 5x with table size of 150M. At the same time, its average relative errors are about 0.2%, which are smaller than that of Sampling since only regression models are used to answer the queries.

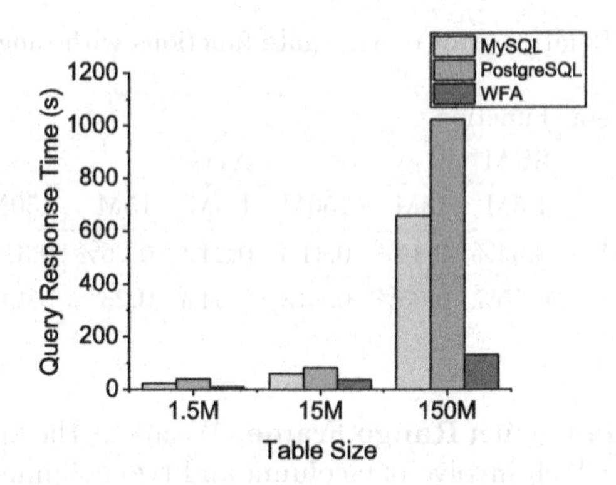

Fig. 6. Response Time of CUME_DIST

Nonaggregate Functions

CUME_DIST The performance of CUME_DIST on different methods is evaluated by following SQL:

```
SELECT cume_dist () over
( ORDER BY o_totalprice ) AS cume FROM orders;
```

Figure 6 shows the corresponding query response times and relative errors of various table sizes for CUME_DIST. When table size is 1.5 and 15 million, WFApprox is about 2x as fast as MySQL and PostgreSQL. And WFApprox achieves much more performance improvement for about 6X when the table size is 150 million. The relative errors are well controlled with an average value of 0.12%.

RANK The evaluating of RANK is based on the following SQL:

```
SELECT * FROM  (SELECT *, rank ()
over (ORDER BY o_totalprice) AS rank FROM orders)
WHERE rank < k;
```

where we set k to 1000, 10000, 50000 and 100000.

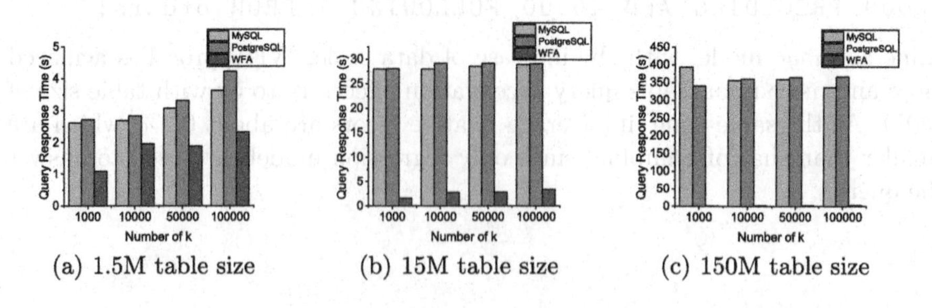

(a) 1.5M table size (b) 15M table size (c) 150M table size

Fig. 7. Response time of RANK

The response times of RANK are shown in Fig. 7. Since WFApprox can skip the sorting step when processing RANK, it brings about significant accelerations of 2x, 10x and 100x for 1.5M, 15M and 150M tables respectively, while the relative errors are below 3.5% except for table size 150M. When table size is 150M, the column *o_totalprice* of *orders* is too dense that the accuracy of KDE in handling RANK degrades rapidly, resulting in relative errors of about 10%. Please note that the errors here only lead to redundant tuples, which can be simply ignored by users.

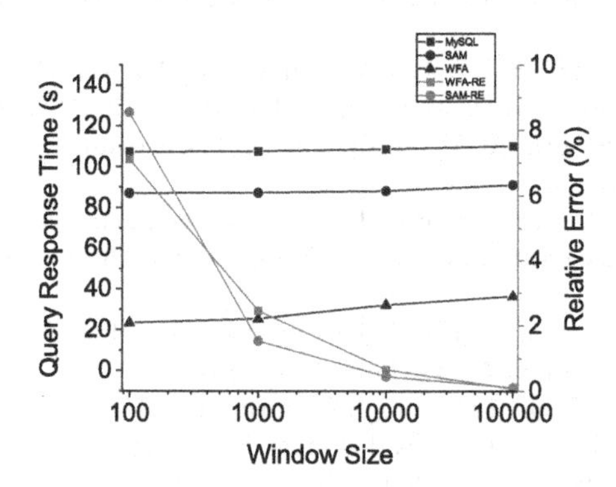

Fig. 8. Influence of window size on response time and relative error

Window Size Effect. Figure 8 shows the effect of window size on query response times and relative errors. Window sizes vary from 100 to 100000. Due to the adoption of increment computing, query response times of all three methods are not sensitive to window sizes. Regardless of window size, WFApprox is 3X faster than Sampling. As window size increases, the relative errors drops significantly, bringing it to below 0.1% when window size is 100000. WFApprox achieves almost same relative errors compared to Sampling with all window sizes.

5 Conclusion

In this paper, we have shown how WFApprox works to approximately process window functions. We choose two ML models including density estimator and regression model, and provide specific calculation methods for different functions using these models in window functions. Additionally, model approximation and incremental computation are proposed to improve computing efficiency. We have implemented our techniques in Python and conducted a series of experiments over TPC-H benchmark by comparing our methods with MySQL, PostgreSQL and Sampling, indicating that WFApprox provides substantial performance gains while producing small errors. As for future work, the providing of error bound and better accuracy-efficiency trade-offs are our main directions.

Acknowledgement. This work was supported by National Key R&D Program of China (No. 2017YFC0803700), NSFC grants (No. 61532021 and 61972155), Shanghai Knowledge Service Platform Project (No. ZF1213) and Zhejiang Lab (No. 2019KB0AB04).

References

1. Acharya, S., Gibbons, P.B., Poosala, V., Ramaswamy, S.: The aqua approximate query answering system. In: ACM Sigmod Record, vol. 28, pp. 574–576. ACM (1999)
2. Agarwal, S., Mozafari, B., Panda, A., Milner, H., Madden, S., Stoica, I.: Blinkdb: queries with bounded errors and bounded response times on very large data. In: Proceedings of the 8th ACM European Conference on Computer Systems, pp. 29–42. ACM (2013)
3. Anagnostopoulos, C., Triantafillou, P.: Query-driven learning for predictive analytics of data subspace cardinality. ACM Trans. Knowl. Discov. Data (TKDD) **11**(4), 47 (2017)
4. Cao, Y., Chan, C.Y., Li, J., Tan, K.L.: Optimization of analytic window functions. Proc. VLDB Endowment **5**(11), 1244–1255 (2012)
5. Chandramouli, B., et al.: Trill: a high-performance incremental query processor for diverse analytics. Proc. VLDB Endowment **8**(4), 401–412 (2014)
6. Hellerstein, J.M., Haas, P.J., Wang, H.J.: Online aggregation. In: ACM Sigmod Record, vol. 26, pp. 171–182. ACM (1997)
7. Jones, E., Oliphant, T., Peterson, P., et al.: Scipy: open source scientific tools for Python (2001)
8. Kandula, S., et al.: Quickr: lazily approximating complex adhoc queries in bigdata clusters. In: Proceedings of the 2016 International Conference on Management of Data, pp. 631–646. ACM (2016)
9. Kraska, T., Beutel, A., Chi, E.H., Dean, J., Polyzotis, N.: The case for learned index structures. In: Proceedings of the 2018 International Conference on Management of Data, pp. 489–504. ACM (2018)
10. Leis, V., Kundhikanjana, K., Kemper, A., Neumann, T.: Efficient processing of window functions in analytical SQL queries. Proc. VLDB Endowment **8**(10), 1058–1069 (2015)
11. Ma, Q., Triantafillou, P.: Dbest: revisiting approximate query processing engines with machine learning models. In: Proceedings of the 2019 International Conference on Management of Data, pp. 1553–1570. ACM (2019)
12. Pansare, N., Borkar, V.R., Jermaine, C., Condie, T.: Online aggregation for large mapreduce jobs. Proc. VLDB Endowment **4**(11), 1135–1145 (2011)
13. Park, Y., Mozafari, B., Sorenson, J., Wang, J.: Verdictdb: universalizing approximate query processing. In: Proceedings of the 2018 International Conference on Management of Data, pp. 1461–1476. ACM (2018)
14. Park, Y., Tajik, A.S., Cafarella, M., Mozafari, B.: Database learning: toward a database that becomes smarter every time. In: Proceedings of the 2017 ACM International Conference on Management of Data, pp. 587–602. ACM (2017)
15. Pedregosa, F., et al.: Scikit-learn: machine learning in Python. Journal of machine learning research **12**, 2825–2830 (2011)
16. Rendle, S.: Scaling factorization machines to relational data. Proc. VLDB Endowment **6**, 337–348 (2013)

17. Schelter, S., Soto, J., Markl, V., Burdick, D., Reinwald, B., Evfimievski, A.: Efficient sample generation for scalable meta learning. In: 2015 IEEE 31st International Conference on Data Engineering, pp. 1191–1202. IEEE (2015)
18. Schleich, M., Olteanu, D., Ciucanu, R.: Learning linear regression models over factorized joins. In: Proceedings of the 2016 International Conference on Management of Data, pp. 3–18. ACM (2016)
19. Shanmugasundaram, J., Fayyad, U., Bradley, P.S., et al.: Compressed data cubes for olap aggregate query approximation on continuous dimensions. In: KDD vol. 99, pp. 223–232. Citeseer (1999)
20. Song, G., Qu, W., Liu, X., Wang, X.: Approximate calculation of window aggregate functions via global random sample. Data Sci. Eng. **3**(1), 40–51 (2018). https://doi.org/10.1007/s41019-018-0060-x
21. Wesley, R., Xu, F.: Incremental computation of common windowed holistic aggregates. Proc. VLDB Endowment **9**(12), 1221–1232 (2016)
22. Wu, S., Ooi, B.C., Tan, K.L.: Continuous sampling for online aggregation over multiple queries. In: Proceedings of the 2010 ACM SIGMOD International Conference on Management of Data, pp. 651–662. ACM (2010)

BiSample: Bidirectional Sampling for Handling Missing Data with Local Differential Privacy

Lin Sun[1(✉)], Xiaojun Ye[1], Jun Zhao[2], Chenhui Lu[1], and Mengmeng Yang[2]

[1] School of Software, Tsinghua University, Beijing, China
{sunl16,luch18}@mails.tsinghua.edu.cn, yexj@tsinghua.edu.cn
[2] Nanyang Technological University, Singapore, Singapore
{junzhao,melody.yang}@ntu.edu.sg

Abstract. Local differential privacy (LDP) has received much interest recently. In existing protocols with LDP guarantees, a user encodes and perturbs his data locally before sharing it to the aggregator. In common practice, however, users would prefer not to answer all the questions due to different privacy-preserving preferences for some questions, which leads to data missing or the loss of data quality. In this paper, we demonstrate a new approach for addressing the challenges of data perturbation with consideration of users' privacy preferences. Specifically, we first propose BiSample: a bidirectional sampling technique value perturbation in the framework of LDP. Then we combine the BiSample mechanism with users' privacy preferences for missing data perturbation. Theoretical analysis and experiments on a set of datasets confirm the effectiveness of the proposed mechanisms.

Keywords: Local differential privacy · Missing data · Randomized response

1 Introduction

With the development of big data technologies, numerous data from users' side are routinely collected and analyzed. In online-investigation systems, statistical information, especially the frequency and mean values can help investigators know about the investigated population. However, users' data are collected at the risk of privacy leakage. Recently, local differential privacy (LDP) [5] has been proposed as a solution to privacy-preserving data collection and analysis since it provides provable privacy protection, regardless of adversaries' background knowledge. Usually, protocols with LDP guarantees can be broken down into an Encode-Perturb-Aggregate paradigm. For single round of data sharing, each user **encodes** his value (or tuple) into a specific data format, and then **perturbs** the encoded value for privacy concerns. At last, all the perturbed data are **aggregated** by an untrusted collector. Mechanisms with LDP guarantees have

Y. Nah et al. (Eds.): DASFAA 2020, LNCS 12112, pp. 88–104, 2020.
https://doi.org/10.1007/978-3-030-59410-7_6

been implemented in many real-world data collecting systems, such as Google's RAPPOR [9] and Microsoft's telemetry data analyzing system [4].

Although the LDP can balance the users' privacy and data utilities, existing solutions assume that the investigated users follow the collecting process truthfully. However, in an investigation system, even though the investigator claims the collection process satisfies LDP, individuals may refuse to confide some specific questions due to following considerations: 1) the provided privacy-preserving level is not as expected, or 2) he just doesn't want to tell anything about the question. For example, if an investigator designed a $\ln 3$-LDP mechanism for personal-related data analyzing, those who think the privacy-preserving is good enough would provide the real value for perturbation, while those who extremely care about their healthy states might evade certain questions such as "Do you have cancer?" because they think the in-built privacy-preserving guarantee by LDP mechanism is not private enough. As existing perturbation solutions demand an input, these individuals would randomly pick an answer (or just answer "No") and use it for perturbation (we call these **fake answers**). In the perturbed space, the fake answers will lead to evasive bias. Thus, applying existing LDP mechanisms without considering users' privacy preferences brings unpredictable estimation errors.

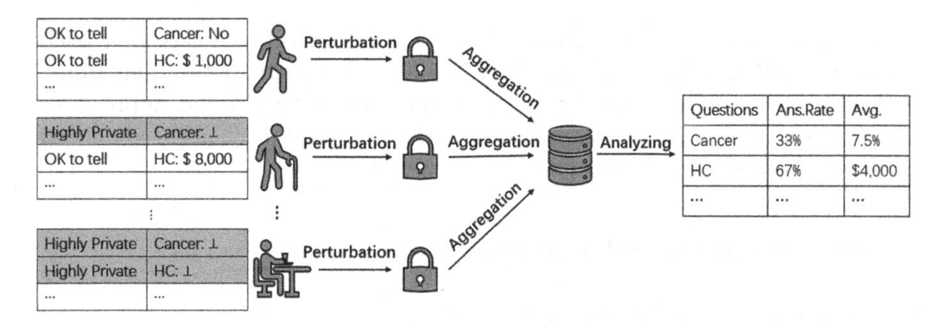

Fig. 1. Missing Data Collecting and Analyzing Framework. For space consideration, HC represents the Hospitalization Cost and Ans. Rate represents the rate of individuals who provide real value. The symbol $\perp$ occurs when users are not willing to provide real value, even in the LDP framework.

In this paper, we consider a "providing **null-value**" procedure to avoid the fake answers when users perturb their data. Figure 1 lists users' privacy preferences and original answers of each investigated question. Instead of sending a fake answer when the provided privacy-preserving level is not as expected, the individual sends a null-value to avoid a biased estimated result. The untrusted aggregator wants to analyze basic statistical information, 1) how many people provide the real value, 2) what is the frequency/mean of the whole investigated population. The privacy-preserving is two-fold. Firstly, users who have higher privacy concerns only provide null-value. Secondly, both the null-value and provided data from users' side are perturbed by a LDP mechanism. The aggregator cannot learn whether the original data is null-value nor what the data is when he receives a perturbed record.

We for the first time consider the influence of users' cooperation on the estimation accuracy. We first propose a bi-directional sampling mechanism called BiSample and use it for numerical value perturbation. Then we extend the BiSample to be capable of the null-value perturbation while still being locally differentially private. In general, this paper presents following contributions:

- For the first time, we consider that users will not provide the true data for perturbation in a collecting and analyzing framework with LDP guarantees. Our proposed data perturbation framework provides new insights into improving data utilities by modeling users' privacy-preserving preferences.
- We propose BiSample, a bi-directional sampling mechanism for data perturbation. Literately, the BiSample mechanism can replace the Harmony [13] solution for mean estimation. Furthermore, we extend the BiSample to be capable of perturbing null-value data. Our mechanism allows to analyze the rate of users who provided true answers and can be used for frequency/mean estimation with LDP guarantees.
- Experimentally, the proposed mechanism achieves lower estimation error than existing mechanisms.

This paper is organized as follows. In Sect. 2, we provide the necessary background of LDP and define the problem of analyzing missing data in our framework. Then we propose the BiSample mechanism in Sect. 3 and apply it for missing data in Sect. 4. The evaluations of the proposed mechanism are shown in Sect. 5. At last, the whole paper is concluded in Section 6.

2 Preliminaries and Problem Definition

2.1 Local Differential Privacy (LDP)

Definition 1 (Local Differential Privacy [2,5–7]**).** *A randomized mechanism $\mathcal{M}(\cdot)$ achieves ϵ-local differential privacy if and only if for every two input tuples t_1, t_2 in the domain of $\mathcal{M}$, and for any output $t^* \in \text{Range}(\mathcal{M})$ that:*

$$\Pr[\mathcal{M}(t_1) = t^*] \leq \exp(\epsilon) \cdot \Pr[\mathcal{M}(t_2) = t^*] \tag{1}$$

Unlike earlier attempts to preserve privacy, such as k-anonymity [16] and l-diversity [12], the LDP retains "plausible deniability" of sensitive information. The LDP has been used in a variety of application areas, such as heavy hitters estimation [1,2,14] and marginal release [3].

The canonical solution towards LDP is the randomized response (RR [8,18]). Specifically, to collect sensitive information from users, e.g., whether the patient is a HIV carrier, RR is used for perturbing actual answers while still guarantees that i) each user's answer provides plausible deniability, ii) the aggregator can get an unbiased estimation over the whole population. Many start-of-the-art mechanisms use RR as a core part to provide privacy guarantees, such as the LDPMiner [14], LoPub [15] and RAPPOR [9]. To handle categorical data with arbitrary number of possible values, the k-RR [10] is proposed. In typical RR, each user shares his answer truthfully with probability p and provide the opposite answer with $1 - p$. To achieve ϵ-LDP, we set $p = \frac{\exp(\epsilon)}{\exp(\epsilon)+1}$.

Let f_r denote the proportion of positive (resp. negative) answers received by the aggregator, the frequency of positive (resp. negative) answers before perturbing can be estimated by:

$$f^* = \frac{p - 1 + f_r}{2p - 1}, \tag{2}$$

then f^* is an unbiased estimator of f.

Recently, the numerical value perturbation under LDP for mean estimation has been addressed in the literature. We briefly introduce the Harmony [13] and Piecewise mechanism [17].

Harmony. Nguyên *et al.* [13] proposed **Harmony** for collecting and analyzing data from smart device users. Shown as Algorithm 1, Harmony contains three steps: discretization, perturbation and adjusting. The discretization (Line 1) is used to generate a discretized value in $\{-1, 1\}$, then Randomized Response is applied to achieve ϵ-LDP (Line 2). At last, to output an unbiased value, the perturbed value is adjusted (Line 3).

Algorithm 1. Harmony [13] for Mean Estimation.

Input: value $v \in [-1, 1]$ and privacy budget ϵ.
Output: discretized value $x^* \in \{-\frac{\exp(\epsilon)+1}{\exp(\epsilon)-1}, \frac{\exp(\epsilon)+1}{\exp(\epsilon)-1}\}$
1: Discretize value to $v^* \in \{-1, 1\}$ by:

$$v^* = \text{Dis}(v) = \begin{cases} -1 & \text{with probability } \frac{1-v}{2}; \\ 1 & \text{with probability } \frac{1+v}{2}. \end{cases}$$

2: Perturb v^* by using the randomized response:

$$v^* = \begin{cases} v^* & \text{with probability } \frac{\exp(\epsilon)}{\exp(\epsilon)+1}; \\ -v^* & \text{with probability } \frac{1}{\exp(\epsilon)+1}. \end{cases}$$

3: Adjusted the perturbed data by:

$$v^* = \frac{\exp(\epsilon) + 1}{\exp(\epsilon) - 1} \cdot v^*.$$

4: **return** v^*.

Piecewise Mechanism. The Piecewise Mechanism (PM) [17] is another perturbation solution for mean estimation. Unlike the Harmony, the output domain of PM is continuous from $-\frac{\exp(\epsilon/2)+1}{\exp(\epsilon/2)-1}$ to $\frac{\exp(\epsilon/2)+1}{\exp(\epsilon/2)-1}$. The PM is used for collecting a single numeric attribute under LDP. Based on PM, [17] also build a Hybrid Mechanism (HM) for mean estimation. The PM and HM obtain higher result accuracy compared to existing methods.

PrivKVM. The PrivKVM mechanism [19] is used to estimate the frequency of keys and the mean of values in Key-Value data with local differential privacy guarantee. To preserve the correlation between keys and values, the PrivKVM uses a interactive protocol with multiple rounds for mean estimation. The estimated mean in PrivKVM is unbiased theoretically when the number of iteration is large enough. The PrivKVM can also be used for missing data perturbation where the key represents whether the data is missing. We will later show that the proposed BiSample mechanism achieves high accuracy in mean estimation.

2.2 Problem Definition

This paper researches the problem of data collecting and analyzing while considering users' privacy preferences in the context of LDP. When it comes to multi-dimensional data, the classical solution is to sample and perturb one item to avoid splitting the privacy budget. For simplicity, we assume that each user u_i only holds one single value v_i (Table 1).

Table 1. Notations.

Symbol	Description
$\mathcal{U} = \{u_1, u_2, ..., u_n\}$	The set of users, where $n = \lvert\mathcal{U}\rvert$
v_i	Value of user u_i, $v_i \in [-1, 1] \cup \{\perp\}$
ϵ_u^i	Privacy demand of u_i
ϵ	Privacy budget of perturbation mechanism
p	$p = \exp(\epsilon)/(\exp(\epsilon) + 1)$

Modeling Users' Privacy Preferences. As detailed in the introduction part, for one single investigating question, different users have different privacy preferences. Without loss of generality, we use ϵ_u^i to describe the privacy-preserving preferences of u_i and we assume that user u_i only collaborates with the data collector when the provided privacy-preserving level is higher than expected (which is $\epsilon \leq \epsilon_u^i$). When the provided privacy-preserving level is not as expected ($\epsilon > \epsilon_u^i$), the user u_i would provide a null-value (represented by $v_i = \perp$) instead of the fake answer for perturbation.

After perturbation, data from users' side are collected by an untrusted aggregator, who wants to learn some statistical information from all users, especially the rate of null-value and the mean value of from users' side.

Definition 2 (Mean of missing data). *For a list of values* $\boldsymbol{v} = \{v_1, v_2, ..., v_n\}$ *where each value* v_i *from user* u_i *is in domain* $[-1,1] \cup \{\perp\}$*, the missing rate (mr) and the mean (m) of* $\boldsymbol{v}$ *are defined as:*

$$mr = \frac{\#\{v_i | v_i = \perp\}}{n}, \quad m = \frac{\sum_{v_i \neq \perp} v_i}{\#\{v_i | v_i \neq \perp\}}. \tag{3}$$

Also, when $\forall i : \epsilon_u^i \geq \epsilon$, the estimation of mean of missing data turns to be the traditional mean estimation problem.

3 BiSample: Bidirectional Sampling Technique

Inspired by the Harmony mechanism that only uses the probability of $\Pr[v = 1]$ for mean estimation, we propose a bidirectional sampling technique, referred to as the *BiSample Mechanism*. The BiSample mechanism takes a value $v \in [-1, 1]$ as input and outputs a perturbed tuple $\langle s, b\rangle$ where s represents the sampling direction and b represents the sampling result of v. Specifically, the BiSample mechanism contains two basic sampling directions, which is defined as:

- **Negative Sampling with LDP**. The negative sampling is used to estimate the frequency of -1 after discretization. The perturbing procedure of negative sampling is:

$$\Pr[b = 1] = (2p - 1) \cdot \Pr[\text{Dis}(v) = -1] + (1 - p). \tag{4}$$

- **Positive Sampling with LDP**. Like negative sampling, the positive sampling is used to estimate the frequency of 1 after discretization. Notably, the typical RR is positive sampling.

$$\Pr[b = 1] = (2p - 1) \cdot \Pr[\text{Dis}(v) = 1] + (1 - p). \tag{5}$$

Assuming the input domain is $[-1, 1]$, Algorithm 2 shows the pseudo-code of BiSample. Without loss of generality, when the input domain is $[L, U]$, the user (i) computes $v' = \frac{2}{U-L} \cdot v + \frac{L+U}{L-U}$, (ii) perturbs v' using the BiSample mechanism, and (iii) shares $\langle s, (\frac{U-L}{2}) \cdot b + \frac{U+L}{2}\rangle$ with the aggregator, where s denotes the sampling method and b is the sampling result of v'. In Algorithm 2, Lines 2–3 show the negative sampling process and Lines 5–6 denote the positive sampling. We prove that the combination of positive and negative sampling satisfies ϵ-LDP.

Algorithm 2. *BiSample*(v, ϵ): Bidirectional Sampling Mechanism

Input: a value $v \in [-1, 1]$, privacy budget ϵ.
1: sample a uniformly variable $s \in \{0, 1\}$ representing the sampling direction.
2: **if** $s = 0$ **then**
3: use **Negative Sampling**: generate a Bernoulli variable b with:

$$\Pr[b = 1] = \frac{1 - \exp(\epsilon)}{1 + \exp(\epsilon)} \cdot \frac{v}{2} + \frac{1}{2}.$$

4: **else**
5: use **Positive Sampling**: generate a Bernoulli variable b with:

$$\Pr[b = 1] = \frac{\exp(\epsilon) - 1}{\exp(\epsilon) + 1} \cdot \frac{v}{2} + \frac{1}{2}.$$

6: **end if**
7: **return** $\langle s, b \rangle$.

Theorem 1. *The BiSample mechanism* $\mathcal{M} = \text{BiSample}(\cdot)$ *guarantees* ϵ*-LDP.*

Proof. For any $t_1, t_2 \in [-1, 1]$ *and output* $o \in \text{Range}(\mathcal{M})$*, we have:*

$$\begin{aligned}
&\ln \max_{t_1, t_2 \in [-1,1], o \in \text{Range}(\mathcal{M})} \frac{\Pr[\mathcal{M}(t_1) = o]}{\Pr[\mathcal{M}(t_2) = o]} \\
&= \ln \max_{t_1, t_2 \in [-1,1], b \in \{0,1\}} \frac{\Pr[\mathcal{M}(t_1) = \langle 0, b \rangle]}{\Pr[\mathcal{M}(t_2) = \langle 0, b \rangle]} \\
&= \ln \frac{\max_{t_1 \in [-1,1]} \Pr[\mathcal{M}(t_1) = \langle 0, 0 \rangle]}{\min_{t_2 \in [-1,1]} \Pr[\mathcal{M}(t_2) = \langle 0, 0 \rangle]} \\
&= \ln \left(\frac{\exp(\epsilon)}{2(\exp(\epsilon) + 1)} \Big/ \frac{1}{2(\exp(\epsilon) + 1)} \right) = \epsilon.
\end{aligned} \tag{6}$$

According to the definition of LDP, the BiSample achieves ϵ*-LDP.*

With the BiSample mechanism, a value v_i in the input domain is perturbed into a two-bit tuple $\mathcal{M}_{\text{BiSample}}(v_i) = \langle s_i, b_i \rangle$. The result is two-fold. First, the s_i indicates whether the sampling mechanism is positive sampling or not. Second, the b_i represents the sampling value with correspond sampling mechanism. For the aggregator, let $\mathcal{R} = \{\langle s_1, b_1 \rangle, \langle s_2, b_2 \rangle, ... \langle s_n, b_n \rangle\}$ be the perturbed data received from all the users and f_{POS} (resp. f_{NEG}) be the aggregated frequency of positive sampling (resp. negative sampling), which is given by:

$$f_{\text{POS}} = \frac{\#\{\langle s_i, b_i \rangle | \langle s_i, b_i \rangle = \langle 1, 1 \rangle, \langle s_i, b_i \rangle \in \mathcal{R}\}}{\#\{\langle s_i, b_i \rangle | s_i = 1, \langle s_i, b_i \rangle \in \mathcal{R}\}}, \tag{7}$$

$$f_{\text{NEG}} = \frac{\#\{\langle s_i, b_i \rangle | \langle s_i, b_i \rangle = \langle 0, 1 \rangle, \langle s_i, b_i \rangle \in \mathcal{R}\}}{\#\{\langle s_i, b_i \rangle | s_i = 0, \langle s_i, b_i \rangle \in \mathcal{R}\}}. \tag{8}$$

Theorem 2. *m^* is an unbiased estimator of $m = \frac{1}{n}\sum_{i:i\in[n]} v_i$, where m^* is given by:*

$$m^* = \frac{1}{2p-1}\left(f_{\mathrm{POS}} - f_{\mathrm{NEG}}\right). \tag{9}$$

Proof. Firstly, the m^* can be represented by:

$$\begin{aligned}\mathbb{E}[m^*] &= \mathbb{E}\left[\frac{1}{2p-1}\left(f_{\mathrm{POS}} - f_{\mathrm{NEG}}\right)\right]\\ &= \mathbb{E}\left[\frac{1}{2p-1}\left((f_{\mathrm{POS}} + p - 1) - (f_{\mathrm{NEG}} + p - 1)\right)\right]\\ &= \left(\mathbb{E}\left[\frac{f_{\mathrm{POS}} + p - 1}{2p-1}\right] - \mathbb{E}\left[\frac{f_{\mathrm{NEG}} + p - 1}{2p-1}\right]\right). \end{aligned} \tag{10}$$

Then, according to Eq. (2), *the $\frac{f_{\mathrm{POS}}+p-1}{2p-1}$ (resp. $\frac{f_{\mathrm{NEG}}+p-1}{2p-1}$) represents the estimated frequency of number* 1 *(resp.* -1*) before perturbation. According to the bidirectional sampling process, we then have:*

$$\begin{aligned}\mathbb{E}[m^*] &= \frac{1}{n}\cdot\mathbb{E}\left[\#\{i|\,\mathrm{Dis}(v_i) = 1\}\right] - \frac{1}{n}\cdot\mathbb{E}\left[\#\{i|\,\mathrm{Dis}(v_i) = -1\}\right]\\ &= \frac{1}{n}\cdot\sum_{i\in[n]} v_i = m. \end{aligned} \tag{11}$$

We then conclude that m^* is unbiased. Also, the variance of BiSample is given by:

$$\begin{aligned}\mathrm{Var}[m^*] &= \mathbb{E}[(m^*)^2] - (\mathbb{E}[m^*])^2\\ &= \left(\frac{\exp(\epsilon)+1}{\exp(\epsilon)-1}\right)^2 - m^2. \end{aligned} \tag{12}$$

Therefore, the worst-case variance of the BiSample equals $\left(\frac{\exp(\epsilon)+1}{\exp(\epsilon)-1}\right)^2$, which is the same as the Harmony solution. Normally, when using perturbation in d-dimensional data with ϵ-LDP guarantee, the maximum difference between the true mean and the estimated mean is bounded with high probability. Shown as Theorem 3, the proof is similar to the one in [13].

Theorem 3. *For $j \in [d]$, let m_j^* denote the estimator of $m_j = \frac{1}{n}\sum_{i\in[n]} v_{i,j}$ by the BiSample mechanism. With at least probability $1-\beta$, we have:*

$$\left|m_j^* - m_j\right| = O\left(\frac{\sqrt{d\cdot\log(d/\beta)}}{\sqrt{n}\cdot\epsilon}\right). \tag{13}$$

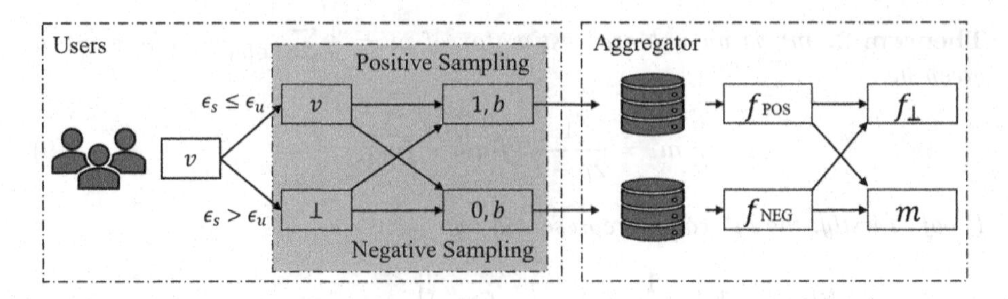

Fig. 2. The BiSample mechanism framework for missing data.

4 BiSample Mechanism for Missing Data Perturbation

The proposed BiSample mechanism uses a bi-directional sampling technique for numerical value perturbation. However, it cannot handle the fake answers. In this section, we consider the **providing null-value procedure** and propose the BiSample-MD framework that extends the BiSample for missing data.

Figure 2 illustrates the BiSample-MD model. We use ϵ_u^i to represent the privacy preference of u_i and use ϵ to represent the privacy budget of the perturbation mechanism provided by the aggregator. Before perturbing value locally, each user uses the $PV(v, \epsilon_u, \epsilon)$ (shown as Algorithm 3) to decide whether using the real value or not. When the privacy-preserving level of perturbation mechanism is higher than user's expectation, the $PV(v_i, \epsilon_u^i, \epsilon)$ returns the real value $v' = v$. Otherwise, the $PV(\cdot)$ returns a null-value $v' = \perp$.

Algorithm 3. $PV(v, \epsilon_u, \epsilon)$: Prepare Value.

Input: user's value $v \in [-1, 1]$ and user's expected privacy budget ϵ_u, system privacy budget ϵ.
1: **if** $\epsilon \leq \epsilon_u$ **then**
2:     **return** v.
3: **else**
4:     **return** $\perp$.
5: **end if**

Then v' is used for perturbing instead of v, the domain of v' is $[-1, 1] \cup \{\perp\}$. We then design the BiSample-MD algorithm for perturbing v'. Even though the input domain is different from that of BiSample, we still design the output domain to be $s \in \{0, 1\}, b \in \{0, 1\}$. The BiSample-MD perturbation process is detailed in Algorithm 4. Like BiSample, the BiSample-MD also contains positive sampling and negative sampling. When $v' = \perp$, both the positive and negative sampling all sample $b = 1$ with probability $1/(\exp(\epsilon)+1)$. The following theorem shows that the BiSample-MD algorithm satisfies ϵ-LDP.

Algorithm 4. *BiSample-MD*$(v, \epsilon_u, \epsilon)$: BiSample for Missing Data.

Input: user's value $v \in [-1, 1]$ and user's expected privacy budget ϵ_u, system privacy budget ϵ.

1: $v' = PV(v, \epsilon_u, \epsilon)$
2: sample a uniformly variable $s \in \{0, 1\}$ representing the sampling direction.
3: **if** $s = 0$ **then**
4: Generate a Bernoulli variable b with:

$$\Pr[b = 1] = \begin{cases} \frac{1-\exp(\epsilon)}{1+\exp(\epsilon)} \cdot \frac{v'}{2} + \frac{1}{2} & \text{if } v' \in [-1, 1]; \\ \frac{1}{\exp(\epsilon)+1} & \text{if } v' = \perp. \end{cases}$$

5: **end if**
6: **if** $s = 1$ **then**
7: Generate a Bernoulli variable b with:

$$\Pr[b = 1] = \begin{cases} \frac{\exp(\epsilon)-1}{\exp(\epsilon)+1} \cdot \frac{v'}{2} + \frac{1}{2} & \text{if } v' \in [-1, 1]; \\ \frac{1}{\exp(\epsilon)+1} & \text{if } v' = \perp. \end{cases}$$

8: **end if**
9: **return** $\langle s, b \rangle$.

Theorem 4. *Algorithm 4 achieves ϵ-LDP.*

Proof. Shown in Theorem 1, it is proven that when $\epsilon_u^i > \epsilon$, the BiSample-MD mechanism is ϵ-LDP. In this way, we only need to consider the situation when the null-value occurs. Without loss of generality, we assume $t_1 = \perp$. we have:

$$\max_{t_2 \in [-1,1], s \in \{0,1\}, b \in \{0,1\}} \left\{ \frac{\Pr[\mathcal{M}(\perp) = \langle s, b \rangle]}{\Pr[\mathcal{M}(t_2) = \langle s, b \rangle]}, \frac{\Pr[\mathcal{M}(t_2) = \langle s, b \rangle]}{\Pr[\mathcal{M}(\perp) = \langle s, b \rangle]} \right\}$$
$$= \max_{t_2 \in [-1,1], b \in \{0,1\}} \frac{\Pr[\mathcal{M}(\perp) = \langle 0, b \rangle]}{\Pr[\mathcal{M}(t_2) = \langle 0, b \rangle]} = \frac{\Pr[\mathcal{M}(\perp) = \langle 0, 0 \rangle]}{\min_{t_2 \in [-1,1]} \Pr[\mathcal{M}(t_2) = \langle 0, 0 \rangle]}.$$

According the perturbation mechanism, the numerator is given by:

$$\Pr[\mathcal{M}(\perp) = \langle 0, 0 \rangle] = \Pr[s = 0] \cdot \Pr[b = 0] = \frac{1}{2} \cdot \frac{\exp(\epsilon)}{1 + \exp(\epsilon)}, \tag{14}$$

while the denominator can be calculated by:

$$\min_{t_2 \in [-1,1]} \Pr[\mathcal{M}(t_2) = \langle 0, 0 \rangle] = \Pr[\mathcal{M}(-1) = \langle 0, 0 \rangle] = \frac{1}{2} \cdot \frac{1}{\exp(\epsilon) + 1}. \tag{15}$$

According to Eq. (14) and Eq. (15), the privacy budget is bounded by ϵ when the value is a null-value. To sum up, the BiSample-MD algorithm is ϵ-LDP.

The perturbed data are then collected by the aggregator. Let s be the sum of values provided truthfully, which is given by $s = \sum_{i:\epsilon_u^i < \epsilon} v_i$ and $f_\perp$ be the fraction of users who provide a null-value for perturbation, which is given by $f_\perp = \#\{i : \epsilon_u^i < \epsilon\}/n$. Then we can estimate s and $f_\perp$ by:

$$s^* = \frac{n}{2p-1} \cdot (f_{\text{POS}} - f_{\text{NEG}}), \qquad f_\perp^* = \frac{1 - f_{\text{POS}} - f_{\text{NEG}}}{2p-1}, \tag{16}$$

where the f_{POS} and f_{NEG} are defined in Eq. 7 and 8. The correctness of the estimation is given by the following theorem.

Theorem 5. *s^* and $f_\perp^*$ are unbiased estimators of s and $f_\perp$.*

Proof. The main intuition behind this theorem is that in the positive sampling process, the perturbed result only contains whether the value is 1 or not. Under such principle, there is no difference between $v = -1$ of $v = \perp$ when using positive sampling. Thus, following the proof of Theorem 2, it is easy to prove that s^* is unbiased. For $f_\perp$ we have:

$$\begin{aligned}
\mathbb{E}[f_\perp^*] &= \mathbb{E}[\frac{1 - f_{\text{POS}} - f_{\text{NEG}}}{2p-1}] \\
&= 1 - \mathbb{E}[\frac{f_{\text{POS}} + p - 1}{2p-1}] - \mathbb{E}[\frac{f_{\text{NEG}} + p - 1}{2p-1}] \\
&= f_\perp.
\end{aligned} \tag{17}$$

Thus, both s^* and $f_\perp^*$ are unbiased.

With the unbiased estimator of the sum of values $\sum_{v_i \neq \perp} v_i$ and the unbiased estimator of the missing rate, we can then estimate the mean by:

$$m^* = \frac{s^*}{n \cdot (1 - f_\perp^*)} = \frac{f_{\text{POS}} - f_{\text{NEG}}}{f_{\text{POS}} + f_{\text{NEG}} + 2p - 2}. \tag{18}$$

5 Experiments

5.1 Experimental Settings

Datasets. To evaluate the proposed mechanisms, we first generated three synthetic datasets: the GAUSS follows Gaussian distribution with location $\mu = 0.5$ and scale $\sigma = 0.1$, the EXP dataset follows Exponential distribution with scale 0.1 and the UNIFORM dataset follows uniform distribution. Each dataset contains 10^5 users. We also use the ADULT dataset [11] for evaluation, in which we extract the Age attribution and regularize each value to $[-1, 1]$ (Table 2).

Table 2. Dataset Description.

Dataset	Distribution	# Instances	Mean value
EXP	Exponential	10^5	−0.831
GAUSS	Gaussian	10^5	0.499
UNIFORM	Uniform	10^5	−0.001
ADULT	–	32561	-0.409

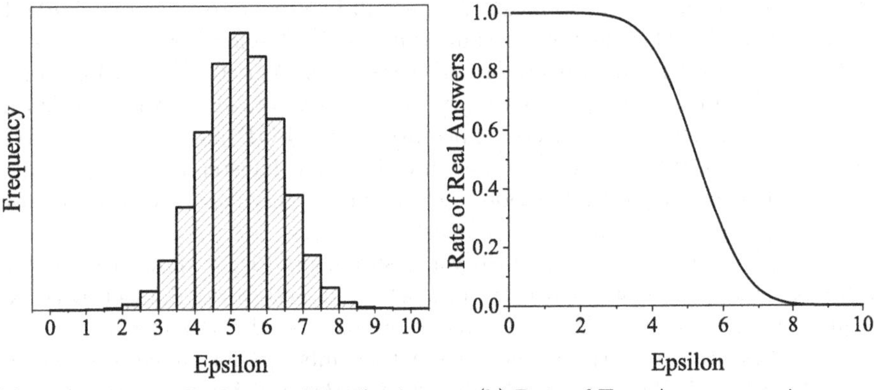

(a) Users' Privacy Preferences Distribution. (b) Rate of True Answers varying ϵ.

Fig. 3. The impact of users' privacy preferences to rate of true answers.

Methodology for Null-Value Perturbation. In terms of mean estimation, we compare the proposed BiSample mechanism with the Harmony [13] and the Piecewise mechanism (PM [17]). For missing data perturbation, we encode the original data to a key-value format. The real value v is represented by $\langle 1, v\rangle$ and the null-value $v = \perp$ is represented by $\langle 0, -\rangle$. We then use PrivKVM [19] for the missing rate estimation and mean estimation. The missing rate is given by $f_\perp = 1 - f_k$, where f_k is the frequency of key given by PrivKVM. We use one real iteration and five virtual iterations.

Utility Metric. All experiments are performed 100 times repeatedly. We evaluate the performance of missing rate (mr) estimation and mean estimation (m) by the average absolute error and variance, which are defined by ($T = 100$):

$$\begin{cases} \mathrm{AE}(mr) = \frac{1}{T}\sum |f_\perp - f_\perp^*|, & \mathrm{AE}(m) = \frac{1}{T}\sum |m - m^*|; \\ \mathrm{Var}(mr) = \frac{1}{T}\sum (f_\perp - f_\perp^*)^2, & \mathrm{Var}(m) = \frac{1}{T}\sum (m - m^*)^2. \end{cases} \tag{19}$$

where $f_\perp$ and m (resp. $f_\perp^*$ and m^*) are the true (resp. estimated) missing rate and the mean value.

5.2 Varying User Behaviors

In this part of experiments, we consider the task of collecting an 1-dimensional value from each user while considering users' privacy preferences. Since no existing solution researched the distribution of users' privacy preferences, shown as Fig. 3(a), we generated Gaussian data with $\mu = 5$ and $\sigma = 1.5$ as the users' preferences distribution. Figure 3(b) plots the rate of users who would truthfully provide the real value for perturbation according to ϵ. Basically, when the privacy budget ϵ is small, the privacy-preserving level provided by the perturbation mechanism is high, so most users would like to share their real value. In contrast, with a high ϵ, few users want to use real value for perturbation as the perturbing process is not privacy-preserving enough. As existing solutions forces an input, we consider two kinds of user behaviors when the privacy-preserving level of LDP is lower than users' expectation: the **TOP** mode and the **RND** mode. In the TOP mode, users always use the value 1 instead of the real value for perturbation. In the RND mode, each user randomly generates a value, uses it for perturbation and shares the perturbed result.

Using the absolute error as utility measurement, Fig. 4 shows the average absolute error over both synthetic datasets and real-world dataset with the change of ϵ, where the TOP and RND represent user behaviors when $\epsilon_u < \epsilon$. For the presentation purpose, methods with similar performance are grouped together in Fig. 4. We first observe that for the PM and Harmony mechanisms, the performance is close to each other. For these two mechanisms, the influence of user behaviors (TOP or RND) is great. In Fig. 4(c), it is a coincidence that the performance of RND-based mechanisms are as good as the BiSample mechanism because for uniformly distributed data. This is explainable as randomly generate a fake answer would not affect the mean value statistically. Moreover, we find out that the absolute error decreases first and increases later with the increase of privacy budget. In the first phase, when the privacy budget is relatively small ($\epsilon < 0.5$), the main estimation errors are caused by perturbation process as the LDP mechanisms provide high-level privacy guarantees. When the privacy budget is big, most of the users would use the null-value for perturbation, which also causes estimation error.

Usually, in conventional settings without consideration of users' privacy preferences, the error of mean estimation becomes smaller with a larger ϵ, as the privacy-preserving level decreases. However, our simulation shows a different opinion. With the increase of privacy budget, the estimation performance would also become poor. As for users, when the privacy budget is too large, they refuse to provide the real value for perturbation because they think their privacy is not well-guaranteed. Also, in this setting, only the PrivKVM and BiSample mechanisms can estimate the missing rate.

5.3 Varying Missing Rate

The prior experiments are based on the assumption that users' privacy preferences follow a Gaussian distribution. The main reason why users' privacy preferences greatly impact the estimation error is that the rate of users who provide

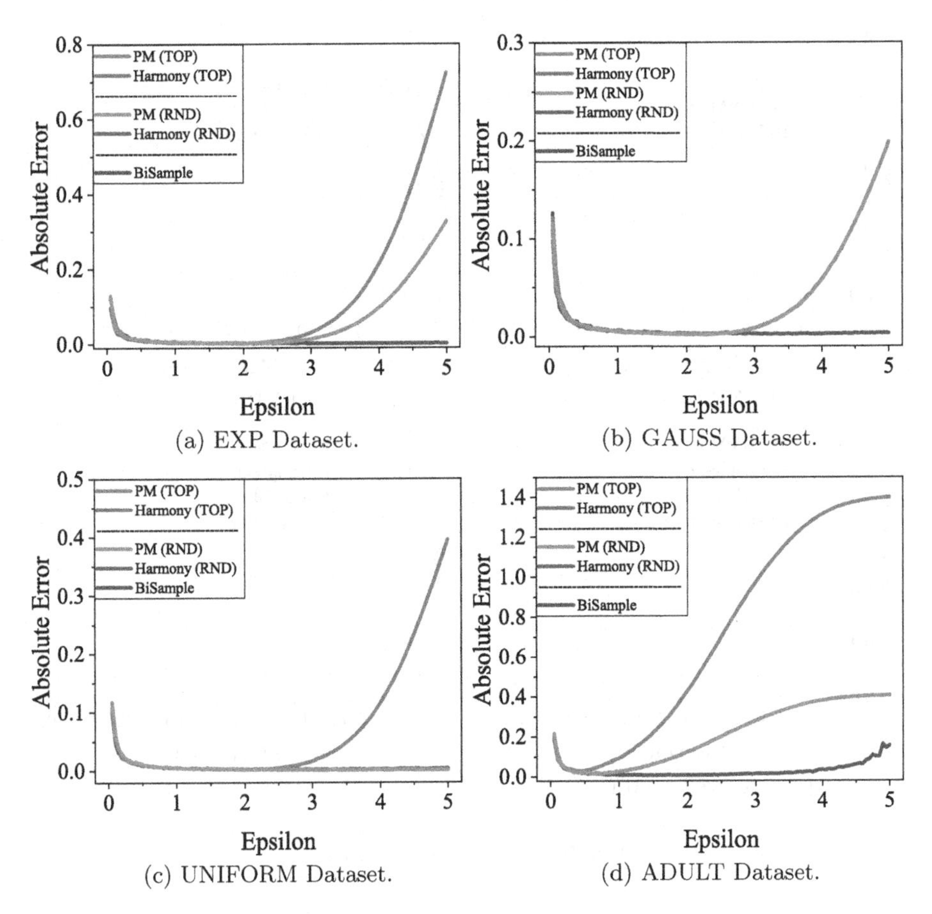

(a) EXP Dataset. (b) GAUSS Dataset.

(c) UNIFORM Dataset. (d) ADULT Dataset.

Fig. 4. Mean estimation on different datasets.

the real value for perturbation changes. Thus in this part of experiment, we directly explore the influence of the missing rate. We fix $\epsilon \in \{0.1, 1\}$ and vary the missing rate to evaluate the estimation error. We only compare the BiSample with PrivKVM as they both can be used for missing rate estimation.

The results are shown in Fig. 5. As expected, for all of the approaches, the utility measurements decrease when the privacy budget increases. When the missing rate is too high, the mean estimation becomes meaningless, as few data can be used. We also observe that for both missing rate estimation (the curve with "−mr") and mean estimation (the curve with "−m"), the proposed BiSample is superior to PrivKVM. We think the reason is that compared with PrivKVM, we only use value for sampling, thus the introduced noise is lower to that of PrivKVM. Another observation is that in each experiment, the missing rate estimation is more accurate than the mean estimation. The reason is that the missing rate estimation only uses typical randomized response, while the

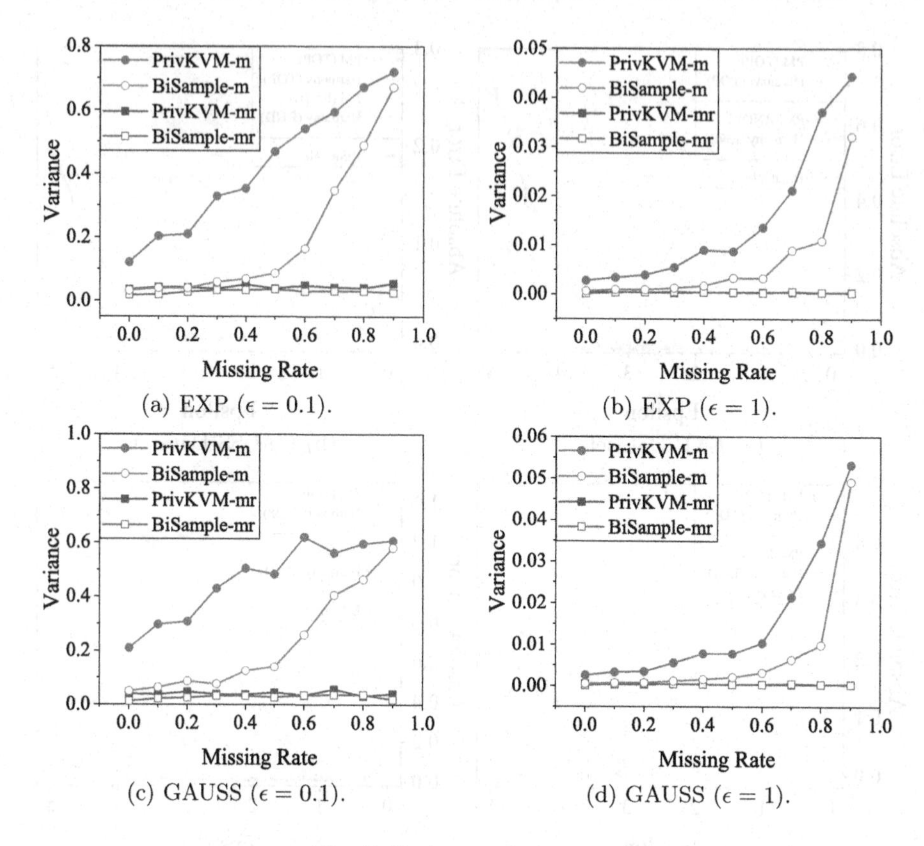

(a) EXP ($\epsilon = 0.1$).

(b) EXP ($\epsilon = 1$).

(c) GAUSS ($\epsilon = 0.1$).

(d) GAUSS ($\epsilon = 1$).

Fig. 5. Variance missing rate.

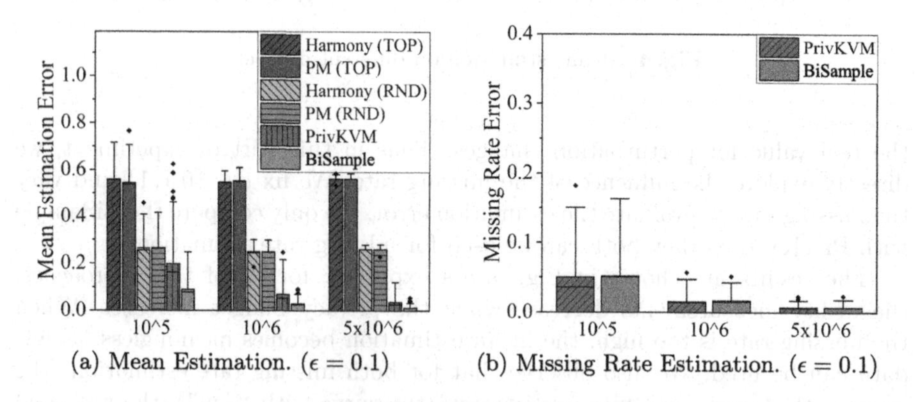

(a) Mean Estimation. ($\epsilon = 0.1$)

(b) Missing Rate Estimation. ($\epsilon = 0.1$)

Fig. 6. Estimation performance varying data size.

perturbation process involves both discretization and perturbation. When $\epsilon = 1$, the variance of missing rate by PrivKVM and BiSample are close.

5.4 Varying Size of Data

We also consider the influence of the size of data on mean estimation and missing rate estimation. We use the absolute error for evaluation. In the mean estimation of Fig. 6(a), we first observe that for PrivKVM and BiSample, the estimation error decreases with a larger size of data. The error of both Harmony and PM is very high because the main inaccuracy is that these two mechanism can not handle the missing data. We observe from Fig. 6(b) that the difference between PrivKVM and BiSample is not obvious in terms of missing rate estimation. Overall, the BiSample outperforms PrivKVM in terms of mean estimation.

6 Conclusion

In this paper, we research the influence of users' privacy-preserving preferences on mean estimation in the framework of LDP. We first propose BiSample, a bidirectional sampling technique for value perturbation. Then users' privacy preferences are considered to avoid fake answers from the user side. Experimental results show that the proposed mechanism can be used for both conventional mean estimation and null-value perturbation with LDP guarantees. As the proposed BiSample mechanism can be used to estimate the rate of users who provide real value under privacy budget ϵ, we plan to study how to set the privacy budget appropriately for the aggregator side in future.

Acknowledgements. This work is supported by National Key Research and Development Program of China (No. 2019QY1402/2016YFB0800901). Jun Zhao's research was supported by Nanyang Technological University (NTU) Startup Grant M4082311.020, Alibaba-NTU Singapore Joint Research Institute (JRI) M4062640. J4I, Singapore Ministry of Education Academic Research Fund Tier 1 RG128/18, RG115/19, and Tier 2 MOE2019-T2-1-176, and NTU-WASP Joint Project M4082443.020.

References

1. Bassily, R., Nissim, K., Stemmer, U., Thakurta, A.G.: Practical locally private heavy hitters. In: Advances in Neural Information Processing Systems, pp. 2288–2296 (2017)
2. Bassily, R., Smith, A.: Local, private, efficient protocols for succinct histograms. In: Proceedings of the Forty-seventh Annual ACM Symposium on Theory of Computing, pp. 127–135. ACM (2015)
3. Cormode, G., Kulkarni, T., Srivastava, D.: Marginal release under local differential privacy. In: Proceedings of the 2018 International Conference on Management of Data, pp. 131–146. ACM (2018)
4. Ding, B., Kulkarni, J., Yekhanin, S.: Collecting telemetry data privately. In: Advances in Neural Information Processing Systems, pp. 3571–3580 (2017)
5. Duchi, J.C., Jordan, M.I., Wainwright, M.J.: Local privacy and statistical minimax rates. In: 2013 IEEE 54th Annual Symposium on Foundations of Computer Science, pp. 429–438. IEEE (2013)

6. Duchi, J.C., Jordan, M.I., Wainwright, M.J.: Privacy aware learning. J. ACM (JACM) **61**(6), 38 (2014)
7. Duchi, J.C., Jordan, M.I., Wainwright, M.J.: Minimax optimal procedures for locally private estimation. J. Am. Stat. Assoc. **113**(521), 182–201 (2018)
8. Dwork, C., Roth, A., et al.: The algorithmic foundations of differential privacy. Found. Trends® Theoret. Comput. Sci. **9**(3–4), 211–407 (2014)
9. Erlingsson, Ú., Pihur, V., Korolova, A.: Rappor: randomized aggregatable privacy-preserving ordinal response. In: Proceedings of the 2014 ACM SIGSAC Conference on Computer and Communications Security, pp. 1054–1067. ACM (2014)
10. Kairouz, P., Oh, S., Viswanath, P.: Extremal mechanisms for local differential privacy. In: Advances in Neural Information Processing Systems, pp. 2879–2887 (2014)
11. Kohavi, R., Becker, B.: UCI repository of machine learning databases: Adult data set (1999). https://archive.ics.uci.edu/ml/datasets/Adult
12. Li, N., Li, T., Venkatasubramanian, S.: t-closeness: Privacy beyond k-anonymity and l-diversity. In: 2007 IEEE 23rd International Conference on Data Engineering, pp. 106–115. IEEE (2007)
13. Nguyên, T.T., Xiao, X., Yang, Y., Hui, S.C., Shin, H., Shin, J.: Collecting and analyzing data from smart device users with local differential privacy. arXiv preprint arXiv:1606.05053 (2016)
14. Qin, Z., Yang, Y., Yu, T., Khalil, I., Xiao, X., Ren, K.: Heavy hitter estimation over set-valued data with local differential privacy. In: Proceedings of the 2016 ACM SIGSAC Conference on Computer and Communications Security, pp. 192–203. ACM (2016)
15. Ren, X., et al.: LoPub: high-dimensional crowdsourced data publication with local differential privacy. IEEE Trans. Inf. Forensics Secur. **13**(9), 2151–2166 (2018)
16. Samarati, P., Sweeney, L.: Protecting privacy when disclosing information: k-anonymity and its enforcement through generalization and suppression. Technical report, SRI International (1998)
17. Wang, N., et al.: Collecting and analyzing multidimensional data with local differential privacy. In: 2019 IEEE 35th International Conference on Data Engineering (ICDE), pp. 638–649. IEEE (2019)
18. Warner, S.L.: Randomized response: a survey technique for eliminating evasive answer bias. J. Am. Stat. Assoc. **60**(309), 63–69 (1965). https://doi.org/10.1080/01621459.1965.10480775. http://www.tandfonline.com/doi/abs/10.1080/01621459.1965.10480775
19. Ye, Q., Hu, H., Meng, X., Zheng, H.: PrivKV: key-value data collection with local differential privacy. In: IEEE Symposium on Security and Privacy (SP), May 2019

PrivGMM: Probability Density Estimation with Local Differential Privacy

Xinrong Diao[1], Wei Yang[1(✉)], Shaowei Wang[2(✉)], Liusheng Huang[1], and Yang Xu[1]

[1] School of Computer Science and Technology, University of Science and Technology of China, Hefei, China
qubit@ustc.edu.cn
[2] Tencent Games, Shenzhen, China
seawellwang@tencent.com

Abstract. Probability density estimation is a fundamental task in data analysis that can estimate the unobservable underlying probability density function from the observed data. However, the data used for density estimation may contain sensitive information, and the public of original data will compromise individuals' privacy. To address this problem, we in this paper propose a private parametric probability density estimation mechanism, called PrivGMM. It provides strong privacy guarantees locally (e.g., on personal computers or mobile phones) and efficiently (i.e., computation cost is small) for users. Meanwhile, it provides an accurate estimation of parameters of the probability density model for data collectors. Specifically, in a local setting, each user adds noise to his/her original data, given the constraint of local differential privacy. On the server side, we employ the Gaussian Mixture Model, which is a popular model to approximate distributions. To reduce the effect of noise, we formulate the parametric estimation problem with a multi-layer latent variables structure, and utilize Expectation-Maximization algorithm to solve the Gaussian Mixture Model. Experiments in real datasets validate that our mechanism outperforms the state-of-the-art methods.

Keywords: Local differential privacy · Probability density estimation · Gaussian Mixture Model

1 Introduction

Currently, with the development of big data analysis, service providers are keen on collecting user data to perform analysis and improve their services. Density estimation is a fundamental task in data analysis and can provide abundant information. For example, the trend in users' long-term credit card spending records could be used by banks to evaluate users' credit ratings.

In some situations, the density distribution can be modeled in simple distributions, such as exponential distributions or power-law distributions. However,

Y. Nah et al. (Eds.): DASFAA 2020, LNCS 12112, pp. 105–121, 2020.
https://doi.org/10.1007/978-3-030-59410-7_7

in realistic cases, most density distributions are complex, and there is no suitable parametric form to describe them. Therefore, in this paper, we consider the Gaussian Mixture Models (GMM) [1,4,19–21], which is a popular parametric density estimation model in statistics. In GMM, all data are assumed to be drawn by randomly sampling from one of K Gaussian distributions $G_1, G_2, ..., G_K$, and the goal of the learning task is to calculate the parameters (weight, mean, and variance) of each Gaussian distribution. In theory, GMM could approximate any distribution.

Density estimation without privacy concerns is well studied in the literature. However, with the perfection of laws and users' increasing awareness of privacy protection, users may want to maintain plausible deniability with their sensitive data. Directly collecting and publishing parameters of the probability density function will sacrifice privacy and no longer satisfy users' real-life needs for data privacy protection.

Some work has been done on privacy preserving GMM. However, most of them [17,22,24] are considered in the centralized model, which means there exists a trusted curator and users will send their unprocessed data to the data collector. The collector uses true data to estimate the parameters of density distribution and add noise to parameters before publishing. In reality, the assumption of a trusted curator will increase privacy leakage risk because collectors may trade users' data for commercial benefit.

Recently, as an answer to privacy protection in local settings, local differential privacy has been proposed to perturb data locally, and it [7] has been implemented by Google [11], Apple [12] and Microsoft [6]. Its main idea is that users perturb data at the client side and send perturbed data to the collector. Compared with central differential privacy, local differential privacy provides privacy protection control to users and therefore provides stronger privacy guarantees.

Although local differential privacy preservation on density estimation has been studied, most of these studies focus on simple distributions such as Gaussian distributions (e.g., [15]) and power-law distributions (e.g., [13]). In reality, most distributions are more complicated and cannot be described with a single simple distribution. Therefore, we need a more universal method to handle complicated distributions under local differential privacy. To the best of our knowledge, no existing work has focused on GMM with local differential privacy. In particular, consider the problem of recovering parameters of GMM, where (i) K Gaussian distributions compose the GMM, and (ii) each sample belongs to which Gaussian distribution is not clear, and (iii) the true data of each user is unknown, we can only obtain perturbed (observed) data, and (iv) we aim to estimate the mean, variance and weight of each Gaussian distribution. If we extend existing local differential privacy-preserving methods to the GMM privacy-preserving problem, such as perturbing data locally and directly using perturbed data to perform estimation work, it will incur low accuracy because the perturbed dataset can only provide a perturbed distribution.

To address the problem discussed above, we propose PrivGMM, a novel mechanism that accurately estimates parameters of GMM while satisfying local differential privacy. In this mechanism, user data are sanitized by adding Gaussian noise, which means that the noise is randomly sampled from a Gaussian distribution. When data are collected in the data collector, we extend the Expectation-Maximization (EM) algorithm and GMM by appending an extra latent variable (noise). The noise latent variable denotes that the mechanism will estimate the data's noise in each iteration, thus obtaining estimated true data to calculate the parameter of true distribution.

The remainder of the paper is organized as follows. Section 2 introduces preliminaries. Section 3 reviews the relevant work. Section 4 presents the construction of the PrivGMM. Section 5 presents and analyzes the experimental performance. Finally, Sect. 6 concludes the work and discusses future works.

Table 1. Notations

Notation	Description
D	Dataset
D'	Neighborhood dataset of D
N	The number of data $N = \|D\|$
y_i	The i-th data in D
K	Number of Gaussian distributions
ϵ	Privacy budget
δ	Approximate leakage probability
$\tilde{\alpha_j}$, $\tilde{\mu_j}$, $\tilde{\sigma_j^2}$	j-th Gaussian distribution parameters

2 Preliminaries

2.1 Local Differential Privacy

The main notations used in this paper are listed in Table 1. The protection model under local differential privacy [7] fully considers the possibility that the data collector steals or divulges user privacy during data collection. In the local model, each user perturbs his own data before sending the data to the collector. More formally, let D denote the database, privacy budget $\epsilon > 0$, $0 < \delta < 1$.

Definition 1 *(Local differential privacy). An algorithm $\Psi : X \rightarrow R$ satisfies (ϵ, δ)-local differential privacy if and only if for any two input elements $t, t' \in X$ and for any output t^*, the following inequality always holds.*

$$Pr[(\Psi(t) = t^*] \leq \exp(\epsilon) \cdot Pr[\Psi(t') = t^*] + \delta. \tag{1}$$

The variant δ is introduced by [9]; this variant ensures that the privacy loss is bounded by ϵ with a probability of at least $1-\delta$. When $\delta > 0$, it is called approximate local differential privacy, and when $\delta = 0$, it is called pure local differential privacy.

Intuitively, local differential privacy guarantees that given the output t^*, the adversary cannot infer that the input of randomized algorithm Ψ is t or t' at a high confidence level. It controls the difference between two single elements' output in one database.

There is a post-processing theorem [10] about local differential privacy that means that if algorithm Ψ is (ϵ, δ)-local differential privacy, then a data analysis of the results of the algorithm cannot increase privacy loss.

Theorem 1 (Postprocessing). *Let $A : R \rightarrow R'$ be a randomized mapping function; let $\Psi : X^m \rightarrow R$ be a (ϵ, δ)-local differential privacy algorithm, then $A \cdot \Psi : X^m \rightarrow R'$ is (ϵ, δ)-local differential privacy.*

2.2 Gaussian Mechanism

To satisfy differential privacy, several fundamental mechanisms exist for different data types. For numerical attributes, a common method for approximating a numerical function $f : D \rightarrow R$ is to add noise, which is calibrated with a sensitivity of f. The sensitivity of f, S_f is defined as the maximum magnitude of $|f(d) - f(d')|$, where d and d' are adjacent inputs. The Gaussian mechanism [25] is a popular technique to implement (ϵ, δ)-differential privacy, which adds noise random sampled from a zero-mean Gaussian distribution to users' data. In particular, the true data of user u_i is noted as x_i, and the perturbed result is y_i, and the Gaussian distribution is defined as

$$y_i = x_i + N(0, \sigma^2). \tag{2}$$

More formally, the application of the Gaussian mechanism needs to satisfy the following constraints:

Theorem 2 (Gaussian Mechanism). *Let $0 < \delta < 0.5$, for $c = \sqrt{ln\frac{2}{\sqrt{16\delta+1}-1}}$, the Gaussian Mechanism with parameter $\sigma = \frac{(c+\sqrt{c^2+\epsilon})S_f}{\epsilon\sqrt{2}}$ is (ϵ, δ)-differentially private.*

2.3 Gaussian Mixture Model and Expectation-Maximization Algorithm

GMM (Gaussian Mixture Model). As a frequently used model of density estimation, GMM is used to accurately quantify density distributions with Gaussian probability density functions, which is a model formed by dividing distributions into several parts based on Gaussian probability density functions. Formally, let D denote a dataset containing N records $y_1, y_2, ..., y_N$, GMM has the following probability distribution model:

$$P(y|\Theta) = \sum_{j=1}^{K} \alpha_j \Phi(y|\Theta_j), \tag{3}$$

with weights obey $\sum_{j=1}^{K} \alpha_j$; $\Phi(y|\Theta_j)$ is Gaussian distribution density and $\Theta_j = (\mu_j, \sigma_j^2)$ are the mean and variance of j-th Gaussian distribution.

We do not know the observed data y_i generated by which Gaussian distribution. To address this problem, we introduce a latent variable β_{ij}. If record y_i generated by j-th Gaussian distribution, then $\beta_{ij} = 1$; otherwise, $\beta_{ij} = 0$. Combined observed data y_i and unobserved data β_{ij}, the log-likelihood function of full data can be written as follows:

$$log(P(y, \beta, z|\Theta)) = \sum_{j=1}^{K} \left\{ n_j log\alpha_j + \sum_{i=1}^{N} \beta_{ij} log[\frac{1}{\sqrt{2\pi}\sigma_j} exp(\frac{(y_i - \mu_j)^2}{2\sigma_j^2})] \right\}, \tag{4}$$

where $n_j = \sum_{i=1}^{N} \beta_{ij}$. We cannot use maximum likelihood estimation to solve this equation because of the presence of summation in the logarithm operation.

EM (Expectation-Maximization) Algorithm. To address this problem, EM algorithm is proposed because it is efficient and powerful. EM algorithm is used to estimate maximum likelihood of the probability model parameter with latent variable. Specifically, it estimates the latent variables (E-step) and update the parameter of the model (M-step). The EM algorithm executes these two steps until convergence or reaches the maximum number of iterations. Here are the update results in each iteration of EM algorithm, and it is the basis of our work. These repeated two steps will converge to a local maximum, as proven by [5].

E-step:

$$\beta_{ij} = \frac{\alpha_j \Phi(y_i|\Theta_j)}{\sum_{j=1}^{K} \alpha_j \Phi(y_i|\Theta_j)}, i = 1, 2, ..., N; j = 1, 2, ..., K \tag{5}$$

M-step:

$$\alpha_j = \frac{\sum_{i=1}^{N} \beta_{ij}}{N}, j = 1, 2, ..., K \tag{6}$$

$$\mu_k = \frac{\sum_{i=1}^{N} \beta_{ij} y_i}{\sum_{i=1}^{N} \beta_{ij}}, j = 1, 2, ..., K \tag{7}$$

$$\sigma_j^2 = \frac{\sum_{i=1}^{N} \beta_{ij} (y_i - \mu_j)^2}{\sum_{i=1}^{N} \beta_{ij}}, j = 1, 2, ..., K \tag{8}$$

3 Related Work

Some works have already been performed on differentially private versions of density estimation, especially in the central setting. We introduce some representative works here. Karwa et al. [18] and Kamath et al. [16] studied the Gaussian distribution under central differential privacy and focused on one-dimensional data distributions and high-dimensional cases respectively. Because the above method can handle only specific and simple distributions, such as the Gaussian distribution, Wu et al. [24] proposed a DPGMM mechanism, which considers GMM under central differential privacy. The key idea is to add Laplace noise to the parameter in each iteration of the EM algorithm. Kamath et al. [17] also studied GMM under central differential privacy and focused on high-dimensional data with well-separated Gaussian distributions.

In local setting, Joseph et al. [14] proposed a mechanism which considers single Gaussian distribution under local differential privacy. They proposed both adaptive two-round solutions and nonadaptive one-round solutions for locally private Gaussian estimation. However, these methods are limited in specific distribution and to the best of our knowledge, there is no specific work about universal parametric density estimation under local differential privacy. In this paper, we study the estimation of GMM parameters. A relevant work is a paper of Duchi et al. [8]. They proposed a nonparametric method to estimate probability density function which is no request for prior knowledge. However, the nonparametric method is not accurate enough and will lose some underlying information of the original distributions. In contrast, our PrivGMM leverage prior knowledge of noise distribution and true data distribution to calibrate the density estimation.

4 PrivGMM

In this section, we present our mechanism PrivGMM, which is a density estimation mechanism with learning parameters of GMM under local differential privacy. In the following, we will present our main idea in Sect. 4.1. Then we propose local privacy preservation in Sect. 4.2. In Sect. 4.3, we show the density estimation aggregation algorithm. Section 4.4 presents the entire algorithm. Finally, we show our theoretical analysis in Sect. 4.5.

4.1 Main Idea

The main idea of PrivGMM includes a random component for each user and a density estimator for server side. Let x_i denote the i-th user's original data, namely, true data. To preserve privacy, we use the Gaussian mechanism in local setting, which means each user randomly sample a noise from the specified Gaussian distribution and adds it to his/her true data. Thus, each user obtains the perturbed data y_i for delivery to the server. On server side, we extend GMM and EM algorithm with multi-layer variables structure. Specifically, after aggregating data from all users, in the E-step of modified algorithm, we append a latent variable z_i to denote the estimate of noise of i-th data in each iteration and calculate its conditional distribution given the current model and observed data and obtain the new Q function. In the M-step of modified algorithm, we calculate the parameters which maximize the Q function. In general, we propose a practical density estimation mechanism under local differential privacy.

4.2 Data Randomization

Given a record x_i, a popular randomization method f is to apply a Gaussian mechanism [25]. Let $y_i = f(x_i)$, and because we consider the data to have a numerical domain $[0, 1]$, we have the sensitivity $S_f = 1$. Each user u_i can obtain his/her perturbed data y_i:

$$y_i = x_i + z_i, \tag{9}$$

where z_i denotes a random variable that follows a Gaussian distribution with zero mean and variance σ^2, with the following probability density:

$$pfd(z_i) = \frac{1}{\sqrt{2\pi}\sigma} exp(\frac{z_i}{2\sigma^2})^2. \tag{10}$$

σ^2 is calculated by the Gaussian Mechanism we described in Sect. 2.2 given privacy budget ϵ, δ and S_f. Through this step, each user can obtain (ϵ, δ)-LDP private view y_i and deliver it to the server side. According to Theorem 1 described in the background, the analysis procedure executed in the collector will not increase the privacy cost. So the whole mechanism satisfies (ϵ, δ)-LDP.

4.3 Parameter Estimation

When the server aggregates all observed data $Y = y_1, y_2, ..., y_N$, it is not a good idea to execute the EM algorithm directly because the noise is large and we do not know the exact value of it, so the noise will disturb estimation results. Therefore, we view the noise of each data as missing data and add one more latent variable z_i to the original model. Therefore, we view observed data y_i and two latent variables z_i and β_{ij} as full data. The value of β_{ij} is 1 or 0. If y_i is from j-th Gaussian, then $\beta_{ij} = 1$. The likelihood function of the full data is updated as follows:

$$\begin{aligned} P(y, \beta, z|\Theta) &= \prod_{i=1}^{N} P(y_i, \beta_{i1}, \beta_{i2}, ..., \beta_{ij}, z_i|\Theta) \\ &= \prod_{j=1}^{K} \prod_{i=1}^{N} [\alpha_j \Phi(y_i - z_i|\Theta_j)]^{\beta_{ij}}. \end{aligned} \tag{11}$$

Next, the log-likelihood function of the full data is updated as:

$$
\begin{aligned}
&logP(y,\beta,z|\Theta)\\
&=\sum_{j=1}^{K}\left\{n_j log\alpha_j+\sum_{i=1}^{N}\beta_{ij}[log(\frac{1}{\sqrt{2\pi}})-log\sigma_j-\frac{1}{2\sigma_j^2}(y_i-z_i-\mu_j)^2]\right\},
\end{aligned}
\tag{12}
$$

where $n_k=\sum_{i=1}^{N}\beta_{ij}$, $\sum_{j=1}^{K}n_j=N$.

E-step. In E-step, we calculate the expected value of Eq. (12) to obtain the Q-function.

$$
\begin{aligned}
&Q(\theta,\theta^{(t)})\\
&=E(logP(y,\beta,z|\theta)|y,\theta^{(t)})\\
&=\sum_{j=1}^{K}\left\{\sum_{i=1}^{N}E(\beta_{ij})log\alpha_j+\sum_{i=1}^{N}E(\beta_{ij})[log(\frac{1}{\sqrt{2\pi}})-log\alpha_j-\frac{1}{2\sigma_j^2}E(y_i-z_i-\mu_j)^2]\right\}.
\end{aligned}
\tag{13}
$$

We analyze this equation and find that there are three additional computation steps before M-step. First, we need to calculate the expected value of β_{ij} given the observed data y and the current model parameter $\theta^{(t)}$. The value of $E(\beta_{ij}|y,\theta)$ equals to the probability of y_i drawn from the j-th perturbed Gaussian distribution component given the current model parameter.

$$
E(\beta_{ij}|y,\theta)=\frac{\alpha_j\Phi(y_i|\mu_j,\sigma_j^2+\sigma^2)}{\sum_{j=1}^{K}\alpha_j\Phi(y_i|\mu_j,\sigma_j^2+\sigma^2)}. \tag{14}
$$

Next, we need to calculate the expected value of z_i and z_i^2. In local settings, we use the Gaussian mechanism to protect numerical data. We add noise z_i, which is randomly sampled from a standard Gaussian distribution with parameter σ^2. We have prior knowledge that the user u_i true data x_i follows one of the K Gaussian distributions $N_1(\mu_j,\sigma_j^2)$. Through the Gaussian mechanism, we know that the noise of z_i is drawn from a Gaussian distribution with zero mean and known variance $N_2(0,\sigma^2)$. Therefore, the observed data y_i also follow a Gaussian distribution $N_3(\mu_j,\sigma_j^2+\sigma^2)$, which is composed of N_1 and N_2. Considering the joint distribution of the noises and the observed data, we can obtain the conditional distribution of noise given the observations where the V is the covariance of noises and observed data.

$$
\begin{aligned}
\mu_{noise|obs}&=\mu_{noise}+V_{noise,obs}V_{obs,obs}^{-1}(obs-\mu_{obs})\\
&=\frac{\sigma^2(y_i-\mu_j)}{\sigma_j^2+\sigma^2},
\end{aligned}
\tag{15}
$$

$$V_{noise|obs} = V_{noise,noise} - V_{noise,obs}V_{obs,obs}^{-1}V_{obs,noise} = \frac{\sigma^2\sigma_j^2}{\sigma^2+\sigma_j^2}. \quad (16)$$

Now, we can use conditional expectation (15) and conditional variance (16) to calculate the expected value of z_i and z_i^2. We denote them as m_{ij} and v_{ij}.

$$m_{ij} = \mu_{noise|obs}, \quad (17)$$

$$v_{ij} = D(z_i|y_i,\theta) + E^2[z_i|y_i,\theta] = \frac{\sigma^2\sigma_j^2}{\sigma^2+\sigma_j^2} + [\frac{\sigma^2(y_i-\mu j)}{\sigma_j^2+\sigma^2}]^2. \quad (18)$$

Substitute Eq. (17) and Eq. (18) into Eq. (13). We obtain the Q function of our EM algorithm.

$$Q(\theta,\theta^{(t)}) = \sum_{j=1}^{K} n_j log\alpha_j + \sum_{i=1}^{N}\beta_{ij}[log\frac{1}{\sqrt{2\pi}} - log\sigma_j - \frac{(y_i-\mu_j)^2 - 2m_{ij}(y_i-\mu_j)+v_{ij}}{2\sigma_j^2}]. \quad (19)$$

M-step. In the maximum step, we calculate the value of θ, which can maximize the Q function. Take the partial derivative of α_j, μ_j and σ_j with the Q function and make the partial derivative equal to zero with limitations that $\sum_{j=1}^{K}\tilde{\alpha_j} = 1$.

$$\tilde{\alpha_j} = \frac{\sum_{i=1}^{N}\beta_{ij}}{N}, \quad (20)$$

$$\tilde{\mu_j} = \frac{\sum_{i=1}^{N}\beta_{ij}(y_i - m_{ij})}{\sum_{i=1}^{N}\beta_{ij}}, \quad (21)$$

$$\tilde{\sigma_j^2} = \frac{\sum_{i=1}^{N}\beta_{ij}[(y_i-\mu_j)^2 - 2m_{ij}(y_i-\mu_j)+v_{ij}]}{\sum_{i=1}^{N}\beta_{ij}}. \quad (22)$$

4.4 PrivGMM

We present our whole mechanism in Algorithm 1 and Algorithm 2, respectively. They correspond the local perturbed procedure and estimation procedure respectively. In Algorithm 1, at lines 1, given privacy budget ϵ, sensitivity S_f and δ we calculate variance of noise. At Line 2, we random sample noise z_i from $N(0,\sigma^2)$. At Line 3, we generate private views y_i of original data x_i. In Algorithm 2, at line 1, we initialize parameters of algorithm. At line 3–10, we execute E-step, calculate latent variables and Q function. And at line 12–16 we execute M-step, calculate parameters in this iteration. At line 17–19, judge whether the algorithm is convergence. Algorithm 2 executes until convergence or reach the maximum number of iterations.

Algorithm 1. Local privacy preservation component of (ϵ,δ)-PrivGMM mechanism.

Input:
User u_i's true data x_i;
Privacy budget ϵ;
Sensitivity S_f;
Failing probability δ
Output:
User u_i's perturbed data, y_i;
1: Calculate $c = \sqrt{ln\frac{2}{\sqrt{16\delta+1}-1}}$.
2: Calculate $\sigma = \frac{(c+\sqrt{c^2+\epsilon})S_f}{\epsilon\sqrt{2}}$.
3: $z_i \sim N(0, \sigma^2)$;
4: $y_i = x_i + z_i$;
5: **return** y_i;

4.5 Theoretical Analysis

Complexity Analysis. Compared with the non-private EM algorithm, PrivGMM appends extra calculation for the latent variable. The time complexity of the nonprivate EM algorithm is $O(NK)$ in each iteration, and the time complexity of calculating the latent variable is also $O(NK)$, so the total time complexity of PrivGMM in each iteration is $O(NK)$.

PrivGMM Maximizes the Likelihood. In this section, we prove that our algorithm maximizes the likelihood of full data (y, z, β) meanwhile maximizing the likelihood of observed data y. Suppose that dataset $y_1, y_2, ..., y_N$, given sample y_i, we assume that there is one distribution function Q_i of latent variables z_i and j. Q_i satisfies that $\sum_j \int_{z_i} Q_i(z_i, j)dz = 1$ and $Q_i(z_i, j) \geq 0$. The likelihood function with observed data can be written as follows:

$$\begin{aligned} logp(y|\theta) &= \sum_{i=1}^{N} logp(y_i|\theta) = \sum_{i=1}^{N} log \sum_j \int_z dzp(y_i, z_i, j|\theta) \\ &= \sum_{i=1}^{N} log \sum_j \int_z dz \frac{p(y_i, z_i, j|\theta)}{Q_i(z_i, j)} Q_i(z_i, j) \\ &\geq \sum_{i=1}^{N} \sum_j \int_z dzQ_i(z_i, j) log \frac{p(y_i, z_i, j|\theta)}{Q_i(z_i, j)} \end{aligned} \tag{23}$$

We get this inequality using Jensen's inequality and this inequality is equal when

$$Q_i(z_i, j) = p(z_i, j|y, \theta). \tag{24}$$

Algorithm 2. Density estimation component of (ϵ,δ)-PrivGMM mechanism.

Input:
N users' observed data, $Y = \{y_1, y_2, ..., y_N\}$;
Number of Gaussian distributions, K ;
Convergence Threshold EPS
Maximum number of iterations T;

Output:
Estimation of K Gaussian distributions' parameter $\theta = \left\{\tilde{\alpha_1}, ..., \tilde{\alpha_K}, \tilde{\mu_1}, ..., \tilde{\mu_K}, \tilde{\sigma_1^2}, ..., \tilde{\sigma_K^2}\right\}$;

1: Initialize parameters;
2: **for** $t = 1$ to T **do**
3: E-step
4: **for** $i = 1$ to N **do**
5: **for** $j = 1$ to K **do**
6: $\hat{\beta_{ij}} = \frac{\alpha_j \Phi(y_i|\mu_j, \sigma_j^2+\sigma^2)}{\sum_{j=1}^{K} \alpha_j \Phi(y_i|\mu_j, \sigma_j^2+\sigma^2)}$
7: $m_{ij} = \frac{\sigma^2(y_i-\mu_j)}{\sigma_j^2+\sigma^2}$
8: $v_{ij} = \frac{\sigma^2\sigma_j^2}{\sigma^2+\sigma_j^2} + [\frac{\sigma^2(y_i-\mu j)}{\sigma_j^2+\sigma^2}]^2$
9: **end for**
10: **end for**
11: M-step
12: **for** j=1 to K **do**
13: $\tilde{\alpha_j} = \frac{\sum_{i=1}^{N} \beta_{ij}}{N}$
14: $\tilde{\mu_j} = \frac{\sum_{i=1}^{N} \beta_{ij}(y_i-m_{ij})}{\sum_{i=1}^{N} \beta_{ij}}$
15: $\tilde{\sigma_j^2} = \frac{\sum_{i=1}^{N} \beta_{ij}[(y_i-\mu_j)^2-2m_{ij}(y_i-\mu_j)+v_{ij}]}{\sum_{i=1}^{N} \beta_{ij}}$
16: **end for**
17: **if** $|\theta^t - \theta^{t-1}| < EPS$: **then**
18: return θ^t
19: **end if**
20: **end for**
21: **return** θ^T;

Combined with this equation, we omit the term which has no influence on optimization and get the function H:

$$
\begin{aligned}
H &= \sum_{ij} \beta_{ij} \int_z dzp(z_i|y_i, j, \theta) logp(y_i, z_i, j|\theta) \\
&= \sum_{ij} \beta_{ij} \int_z dzp(z_i|y_i, j, \theta) logp(z_i|y)p(y|j, \theta)p(j|\theta) \\
&= \sum_{ij} \beta_{ij} \int_z dzp(z_i|y_i, j, \theta)[log(\frac{1}{\sqrt{2\pi}}) - log\alpha_j - \frac{1}{2\sigma_j^2}(y_i - z_i - \mu_j)^2]
\end{aligned}
\tag{25}
$$

Comparing this function with the Q function of our mechanism Eq. (19), we can observe that they both calculate the expected values of z_i and z_i^2 given observed data y_i and the parameters of the current model in E-step. In M-step, as the log likelihood of our model increases, the log likelihood of model with observed data also increases. Therefore, the likelihood of observed data increases monotonically and our algorithm will reach a local maxima of the likelihood.

5 Experiments

In this section, we present the performance of our method. We implement our mechanism in Python and performed all experiments on a desktop computer with Intel Core i5-3470 3.20 GHz CPU.

5.1 Experimental Settings

General Settings. We evaluate PrivGMM on two real datasets from the UCI Machine Learning Repository. The Tamilnadu Electricity Board Hourly Readings dataset and the Anuran Calls (MFCCs) dataset, are real numerical datasets and used frequently in their fields. The number of records in Tamilnadu dataset is 45871 and the number of records in MFCCs dataset is 7195. We choose one dimension from the two datasets respectively. For all datasets, we normalize the domain of attributes into [0, 1]. In experiments, we set the number of Gaussian distributions K ranges from 1 to 7, $\delta = 0.01$, and the privacy budget ϵ ranges from 1 to 10.

Initialization. The EM algorithm can ensure that the likelihood function will converge to a local maximum but it cannot ensure likelihood will convergence to a global maximum, as we prove earlier. Thus, initialization has a strong effect on the results. Some research has studied the initialization of GMM [2,3]. However, they are not applicable to PrivGMM because the true data are unavailable to the data collector. Therefore, we use a simple method to initialize our parameters. Given K, we assume each Gaussian distribution has same weight. We divide the data domain into K segments and initialize the mean as each segment's midpoint. We assume that each Gaussian distribution can cover each segment but does not overlap strongly, so we set the following parameters:

$$alpha_j = \frac{1}{K}, \tag{26}$$

$$\mu_j = \frac{2j+1}{2K}, j = 0, 1, ..., K-1, \tag{27}$$

$$\sigma_j^2 = (\frac{1}{4K})^2. \tag{28}$$

Metrics. We use the metric to measure the accuracy of our density estimation function, which is popular in the literature, Integrated Square Error (ISE) [23].

ISE measures the absolute error of density estimation with respect to the real probability density function. Specifically, $\widehat{f}$ is the estimation function, and f is the real density function. We obtain the following error measure method:

$$ISE = ||\widehat{f} - f||_2^2 = \int_x (\widehat{f}(x) - f(x))^2 dx. \tag{29}$$

5.2 Results

In this section, we compare PrivGMM with a state-of-the-art method in [8](denoted as NDM) , and we evaluate the effect of parameters K in PrivGMM.

Compare with *NDM*. In the local differential privacy preserving density estimation field, there is not much research in this field. There is a relevant work [8] denoted as NDM. NDM is based on the nonparametric density estimation method, so the result of this mechanism is rough and cannot tell the underlying knowledge of the target distribution. From Fig. 1 and Fig. 2, we can observe that our mechanism PrivGMM performs much better than NDM in both the Tamilnadu dataset and the MFCCs dataset. NDM has many restrictions for objective function so the reason of bad performance of NDM is maybe these two real datasets' distribution do not meet the requirements of NDM. For presenting our mechanism results more concrete, we present our results in Fig. 3(a) (b) separately. From Fig. 3, we can observe that as the ϵ increases, PrivGMM performs better.

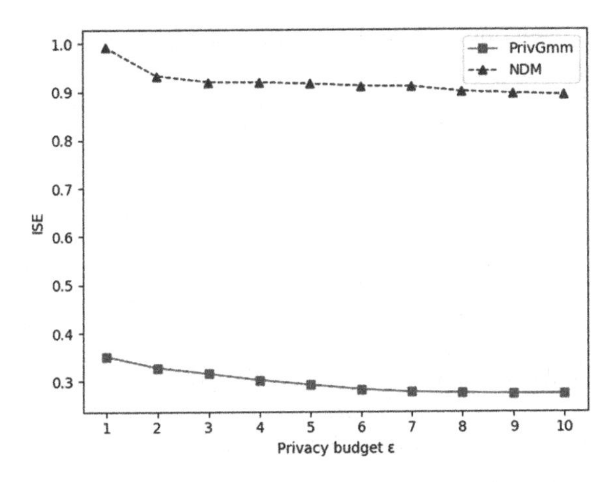

Fig. 1. Effect of privacy budget in Tamilnadu dataset.

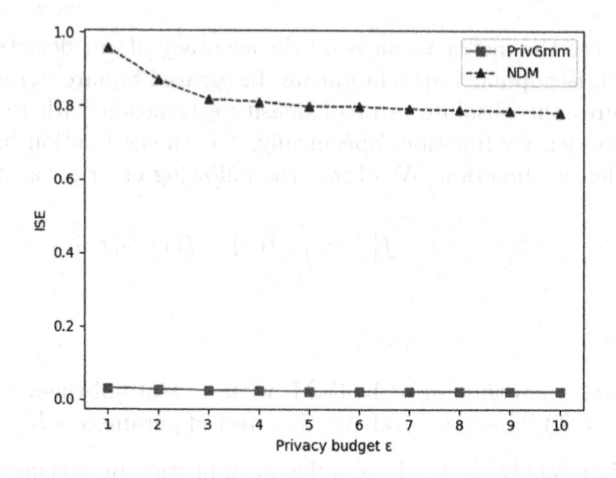

Fig. 2. Effect of privacy budget in MFCCs dataset.

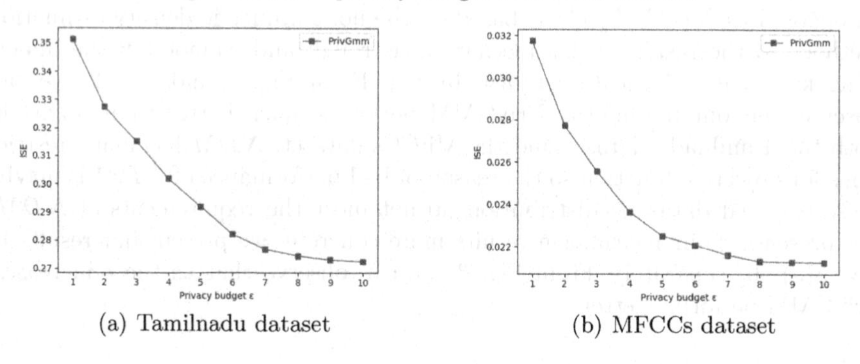

(a) Tamilnadu dataset (b) MFCCs dataset

Fig. 3. Concrete present of PrivGMM with effect of privacy budget in real datasets

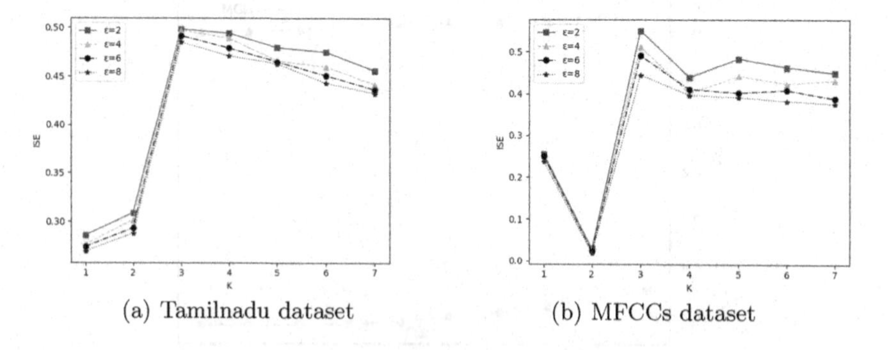

(a) Tamilnadu dataset (b) MFCCs dataset

Fig. 4. Effect of K

Evaluation of PrivGMM

Effect of Number of Gaussian Distributions (K). From Fig. 4(a), we can see that in Tamilnadu dataset, the ISE increases at first and starts decreasing when $K = 3$. From Fig. 4(b), we can observe that in MFCCs dataset, the ISE first decreases and increases when $K = 3$ and performs best when K equals to 2. The reason is that $K = 2$ is maybe a good fit to MFCCs dataset. Therefore, we can conclude that the value of K has great effect to our results. This phenomenon presents the importance of the choice of K.

6 Conclusion

In this paper, we presented PrivGMM, which is an effective and accurate privacy-preserving mechanism to estimate density under local differential privacy. We employed the Gaussian mechanism to sanitize numerical data independently and rigorously. On the server side, we added a noise latent variable to each iteration to estimate density accurately. Through theoretical analysis, we proved the effectiveness and accuracy of our mechanism and its robustness in various parameter settings. Experimental results in both synthetic datasets and real datasets showed that our PrivGMM outperforms the state-of-the-art methods.

In the future, we plan to extend the application of PrivGMM to high-dimensional data. We also plan to analyze local differential privacy in other statistical inference tasks, such as logistic regressions and Markov chains.

Acknowledgments. This work was supported by the Anhui Initiative in Quantum Information Technologies (No. AHY150300). Wei Yang and Shaowei Wang are both corresponding authors.

References

1. Bhaskara, A., Charikar, M., Moitra, A., Vijayaraghavan, A.: Smoothed analysis of tensor decompositions. In: Proceedings of the Forty-Sixth Annual ACM Symposium on Theory of Computing, pp. 594–603. ACM (2014)
2. Biernacki, C., Celeux, G., Govaert, G.: Choosing starting values for the EM algorithm for getting the highest likelihood in multivariate Gaussian mixture models. Comput. Stat. Data Anal. **41**(3–4), 561–575 (2003)
3. Blömer, J., Bujna, K.: Adaptive seeding for gaussian mixture models. In: Bailey, J., Khan, L., Washio, T., Dobbie, G., Huang, J.Z., Wang, R. (eds.) PAKDD 2016. LNCS (LNAI), vol. 9652, pp. 296–308. Springer, Cham (2016). https://doi.org/10.1007/978-3-319-31750-2_24
4. Chaudhuri, K., Rao, S.: Beyond Gaussians: spectral methods for learning mixtures of heavy-tailed distributions. In: COLT, vol. 4, p. 1 (2008)
5. Dempster, A.P.: Maximum likelihood estimation from incomplete data via the EM algorithm. J. R. Stat. Soc. Ser. B (Stat. Methodol.) **39**, 1–38 (1977)
6. Ding, B., Kulkarni, J., Yekhanin, S.: Collecting telemetry data privately. In: Advances in Neural Information Processing Systems, pp. 3571–3580 (2017)

7. Duchi, J.C., Jordan, M.I., Wainwright, M.J.: Local privacy and statistical minimax rates. In: 2013 IEEE 54th Annual Symposium on Foundations of Computer Science, pp. 429–438. IEEE (2013)
8. Duchi, J.C., Jordan, M.I., Wainwright, M.J.: Minimax optimal procedures for locally private estimation. J. Am. Stat. Assoc. **113**(521), 182–201 (2018)
9. Dwork, C., Kenthapadi, K., McSherry, F., Mironov, I., Naor, M.: Our data, ourselves: privacy via distributed noise generation. In: Vaudenay, S. (ed.) EUROCRYPT 2006. LNCS, vol. 4004, pp. 486–503. Springer, Heidelberg (2006). https://doi.org/10.1007/11761679_29
10. Dwork, C., McSherry, F., Nissim, K., Smith, A.: Calibrating noise to sensitivity in private data analysis. In: Halevi, S., Rabin, T. (eds.) TCC 2006. LNCS, vol. 3876, pp. 265–284. Springer, Heidelberg (2006). https://doi.org/10.1007/11681878_14
11. Erlingsson, Ú., Pihur, V., Korolova, A.: RAPPOR: randomized aggregatable privacy-preserving ordinal response. In: Proceedings of the 2014 ACM SIGSAC Conference on Computer and Communications Security, pp. 1054–1067. ACM (2014)
12. Greenberg, A.: Apple's differential privacy is about collecting your data-but not your data. Wired, June 13 (2016)
13. Jia, J., Gong, N.Z.: Calibrate: frequency estimation and heavy hitter identification with local differential privacy via incorporating prior knowledge. In: IEEE INFOCOM 2019-IEEE Conference on Computer Communications, pp. 2008–2016. IEEE (2019)
14. Joseph, M., Kulkarni, J., Mao, J., Wu, S.Z.: Locally private Gaussian estimation. In: Advances in Neural Information Processing Systems, pp. 2980–2989 (2019)
15. Joseph, M., Kulkarni, J., Mao, J., Wu, Z.S.: Locally private Gaussian estimation. arXiv preprint arXiv:1811.08382 (2018)
16. Kamath, G., Li, J., Singhal, V., Ullman, J.: Privately learning high-dimensional distributions. arXiv preprint arXiv:1805.00216 (2018)
17. Kamath, G., Sheffet, O., Singhal, V., Ullman, J.: Differentially private algorithms for learning mixtures of separated Gaussians. In: Advances in Neural Information Processing Systems, pp. 168–180 (2019)
18. Karwa, V., Vadhan, S.: Finite sample differentially private confidence intervals. arXiv preprint arXiv:1711.03908 (2017)
19. Kothari, P.K., Steinhardt, J., Steurer, D.: Robust moment estimation and improved clustering via sum of squares. In: Proceedings of the 50th Annual ACM SIGACT Symposium on Theory of Computing, pp. 1035–1046. ACM (2018)
20. Kumar, A., Kannan, R.: Clustering with spectral norm and the k-means algorithm. In: 2010 IEEE 51st Annual Symposium on Foundations of Computer Science, pp. 299–308. IEEE (2010)
21. Moitra, A., Valiant, G.: Settling the polynomial learnability of mixtures of Gaussians. In: 2010 IEEE 51st Annual Symposium on Foundations of Computer Science, pp. 93–102. IEEE (2010)
22. Nissim, K., Raskhodnikova, S., Smith, A.: Smooth sensitivity and sampling in private data analysis. In: Proceedings of the Thirty-Ninth Annual ACM Symposium on Theory of Computing, pp. 75–84. ACM (2007)
23. Scott, D.W.: Parametric statistical modeling by minimum integrated square error. Technometrics **43**(3), 274–285 (2001)

24. Wu, Y., Wu, Y., Peng, H., Zeng, J., Chen, H., Li, C.: Differentially private density estimation via Gaussian mixtures model. In: 2016 IEEE/ACM 24th International Symposium on Quality of Service (IWQoS), pp. 1–6. IEEE (2016)
25. Zhao, J., et al.: Reviewing and improving the Gaussian mechanism for differential privacy. arXiv preprint arXiv:1911.12060 (2019)

A Progressive Approach for Computing the Earth Mover's Distance

Jiacheng Wu, Yong Zhang(✉), Yu Chen, and Chunxiao Xing

BNRist, Department of Computer Science and Technology, RIIT,
Institute of Internet Industry, Tsinghua University, Beijing, China
{wu-jc18,y-c19}@mails.tsinghua.edu.cn,
{zhangyong05,xingcx}@tsinghua.edu.cn

Abstract. Earth Mover's Distance (EMD) is defined as the minimum cost to transfer the components from one histogram to the other. As a robust similarity measurement, EMD has been widely adopted in many real world applications, like computer vision, machine learning and video identification. Since the time complexity of computing EMD is rather high, it is essential to devise effective techniques to boost the performance of EMD-based similarity search. In this paper, we focus on deducing a tighter lower bound of EMD, which still remains the bottleneck of applying EMD in real application scenarios. We devise an efficient approach to incrementally compute the EMD based on the primal-dual techniques from linear programming. Besides, we further propose progressive pruning techniques to eliminate the dissimilar results as well as enable early termination of the computation. We conduct extensive experiments on three real world datasets. The results show that our method achieves an order of magnitude performance gain than state-of-the-art approaches.

Keywords: Similarity search · Earth Mover's Distance · Primal-dual

1 Introduction

Earth Mover's Distance (EMD) is an important similarity metric to measure the dissimilarity between objects represented by histograms. For any two histogram tuples, their EMD is defined as the minimum cost to transform one histogram into the other. Compared with other similarity metrics for histograms such as Euclidean Distance, EMD is more robust in identifying dissimilarity between objects since it closely follows the human's perception of differences. Consequently, EMD is widely used in real-world applications from many domains, including image retrieval [5,6], computer vision [7,8], video identification [11] and multimedia database [2].

In the database community, EMD-based similarity search has been extensively studied in the past decades. Given a query and a set of histogram records,

This work was supported by NSFC (91646202), National Key R&D Program of China (2018YFB1404401, 2018YFB1402701).

Y. Nah et al. (Eds.): DASFAA 2020, LNCS 12112, pp. 122–138, 2020.
https://doi.org/10.1007/978-3-030-59410-7_8

there are two variants of EMD-based similarity search: threshold-based and kNN search. In this paper, we focus on the kNN search problem, which identifies the k records that has the minimum EMDs to given query. In general, EMD can be computed by solving an instance of transportation problem, which can be formulated as a minimum cost flow problem[1], e.g. the network simplex algorithm. Due to the high computational cost of EMD, it is necessary to propose powerful pruning and bounding techniques to filter out unpromising records.

Some early studies utilize other similarity metrics, such as L_p Norm distance and Euclidean distance, as the lower bound of EMD [2,9,15,17]. Then they devise pruning and indexing techniques based on such similarity metrics. However, as [10] pointed out, the effectiveness of a lower bound for EMD largely depends on various factors and the process of EMD computation still remains the bottleneck of the whole similarity search process due to its extreme time complexity. Thus, this category of studies aims at tightening the lower bounds based on the process of EMD computation. Actually, they are orthogonal to and can be easily integrated with the state-of-the-art filtering techniques. While some improvements are proposed on feature histograms, not much work is known to study the computation based on its dual problems. Compared with the previous approaches, our method based on dual problem has fewer computations during each iteration and is also cache-friendly due to our contiguous access patterns.

In this paper, we propose an effective approach to support EMD-based similarity search by leveraging the primal-dual theory [4] in linear programming. Since any solution for dual problem is correlated with a lower bound for primal problem, we propose an incremental method to progressively find better feasible solutions for dual problem during the whole process of computation. In addition, we boost the search performance by terminating the calculations of EMD timely. Moreover, we deduce a tighter lower bound with the help of a cost-based approach to accelerate the convergence of the kNN search process.

The contribution of this paper is summarized as following:

- We devise incremental filling algorithms based on the primal-dual techniques from linear programming.
- We utilize a cost-based method to maximize the increment for lower bounds to accelerate the convergence of calculation.
- We propose a progressively pruning technique which can terminate the calculation of EMD early.
- We conduct extensive experiments on three real world datasets. The results show that our method achieves an order of magnitude performance gain than state-of-the-art approaches.

The rest of this paper is organized as following: Section 2 formulates the problem, introduces necessary background knowledge and presents the overall framework. Section 3 presents a cost-based method for candidate generation.

[1] https://en.wikipedia.org/wiki/Earth_mover%27s_distance.

Section 4 introduces refinement and pruning techniques. Section 5 reports experimental results. Section 6 surveys the related work. Finally Sect. 7 concludes the paper.

2 Preliminaries

In this section, we revisit the definition of EMD and the similarity search problem based on EMD. We also revise the common filter-and-refinement framework for the similarity search. Then we present its dual problem and finally provide some new insights as the cornerstone of our proposed techniques.

2.1 Primal-Dual Theory of Computing EMD

EMD is used to evaluate the difference between two n-dimensional histograms, which can be calculated as the minimal effort to transfer from the one to the another. Formally, given two histograms $\boldsymbol{p} = (p_1, \cdots, p_i, \cdots, p_n)$ and $\boldsymbol{q} = (q_1, \cdots, q_j, \cdots, q_n)$ and a cost matrix $\boldsymbol{C}$ where c_{ij} is the cost to transfer from p_i to q_j, the EMD is defined as the optimum to satisfy the following requirements:

$$\begin{aligned} EMD(\boldsymbol{p},\boldsymbol{q}) &= \min_{f_{ij}} \sum_{i=1}^{n}\sum_{j=1}^{n} c_{ij} \cdot f_{ij} \\ \text{such that } \forall i &: \sum_{j=1}^{n} f_{ij} = p_i \ , \ \forall j : \sum_{i=1}^{n} f_{ij} = q_j \ , \ \forall i,j : f_{ij} \geq 0 \end{aligned} \tag{1}$$

Actually, above formulation implicitly requires that the sum of the components in both $\boldsymbol{p}$ and $\boldsymbol{q}$ should be equal. Following previous studies [9,10], we normalize all histograms to guarantee $\sum_{i=1}^{n} p_i = 1$.

Furthermore, EMD can be considered as a linear programming optimization problem which is called the *Primal Problem* and its *Dual Form* can be expressed as Eq. (2):

$$\begin{aligned} DualEMD(\boldsymbol{p},\boldsymbol{q}) &= \max_{\phi_i,\pi_j} \sum_{i=1}^{n} \phi_i \cdot p_i + \sum_{j=1}^{n} \pi_j \cdot q_j \\ \text{such that: } & \forall i,j : \phi_i + \pi_j \leq c_{ij} \ , \ \forall i : \phi_i \in \mathbb{R} \ , \ \forall j : \pi_j \in \mathbb{R} \end{aligned} \tag{2}$$

We denote the f_{ij} and the pair $\langle \phi_i, \pi_j \rangle$ that satisfy Eqs. (1) and (2) as the *feasible solutions* for the primal problem and dual problem, respectively.

By the primal-dual theory [4], we will get the following inequalities for all feasible solutions for primal and dual problems:

$$\sum_{i=1}^{n} \phi_i \cdot p_i + \sum_{j=1}^{n} \pi_j \cdot q_j \leq DualEMD(\boldsymbol{p},\boldsymbol{q}) = EMD(\boldsymbol{p},\boldsymbol{q}) \leq \sum_{i=1}^{n}\sum_{j=1}^{n} c_{ij} \cdot f_{ij} \tag{3}$$

Therefore, we can deduce a lower bound for $EMD(\boldsymbol{p},\boldsymbol{q})$ for each feasible solution of dual problems, which is a start point of our proposed techniques (Table 1).

Table 1. A concrete example of EMD

Cost Matrix

C	p_1	p_2	p_3	p_4
q_1	0.0	0.2	0.6	1.0
q_2	0.2	0.0	0.3	0.7
q_3	0.6	0.3	0.0	0.8
q_4	1.0	0.7	0.8	0.0

Dataset and Query

$\boldsymbol{p}^1 = \{0.5, 0.2, 0.1, 0.2\}$
$\boldsymbol{p}^2 = \{0.1, 0.4, 0.0, 0.5\}$
$\boldsymbol{p}^3 = \{0.2, 0.3, 0.5, 0.0\}$
$\boldsymbol{p}^4 = \{0.1, 0.4, 0.3, 0.2\}$
$\boldsymbol{q} = \{0.1, 0.3, 0.2, 0.4\}$

2.2 Filter-and-Refinement Framework

Next we formulate the problem of EMD-based similarity search. In this paper, we use the problem of kNN similarity search based on EMD to illustrate the proposed techniques, which is formally stated in Definition 1. Of course our proposed techniques can also be adopted in the threshold-based search problem.

Definition 1. *Given a cost matrix C, a dataset $\mathscr{D}$ of histograms and a query histogram q, the k-nearest neighbor (kNN) query aims to find the subset $\mathscr{S} = \{p | \forall p' \in \mathscr{D} : EMD(p, q) \leq EMD(p', q)\}$ where $|\mathscr{S}| = k$.*

Example 1. We consider the toy example shown in Fig. 1. It includes the cost matrix $\boldsymbol{C}$ and the dataset with 4 histogram tuples, also for query q. The EMDs between q and each record are 0.26, 0.11, 0.33 and 0.15, respectively. If we specify $k = 2$, then the result of kNN query is $\{\boldsymbol{p}^2, \boldsymbol{p}^4\}$.

According to previous studies [2,9,15], EMD-based similarity search can be evaluated based on a filter-and-refinement framework. The EMD between the query q and every histogram $p \in D$ is estimated with the help of lower bound filtering techniques, such as the normal distribution index [9], the reduced dimension lower bound [15], and independent minimization [2]. In general, these filters are applied in an order starting from quick-and-dirty ones to slow-and-accurate ones. For candidates p that cannot be pruned by the filters, the true value of $EMD(q, p)$ is calculated by a black-box computation module, such as SSP or transportation simplex [4]. Our method focuses on the step of verifying EMD and can be seamlessly integrated with the above filtering techniques.

2.3 Incremental Filling Problem

We can solve the above dual problem by incrementally filling the matrix for EMD computation. Such a value matrix ($\boldsymbol{V}$) demonstrates the whole process of calculating EMD. Here each row corresponds to ϕ_i while each column corresponds to π_j. The value of a cell in the ith row and j^{th} column is denoted as $v_{ij} = c_{ij} - \phi_i - \pi_j$. In the process of computing EMD, we will have $v_{ij} = 0$ when an equation $\phi_i + \pi_j = c_{ij}$ holds. For other unsatisfied cases, v_{ij} should be greater than 0 by the constraints. Therefore, the feasible solution of the dual problem can be considered as the feasible patterns whose all cells are no smaller than 0, which is formally stated in Definition 2.

Definition 2. *The* feasible solution *is a matrix in which all cells* v_{ij} *are no smaller than* 0, *i.e.* $\forall i, j,\ v_{ij} = c_{ij} - \phi_i - \pi_j \geq 0$.

A feasible solution can also represented by a pair $\boldsymbol{V} = \langle \boldsymbol{\phi}, \boldsymbol{\pi} \rangle$ without disambiguation. Our target is to find the *best solution* $\langle \boldsymbol{\phi}, \boldsymbol{\pi} \rangle$, which is the feasible solution maximizing the objective function $\mathcal{T}$ shown in the following Eq. (4).

$$\mathcal{T}_{\boldsymbol{p},\boldsymbol{q}}(\boldsymbol{\phi}, \boldsymbol{\pi}) = \sum_i \phi_i \cdot p_i + \sum_j \pi_j \cdot q_j \tag{4}$$

Sometimes the subscript can be ignored when it is obvious in the context.

Table 2. Example of feasible solution

The Given Cost Matrix

$\boldsymbol{C}$	p_1	p_2	p_3	p_4
q_1	0.0	0.2	0.6	1.0
q_2	0.2	0.0	0.3	0.7
q_3	0.6	0.3	0.0	0.8
q_4	1.0	0.7	0.8	0.0

The Feasible Solution

	π_1 0.0	π_2 0.0	π_3 0.0	π_4 −0.7
ϕ_1 0.0	0.0	0.2	0.6	1.7
ϕ_2 0.0	0.2	0.0	0.3	1.4
ϕ_3 0.0	0.6	0.3	0.0	1.5
ϕ_4 0.7	0.3	0.0	0.1	0.0

Example 2. Given $\boldsymbol{p}^2 = \{0.1, 0.4, 0.0, 0.5\}$ and $\boldsymbol{q} = \{0.1, 0.3, 0.2, 0.4\}$, We show an example of feasible solution $\boldsymbol{\phi} = \{0.0, 0.0, 0, 0.7\}$ and $\boldsymbol{\pi} = \{0.0, 0.0, 0.0, -0.7\}$ for the incremental filling problem in Table 2. Each cell is the $v_{ij} = c_{ij} - \phi_i - \pi_j$. The current value of objective function $\mathcal{T}_{\boldsymbol{p}^2,\boldsymbol{q}}(\boldsymbol{\phi}, \boldsymbol{\pi})$ is 0.21, which is a lower bound of $EMD(\boldsymbol{p}^2, \boldsymbol{q})$.

Actually, we can only modify all ϕ_i and π_j to adjust the cells v_{ij}. Moreover, we find some cells are dependent with other cells. For instance, the cells $v_{11}, v_{12}, v_{21}, v_{22}$ are correlated under the given cost matrix since $v_{11} + v_{22} - v_{12} - v_{21} = c_{11} + c_{22} - c_{12} - c_{21} = Constant$. Therefore, we cannot arbitrarily assign value for any cell without taking care of such dependency. Based on above observations, we therefore propose some definitions to depict the dependency of cells.

To show the dependency of cells, we represent it by vector forms: The *vector representation* $\boldsymbol{r}_{ij}$ of cell v_{ij} is the vector with $2n$ components that only have 1 on the position i and $n + j$ while 0 on the other positions. Given k cells in the matrix, we call them *linear dependent* if and only if there exist $2m (< k)$ cells with their vector representation $\boldsymbol{r}^i (i = 1, \cdots, 2m)$ and corresponding η_i $(i = 1, \cdots, 2m)$, half of which are -1 and half are 1, leading to Eq. (5):

$$\sum_{i=1}^{2m} \eta_i \boldsymbol{r}^i = \boldsymbol{0} \tag{5}$$

Otherwise, these k cells are *linearly independent.* Besides, a cell v is linearly independent with other cells if and only if they are linearly independent and the cell v is one of above $2m$ cells.

Example 3. In the above situation where dimension $n = 4$, we revisit these special 4 cells v_{11}, v_{12}, v_{21}, v_{22}. Their vector representations are $\boldsymbol{r}^1 = (1,0,0,0,1,0,0,0)$, $\boldsymbol{r}^2 = (1,0,0,0,0,1,0,0)$, $\boldsymbol{r}^3 = (0,1,0,0,1,0,0,0)$ and $\boldsymbol{r}^4 = (0,1,0,0,0,1,0,0)$. These cells are linearly dependent since there exist $\eta_1 = 1, \eta_2 = -1, \eta_3 = 1, \eta_4 = -1$, which makes all components of $\sum_{i=1}^{4} \eta_i \cdots \boldsymbol{r}^i$ are $\mathbf{0}$. Obviously, we are easy to find that cells v_{11}, v_{12}, v_{13} are linearly independent.

Based on the notion of linearly dependent, we then only need to take care of those linearly independent cells. Specifically, the best feasible solution must contains $2n-1$ linearly independent cells with value 0, which is denoted as *0-cell.* In this case, all 0-cells must be linearly independent in any feasible solutions. Consequently, for any $2m$ linearly dependent cells, their cost c_i and corresponding η_i satisfy $\sum c_i \cdot \eta_i \neq 0$. If the cost matrix does not satisfy it, we can use *Perturbation Method* [3] to perform preprocessing by adding different infinitesimals to the cost of cells in distinct position.

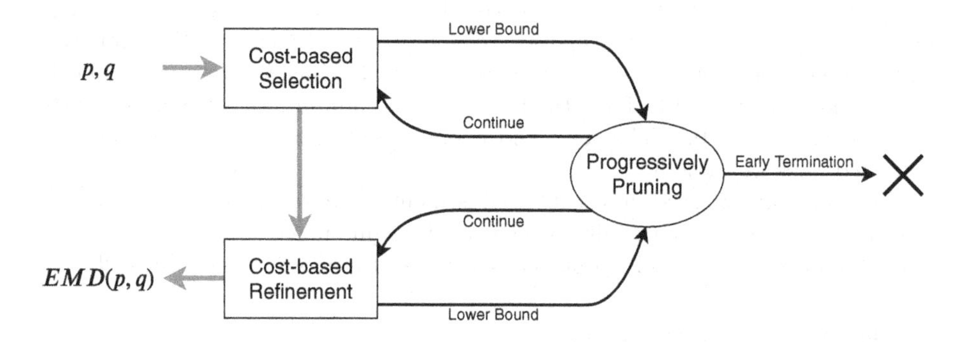

Fig. 1. The overall procedure of our method

Based on the previous discussion, we need to assign $2n-1$ 0-cells in our table. First of all, we assign each $\phi_i = \min_j c_{ij}$ and $\pi_j = 0$. Therefore, each column must have at least one 0-cell. Then we need to choose the remaining 0-cells and make the feasible solution to become the best solution. To this end, we propose a 2-stage framework as is shown in Fig. 1:

- Cost-Based Selection (in Sect. 3): With the help of cost model, we generate a feasible solution which therefore is a candidate of best solution.
- Cost-Based Refinement (in Sect. 4): We refine the candidates of best solution by a similar cost model and finally get the results of EMD.

Besides, we also propose a progressively pruning technique to accelerate the above two stages as described in Sect. 4.2.

3 Cost-Based Selection

In this section, we illustrate the details of the cost-based selection stage. In each iteration, we first use the cost model to select the best one among multiple feasible cells. After that, we show how to modify ϕ and π in order to select cells that will be turned to 0-cells while keeping current 0-cells unchanged. Then after $n-1$ iterations, we get a feasible solution which is a candidate of best solution.

In above steps, we need to identify the **feasible cells**, which are non-zero cells that can be turned to a 0-cell by modifying ϕ and π, while keeping current 0-cells unchanged to guarantee that the modified $\langle \phi, \pi \rangle$ is still a feasible solution.

Note that after modifying ϕ and π, preserving current 0-cells is also a requirement for the adjustment on ϕ and π. We call such adjustment *feasible adjustment* which is formally defined in Definition 3.

Definition 3. *The feasible adjustment $\mathscr{L}$ is an adjustment on ϕ and π that preserves the existed 0-cells.*

Here we only consider the simple feasible adjustment that cannot be combined from other feasible adjustments whose influenced ϕ and π are different.

Example 4. In Table 3 starting from column 4, we first add 0.1 to ϕ_4. Then in order to preserve the 0-cell v_{44}, we need subtract 0.1 from π_4. However, the addition on ϕ_4 also influences the 0-cell v_{42} in row 4. Thus, to preserve the 0-cell v_{42}, we do $\pi_2 = \pi_2 - 0.1$. Then the 0-cell v_{22} on column 2 is influenced, which requires $\phi_2 = \phi_2 + 0.1$. The rows and columns influenced are highlighted use these red lines in Table 3.

The cells and ϕ, π after feasible adjustment is shown in Table 4. The red cells are those cells that are influenced after the adjustment, which we will study later. We find that the cell v_{43} becomes a 0-gird, therefore is a feasible cell.

Table 3. Before feasible adjustment

	π_1 0.0	π_2 0.0	π_3 0.0	π_4 −0.7
ϕ_1 0.0	0.0	0.2	0.6	1.7
ϕ_2 0.0	0.2	0.0	0.3	1.4
ϕ_3 0.0	0.6	0.3	0.0	1.5
ϕ_4 0.7	0.3	0.0	0.1	0.0

Table 4. After feasible adjustment

	π_1 0.0	π_2 −0.1	π_3 0.0	π_4 −0.8
ϕ_1 0.0	0.0	0.3	0.6	1.8
ϕ_2 0.1	0.1	0.0	0.2	1.4
ϕ_3 0.0	0.6	0.4	0.0	1.6
ϕ_4 0.8	0.2	0.0	0.0	0.0

Based on above example, we find that a feasible adjustment can be determined by simply adjusting few influenced rows and columns. Moreover, the rows and columns affected by the feasible adjustment can be identified by the following procedure: 1) Starting from a 0-cell v_{ij}, we mark i rows and j columns as influenced, since both ϕ_i and π_j are influenced. 2) If other 0-cells are in row i or column j, then we mark the rows and columns of these other 0-cells as influenced. 3) Repeat Step 2 until we cannot get any more influenced rows or columns.

Next we introduce more detailed ideas. A feasible adjustment can be represented by Eq. (6):

$$\mathscr{L}^{\Delta} = \{\mu_1, \cdots, \mu_s\} \cup \{\nu_1, \cdots, \nu_t\} = \mathscr{L}_{\mu}^{\Delta} \cup \mathscr{L}_{\nu}^{\Delta} \tag{6}$$

where Δ (can be either positive or negative) is the **adjustment amount** added to ϕ and subtracted from π. μ and ν are the row number and column number which are influenced by the feasible adjustment, respectively. The $\mathscr{L}_{\mu}^{\Delta}$ and $\mathscr{L}_{\nu}^{\Delta}$ are the sets that contain all μ and ν, respectively.

Then we show how to find the feasible cells under the feasible adjustment. Noticed that the feasible adjustment can help us preserve the existed 0-cell. If we want to let a non-0-cell v_{ij} be a 0-cell, we can apply some changes on row i or column j, and then do the feasible adjustment. However, though current 0-cells are preserved, the result $\langle \boldsymbol{\phi}, \boldsymbol{\pi} \rangle$ may not be a feasible solution due to some negative cells. For instance, in Example 4, v_{21} can be a 0-cell if $\Delta = 0.2$. However, v_{43} will be negative. Besides, the adjustment will not influence the cells that are linearly dependent with the current 0-cell, whose rows and columns are both influenced in the adjustment.

Therefore, given a feasible adjustment $\mathscr{L}^{\Delta} = \mathscr{L}_{\mu}^{\Delta} \cup \mathscr{L}_{\nu}^{\Delta}$ with arbitrarily specified Δ, the corresponding feasible cell v^f should satisfy either of the two cases in Eq. (7)

$$v^f = \arg\min\{v_{ij} | i \in \mathscr{L}_{\mu}^{\Delta} \wedge j \notin \mathscr{L}_{\nu}^{\Delta}\} \text{ or } v^f = \arg\min\{v_{ij} | i \notin \mathscr{L}_{\mu} \wedge j \in \mathscr{L}_{\nu}\} \tag{7}$$

The feasible cell must be the cell with minimum v_{ij} whose row is influenced by the adjustment but column is not or vice versa. The minimum property of feasible cell guarantees that after the adjustment from a feasible solution, it can still generate a feasible solution. If the v^f is the first case in Eq. (7), then the $\Delta = v^f$; while if the v^f is the second case, $\Delta = -v^f$ since we need add the value to influenced column π and subtract the value from ϕ.

Therefore, given a feasible cell $\pm v^f$[2], we can have its feasible adjustment $\mathscr{L}^{\pm v^f}$. Then if we do the feasible adjustment $\mathscr{L}^{\pm v^f} = \{\mu_1, \cdots, \mu_s\} \cup \{\nu_1, \cdots, \nu_t\}$, the objective function $\mathcal{T}_{\boldsymbol{p},\boldsymbol{q}}(\boldsymbol{\phi}, \boldsymbol{\pi})$ will change. We then consider the change amount of the objective function by this given feasible cell as the **cost of the feasible cell** $\mathcal{C}(\pm v^f)$, which is $(\pm v^f) \times (\sum_{i=1}^{s} p_{\mu_i} - \sum_{j=1}^{t} q_{\nu_j})$.

Next we select the optimal feasible cell with maximum cost among all possible feasible cells. For example in Table 3, the optimal cell is brown cell v_{43} after some simple calculations. Then we do the feasible adjustment based on the optimal feasible cell. After we repeat the above procedure by $n-1$ times, we have $2n-1$ linearly independent 0-cells, which is one of the candidate best feasible solutions. Such process can be summarized in Algorithm 1:

We first do the initialization for $\boldsymbol{\phi}, \boldsymbol{\pi}$ (line 2), and have n 0-cells. Then in each iteration, we recompute all values v_{ij} (line 4) and collect the information of

[2] We use the sign to distinguish the two types of a feasible cell. We then consider a feasible cell which has two different types in two (and at most two) different adjustments as two different feasible cells.

Algorithm 1: Cost-Based Selection

Input: q: The query; p : The record; C: The EMD cost matrix
Output: ϕ, π: The feasible solution

1 **begin**
2 Intialize ϕ, π, $\forall i : \phi_i = min_j c_{ij}$; $\forall j : \pi_j = 0$;
3 **for** *iter = 1 to n-1* **do**
4 Compute all cells v_{ij} using C, ϕ, π
5 $\mathcal{K} = \emptyset$
6 **foreach** *feasible cells* $\pm v^f$ **do**
7 Obtain its feasible adjustment $\mathscr{L}^{\pm v^f}$
8 Calculate its Cost $\mathcal{C}(\pm v^f)$
9 Add $\langle \pm v^f, \mathscr{L}^{\pm v^f}, \mathcal{C}(\pm v^f) \rangle$ into $\mathcal{K}$
10 Obtain the tuple $\langle \pm v^f_{opt}, \mathscr{L}^{\pm v^f}_{opt}, \mathcal{C}(v^f_{opt}) \rangle \in \mathcal{K}$ with maximum $\mathcal{C}$
11 $\phi_i = \phi_i + (\pm v^f_{opt})$ if $i \in \mathscr{L}^{\pm v^f}_{opt}$ for all i
12 $\pi_j = \pi_j - (\pm v^f_{opt})$ if $j \in \mathscr{L}^{\pm v^f}_{opt}$ for all j
13 **return** $\langle \phi, \pi \rangle$.
14 **end**

each feasible cells into $\mathcal{K}$ (line 6–9), such as its feasible adjustment and its cost. We then pick the feasible cell with maximum cost from $\mathcal{K}$ (line 10). Finally, for the best adjustment $\mathscr{L}'$, we update ϕ_i and π_j by the optimal feasible cell and its feasible adjustment (line 11–12).

4 Refinement

4.1 Cost-Based Refinement

The result of above selection algorithm just returns a candidate feasible solution instead of the final results. In this section, we further propose an algorithm to refine the candidate feasible solution and obtain the final results.

The basic idea is shown as follows: since the best solution also has $2n - 1$ linearly independent 0-cells, then we just find a replaceable non-0-cell to replace a current 0-cell in each iteration. Here the **replaceable cell** are those non-0-cells which can be 0-cells by modifying ϕ and π, preserving other current 0-cells except for the given one, and ensuring that the modified $\langle \phi, \pi \rangle$ is still a feasible solution. Besides, the **replaceable adjustment** $\widetilde{\mathscr{L}}$ is an adjustment on ϕ and π that preserves the existed 0-cells except for the given one. In addition, this replaceable adjustment should make the give 0-cell be a cell with positive value. We can guarantee each replaceable adjustment will increase the objective function $\mathcal{T}$ by utilizing the similar cost-model until reaching the convergence point, which is the precise value of EMD we need.

Example 5. In Table 5, we show a replaceable adjustment for the given 0-cell v_{42}. In order to let v_{42} positive, we subtract 0.2 from π_2. However, this effects

Table 5. Before replaceable adjustment

	π_1 −0.3	π_2 −0.1	π_3 0.0	π_4 −0.8
~~ϕ_1 0.3~~	~~0.0~~	~~0.0~~	~~0.3~~	~~1.5~~
~~ϕ_2 0.1~~	~~0.4~~	~~0.0~~	~~**0.2**~~	~~1.4~~
ϕ_3 0.0	0.9	0.4	0.0	1.6
ϕ_4 0.8	0.5	*0.0*	0.0	0.0

Table 6. After replaceable adjustment

	π_1 −0.5	π_2 −0.3	π_3 0.0	π_4 −0.8
ϕ_1 0.5	0.0	0.0	0.1	1.3
ϕ_2 0.3	0.4	0.0	**0.0**	1.2
ϕ_3 0.0	1.1	0.6	0.0	1.6
ϕ_4 0.8	0.7	**0.2**	0.0	0.0

the 0-cells v_{12} and v_{22}. In order to preserve these 0-cells, we adjust $\phi_1 = \phi_1 + 0.2$ and $\phi_2 = \phi_2 + 0.2$. Also, 0-cell v_{11} in first column becomes negative, then $\pi_1 = \pi_1 = 0.2$. The state after the replaceable adjustment is shown in Table 6.

Similar to the selection algorithm, the replaceable adjustment is also only determined by the adjusting amount Δ and the influenced rows and columns. We can get the influenced rows and columns in a same procedure with the feasible adjustment except in the first step: we just obtain the influenced column j without row i for the given started 0-cell. Thus, we can use the similar representation $\widetilde{\mathscr{L}^{\Delta}} = \{\widetilde{\mu}_1, \cdots, \widetilde{\mu}_s\} \cup \{\widetilde{\nu}_1, \cdots, \widetilde{\nu}_t\} = \widetilde{\mathscr{L}^{\Delta}_{\mu}} \cup \widetilde{\mathscr{L}^{\Delta}_{\nu}}$, where Δ is the amount transported from π $(-\Delta)$ to ϕ $(+\Delta)$ which is specified by each $\widetilde{\mu}$ and $\widetilde{\nu}$. In addition, Δ should be positive since the replaceable adjustment requires that the given 0-cells should have a positive value after the adjustment.

Now, given a 0-cell and its replaceable adjustment $\widetilde{\mathscr{L}^{\Delta}}$, the corresponding replaceable cells v^r should satisfy Eq. (8)

$$v^r = \arg\min\{v_{ij} | i \in \widetilde{\mathscr{L}^{\Delta}_{\mu}} \wedge j \notin \widetilde{\mathscr{L}^{\Delta}_{\nu}}\} \tag{8}$$

After applying the replaceable adjustment with $\Delta = v^r$, the cell v^r becomes 0-cell, and the given 0-cell has value v^r. Other 0-cells stay the same due to the replaceable adjustment itself, and the modified $\langle \boldsymbol{\phi}, \boldsymbol{\pi} \rangle$ is still a feasible solution due to the minimum property of replaceable cells. We then denote the relevant replaceable adjustment as $\widetilde{\mathscr{L}^{v^r}}$.

Then we define the cost based on the replaceable adjustment and the replaceable cells as Definition 4.

Definition 4. *With a given 0-cell should be replaced, we have the replaceable adjustment* $\widetilde{\mathscr{L}^{v^r}} = \{\widetilde{\mu}_1, \cdots, \widetilde{\mu}_s\} \cup \{\widetilde{\nu}_1, \cdots, \widetilde{\nu}_t\}$ *and the replaceable cell* v^r*, then the* ***cost of replaceable cell*** *is* $\widetilde{C}(v^r) = v^r \times (\sum_{i=1}^{s} p_{\widetilde{\mu}_i} - \sum_{j=1}^{t} q_{\widetilde{\nu}_j})$.

Each 0-cell in the current feasible solution will have its corresponding replaceable cells and adjustment, among all of them we then choose the optimal cell v^r_{opt} with maximum cost $\widetilde{C}$. Then we apply the relevant replaceable adjustment on current feasible solution. When the maximum cost is non-positive, we stop the whole refinement stage.

Algorithm 2: Cost-Based Refinement

Input: $\boldsymbol{q}$: The query; $\boldsymbol{p}$: The records; $\boldsymbol{C}$: The EMD cost matrix, $\boldsymbol{\phi}, \boldsymbol{\pi}$: The feasible solution from Selection Algorithm

Output: $\boldsymbol{\phi}$, $\boldsymbol{\pi}$: The best solution, res: the precise value of $EMD(\boldsymbol{p}, \boldsymbol{q})$

1 **begin**
2   **while** *True* **do**
3   Compute the cell's value v_{ij} using $\boldsymbol{C}$
4     $\mathcal{K} = \emptyset$
5   **foreach** *0-cell* **do**
6   Obtain the replaceable adjustment $\widetilde{\mathscr{L}}^{v^r}$ and replaceable cell v^r
7   Obtain the cost of replaceable cell $\widetilde{\mathcal{C}}(v^r)$
8   **if** $\widetilde{\mathcal{C}}(v^r) > 0$ **then**
9   Add $\langle v^r, \widetilde{\mathscr{L}}^{v^r}, \widetilde{\mathcal{C}}(v^r)\rangle$ into $\mathcal{K}$
10   **if** $\mathcal{K} == \emptyset$ **then**
11   **break**
12   Obtain the tuple $\langle v^r_{opt}, \widetilde{\mathscr{L}}^{v^r}_{opt}, \widetilde{\mathcal{C}}(v^r_{opt})\rangle \in \mathcal{K}$ with maximum $\widetilde{\mathcal{C}}$
13     $\phi_i = \phi_i + v^r_{opt}$ if $i \in \widetilde{\mathscr{L}}^{v^r}_{opt}$ for all i
14     $\pi_j = \pi_j - v^r_{opt}$ if $l^t_j \in \widetilde{\mathscr{L}}^{v^r}_{opt}$ for all j
15   $res = \mathcal{T}_{\boldsymbol{p},\boldsymbol{q}}(\boldsymbol{\phi}, \boldsymbol{\pi})$
16   **return** $\boldsymbol{\phi}, \boldsymbol{\pi}, res$.
17 **end**

Example 6. In Table 6, after replaceable adjustment, we find the current state satisfies the stop condition, and the objective function $\mathcal{T}_{\boldsymbol{p}^2,\boldsymbol{q}}(\boldsymbol{\phi}, \boldsymbol{\pi})$ is $0.5 * 0.1 + 0.3 * 0.4 + 0.0 * 0.0 + 0.8 * 0.5 + (-0.5) * 0.1 + (-0.3) * 0.3 + 0.0 * 0.2 + (-0.8) * 0.4 = 0.11$, which is the precise value of EMD between $\boldsymbol{p}^2$ and $\boldsymbol{q}$.

Algorithm 2 illustrates the whole refinement process: We accept the feasible solution represented by $\boldsymbol{\phi}$, $\boldsymbol{\pi}$ returned from Algorithm 1. In each iteration, we first refresh the cell values based on current $\boldsymbol{\phi}$ and $\boldsymbol{\pi}$, then we traverse each 0-cell, and obtain the corresponding replaceable adjustment and cell whose cost is positive (line 5–9). We choose the optimal replaceable cell with maximum cost from the above collection (line 12). Next, we adjust the $\boldsymbol{\phi}$ and $\boldsymbol{\pi}$ (line 13–14), i.e. subtracting $v^{opt}_{\mathscr{L}^h}$ from ϕ_i and adding $v^{opt}_{\mathscr{L}^h}$ to π_j which is influenced in the replaceable adjustment. We repeat the above procedure until we cannot find any replaceable cell with positive cost (line 10). Finally, we calculate the precise EMD value based on the $\boldsymbol{\phi}$ and $\boldsymbol{\pi}$ (line 15).

4.2 Progressively Pruning

We also propose a progressively pruning technique, which might enable early termination of the refinement process.

After each iteration, we get the feasible solution represented by $\boldsymbol{\phi}$ and $\boldsymbol{\pi}$. Based on the Eq. (3), the feasible solution results in a lower bound whose value is $\sum_i \phi_i \cdot p_i + \sum_j \pi_j \cdot q_j$. Since our goal is to identify the histograms whose EMD is within a given threshold τ[3], we assert that the current record can be safely pruned if Eq. (9) holds.

$$\mathcal{T}_{\boldsymbol{p},\boldsymbol{q}}(\boldsymbol{\phi}, \boldsymbol{\pi}) = \sum_i \phi_i \cdot p_i + \sum_j \pi_j \cdot q_j \geq \tau \tag{9}$$

Then the pruning procedure can be described as follows: we check whether Eq. 9 holds or not and then perform pruning upon the requirement after every e iterations, which is a tunable parameter. Above pruning procedure can be integrated into the refinement Algorithm 2 after line 14. Similarly, it can also be integrated into the cost-based selection algorithm (inserted into Algorithm 1 after line 12).

5 Evaluation

5.1 Experiment Setup

The evaluation is conducted based on the methodology of the paper [10]. We evaluate our proposed techniques on three real world datasets: ImageNet, Flickr and Friends. ImageNet is taken from [9], while Flickr and Friends are both taken from [10]. Since these datasets are not public, we just use the same ways to acquire from the websites and process.

The details of datasets are shown in Table 7.

Table 7. Statistics of datasets

Dataset	# Objs	# Bins	Description
ImageNet	1.33M	1024 (32 × 32)	Images from ImageNet[a] project.
Flickr	2.01M	100 (10 × 10)	Images crawled from Flickr[b].
Friends	0.32M	768 (24 × 32)	Images captured per second from "Friends"

[a]http://www.image-net.org/
[b]http://www.flickr.com

In the evaluation, we compare our method named CIF (**C**ost-based **I**cremental **F**illing algorithm) which is the combination of our proposed techniques in the previous sections, with a set of EMD computation methods when used as a black-box module of the filter-and-refinement framework. In all cases, the filtering of candidates is done by applying the state-of-the-art filtering techniques, such as the normal distribution index, full projection and independent minimization which is all mentioned in Subsect. 2.2. For the candidates that pass the filter phase, we compare the application of the following three EMD computation methods: (1) TS (**T**ransportation **S**implex [1]) is a optimization algorithm

[3] For KNN query it can be regarded as the maximum EMD from current k results.

for linear programming. (2) SSP (**S**uccessive **S**hortest **P**ath [4]) is a network-flow based algorithm to find the minimum-cost flow in a bipartite graph what EMD can be transformed to. (3) SIA (**S**implified graph **I**ncremental **A**lgorithm) is in conjunction with PB (**P**rogressive **B**ounding) proposed in [10].

All the algorithms, including our proposed method and those state-of-the-art ones, are implemented in C++. The experiments are run on a 1.80 GHz octa-core machine running Ubuntu 18.04, with 16 GBytes of main memory. We run all the algorithms 5 times and report the average results. Among all experiments, we use the default parameter: the length of iteration for pruning $e = 1$, whose selection will be discussed later.

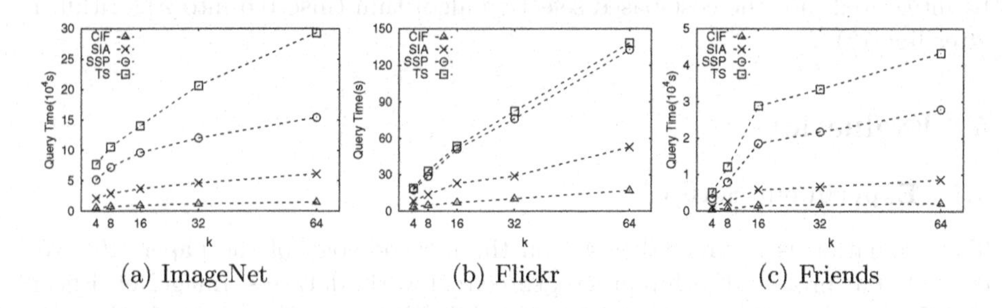

(a) ImageNet (b) Flickr (c) Friends

Fig. 2. Compare with state-of-the-art methods

5.2 Compare with State-of-the-Art Methods

First we compare CIF with state-of-the-art methods. For all baseline methods, we try our best to tune their parameters according to the descriptions in previous studies. The results are shown in Fig. 2. We can see that CIF achieves the best result on all datasets. On Friends ($k = 8$), the join time of CIF is 744 s, while SIA, SSP and TS use 2579, 7890 and 12069 s respectively.

CIF is better than SIA because of fewer computations in each incremental iteration and the more cache-friendly access patterns and storage means. Since our main data structure is the set of cells which can be stored in a contiguous memory while SIA needs to store the bi-graph, which is not cache-efficient for the access and store. Moreover, CIF beats both SSP and TS since we utilize the progressively pruning and the cost model to avoid unnecessary computations and therefore accelerate the whole calculation of EMD.

5.3 Effect of Proposed Techniques

Then, we evaluate the effect of proposed techniques. We implement several methods combining different proposed techniques. In Fig. 3, **Select** is short for cost-based selection in Sect. 3; **Refine** is short for cost-based refinement in Sect. 4 and **Prune** is short for progressively pruning in Sect. 4.2. Therefore, **Select+Refine** means not to use the progressively pruning while **Refine+Prune** means that

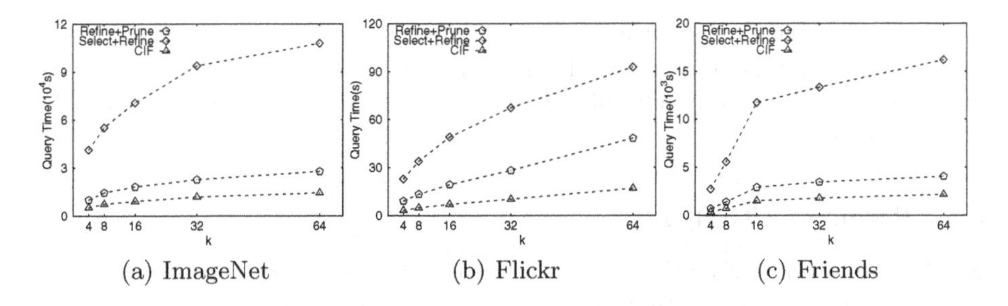

Fig. 3. Effect of proposed techniques

we start the refinement from a state with $2n - 1$ 0-cells obtained by anrbitrarily feasible solutions with $2n - 1$ linearly independent 0-cells, which is not optimal in the cost-based selection. In addition, it is no meaning to say **Select+Prune** since it even cannot be a calculation for EMD.

The results are shown in Fig. 3. CIF achieves a large performance improvement over **Select+Refine** under all the settings. This is because that, without the pruning techniques, we need to do the whole computational procedure of EMD, which costs much more time just like what the traditional EMD computation algorithms do. And it is also possible to prune the records after very limited iterations for current records when we get a tighter threshold after accessing enough records. Also, CIF outperforms **Refine+Prune** since choosing the optimal cell in the cost-based selection algorithm accelerates the convergence of refinement procedure. Sometimes, the upper bound obtained after the cost-based selection is tight enough for pruning.

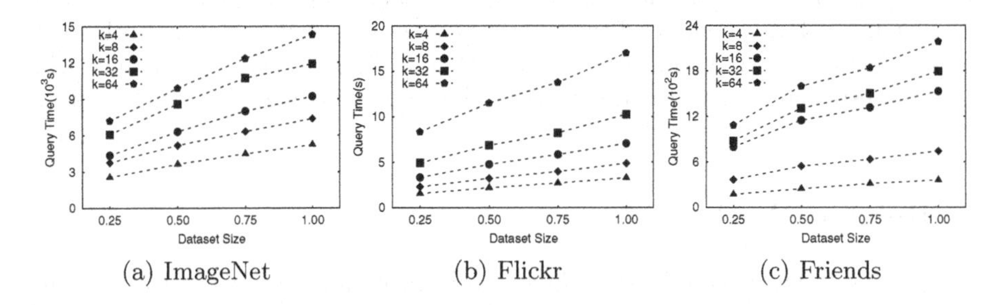

Fig. 4. Scalability: varying dataset size

5.4 Scalability

We conduct experiments to test effects of scaling up on the three real datasets. The results of scaling up are shown in Fig. 4. We vary the data size from 25% to 100% of the dataset. For example, on the ImageNet dataset ($k = 4$), the time costs of query with 25%, 50%, 75%, 100% dataset are 2556, 3643, 4488, 5233 s respectively.

We notice that the performance is less influenced by varying the dataset size comparing to the main factor k. This is a merit in the real application, since we always retrieve a fixed number of records even though the database size may be arbitrarily large.

5.5 Parameter Tuning

Finally, we evaluate the effect of parameter e proposed in progressively pruning. This parameter specifies that the pruning technique takes effect only after e iterations in both selection and refinement algorithms. The results are show in Fig. 5.

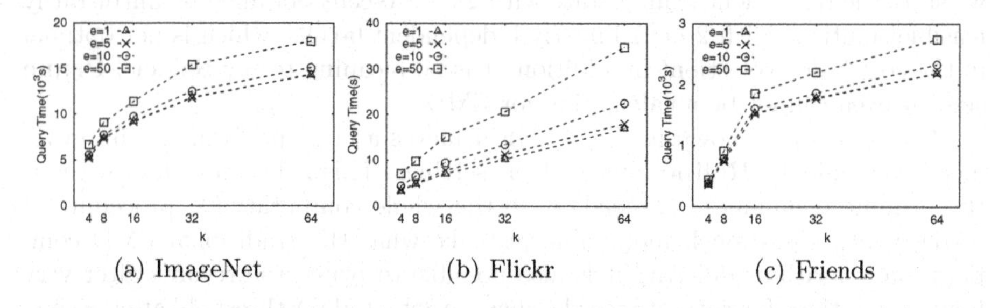

Fig. 5. Effect of parameter e in progressively pruning

We find that the performance is not sensitive to e when e is small especially for high dimensional datasets ImageNet and Friends. This is because the cost of checking pruning conditions is only $O(n)$ which can be ignored when the dimension n is large, compared to the cost $O(n^2)$ for each iteration in both cost-based selection and refinement. However, when e is larger, the performance will decrease, especially for Flickr due to the delayed pruning and extra unnecessary refinement.

6 Related Work

Earth Mover's Distance (EMD) is firstly introduced and applied in the domain of image processing [7]. It acts as an effective similarity measure for a variety of applications that can be represented by signatures denoting each object, or by a histogram consisting of shared features in the feature space. Examples of data types supported by EMD include probabilistic, image, video, document and so on.

Existing approaches in the database community employ a filter-and-verification framework to support the problem of EMD-based similarity search. The idea of filter-and-verification has been widely adopted in the research of string similarity queries [13,14,16,18–20]. Some studies use other similarity metrics as the lower bound of EMD and devise filtering and indexing techniques.

Assent et al. [2] proposed several lower bounds of EMD using the L-p Norm distance while Wichterich et al. [15] shown a dimension reduction technique to estimate the lower bound of EMD. Xu et al. [17] built a B^+-Tree to support and accelerate the similarity search for EMD. Ruttenberg et al. [9] proposed a projection based lower bound and built a high dimensional index to accelerate query processing. Above methods just rely on other similarity metrics, which ignore the process of computing EMD. Therefore, the bound is rather loose and the filter power is poor. There are also some other studies related to the computation of EMD. Uysal et al. [12] devised the IM-Sig algorithm estimates the lower bound of EMD by relaxing the constraints. Tang et al. [10] proposed a progressive bound for verifying the EMD as well as improving the performance for kNN query.

7 Conclusion

In this paper, we study the problem of EMD-based similarity search. To do better cache-friendly calculation, we utilize the primal-dual theory from the linear programming to incrementally find tighter lower bounds during the whole procedure of computation. Such progressively bounds can be used to stop the whole computation timely when some bounds are tight enough. We further develop a cost-based selection to maximize the increment for lower bounds, accelerating the convergence of calculation. Experimental results on real world datasets demonstrate the superiority of our proposed method.

References

1. Ahuja, R.K., Magnanti, T.L., Orlin, J.B.: Network Flows - Theory, Algorithms and Applications. Prentice Hall, Englewood Cliffs (1993)
2. Assent, I., Wenning, A., Seidl, T.: Approximation techniques for indexing the earth mover's distance in multimedia databases. In: ICDE, p. 11 (2006)
3. Charnes, A.: Optimality and degeneracy in linear programming. Econometrica (pre-1986) **20**(2), 160 (1952)
4. Hillier, F.S., Lieberman, G.J.: Introduction to Mathematical Programming. McGraw-Hill, New York (1995)
5. Jang, M., Kim, S., Faloutsos, C., Park, S.: A linear-time approximation of the earth mover's distance. In: CIKM, pp. 505–514 (2011)
6. Pele, O., Werman, M.: Fast and robust earth mover's distances. In: ICCV, pp. 460–467 (2009)
7. Rubner, Y., Tomasi, C., Guibas, L.J.: A metric for distributions with applications to image databases. In: ICCV, pp. 59–66 (1998)
8. Rubner, Y., Tomasi, C., Guibas, L.J.: The earth mover's distance as a metric for image retrieval. Int. J. Comput. Vis. **40**(2), 99–121 (2000)
9. Ruttenberg, B.E., Singh, A.K.: Indexing the earth mover's distance using normal distributions. PVLDB **5**(3), 205–216 (2011)
10. Tang, Y., U, L.H., Cai, Y., Mamoulis, N., Cheng, R.: Earth mover's distance based similarity search at scale. PVLDB **7**(4), 313–324 (2013)

11. Uysal, M.S., Beecks, C., Sabinasz, D., Schmücking, J., Seidl, T.: Efficient query processing using the earth's mover distance in video databases. In: EDBT, pp. 389–400 (2016)
12. Uysal, M.S., Beecks, C., Schmücking, J., Seidl, T.: Efficient filter approximation using the earth mover's distance in very large multimedia databases with feature signatures. In: CIKM, pp. 979–988 (2014)
13. Wang, J., Lin, C., Li, M., Zaniolo, C.: An efficient sliding window approach for approximate entity extraction with synonyms. In: EDBT, pp. 109–120 (2019)
14. Wang, J., Lin, C., Zaniolo, C.: MF-join: efficient fuzzy string similarity join with multi-level filtering. In: ICDE, pp. 386–397 (2019)
15. Wichterich, M., Assent, I., Kranen, P., Seidl, T.: Efficient EMD-based similarity search in multimedia databases via flexible dimensionality reduction. In: SIGMOD, pp. 199–212 (2008)
16. Wu, J., Zhang, Y., Wang, J., Lin, C., Fu, Y., Xing, C.: Scalable metric similarity join using MapReduce. In: ICDE, pp. 1662–1665 (2019)
17. Xu, J., Zhang, Z., Tung, A.K.H., Yu, G.: Efficient and effective similarity search over probabilistic data based on earth mover's distance. PVLDB **3**(1), 758–769 (2010)
18. Yang, J., Zhang, Y., Zhou, X., Wang, J., Hu, H., Xing, C.: A hierarchical framework for top-k location-aware error-tolerant keyword search. In: ICDE, pp. 986–997 (2019)
19. Zhang, Y., Li, X., Wang, J., Zhang, Y., Xing, C., Yuan, X.: An efficient framework for exact set similarity search using tree structure indexes. In: ICDE, pp. 759–770 (2017)
20. Zhang, Y., Wu, J., Wang, J., Xing, C.: A transformation-based framework for KNN set similarity search. IEEE Trans. Knowl. Data Eng. **32**(3), 409–423 (2020)

Predictive Transaction Scheduling for Alleviating Lock Thrashing

Donghui Wang, Peng Cai[✉], Weining Qian, and Aoying Zhou

School of Data Science and Engineering, East China Normal University, Shanghai 200062, People's Republic of China
donghuiwang@stu.ecnu.edu.cn,
{pcai,wnqian,ayzhou}@dase.ecnu.edu.cn

Abstract. To improve the performance for high-contention workloads, modern main-memory database systems seek to design efficient lock managers. However, OLTP engines adopt the classic FCFS strategy to process operations, where the generated execution order does not take current and future conflicts into consideration. In this case, lock dependencies will happen more frequently and thus resulting in high transaction waiting time, referred to as lock thrashing, which is proved to be the main bottleneck of lock-based concurrency control mechanisms. In this paper, we present a transaction scheduler that generates efficient execution order to alleviate the lock thrashing issue. To proactively resolve conflicts, LOTAS predicts which data will be accessed by following operations through building Markov-based prediction graphs. Then LOTAS uses the information to schedule transactions by judging whether a transaction needs to be deferred to acquire locks. Experimental results demonstrate that LOTAS can significantly reduce the lock waiting time and improves the throughput up to 4.8x than the classic FCFS strategy under highly contended workloads.

Keywords: High conflicting · Concurrency control · Lock thrashing · Transaction scheduling · Prediction

1 Introduction

Main memory database systems hold the entire working dataset into memory. As a result, the transaction processing performance is no longer dominated by the slow disk I/O. The study shows that about 16–25% of transaction time is spent on the lock manager in a main memory database system even with a single core [9]. Thus, modern main memory database systems are devoted to design more lightweight and efficient lock managers. For example, all lock information is removed from the centralized lock data structure, and co-located with the raw data (e.g., an additional field in the header of a tuple is used to represent its current locking state) [20,25,29].

However, even with improved implementations of lock managers, the performance of transaction processing still suffers due to *lock thrashing* especially

Y. Nah et al. (Eds.): DASFAA 2020, LNCS 12112, pp. 139–156, 2020.
https://doi.org/10.1007/978-3-030-59410-7_9

under highly contended workloads [2,29]. In this scenario, the chains of lock dependencies are easily to get very long, resulting in longer and longer transaction waiting time. As shown in Fig. 1, transactions waiting in the dependency chains can only execute one by one in order, which results in less concurrent execution. In this paper, our intuition is that the lock thrashing problem can be proactively alleviated through effective scheduling.

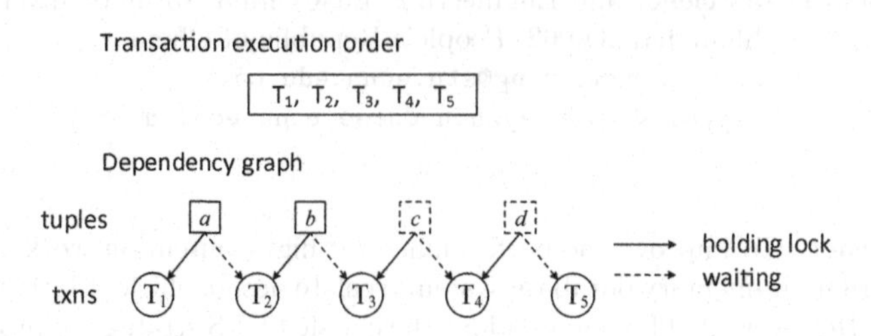

Fig. 1. Example of lock thrashing. T_5 waits on T_4; T_4 waits on T_3 and so on.

Under the one-shot transaction model, it is easy to schedule a conflict-free execution order because the full read and write sets of each transaction can be obtained before running a batch of transactions. For example, Calvin [23] generates a deterministic order for a batch of transactions by locking all tuples in the read/write set at the very start of executing the transaction. It aggressively avoids all contentions between running transactions.

When the read/write set of a transaction cannot be known at the very beginning, Calvin performs additional read operations to obtain the read and write set. However, this mechanism is only suitable for stored-procedure transactions, where transaction logics and input parameters are known before running the transaction. According to a recent survey of DBAs on how applications use databases [16], its results reported that 54% responses never or rarely used stored procedures and 80% interviewed said less than 50% transactions running on their databases were executed as stored procedures. Transaction logics in many real-world OLTP applications are still implemented in the application side, and interact with the database by using APIs because of the scalability and flexibility of the multi-tier architecture. In this scenario, clients or middlewares execute a transaction by repeatedly sending statements to a database server one by one and then waiting the response until this transaction is committed or aborted. Therefore, it is unfeasible to obtain the full read and write set at the start of a transaction.

In this paper, we present LOTAS, a lock thrashing aware transaction scheduling mechanism which relaxes the assumption of transaction execution model. By analyzing transactional logs offline, LOTAS collects and groups a set of statement template sequences where each sequence denotes a transaction, and then mines the correlation of parameters in different statements for each group. As a result, LOTAS can predict which data next operations of a started transaction will access at runtime even the transaction is not running with the stored procedure. Before acquiring a lock, LOTAS detects whether current operation and subsequent operations of a started transaction will exacerbate lock thrashings across live transactions and proactively makes scheduling decisions. For example, deferring to acquire this lock.

The paper is organized as follows. Section 2 gives motivational examples of proactive scheduling to resolve lock thrashing problem and sketches the overall framework of LOTAS. Section 3 shows several preliminary works of LOTAS, including how to build Markov-based prediction model and how to track hot records. In Sect. 4, we describe the transaction schedule mechanism and the details about our implementation. Section 5 presents the results of performance evaluation. Finally, in Sect. 6, we give an overview of related works and Sect. 7 summarizes and concludes the paper.

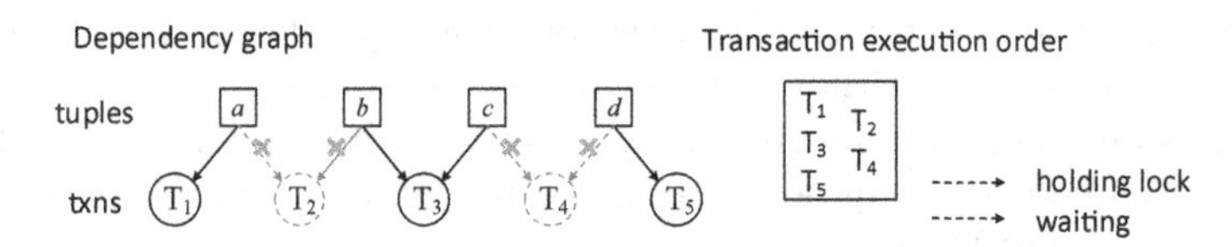

Fig. 2. Motivational example that uses predictive information to make scheduling.

2 LOTAS Overview

Motivation Example of LOTAS — In Fig. 2, we present a motivation example to show how proactive scheduling can avoid lock thrashing problem and achieve more concurrency. Recalling the example in Sect. 1, Fig. 1 shows the original execution under the FCFS strategy, which creates a long dependency chain. As a result, transaction T_5 can not be processed until the completion of T_4 while T_4 depends on T_3; T_3 waits on T_2, and so on.

Figure 2 shows a scheduled execution that avoid lock thrashing. For example, when T_2 attempts to acquire lock on b, it finds that its next wanted lock on a has been held by transaction T_1. Without scheduling, in this case, transaction T_2 will obtain the lock on b and is added to the waiting list of record a, waiting for a to become available (like the behavior in Fig. 1). However, such behavior will cause that T_2 is blocked by T_1 and T_3 is blocked by T_2. When the locked tuple is popular, the blocking will be heavier and result in more severe performance degradation. To avoid been blocked by T_1 and block other transactions that try to lock b, LOTAS makes T_2 voluntarily give up locking b. As a result, delaying T_2 would allow T_1 and T_3 can be executed concurrently and be finished as soon as possible and thus the lock holding time can be reduced.

This example motivates us to use the predictive information to generate an effective scheduled execution order under high contentions. The foundation of scheduling decision is based on that the records accessed by the forthcoming operations are known in advance. However, we have no assumption of one-shot transaction model.

Illustrating Example for Operation Prediction — From the observation of transactional logs, a transaction usually executes a sequence of parameterized SQL statements repeatedly. Most of the time, the parameters in next transaction operation can be inferred from their previous operations. Figure 3 presents an illustrating example of the main OLTP transaction PAYMENT in TPC-C workload, in which tuples on table WAREHOUSE and DISTRICT are popular and usually causing heavy lock waiting across concurrent transactions. The first operation selects data from table warehouse with the parameter *w_id*, and the second operation selects data from table district with parameter *d_w_id* and *d_id*, where the *d_w_id* is actually the same with the parameter *w_id* in the first operation. It can be observed that, the third operation updates warehouse by using the same parameter *w_id* with the first operation. Similarly, parameters in the fourth operation can also be extracted from the previous operations. Therefore, both of them are predictable. Although transactions are not executed with one-shot model, subsequent operations are still associated with previous executed operations. To this end, we propose to build the prediction graph for each type of transaction by using Markov model.

1. **SELECT** WAREHOUSE **WHERE** w_id = ***w_id***
2. **SELECT** DISTRICT **WHERE** w_id = ***w_id*** and d_id = ***d_id***
3. **UPDATE** WAREHOUSE **WHERE** w_id = ***w_id***
4. **UPDATE** DISTRICT **WHERE** w_id = ***w_id*** and d_id = ***d_id***
5. **SELECT** CUSTOMER **WHERE** c_w_id = ***c_w_id*** and c_d_id = ***c_d_id*** and c_id = ***c_id***
6. **UPDATE** CUSTOMER **WHERE** c_w_id = ***c_w_id*** and c_d_id = ***c_d_id*** and c_id = ***c_id***
7. **INSERT** HISTORY **VALUES** (***h_id, h_value***)

Fig. 3. The statement sequence of PAYMENT.

LOTAS Framework — The framework of LOTAS is shown in Fig. 4, which is consisted of three main components. The first part (left side in Fig. 4) is to reverse-engineer transaction logics based on offline transactional logs, and build the prediction model for each kind of transaction. The second component (box in the right-top) is to maintain the temperature statistics to identify hot records and trigger scheduling for lock requests on these records. The third part (middle side) demonstrates the main process of LOTAS, including three steps. First, LOTAS loads the predicted Markov prediction graph for a started transaction after the first few statements have been executed; second, LOTAS makes prediction for this transaction and fills the *pred-set* stored in the transaction context; third, before invoking the function acquire_lock of the lock manager, LOTAS judges whether the current lock request need to be deferred to get the lock according to the tracked lock thrashing information.

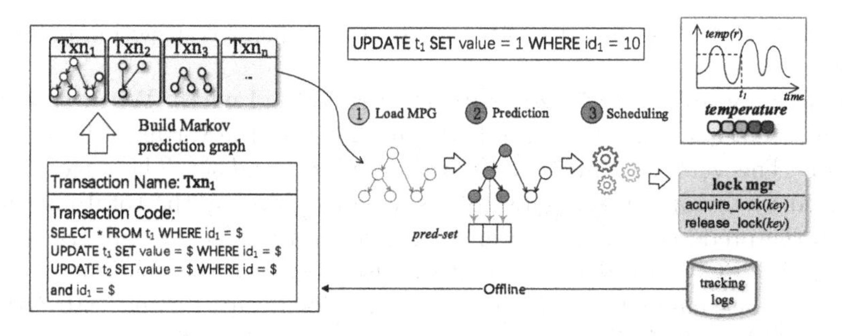

Fig. 4. Framework of LOTAS. Markov prediction graph is built offline. Scheduling is invoked before acquiring the lock.

3 Preliminary

This section presents the preliminary works of LOTAS, including how to make prediction through the Markov model, and how to identify hot records.

3.1 Building Markov Prediction Model

Problem Specification — The scheduling algorithm requires to exactly predict which data will be accessed, which can be inferred by the transaction access pattern and the parameters of previously executed statements. Markov model, as a widely adopted solution for the sequence prediction problem, takes small costs on the computation, which is also suitable for our problem. In this work, we define the Markov prediction model of a transaction $\mathcal{M}_\ell$ as an directed acyclic graph of the transaction execution path. The vertex $v_i \in V(\mathcal{M}_\ell)$ represents an unique statement template, which contains (1) the identifier of the template, (2) the set of involved parameters (e.g., *param-set*), (3) the set of predicted keys of subsequent operations after v_i (e.g., *pred-set*). The outgoing edges from a vertex v_i represent the probability distribution that a transaction from v_i's state to one of its subsequent states. The Markov prediction graph is generated in two phases. In the first phase, we construct the transaction access graph according to the trace log. In the second phase, we mine parameter dependency relationships between vertices and fill the *pred-set* for each vertex.

Extracting Transaction Access Patterns — In the first step, transaction patterns are extracted from the tracking logs produced by the database. Each transaction log contains a sequence of statements, which can be ended with COMMIT or ABORT. We first converts all raw statement sequences appeared in transactional logs into a collection of template sequences by replacing each parameter in the query with a wildcard character (e.g., '$'). Then, the collected template sequences are divided into different clusters. This step helps to find frequent transaction patterns. There are lots of ways to complete this step. We adopt a plain and effective clustering approach based on similarity calculated by *Levenshtein distance*. If the count of similar sequences in a cluster is greater than a threshold, this sequence is regarded as a frequent transaction pattern in current workload.

An access graph is initialized with no vertices and edges. When traversing the template sequence, a vertex is built and hung on the vertex of its previous template. The vertex is initialized with an empty *pred-set*. The edge value is calculated by dividing the number of visits to the edge by the total number of visits to the predecessor vertex. Figure 5(a) shows the access graph transaction PAYMENT in TPC-C workload. 60% of the users retrieve the customer's profile based on its last name, thus the weight of edge UpdateDist → GetCustbyName is 0.6. Looping logic is similar to the branching logic, where a state v_i has a certain probability of transiting to a previous state.

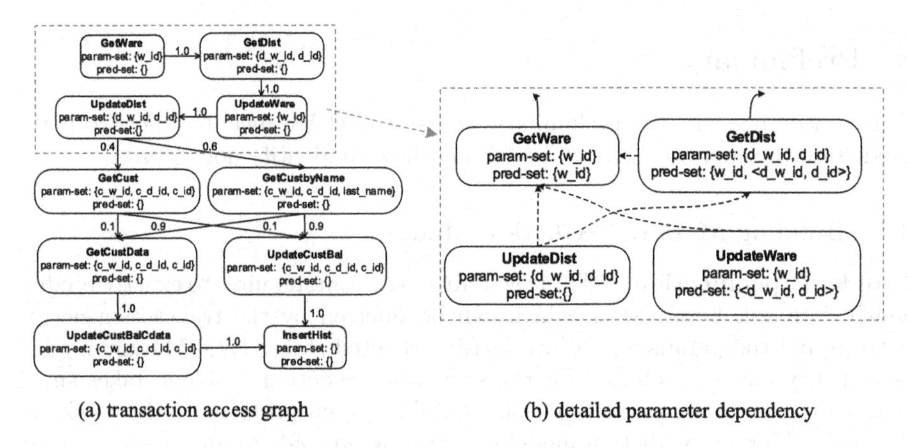

(a) transaction access graph (b) detailed parameter dependency

Fig. 5. Markov prediction graph for PAYMENT.

Finding Parameter Dependency — After constructing the access graph of the Markov prediction model, we mine the parameter dependency relationship in this phase and fill the *pred-set* for each vertices. Thus we could quickly perform predictions online without expensive computations.

Predicting what tuples following operation of a start transaction are going to access is not an easy task, since a parameter in a request can be totally independent, which means the parameter value is randomly inputted by clients. We consider an operation to be predictable if each of its parameters either depends on the previous operations or is a *constant*. A dependent parameter can be *variable-dependent or result-dependent.*

The value of a variable-dependent parameter is determined by the parameters of previous requests. We use *mapping functions* to find variable-dependencies. A mapping function between p_1 and p_2 can help to obtain the value of p_2 by giving the value of p_1. Basically, the mapping function can be linear for numerical values or non-linear for non-numerical values. The variable-dependencies among parameters can be mined by using liner regression analysis. While the mapping functions of non-numerical parameters can be trained as a classification task by using parameter values from previous operations as inputs, and parameter values of follow-up operations as outputs. A result-dependent parameter depends on the retrieval result of previous operations. For example, in the PAYMENT, if we get customer information by its last name, the parameter *c_id* in the following operations comes from the result of GetCustbyName. Basically, exploring dependencies among result-dependent parameters needs to record retrieval result sets of all attributions of SELECT statements. Then the method to explore dependency functions between result sets and requests' parameters is similar to variable-dependent parameters.

To illustrate this more specifically, consider the example of PAYMENT again, Fig. 5(b) shows the detailed parameter dependency of the first four statements. A dashed-line arrow from v_i to v_j represents that the parameter in v_i depends on v_j, while the solid arrow means that one of a parameter of this vertex depends on client inputs. We traverse the transaction access graph and for each vertex v_i, and find the vertices set $v_1^i, v_2^i, ..., v_k^i$ that are connected with v_i by the dashed-line arrows. For each vertex in $v_1^i, v_2^i, ..., v_k^i$, we check whether v_i is its last dependent node. If so, its *params-set* will be appended into the *pred-set* of v_i. The next vertex inherits the *pred-set* from its previous vertices and removes the key that has been executed.

Discussion — It is clear that the predictable parameter information is limited especially when most parameters of a transaction are independent. However, we investigated several OLTP benchmarks including AuctionMark, SEATs, Smallbank, TPC-C and swingBench, which contain 37 transactions and 505 requests. Through mining the prediction information on all parameters in requests, we found that about 24% of them are independent variables which are totally randomly inputted by clients. The remaining 76% parameters are predictable, where the variable-dependent parameters account for more than 61%. We believe that predicting independent parameters in requests is still an open question of worth resolving in the future works especially with the progress on incorporating prediction machine learning models into DBMS.

Prediction Process — The Markov prediction graphs are built offline and can be stored in the persistent storage. When the database is running, it first loads prediction graphs into its own workspace. When an operation coming, if it is a new-started transaction, the scheduler first needs to find its corresponding Markov prediction graph. We extract the template of this operation and calculate the hashing value to match all graphs. If there is only one model matches, the Markov predication graph is found. If more than one matches are found, more operation information is needed to match the exact Markov graphs. If there is a Markov prediction graph already in the transactional context, we find the vertex in the graph that matches this operation. Then the *pred-set* of this vertex is filled with existing parameter values to obtain the exact set of predicted keys.

3.2 Tracking Hot Records

If the lock manager within the database is designed to track waiting transactions on the records, it is easily to identify whether a tuple is hot or not through judging the number of waiting transactions on it. However, when the concurrency control adopts NO WAIT mechanism [29] to resolve lock failures, there has no statistics on waiting transactions. In this case, hot records cannot be identified through the lock manager.

The hot record tracking mechanism in LOTAS is independent to the lock manager. It tracks popular data at runtime and begins to heat up the temperature of the data which is accessed frequently in a period of time. If the access frequency decreases, its temperature is cooled down. We use a 64-bit field *temp-val* in the header of the record to maintain the temperature statistic. Since frequent updates to *temp-val* potentially hurt performance, temperature update operations are online sampled with a low probability (e.g., 1%) for updating temperature statistics.

With the passage of time, the temperature is gradually cooling. Inspired by Newton's law of cooling, the temperature has an exponential decay, which is determined by the cooling factor and the time gap between current time and last update time. The following formula is used for computing temperature values.

$$temp(r) = \begin{cases} temp(r) * (epochGap + 1)^{c_\theta} + h_\Delta, \\ \qquad\qquad\qquad \text{if } curTime - lastTime \geq epoch \\ \min\left(temp(r) + h_\Delta, \mathcal{H}_{up}\right), \\ \qquad\qquad\qquad \text{if } curTime - lastTime < epoch \end{cases} \tag{1}$$

For a sampled update operation on record r, the formula first checks whether the current time is in the same epoch with the last cooling epoch. If the time gap is greater than an epoch, $temp(r)$ should be decayed with a decay coefficient which is decided by epoch difference and cooling factor, i.e., *epochGap* and c_θ ($c_\theta < 0$). Besides, an increment (h_Δ) is added to $temp(r)$. If the time gap between two adjacent updates is less than an epoch, which means that these two heating-up operations are so close that it's not necessary to decay the temperature. In this condition, $temp(r)$ only needs to be added with the heating factor.

When r is already hot, its temperature value can not continue to increase in an unlimited way. Otherwise, two hot records may have a big difference in their heat values, and it's difficult to cool down at an unified speed especially when both of them become cold in the next epoch. To resolve this problem, we set an upper bound to $temp(r)$, denoted as $\mathcal{H}_{up}$. Thus, heat values are guaranteed to be smaller than $\mathcal{H}_{up}$ so that stale, hot records can be filtered out by a fast and unified cooling-down speed.

The record is regarded as a hot record if its temperature exceeds to the heat threshold, which is predefined as $\mathcal{H}$. Some expired hot data may not be updated in recent epochs, so their temperatures will not be triggered to be refreshed and remain high. Thus there is a timed garbage collection function to remove obsolete hot records.

4 Scheduling Algorithm

In this section, we propose a lightweight scheduling algorithm. It produces little overhead in both identifying and avoiding lock thrashing problems. In the following, we firstly present related data structures, and then describe the scheduling algorithm.

A transaction maintains the following fields in its local running context:

- *pred-set* is used to maintain the predicted hot locks this transaction is going to acquire;
- *hold-hot-locks* is to track held hot locks of this transaction.

Here, *pred-set* is filled by the prediction process as described in Sect. 3.1. *hold-hot-locks* stores the key of all acquired hot locks of a started transaction. Maintaining holding locks in the execution context is common in the implementation of concurrency control mechanism, which records the locks that need to be released.

A record stores the following information in its header for LOTAS:

- *lock-state*, a 64-bit variable, representing the state of the lock;
- *reg-lock*, a 64-bit variable, encoding the state of the registered lock;
- *temp-val*, a 64-bit variable, recording the temperature of this record.

The value of *temp-val* is updated according to the formula in Sect. 3.2. The first bit of *lock-state* tells whether the lock is in write mode, i.e. 1 means in write mode, 0 in read mode. If the lock is in write mode, the last 63 bits refer to the identifier of the transaction that are holding its write lock. Otherwise, the rest bits represent how many transactions are holding the read lock on it. The information in *lock-state* is used to lightweightly pre-determine whether a request could obtain the lock on this tuple. If the database uses per-tuple lock, i.e. the lock information is co-located with each tuple, thus *lock-state* is omitted to eliminate redundancy.

The field *reg-lock* is used to encode the registered lock on this tuple. The encoding is the same as *lock-state*, which also has read lock and write lock. When a transaction obtained a hot lock, it will register the lock in its *pred-set* by updating the *reg-lock* on them to notify that this lock has been pre-locked by a started transaction that has held other hot locks.

The scheduling decisions are made before lock acquisition. Hence, LOTAS works before invoking acquire_lock of the lock manager. The function need_deferral in Algorithm 1 gives the main rules of our scheduling algorithm. Before transaction t_x attempts to acquire a lt lock on record r, LOTAS judges whether current request needs to be deferred to execute according to the following cases:

- Case 1: if the request lock is not hot or t_x has acquired one or several hot locks, it is clear that, completing transactions that hold hot locks as early as possible could reduce the probability of contentions. Thus this request does not need to be deferred (line 1–2).

- Case 2: if the request lock on r can not be compatible with *r.reg-lock*, which means that continue holding locks on r will block the execution of other transactions that have acquired other hot locks, and thus deferring this transaction could reduce the risk of long blocking chains (line 3–4).
- Case 3: in the cases that one of the predicted keys r_i refers to a hot tuple and lt can not be compatible with r_i*.lock-state* or r_i*.reg-lock*, which means that this transaction would be blocked on r_i definitely, thus this request is preferred to be delayed to avoid blocking started transactions' execution on r (line 5–9).

compatible(*lock-state*, lt, t_x) — If the first bit of *lock-state* is 0 and lt is not a write type, then it is compatible. Otherwise, it is incompatible. If the first bit of *lock-state* is 1, which means that it is in WRITE mode, and if lt is a read type, it is incompatible. If lt is a write type, unless the rest bits of *lock-state* equals to the identifier of t_x, it is incompatible.

After acquiring the lock on a hot tuple r, *r.lock-state* are updated and the locks on *pred-set* are registered to *r.reg-lock*. If lt is a WRITE lock, *r.lock-state* is set to $t_x.id$ xor $mask = 0x80000000$, whose first bit is 1 and the rest are zeros. Otherwise, *r.lock-state* is added by 1 if the lock type is READ. The lock registering operation on *reg-lock* is similar to that on *lock-state.*

Discussion — Even through the extra informations only occupy three 64 bits, the storage cost of adding them for each tuple is not negligible especially when the original tuple just occupies a small number of bits (e.g., the integer). Moreover, modifying a tuple's header and adding extra information is too invasive for the existing DBMSs. For these reasons, LOTAS has an alternative solution. That is tracking information is not co-located in the header of the tuple. Instead, there are three individual data structures to track the information required by LOTAS. Each can be implemented by a lock-free hash table. Additionally, to reduce the storage cost, only hot tuples are involved in these structures. As a result the storage cost can be calculated to $\mathcal{O}(m)$, where m is the number of hot tuples in the current workload. With a little effort, LOTAS can be integrated in any existing DBMS.

5 Evaluation

In this section, we evaluate the performance of LOTAS for answering the following questions. The first question is whether LOTAS could improve the performance under the highly contended workloads, and how much performance gain obtained by LOTAS compared with classical FCFS strategy. Another question is how much prediction and scheduling overheads produced by LOTAS. The final question is that how LOTAS performs under a variety of workloads. For example, the workload with different contention rates and workload that contains a lot of unpredictable transactions.

Experiment Setup — We have implemented LOTAS in the open source main memory database prototype DBx1000 [1]. The default concurrency control and

Algorithm 1: Judge whether a transaction needs to defer to invoke *acquire_lock*

```
   function need_deferral
      Input: Transaction t_x, Record r, Lock type lt
 1    if r.temp-val < H ∨ t_x.hold-hot-lock then
 2        return false;
 3    else if !compatible(r.reg-lock, lt) then
 4        return true;
 5    else
 6        foreach r_i in t_x.pred-set do
 7            if r_i.temp-val ≥ H then
 8                if !compatible(r_i.lock-state, lt) or !compatible(r_i.reg-lock, lt)
                  then
 9                    retrun true;
10    return false;
```

lock manager is based on the two-phase locking with an improved implementation [29], in which the lock information is maintained in a per-tuple fashion where each transaction only latches the tuples that it needs instead of having a centralized lock table. Besides, the *waits-for* graph used in the deadlock detection is maintained in a decentralized data structure, which is partitioned across cores and making the deadlock detector completely lock-free.

The database server is equipped with two 2.00 GHz E5-2620 processors (each with 8 physical cores) and 192 GB DRAM, running CentOS 7.4 64 bit. With hyper-threading, there are a total of 32 logical cores at the OS level.

Benchmark — In the following experiments, we use two benchmarks to test the performance of LOTAS. TPC-C: a standard online transaction processing (OLTP) benchmark. The transaction parameters are generated according to the TPC-C specification. Micro-Benchmark: a customized benchmark based on the YCSB. We build it to evaluate LOTAS under workloads with different skewness and various types of transactions, i.e., predictable and unpredictable transactions. The scheme of Micro-Benchmark contains a single table (usertable) which has one primary key (INT64) and 9 columns (VARCHAR). The usertable is initialized to consist of 10 million records. The default transaction in Micro-Benchmark first reads two serial items, and then updates the value of these two items. Thus it is predictable. We also modify the transaction logic to make it unpredictable and test the performance of LOTAS. The input parameters are generated according to the Zipfian distribution and the degree of skewness is controlled by the parameter θ.

We take the widely used FCFS as the baseline, where transactions are executed in their arrival orders. Beyond that, all other processes adopted by LOTAS and FCFS stay the same. For each experiment, we run 10 million transactions three times, and the results are averaged over the three runnings. Besides, each experiment has a 10-s warm-up.

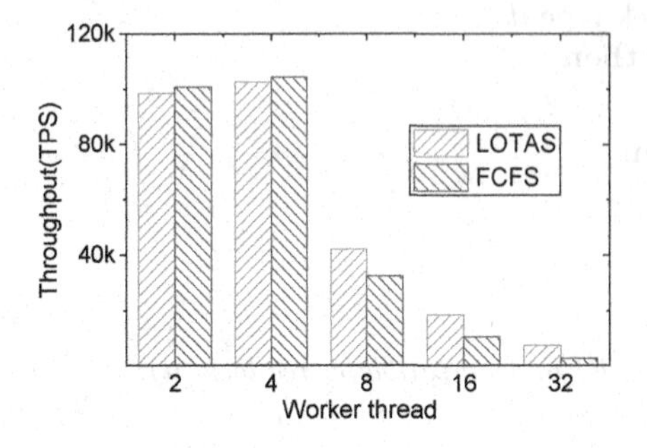

Fig. 6. Throughput of TPC-C (ware = 4).

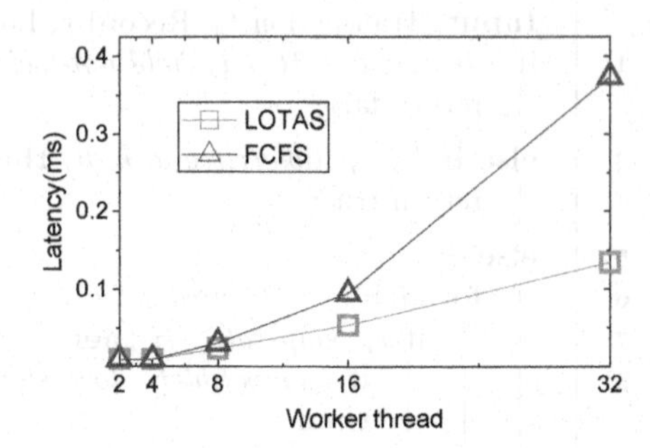

Fig. 7. Latency of TPC-C (ware = 4).

5.1 Overall Performance

We first compare the overall performance of FCFS and LOTAS under the TPC-C workload. In this experiment, the database is populated with 4 warehouses. We observe the throughput and average transaction latency by varying the number of worker thread from 2 to 32.

Experimental results are shown in Fig. 6 and 7. We can observe that the throughput of both LOTAS and FCFS is decreasing with increasing worker threads. When the number of worker thread is less than 8, contention is rarely in the workload since one warehouse would not be concurrently accessed by two worker threads. As a result, LOTAS performs similarly to FCFS. Once the number of worker thread exceeds the number of warehouse, lock contention is heavy and the lock thrashing problem emerges, making transactions suffers from long lock waiting time. In this scenario, LOTAS could achieve 1.3–2.8x higher throughput than FCFS. As for latency, the average transaction latency of LOTAS is 22.5%–64.2% of FCFS. This is mainly due to the fact that LOTAS identifies the lock thrashing problem and proactively defer some transactions, thus reducing the lock waiting time of transactions and achieves more concurrent execution.

5.2 Impact of Contention Rate

To investigate the impact of skewness, we use TPC-C and Micro-Benchmark to compare LOTAS and FCFS under the workloads at different rates of contention. First, the throughput of TPC-C with various number of warehouse are tested. As warehouse records are accessed by the uniform distribution in TPC-C, thus the fewer warehouses are used in the workload, the more contentions will be

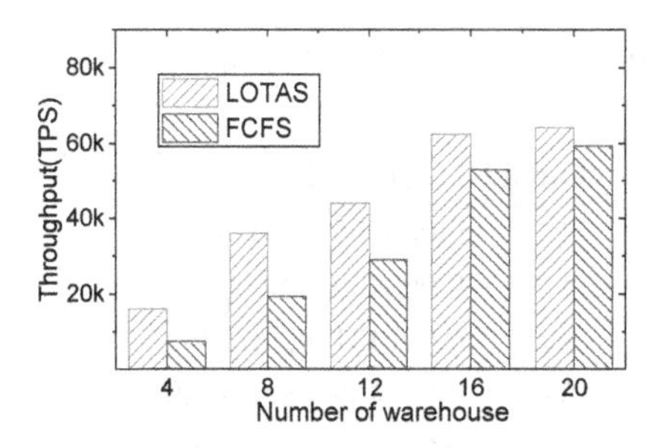

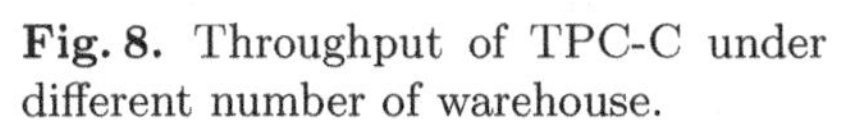
Fig. 8. Throughput of TPC-C under different number of warehouse.

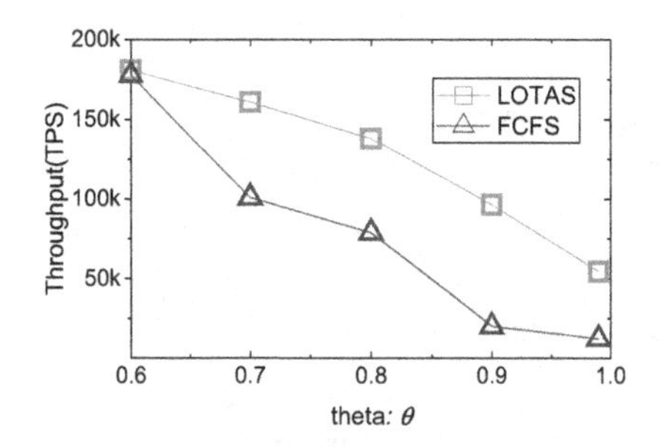

Fig. 9. Throughput of MicroBenchmark under different levels of skewness.

introduced. The number of worker thread is fixed to 32. Result is shown in Fig. 8. In general, LOTAS performs better than FCFS. And the less warehouses involved, LOTAS improves more (about 1.9x than FCFS with 4 warehouses).

We also evaluate LOTAS by varying the parameter θ of Zipfian distribution from 0.6 to 0.99. The number of worker thread is fixed to 32. Figure 9 illustrates the result. In general, the performance trend is similar to that in TPC-C. We find that LOTAS could achieve 1.6x–4.8x higher throughput than FCFS, which is more significant than that in TPC-C. The reason is that when the parameter θ is higher than 0.9, the contention rate is extremely higher than TPC-C. More and more transactions will access fewer records. Besides, transaction logic in Micro-Benchmark is simpler. Therefore, the percentage of lock waiting time in the whole transaction execution time is relatively higher and thus the performance improvements LOTAS can make compared to FCFS under Micro-Benchmark is more notable than that under TPC-C.

5.3 Overhead Analysis

In this experiment, we estimate the overhead brought by LOTAS by calculating the proportion of time spent on prediction phase and the scheduling phase to total execution time. The experimental settings are the same as Sect. 5.1. We divide the total execution time into four parts: Lock-mgr (time used by lock manager, including deadlock detection, lock waiting time), Schedule (time mainly used by the scheduling algorithm before lock acquisition), Pred (time used in predicting subsequent keys) and Useful (time used by other useful works to execute transactions, for example, index reading, computation). Results are shown in Fig. 10.

It can be seen that, when the database is running with 32 worker threads, LOTAS spends about 17.4% of the total execution time in Pred and Schedule. While in the same time, the time spent in Lock-mgr is greatly reduced (from 79% to 0.074%). This is mainly because of the significant decrease of lock waiting time (as shown in Table 1). LOTAS makes transactions that may intensify the risk of lock thrashing deferring to acquire locks, enabling live transactions to finish as soon as possible.

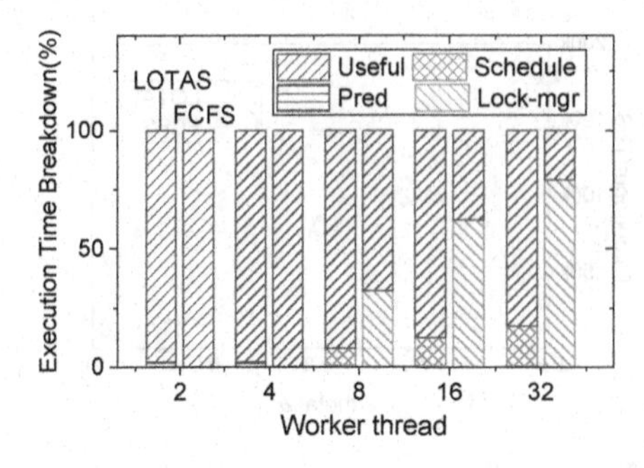

Fig. 10. Execution time breakdown.

Table 1. Average lock waiting time (us)

Thread	LOTAS	FCFS
2	0.000155	0.000384
4	0.000400	0.000596
8	0.007461	9.887786
16	0.037435	58.972930
32	0.099059	295.960092

Furthermore, under the workload with less or no contentions, the performance gains from scheduling algorithm are greatly reduced. With 2 and 4 workers, LOTAS spends 2% time in filling *pred-set*. While in these cases, there has no hot tuples. The scheduling algorithm would not be triggered actually. As the Markov prediction graphs are built and analyzed offline, it has no direct impact on transaction performance. Therefore, under low-contended workloads, the overhead produced by LOTAS is negligible.

5.4 Impact of Unpredictable Transactions

This experiments evaluate the performance of LOTAS under the high-contention workload where some of the transactions are *unpredictable*. Unlike the predictable transactions in Micro-Benchmark, all parameters in the unpredictable transactions comes from the client. The percentage of unpredictable transactions is varied from 0% to 100%. The number of worker thread is fixed in 32 and the parameter θ is set to 0.9 in this experiment. Figure 11 shows the results. With the increasing ratio of unpredictable transaction, the throughput achievements brought by LOTAS are getting smaller and smaller. This is expected since more and more lock thrashings among unpredictable transactions cannot be identified by LOTAS. Even so, when there are 60% contended transactions can not be

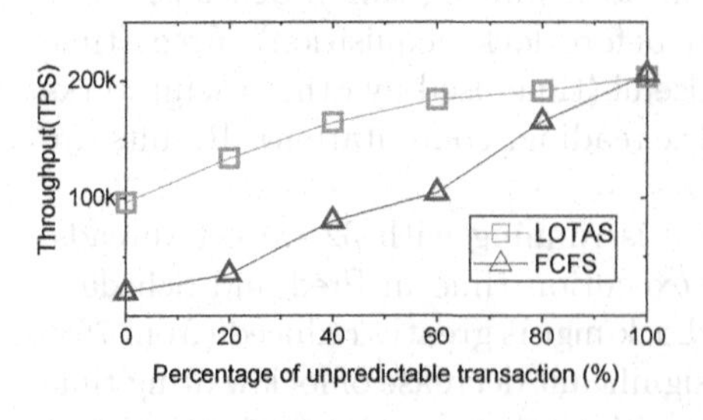

Fig. 11. Impact of the ratio of unpredictable transactions.

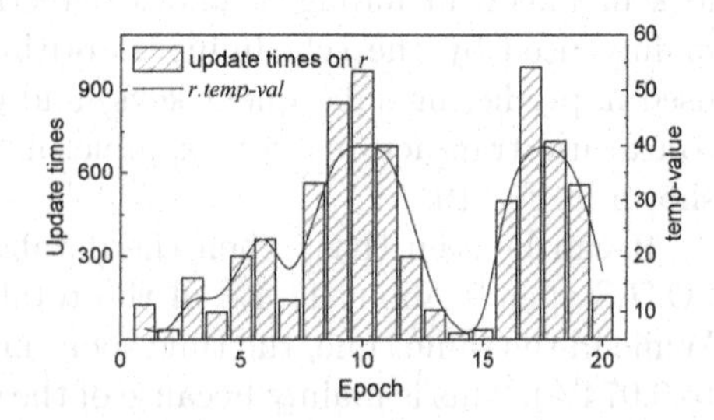

Fig. 12. The effectiveness of hot records tracking mechanism.

predicted, LOTAS still could deliver a 1.8x higher throughput according to the identified lock thrashings across predictable transactions.

5.5 Effectiveness of Hot Record Tracking Mechanism

We conduct an individual experiment for the hot record tracking mechanism. The parameter c_θ in formula 1 is set -0.5, epoch is set to 0.1 ms, h_Δ is set to 1, and $\mathcal{H}_{up}$ is set to 40. We change the update times on the record r and sample the value of *r.temp-val* every epoch. Result shown in Fig. 12 illustrates that the value of *r.temp-val* calculated by our proposed temperature computation formula can reflect the hotness of records and is able to rapidly adjust the temperature value when the hotness changes.

6 Related Works

Database Scheduling — The FCFS policy is commonly used in commercial DBMSs like MySQL, SQL Server and DB2. Many recent researches proposed to schedule transaction in a serializable order before execution based on the assumption of one-shot transaction [15,23,30]. The intelligent scheduling [30] proposed to schedule conflicting transactions into the same queue according to the calculated conflict possibility by comparing the full read/write sets of current transactions in each queue. STRIFE [18] presented an algorithm to schedule a batch of transactions into clusters that do not have cross-cluster conflicts and a small residual cluster. Transactions under different clusters can be executed in parallel without contention.

There are many scheduling works focusing on preempting locking [10,13,14, 24]. Boyu et al. [24] proposed a lock scheduling mechanism, which allows the transaction that blocks most transactions has highest priority to acquire the lock at the lock release phase. Quro [28] reorders transaction codes from the application side. These techniques are compatible with LOTAS, which enable scheduling to improve the performance at multiple stages.

Transaction scheduling has also been studied in the execution of OCC (e.g., abort transactions that will cause living transactions to abort [8]). The recent Bailu et al. [3] proposed to reorder a batch of transaction reads and writes at the storage layer and the validation phase. The idea of reordering operations in a batch is somewhat similar to LOTAS, which aims to proactively avoid exacerbating contention among live transactions.

Improvements over High-Contention Workloads — For pessimistic CC schemes (e.g. 2PL), many works focused on designing a lightweight and efficient lock manager [6,11,20,29]. Kun Ren et al. proposed to partition transaction processing functionality across CPU cores and each thread is responsible for handling conflicts over some specific records [19]. For optimistic CC, Tianzheng et al. proposed a mostly-optimistic concurrency control to enhance OCC under highly contended dynamic workloads by using a mix of optimistic and pessimistic concurrency control protocols [25]. Transaction chopping technique [22] is also widely used to deal with high contention issues [4,15,26,27].

Workload Prediction — Prediction has many applications in transaction or query processing within the database [5,7,12,17,21]. Ganapathi et al. [5] used kernel canonical correlation analysis to predict performance metrics (e.g., running time, resource usage). QB5000 [12] forecasts the query arrival rate by using linear regression and recurrent neural network in order that optimal indexes could be chosen for the target workload in real-time. Andy Pavlo et al. [17] adopted the Markov model to forecast the future behaviors (e.g., access which partition, abort or commit) of started transactions according to current executed operations. The predicted information is used to choose the optimal partition to run this transaction or disable undo logging for non-aborting transactions. PROMISE [21] also used Markov model to learn OLAP query patterns, and then it pre-fetches the data for the following queries. Therefore, in LOTAS, we adopt Markov model to predict subsequent operations about what tuples they are going to access, which is a decent fit.

7 Conclusions

In this paper, we presented LOTAS, a lock thrashing aware transaction scheduling mechanism for highly contended workloads. We propose to predict subsequent data of a running transaction according to the previous operations and parameters through building the Markov-based prediction model. At runtime, according to the identification of current and future possible lock thrashings, LOTAS decides whether a transaction requires to be postponed to acquire locks. Experimental results demonstrate that, under high contention workloads, LOTAS could greatly reduce lock waiting time and gain 1.3x–4.8x improvements in throughput than the classical FCFS scheduling. Also, the time overhead produced by LOTAS is negligible even under low-contended workloads.

Acknowledgement. This work is partially supported by National Key R&D Program of China (2018YFB1003404), and NSFC under grant numbers 61972149, and ECNU Academic Innovation Promotion Program for Excellent Doctoral Students. The corresponding author is Peng Cai. We thank anonymous reviewers for their very helpful comments.

References

1. Dbx1000. https://github.com/yxymit/DBx1000
2. Bernstein, P.A., Newcomer, E.: Principles of Transaction Processing for Systems Professionals. Morgan Kaufmann, Burlington (1996)
3. Ding, B., Kot, L., Gehrke, J.: Improving optimistic concurrency control through transaction batching and operation reordering. PVLDB **12**(2), 169–182 (2018)
4. Faleiro, J.M., Abadi, D., Hellerstein, J.M.: High performance transactions via early write visibility. PVLDB **10**(5), 613–624 (2017)

5. Ganapathi, A., Kuno, H.A., et al.: Predicting multiple metrics for queries: better decisions enabled by machine learning. In: ICDE, pp. 592–603 (2009)
6. Guo, J., Chu, J., Cai, P., Zhou, M., Zhou, A.: Low-overhead paxos replication. Data Sci. Eng. **2**(2), 169–177 (2017). https://doi.org/10.1007/s41019-017-0039-z
7. Gupta, C., Mehta, A., Dayal, U.: PQR: predicting query execution times for autonomous workload management. In: ICAC, pp. 13–22 (2008)
8. Haritsa, J.R., Carey, M.J., Livny, M.: Dynamic real-time optimistic concurrency control. In: RTSS, pp. 94–103 (1990)
9. Harizopoulos, S., Abadi, D.J., Madden, S., Stonebraker, M.: OLTP through the looking glass, and what we found there. In: SIGMOD, pp. 981–992 (2008)
10. Huang, J., Mozafari, B., et al.: A top-down approach to achieving performance predictability in database systems. In: SIGMOD, pp. 745–758 (2017)
11. Jung, H., Han, H., Fekete, A.D., Heiser, G., Yeom, H.Y.: A scalable lock manager for multicores. In: SIGMOD, pp. 73–84 (2013)
12. Ma, L., Aken, D.V., Hefny, A., et al.: Query-based workload forecasting for self-driving database management systems. In: SIGMOD, pp. 631–645 (2018)
13. McWherter, D.T., Schroeder, B., Ailamaki, A., et al.: Priority mechanisms for OLTP and transactional web applications. In: ICDE, pp. 535–546 (2004)
14. McWherter, D.T., Schroeder, B., Ailamaki, A., et al.: Improving preemptive prioritization via statistical characterization of OLTP locking. In: ICDE, pp. 446–457 (2005)
15. Mu, S., Cui, Y., Zhang, Y., Lloyd, W., Li, J.: Extracting more concurrency from distributed transactions. In: OSDI, pp. 479–494 (2014)
16. Pavlo, A.: What are we doing with our lives?: nobody cares about our concurrency control research. In: SIGMOD, p. 3 (2017)
17. Pavlo, A., Jones, E.P.C., Zdonik, S.B.: On predictive modeling for optimizing transaction execution in parallel OLTP systems. PVLDB **5**(2), 85–96 (2011)
18. Prasaad, G., Cheung, A., Suciu, D.: Improving high contention OLTP performance via transaction scheduling. CoRR, abs/1810.01997 (2018)
19. Ren, K., Faleiro, J.M., Abadi, D.J.: Design principles for scaling multi-core OLTP under high contention. In: SIGMOD, pp. 1583–1598 (2016)
20. Ren, K., Thomson, A., Abadi, D.J.: Lightweight locking for main memory database systems. PVLDB **6**(2), 145–156 (2012)
21. Sapia, C.: PROMISE: predicting query behavior to enable predictive caching strategies for OLAP systems. In: Kambayashi, Y., Mohania, M., Tjoa, A.M. (eds.) DaWaK 2000. LNCS, vol. 1874, pp. 224–233. Springer, Heidelberg (2000). https://doi.org/10.1007/3-540-44466-1_22
22. Shasha, D.E., Llirbat, F., Simon, E., et al.: Transaction chopping: algorithms and performance studies. ACM Trans. Database Syst. **20**(3), 325–363 (1995)
23. Thomson, A., Diamond, T., Weng, S., et al.: Calvin: fast distributed transactions for partitioned database systems. In: SIGMOD, pp. 1–12 (2012)
24. Tian, B., Huang, J., Mozafari, B., Schoenebeck, G.: Contention-aware lock scheduling for transactional databases. PVLDB **11**(5), 648–662 (2018)
25. Wang, T., Kimura, H.: Mostly-optimistic concurrency control for highly contended dynamic workloads on a thousand cores. PVLDB **10**(2), 49–60 (2016)
26. Wang, Z., Mu, S., Cui, Y., Yi, H., Chen, H., Li, J.: Scaling multicore databases via constrained parallel execution. In: SIGMOD, pp. 1643–1658 (2016)
27. Xie, C., Su, C., Littley, C., Alvisi, L., Kapritsos, M., Wang, Y.: High-performance ACID via modular concurrency control. In: SOSP, pp. 279–294 (2015)
28. Yan, C., Cheung, A.: Leveraging lock contention to improve OLTP application performance. PVLDB **9**(5), 444–455 (2016)

29. Yu, X., Bezerra, G., Pavlo, A., et al.: Staring into the abyss: an evaluation of concurrency control with one thousand cores. PVLDB **8**(3), 209–220 (2014)
30. Zhang, T., Tomasic, A., Sheng, Y., Pavlo, A.: Performance of OLTP via intelligent scheduling. In: ICDE, pp. 1288–1291 (2018)

Searchable Symmetric Encryption with Tunable Leakage Using Multiple Servers

Xiangfu Song[1], Dong Yin[1], Han Jiang[2], and Qiuliang Xu[2](✉)

[1] School of Computer Science and Technology, Shandong University, Jinan, China
bintasong@gmail.com, dybean1994@gmail.com
[2] School of Software, Shandong University, Jinan, China
{jianghan,xql}@sdu.edu.cn

Abstract. Searchable symmetric encryption has been a promising primitive as it enables a cloud user to search over outsourced encrypted data efficiently by only leaking small amount of controllable leakage. However, recent leakage-abuse attacks demonstrate that those stand leakage profiles can be exploited to perform severe attacks – the attacker can recover query or document with high probability. Ideal defending methods by leveraging heavy cryptographic primitives, e.g. Oblivious RAM, Multiparty Computation, are still too inefficient for practice nowadays.

In this paper, we investigate another approach for countering leakage-abuse attacks. Our idea is to design SSE with tunable leakage, which provides a configurable way for trade-off between privacy and efficiency. Another idea is to share the leakage among multiple non-collude servers, thus a single server can only learn partial, rather than the whole leakage. Following the ideas, we proposed two SSE schemes. The first scheme uses two servers and is static, which serves as the first step to emphasize our design methodology. Then we propose a dynamic SSE scheme, by additionally use a third server to hold dynamic updates. We demonstrate that the leakage for the third server is only *partial update history*, a newly defined leakage notion that leaks limited information rather than the whole update history. Our schemes provide stronger security that hides search/access pattern in a tunable way as well as maintains forward and backward privacy. We also report the performance of our constructions, which shows that both schemes are efficient.

Keywords: Leakage-abuse attack · Searchable encryption · Tunable leakage · Multiple servers

1 Introduction

Searchable Symmetric Encryption (SSE) scheme enables the client to securely outsource its encrypted data to cloud server(s) while maintaining the capability of private query from the client. The main task of the SSE design is to

This work is supported by National Natural Science Foundation of China (Grant No. 61572294, 61632020).

Y. Nah et al. (Eds.): DASFAA 2020, LNCS 12112, pp. 157–177, 2020.
https://doi.org/10.1007/978-3-030-59410-7_10

build a secure searchable index that assists the server to perform search on the encrypted data. In recent years, many progress [3–5,8,9,16,18,20,23–25,35] made to improve its efficiency, functionality and security.

Most SSE schemes trade efficiency with the price of some small controllable leakage. Those leakage, which is formally defined in the seminal work [15], mainly includes *search pattern* (whether a query match with a previous query or not), *access pattern* (all matched document identifiers from a query) and *volume pattern* (the result size of a query). For a long time, above leakages are deemed to be small and regarded as "common leakage" for almost all SSE schemes.

Unfortunately, the seemingly acceptable leakage can be exploited to perform severe attacks. In general, all those attacks aim to recover query content with high probability. In particular, the IKK Attack [22] exploits access pattern leakage assuming the background information of underlying data collection. The Counter Attack [7], which can recover queries by exploiting access pattern and volume pattern leakage. More recently, Zhang et al. [37] proposed file-injection attack to encrypted e-mail service by exploiting access pattern and update leakage, which significantly improves recovery rate. All above attacks indict that those so-called *small* leakage was poorly understood in the past.

To defend against leakage-abuse attacks, we can leverage generic cryptographic primitives. The Oblivious RAM (ORAM) simulation can be used to eliminate the search and access pattern leakage, and volume pattern leakage can be trivially suppressed by padding all search result to a maximize size. However, those approaches significantly increase the complexity of SSE scheme. Specifically, the ORAM simulation will incur a $O(\log N)$ (N is the storage size) communication blow-up per memory access. The native padding approach will result worst linear search time and quadratic index size.

After many years' trials, the inherent leakage of SSE is quite hard to totally eliminate without incurring a huge overhead, even slight amount of leakage reduction is quite challenging. It would be desirable to seek some efficient configurable leakage reduction approaches, meanwhile maintaining practical efficiency.

1.1 Overview and Contribution

In this paper, we investigate how to design efficient SSE with tunable leakage. Specifically, we want to obtain SSE scheme with adjustable leakage in search pattern, access pattern and volume pattern.

Our first intuitive idea is to partition the index into two parts, separately put them on two non-colluded servers, and move the index around servers during search to hide search pattern. By this approach, we can break query linkability on search token level and only leave a *one-time* search pattern for the server being queried. However, things are more complex than we expect when concerning the whole leakage of a SSE scheme. The reason is that the server may learn information from access pattern either from the search result or its associated volume information. Therefore, we must handle those leakage carefully otherwise the server may still deduce which keyword is being queried, given some background information of the underlying dataset.

To handle the query result leakage, we firstly require the search protocol to be result-hiding, i.e., the searched result remains encrypted until the client performs local decryption. Secondly, we need to suppress the volume pattern leakage. To totally eliminate volume pattern, one trivial and inefficient approach is to pad all search result to a maximize size, but this approach bring too much overhead to communication and storage. We take a more efficient way to handle the leakage, the idea is to group the keywords, and pad the volume of keywords in same group to a common max size, which are also used in some recent works [4,16]. This provides a tunable notion between security and efficiency, and makes SSE scheme more configurable depending on application scenarios and security level. By the grouping technique, the server cannot distinguish which keyword within a group is being queried as they all have the same leakage profile.

For our dynamic scheme, we also consider forward and backward privacy. We design efficient token generation method and carefully integrate it with above leakage-reduction mechanism. We also analyse the additional leakage of our dynamic scheme and formally capture it by a *partical update history*, a leakage that is smaller than the whole update history.

On Feasibility of Using Non-colluded Servers. Non-collusion assumption can be realized by using two conflict-of-interest service providers, e.g., Amazon AWS and Microsoft Azure. The assumption is believed to be reasonable in such case and many cryptographic protocol primitives were proposed based on that since the very beginning of modern cryptography. Just name a few, Private Information Retrieval (PIR) [10], Multi-party Computation, and more recent schemes such as Outsource Computation, Privacy Preserving Machine Learning as well as Searchable Encryption [11,14,20,21]. Moreover, we stress that unlike many SSE schemes under multi-servers setting, servers in our scheme do not need to communicate with each other. A server may even be unaware of the existence of others, let alone collude with them.

Our contribution are summarized as follows:

- We propose two SSE schemes with tunable leakage. Our first scheme is only static, which serves as the first step to demonstrate our idea. The static scheme leverages two servers, with a novel index structure to hide information in a tunable way. Then we propose a dynamic SSE to support fully dynamic update, by leveraging a third server that only learns *partial update history* (a security notion we will be explained later) while maintaining the same leakage for the first two servers.
- Our schemes provide a tunable leakage in search pattern, access pattern and size pattern. For our dynamic scheme, we additionally require it to be with forward privacy and backward privacy, a stronger security notion for dynamic SSE schemes. We carefully analyze the leakage for our two schemes, and capture them with well-defined leakage functions. A formal proof is provided to argue the security of our construction.
- Our schemes only involve efficient symmetric primitives such as pseudorandom function (PRF), hash function, which is simply and efficient. We implement our scheme and the experiment shows that both constructions are efficient.

2 Related Works

All existing leakage-abuse attacks leverages leakage from query, and some of them assume the attacker knows some background information. In following part, we will give a brief review on the leveraged leakage and basic ideas behind, as well as existing defending countermeasures.

2.1 The Attacks

The IKK Attack. The IKK attack was proposed by Islam et al. [22]. The attack exploits access pattern leakage to recover search query, with some background information on the underlying database, the server knows all possible keywords and their co-occurrence probability information (i.e., the probability of any two keywords appear in a document). Specifically, the adversary builds an $m \times m$ matrix M in which each element is the co-existing probability of possible keyword pair. Then after observing n search queries, the adversary can learn if a document matches two queries, then it can construct a $n \times n$ matrix $\hat{M}$ with their co-occurrence rate for each query. By matching between M and $\hat{M}$ as shown from [22], the adversary can recover search queries efficiently.

Count Attack. By extending IKK attack, Cash et al. [7] proposed count attack to improve efficiency and recovery rate, by further exploiting volume leakage during search query. As usual, the adversary is assumed to know the background information of the underlying dataset *plus* the size patterns of all keywords. The idea of count attack is quite straightforward. Specifically, if the number of matching document is unique, then the adversary can immediately match the query with a known keyword by that unique volume pattern, otherwise, any keyword with that volume pattern will be candidate. Now the adversary can leverage the idea of IKK attack to further recovery the query by the access pattern leakage. The count attack is more efficient and accurate than IKK attack, and it can recover almost all queries efficiently for Enron dataset.

File-injection Attack. The File-injection attack was initially proposed by Zhang et al. [37] for encrypted e-mail system. The attack assumes the adversary can launch specific document injection into the encrypted database (e.g., sending e-mail to client and wait the client to update those dedicate file into the system) and later can recover query content by observing the leakage from queries. Two versions of attack was provided in [37]. The static attack only exploits access pattern leakage and the adaptive version further leverages update leakage to improve recover rate. To date, the adaptive file-injection attack can be resolved by forward-private SSE [3,17,25,35] in which no information on the keyword is leaked out during update query. Nowadays, forward privacy can be obtained efficiently by only using symmetric primitives and is almost for free compared to non-forward-private SSE schemes. However, the static version of the attack is still hard to resolve as most SSE schemes leaks access pattern leakage during search, either from searching the index or accessing the fetched files.

Search Pattern Based Attack. The first attack leveraging search pattern leakage, as far as we know, is from Liu et al. [29]. The attack assumes the adversary can obtain some background information, such as search trend from Google Trend, domain-specific information and user search habit. The attacker records all search pattern over an interval of time, and then she can find the best match between background information with the search pattern, e.g., by applying some similarity measurement. Recently, some attacks for range and k-NN query by exploiting search pattern leakage were proposed [26,30]. Those attacks aim to improve recovery rate without pre-supposed query distribution while prior attacks require the query is issued uniformly.

2.2 Prior Countermeasures

Kamara et al. [24] studied how to suppress search pattern leakage. They proposed compilers that make any dynamic SSE scheme rebuildable efficiently and further transforms any rebuildable scheme to hide search pattern leakage, based on a non-black-box ORAM-like simulation. They claimed that for certain cases the compiled scheme is more efficient than black-box ORAM simulation. Huang et al. [21] proposed search-pattern-oblivious index structure based on two-server model. The idea is similar to our's, but the storage overhead of [21] is $O(nm)$ where n denotes sizes of keyword set and m denotes number of documents, which is too high for large document set.

Reducing Search/Access Pattern Leakage. Towards access pattern leakage, the most effective approach is by leveraging ORAM or PIR [16,20]. Demertzis et al. [16] recently proposed several searchable encryption with tunable leakage. The idea is to use grouping technique [4] to hide volume pattern and then a ORAM to store padded indexes. Hoang et al. [20] designed an oblivious searchable encryption scheme by combining PIR and write-only ORAM [2] based on bit-string index. The idea is to use oblivious-write mechanism to perform update, thus hides update leakage, and use PIR during search to hide access pattern. One noticeable property of bit-string index is that the size pattern can be hided naturally, so prior padding procedure is not required. But as we shown previously, the storage overhead of bit-string index do not scale to large database.

Reducing Volume Leakage. Some recent schemes also focus on reducing volume pattern, a leakage that is actually more difficult to resolve. Even by using ORAM, the number of storage access is still exposed. One of the native approach is to pad all result set to a max size, but this approach is too unfriendly to storage complexity and search efficiency. Bost et al. [4] made a balance between security and efficiency, specifically, in their scheme keywords are divided into group based on their result size, and then each keyword is padded to the max result size of its group. Kamara and Moataz [23] proposed to use a computational security notion for volume pattern, they designed a pseudo-random transform takes an input set and generates a new one such that the size is generated using a pseudo-random function. However, query correctness cannot be guaranteed because when truncation occurs, several values are removed from the new result set. Those schemes, either require more storage overhead as a result of padding mechanism, or cannot guarantee search correctness, significantly impede their practical deployment. Very recently, Patel et al. [33] proposed an efficient scheme without padding or pseudorandom transform mechanism, thus achieved un-lossy volume leakage reduction with optimal $O(N)$ storage complexity. The basic idea is to use a Pseudorandom Function (PRF) with a small output domain, and place each index by Cuckoo Hashing [32]. Their transform works for static setting, and it is not clear how to efficiently extent their idea to dynamic SSE schemes.

Forward and Backward Private SSE Schemes. To handle information leakage from update phrase, most SSE schemes with forward and backward privacy were proposed. The forward privacy, which is the key to resolve adaptive version of file-injection attack, was firstly defined in [36]. It is until in [3], Bost proposed an efficient construction based on trapdoor permutations, which achieves asymptotically optimal search and update complexity. However, the update is not efficient enough as the involvement of public key primitives (i.e., only about 4500 updates per second as reported in the paper). After that, many symmetric-key-only forward private SSE schemes were proposed [3,25,35], those works significantly improves the performance of forward private SSE schemes.

There are also recent schemes that focuses on backward privacy for DSSE schemes. In general, a DSSE scheme with backward privacy requires that a honest-but-curious server cannot learn that a document was ever added into the system and then deleted, between any two adjacent search requests. The construction in [36] support certain level of backward privacy, but the security notion was not formally defined at that time. Bost et al. [5] defined three levels of backward privacy from the most secure to the least, and proposed three constructions correspondingly. Recently, SSE schemes with better efficiency-security trade-off are proposed [18,27]. Nowadays, forward private SSE schemes can be obtained without much efficiency degradation compared with their non-forward-private versions.

Countermeasures Leveraging Secure Hardware. Another approach is by leveraging secure hardware, e.g., intel SGX [12]. Recently, many searchable encryption schemes [1,13,31,34] based on secure hardware were proposed. However, it is still a concern since secure hardware may not resist with many kinds of side-channel attack [6,19,28].

3 Preliminary

3.1 Negligible Function

A function $\nu : \mathbb{N} \rightarrow \mathbb{R}$ is negligible in λ if for every positive polynomial p, $\nu(\lambda) < 1/p(\lambda)$ for sufficiently large λ.

3.2 Searchable Encryption

Let a database $\mathsf{DB} = \{(\mathsf{ind}_i, \mathsf{W}_i)\}_{i=1}^{D}$ be a D-vector of identifier/keyword-set pairs, where $\mathsf{ind}_i \in \{0,1\}^l$ is a document identifier and $\mathsf{W}_i \subseteq \mathcal{P}(\{0,1\}^*)$. The universe of keywords of the database DB is $\mathsf{W} = \cup_{i=1}^{D}\mathsf{W}_i$. We use $N = \sum_{i=1}^{D}|\mathsf{W}_i|$ to denote the number of document/keyword pairs and $\mathsf{DB}(w) = \{\mathsf{ind}_i | w \in \mathsf{W}_i\}$ to denote the set of documents that contain the keyword w. A *dynamic searchable symmetric encryption* (DSSE) scheme $\Pi = \{\mathsf{Setup}, \mathsf{Search}, \mathsf{Update}\}$ consists of three protocols ran by the client and the server(s):

- $((K,\sigma); \mathsf{EDB}) \leftarrow \mathsf{Setup}(\lambda, \mathsf{DB}; \perp)$: It takes a security parameter λ and a database DB as inputs and outputs (K,σ) to the client and EDB to the server, where K is a secret key, σ is the client's state, and EDB is the encrypted database.
- $((\sigma', \mathsf{DB}(w)); \mathsf{EDB}') \leftarrow \mathsf{Search}(K,\sigma,w; \mathsf{EDB})$: The client's input consists of its secret key K, the state σ and a keyword w, the server's input is the encrypted database EDB. The client's output includes a possibly updated state σ' and $\mathsf{DB}(w)$, i.e. the set of the identifiers of the documents that contain the keyword w. The server's output is the possibly updated encrypted database EDB'.
- $(\sigma'; \mathsf{EDB}') \leftarrow \mathsf{Update}(K,\sigma, ind, w, op; \mathsf{EDB})$: The client's input is the secret key K, the state σ, a document identifier ind, a keyword w and an operation type op. The server's input is EDB. The operation op is taken from the set {add, del}, which means the client wants to add or delete a document/keyword pair. The client updates its state and the server updates EDB as requested by the client.

3.3 Security Definition

All existing searchable encryption schemes leak more or less some information to the server, as a trade-off to gain efficiency. Thus, The security of searchable encryption is defined in the sense that no more information is leaked than allowed. This is captured by using the simulation paradigm and providing the simulator a set of predefined leakage functions $\mathcal{L} = \{\mathcal{L}_{\text{Setup}}, \mathcal{L}_{\text{Search}}, \mathcal{L}_{\text{Update}}\}$.

Definition 1 (Adaptively Secure Searchable Encryption). *Let $\Pi = \{\mathsf{Setup}, \mathsf{Search}, \mathsf{Update}\}$ be a searchable encryption scheme, $\mathcal{A}$ be an adversary, $\mathcal{S}$ be a simulator parameterized with leakage function $\mathcal{L} = \{\mathcal{L}_{\text{Setup}}, \mathcal{L}_{\text{Search}}, \mathcal{L}_{\text{Update}}\}$. We define the following two probabilistic experiments:*

- $\boldsymbol{Real}_{\mathcal{A}}^{\Pi}(\lambda)$: *$\mathcal{A}$ chooses a database DB, the experiment runs $\mathsf{Setup}(\lambda, \mathsf{DB}; \perp)$ and returns EDB to $\mathcal{A}$. Then, the adversary adaptively chooses queries q_i. If q_i is a search query then the experiment answers the query by running $((\sigma_{i+1}, \mathsf{DB}_i(w_i)), \mathsf{EDB}_{i+1})) \leftarrow \mathsf{Search}(K,\sigma_i,q_i; \mathsf{EDB}_i)$. If q_i is an update query, then the experiment answers the query by running $(\sigma_{i+1}, \mathsf{EDB}_{i+1}) \leftarrow \mathsf{Update}(K,\sigma_i,q_i; \mathsf{EDB}_i)$. Finally, the adversary $\mathcal{A}$ outputs a bit $b \in \{0,1\}$.*

- $\boldsymbol{Ideal}^{\Pi}_{\mathcal{A},\mathcal{S}}(\lambda)$: *$\mathcal{A}$ chooses a database DB. Given the leakage function $\mathcal{L}_{\mathsf{Setup}}(DB)$, the simulator $\mathcal{S}$ generates an encrypted database EDB $\leftarrow \mathcal{S}(\mathcal{L}_{\mathsf{Setup}}(DB))$ and returns it to $\mathcal{A}$. Then, the adversary adaptively chooses queries q_i. If q_i is a search query, the simulator answers the query by running $\mathcal{S}(\mathcal{L}_{\mathsf{Search}}(q_i))$. If q_i is an update query, the simulator answers the query by running $\mathcal{S}(\mathcal{L}_{\mathsf{Update}}(q_i))$. Finally, the adversary $\mathcal{A}$ outputs a bit $b \in \{0,1\}$.*

We say Π is an $\mathcal{L}$-adaptively-secure searchable encryption scheme if for any probabilistic, polynomial-time (PPT) adversary $\mathcal{A}$, there exists a PPT simulator $\mathcal{S}$ such that:

$$|\Pr(\boldsymbol{Real}^{\Pi}_{\mathcal{A}}(\lambda) = 1) - \Pr(\boldsymbol{Ideal}^{\Pi}_{\mathcal{A},\mathcal{S}}(\lambda) = 1)| \leq \boldsymbol{negl}(\lambda)$$

4 The Construction

In this section, we will present our two SSE constructions with tunable leakage using multi-servers, and analyse their concert leakage. Our first construction is only static, which serves as a first step to express our idea of designing SSE with tunable leakage. However, when applying the idea to support fully dynamic update, a third server is additionally added to support fully dynamic update. By this approach, the leakage of first two servers remains as the same, which keeps good property of the static scheme and also simplifies the security proof. We further analyze the leakage of the third server, and propose a new security notion call *partial update history* to capture it, the leakage is limited compared with original full update history.

Padding Algorithm. Before processing to generating encrypted database, the client will first do a padding algorithm to the raw index data. The padding algorithm is quite straightforward. The client chooses a group parameter G, α where G is the minimum padding size and α is the number of group. Each keyword in Group $i \in [1, \alpha]$ will be padded to with equal size G^i, e.g., if $G = 10$, $i = 2$, then for keyword w in group 2, $\mathsf{DB}(w)$ will be padded to be $\mathsf{PDB}(w)$ such that $|\mathsf{PDB}(w)| = 100$. Finding the group of a keyword is efficient – the client just compute i such that $G^{i-1} < |\mathsf{DB(w)}| \leq G^i$.

Additionally, we require dummy identifiers be distinguishable from those valid ones, which allows the client to filter dummy identifiers when obtaining search result. This is easy to achieve – by putting one single bit into a identifier to denote dummy or not.

4.1 The Static Scheme

In the following pseudo codes, we use pseudo-random function F, keyed hash function H to generate tokens needed during the protocol, a hash function h, and $\mathsf{SE.Enc}(\cdot)$ and $\mathsf{SE.Dec}(\cdot)$ to denote the encryption and decryption algorithm of an $\mathsf{IND\text{-}CPA}$ secure symmetric encryption scheme SE (e.g., AES).

Overview. The main idea of the static construction is to leverage two servers combining with a adjustable padding scheme to provide tunable SSE leakage. Before the searchable index is outsourced to servers, the client will apply the padding algorithm on the database DB to get its padded version PDB. The client then encrypts the padded index like normal SSE schemes and outsources to servers. Some necessary information will be stored on client side, which is used to perform subsequent search query. Specifically, we will let the client store (b, s_w, u_w) for each keyword in which b denotes the current server that holds the index associated with w, s_w denotes counter of search queries that ever performed to w and u_w denotes total number of identifiers associate with w. As we will show later, those information plays a core role in the design of our schemes.

```
Setup(λ, DB; ⊥)
  Client:
 1: k_s, k_u ←$ {0,1}^λ
 2: Σ, EDB_0, EDB_1 ← empty map
 3: PDB ← Padding(DB)
 4: for w ∈ W do
 5:     u_w, s_w ← 0, b ←$ {0,1}
 6:     k_w ← F(k_s, h(w)||s_w)
 7:     for ind ∈ PDB(w) do
 8:         e ← H(k_w, u_w)
 9:         v ← SE.Enc(k_u, ind)
10:         u_w ← u_w + 1
11:         EDB_b[e] ← v
12:     end for
13:     Σ[w] ← (b, s_w, u_w)
14: end for
15: Client locally stores Σ
16: Send EDB_b to server S_b for b ∈ {0,1}

Search(k_s, Σ, w; EDB_0, EDB_1)
  Client:
17: (b, s_w, u_w) ← Σ[w]
18: k_w ← F(k_s, h(w)||s_w)
19: Send (k_w, u_w) to server S_b
  Server S_b:
20: R ← {}
21: for i ← 0 to u_w do            ▷ can be parallel
22:     e ← H(k_w, i)
23:     R ← R ∪ {EDB_b[e]}
24: end for
25: Send R to client
26: Delete R from EDB_b
  Client:
27: D, EM ← {}, u_w ← 0
28: k_w ← F(k_s, h(w)||s_w + 1)
29: for c ∈ R do
30:     ind ← SE.Dec(k_u, c)
31:     D ← D ∪ {ind} if ind is not dummy
32:     e ← H(k_w, u_w)
33:     v ← SE.Enc(k_u, ind)
34:     EM ← {(e, v)}
35:     u_w ← u_w + 1
36: end for
37: Σ[w] ← (b ⊕ 1, s_w + 1, u_w)
38: Send EM to server S_{b⊕1}
39: Output D as search result
  Server S_{b⊕1}:
40: Insert EM into EDB_{b⊕1}
```

Fig. 1. Pseudocode of the static scheme

Setup. The client firstly generates two random keys k_s, k_u kept private from servers, and a map Σ to store necessary state information of each keyword. The map Σ will later be used to perform query and shuffle procedure. The client also initializes two empty map EDB_0, EDB_1 for each server, which will later be outsourced to server S_0 and S_1, respectively.

Given the index DB, the client will first pad the initial DB by the Padding algorithm, then encrypts the padded index. Specifically, for each keyword w, all its result identifiers PDB(w) will be randomly put on S_0 or S_1. The client generates an initial key k_w for keyword w, and use k_w to derive entry reference e for each encrypted value v, which is then updated into EDB_b. The client will also update the initial state information of w into Σ, and locally stores the state map. In the end, the client sends EDB_b to server S_b and finalizes the setup phrase.

Search. When searching a keyword w, the client firstly fetches w's state information from Σ, then generates search token k_w correspondingly using a PRF F, with the state information s_w as the part of input. The intuition is that after each search, s_w is increased, then k_w changed correspondingly. Therefore, search tokens for the same keyword is issued in an unlinkable way, which protects search pattern leakage on the search token level (Fig. 1).

The client then send (k_w, u_w) to server S_b that holds the searched index. S_b then finds all entries for the search query and return encrypted values back to the client. After obtaining those ciphertexts, the client decrypts them locally and learns whether an identifier is dummy or not because the padding algorithm have already marked them in the first place. In the end, the client re-encrypts prior identifiers (with dummy ones) and uploads them to other server. The client also updates local state correspondingly, which will enable future search towards the same keyword. The server $S_{b\oplus 1}$ can just insert the received values into its database.

4.2 The Dynamic Scheme

The additional property for dynamic SSE is allowing dynamically update to the remote encrypted database. Therefore, we must handle the leakage during update phrase properly, which means we will empower the dynamic scheme with forward privacy. As we will show later, our dynamic scheme leaks less information than most existing SSE schemes with forward/backward privacy. Before describing the detail, let us firstly demonstrate the leakage that the server may learn.

Update History and New Definition. Compared with static SSE, there are two kinds of extra (possible) leakage for dynamic SSE. Informally, the first kind is from update queries. For many existing SSE schemes, an update query will immediately leak its relation with a previous search and/or update query. The leakage can be reduced with so-called forward privacy, which ensures that any update query does not leak any information about the updated keyword. In our dynamic SSE scheme, we give a simple mechanism to enable forward privacy efficiently.

A more annoying leakage is *update history* from search queries. In particular, when searching a keyword, even the server is unaware of underlying searched contents, the server may learn all timestamps when previous updates issued to the keyword. The leakage is formally defined as UpHist(w). Ideally, we would like a search protocol leaks no information, i.e., $\mathcal{L}^b_{\mathsf{search}}(w) = (\perp)$. However, it is too high to achieve unless using heavy cryptographic primitives, e.g., PIR or

ORAM (to eliminate history information) plus worst-case padding (to eliminate size leakage).

Therefore, for practical reason, we would like a weaker security notion that allows the adversary to only learn *partial update history*. Let us demonstrate the differences between full and partial update history with an example. Consider the following sequence: $(\mathsf{add}, \mathsf{ind}_1, \{w_1, w_2\})$, $(\mathsf{add}, \mathsf{ind}_2, \{w_1, w_3\})$, $(\mathsf{delete}, \mathsf{ind}_3, \{w_1, w_2\})$, $(\mathsf{search}, \{w_1\})$, $(\mathsf{add}, \mathsf{ind}_4, \{w_1\})$, $(\mathsf{search}, \{w_1\})$. At time 6, the first notion for keyword w_1 leaks $\mathsf{UpHist}(w) = \{1, 2, 3, 5\}$ while the second only leaks $\mathsf{PartUpHist}(w) = \{5\}$.

Dynamic SSE from Three Servers. Following our leakage definition, in our dynamic scheme we additionally require a third server to hold dynamic updates. The update protocol is initiated by a forward-private update mechanism, and the only leakage is from search protocol, but the leakage is limited – only partial update history rather than the whole update history. Meanwhile, the functionalities and leakage of first two servers remain unchanged compared to the static scheme.

The setup protocol remains almost same as the static scheme with only minor differences – the client additionally stores a partial update counter p_w for each keyword w, which enables the client to pseudo-randomly generate fresh key for partial updates hold by the third server. In the following pseudo code, we only present the search and update part of our dynamic scheme.

Update. As we described above, all updates will be upload to the third server S_2. To ensure that the server only learns partial update history (i.e., updates timestamps after the last search for a keyword), we use an update oblivious mechanism for the update protocol. The main idea behind is that after each search, the client will change to use a fresh key for future updates. The fresh key is hidden from server within current round and will only be revealed until next search query for the same keyword. By using the revealed key, the server can learn all updates perform in current round, but not those updates to the same keyword in previous rounds. This is the main idea to only reveal partial update history to the third server (Fig. 2).

More specifically, in the update protocol, we use the existing state information in $\boldsymbol{\Sigma}$ to generate the fresh key. The client use a PRF F to generate the fresh secret of current round by using counter s_w, and use keyed hash function H on u_w to generate individual random entry within current round. When searching keyword w, the client will reveal the corresponding key k_p, the server can use the revealed key to find all updates in *last* round, but not for the whole updates from the very beginning. After that, s_w will be increased, and the key for future updates will again be hidden from the server (because the new key will be another PRF output). Note that the search token for each rounds is unpredictable as the key of PRF is only known to the client.

Search. The search protocol is quite similiar to our static scheme. Note that the index will contain two parts – the main part is located either on S_0 or S_1 and partial part on S_2. Same as our static scheme, the client will firstly find the server

```
Update(k_s, Σ, (op, w, ind); EDB_2)
  Client:
  1: (b, s_w, u_w, p_w) ← Σ[w]
  2: k_p ← F(k_s, h(w)||s_w||2)
  3: e ← H(k_p, p_w)
  4: v ← SE.Enc(k_u, op||ind)
  5: Send (e, v) to server S_2
  Server S_2:
  6: EDB_2[e] ← v

Search(k_s, Σ, w; EDB_0, EDB_1)
  Client:
  1: (b, s_w, u_w, p_w) ← Σ[w]
  2: k_w ← F(k_s, h(w)||s_w||b)
  3: k_p ← F(k_s, h(w)||s_w||2)
  4: Send (k_w, u_w) to server S_b
  5: Send (k_p, p_w) to server S_2
  Server S_b:
  6: R ← {}
  7: for i ← 0 to u_w do                  ▷ can be parallel
  8:     e ← H(k_w, i)
  9:     R ← R ∪ {EDB_b[e]}
  10: end for
  11: Send R back to client
  12: Delete R from EDB_b
  Server S_2:                           ▷ S_2 is the third server
  13: P ← {}
  7: for i ← 0 to p_w do                  ▷ can be parallel
  8:     e ← H(k_p, i)
  9:     P ← P ∪ {EDB_2[e]}
  10: end for
  11: Send P back to client
  12: Delete P from EDB_2
  Client:
  13: D, EM ← {}
  14: for c ∈ R ∪ P do
  15:     (op, ind) ← SE.Dec(k_u, c)
  16:     if op = "add" ∧ ind is valid then
  17:         D ← D ∪ {ind}
  18:     else
  19:         D ← D − {ind}
  20:     end if
  21: end for
  22: Output D as search result
  23: PD ← Padding(D)
  24: for ind ∈ PD do
  25:     e ← H(k_s, h(w)||s_w + 1||b ⊕ 1)
  26:     v ← SE.Enc(k_u, "add"||ind)
  27:     EM[e] ← v
  28: end for
  29: Σ[w] ← (b ⊕ 1, s_w + 1, |PD|, 0)
  30: Send EM to server S_{b⊕1}
  Server S_{b⊕1}:
  31: Insert EM into EDB_{b⊕1}
```

Fig. 2. Pseudocode of the dynamic scheme

S_b that holds the main part of the index for the searched keyword, and receive back the encrypted result. For those partial updates (updates in last round for the searched keyword) on server S_2, the client will reveal the round key k_p for those partial updates, by the key k_p, S_2 can find all partial updates and send corresponding encrypted values back to the client.

The client then performs local decryption and necessary deletion to remove deleted document identifiers. After that, the client should re-encrypt the result and upload to server $S_{b\oplus 1}$. Note that we want to hide the actual result size from server $S_{b\oplus 1}$, so the client should pad the result to the closest group size. By doing that, the server $S_{b\oplus 1}$ receiving the uploaded encrypted result is unaware the size of partial updates, in a way that we can kept the leakage unchanged for server S_0 and S_1 compared with the static scheme.

I/O Optimization. Recently, there are many works demonstrate that when the database is big enough, I/O will dominate the efficiency. Cash et al. [9] show that any SSE schemes cannot achieve optimal storage, read efficiency and locality simultaneously. The theorem certainly applies to our schemes. Intuitively, under the constraints of optimal storage and read efficiency, the encrypted indexes must be randomly stored among the space of storage, which may incur huge I/O latency during search due to random storage access.

We observe that the leakage during search can be leveraged to further optimize I/O efficiency, similar trick can also be found in some recent works [5,35]. More specifically, after each search query, the result size (grouping information) is already leaked out. Therefore, when the client upload those re-encrypted results on the other server, the server can just put them in continuous storage address, to improve I/O efficiency for future search query.

4.3 Leakage Analysis

Leakage of Static Scheme. Informally, setup leakage only contains the size of encrypted database and nothing else. As for search protocol, the queried server S_b learns the size of search result (or equivalently, the group information gID), and the time that they are updated into the encrypted database. Though, the leakage is only one-time and will soon be moved to the other server. The one-time leakage is very limited. In particular, suppose the total number of group is α, then our static construction only leaks $\log \alpha$ bits information of search/volume pattern. Moreover, due to the result-hiding property of search protocol, the result identifiers are hided from the server, thus limited information of access pattern is leaked.

The leakage of update and search protocol for server S_b, $b \in \{0, 1\}$ can be formally defined as follows, Here $\mathsf{LastUpHist}(w)$ contains when the searched keyword is added to the database, but the leakage is only one-time.

$$\mathcal{L}^b_{\text{setup}}(\mathsf{DB}) = \{N_b\}$$
$$\mathcal{L}^b_{\text{search}}(w) = \{b, \text{gID}, \mathsf{LastUpHist}(w)\}$$

Theorem 1. *Let F be a pseudorandom function, H be a hash function modeled as random oracle, and SKE be a $\mathsf{IND\text{-}CPA}$ secure symmetric key encryption scheme. Define leakage $\mathcal{L}_{\text{static}} = (\mathcal{L}^b_{\text{setup}}, \mathcal{L}^b_{\text{search}})$ where $b \in \{0, 1\}$, then the above scheme is a $\mathcal{L}_{\text{static}}$-adaptively-secure SSE scheme.*

Leakage of Dynamic Scheme. In our dynamic SSE scheme, the first two server works exactly same as the static scheme, i.e., searching and receiving encrypted result that are properly padded. Therefore, the leakage for the first two servers S_0, S_1 remain the same as our static scheme. The only additional leakage is from the third server S_2, in which the server learns partial updates history during search. Nevertheless, the server S_2 still cannot match updates performed in different rounds to the same keyword together, due to our re-keying technique. Therefore, the partial update history on S_2 is limited. Regarding update protocol, the third server cannot learn any valuable information during update, thus we can obtain forward privacy. Formally, we have the following leakage profile for the third server S_2.

$$\mathcal{L}^2_{\text{setup}}(\mathsf{DB}) = \{\perp\}$$

$$\mathcal{L}^2_{\text{update}}(w, ind, op) = \{\bot\}$$
$$\mathcal{L}^2_{\text{search}}(w) = \{\mathsf{partUpHist}(w)\}$$

Theorem 2. *Let F be a pseudorandom function, H be a hash function modeled as random oracle, and SKE be a IND-CPA secure symmetric key encryption scheme. Define leakage $\mathcal{L}_{\text{dynamic}} = (\mathcal{L}^b_{\text{setup}}, \mathcal{L}^b_{\text{search}})$ where $b \in \{0, 1, 2\}$, then the above scheme is a $\mathcal{L}_{\text{dynamic}}$-adaptively-secure SSE scheme.*

All proofs can be found in Appendix.

5 Experiment and Performance Evaluation

5.1 Implementation and Experiment Setting

We implemented our schemes in C++, with OpenSSL (version 1.0.2t) as the cryptographic library, the underlying database is rocksDB and gRPC for communication. The evaluation is performed on Ubuntu 16.04 equipped with 4 cores 3.40 GHz Intel Core i5-7500 CPU, 8 GB RAM, and 1 TB hard disk.

5.2 Performance Result

Padding Ratio. Regarding the padding algorithm. Intuitively, the less the number of groups is, the securer the scheme is, and correspondingly, more dummy identifiers are needed. We first report the price paid from the padding algorithm. In particular, we show the relation between dummy ratio r (i.e., r = # padded index / # raw index) and group parameters G. We assume the distribution of result size of the raw database follows Zipf distribution, which is a reasonable observation for real-world document, and is also adapted in recent SSE schemes [24,33]. Our setting is similar to [24], specifically, we parameterize Zipf distribution $Z_{n,s}$ with probability mass function $f_{n,s} = r^{-s}/H_{n,s}$ where $H_{n,s} = \sum_{i=1}^{n} i^{-s}$. For i-th keyword, its result length is $T/(i \cdot H_{n,s})$ where $s = 1$ and $T = \alpha \cdot n$ in which α is an empirical parameter to adjust maximal result length of the document set.

We choose the group parameter G from 2^1 to 2^6. The price of padding is measured by the padding ratio, i.e., the extra storage overhead due to padding. As we can see in Fig. 3, the padding ratio r is almost linear with group parameter G. The result is as excepted – the number of group is decreased with increasing of G, then more dummy index is required, which is the price paid for better security.

Search Efficiency. We measure search efficiency for three synthetic database with database size range from 1e6 to 1e8. We first measure search performance of the first-time search (i.e., the first time search for a keyword after setup). For this case, the searched index is randomly resenting among server's storage, each search will incur a random I/O access. As we can see in Fig. 4, the search

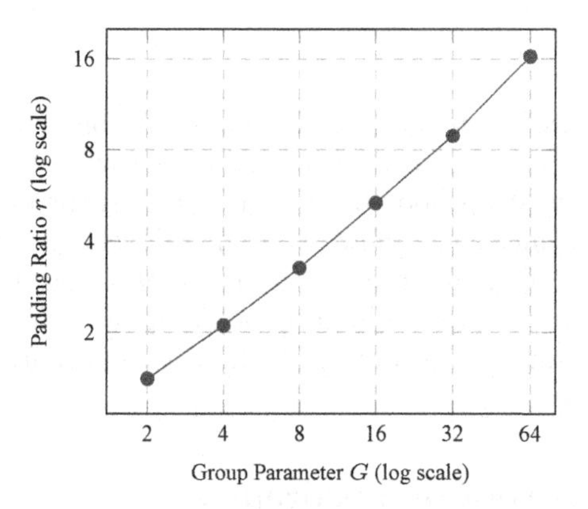

Fig. 3. Padding price for different group parameter G

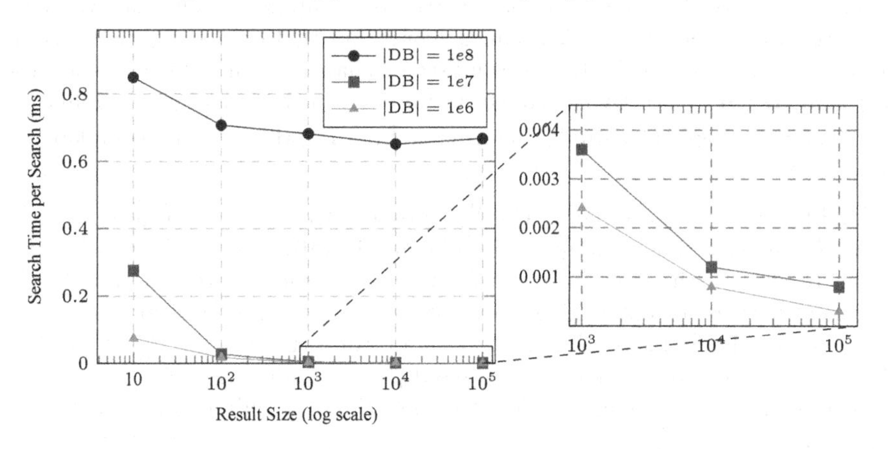

Fig. 4. Search efficiency for server-side search operation

time for per search is decreasing when result size is increasing, as the overhead is atomized. However, when the database is large, i.e., $|\mathrm{DB}| = 1e8$, search efficiency is sharply decrease. The reason is that for $|\mathrm{DB}| = 1e8$, the database cannot be loaded into RAM of the holding machine, so random disk access incur a huge overhead.

We then test our I/O optimization technique for non-first-time search, and compare it with the first-time search. The optimization can highly improve search efficiency. In particular, for a result size of 100000 on $|\mathrm{DB}| = 1e8$, the optimized search time per entry is only 0.002 ms, 300 times faster than its first-time search. Due to page limits, we will put more performance information in full version of this paper.

6 Conclusion

In this paper, we investigate how to design searchable symmetric encryption with tunable leakage, which provide a configurable way for the trade-off between security and efficiency. We proposed two SSE schemes, the first scheme uses two servers and is static, which shows how to obtain SSE scheme with tunable leakage by using two servers. To make the scheme for supporting dynamic update, we additionally use a third server to hold dynamic updates. We demonstrate that the leakage for the third server only leaks limited information rather then the whole update history.

Appendix A Proof of Theorem 1

Proof. We prove the security of our scheme by ideal/real simulation. We start from $\mathbf{Real}^{\Pi}_{\mathcal{A}}(\lambda)$ and construct a sequence of games that differs slightly from the previous game and show they are indistinguishable. Eventually we reach the last game $\mathbf{Ideal}^{\Pi}_{\mathcal{A},\mathcal{S}}(\lambda)$. By the transitive property of the indistinguishability, we conclude that $\mathbf{Real}^{\Pi}_{\mathcal{A}}(\lambda)$ is indistinguishable from $\mathbf{Ideal}^{\Pi}_{\mathcal{A},\mathcal{S}}(\lambda)$ and complete our proof.

Hybrid G_1: G_1 is the same as $\mathbf{Real}^{\Pi}_{\mathcal{A}}(\lambda)$ except that instead of generating t_w using F, the experiment maintain a mapping **Token** to store $(h(w)||s_w||b, k_w)$ pairs. In the search protocol, when k_w is needed, the experiment first checks whether there is an entry in **Token** for $h(w)||s_w||b$, if so returns the entry; otherwise randomly picks a k_w in $\{0,1\}^{\ell_k}$ and stores the $(h(w)||s_w||b, k_w)$ pair in **Token**. It's trivial to see that G_1 and $\mathbf{Real}^{\Pi}_{\mathcal{A}}(\lambda)$ are indistinguishable, otherwise we can distinguish a pseudo-random function F and a truly random function.

Hybrid G_2: G_2 and G_1 is the same except that in G_2 the experiment will maintain a map **E**. Instead of producing e by calling $H(k_w, u_w)$, in Setup protocol, the experiment will replace it with the following procedure:

$$\mathbf{E}[k_w||u_w] \leftarrow \{0,1\}^{\ell}$$

The intuition is to firstly sampling random string during setup, and program H correspondingly to maintain consistency. Now in search protocol, the entry e is generated with the following procedure:

$$\mathbf{H}[k_w||u_w] \leftarrow \mathbf{E}[k_w||u_w]$$

G_1 and G_2 behaves exactly the same except that in G_2, with some probability inconsistency in random oracle query results can be observed. If the adversary queries $\mathbf{H}$ with $k_w||u_w$ before the next search query, it will get a value e' such that with a overwhelming probability $e' \neq e$ because $\mathbf{H}[k_w||u_w]$ has not been updated and a random string e' is chosen by the oracle in this case. If the adversary queries $\mathbf{H}$ with $k_w||u_w$ again after the next search query, e will be updated to the $\mathbf{H}$ and the query result will be e. If the inconsistency is observed (we denote this event as **Bad**), the adversary knows it is in G_2. We have:

$$\Pr[G_1 = 1] - \Pr[G_2 = 1] \leq \Pr[\mathbf{Bad}]$$

The event **Bad** can only happen if the adversary can query the oracle with $k_w||u_w$. Since k_w is pseudorandom output of PRF F and is unknown to the adversary before search, the probability of the adversary choosing k_w is $2^{-\lambda}$. A PPT adversary can make at most $q = \mathbf{poly}(\lambda)$ guesses, then $\Pr[\mathbf{Bad}] \leq \frac{q}{2^\lambda}$, which is also negligible.

Hybrid G_3: G_3 and G_2 is the same except that in G_3 we change the way of generating encrypted value v. Note that in G_2, the value v is generated by a IND-CPA secure symmetric key encryption scheme SE, therefore the ciphertext is semantically secure. In G_3, the experiment simply choose a random value with the same length as SE's ciphertext whenever the SE.Enc is called. After above replacement, some code are not necessary, so we do cleanup for readability. The full pseudocode of G_4 can be found in Fig. 5.

```
Setup(λ, DB; ⊥)
    Client:
 1: Σ, EDB_0, EDB_1 ← empty map
 2: PDB ← Padding(DB)
 3: for w ∈ W do
 4:     u_w, s_w ← 0, b ←$ {0,1}
 5:     if Token[w||s_w||b] = ⊥ then
 6:         Token[w||s_w||b] ←$ {0,1}^λ
 7:     end if
 8:     k_w ← Token[w||s_w||b]
 9:     for ind ∈ PDB(w) do
10:         E[k_w||u_w] ← {0,1}^ℓ
11:         v ← {0,1}^|v|
12:         u_w ← u_w + 1
13:         EDB_b[e] ← v
14:     end for
15:     Σ[w] ← (b, s_w, u_w)
16: end for
17: Client locally store Σ
18: Send EDB_b to server S_b for b ∈ {0,1}

Search(k_s, Σ, w; EDB_0, EDB_1)
    Client:
17: (b, s_w, u_w) ← Σ[w]
18: k_w ← Token[w||s_w||b]
19: Send (k_w, u_w) to server S_b
    Server S_b:
20: R ← {}
21: for i ← 0 to u_w do
22:     Programing H(k_w, i) ← E[k_w||i]
23: end for
24: for i ← 0 to u_w do            ▷ can be parallel
25:     e ← H(k_w, i)
26:     R ← R ∪ {EDB_b[e]}
27: end for
28: Send R back to client
    Client:
29: D, EM ← {}, u_w ← 0
30: for c ∈ R do
31:     e ← {0,1}^ℓ
32:     v ← {0,1}^|v|
33:     EM ← {(e, v)}
34:     u_w ← u_w + 1
35: end for
36: Σ[w] ← (b ⊕ 1, s_w + 1, u_w)
37: Send EM to server S_{b⊕1}
38: Output D as search result
    Server S_{b⊕1}:
39: Insert EM into EDB_{b⊕1}
```

Fig. 5. Pseudocode of G_3

The asversary cannot distinguish between G_2 and G_3 because of the semantic security of SE, hence we have:

$$\Pr[G_2 = 1] - \Pr[G_3 = 1] \leq \mathbf{negl}(\lambda)$$

Hybrid G_4: In G_4, we perform further simplification for the pseudocode. Firstly, the experiment does not need to generate (e, v) pair for each keyword separately. The reason is that in G_3, both e and v are generated by random sampling, so randomly generating N_b pairs of string in G_4 is not distinguishable from the prior approach.

Similarly, when receiving search query, if the query will touch storage in EDB_b that is from setup, then the experiment just randomly chooses u_w non-touched entries, and then programs the random oracle; Otherwise, if the search query is for a keyword that has been search previously, then the experiment will program the random oracle, in a way that make the server return those encrypted values that is previously uploaded by client from a shuffle phrase – the experiment can do above decision by simply checking if s_w equals to 0. For other parts, G_4 behaves the same as G_3

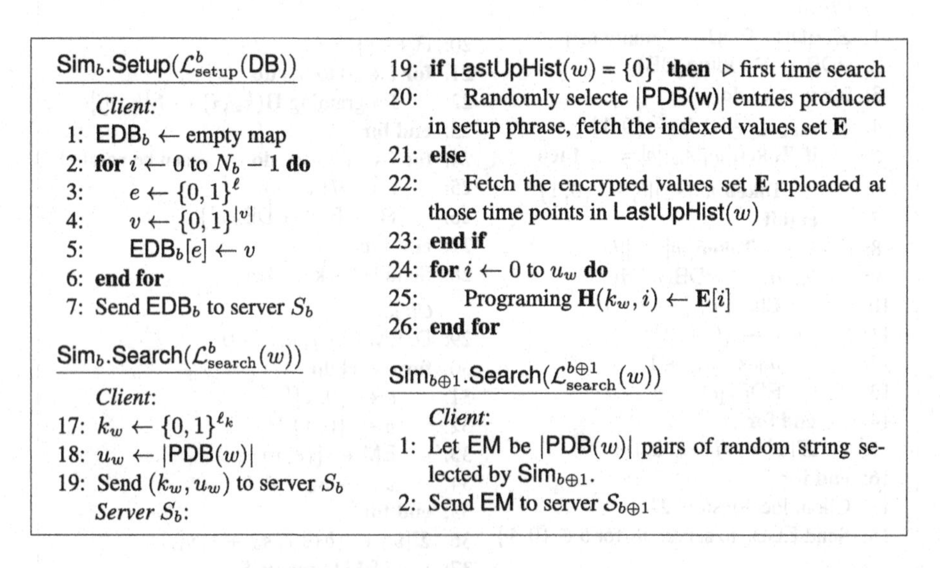

Fig. 6. Pseudocode of simulator $\mathcal{S}$

Ideal$^{\Pi}_{\mathcal{A},Sim}(\lambda)$: In ideal world, the simulator needs to simulate all views of the server just using the allowed leakage profile. The simulation is quite simple. In setup protocol, the simulator can just randomly generate N_b pair of random strings and upload to server S_b to completes the simulation. The tricky part is about the simulation of search protocol. Note that in a our search protocol, the client will firstly query server S_b for result, then re-encrypt the result and upload to the other server $S_{b\oplus 1}$. To complete the simulation, in ideal world there are two simulators for S_0 and S_1 separately. The first simulator Sim_b will simulator the view for S_b and simulator $\mathsf{Sim}_{b\oplus 1}$ will simulate the view of $S_{b\oplus 1}$ (the intuition is each simulator can independently complete simulation without sharing information, which is consistent with the non-collusion assumption).

For the simulation for server S_b, simulator Sim_b will randomly generate k_w, get the group size u_w from the leakage function, and send them to server S_b. After that, the simulator should program the random oracle to make some entries match with the impending search operation. Specifically, the simulator get leakage $\mathsf{LastUpHist}(w)$ from the leakage function. If $\mathsf{LastUpHist}(w)$ only contains 0, then the simulator knows the keyword is never searched before, then it can just randomly select $|PDB(w)|$ untouched storage entries that are produced from setup protocol (not includes those that are uploaded due to search re-shuffle!); if not, then the simulator knows the keyword has been searched before, and then it can get the re-shuffle time and fetch all required entries. Then Sim_b program random oracle and complete the simulation. The simulation of $\mathsf{Sim}_{b\oplus 1}$ is simple, it just generate $|\mathsf{PDB}(w)|$ random (e, v) pairs and upload to server $S_{b\oplus 1}$. For other parts, $\mathbf{Ideal}^{\Pi}_{\mathcal{A},Sim}(\lambda)$ behaves the same as G_4.

All summed up, we have:

$$|\Pr[\mathbf{Real}^{\Pi}_{\mathcal{A}}(\lambda) = 1] - \Pr[\mathbf{Ideal}^{\Pi}_{\mathcal{A},Sim}(\lambda) = 1]| \leq \mathbf{negl}(\lambda)$$

Appendix B Proof of Theorem 2

The proof of Theorem 2 is almost the same with the proof of Theorem 1, the only difference is to handle the partial update in the third server. Note that the $\mathsf{partUpHist(w)}$ enables the simulator to learn all update timestamps for a temporary keyword that the client updated previously, then the simulator can use similar technique to program the random oracle like Fig. 6. Therefore, we decide not to rewrite the proof here.

References

1. Ahmad, A., Kim, K., Sarfaraz, M.I., Lee, B.: OBLIVIATE: a data oblivious filesystem for intel SGX. In: NDSS (2018)
2. Blass, E.O., Mayberry, T., Noubir, G., Onarlioglu, K.: Toward robust hidden volumes using write-only oblivious ram. In: ACM CCS, pp. 203–214. ACM (2014)
3. Bost, R.: $\Sigma o\varphi o\varsigma$: forward secure searchable encryption. In: ACM CCS, pp. 1143–1154 (2016)
4. Bost, R., Fouque, P.: Thwarting leakage abuse attacks against searchable encryption - a formal approach and applications to database padding. IACR Cryptology ePrint Archive (2017)
5. Bost, R., Minaud, B., Ohrimenko, O.: Forward and backward private searchable encryption from constrained cryptographic primitives. In: ACM CCS, pp. 1465–1482 (2017)

6. Brasser, F., Müller, U., Dmitrienko, A., Kostiainen, K., Capkun, S., Sadeghi, A.R.: Software grand exposure: {SGX} cache attacks are practical. In: {USENIX} ({WOOT}) (2017)
7. Cash, D., Grubbs, P., Perry, J., Ristenpart, T.: Leakage-abuse attacks against searchable encryption. In: ACM CCS, pp. 668–679 (2015)
8. Cash, D., Jarecki, S., Jutla, C., Krawczyk, H., Roşu, M.-C., Steiner, M.: Highly-scalable searchable symmetric encryption with support for boolean queries. In: Canetti, R., Garay, J.A. (eds.) CRYPTO 2013. LNCS, vol. 8042, pp. 353–373. Springer, Heidelberg (2013). https://doi.org/10.1007/978-3-642-40041-4_20
9. Cash, D., Tessaro, S.: The locality of searchable symmetric encryption. In: Nguyen, P.Q., Oswald, E. (eds.) EUROCRYPT 2014. LNCS, vol. 8441, pp. 351–368. Springer, Heidelberg (2014). https://doi.org/10.1007/978-3-642-55220-5_20
10. Chor, B., Goldreich, O., Kushilevitz, E., Sudan, M.: Private information retrieval. In: FOCS, pp. 41–50. IEEE (1995)
11. Chow, S.S., Lee, J.H., Subramanian, L.: Two-party computation model for privacy-preserving queries over distributed databases. In: NDSS (2009)
12. Costan, V., Devadas, S.: Intel SGX explained. IACR Cryptology ePrint Archive 2016(086), pp. 1–118 (2016)
13. Cui, S., Belguith, S., Zhang, M., Asghar, M.R., Russello, G.: Preserving access pattern privacy in SGX-assisted encrypted search. In: ICCCN, pp. 1–9. IEEE (2018)
14. Cui, S., Song, X., Asghar, M.R., Galbraith, S.D., Russello, G.: Privacy-preserving searchable databases with controllable leakage. arXiv preprint arXiv:1909.11624 (2019)
15. Curtmola, R., Garay, J., Kamara, S., Ostrovsky, R.: Searchable symmetric encryption: improved definitions and efficient constructions. J. Comput. Secur. **19**(5), 895–934 (2011)
16. Demertzis, I., Papadopoulos, D., Papamanthou, C., Shintre, S.: SEAL: attack mitigation for encrypted databases via adjustable leakage. IACR Cryptology ePrint Archive (2019)
17. Etemad, M., Küpçü, A., Papamanthou, C., Evans, D.: Efficient dynamic searchable encryption with forward privacy. PoPETs **2018**(1), 5–20 (2018)
18. Ghareh Chamani, J., Papadopoulos, D., Papamanthou, C., Jalili, R.: New constructions for forward and backward private symmetric searchable encryption. In: ACM CCS, pp. 1038–1055. ACM (2018)
19. Götzfried, J., Eckert, M., Schinzel, S., Müller, T.: Cache attacks on intel SGX. In: Proceedings of the 10th European Workshop on Systems Security, p. 2. ACM (2017)
20. Hoang, T., Yavuz, A.A., Durak, F.B., Guajardo, J.: Oblivious dynamic searchable encryption on distributed cloud systems. In: Kerschbaum, F., Paraboschi, S. (eds.) DBSec 2018. LNCS, vol. 10980, pp. 113–130. Springer, Cham (2018). https://doi.org/10.1007/978-3-319-95729-6_8
21. Hoang, T., Yavuz, A.A., Guajardo, J.: Practical and secure dynamic searchable encryption via oblivious access on distributed data structure. In: ACSAC, pp. 302–313 (2016)
22. Islam, M.S., Kuzu, M., Kantarcioglu, M.: Access pattern disclosure on searchable encryption: ramification, attack and mitigation. In: NDSS (2012)
23. Kamara, S., Moataz, T.: Computationally volume-hiding structured encryption. In: Ishai, Y., Rijmen, V. (eds.) EUROCRYPT 2019. LNCS, vol. 11477, pp. 183–213. Springer, Cham (2019). https://doi.org/10.1007/978-3-030-17656-3_7

24. Kamara, S., Moataz, T., Ohrimenko, O.: Structured encryption and leakage suppression. In: Shacham, H., Boldyreva, A. (eds.) CRYPTO 2018. LNCS, vol. 10991, pp. 339–370. Springer, Cham (2018). https://doi.org/10.1007/978-3-319-96884-1_12
25. Kim, K.S., Kim, M., Lee, D., Park, J.H., Kim, W.: Forward secure dynamic searchable symmetric encryption with efficient updates. In: ACM CCS, pp. 1449–1463 (2017)
26. Kornaropoulos, E.M., Papamanthou, C., Tamassia, R.: The state of the uniform: attacks on encrypted databases beyond the uniform query distribution. IACR Cryptology ePrint Archive, p. 441 (2019)
27. Lai, R.W.F., Chow, S.S.M.: Forward-secure searchable encryption on labeled bipartite graphs. In: ACNS, pp. 478–497 (2017)
28. Lee, J., et al.: Hacking in darkness: return-oriented programming against secure enclaves. In: USENIX Security, pp. 523–539 (2017)
29. Liu, C., Zhu, L., Wang, M., Tan, Y.A.: Search pattern leakage in searchable encryption: attacks and new construction. Inf. Sci. **265**, 176–188 (2014)
30. Markatou, E.A., Tamassia, R.: Full database reconstruction with access and search pattern leakage. IACR Cryptology ePrint Archive, p. 395 (2019)
31. Mishra, P., Poddar, R., Chen, J., Chiesa, A., Popa, R.A.: Oblix: an efficient oblivious search index. In: IEEE SP, pp. 279–296. IEEE (2018)
32. Pagh, R., Rodler, F.F.: Cuckoo hashing. J. Algorithms **51**(2), 122–144 (2004)
33. Patel, S., Persiano, G., Yeo, K., Yung, M.: Mitigating leakage in secure cloud-hosted data structures: volume-hiding for multi-maps via hashing. In: ACM CCS, pp. 79–93. ACM (2019)
34. Sasy, S., Gorbunov, S., Fletcher, C.W.: Zerotrace: oblivious memory primitives from intel SGX. IACR Cryptology ePrint Archive 2017, 549 (2017)
35. Song, X., Dong, C., Yuan, D., Xu, Q., Zhao, M.: Forward private searchable symmetric encryption with optimized I/O efficiency. IEEE Trans. Dependable Secure Comput. **17**(5), 912–927 (2020). https://doi.org/10.1109/TDSC.2018.2822294
36. Stefanov, E., Papamanthou, C., Shi, E.: Practical dynamic searchable encryption with small leakage. In: NDSS (2014)
37. Zhang, Y., Katz, J., Papamanthou, C.: All your queries are belong to us: the power of file-injection attacks on searchable encryption. In: USENIX Security, pp. 707–720 (2016)

Completely Unsupervised Cross-Modal Hashing

Jiasheng Duan(✉), Pengfei Zhang, and Zi Huang

School of Information Technology and Electrical Engineering,
The University of Queensland, Brisbane, Australia
j.duan@uqconnect.edu.au, mima.zpf@gmail.com, huang@itee.uq.edu.au

Abstract. Cross-modal hashing is an effective and practical way for large-scale multimedia retrieval. Unsupervised hashing, which is a strong candidate for cross-modal hashing, has received more attention due to its easy unlabeled data collection. However, although there has been a rich line of such work in academia, they are hindered by a common disadvantage that the training data must exist in pairs to connect different modalities (*e.g.*, a pair of an image and a text, which have the same semantic information), namely, the learning cannot perform with no pair-wise information available. To overcome this limitation, we explore to design a Completely Unsupervised Cross-Modal Hashing (CUCMH) approach with none but numeric features available, *i.e.*, with neither class labels nor pair-wise information. To the best of our knowledge, this is the first work discussing this issue, for which, a novel dual-branch generative adversarial network is proposed. We also introduce the concept that the representation of multimedia data can be separated into content and style manner. The modality representation codes are employed to improve the effectiveness of the generative adversarial learning. Extensive experiments demonstrate the outperformance of CUCMH in completely unsupervised cross-modal hashing tasks and the effectiveness of the method integrating modality representation with semantic information in representation learning.

Keywords: Cross-modal hashing · Completed unsupervised · Modality representation

1 Introduction

There has been a dramatic explosion of multimedia data on the Internet during the last decade. Data are always collected from multiple sources and exist in heterogeneous modalities, such as image, text, audio, video, and etc. This phenomenon gives rise to an increasing requirement on effective multimedia interaction technology. In this situation, cross-modal retrieval has been an area of active research. In general, a cross-modal retrieval system takes queries in one modality (*e.g.*, images) to search data in other modalities (*e.g.*, texts) with certain similarity metrics. Due to the heterogeneity of multimedia, *i.e.*, multi-modal data

Y. Nah et al. (Eds.): DASFAA 2020, LNCS 12112, pp. 178–194, 2020.
https://doi.org/10.1007/978-3-030-59410-7_11

residing in different feature spaces, the key challenge of a cross-modal retrieval system is how to model the similarity relationships across modalities, so that data in various modalities can be measured directly.

Due to the effectiveness and practicability, cross-modal hashing has received increasing attention in the community in recent years. In the context of cross-modal retrieval, hashing is the technique that aims to project high-dimensional data in heterogeneous modalities into a common low-dimensional discrete Hamming space, meanwhile the underlying inter-modal and intra-modal correlations are well preserved. A rich line of existing work of this area have been proposed, which are roughly categorized into unsupervised hashing and supervised hashing. Limited by the available training datasets, research work of this topic mostly focus on bi-modal (images and text) hashing.

Supervised cross-modal hashing [2,10,12,17,31–34,36] typically adopts guidance information, *i.e.*, class labels, to define the similarity relationships among heterogeneous data. Compared with supervised learning, unsupervised cross-modal hashing [4,14,15,19,22,26,27,30,35,37] learns similarity relationships based on the data distribution or structure property because no class labels available. It is obviously easier to cater unlimited amount of unlabeled training data, which makes unsupervised learning strategy more practical and allows it to be a strong candidate of cross-modal hashing methods.

However, almost all the existing unsupervised methods rely on the pair-wise information, namely, images and texts exist in pairs and a pair of data have the same semantic information, *i.e.*, class labels, in a single dataset. The approaches build the correspondence between different modalities through the pair-wise information. However, this common practice has become a obvious disadvantage hindering the research, because the learning cannot be separated with pair-wise information. This means that the training data used must exist in aligned pairs, which results in a significant limitation in data collection. To overcome this challenge, we explore to design a Completely Unsupervised generative adversarial Cross-Modal Hashing (CUCMH) approach with none but numeric features available, *i.e.*, with neither class labels nor pair-wise information. The main idea is to build a dual-branch Generative Adversarial Network (GAN) to produce the fake instances of modality 1 from the generated binary codes of modality 2, then re-generate the hash codes in modality 1 from these fake instances, and vice versa. Ideally, the re-generated hash codes should be the same as the first generated ones. However, the lack of information from other modalities limits the effectiveness of this learning strategy, so we strive to find a way to bridge the gap between modalities.

We are inspired by the idea of style transfer, which has been hotly discussed in neural language processing and computer vision, such as [6,7,38]. A style transfer method aims to transfer the images or texts into different styles with the semantic content preserved, *e.g.*, transfer a paragraph in the official news into another paragraph in funny story style but with the same semantic information. Hence, arguably, the expression of multimedia data can be separated into content and style representation. The former expresses the semantic content and the latter represents the data 'appearance'. In other words, the combination of a certain semantic content with different modality representation can be encoded into the same semantic binary code. To draw an analogy, a person is always himself even if he wears different clothes. Under this condition, we exploit the representation of the modality-specific "appearance" as a bridge to surmount the difficulty of no information transferred between different modalities. Specifically, the architecture of CUCMH is composed of two branches for images and texts, respectively. Each branch is built with two encoders and a GAN. The encoders encode image and text features into binary codes for semantic content and short codes for modality representation, then the two types of codes are concatenated across the modalities and be transformed to dummy data samples through the GAN. Finally the dummy data are encoded into binary codes by the semantic encoders, which are compared with the original binary codes. In this way, the two modalities are connected through the combination of two types of codes. The details of our framework will be described in Sect. 3 and the architecture is illustrated in Fig. 1. The main contributions of this paper are summarized as follows:

- To the best of our knowledge, this is the first work that discusses unsupervised cross-modal hashing with no pair-wise information. We propose a novel dual-branch generative adversarial network to overcome the major challenge – there is no correspondence between different modalities.
- We introduce the concept that the representation of multimedia data can be separated into content manner and style manner. With this idea, we leverage the cross-modal concatenated codes to solve the problem – the lack of information transferred between modalities when pair-wise information unavailable in unsupervised learning.
- Extensive experiments demonstrate the desirable performance of CUCMH in completely unsupervised cross-modal hashing tasks and the effectiveness of the method that integrates modality representation with semantic information in representation learning.

The rest of this paper is organized as follows: in Sect. 2, we discuss the related previous work; in Sect. 3, we explicitly introduce the proposed approach CUCMH; in Sect. 4, the extensive experiments will be discussed; and Sect. 5 is the conclusion of the whole paper.

2 Related Work

In recent years, cross-media analysis has been extensively studied to explore the correlations between different modalities. Particularly, cross-modal retrieval has become more and more active in the community.

A common practice of cross-modal retrieval is to learn a shared real-value latent subspace where different modal data are concatenated into the same measurements for searching and ranking [1,5,16,20,21,23,24,28]. However, these real-valued methods are unfeasible to be adapted to large-scale multi-modal

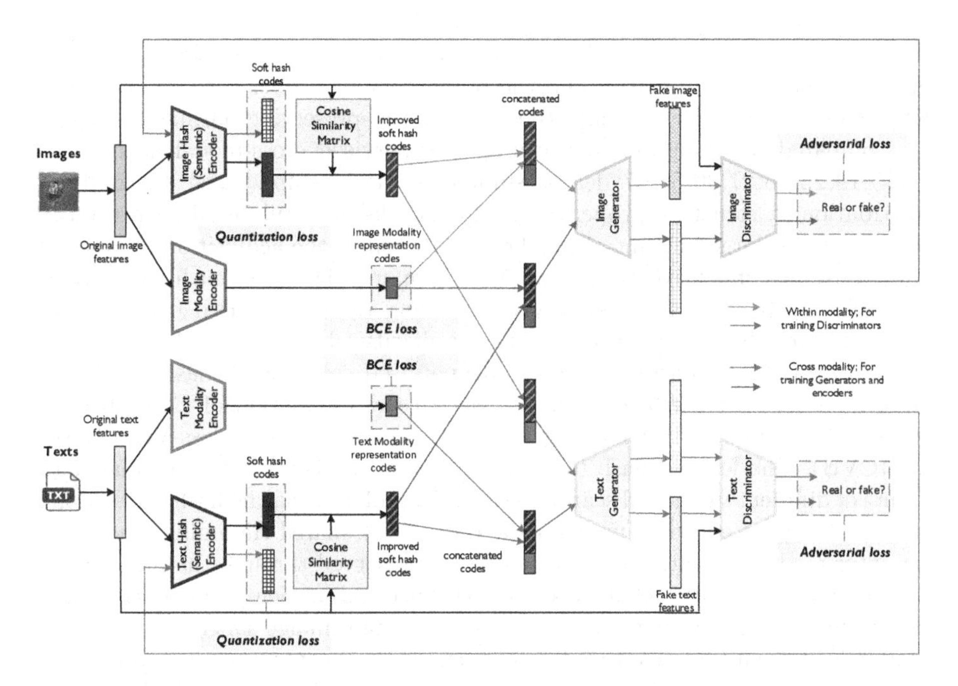

Fig. 1. The overview of Completely Unsupervised Cross-Modal Hashing.

retrieval tasks because of the inevitable cumbersome computational and space complexities. Fortunately, binary representation learning, known as hashing, utilizes efficient xor operations to calculate Hamming distance rather than Euclidean distance, which greatly reduces the computation burdens on multimedia analysis. A rich line of existing work of cross-modal hashing have been proposed, which are roughly categorized into unsupervised hashing and supervised hashing.

Supervised cross-modal hashing [2,10,12,17,31–34,36] typically adopts guidence information, *i.e.*, class labels, to define the similarity relationships among heterogeneous data. Zhang et al. [34] employ semantic correlation maximization (SCM) to integrate semantic labels into the hashing learning procedure for large-scale data modelling, and reconstruct the multi-modal similarity matrix through the learned hash codes to speed up the training process. Xu et al. [32] describe an efficient discrete optimization algorithm (DCH) to iteratively generate the binary codes bit by bit, which optimizes the hashing procedure with the proposed linear classification framework. Bronstein et al. [2] made the first attempt to learn hash functions for the multi-modal data with the information of cross-modal pairs. They propose the method, called cross-modal similarity sensitive hashing (CMSSH), which learns two series of hash functions to generate similar binary codes for pairs of relevant data points with different modalities and dissimilar codes for unrelated pairs. Jiang et al. [12] propose an end-to-end cross-modal hashing framework, dubbed deep cross-modal hashing (DCMH), to complete feature learning and hash function learning process. Lin et al. [17] integrate the semantic matrix and the probability distribution to minimize the KL-divergence in order to complete the semantics-preserving hashing (SePH).

Compared with supervised learning, unsupervised cross-modal hashing [4, 14,15,19,22,26,27,30,35,37] learns similarity relationships based on the data distribution or structure property and no class labels available. It is obviously easier to cater large amount of unlabeled data, since the acquisition of labelled data consumes expensive human labors. Therefore, there are much more training data can be used for unsupervised learning than supervised manner. Sun et al. [27] are the first to introduce the similarity idea, Canonical Correlation Analysis (CCA) [9], into the area of unsupervised cross-modal hashing, then a large number of research work emergence on the foundation of CCA. Kumar et al. [14] adapt spectral hashing for single modal [29] to Cross-View Hashing (CVH). The learned hash functions generate similar codes for similar data points of different views and search cross-view similarity via the hash codes. Song et al. [26] suggest encoding multi-modal data into a shared Hamming space with the inter-media and intra-media consistency preserved through the inter-media hashing method (IMH). Zhou et al. [37] designed Latent Semantic Sparse Hashing (LSSH), which leverage Sparse Coding to obtain image salient features and Matrix Factorization to capture the text semantic concepts, then integrate them into a common space. Ding et al. [4] proposed a Collective Matrix Factorization Hashing (CMFH) based on the assumption that a group of an identical hash codes, which learned with the collaboration of collective matrix factorization and a latent factor model, can represent data of all modalities. Wu et al. [30] integrate feature learning and binarization, and preserve discrete constraints when learning the hash functions.

Generative Adversarial Networks (GANs) [8] have been a candidate for unsupervised cross-modal hashing, since the related research became booming in the community. GANs is built with a generative network (G) and a discriminative network (D), which contest with each other. G generates candidates while the D evaluates them, in which way the model learns data distribution and discriminative feature representation. Li et al. [15] propose the Unsupervised coupled Cycle generative adversarial Hashing networks (UCH), which builds an outer-cycle and an inner-cycle network, to simultaneously optimize the representation learning for common data features and hash codes. Zhang et al. [35] design a model with GANs to generate pairs of images and texts with the modality-correlation-preserved graph manifold structure for learning hash functions.

However, nearly all the existing unsupervised methods rely on the pair-wise information to build the correspondence between various modalities, which seriously limits the availability of multimedia data in terms of training. Therefore, we explore to overcome this limitation in this paper.

3 Proposed Approach

In this section, we describe the Completely Unsupervised Modality-bridge Generative Adversarial Cross-Modal Hashing (CUCMH) in details. The Generative

Adversarial Networks (GANs) will be first introduced as the base algorithm, and then we will discuss the proposed framework architecture, the objective functions, and the learning and testing process. Without loss of generality, this paper focuses on bi-modal (images and text) hashing, and our method can be easily extended to handle multiple-modal hashing.

3.1 Preliminaries

Generative Adversarial Networks (GANs) serve as the basis of CUCMH. As described in Sect. 2, GANs play a turn-wise min-max game to learn the representation of data distribution and discriminative features. It contains a generator (G) and a discriminator (D). G learns to generate dummy data samples to fool D, while D is trained to distinguish the fake candidates from the real data. This "game" is played with the following function:

$$\min_{G}\max_{D}\mathbf{K}(D,G)\sum_{i=1}^{n}\mathbb{E}_{x_i\in p_{data}(x)}[logD(x_i)]+\mathbb{E}_{z\in p_z(z)}[log(1-D(G(z))] \quad (1)$$

where x_i represents a data sample; p_{data} and p_z denote the distribution of the real data and the fake data samples, respectively; z is a noise point. GAN aims to cast p_z to p_{data}, namely, G learns to generate a distribution from z, which has the high likeness with the real data. In the learning process, G and D are alternatively fixed when the other is trained.

3.2 Problem Definition

Given a cross-modal dataset that consists of a image set $V=\{v_i\}_{i=1}^{n}$ and a text set $T=\{t_j\}_{j=1}^{n}$, v_i and t_j are the i-th raw image data instance and j-th text data sample. The purpose of CUCMH is to learn two sets of hash functions E_v^H and E_t^H for images and texts, respectively, which are able to generate reliable common semantic hash codes $B_\Delta\in\{+1,-1\}^{C\times n}$ of the two modalities, where C denotes the length of the binary codes. In CUCMH, we denote B, the output of the binary code encoder, as the soft representation of B_Δ, *i.e.*, B_Δ are generated by applying a sign function to B:

$$B_\Delta=sign(B) \quad (2)$$

3.3 Proposed CUCMH

Figure 1 presents the overview of the proposed CUCMH, which contains two branches for images and texts, respectively. Each branch consists of 4 parts, including a semantic encoder E^H, a modality expression encoder E^M, a generator G and a discriminator D. These components of the two branches are denoted using E_v^H, E_t^H, E_v^M, E_t^M, G_v, G_t, D_v and D_t. We first extract the features of raw image and text data with pre-trained networks. Specifically, raw images are represented with the features from deep CNNs, and the BOW vectors of texts go through a text CNN model [13] to generate embeddings of continuous values. We save these feature vectors as two sets, F_v and F_t. Then the feature vectors are sent into the two branches in mini-batches simultaneously.

Firstly, F_v passes through E_v^H and E_v^M simultaneously to produce soft semantic hash codes B_v and modality representation codes M_v. Likewise, we gain B_t and M_t by passing F_t through E_t^H and E_t^M. From this process, we get $B_v = \{b_{v_i}\}_{i=1}^n$, $B_t = \{b_{t_j}\}_{j=1}^n$, $M_v = \{m_{v_i}\}_{i=1}^n$ and $M_t = \{m_{t_j}\}_{j=1}^n$. That is

$$B_v = E_v^H(F_v, W_v^H) \tag{3}$$

$$M_v = E_v^M(F_v, W_v^M) \tag{4}$$

$$B_t = E_t^H(F_t, W_t^H) \tag{5}$$

$$M_t = E_t^M(F_t, W_t^M) \tag{6}$$

where E_v^H, E_v^M, E_t^H and E_t^M are networks that consist of several linear layers followed by activation layers. Each activation layer is also a output layer with the Tanh function to relax the vectors to soft representation of binary codes. W_v^H, W_v^M, W_t^H and W_t^M are the network parameters. The details will be depicted in the experiment section.

As discussed in Sect. 1, the underlying inter-modal and intra-modal correlations should be well preserved in a reliable cross-modal hashing method. Therefore, for preserving the intra-modal correlations, *i.e.*, the similarity structure within a modality, we normalize F_v and F_t to $\hat{F}_v$ and $\hat{F}_t$, which are in the unit $l2$-norm each row, and create the cosine similarity matrices of the normalized feature vectors for the two modalities, namely $S_v = \hat{F}_v\hat{F}_v^T$ and $S_t = \hat{F}_t\hat{F}_t^T$. Then, B_v and B_t are updated to B_{v_s} and B_{t_s} by multiplying the similarity matrices, namely,

$$B_{v\_s} = S_v B_v \tag{7}$$

$$B_{t\_s} = S_t B_t \tag{8}$$

In this way, the neighborhood structures within the individual modalities are kept into B_{v_s} and B_{t_s}. Besides, the effective information of the neighbors in the original data distribution is integrated into the soft hash codes to enhance their semantic representation.

In order to train D_v and D_t, we combine B_{v_s} with M_v and combine B_{t_s} with M_t by concatenating them, which is denoted as

$$U_{vv} = B_{v\_s} \oplus M_v \tag{9}$$

$$U_{tt} = B_{t\_s} \oplus M_t \tag{10}$$

Then, G_v is fixed and generates fake image feature vectors from U_{vv} and D_v learns to judge the quality of the generated data through comparing the generated vectors with F_v. Meanwhile, D_t is trained in the same way.

After the parameters of discriminators W_v^D and W_t^D of D_v and D_t are updated, we train the generators. Firstly, we calculate

$$U_{vt} = B_{v\_s} \oplus M_t \tag{11}$$

$$U_{tv} = B_{t\_s} \oplus M_v \tag{12}$$

Here, U_{vt} preserves the semantic information of F_v but has the "appearance" of text data, since the semantic content is stored in B_{v_s} and M_t provides the expression style. Likewise, U_{tv} has the text semantic content and the image "appearance". We denote $U_{vt} = \{u_{vt_i}\}_{i=1}^n$ and $U_{tv} = \{u_{tv_j}\}_{j=1}^n$.

We send U_{vt} into G_t to produce dummy text feature vectors represented as F_t^d, then F_t^d passes through E_t^H to generate soft hash codes $B_t^{'}$. Ideally, $B_t^{'}$ should be the same as B_v since they have the same semantic content. Analogously, U_{tv} goes through G_v to generate fake image feature vectors F_v^d, which is then encoded to $B_v^{'}$ by E_v^H. Through this process, we gain the dummy data samples, $F_v^d = \{f_{v_i}^d\}_{i=1}^n$, $F_t^d = \{f_{t_j}^d\}_{j=1}^n$, and the re-generated soft binary codes, $B_v^{'} = \{b_{v_i}^{'}\}_{i=1}^n$, $B_t^{'} = \{b_{t_j}^{'}\}_{j=1}^n$. The formulations are written as follows:

$$F_t^d = G_t(U_{vt}, W_t^G) \tag{13}$$

$$B_t^{'} = E_t^H(F_t^d, W_t^H) \tag{14}$$

$$F_v^d = G_v(U_{tv}, W_v^G) \tag{15}$$

$$B_v^{'} = E_v^H(F_v^d, W_v^H) \tag{16}$$

where W_t^G, W_t^H, W_v^G, W_v^H represent the model parameters. This procedure allows data in different modalities to be combined together, which solves the problem of no correspondence between modalities when no pair-wise information provided. In addition, the cross-generation process forces the two branches to interact with each other so that the inter-modal relationships are preserved during learning.

3.4 Objective Functions

As discussed above, the architecture of CUCMH is built based on two GANs. The two discriminators D_v and D_t provides adversarial loss for the two generators G_v and G_t. The loss functions are defined following the way described in Sect. 3.1:

$$\begin{aligned} \mathcal{L}_{adv}^V = \min_{G_v} \max_{D_v} \mathbf{K}(D_v, G_v) \sum_{i=1}^{n} \mathbb{E}_{f_i \in p_{F_v}(f)}[logD(f_i)] \\ + \mathbb{E}_{u_{*v} \in p_{u_{*v}}(u_{*v})}[log(1 - D(G(u_{*v}))] \end{aligned} \tag{17}$$

$$\mathcal{L}_{adv}^{T} = \min_{G_t} \max_{D_t} \mathbf{K}(D_t, G_t) \sum_{i=1}^{n} \mathbb{E}_{f_i \in p_{F_t}(f)}[log D(f_i)] \\ + \mathbb{E}_{u_{*t} \in p_{u_{*t}}(u_{*t})}[log(1 - D(G(u_{*t}))] \tag{18}$$

where u_{*v} represents the reconstructed soft hash code concatenated with image modality expression and $*$ can be either v or t as described in Sect. 3.3. Correspondingly, u_{*t} is the concatenated soft binary code with text modality representation.

Furthermore, For each branch, CUCMH uses a reconstruction loss to alleviate the gap between the soft semantic binary codes generated in two times, namely, to pull B_v generated from F_v and $B_v^{'}$ generated from F_v^d through E_v^H as close as possible. Likewise, B_t generated from F_t and $B_t^{'}$ generated from F_t^d through E_t^H are forced to get closer. We employ the quantization loss to achieve this purpose. The loss function is formulated as

$$\mathcal{L}_{rec}^{V} = \left\| B_v - B_v^{'} \right\|_2^2 \tag{19}$$

$$\mathcal{L}_{rec}^{T} = \left\| B_t - B_t^{'} \right\|_2^2 \tag{20}$$

where $\|\cdot\|_2$ indicates $l2$ norm.

In addition, to improve the representation capacity of the modality expression encoders E_v^M and E_t^M, we set a modality classification loss. Specifically, a shared classifier is designed for both of the encoders to differentiate the modality of the outputs, and the prediction is measured by the binary cross-entropy loss, which are

$$\mathcal{L}_{m}^{V} = -(p_v log(p_t) + (1 - p_v) log(1 - p_t)) \tag{21}$$

$$\mathcal{L}_{m}^{T} = -(p_t log(p_v) + (1 - p_t) log(1 - p_v)) \tag{22}$$

Here, p_v and p_t are the predicted probability of the modality labels. The final objective functions of CUCMH, *i.e.*, the total loss of the image branch $\mathcal{L}_{total}^{V}$, is composed of $\mathcal{L}_{adv}^{V}$, $\mathcal{L}_{rec}^{V}$ and $\mathcal{L}_{m}^{V}$. That is

$$\mathcal{L}_{total}^{V} = \alpha_1 \mathcal{L}_{adv}^{V} + \alpha_2 \mathcal{L}_{rec}^{V} + \alpha_3 \mathcal{L}_{m}^{V} \tag{23}$$

and the total loss of the text branch $\mathcal{L}_{total}^{T}$ is written as

$$\mathcal{L}_{total}^{T} = \alpha_1 \mathcal{L}_{adv}^{T} + \alpha_2 \mathcal{L}_{rec}^{T} + \alpha_3 \mathcal{L}_{m}^{T} \tag{24}$$

where α_1, α_2 and α_3 are trade-off hyper-parameters.

3.5 Learning

CUCMH leverages the alternating learning strategy through the network training. We optimize the generators and discriminators separately. As shown in Algorithm 1, we first randomly sample a mini-batch of F_v and F_t, which are denoted as F_{v_batch} and F_{t_batch}. They pass through E_v^H, E_v^M and G_v with S_v, as well as E_t^H, E_t^M and G_t with S_t to get fake data samples. In this process, the parameters of encoders and generators are fixed and we train D_v and D_t using the detached $F^d_{v\_batch}$ and $F^d_{t\_batch}$, then update D_v and D_t parameters. After this, we re-pass F_{v_batch} and F_{t_batch} through encoders and generators to optimize these components with the parameters of D_v and D_t fixed.

Algorithm 1 CUCMH Learning.

1: **Inputs:**
  Image feature vectors F_v and text feature vectors F_t
2: **Outputs:**
  Binary code matrix B_Δ;
  Parameters W_v^H, W_v^M, W_t^H, W_t^M, W_v^G, W_t^G, W_v^D and W_t^D of the networks;
3: **Initialize:**
  Hyper-parameters: α_1, α_2 and α_3;
  Hash code length C;
  Mini-batch size and learning rate;
4: **for** k epochs **do**
5:   **for** t time steps **do**
6:     Randomly sample a mini-batch from F_v and F_t, respectively, to form F_{v_batch} and F_{t_batch}.
7:     Pass F_{v_batch} through E_v^H and E_v^M, calculate B_{v_batch} and M_{v_batch} by forward propagation; pass F_{t_batch} through E_t^H and E_t^M, calculate B_{t_batch} and M_{t_batch} by forward propagation.
8:     Build S_v and S_t; calculate $B_{v\_s\_batch}$ and $B_{t\_s\_batch}$ according to (7) and (8).
9:     Calculate U_{vv_batch} and U_{tt_batch} according to (9) and (10).
10:     Forward propagate the detached U_{vv_batch} and U_{tt_batch} through G_v and G_t to get fake image and text feature vectors; D_v and D_t distinguish F_v and F_t from fake features and calculate the loss.
11:     Update W_v^D and W_t^D with BP algorithm.
12:     Re-generate $B_{v\_s\_batch}$ and $B_{t\_s\_batch}$ in the same way as 7 and 8.
13:     Calculate U_{vt_batch} and U_{tv_batch} according to (11) and (12).
14:     Forward propagate U_{tv_batch} and U_{vt_batch} through G_v and G_t to get F_v^d and F_t^d.
15:     Forward propagate F_v^d and F_t^d through E_v^H and E_t^H to get $B_v^{'}$ and $B_t^{'}$.
16:     Calculate loss according to (23) and (24); Update W_v^H, W_v^M, W_t^H, W_t^M, W_v^G and W_t^G with BP algorithm.
17:   **end for**
18: **end for**
19: Optimize B_Δ according to (2).

3.6 Testing

In testing, given a query feature vector q_v or q_t, CUCMH passes it through the well-trained corresponding Hashing encoder, *i.e.*, E_v^H or E_t^H, to compute the soft hash code and then signs it to the binary code. Meanwhile, hash codes of the instances in database are calculated in the same way. Through calculating the Hamming distance and ranking, we gain the most similar data entities in the other modality of q_v or q_t.

4 Experiments

The experiments are performed on two popular datasets: MIRFlickr dataset [11] and NUS-WIDE dataset [3]. Since this is the very first work under the setting of unsupervised learning with no pair-wise information available, there is no such previous work to compare. Therefore, we compare CUCMH with 2 supervised learning methods and 3 pair-wise information guided unsupervised algorithms. In addition, we perform the ablation study to evaluate the effectiveness of modality representation. In this section, two datasets will be introduced at first, and the settings of experiments will be described in details, which includes the algorithms compared and the evaluation metrics. Finally, we will analyze the experiment results to dig deeply into the proposed method.

4.1 Datasets and Features

MIRFlickr 25K dataset [11] is a large multimedia dataset with 25,000 images retrieved from Flickr. Each image has multiple textual tags and belongs to at least one of the total 24 categories. We use 5% of the total data instances as the testing set and the rest as the retrieved database. For the supervised methods compared, 5000 images and corresponding tags are provided for training. The images are represented with 4,096-D deep feature vectors from the pre-trained 19-layer VGGNet [25] and texts are used as 1,000-D BOW vectors.

NUS − WIDE dataset [3] totally contains 269,498 images collected from Flickr, and each image is annotated with several corresponding textual tags. The data are categorised into 81 concepts and each data instance belongs to one or multiple concepts. We select 10 most common concepts as class labels and the corresponding images with text tags form the experiment dataset. 2000 data instances are randomly selected as the testing set and the remaining act as the training set. We utilize 5000 images and corresponding tags for the supervised approaches training. The images are extracted 4,096-D deep feature vectors from a 19-layer VGGNet [25] for the experiments and text are 1,392-D Bag-of-Visual-Words vectors.

4.2 Experiment Settings

Evaluation Metrics. Hamming ranking and hash lookup are two popular retrieval procedures in Hashing-based retrieval [18]. We evaluate these two procedures of CUCMH and the baselines. In Hamming ranking procedure, the data entities in the retrieval set are ranked by the Hamming distances to the given query. We adopt the widely used metric mean average precision (mAP) [18] to measure the accuracy of the Hamming ranking procedure. In hash lookup procedure, all candidate points in the retrieval set within a certain Hamming radius away from the query are returned. The Precision-Recall Curve is used to measure the performance of the hash lookup.

4.3 Baselines

To demonstrate the effectiveness of CUCMH, we conduct experiments to compare the performance with the existing work. We compare CUCMH with 2 supervised learning methods (including CMSSH and SCM) and 3 pair-wise information guided unsupervised algorithms (including CVH, PDH, and CCQ). We describe these baselines briefly as follows.

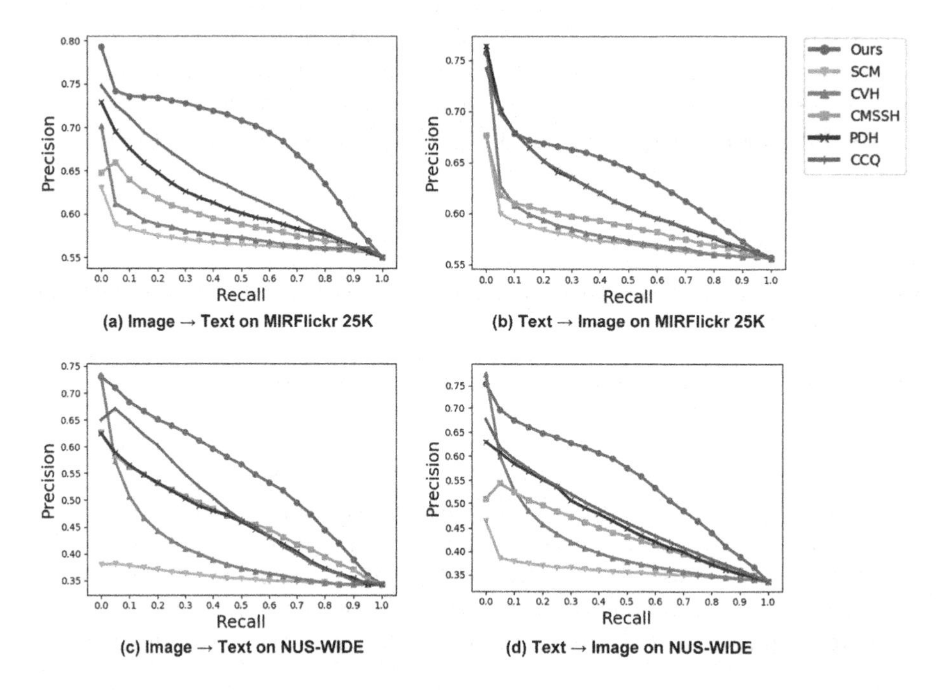

Fig. 2. Precision-recall curves on MIRFlickr (a, b) and NUS-WIDE (c, d) dataset. Thecode length is 64 bits.

CMSSH [2] solves a binary classification problem with learning a mapping which can be efficiently learned using boosting algorithms. SCM [34] proposes a semantic correlation maximization method based on semantic labels for improving the efficiency of cross-modal modal hashing. CVH [14] adopts the spectral hashing for single modal to cross-view hashing tasks. It preserves both intra and inter view correlations with a generalized eigenvalue formulation. PDH [22] embeds proximity of original data and designs an objective function, which is optimized based on block coordinate descent algorithm, to maintain the predictability of the the generated binary code. CCQ [19] proposes a seamless latent semantic analysis (LSA) framework with multimodal correlation and composite quantization integrated, so that data in different modalities are encoded into an isomorphic latent space and then into the common binary space.

Table 1. The mAP scores of two retrieval tasks on MIRFlickr 25K dataset with different lengths of hash codes.

Methods	image→text				text→image			
	16-bits	32-bits	64-bits	128-bits	16-bits	32-bits	64-bits	128-bits
CVH	0.602	0.587	0.578	0.572	0.607	0.591	0.581	0.574
PDH	0.623	0.624	0.621	0.626	0.627	0.628	0.628	0.629
CCQ	0.637	0.639	0.639	0.638	0.628	0.628	0.622	0.618
CMSSH	0.611	0.602	0.599	0.591	0.612	0.604	0.592	0.585
SCM	0.585	0.576	0.570	0.566	0.585	0.584	0.574	0.568
Baseline	0.609	0.617	0.626	0.637	0.610	0.625	0.632	0.632
CUCMH(ours)	0.637	0.641	0.643	0.641	0.631	0.640	0.642	0.639

Table 2. The mAP scores of two retrieval tasks on NUS-WIDE dataset with different lengths of hash codes.

Methods	image→text				text→image			
	16-bits	32-bits	64-bits	128-bits	16-bits	32-bits	64-bits	128-bits
CVH	0.458	0.432	0.410	0.392	0.474	0.445	0.419	0.398
PDH	0.475	0.484	0.480	0.490	0.489	0.512	0.507	0.517
CCQ	0.504	0.505	0.506	0.505	0.499	0.496	0.492	0.488
CMSSH	0.512	0.470	0.479	0.466	0.519	0.498	0.456	0.488
SCM	0.389	0.376	0.368	0.360	0.388	0.372	0.360	0.353
Baseline	0.519	0.538	0.547	0.545	0.508	0.525	0.530	0.522
CUCMH(ours)	0.535	0.550	0.553	0.548	0.537	0.542	0.537	0.531

4.4 Implementation Details

Module Architecture. As discussed in Sect. 3, the CUCMH consists of two branches and each branch contains a semantic encoder, a modality encoder, a generator and a discriminator. All components are constructed with multiple fully-connected layers followed by activation functions. For instance, the transformation of the vector dimension through E_v^H is $4,096 \rightarrow 2,048 \rightarrow 512 \rightarrow 256 \rightarrow 64$ and that through G_v is $(64+5) \rightarrow 256 \rightarrow 512 \rightarrow 2,048 \rightarrow 4,096$, where 64 is the length of the hash code and 5 is the length of the modality code. The inputs of these whole two branches, *i.e.*, the inputs of the semantic encoders and the modality encoders, are 4,096D embeddings. The dimension of semantic encoder outputs is the length of hash codes, *i.e.*, 16, 32, 64 and 128. We conduct an experiment to explore the influence from the length of modality codes, *i.e.*, the dimension of the modality encoder outputs, and found that when the code length is set to 5, our model can achieve the best performance. The inputs of the discriminators and the outputs of the generators are in 4,096-D.

Parameter Settings. The loss functions for both the image and text branch are composed with 3 components as summarized in (23) and (24). Through performing a series of experiments, we set $\alpha_1 = 1$, $\alpha_2 = 1$, and $\alpha_3 = 0.5$ to achieve the best testing results.

4.5 Experiment Results and Analysis

Comparisons with Other Work. Since there is no previous work under the same setting, we compare our method with 2 supervised learning methods and 3 pair-wise information guided algorithms. Nevertheless, CUCMH can achieve superior retrieval performance in comparison with these work. The mAP results of all the algorithms on different hash code lengths for two datasets are listed in Table 1 and Table 2. The Precision-Recall curves (with 64-bit hash codes) are shown in Fig. 2. According to the experiment results, CUCMH attains desirable performance.

In the mAP results, it can be seen that CUCMH outperforms all competitors. On MIRFlickr-25K dataset, CUCMH achieves the best mAP of 0.643 on image $\rightarrow$ text and 0.642 on text $\rightarrow$ image task in 64-bits. On NUS-WIDE dataset, CUCMH performs the best on image $\rightarrow$ text with the mAP score of 0.553 in 64-bits and on text $\rightarrow$ image with the mAP score of 0.542 in 32-bits. In comparison with other methods, the performance of CUCMH is desirable. Take the example of the results on MIRFlickr-25K, on image $\rightarrow$ text, compared with CCQ [19], which performs the best among the unsupervised competitors, CUCMH increases the mAP score by around 0.004, and the mAP of CUCMH is around 0.04 higher than the supervised methods in 64-bits; on NUS-WIDE dataset, the mAP score of our method is more than 0.4 higher on both image $\rightarrow$ text task in 64-bits and text $\rightarrow$ image task in 32 bits.

From the Precision-Recall curves, it can be observed that CUCMH is at the top of the list of all competitors, which means that our method has a better precision-recall balance for the lookup procedural than other work. For example, in the image → text task on NUS-WIDE, with the same recall at 0.3, the precision of CUCMH reaches 0.66, which is around 0.035 higher than the second best method CCQ. This indicates that CUCMH can find the same number of relevant data candidates by retrieving less data in the database than other methods.

Effect of Modality Representation. To further explore the effectiveness of the Modality Representation in CUCMH, we design a baseline method without the Modality Encoder. This model is implemented as the same as CUCMH, excluding only that the inputs of the generators are the improved semantic codes instead of the concatenated codes (*i.e.*, the combination of semantic codes and modality codes). Table 1 and Table 2 show the results of the comparison between this baseline and CUCMH. It is can be seen that CUCMH outperforms the baseline in all cases. This demonstrates that the modality representation is able to improve the connection between different modalities and promote the learning of hash functions.

5 Conclusion

In this paper, we propose a novel dual-branch generative adversarial network to tackle the challenge from unsupervised cross-modal hashing with no pairwise information. We believe this is the first work to discuss this issue. We also introduce the concept that the representation of multimedia data can be separated into content and style, and employ the modality representation codes to improve the effectiveness of the cross-modal generative adversarial learning. The experiments demonstrate the desirable performance of CUCMH in completely unsupervised cross-modal hashing tasks and the effectiveness of the method of integrating modality representation with semantic information in representation learning.

Acknowledgement. This work was partially supported by Australian Research Council Discovery Project (ARC DP190102353).

References

1. Andrew, G., Arora, R., Bilmes, J.A., Livescu, K.: Deep canonical correlation analysis. In: ICML (2013)
2. Bronstein, M.M., Bronstein, A.M., Michel, F., Paragios, N.: Data fusion through cross-modality metric learning using similarity-sensitive hashing. In: CVPR (2010)
3. Chua, T., Tang, J., Hong, R., Li, H., Luo, Z., Zheng, Y.: NUS-WIDE: a real-world web image database from national university of Singapore. In: CIVR (2009)
4. Ding, G., Guo, Y., Zhou, J.: Collective matrix factorization hashing for multimodal data. In: CVPR (2014)

5. Frome, A., et al.: Devise: a deep visual-semantic embedding model. In: NeurIPS (2013)
6. Fu, Z., Tan, X., Peng, N., Zhao, D., Yan, R.: Style transfer in text: exploration and evaluation. In: AAAI (2018)
7. Gatys, L.A., Ecker, A.S., Bethge, M.: Image style transfer using convolutional neural networks. In: CVPR (2016)
8. Goodfellow, I.J., et al.: Generative adversarial nets. In: NeurIPS (2014)
9. Hardoon, D.R., Szedmák, S., Shawe-Taylor, J.: Canonical correlation analysis: an overview with application to learning methods. Neural Comput. **16**(12), 2639–2664 (2004)
10. Hu, Y., Jin, Z., Ren, H., Cai, D., He, X.: Iterative multi-view hashing for cross media indexing. In: ACMMM (2014)
11. Huiskes, M.J., Lew, M.S.: The MIR flickr retrieval evaluation. In: SIGMM (2008)
12. Jiang, Q.Y., Li, W.J.: Deep cross-modal hashing. In: CVPR (2017)
13. Kim, Y.: Convolutional neural networks for sentence classification. In: EMNLP (2014)
14. Kumar, S., Udupa, R.: Learning hash functions for cross-view similarity search. In: IJCAI (2011)
15. Li, C., Deng, C., Wang, L., Xie, D., Liu, X.: Coupled cyclegan: unsupervised hashing network for cross-modal retrieval. In: AAAI (2019)
16. Li, D., Dimitrova, N., Li, M., Sethi, I.K.: Multimedia content processing through cross-modal association. In: ACMMM (2003)
17. Lin, Z., Ding, G., Hu, M., Wang, J.: Semantics-preserving hashing for cross-view retrieval. In: CVPR (2015)
18. Liu, W., Mu, C., Kumar, S., Chang, S.: Discrete graph hashing. In: NeurIPS (2014)
19. Long, M., Cao, Y., Wang, J., Yu, P.S.: Composite correlation quantization for efficient multimodal retrieval. In: SIGIR (2016)
20. Ngiam, J., Khosla, A., Kim, M., Nam, J., Lee, H., Ng, A.Y.: Multimodal deep learning. In: ICML (2011)
21. Rasiwasia, N., et al.: A new approach to cross-modal multimedia retrieval. In: ACMMM (2010)
22. Rastegari, M., Choi, J., Fakhraei, S., Hal III, H., Davis, L.S.: Predictable dual-view hashing. In: ICML (2013)
23. Rosipal, R., Krämer, N.: Overview and recent advances in partial least squares. In: SLSFS (2005)
24. Sharma, A., Kumar, A., Daumé, H., Jacobs, D.W.: Generalized multiview analysis: a discriminative latent space. In: CVPR (2012)
25. Simonyan, K., Zisserman, A.: Very deep convolutional networks for large-scale image recognition. In: ICLR (2015)
26. Song, J., Yang, Y., Yang, Y., Huang, Z., Shen, H.T.: Inter-media hashing for large-scale retrieval from heterogeneous data sources. In: SIGMOD (2013)
27. Sun, L., Ji, S., Ye, J.: A least squares formulation for canonical correlation analysis. In: ICML (2008)
28. Tenenbaum, J.B., Freeman, W.T.: Separating style and content with bilinear models. Neural Comput. **12**, 1247–1283 (2000)
29. Weiss, Y., Torralba, A., Fergus, R.: Spectral hashing. In: NeurIPS (2008)
30. Wu, G., et al.: Unsupervised deep hashing via binary latent factor models for large-scale cross-modal retrieval. In: IJCAI (2018)
31. Wu, L., Wang, Y., Shao, L.: Cycle-consistent deep generative hashing for cross-modal retrieval. IEEE Trans. Image Process. **28**(4), 1602–1612 (2019)

32. Xu, X., Shen, F., Yang, Y., Shen, H.T., Li, X.: Learning discriminative binary codes for large-scale cross-modal retrieval. IEEE Trans. Image Process. **26**, 2494–2507 (2017)
33. Ye, Z., Peng, Y.: Multi-scale correlation for sequential cross-modal hashing learning. In: ACMMM (2018)
34. Zhang, D., Li, W.J.: Large-scale supervised multimodal hashing with semantic correlation maximization. In: AAAI (2014)
35. Zhang, J., Peng, Y., Yuan, M.: Unsupervised generative adversarial cross-modal hashing. In: AAAI (2018)
36. Zhen, Y., Yeung, D.: Co-regularized hashing for multimodal data. In: NeurIPS (2012)
37. Zhou, J., Ding, G., Guo, Y.: Latent semantic sparse hashing for cross-modal similarity search. In: SIGIR (2014)
38. Zhu, J., Park, T., Isola, P., Efros, A.A.: Unpaired image-to-image translation using cycle-consistent adversarial networks. In: ICCV (2017)

High Performance Design for Redis with Fast Event-Driven RDMA RPCs

Xuecheng Qi, Huiqi Hu(✉), Xing Wei, Chengcheng Huang, Xuan Zhou, and Aoying Zhou

School of Data Science and Engineering, East China Normal University, Shanghai, China
{xcqi,simba_wei,cchuang}@stu.ecnu.edu.cn,
{hqhu,xzhou,ayzhou}@dase.ecnu.edu.cn

Abstract. Redis is a popular key-value store built upon socket interface that remains heavy memory copy overhead within the kernel and considerable CPU overhead to maintain socket connections. The adoption of Remote Direct Memory Access (RDMA) that incorporates outstanding features such as low-latency, high-throughput, and CPU-bypass make it practical to solve the issues. However, RDMA is not readily suitable for integrating into existing key-value stores. It has a low-level programming abstraction and the original design of existing systems is a hurdle to exploit RDMA's performance benefits. RPCs can provide simple abstract programming interfaces that make it easy to be integrated into existing systems. This paper proposes a fast event-driven RDMA RPC framework named FeRR to promote the performance of Redis. First, we describe our design of FeRR that is based on one-sided RDMA verbs. Second, we propose an efficient request notification mechanism using event-driven model that can decrease the CPU consumption of polling requests. Finally, we introduce a parallel task engine to eschew the bottleneck of the single-threaded execution framework in Redis. Comprehensive experiment shows that our design achieves orders-of-magnitude better throughput than Redis - up to 2.78 million operations per second and ultra-low latency - down to 10 μs per operator on a single machine.

Keywords: RDMA · FeRR · Redis

1 Introduction

Existing key-value stores like Redis [1] use the conventional socket I/O interface, which is easy to develop and compatible with commodity NIC. Because of large CPU copy overhead that network data package should be copied between user space to kernel space, about 70% time is spent on receiving queries and sending responses over TCP [2]. Furthermore, low-speed NIC confines the performance of key-value stores. Remote Direct Memory Access (RDMA) can provide high-throughput and low-latency network communication. There has been increasing

Y. Nah et al. (Eds.): DASFAA 2020, LNCS 12112, pp. 195–210, 2020.
https://doi.org/10.1007/978-3-030-59410-7_12

interest in recent years to build high performance key-value stores using RDMA-capable network [3–8].

In the above works, many design considerations in RDMA suit are confined. Pilaf [4] and FaRM [5] use one-sided RDMA read to traverse remote data structures but suffer from multiple round trips across the network. As we demonstrate later, RDMA read cannot saturate the peak performance of the RDMA hardware. Herd [7] transmits clients' requests to server memory using RDMA write and polling server memory for incoming requests using its CPU. However, this request notification approach is inefficient and bings considerable CPU overhead.

Although RDMA-capable network is a treasure, integrating it into existing commercial systems is disruptive. These prototype systems type the code to a specific interconnect API to support RDMA communication. But for existing systems, this can lead to significant code changes. In terms of simplicity, RPCs reduce the software complexity required to design key-value stores compared to one-sided RDMA-based systems.

To solve these issues, we propose FeRR, a fast event-driven RDMA RPC framework that delivers low latency and high throughput. FeRR has fully considered network primitives of RDMA hardware and possesses an efficient event-driven request notification mechanism. FeRR provides simple abstract programming interfaces that make it easy to be integrated into existing systems. Furthermore, we have designed and implemented a new novel branch of Redis over FeRR, FeRR-driven Redis.

Since FeRR can improve the performance of Redis, several new issues have emerged. Redis Serialization Protocol (RESP) is not suitable for RDMA verbs but is necessary for multiple data type support. Recent works [4,7] need no serialization protocol, they set only one data type of $<key, value>$ as $<string, string>$. As a remedy, we take a less disruptive approach to optimize the existing RESP to relieve the memory copy overhead, meanwhile maintain the support of multiple data types.

When network is not the bottleneck, the single thread framework of Redis is the new issue. To address this problem, we design a parallel task engine for FeRR-driven Redis. The major part of this engine is a cuckoo hashing [9] table with optimistic locking. Cuckoo hashing is friendly for read-intensive workloads while many workloads of key-value stores are predominately reads, with few writes [10]. Taking constant time in the worst case, cuckoo hashing has great lookup efficiency. Meanwhile, optimistic locking outperforms pessimistic locking for read-intensive workloads.

Overall, this paper makes three key contributions:

(1) We discuss our design considerations based on a sufficient analysis of the performance of RDMA and disadvantages of previous works.
(2) We propose a fast event-driven RDMA framework name FeRR that delivers extremely high throughput and low latency.
(3) We design and implement a high-performance version of Redis named FeRR-driven Redis, including an optimized serialization protocol for low-latency transmission, and a parallel task engine for high-throughput execution.

We implemented these designs on Redis v3.0, branching a new branch of Redis. We conducted experiments on ConnectX-3 RNIC to evaluate the performance of our design. Experiments show that FeRR-driven Redis achieves throughput up to 2.78 million operations per second and latency down to 10 µs per operation, which are over 37.6× higher throughput and only 15% latency of Redis on IPoIB.

2 Preliminary

2.1 RDMA

RDMA enables zero-transfer, low round-trip latency, and low CPU overhead. RDMA capable networks can provide 56 Gbps of bandwidth and 2 µs round-trip latency with Mellanox ConnectX-3 RNIC. In this section, we provide an overview of RDMA features: verb types, transport modes that are used in the rest of the paper.

Message Verbs: SEND and RECV, provide user-level two-sided message passing that involves remote machine's CPU. Before a SEND operation, a pre-posted RECV work request should be specified by the remote machine where the request is written to. Then, responder could get the request by polling the completion queue. Compared to RDMA verbs, message verbs have higher latency and lower throughput.

RDMA Verbs: WRITE and READ are one-sided operations, namely, allow full responder's CPU bypass that client can write or read directly the memory of responder without its CPU involved. The new type of message passing technique can relieve the overhead of responder's CPU since responder is unaware of client's operations. Furthermore, one-sided operation has the lowest latency and highest throughput.

Transports Mode: RDMA transports are divided into reliable or unreliable, and connected or unconnected. Reliable transports guarantee that messages are delivered in order and without corruption, while unreliable has no such guarantee. InfiniBand uses a credit-based flow control in the link layer to prevent loss of data, and CRC to ensure data integrity [11]. Thus, the packet losses of unreliable transports are extremely rare. The difference between connected and unconnected transports is the number of connected queue pairs (QP). Connected transports need that one QP sends/receives with exactly one QP, and unconnected transports allow one QP sends/receives with any QP. RDMA verbs support two types of connected transports: Reliable Connection (RC) and Unreliable Connection (UC). Unreliable Datagram (UD) supports only SEND operations.

2.2 Redis

The existing open-source Redis implementation is designed using traditional Unix socket interface. While having an only single thread to handle socket connections, Redis has built an I/O multiplexing event model, such as epoll/select,

Table 1. Redis serialization protocol

Data type	Label byte	Example
Simple String	+	"$+ok\backslash r\backslash n$"
Errors	−	"$-Error\ message\backslash r\backslash n$"
Integers	:	"$: 1\backslash r\backslash n$"
Bulk String	\$	"$\$6\backslash r\backslash nfoobar\backslash r\backslash n$"
Arrays	*	"$*3\backslash r\backslash n : 1\backslash r\backslash n : 2\backslash r\backslash n : 3\backslash r\backslash n$"
Example:	$*3\backslash r\backslash n\$3\backslash r\backslash nSET\backslash r\backslash n\$3\backslash r\backslash nkey\backslash r\backslash n\$5\backslash r\backslash nHello\backslash r\backslash n$	

that has the ability to handle hundreds of connections and achieve good performance. In Redis, I/O multiplexing event model considers operations as file events(except time event). A connection, a command or a reply is constructed as a file event and inserted into an event set which is monitored by epoll() in Linux. Then Linux kernel epolls a fired file event for a corresponding function to process when its socket is ready to read or write.

Different from Memcached [12], multiple data types are supported in data transport of Redis, such as string, integer, array. As Table 1 shows, RESP attaches one byte to different types of data to distinguish them. Taking a set command **"SET key Hello"** as an example, it should be encoded in client-side as "Example" in Table 1 that means a size of 3 arrays including 3 bulk strings with respective length 3, 3, 5. And server decodes the encoded buffer to a set command to process. RESP remains several times of memory copy for encoding and decoding.

2.3 Related Work

RDMA-Optimized Communication: The HPC community engages in taking advantage of RDMA with Infiniband to improve the MPI communication performance, such as MVAPICH2 [13] OpenMPI [14]. They provide RDMA-based user-level libraries that support the MPI interface. These works [3,15,16] utilize RDMA to improve the throughput and reduce the CPU overhead of communication of the existing systems Hadoop, HBase, Memcached. However, Most of them only use SEND/RECV verbs as a fast alternative to socket-based communication despite leveraging one-sided RDMA primitives. Memcached-RDMA [3] fixes message verbs and RDMA verbs to build a communication library. For put operations, the client sends the server a local memory location using SEND verb and the server reads the key-value pair via RDMA read. Get operations involve SEND verb and RDMA write. The server writes the data using RDMA write into the allocated memory address pre-sent by the client.

User-Level Communication: Other than the usage of RDMA capable network, MICA [17] utilizes the Intel DPDK library [18] to build the key-value store on classical Ethernet. The Intel DPDK library supports zero-copy technology that eliminates CPU interrupt and memory copy overhead. Specifically,

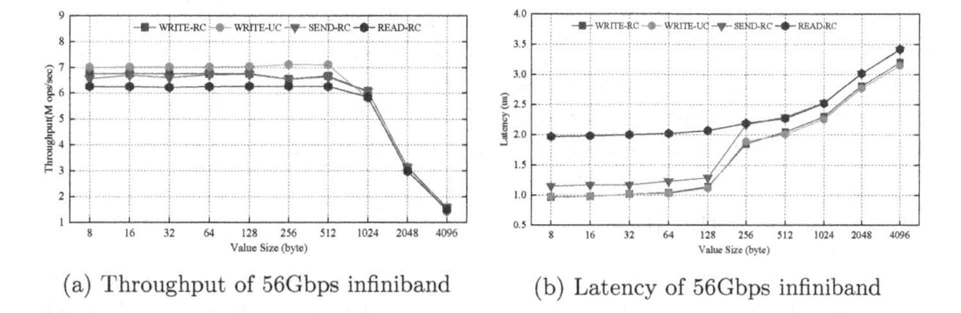

(a) Throughput of 56Gbps infiniband

(b) Latency of 56Gbps infiniband

Fig. 1. Throughput and latency under different verbs type and transport modes

MICA combines exclusive partitions with CPU cores to avoid synchronization but introduces core imbalance for skewed workloads. KV-Direct [8] introduces a new key-value store built on a programmable NIC. It offloads key-value processing on CPU to the FPGA-based programmable NIC to extend RDMA primitives from memory operations to key-value operations and enable remote direct access to the host memory. As a result, systems on user-level communication can achieve matched throughput to RDMA based solutions, nonetheless with higher latency.

RDMA-Based Key-Value Store: RAMCloud [19] takes advantage of message verbs of RDMA to build an in-memory, persistent key-value store. Pilaf [4] lets clients directly execute get operations at client side using multiple RDMA read operations and introduces a self-verification data structure to detect read-write races between client-side get operations and server-side put operations. Meanwhile, HydraDB [6] uses RDMA read to operate get as well and implement the client cache to promote system performance. HERD [7] points out that RDMA write has better throughput and lower latency than RDMA read. For this, HERD building a high-performance key-value store using one-sided RDMA write for sending requests and use SEND over UC for replying responses because datagram transport can scale better for applications with a one-to-many topology. FaRM [5] exploits RDMA verbs to implement global shared address among distributed nodes.

3 Design Consideration

3.1 Network Primitives Choice

Since we have introduced FeRR, the first challenge is that how to choose networking primitives of RDMA to saturate the peak performance of RNIC. We conduct an experiment to measure the throughput and latency of ConnectX-3 56 Gbps InfiniBand under different verbs and transport modes. As Fig. 1 shows, RDMA write over UC have the best performance in both throughput and latency. As to verbs, message verbs require involvement of the CPU at both the sender and receiver, but RDMA write bypass the remote CPU to operate directly on remote memory. Meanwhile, RDMA write outperforms RDMA read as well, because the responder does not need to send packets back, RNIC performs less processing. Performance of RDMA write over UC and RC are nearly identical, using UC is still beneficial that it requires less processing at RNIC to save capacity. Therefore, we select RDMA write over UC to as network primitives of FeRR.

3.2 Request Notification

Another issue is how to build an efficient request notification mechanism. As well-known, RDMA verbs allow CPU bypass that can greatly relieve the overhead of the server's CPU since the server is unaware of the client's operations. Therefore, the server needs to scan the message pool continuously to get new requests, which leads to high CPU overhead. Also, with increasing number of clients, latency is increased much. Furthermore, Redis server's CPU is easier to become a bottleneck due to complex code path. Event-driven way is very attractive that it relieves CPU consumption and is effective for request notification. Therefore, we design an event-driven request notification mechanism based on RDMA-write-with-imm. It is the same as RDMA write, but a 32-bit immediate data will be sent to remote peer's receive (CQ) to notify the remote CPU of the completion of write. Meanwhile, this RDMA primitive retains the property of high throughput and low latency of RDMA write.

4 FeRR

In this section, we explain the design of FeRR, and how we reap the benefits of RDMA. We first motivate a well-tuned RPC architecture, which incorporates RDMA-write-with-imm as network primitive. Then we describe an event-driven mechanism that can efficiently poll requests while saving CPU capacity.

4.1 Architecture

In the architecture of FeRR, each client connects the server with two QPs where one takes charge of the outbound verbs and the other for the inbound verbs. As Fig. 2 shows, a client maintains two buffers that one for sending requests and the other for receiving responses. Server has a message pool for temporary storage when exchanging messages. After establishing the connection with server, a client is given a clientID and allocated a pair of correspondingly buffers in the pool for communication. By doing so, the server can easily manage the allocation and displacement of buffers in the message pool.

FeRR is fundamentally built on RDMA-write-with-imm that can notify remote machine immediately. RDMA-write-with-imm consumes a receive request form the QP of remote side and the request is processing immediately after being polled from the CQ. Besides, it is able to carry a 32-bit immediate data in the message. We attach the client's identifier in the immediate data filed including a clientID and an offset of the client's receive buffer. The server can directly locate

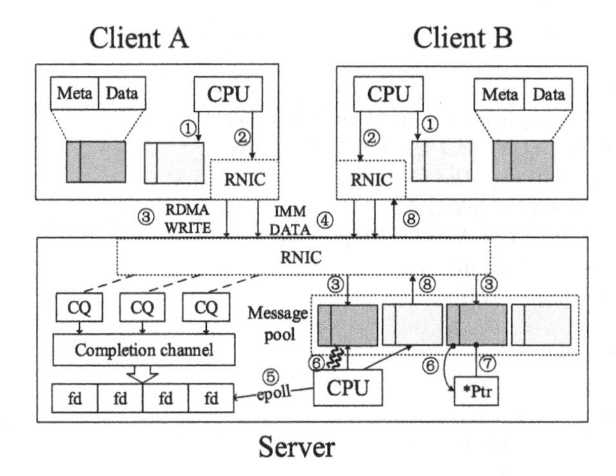

Fig. 2. Architecture of FeRR

the request in the message pool with the identifier instead of scanning the whole pool.

Figure 2 illustrates the process of a FeRR call. Before the call, the server and clients exchange their pre-allocated memory space addresses and keys via socket connection. A client creates and serializes a new request into the request buffer ①, then calls the write API of FeRR ②. FeRR uses RDMA-write-with-imm to write the memory-object of request into the pre-exchanged memory space in the message pool of server ③. FeRR uses the IMM value to include the clientID and the offset that indicates which buffer in the message pool ④. In the server, a polling thread epolls entries from CQs via completion channel ⑤. The server read CQE to get offset and clientID to know the request is in which buffer and belongs to which client. The user thread decodes the request and constructs the corresponding pointer to reference the memory object ⑥. FeRR returns the pointer to a worker thread to handle the request ⑦. Finally, response is written back to the client ⑧.

4.2 Event-Driven Request Notification

In FeRR, we design an event-driven mechanism for efficient request notification. For the conventional way, the server needs polling the CQ ceaselessly to learn whether there is a work completion. Reading the work completions using events decreases the CPU consumption because it eliminates the need to perform constant polling the CQ. A completion channel can monitor several CQs and notify the thread once a work completion yields. FeRR registers fixed number (can be configured) of CQs to one completion channel and epolls the coming completion event. After the request notification, the worker thread polls the work completion from the CQ. Finally, the request is got and processed by the thread. This is the fundamental procedure of event-driven request notification in FeRR.

Algorithm 1: Event-driven request notification

```
1  Function Binding():
2      channel ← initialize a Completion Channel;
3      for i = 0 to CQ_NUM_PER_CHANNEL do
4          cq[i] ← initialize the i_th CQs;
5          cq[i].channel ← channel ;
6          for j = 0 to QP_NUM_PER_CQ do
7              qp[j] ← initialize the j_th QPs;
8              qp[j].cq ← cq[i] ;

9  Function f_epoll():
10     while true do
11         ret ← epoll(channel_fd);
12         if ret = true then
13             cq ← ibv_get_cq_event(channel);
14             while cq has any Work Completion do
15                 wc ← ibv_poll_cq(cq);
16                 notify corresponding worker thread to process the request;
```

The native approach has a major bottleneck that is inefficient. Upon polling a work completion, the worker thread needs to switch for polling to process the request. As one CQ is always shared by multiple QPs, there are probably more than one work completions in the CQ. To solve the bottleneck, we exploit specific polling thread to take charge of request notification. As Algorithm 1 shows, FeRR first initializes a completion channel "channel" which is used for monitoring CQs (line 2). Then, FeRR initializes CQs and registers a fixed number of them to the "channel" (line 3–5). Next, QPs are created and a fixed number of them are bound to one CQ (line 6–8). In *f_epoll*() function, FeRR calls *epoll*() to wait for completion event and gets the CQ context (line 10–13). After that, Polling threads call *ibv_poll_cq*() to poll all of the completions from the CQ and notify corresponding worker threads to process the requests (14–16).

4.3 APIs of FeRR

FeRR supports three kinds of interface: RPC constructor(f_init())), memory management(f_reg(), f_free()) and messaging interface(f_send(), f_recv()) for user-level applications. We select these semantics because they are simple to RDMA-driven application programmers. Table 2 lists major APIs of FeRR.

5 FeRR-Dirven Redis

In this section, we design and implement FeRR into Redis that proposes a new branch of Redis, FeRR-Redis. Then we propose a subtle serialization protocol

Table 2. Major APIs of FeRR

Interface	Description
status → f_init()	Initialize resources for FeRR
addrID → f_reg()	Register memory space into RNIC
status → f_free()	Free the memory space registered
status → f_send()	FeRR sends data to remote machine
status → f_recv()	Receives next FeRR send
status → f_epoll()	Epoll WQEs from the completion channel

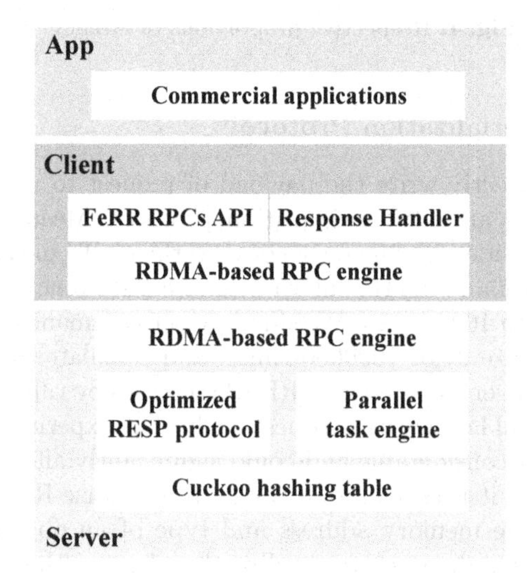

Fig. 3. Overview framework of FeRR-Redis

to replace the origin RESP, which not only reserves the supporting of multiple value types but also solves the memory copy issue. Finally, we point out that single-threaded task engine is a new bottleneck instead of socket-based network and propose a parallel task engine for Redis.

5.1 Overview

Figure 3 illustrates the overview framework of FeRR-Redis that contains two parts: (1) Client-side layer: Client side provides APIs of FeRR to the application layer and is in charge of handling responses. (2) Server-side layer: The main part of this layer is a parallel task engine. To embrace Redis with multi-threaded execution, we implement a concurrent cuckoo hashing table with optimistic locking. Besides, we propose an optimized serialization protocol of Redis to substitute the original one.

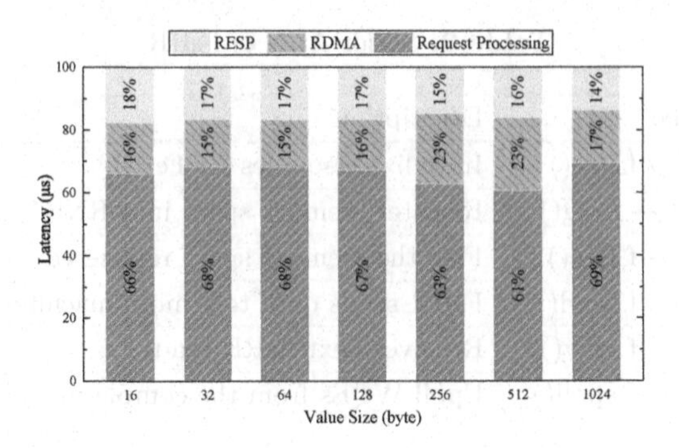

Fig. 4. Respective proportions of latency

5.2 Optimized Serialization Protocol

RDMA write can directly write the payload of request to responder's memory region, neither needs an extra serialization protocol. Meanwhile, serialization protocol leads to considerable memory copy overhead. To measure the overhead of RESP, we conduct an experiment to evaluate its latency on different value sizes from 16 bytes to 1024 bytes. We simulate set commands and use RESP to serialize and deserialize it for 1000000 times, and calculate the average latency. As Fig. 4 shows, the average latency of RESP increases by value size and occupies about 17% of the total latency of a request-reply. The experiment shows that the overhead of RESP is considerable and could significantly affect the performance of FeRR-Redis, while it is crucial and worthy to optimize RESP.

Supposed that the memory address and type of an object are known, it's easy to new a corresponding pointer to directly reference the object that can be manipulated by this pointer. Likewise, serialization protocol is a facile thing if we can use a request pointer to reference the request object. We consider that RDMA supports memory-object access that an object in client can directly be copied to the memory of the server, like a memory copy operation. The remaining issue is how to exactly know the struct type of this request to construct the pointer.

As Fig. 2 shows, we divide the request buffer into two parts: metadata part and data part, the former is used to store encoding bytes to distinguish the struct type of the command, the latter is used to store the command. Taking a request as an example, the data part stores the command *SET key Hello*, while the encoding bytes are stored in the metadata part. The entire request should be "$*3\$3\$3\$5SETkeyHello$". Then the client writes the request buffer into server memory region. Upon getting the request, the server decodes the metadata part of the request to get the encoding bytes of the struct. It is easy to realize a 3-element string struct and constructs a corresponding struct pointer to reference the command. Then the server processes the command by manipulating this pointer. For multiple data types, encoding bytes distinguish them using different byte in the metadata part. The optimized serialization protocol is a simple and effective approach that relieves the default memory copy overhead.

5.3 Parallel Task Engine

In order to support multi-threaded execution, we replace the chained hash table used in Redis that is not friendly to concurrent access. A chained hash table uses a linked list to deal with hash conflict where the new hash table entry is linked to the last entry of the conflicting bucket. Chaining is efficient for inserting or deleting single keys. However, a lookup may require scanning the entire chain that is time-consuming.

Cuckoo hashing table is an attractive choice where each lookup requires only k memory reads on average with k hash functions for get operations. In many application scenarios, 95% of the operations are get operations [10]. It makes cuckoo hashing more attractive that is designed for read-intensive workloads of Redis. Second, we should cope with concurrency control challenge. There are serveral ways like mutexes, optimistic locking [20], or lock-free data structures to ensure consistency among multiple threads. For read-intensive workloads, optimistic locking outperforms pessimistic locking because of no overhead of context switch. Furthermore, lock-free data structures are too hard to code correctly and integrate into Redis, so we minimize the synchronization cost with efficient optimistic locking scheme.

Concurrent Hash Index: We use the 3-way cuckoo hashing table to replace the original chaining hash table. The 3-way cuckoo hashing means the hash schema has 3 orthogonal hash functions. When a new key comes, it chooses one of the 3 locations that are empty to insert. If all possible locations for the new key to insert are full, resident key-value pair is kicked to an alternative location to make space for the new key. There may happen cascaded kicking until each key-value pair is in the proper location. The hash table is resizing when the number of kicks exceeds the limit or when a cycle is detected.

This data structure is space-efficient that pointers are eliminated from each key-value object of the chained hashing table. Cuckoo hashing increases space efficiency by around 40% over the default [20]. Meanwhile, cuckoo hashing is friendly to lookup operations that needs few hash entry traversals per read. At a fill ratio of 95%, the average probes in 3-way Cuckoo hashing are 1.5. In the worst case, the max lookup time of 3-way cuckoo hashing is 3 while chained hashing needs to search the whole hash chain that takes much more time.

Supporting Concurrent Access: To solve the consistency issue, we adopt the optimistic locking schema that performs reads optimistically without satisfying get operations when generating put. We add a 32-bit version field at the start of each hash table entry. The initial version number is zero, and the maximum is 2^32-1. When nearly reaching the maximum version number, it is about to be made zero by a successful put operation.

For a get operation, it first read the key version: if this version is odd, there must be a concurrent put operator on the same key, and it should wait and retry; otherwise, the version is even, it reads the key-value pair immediately. Upon completion of entry fetched from the hash table, it reads the version again and checks if the current version number is equal to the initial version number. If the check fails, the get operation retries.

We use a compare-swap (CAS) operation instruction to allows multiple put operations to access the same entry. Before a put operation displaces the original key-value, it first reads the relevant version and waits until the initial version number is even. Then the writer increases the relevant version by one using a CAS operation. If succeeds, the odd version number indicates other operations to wait for an on-going update for the entry. Upon completion, it increases the version number by one again. As a result, the key version increases 2 and keeps even after each displacement.

6 Evaluation

In this section, we analyze the overall performances of our high-performance design of FeRR-Redis, then the benefits from each mechanism design.

6.1 Experimental Setup

Hardware and Configuration: Our experiments run on a cluster of five machines, one for server and the other four for clients. Each machine is equipped with an Intel(R) Xeon(R) Silver 4110 CPU @ 2.10 GHz processor. 192 GB memory is installed on each node. All nodes are connected through InfiniBand FDR using Mellanox ConnectX-3 56 Gb/sec. On each machine, we run Centos7.5.

Comparison Target: We compare FeRR-Redis against Redis on IPoIB, FeRR-Redis with single-threaded execution. We run a FeRR-Redis server on one physical machine, while clients are distributed among four remaining machines. All of the Redis implementations disable logging function. We intend to find out the importance of RDMA kernel bypass and parallel task engine of FeRR-Redis. Furthermore, the performance gap between them demonstrates this idea.

YCSB Workloads: In addition to get workload and put workload of redis-benchmark, we use two types of YCSB workloads: read-intensive (90% GET,10% PUT) and write-intensive (50% GET, 50% PUT). Our workloads are uniform workloads, the keys are chosen uniformly at random. This workload is generated off-line using YCSB [21].

6.2 Microbenchmark

We integrate FeRR into redis-benchmark tool in Redis project to supprt RDMA-capable network. We run it to measure the throughput and latency of different workloads.

Latency: Figure 5 shows the latency of get and put operations of FeRR-Redis and Redis (IPoIB) with 16 clients. With few clients, server can process operations

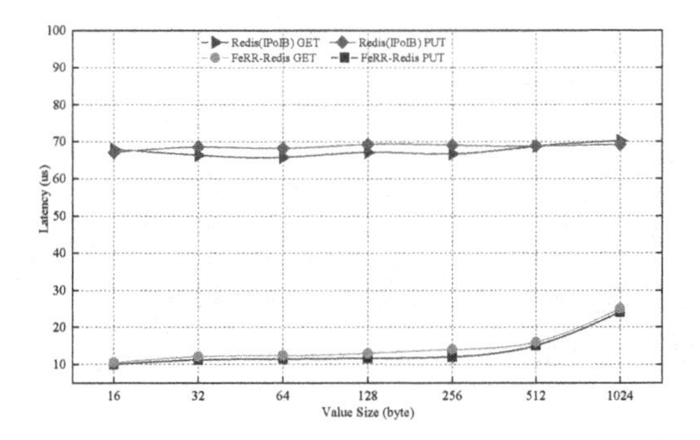

Fig. 5. Latency of FeRR-Redis compared to Redis (IPoIB)

as fast as possible. For small key-values, FeRR-Redis achieves 10.0 µs latency for get operations and 10.4 µs latency for operations. For large key-values (1024bytes values), the latencies of get and put are 24.2 µs and 25.1 µs.

With value size increasing, the average latencies of Redis on IBoIP have a little change. This is because of much time spent to transmit network packets over socket interface and generate serialization and deserialization. No matter for small or large values, the latencies of get and put are more than 60 µs. Generally, the average latencies of FeRR-Redis beat that of Redis on IPoIB by 3×–7×.

Throughput: Figure 6 shows the throughput of FeRR-Redis, single-threaded FeRR-Redis and Redis on IPoIB under different read/write ratio workloads. The throughput of FeRR-Redis is achieved with 16 concurrent clients, compared to the single-threaded FeRR-Redis and Redis on IPoIB running 16 concurrent clients as well. For small key-values (16bytes values), FeRR-Redis can achieve 2.78 million operations per second, compared to 570 Kops/sec for the single-threaded FeRR-Redis and 74 Kops/sec for Redis on IPoIB. For get operations, FeRR-Redis can achieve throughput improvement 4.7× that of FeRR-Redis with single-threaded execution and 37.6× that of Redis (IPoIB), while 4.7× that of FeRR-Redis with single-threaded execution and 29.5× that of Redis (IPoIB) for put operations.

For larger key-values, FeRR-Redis keeps great performance. The throughputs of get and put operations slightly go down while their usage of Infiniband card's performance increases. However, FeRR-Redis cannot saturate the Infiniband card's performance because Redis has an extremely long code path and much exception handling that incurs much CPU overhead. Specifically, for 1024-byte value size, FeRR-Redis achieves 1.28 million get per second and 1.25 million put per second.

Because IBoIP remains CPU copy and kernel involved, Redis on IPoIB is bottlenecked by the poor performance of IPoIB and its performance is the same when running on 10 Gbps Ethernet. When the network is not the bottleneck, the single CPU is the new bottleneck. The single-threaded FeRR-Redis's throughput is restricted and unable to saturate the Infiniband card's performance. As a

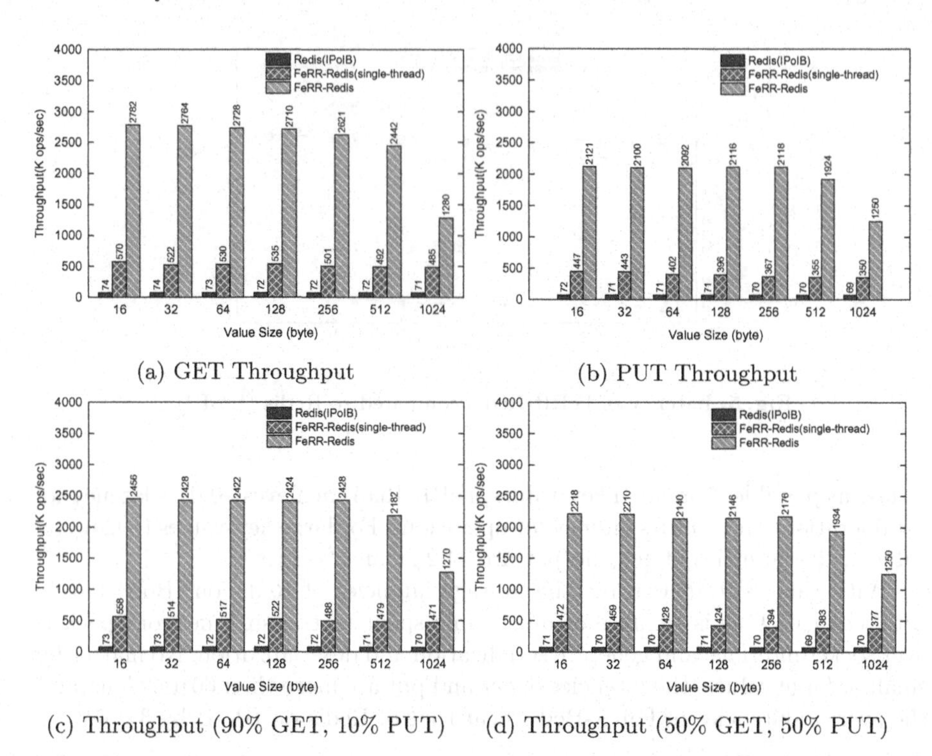

(a) GET Throughput (b) PUT Throughput

(c) Throughput (90% GET, 10% PUT) (d) Throughput (50% GET, 50% PUT)

Fig. 6. Throughput achieved for FeRR-Redis, FeRR-Redis (single-thread), and Redis (IPoIB)

conclusion, we precisely found out the main bottlenecks, and have crafted high-performance design with corresponding approaches to boost the performance of Redis.

As YCSB workloads, FeRR-Redis is outstanding for the read-intensive workload (95% GET, 10% PUT) that achieves peak throughput of 2.45 million operations per second, which is slightly lower than that of get workload. For write-intensive workload (50% GET, 50% PUT), FeRR-Redis achieves almost the same throughput as that of put workload, because read- write contention incurs much overhead. As a result, the data structure of multi-threaded framework is suitable for read-intensive workload.

Optimized Latency: In our previous analysis, RESP consumes several μs to execute, which is almost 15% of the whole latency. Our optimization aims to eliminate memory copy during serialization and deserialization process. Figure 7 shows that FeRR-Redis get latency with optimized serialization protocol decrease by 2–3 μs compared to the default RESP while a native round-trip of RDMA write for 16 byte is only 2 μ. The result demonstrates our optimization for RESP is effective.

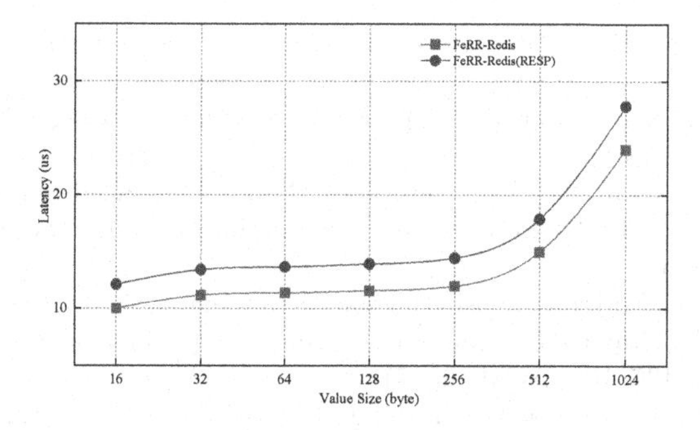

Fig. 7. FeRR-Redis latency with RESP and optimized serialization protocol

7 Conclusion

In this paper, we propose a high-performance design of Redis leveraging RDMA, FeRR-Redis. We solve three key issues of integrating RDMA into Redis: First, we design a fast event-driven RPC framework that is closely coupled with RDMA primitives with high-level programming interface. Second, we optimize RESP to relieve memory copy overhead and support multiple data types via transmission. Finally, we exploit an efficient parallel task engine that embraces Redis with multi-core processing. Evaluations show that FeRR-Redis effectively explores hardware benefits, and achieves orders-of-magnitude better throughput and ultra-low latency than Redis.

Acknowledgement. This work was supported by National Key R&D Program of China (2018YFB1003303), National Science Foundation of China under grant number 61772202, Youth Program of National Science Foundation of China under grant number 61702189, and Youth Science and Technology-Yang Fan Program of Shanghai under Grant Number 17YF1427800.

References

1. Redis homepage. https://redis.io/
2. Metreveli, Z., Zeldovich, N., Kaashoek, M.F.: Cphash: a cache-partitioned hash table, vol. 47. ACM (2012)
3. Jose, J., et al.: Memcached design on high performance RDMA capable interconnects. In: 2011 International Conference on Parallel Processing, pp. 743–752. IEEE (2011)
4. Mitchell, C., Geng, Y., Li, J.: Using one-sided RDMA reads to build a fast, CPU-efficient key-value store. In: USENIX Annual Technical Conference, pp. 103–114 (2013)
5. Dragojević, A., Narayanan, D., Castro, M., Hodson, O.: FaRM: fast remote memory. In: 11th USENIX Symposium on Networked Systems Design and Implementation, pp. 401–414 (2014)

6. Wang, Y., et al.: HydraDB: a resilient RDMA-driven key-value middleware for in-memory cluster computing. In: SC 2015: Proceedings of the International Conference for High Performance Computing, Networking, Storage and Analysis, pp. 1–11. IEEE (2015)
7. Kalia, A., Kaminsky, M., Andersen, D.G.: Using RDMA efficiently for key-value services. In: ACM SIGCOMM Computer Communication Review, vol. 44, pp. 295–306. ACM (2014)
8. Li, B., et al.: KV-direct: high-performance in-memory key-value store with programmable NIC. In: Proceedings of the 26th Symposium on Operating Systems Principles, pp. 137–152. ACM (2017)
9. Pagh, R., Rodler, F.F.: Cuckoo hashing. J. Algorithms **51**(2), 122–144 (2004)
10. Atikoglu, B., Xu, Y., Frachtenberg, E., Jiang, S., Paleczny, M.: Workload analysis of a large-scale key-value store. In: ACM SIGMETRICS Performance Evaluation Review, vol. 40, pp. 53–64. ACM (2012)
11. Connect-IB: Architecture for scalable high performance computing whitepaper. https://www.mellanox.com/pdf/whitepapers/IB_Intro_WP_190.pdf/
12. Memcached homepage. https://memcached.org/
13. Liu, J., et al.: Design and implementation of MPICH2 over InfiniBand with RDMA support. In: 18th International Parallel and Distributed Processing Symposium, Proceedings, p. 16. IEEE (2004)
14. Shipman, G.M., Woodall, T.S., Graham, R.L., Maccabe, A.B., Bridges, P.G.: Infiniband scalability in open MPI. In: Proceedings 20th IEEE International Parallel & Distributed Processing Symposium, pp. 10-pp. IEEE (2006)
15. Lu, X., et al.: High-performance design of Hadoop RPC with RDMA over InfiniBand. In: 2013 42nd International Conference on Parallel Processing, pp. 641–650. IEEE (2013)
16. Huang, J., et al.: High-performance design of HBase with RDMA over InfiniBand. In: 2012 IEEE 26th International Parallel and Distributed Processing Symposium, pp. 774–785. IEEE (2012)
17. Lim, H., Han, D., Andersen, D.G., Kaminsky, M.: MICA: a holistic approach to fast in-memory key-value storage. In: 11th USENIX Symposium on Networked Systems Design and Implementation, pp. 429–444 (2014)
18. Intel DPDK homepage. https://dpdk.org/
19. Ongaro, D., Rumble, S.M., Stutsman, R., Ousterhout, J., Rosenblum, M.: Fast crash recovery in RAMCloud. In: Proceedings of the Twenty-Third ACM Symposium on Operating Systems Principles, pp. 29–41. ACM (2011)
20. Fan, B., Andersen, D.G., Kaminsky, M.: MemC3: compact and concurrent memcache with dumber caching and smarter hashing. Presented as part of the 10th USENIX Symposium on Networked Systems Design and Implementation, pp. 371–384 (2013)
21. Cooper, B.F., Silberstein, A., Tam, E., Ramakrishnan, R., Sears, R.: Benchmarking cloud serving systems with YCSB. In: Proceedings of the 1st ACM Symposium on Cloud Computing, pp. 143–154. ACM (2010)

Efficient Source Selection for Error Detection via Matching Dependencies

Lingli Li[1(✉)], Sheng Zheng[1], Jingwen Cai[1], and Jinbao Li[2(✉)]

[1] Department of Computer Science and Technology, Heilongjiang University, Harbin, China
lilingli@hlju.edu.cn, zhengsheng9601@163.com, bangurpan@163.com
[2] Shandong Artificial Intelligence Institute, Qilu University of Technology, Jinan, China
Lijinb@sdas.org

Abstract. Data dependencies have been widely used in error detection. However, errors might not be detected when the target data set is sparse and no conflicts occur. With a rapid increase in the number of data sources available for consumption, we consider how to apply both external data sources and matching dependencies (a general form of FD) to detect more potential errors in target data. However, accessing all the sources for error detection is impractical when the number of sources is large. In this demonstration, we present an efficient source selection algorithm that can select a proper subset of sources for error detection. A key challenge of this approach is how to estimate the gain of each source without accessing their datasets. To address the above problem, we develop a two-level signature mechanism for estimating the gain of each source. Empirical results on both real and synthetic data show high performance on both the effectiveness and efficiency of our algorithm.

Keywords: Data quality · Consistency · Source selection · LSH · Minhash

1 Introduction

Consistency is one of the central criteria for data quality. Inconsistencies in a database can be captured by the violations of integrity constraints, such as functional dependencies (FDs) and matching dependencies (MDs) [7,14]. In concrete terms, given a database D and a set of integrity constraints Σ, tuples in D that violate rules in Σ are inconsistent and need to be repaired. However, D might still have errors when there are no violations in D. To address this problem, we can compare D with other data sets to identify more inconsistencies and capture more errors. Such an example is described below to illustrate this issue.

Example 1. Consider S_0, S_1 and S_2 in Fig. 1. S_0 is the data for error detection called the *target data*. S_1 and S_2 are two data sources. S_0 has three errors, i.e. x_1[city], x_2[phone] and x_3[state] (in italics and underlined). We assume that the

Y. Nah et al. (Eds.): DASFAA 2020, LNCS 12112, pp. 211–227, 2020.
https://doi.org/10.1007/978-3-030-59410-7_13

schemas of $S_0 - S_2$ have been mapped by existing schema mapping techniques [2]. A set of integrity constraints shown in Fig. 1(a) are used for error detection, in which ϕ_1, ϕ_2 are FDs, and ϕ_3 is an MD. ϕ_1, ϕ_2 state that if two tuples have identical zip values, their cities and states should be identical. ϕ_3 states that if two tuples have similar values on both name and street based on some predefined fuzzy match operators, they should have identical phone numbers.

We assume S_0 satisfies all the rules, S_1 satisfies ϕ_1–ϕ_2, and S_2 satisfies ϕ_3. Clearly, since there are no violations of rules in S_0, none of the above errors can be detected. However, if we compare S_0 with S_1, some errors can be discovered. For instance, x_1[city] is a candidate error based on S_1 and ϕ_1. This is because x_1[zip] $=$ y_1[zip] $=$ 05211 and x_1[city] $\neq$ y_1[city] which violates ϕ_1. Similarly, x_3[state] is also a candidate error based on S_1 and ϕ_2. Finally, let us look at the matching dependency ϕ_3. Suppose x_i [name] $\approx_1$ z_i[name] and x_i [street] $\approx_2$ z_i[street] holds for each $i \in [1, 3]$. x_2[phone] is a candidate error based on S_2 and ϕ_3 since x_2 [name] $\approx_1$ z_2[name] $\wedge$ x_2 [street] $\approx_2$ z_2[street] while x_2 [phone] $\neq$ z_2[phone].

id	expression
ϕ_1	zip→ city
ϕ_2	zip→ state
ϕ_3	name$_{\approx_1}$ $\wedge$ street$_{\approx_2}$ → phone

(a) Dependencies

id	phone	name	street	city	state	zip
x_1	949-1212	Smith, Alice	17 bridge	*Midvile*	AZ	05211
x_2	*555-4093*	Jones, Bob	5 valley	Centre	NY	10012
x_3	212-6040	Black, Carol	9 mountain	Davis	*AA*	07912

(b) Target data S_0 for error detection

id	zip	city	state
y_1	05211	Midville	AZ
y_2	10012	Centre	NY
y_3	07912	Davis	CA

(c) Data source S_1

id	phone	name	street	city
z_1	949-1212	Alice Smith	17 bridge	Midville
z_2	555-8195	Bob Jones	5 valley rd	Centre
z_3	212-6040	Carol Blake	9 mountain	Davis

(d) Data source S_2

Fig. 1. Motivating example

From this example, it can be observed that, given a target data, involving both extra data sources and MDs (since FD is a special form of MD) help us detect more potential errors. The candidate errors will be verified and cleaned by a further step, i.e., posing queries on multiple data sources for truth discovery, which has been widely studied [17]. This paper focuses on how to select proper sources to detect candidate errors in target data.

Error detection crossing data sets brings not only opportunities, but also challenges. First, most of the data sources might be irrelevant to our target data or redundant. In order to discover the most useful sources, we need to study how to evaluate the gain of data sources under a given set of MDs for error detection. Second, accessing data sources requires high costs of both time and money [6]. Thus, we need to study how to balance between gain and cost. Last but not least, it is not worthwhile to access sources for source selection. It brings the big challenge of how to conduct selection from a large number of sources without accessing them? Aiming at the aforementioned challenges, this paper makes the following main contributions.

1. We formulate the problem of source selection for error detection via MDs, called SSED. As far as we know, this is the first study on how to select sources for error detection via MDs. We show that the SSED problem is NP-hard (Sect. 3).
2. A greedy framework algorithm for SSED is developed (Sect. 4). To avoid accessing data sources, we develop a coverage estimation method based on a two-level signature (Sect. 5).
3. Using real-life and synthetic data, the effectiveness and efficiency of our proposed algorithm are experimentally verified (Sect. 6).

2 Background

Let S_0 be the *target* data for error detection and $\mathbb{S} = \{S_i | 1 \leq i \leq m\}$ be the set of *sources*, where each S_i $(0 \leq i \leq m)$ includes a dataset D_i with schema R_i and a set of MDs Σ_i. Note that, FDs can also be included in our set, since MD is a general form of FD. We assume that the schemas of the sources have been mapped by existing schema mapping techniques [2].

2.1 Matching Dependencies

We denote the domain of an attribute X by $dom(X)$. If X consists of a sequence of attributes $X = A_1 A_2 \cdots A_k$, then $dom(X) = dom(A_1) \times dom(A_2) \times \cdots \times dom(A_k)$. A **fuzzy match operator** $\approx_i$ on an attribute A_i is defined on the domain of A_i.

$$\approx_i\colon \mathsf{dom}(A_i) \times \mathsf{dom}(A_i) \rightarrow \{\mathsf{true}, \mathsf{false}\}.$$

It indicates true if two values are similar. This operator is defined by a similarity function f and a threshold θ. For instance, consider $\approx_1$: (Jaccard, 0.8). It means that $t[A_1] \approx_1 t'[A_1]$ iff the Jaccard similarity between $t[A_1]$ and $t'[A_2]$ is at least

θ. We now give the definition of MD. An MD ϕ defined on schemas (R_1, R_2) is an expression of the form,

$$\wedge_{j\in[1,k]}(R_1[X[j]] \approx_j R_2[X[j]]) \to R_1[Y] = R_2[Y],$$

where X is an attribute list, Y is a single attribute in both R_1 and R_2, and $\approx_j$ is the match operator for $X[j]$. We say that the instance (D_1, D_2) satisfies MD ϕ if for any tuples $t_1 \in D_1$ and $t_2 \in D_2$, the following property holds: if $\wedge_{j\in[1,k]}(t_1[X[j]] \approx_j t_2[X[j]])$, then $t_1[Y] = t_2[Y]$. For simplicity, we rewrite $\wedge_{j\in[1,k]}(t_1[X[j]] \approx_j t_2[X[j]])$ as $t_1[X] \approx t_2[X]$. We assume that the fuzzy match operators are predefined.

2.2 Locality-Sensitive Functions

The key idea of locality-sensitive hashing (LSH) [1] is to hash the items in a way that the probability of collision is much higher for items which are close to each other than for those which are far apart.

We first consider functions that take two items and output a decision about whether these items should be a candidate pair. A collection of functions of this form will be called a *family* of functions. Let $d_1 < d_2$ be two distances according to some distance measure d. A family F of functions is said to be (d_1, d_2, p_1, p_2)-*sensitive* if for every f in F:

1. If $d(x, y) \leq d_1$, then $\Pr(f(x) = f(y)) \geq p_1$.
2. If $d(x, y) \geq d_2$, then $\Pr(f(x) = f(y)) \leq p_2$.

Suppose we are given a (d_1, d_2, p_1, p_2)-sensitive family $\mathbf{F}$. We can construct a new family $\mathbf{F}^{r\wedge}$ by the r-way AND-construction on F. We can assert that $\mathbf{F}^{r\wedge}$ is a $(d_1, d_2, (p_1)^r, (p_2)^r)$-sensitive family. Notice that the AND-construction lowers all probabilities. If we choose r judiciously, we can make the small probability p_2 get very close to 0.

Different distance measures have different locality-sensitive families of hash functions. Please read [1] for more examples on this field.

3 Problem Definition

In order to formalize the source selection problem, we first introduce the notation of pattern. Given a tuple $t \in S_i$ and MD $\phi \in \Sigma_i$, the *pattern* of t under ϕ, denoted as $P_{\phi,t}$, is defined as $P_{\phi,t} = t[\mathsf{LHS}(\phi)]$, where $\mathsf{LHS}(\phi)$ is the left-hand side of ϕ, and $t[\mathsf{LHS}(\phi)]$ is the $\mathsf{LHS}(\phi)$ value of t. For example, for the tuple x_1 in Fig. 1(b) and MD ϕ_3 in Fig. 1(a), the pattern P_{ϕ_3,x_1} is ["Smith, Alice", "17 bridge"]. We can reasonably assume that both the target data and the sources are consistent under their dependencies. Otherwise, the dependencies can be initially used to detect and fix the errors. Accordingly, the *pattern set* of S_i on MD $\phi \in \Sigma_i$, denoted by P_{ϕ,S_i}, is defined as

$$P_{\phi,S_i} = \{t[\mathsf{LHS}(\phi)] | t \in S_i\} \tag{1}$$

Similarly, for a source set $\mathcal{S}$, we have

$$P_{\phi,\mathcal{S}} = \cup_{S_i \in \mathcal{S}} P_{\phi,S_i} \tag{2}$$

It is easy to find that using data sources and MDs to detect potential errors can be transformed into a pattern matching problem, as the following proposition described.

Proposition 1. *Given tuple $s \in S_0$ and MD $\phi \in \Sigma_0$, s can be* checked *by the source set $\mathcal{S}$ based on ϕ iff $\exists$ tuple t such that $P_{\phi,s} \approx P_{\phi,t}$, where $t \in S_i$, $S_i \in \mathcal{S}$ and $\phi \in \Sigma_i$. And we say pattern $P_{\phi,s}$ is* covered *by $\mathcal{S}$, denoted by $P_{\phi,s} \tilde{\in} P_{\phi,\mathcal{S}}$. Clearly, for such a tuple t, if $s[RHS(\phi)] \neq t[RHS(\phi)]$, $s[RHS(\phi)]$ is a candidate error detected by $\mathcal{S}$ under ϕ, where $RHS(\phi)$ is the right-hand side of ϕ.*

From Proposition 1, the more the patterns in S_0 are covered by $\mathcal{S}$, the more data in S_0 can be determined whether or not correct. Thus, we define the number of patterns in the target data covered by sources, called *coverage*, as the gain model (Definition 1).

Definition 1. *Given a target data S_0 and a source set $\mathcal{S}$, we call the number of patterns in P_{ϕ,S_0} covered by $P_{\phi,\mathcal{S}}$ as the* **coverage** *of $\mathcal{S}$ under ϕ, denoted as $cov_\phi(\mathcal{S})$. Formally,*

$$cov_\phi(\mathcal{S}) = |\{p | p \in P_{\phi,S_0}, p \tilde{\in} P_{\phi,\mathcal{S}}\}| \tag{3}$$

Thus $cov_\phi(\mathcal{S})$ represents the number of patterns in the target data that can be verified by $\mathcal{S}$ based on ϕ. We further define the coverage of $\mathcal{S}$ under the MD set Σ_0 as the sum of the coverage under each MD,

$$cov(\mathcal{S}) = \sum_{\phi \in \Sigma_0} cov_\phi(\mathcal{S}) \tag{4}$$

Ideally, we wish to maximize the gain, i.e. $cov(\mathcal{S})$, and minimize the cost, i.e., the number of sources in $\mathcal{S}$. However, it is impractical to achieve both goals. Thus, this paper attempts to find a subset of sources that maximizes the coverage with the number of sources no more than a threshold. We call this problem as Source Selection for Error Detection (SSED). The formal definition is shown below.

Problem 1. *Given a target data S_0, a source set $\mathbb{S}$ and a positive integer K, the* ***Source Selection for Error Detection (SSED)*** *is to find a subset $\mathcal{S}$ of $\mathbb{S}$, such that $|\mathcal{S}| \leq K$, and $cov(\mathcal{S})$ is maximized.*

Theorem 1. *SSED is NP-hard.*

The proof is by reduction from the classic unweighted maximum coverage problem [13]. The detail is omitted due to space limitation. □

Algorithm 1. exaGreedy

Require: S_0, $\mathbb{S}$, K
Ensure: a subset $\mathcal{S}$ of $\mathbb{S}$ with $|\mathcal{S}| \leq K$;
1: $\mathcal{S} \leftarrow \emptyset$;
2: **while** $|\mathcal{S}| < K$ **do**
3: **for each** $S_i \in \mathbb{S}$ **do**
4: **for each** MD $\phi \in \Sigma_0 \cap \Sigma_i$ **do**
5: $cov(\{\mathcal{S}, S_i\}) += \texttt{exaCov}(\mathcal{S}, S_i, S_0, \phi)$;
6: $S_{opt} \leftarrow \arg\max_{S_i \in \mathbb{S}} cov(\{\mathcal{S}, S_i\}) - cov(\mathcal{S})$;
7: add S_{opt} into $\mathcal{S}$;
8: remove S_{opt} from $\mathbb{S}$;
9: **return** $\mathcal{S}$;

4 Algorithm for SSED

Due to the NP-hardness of the problem, we devise a greedy approximation algorithm for SSED (shown in Algorithm 1), denoted by exaGreedy. In the main loop, for each unselected source S_i, the coverage of adding S_i into the selected sources $\mathcal{S}$ is computed (Line 3–5); and a source that provides the largest marginal gain is selected (Line 6).

Theorem 2. *Algorithm 1 has an approximation ratio of* $(1-1/\mathrm{e})$ *of the optimal solution.*

Proof Sketch. Since the coverage function is submodular and monotone, this algorithm has $1 - 1/e$ approximation ratio based on the submodular theory [18]. □

Analysis. Given MD ϕ and source S_i, we have to perform the similarity join between S_0 and S_i, thus the expected time of exaGreedy is very high. Moreover, to compute the accurate coverage, all the data sources have to be accessed, which is unacceptable. To address this problem, we build a signature $\text{SIG}(P_{\phi,S_i})$ for each pattern set P_{ϕ,S_i} and use these signatures for coverage estimation. We attempt to achieve the following three goals: (1) The signatures for all the sources are small enough that can be fit in main memory. (2) The signatures can be used to estimate coverages without accessing the original data sets. (3) The estimation process should be both efficient and accurate.

To this end, we devise a signature-based greedy approximation algorithm for SSED (shown in Algorithm 2), denoted by mhGreedy. The input of mhGreedy is the signatures for each source generated beforehand in an offline process. The framework of mhGreedy is generally the same as exaGreedy. There are only two differences: (1) each time when we consider a new source set $\mathcal{S}' = \{\mathcal{S}, S_i\}$, we need to compute its signatures by `UniSig` in the online fashion (lines 6–8, lines 13–14); and (2) coverage needs to be estimated by `CovEst` (line 9) based on their signatures. These implementations will be discussed in detail in the following section.

Algorithm 2. mhGreedy

Require: $\{\text{SIG}(P_{\phi,S_i})|i \in [0, \leq |\mathbb{S}|], \phi \in \Sigma_i\}$, K
Ensure: a subset $\mathcal{S}$ of $\mathbb{S}$ with $|\mathcal{S}| \leq K$;
1: $\mathcal{S} \leftarrow \emptyset$;
2: **while** $|\mathcal{S}| < K$ **do**
3: **for** each $S_i \in \mathbb{S}$ **do**
4: $\mathcal{S}' \leftarrow \{\mathcal{S}, S_i\}$;
5: **for** each MD $\phi \in \Sigma_0$ **do**
6: $\text{SIG}(P_{\phi,\mathcal{S}'}) \leftarrow \text{SIG}(P_{\phi,\mathcal{S}})$;
7: **if** $\phi \in \Sigma_i$ **then**
8: $\text{SIG}(P_{\phi,\mathcal{S}'}) \leftarrow$`UniSig`$(\text{SIG}(P_{\phi,\mathcal{S}}), \text{SIG}(P_{\phi,S_i}))$;
9: $\hat{cov}_\phi(\mathcal{S}') =$ `CovEst`$(\text{SIG}(P_{\phi,\mathcal{S}'}), \text{SIG}(P_{\phi,S_0}))$;
10: $\hat{cov}(\mathcal{S}') \leftarrow \sum_{\phi \in \Sigma_0 \cap \Sigma_i} \hat{cov}_\phi(\mathcal{S}')$;
11: $S_{opt} \leftarrow \arg\max_{S_i \in \mathbb{S}} \hat{cov}(\{\mathcal{S}, S_i\}) - \hat{cov}(\mathcal{S})$;
12: add S_{opt} into $\mathcal{S}$;
13: **for** each $\phi \in \Sigma_0 \cap \Sigma_{\text{opt}}$ **do**
14: $\text{SIG}(P_{\phi,\mathcal{S}}) \leftarrow$`UniSig`$(\text{SIG}(P_{\phi,\mathcal{S}}), \text{SIG}(P_{\phi,S_{opt}}))$;
15: remove S_{opt} from $\mathbb{S}$;
16: **return** $\mathcal{S}$;

5 Coverage Estimation

In this section, we discuss details of signature generation in Sect. 5.1 and coverage estimation in Sect. 5.2.

5.1 Signature Generation

We present a two-level signature mechanism for coverage estimation. First of all, we assume that MD ϕ has the following form:

$$\phi: \; A_{1\approx_1} \wedge \cdots A_{l\approx_l} \wedge A_{l+1=} \wedge \cdots A_{l+l'=} \rightarrow B,$$

in which the match operators for A_1–A_l ($l > 0$) are defined by a similarity metric and a threshold while the match operators for A_{l+1}–$A_{l+l'}$ ($l' \geq 0$) are defined by the equality(=).

The main idea of signature generation is as follows: (1) We first generate the 1st-level signature for each pattern using locality-sensitive hashing, called *lsh-signature*. The goal is to make the coverage of signatures, i.e., the number of matched signatures between source and target, be a scaled version of coverage. To achieve this, we adopt the AND-construction for each locality-sensitive hash family to make probability of collisions between nonmatching patterns get very close to 0, while the probability of collisions between matching patterns stay away from 0. (2) Due to the high costs of both time and space, we next construct a minhash signature for each lsh-signature set, called *mh-signature*, to estimate the coverage of lsh-signatures efficiently and effectively by a much smaller data structure.

First Level: lsh-Signature. Given pattern $P_{\phi,t}$, we construct its lsh-signature, denoted as $\mathsf{lsh}_\phi(t) = [e_1, \cdots, e_{l+l'}]$, as follows. For each attribute A_j, if A_j is with the exact match operator (=), we let $e_j = t[A_j]$; otherwise, suppose f_j is the similarity function defined on A_j by ϕ. Let $\mathbf{F}_j$ denote the locality-sensitive family of f_j, and $\mathbf{F}_j^{h_j\wedge}$ denote the h_j-way AND-construction on $\mathbf{F}_j$. One natural question is how to determine the proper values of all the h_js. We defer the discussion of this problem to Sect. 5.3. The formal definition of lsh-signature is given as below.

Definition 2. *The* ***lsh-signature*** *for a tuple t under* ϕ*, is*

$$lsh_\phi(t) = [lsh_1(t[A_1]), \cdots, lsh_l(t[A_l]), t[A_{l+1}], \cdots, t[A_{l+l'}]],$$

where $lsh_j(t[A_j]) = [f_1(t[A_j]), \cdots, f_{h_j}(t[A_j])]$ *and* f_1*–*f_{h_j} *are* h_j *functions randomly selected from* $\mathbf{F}_j^{h_j\wedge}$*, called the lsh-signature of* $t[A_j]$*.*

Accordingly, the ***lsh-signature set*** *for the source set* $\mathcal{S}$ *under* ϕ *is,*

$$lsh_\phi(\mathcal{S}) = \cup_{S_i \in \mathcal{S}} lsh_\phi(S_i) = \{lsh_\phi(t) | t \in S_i, S_i \in \mathcal{S}\}.$$

To obtain an accurate estimation, it is necessary to generate multiple (say c) independent lsh-signatures for each pattern $P_{\phi,t}$, denoted by $\mathsf{lsh}_\phi^1(t)$–$\mathsf{lsh}_\phi^c(t)$. As a consequence, there are c lsh-signature sets for each pattern set P_{ϕ,S_i}, denoted as $\mathsf{lsh}_\phi^1(S_i)$–$\mathsf{lsh}_\phi^c(S_i)$.

Since the number of lsh-signatures is linear to the dataset size, they might be too large to fit in main memory for large scale datasets. Moreover, the time complexity of coverage estimation would also grow with the number of lsh-signatures. To this end, we present a 2nd-level signature for both the time and space issues.

Second level: mh-Signature. For each lsh-signature set $\mathsf{lsh}_\phi^j(S_i)$, we use its minhash values to be the 2nd-level signature, called the *mh-signature*, since minhashing [5] is a widely-used method for quickly estimating Jaccard similarity and intersection set size.

Definition 3. *The* ***mh-signature*** *for each pattern set* P_{ϕ,S_i}*, denoted by* $mh(lsh_\phi(S_i))$*, is defined as the concatenation of k independent minhash values for each* $lsh_\phi^j(S_i)$*, that is*

$$mh(lsh_\phi(S_i)) = [[mh_1(lsh_\phi^1(S_i)), \cdots, mh_k(lsh_\phi^1(S_i))], \cdots, [mh_1(lsh_\phi^c(S_i)), \cdots, mh_k(lsh_\phi^c(S_i))]].$$

Clearly, the size of mh-signature $\mathsf{mh}(\mathsf{lsh}_\phi(P_{\phi,S_i}))$ is bound to $O(ck)$, which is very small and is independent of the dataset size. Therefore, we can reasonably assume that all the mh-signatures can be stored in main memory.

Implementation of `UniSig` in Algorithm 2. After generating all the mh-signatures for each single source offline, we need to further consider how to generate mh-signatures for multiple sources online. Specifically, given the mh-signatures of source set $\mathcal{S}$ and source S_i under ϕ, how to generate the mh-signature of $\{\mathcal{S}, S_i\}$ under ϕ in an online fashion?

A brute-force method for computing the minhash values of a union set is to initially obtain the union set. Clearly, this method is too inefficient since it requires to store all the lsh-signatures. Fortunately, minhashing has a good property (shown in Theorem 3[1]) that the minhash value of a union set can be obtained based on the minhash value of each set.

Theorem 3. $mh(A_1 \cup A_2) = \min(mh(A_1), mh(A_2))$.

Based on the previous theorem, the mh-signature of $\mathsf{lsh}_\phi(\{\mathcal{S}, S_i\})$ can be easily computed. In particular, given $\mathsf{mh}(\mathsf{lsh}_\phi(\mathcal{S}))$ and $\mathsf{mh}(\mathsf{lsh}_\phi(S_i))$, the mh-signature $\mathsf{mh}(\mathsf{lsh}_\phi(\{\mathcal{S}, S_i\}))$ can be computed by the following equation. For each $j \in [1, k]$,

$$\mathsf{mh}(\mathsf{lsh}_\phi(\{\mathcal{S}, S_i\}))[j] = \min(\mathsf{mh}(\mathsf{lsh}_\phi(\mathcal{S}))[j], \mathsf{mh}(\mathsf{lsh}_\phi(S_i))[j]).$$

5.2 Coverage Estimation

The framework for coverage estimation is shown in Algorithm 3. Since an mh-signature consists of c groups of independent values, an estimated coverage can be obtained based on each of them (lines 1–5). Therefore, we use the average of these c estimated coverages as the final estimation (line 6). Clearly, the time complexity of `CovEst` is $O(ck)$, which is irrelevant to the corresponding dataset size.

Algorithm 3. CovEst

Require: $\mathsf{mh}(\mathsf{lsh}_\phi(\mathcal{S}))$, $\mathsf{mh}(\mathsf{lsh}_\phi(S_O))$, n_ϕ //n_ϕ: the size of the domain of the LHS of ϕ
Ensure: $\hat{cov}_\phi(\mathcal{S})$
1: **for** each j in $[1, c]$ **do**
2: $\quad \hat{s_j} = \frac{1}{k}|\{l|\mathsf{mh}(\mathsf{lsh}_\phi(\mathcal{S})[j][l] = \mathsf{mh}(\mathsf{lsh}_\phi(S_O))[j][l]\}|_0$;
3: $\quad \hat{a_j} = n_\phi k / \sum_{1 \le j \le k} \mathsf{mh}(\mathsf{lsh}_\phi(S_O))[j][l]$;
4: $\quad \hat{b_j} = n_\phi k / \sum_{1 \le j \le k} \mathsf{mh}(\mathsf{lsh}_\phi(\mathcal{S}))[j][l]$;
5: $\quad \hat{y_j} = \frac{\hat{s_j}}{1+\hat{s_j}}(\hat{a_j} + \hat{b_j})$;
6: $\bar{y} = \frac{1}{c} \sum_{1 \le j \le c} \hat{y_j}$;
7: **return** $\bar{y}$;

Why does this estimation work? In the above algorithm, $\hat{y_j}$ is the estimated value of $|\mathsf{lsh}^j_\phi(\mathcal{S}) \cap \mathsf{lsh}^j_\phi(S_O)|$. This equation can be easily derived by the formula of Jaccard similarity and the property of minhashing, i.e., $\Pr(\mathsf{mh}(A) = \mathsf{mh}(B)) = \text{Jaccard}(A, B)$. Accordingly, $\bar{y}$ is the estimation of the expectation $\mathbb{E}[|\mathsf{lsh}_\phi(\mathcal{S}) \cap \mathsf{lsh}_\phi(S_O)|]$. If we choose the h_js judiciously, we can make $\mathbb{E}[|\mathsf{lsh}_\phi(\mathcal{S}) \cap \mathsf{lsh}_\phi(S_O)|] \propto cov_\phi(\mathcal{S})$(will be explained soon in the next subsection). In another word, our estimation $\bar{y}$ is an approximate scaled-down value of coverage. Consequently, our source selection algorithm embedded with this coverage estimation method would choose the source with the maximal gain with a high probability.

[1] https://es.wikipedia.org/wiki/MinHash.

5.3 Determination of h_js

Recall that for each attribute A_j in ϕ, we need to construct h_j-way AND-construction on $\mathbf{F}_j$. In this subsection, we will discuss how to determine the proper values of h_js. For simplicity, we denote all the h_js as a vector $\mathbf{h} = [h_1, \cdots, h_l]$. We let $\mu_{\phi,\mathcal{S}}$ denote the expected size of the matched lsh-signatures between source set $\mathcal{S}$ and target S_0, that is $\mu_{\phi,\mathcal{S}} = \mathbb{E}[|\mathsf{lsh}_\phi(\mathcal{S}) \cap \mathsf{lsh}_\phi(S_0)|]$. In order to assure the greedy source selection algorithm can choose the source with the maximal coverage gain in each iteration, the goal of the $\mathbf{h}$ determination is to ensure $\mu_{\phi,\mathcal{S}}$ is linear to the real coverage, as described below,

$$\mu_{\phi,\mathcal{S}} = d_\phi * cov_\phi(\mathcal{S}),$$

where d_ϕ is a coefficient determined by ϕ.

To solve this problem, we first study the relationship between $\mu_{\phi,\mathcal{S}}$ and $\mathbf{h}$. The following lemma shows that $\mu_{\phi,\mathcal{S}}$ is indirectly determined by $\mathbf{h}$, since $\mu_{\phi,\mathcal{S}}$ is determined by a_ϕ and b_ϕ, and both a_ϕ and b_ϕ are determined by $\mathbf{h}$. The proof is omitted due to space limitation.

Lemma 1. *Let n_1 denote the size of pattern set P_{ϕ,S_0}; n_2 denote the size of pattern set $P_{\phi,\mathcal{S}}$; α_j and β_j denote the expected probability of matching and non-matching pattern pair that have identical lsh-signatures on attribute A_j where $j \in [1, l]$ respectively. We suppose similarities on different attributes are independent of each other. $\mu_{\phi,\mathcal{S}}$ can be represented as follows,*

$$cov_\phi(\mathcal{S})(1 - (1 - a_\phi)(1 - b_\phi)^{n_2 - 1}) + (n_1 - cov_\phi(\mathcal{S}))(1 - (1 - b_\phi)^{n_2}),$$

where a_ϕ and b_ϕ denote the expected probability of matching and non-matching pattern pair that have identical locality-sensitive function values respectively, which are represented as follows,

$$a_\phi = \mathbb{E}(\Pr(lsh_\phi(t_1) = lsh_\phi(t_2) | t_1 \approx t_2)) = \alpha_1^{h_1} \alpha_2^{h_2} \cdots \alpha_l^{h_l},$$

$$b_\phi = \mathbb{E}(\Pr(lsh_\phi(t_1) = lsh_\phi(t_2) | t_1 \not\approx t_2)) = \beta_1^{h_1} \beta_2^{h_2} \cdots \beta_l^{h_l}. \qquad \square$$

Lemma 2 demonstrates that to ensure $\mu_{\phi,\mathcal{S}} \propto cov_\phi(\mathcal{S})$, b_ϕ should be small enough, which means all the h_js need to be large (by Lemma 1). The proof is omitted due to space limitation.

Lemma 2. *When $b_\phi * |P_{\phi,\mathcal{S}}| \to 0$, we have*

$$\mu_{\phi,\mathcal{S}} = a_\phi * cov_\phi(\mathcal{S}).$$

However, when h_j is too large, the probability of collisions between matching patterns approaches 0. As a consequence, the estimated coverage, i.e., the coverage of signatures, would be 0. Motivated by this, we attempt to minimize $\|h\|_1$ under the constraint that $b_\phi * |P_{\phi,\mathcal{S}}| \to 0$. Now we formulate this problem.

Problem 2. *Given ϵ, the* ***determination of*** $\mathbf{h}$ *under MD ϕ is to find the vector $\mathbf{h} = [h_1, \cdots, h_l]$ such that $||h||_1$ is minimal, while*

$$\beta_1^{h_1} \beta_2^{h_2} \cdots \beta_l^{h_l} n \leq \epsilon, \tag{5}$$

where n is the domain size of $P_{\phi,t}$ as the upper bound of any $|P_{\phi,\mathcal{S}}|$, and the values of β_1–β_l can be obtained by sampling.

The following theorem shows an easy way to find the optimal solution for the signature setting problem.

Theorem 4. *Given ϵ and β_1–β_l, $\mathbf{h} = [\lceil \frac{\log(\epsilon/n)}{\log \beta} \rceil, 0, \cdots, 0]$ is the optimal solution for the determination of* $\mathbf{h}$ *problem, where $\beta = \min(\beta_1, \beta_2, \cdots, \beta_l)$.*

The detail is omitted due to space limitation. According to Theorem 4, the property $\mu_{\phi,\mathcal{S}} = a_\phi * cov_\phi(\mathcal{S})$ holds when we set $\mathbf{h} = [\lceil \frac{\log(\epsilon/n)}{\log \beta} \rceil, 0, \cdots, 0]$.

6 Experimental Results

In this section, we report an extensive experimental evaluation. The goals of our study are to investigate the performance of our source selection algorithm in two aspects: the accuracy and efficiency of mhGreedy.

6.1 Experimental Setting

Algorithms. In order to verify our approach, mhGreedy, we implement two baseline algorithms for comparison: (1) exaGreedy (the upper bound): the greedy source selection algorithm based on exact coverage information obtained by accessing to all the sources; and (2) Random (the lower bound): randomly selects K sources from the source set $\mathbb{S}$. All experiments are implemented in Java and executed on a PC with Windows 10, 8 GB RAM and a 2.9 GHz Intel i7-7500U CPU.

Note we did not compare our approach with the optimal source selection algorithm since the optimal algorithm takes a prohibitive amount of time due to its running time exponential in K. From our experiments, the optimal source selection algorithm takes more than 27 h on the synthetic data set with the default setting shown in Table 2.

Table 1. Datasets and dependencies

Dataset	Dependencies
Rest.	$\text{name}_{(\text{Jaccard},0.87)} \wedge \text{addr}_{(\text{Jaccard},0.81)} \rightarrow \text{class}$
Cora	$\text{title}_{(\text{Jaccard},0.89)} \wedge \text{author}_{(\text{Jaccard},0.66)} \rightarrow \text{class}$
Syn	$\text{name}_{(\text{Jaccard},0.86)} \wedge \text{addr}_{(\text{Jaccard},0.81)} \rightarrow \text{blocking\#}$

Datasets. We conducted our experiments over two real data sets: Rest [4], and Cora [4]. In addition, to investigate the impact of the parameters and the scalability of our algorithms, we evaluated the performance of mhGreedy on one synthetic datasets: Syn, that yielded more sources and more tuples. The dependencies for all the datasets are shown in Table 1. The default setting of parameters for Syn is shown in Table 2. The sizes of all the synthetic sources are with the Zipf distribution.

(1) *Rest*: a real data set containing 864 restaurant records from the Fodor's and Zagat's restaurant guides containing 112 duplicates, where all the records from Fodor are chosen to be the target data, and records from Zagat are randomly chosen to generate 100 sources.
(2) *Cora*: a real data set that contains 1295 citation records and 112 duplicates, where 1295 records that do not contain duplicates are chosen to be the target data, and other records are randomly chosen to generate 100 sources.
(3) *Syn*: synthetic data sets generated using the Febrl data generator [3], where original records are all chosen to be the target data, while the duplications of these records are randomly chosen to generate data sources. Table 2 shows the parameters used for generating the data sets and the default settings for the parameters.

Table 2. Parameters for Syn

Param.	description	Val.
n_o	# of orignal records	1000
n_d	# of duplicates	1000
d	# of duplicates per original record	1
m	# of modifications within one attribute	2
#Source	# of sources	100

6.2 Comparison

We compared the effectiveness of the methods with K varying from 1 to 10 on two real data sets. The comparison results are reported in Fig. 2(a), (b). mhGreedy achieves coverage comparable to the baseline, exaGreedy, in which the approximation ratio is around 99%. Sometimes, the coverage of mhGreedy is even larger than exaGreedy, this is because the latter one is not the optimal selection strategy, and some estimation error might lead to better results.

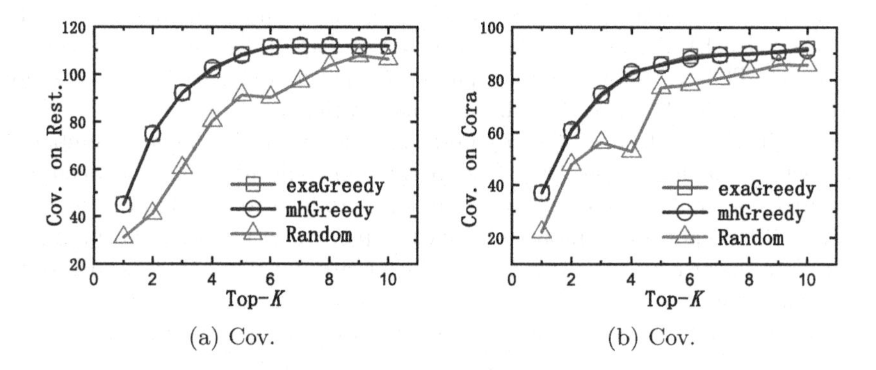

(a) Cov. (b) Cov.

Fig. 2. Comparison results

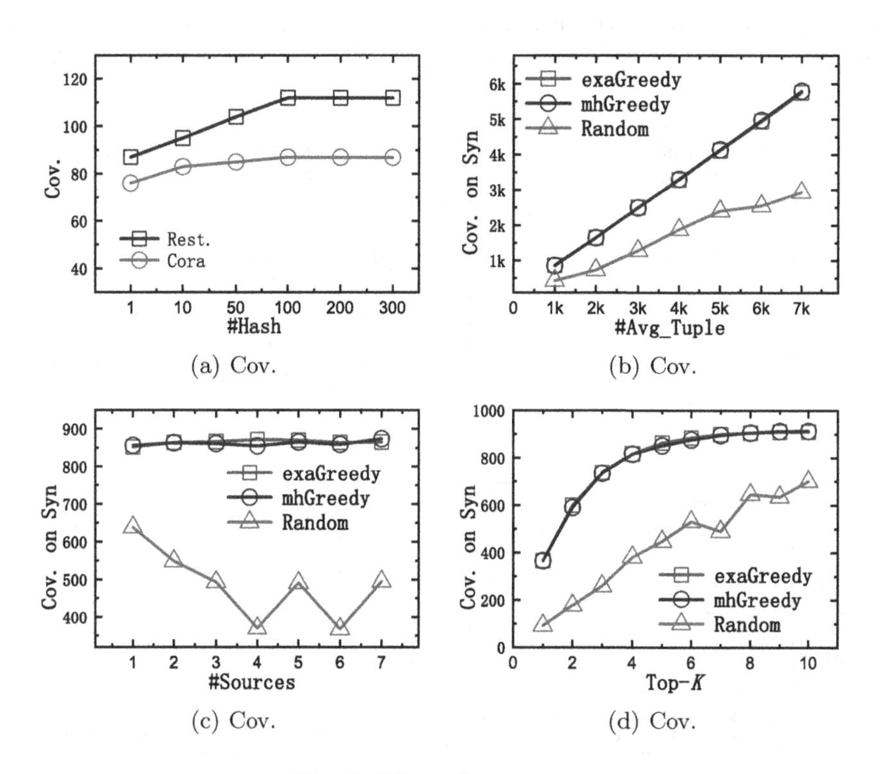

(a) Cov. (b) Cov.

(c) Cov. (d) Cov.

Fig. 3. Effect of parameters

6.3 The Impact of Parameters

We evaluated the impact of the number of hash functions used for mh-signature (#Hash), the average data size (#Avg_Tuple), the number of sources (#Sources), TOP-K on the performance of our algorithms. The results are reported in Fig. 3. We have the following observations.

Figure 3(a) shows that mhGreedy performs quite well with a small number of hash functions (no more than 100). Due to the significant differences between real data sources, a small number of hash functions can lead to a good performance.

Figure 3(b) reports the impact of #Avg_Tuple. However, the performance of mhGreedy is insensitive to #Avg_Tuple, which is around 100%. This is consistent with our analysis.

Figure 3(c) and (d) report the impact of #Source and K on the synthetic data sets respectively. From the results, the effectiveness of mhGreedy is insensitive to both #Source and K, which also coincides with our analysis.

6.4 Efficiency Results

We investigated the runtime performance of mhGreedy. Figure 4(a) reports the runtime of algorithms with various data sizes. We observe that the runtime of exaGreedy grows quadratically with the data size, while the runtime

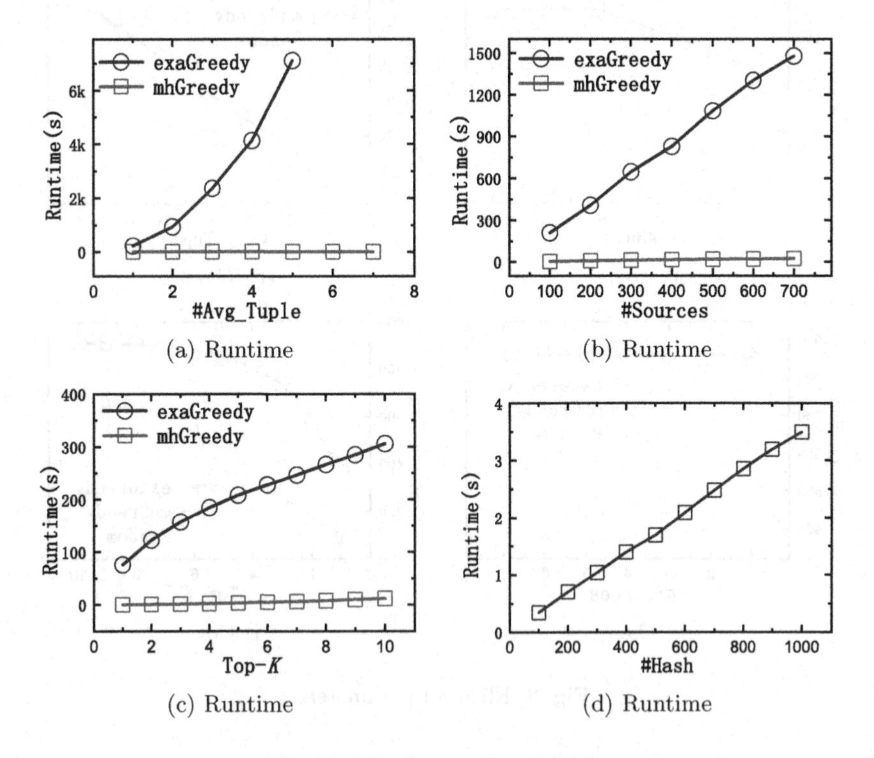

(a) Runtime

(b) Runtime

(c) Runtime

(d) Runtime

Fig. 4. Efficiency and scalability

of mhGreedy is very stable, which is around 3.3s. These results coincide with our analysis in Sect. 5. In Fig. 4(b), we varied the number of sources from 100 to 700. The runtime of all the algorithms grows linearly with #Sources. However, mhGreedy outperforms exaGreedy significantly (>50 times faster than exaGreedy). In Fig. 4(c), we varied TOP-K from 1 to 10. The runtime of exaGreedy is still linear to TOP-K, while the runtime of mhGreedy grows slowly with K and outperforms exaGreedy significantly. Such results show the benefit of mhGreedy regarding both efficiency and scalability. Figure 4(d) shows the runtime of mhGreedy with varying #Hash from 100 to 1000. The runtime grows approximately linearly as #Hash increases. These results verify the scalability of our methods.

Summary. Based on the experimental results, we draw the following conclusions. (a) The sources selected by mhGreedy is comparable to the baseline algorithm, exaGreedy; (b) mhGreedy outperforms exaGreedy both on efficiency and scalability significantly; (c) The effectiveness of mhGreedy is insensitive to most of the parameters; and (d) mhGreedy scales well on both the data size and the number of sources.

7 Related Work

Our work relates to the areas of *error detection and repairing*, *source selection*, *data fusion* and *similarity join size estimation.*

Error detection and repairing [12,22,23] is the task of our paper which has been widely studied due to its importance [8] studied the problem of CFD violations detection for centralized data. [9] and [10] studied the problem of detecting FD and CFD violations in a distributed database. Most of the existing studies conducted error detection and repair on only local data set.

Source selection [6,21] has been recently studied. Most previous work of source selection focus on choosing proper sources for query processing or integration while the objective of SSED is to detect MD violations. [16] studied the problem of source selection for error detection. This approach is based on FDs which are not as general as MD. In addition, it can not be applied for MD since only exact match is supported.

Data fusion [17,19] aims at finding true values of conflicting data items. Various fusion models have been proposed. The most basic and simple fusion method is to take the value provided by the majority vote. Advanced methods assign weights to sources according to their reliability. SSED can be viewed as a preprocessing of data fusion.

Many algorithms [11,15,20] have been proposed for the similarity join size estimation problem which focus on estimating the number of matching pairs for set similarity selection queries. Most of the methods are based on constructing priori samples or indexes for the given sets. However, they can hardly be applied in our coverage estimation problem, since we need to iteratively adding sources into a selected source set and incrementally estimate the updated coverage.

8 Conclusion

We have formulated the problem of how to select sources to detect candidate errors in target data via MDs. In order to support efficient gain estimation of each source, we presented a two-level signature mechanism by applying LSH for fuzzy match and minhash for coverage estimation, where the size of signatures is independent of the data size. The experiments show that our approach finds solutions that are competitive with the near-optimal greedy algorithm and achieves a better performance regarding both efficiency and scalability without accessing data sources.

Acknowledgements. This work is supported by NFSC 61602159; the Heilongjiang Province Natural Science Foundation YQ2019F016, ZD2019F003; and the Fundamental Research Foundation of Universities in Heilongjiang Province for Youth Innovation Team under Grant RCYJTD201805.

References

1. Andoni, A., Indyk, P.: Near-optimal hashing algorithms for approximate nearest neighbor in high dimensions. Commun. ACM **51**(1), 117–122 (2008)
2. Bellahsene, Z., Bonifati, A., Rahm, E.: Schema Matching and Mapping. Data-Centric Systems and Applications. Springer, Heidelberg (2011). https://doi.org/10.1007/978-3-642-16518-4
3. Christen, P.: Febrl - an open source data cleaning, deduplication and record linkage system with a graphical user interface. In: SIGKDD (2008)
4. Christen, P.: A survey of indexing techniques for scalable record linkage and deduplication. TKDE **24**(9), 1537–1555 (2012)
5. Cohen, E., et al.: Finding interesting associations without support pruning. TKDE **13**(1), 64–78 (2001)
6. Dong, X.L., Saha, B., Srivastava, D.: Less is more: selecting sources wisely for integration. VLDB **6**(2), 37–48 (2012)
7. Fan, W.: Dependencies revisited for improving data quality. In: PODS, pp. 159–170 (2008)
8. Fan, W., Geerts, F., Jia, X., Kementsietsidis, A.: Conditional functional dependencies for capturing data inconsistencies. TODS **33**(2), 6 (2008)
9. Fan, W., Geerts, F., Ma S., Müller, H.: Detecting inconsistencies in distributed data. In: ICDE, pp. 64–75. IEEE (2010)
10. Fan, W., Li, J., Tang, N., et al.: Incremental detection of inconsistencies in distributed data. TKDE **26**(6), 1367–1383 (2014)
11. Hadjieleftheriou, M., Yu, X., Koudas, N., Srivastava, D.: Hashed samples: selectivity estimators for set similarity selection queries. PVLDB **1**(1), 201–212 (2008)
12. He, C., Tan, Z.T., Chen, Q., Sha, C.: Repair diversification: a new approach for data repairing. Inf. Sci. **346–347**, 90–105 (2016)
13. Hochbaum, D.S.: Approximating covering and packing problems: set cover, vertex cover, independent set, and related problems. In: Approximation Algorithms for NP-Hard Problems, pp. 94–143. PWS Publishing Co. (1996)
14. Jin, C., Lall, A., Jun, X., Zhang, Z., Zhou, A.: Distributed error estimation of functional dependency. Inf. Sci. **345**, 156–176 (2016)

15. Lee, H., Ng, R.T., Shim, K.: Similarity join size estimation using locality sensitive hashing. PVLDB **4**(6), 338–349 (2011)
16. Li, L., Feng, X., Shao, H., Li, J.: Source selection for inconsistency detection, pp. 370–385 (2018)
17. Li, X., Dong, X.L., Lyons, K., Meng, W., Srivastava, D.: Truth finding on the deep web: is the problem solved? VLDB **6**(2), 97–108 (2012)
18. Nemhauser, G.L., Wolsey, L.A., Fisher, M.L.: An analysis of approximations for maximizing submodular set functions—I. Math. Program. **14**(1), 265–294 (1978). https://doi.org/10.1007/BF01588971
19. Pochampally, R., Sarma, A.D., Dong, X.L., Meliou, A., Srivastava, D.: Fusing data with correlations. In: SIGMOD, pp. 433–444. ACM (2014)
20. Rafiei, D., Deng, F.: Similarity join and self-join size estimation in a streaming environment. CoRR, abs/1806.03313 (2018)
21. Rekatsinas, T., Dong, X.L., Getoor, L., Srivastava., D.: Finding quality in quantity: the challenge of discovering valuable sources for integration. In: CIDR (2015)
22. Ye, C., Li, Q., Zhang, H., Wang, H., Gao, J., Li, J.: AutoRepair: an automatic repairing approach over multi-source data. Knowl. Inf. Syst. **61**(1), 227–257 (2019)
23. Zheng, Z., Milani, M., Chiang, F.: CurrentClean: spatio-temporal cleaning of stale data, pp. 172–183 (2019)

Dependency Preserved Raft for Transactions

Zihao Zhang[1], Huiqi Hu[1(✉)], Yang Yu[1], Weining Qian[1], and Ke Shu[2]

[1] School of Data Science and Engineering, East China Normal University, Shanghai, China
{zach_zhang,yuyang}@stu.ecnu.edu.cn, {hqhu,wnqian}@dase.ecnu.edu.cn

[2] PingCAP Ltd., Beijing, China
shuke@pingcap.com

Abstract. Modern databases are commonly deployed on multiple commercial machines with quorum-based replication to provide high availability and guarantee strong consistency. A widely adopted consensus protocol is Raft because it is easy to understand and implement. However, Raft's strict serialization limits the concurrency of the system, making it unable to reflect the capability of high concurrent transaction processing brought by new hardware and concurrency control technologies. Upon realizing this, the work targets on improving the parallelism of replication. We propose a variant of Raft protocol named DP-Raft to support parallel replication of database logs so that it can match the speed of transaction execution. Our key contributions are: (1) we define the rules for using log dependencies to commit and apply logs out of order; (2) DP-Raft is proposed for replicating logs in parallel. DP-Raft preserves log dependencies to ensure the safety of parallel replication and with some data structures to reduce the cost of state maintenance; (3) experiments on YCSB benchmark show that our method can improve throughput and reduce latency of transaction processing in database systems than existing Raft-based solutions.

Keywords: Consensus protocol · Log replication · Log dependency

1 Introduction

To solve a variety of fault tolerance problems and ensure the availability of distributed database systems, multiple replicas are required. How to keep consistency among replicas is a challenge [10,18]. The traditional approach adopts primary-backup replication [17] to replicate writes from primary replica node to backup nodes. To achieve strong consistency, the replication process is synchronous, which sacrifices performance. In contrast, to achieve better performance, asynchronous replication is adopted, which does not guarantee consistency between primary and backups. Therefore, primary-backup replication requires a compromise between strong consistency and high performance.

Recently, many transactional systems require both high-performance and strong consistency to build mission-critical applications. A common way is to

Y. Nah et al. (Eds.): DASFAA 2020, LNCS 12112, pp. 228–245, 2020.
https://doi.org/10.1007/978-3-030-59410-7_14

use quorum-based replication to maintain a consistent state machine, which is a database instance in each replica. By storing a series of database logs and executing the logs in sequence, the same state can be ensured among replicas. Quorum-based replication only requires replicating to majority nodes, instead of waiting for all backup nodes' response like in primary-backup replication so that it can achieve high performance. Besides, using consensus protocols can guarantee strong consistency, and tolerate up to $\lceil N/2 \rceil - 1$ failures. Therefore, quorum-based replication makes a good balance on performance and consistency. Consensus protocols such as Paxos [11] and Raft [14] are used to keep the log consistent in quorum-based replication. Paxos is a classical algorithm to ensure distributed consensus, but due to the high complexity, many works devoted to simplifying it so that it is easy to understand and implement. Raft is one of them, it can achieve strong consistency as Paxos, and much easier to be implemented. As a result, Raft is widely used in commercial systems like etcd [1], TiDB [2] and PolarFS [5] in recent years.

Unfortunately, to make Raft is an understandable protocol, there are some restrictions that make Raft highly serializable. Logging in Raft is in a sequential order, which means logs should be acknowledged, committed, and applied in sequence, and log hole is not allowed, this dramatically limits the concurrency of systems. Nowadays, multi-core CPU and large memory are commonly used to improve the performance of database. To make better use of hardware resources, concurrency control strategy is adopted to improve the concurrency of transaction execution. Furthermore, in production systems, logs usually be replicated in a pipeline way [13]. After leader sends a log entry to followers, it can send the next one instead of waiting for the response of the last entry. Moreover, when there are multiple connections between leader and followers, follower may receive out-of-order logs. In other words, some logs may reach follower earlier than those in front of them. All of these strategies are to improve system's concurrency, but with the restriction of Raft, the acknowledgment in follower must in order, which causes the serial commit of transactions. This is even worse when the network is poor. If some connections are blocked, followers cannot receive log entries and reply to leader. As a result, leader may also be stuck when majority followers are blocked, causing the system to be unavailable. So even with high concurrency in transaction execution, it still needs to wait for replicating logs serially.

To improve the performance of log replication with Raft, so that it can match the speed of transaction execution, we need to modify Raft to replicate logs in parallel. Recently, a variant protocol of Raft named *ParallelRaft* is used in PolarFS [5], it can replicate logs out-of-order. It is designed for file system and has no guarantee on the order of completion for concurrent commands. Therefore, it does not satisfy transaction's commit semantics and cannot be used in OLTP database systems directly. What's more, it adds a *look behind buffer* into log entry which contains the LBA (Logical Block Address) modified by the previous N log entries, N is the maximum size of a log hole permitted. With this information, it can check if an entry's write set conflicts with the previous to decide if the entry can be applied out-of-order. If the N is too large, it will greatly increase the size of log entry and waste bandwidth.

In this paper, we propose a new variant called *DP-Raft* (dependency preserved Raft) for transactions to support concurrent log replication which will gain high performance in OLTP systems. We redesign Raft to overcome the shortcomings of low parallelism so that it can replicate logs in parallel. In the meantime, we preserve the dependency of log entries to support commit semantics of transactions and make sure the correctness of log applying. Our contributions can be summarized as follows:

- Based on the analysis of log dependencies, we form the requirements to satisfy transaction commit semantics and apply safely in parallel replication. In short, commit order must follow the order of RAW dependency; and to apply logs to a consistent state machine, WAW dependency must be tracked.
- We design DP-Raft to preserve dependencies between log entries and use dependency information to acknowledge, commit and apply log entries out-of-order, which significantly improves Raft's parallelism.
- Experiments show that our new approach achieves a better performance than Raft and ParallelRaft.

The paper is organized as follows: Section 2 gives some related works. Section 3 analyzes if logs can be committed and applied with dependencies. In Sect. 4, we describe DP-Raft in detail. Experimental results are presented in Sect. 5. Finally, we conclude our work in Sect. 6.

2 Related Work

Lamport proposed Paxos [11] to guarantee the consistency in distributed systems, which is a general-purpose consensus protocol for asynchronous environments and has been proven to ensure distributed consensus in theory. Paxos has no primary node, so each request is a new Paxos instance, and must go through two phases to be accepted in the majority. To resolve the problem that Paxos unable to serve consecutive requests which is required in practical, many works devoted to Paxos variants. Multi-Paxos [12] remove the proposal phase to reduce the round number for every single request by electing a leader, who can serve requests continuously. As a result, Multi-Paxos is widely used in distributed database systems like Google's Spanner [7] and MegaStore [4], Tencent's PaxosStore [19] and IBM's Spinnaker [16].

Despite this, Paxos is still difficult to understand and implement correctly. Raft [14] is a consensus protocol designed for understandability. It separates the protocol into leader election and log replication. Leader in charges of processing requests and replicating the results to followers. Raft introduces a constraint that the logs must be processed in serial. This constraint makes Raft easy to understand but also limits the performance of concurrent replication. Because of the simplicity of implementation, many systems such as etcd [1], TiDB [2] and PolarFS [5] use Raft to provide high availability.

As mentioned above, the strict serialization of Raft limits systems' concurrency. Many researchers are working on it to improve Raft's performance.

PloarFS [5] proposed ParallelRaft which supports concurrent log replication. But it is designed for file systems, so out-of-order replication is implemented by recording the write block address of the previous N entries in each log entry. Only checking conflict of write is not enough for commit semantics of transaction, and maintain the write set of N entries in each entry is too expensive for memory and network if we want to get higher concurrency by increasing parameter N. TiDB [2] optimizes Raft in another way, it divides data into partitions, and each partition is a single Raft group to replicate. By using multi-group Raft, it balances the workload in multiple servers, which improves the concurrency of the whole system. Vaibhav et al. [3] discussed the issue of consistently reading on followers so that followers can share the workload of leader to avoid leader becoming a performance bottleneck. DARE [15] gave another way to speed up replication with hardware. It uses RDMA to redesign the Raft like protocol to achieve high performance.

3 Parallel Committing and Applying

The purpose of keeping the restriction of log order in Raft is to make sure of the correctness of apply and finally reach a consistent state among different servers. But in a transactional database system, it is not necessary to guarantee this strict constraint. In database systems, we use Raft to replicate the log entry which records the results of a transaction when the transaction enters commit phase. After the transaction log is committed, the commit of transaction is complete. We keep the log order the same as transaction commit order so that replay log we can achieve the same state. But when transactions do not have a dependency, which means that they do not read or write on the same data, they can be executed in any order. Raft can benefit from the out-of-order execution between such non-dependent transactions, the log entries of these transactions also have no dependencies, so there is no need to strictly acknowledge, commit, and apply them in order. But to ensure the correctness of the state, it is still necessary to keep the order between dependent logs. To do this, we need to analyze how to commit and apply in parallel with dependency.

To correctly commit and apply logs to a consistent state, the relative order among log entries of conflicting transactions must comply with the transactions commit order (the commit order of transactions can be generated by any concurrency control mechanism). It indicates that dependency should be tracked during log processing. We will demonstrate how to use log dependency to guarantee the correctness of Raft in parallel. For the sake of illustration, we use T_i to represent a transaction and L_i as the log entry of transaction T_i. For each transaction, we generate one log entry to record the results of the transaction.

Figure 1 shows four dependencies that may exist between two transactions. In each case of dependency, the logs can be acknowledged by followers in an out-of-order way, but whether it can be committed or applied is determined by the type of dependency.

Dependency	Log Order	Can commit in parallel?	Can apply in parallel?
$T_1 \xrightarrow{WAW} T_2$	$L_1 < L_2$	yes	no
$T_3 \xrightarrow{RAW} T_4$	$L_3 < L_4$	no	yes
$T_5 \xrightarrow{WAR} T_6$	$L_5 < L_6$	yes	yes
$T_7 \xrightarrow{Null} T_8$	$L_7 < L_8$	yes	yes

Fig. 1. Four dependencies may exist between transactions. For the correctness of state machine, the order of committing must track RAW dependencies and the order of applying must track WAW dependencies.

Write after write dependency (WAW): For T_1 and T_2 with WAW dependency, it means that T_2 writes the data which has been modified by T_1. In this scenario, the log entries cannot be applied out-of-order. This is because according to the execution order, the results in L_2 must be written after L_1. If we apply L_2 first, then L_1 will overwrite L_2's update results, which causes an incorrect state. If both logs have been committed, simply applying them in order will reach a consistent state, regardless of the order of commit. And if someone commits failed, it can be ignored and the applied state is consistent as if the failed log entry has never existed. So L_1 and L_2 can commit in parallel.

Read after write dependency (RAW): For T_3 and T_4 with RAW dependency, T_4 reads the result that T_3 wrote. In this scenario, L_4 cannot commit before L_3. If L_4 commits first and then L_3 commits failed, T_4 will read a data that never exists, it is called *DirtyRead*, which is not allowed in transaction processing. When both T_3 and T_4 have been committed in sequence, because they do not update the same data, they can be applied in parallel, which will not cause an inconsistent state.

Write after read dependency (WAR): When T_6 updates the same data that T_5 has read before, there is a WAR dependency between them. In this case, L_5 and L_6 can commit and apply in parallel. Because T_5 has read what it should read, so T_6's update will not affect the result of T_5, they can commit without order. And with the different data they update, they can also be applied in any order. So the WAR dependency need not to track for DP-Raft.

No dependency (Null): If two transactions do not read or write on the same data, as shown in scenario (4), there is no dependency between them, and the log can be committed and applied out-of-order in DP-Raft.

Based on the above analysis, only WAW and RAW dependency of log entries need to be tracked for DP-Raft to commit and apply. When RAW dependency exists, log entries should be committed in sequence, and for WAW dependency, log entries should be applied in order.

4 Dependency Preserved Raft

This section proposes a variant of Raft algorithm for transactions, named as *DP-Raft*, which is independent of concurrency control strategy. It preserves the dependency of transactions in their log entries and uses dependency to decide that whether logs can be committed or applied without order. Because DP-Raft makes the log replication process more parallel, it is suitable for multiple connections between servers and can effectively utilize the network bandwidth. Therefore, it will significantly reduce the latency of log replication.

In the following, we first present related data structures and how to analyze log dependency, then the design of *log replication* and *leader election* algorithms will be introduced.

4.1 Structures of DP-Raft

Log Buffer. DP-Raft uses a ring buffer to store log entries, as shown in Fig. 2, every node has such a log buffer. DP-Raft allows log hole to exist in the log buffer. On followers, log hole means that an entry is empty, but there are non-empty log entries before and after it; on the leader, log holes represent uncommitted entries that exist between committed log entries. Section 3 has mentioned that to ensure the correctness of committing and applying in DP-Raft with log hole, log dependency should be tracked. But if we allow hole log entries to exist without limits, it will cause two problems: (1) According to the strategy that will be introduced in Sect. 4.2, we should not only maintain the write set of hole log entries but also all entries that depend on them. So if we do not limit the range that hole entry exists, the write set will be too large to maintain; (2) Because of the existence of hole entries, we can not use *commitIndex* to represent the status of all entries before this index like Raft. The index of last committed log entry cannot indicate the status of the previous log entries, so we should record all entries' status so that we can know which is committed. This is also difficult to maintain if hole entries appear in a large range.

To solve the problems mentioned above, we introduce a sliding window whose size can be modified. We only allow hole entries in the sliding window, which will limit the range that hole entries may exist. As a result, the maintenance

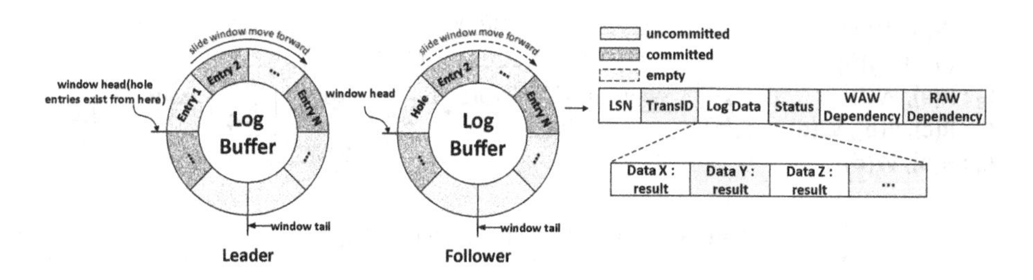

Fig. 2. The structures of log buffer and log entries in DP-Raft. The left one represents the log buffer in leader, and the right one is in follower.

of the write set and entries's status will be simple. Figure 2 shows the window head in leader represents the first uncommitted entry which is also the start point that hole entries can exist in followers. The window head is the same as *commitIndex* in Raft, all entries before window head are committed. The head can move forward when one entry and all the previous are committed. If leader receives acknowledgment of an entry from majority followers, this entry can commit even if someone is uncommitted before it, like Entry 2 in leader's buffer. Reason for Entry 1 not committed may be that more than half followers have not received the replicating request of Entry 1, so the slot of Entry 1 is empty in follower now. But follower receives Entry 2 first, it stores the entry and sends an acknowledgment back to leader, then the slot of Entry 1 will be a log hole. Section 4.3 will explain the replication process in detail.

Comparing to the strategy that allows holes without limitation, we narrow the range of maintaining the write set and entries' status to a sliding window, which significantly reduces the difficulty.

Log Entry. Each log entry has a unique *log sequence number* (*LSN*), which is monotonically increasing and also used as log entry index for DP-Raft. When a transaction is received, leader will process the operations in the transaction and pre-applies the results to memory. Once the transaction enters the commit phase, an LSN will be acquired, and then generates a log entry to reflect the results of the transaction. Log entry format is shown in Fig. 2. We adopt the common *value log* like in [8], so *Log Data* stores the results of the transaction, indicating which data has been modified and its new value. In addition to *Log Data*, a log entry also includes LSN and the transaction ID. *Status* indicates if the entry is committed and applied. Moreover, in order to track log dependency, log entry also stores indexes of WAW and RAW dependent log entries at *WAW* and *RAW Dependency*. With this information, leader can decide if the entry can be committed and followers can decide if it can be applied to the state machine. Compared to ParallelRaft [5] which stores the write set of the previous N entries, storing dependent entries' index will reduce the size of log entries and can satisfy transaction's commit semantics when committing in parallel.

4.2 Dependency Analysis

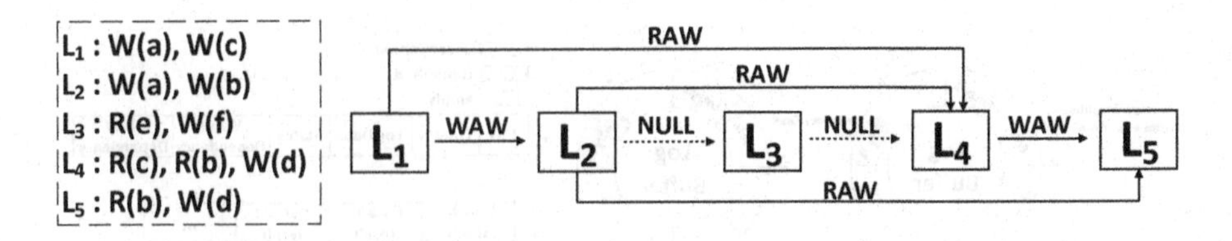

Fig. 3. Transaction dependency graph.

KuaFu [9] has a similar idea about enabling logs to be applied concurrently to reduce the gap between primary and backups by constructing a dependency graph of logs, which will introduce overheads for database systems. While DP-Raft simply analyzes dependency and records dependency in each entry instead of maintaining a dependency graph. Here is an example of log dependency in Fig. 3. The read and write sets of five entries are shown on the left side. Because L_1 and L_2 both write data a, so there is a WAW dependency between L_1 and L_2. For L_1, L_2 and L_4, L_4 read data b and c, which are updated by L_1 and L_2, so L_4 has a RAW dependency with L_1 and L_2. Other log entries' dependencies are also analyzed in the same way.

Once a transaction enters the commit phase, leader will generate a log entry and add it into a slot in log buffer (once an entry is added into log, it will never be overwritten, so DP-Raft guarantees the *Leader Append Only* property). During this phase, log dependency is analyzed. DP-Raft stores the write set of hole entries and entries depend on them in the sliding window. So when L_5 enters commit phase, the write sets of L_1, L_2 and L_4 are stored. Because L_5 reads data b, we find b was last modified by L_2, so it has a RAW dependency with L_2. And d in L_5's write set was updated by L_4 recently, so it connects with L_4 on WAW dependency. If the read or write set of L_5 is overlapping with an entry which is ahead window head, we will ignore it, because the entry is committed so out-of-order commit and apply will not happen. It should be noted that for an entry that does not depend on both hole entries and those who have dependencies with holes, the write set of the entry is no need to be recorded, like L_3.

4.3 Log Replication

After leader adds an entry into its local buffer, it can replicate the entry to followers asynchronously. Log replication in DP-Raft is quite different from Raft [14]. When leader of Raft replicates log entries, the *commitIndex* will be sent together to inform followers of the current commit status of entries. But DP-Raft does not have *commitIndex*, it needs to store the status of all entries in the sliding window. To simplify the cumbersome maintenance, DP-Raft introduces a *commitFlag* which is a bitmap, wherein each bit of the flag represents the status of an entry in the sliding window, with 1 for committed, and 0 for uncommitted. When a replicating request is sent, *commitFlag* and *windowHead* are attached to the request to replace *commitIndex* of Raft. The entire process of log replication will be divided into three parts.

Followers Acknowledge in Parallel. In Raft, when a follower receives a request to append entries, it must ensure the consistency of the log in leader and itself by checking if the last entry is the same. Only when the checking is successful, it can accept the new entries. If the previous entry is missing, follower needs to wait until the previous one is acknowledged. The checking ensures the *Log Matching* property, which is one of the properties to guarantee the safety of Raft. But in DP-Raft, this constraint is removed, it is no need to wait even if some holes exist before. When follower receives new entries from leader, it just adds them into its log buffer based on the entry's LSN and sends an acknowledgment back. In DP-Raft, follower can acknowledge entries in parallel, thus

Algorithm 1: Commit and Apply in Parallel

```
Input: the entry to be committed or applied
/* leader commits in parallel                                         */
1  Function ParallelCommit()
2      if entry.replicated_count ≥ majority then
3          for dependent_entry ∈ entry.RAWDependency do
4              if dependent_entry.status == commited or applied then
5                  can_commit ← true;
6              else
7                  can_commit ← false;
8                  break;
9          if can_commit then
10             entry.status ← committed;
11             new_flag ← commitFlag — 1<<(entry.LSN - windowHead);
12             atomic-cas(&commitFlag, commitFlag, new_flag);
/* followers apply in parallel                                        */
13 Function ParallelApply()
14     if entry.status == committed then
15         for dependent_entry ∈ entry.WAWDependency do
16             if dependent_entry.status == applied then
17                 can_apply ← true;
18             else
19                 can_apply ← false;
20                 break;
21         if can_apply then
22             DoApply();
23             entry.status ← applied;
```

reducing the extra waiting time even when the network is jittery and entries are received out-of-order, so the latency of log replication is significantly reduced.

Even without checking when the previous entry is missing, the *Log Matching* property is still ensured by DP-Raft. If two entries have the same index and term, the two entries and all entries before their index are considered as the same. So Raft ensures this property by comparing the *prevTerm* and *pervIndex* in request to follower's local last entry when acknowledges the replicated entries from leader. In contrast to this, DP-Raft relaxes this constraint. It skips the check phase and saves the entry directly if the previous one is missing, but the previous entry's information is also stored. When the previous entry arrives later, it will be compared with the previous information stored before to see if matches. If not, the entry located behind will be discarded. As a result, follower cannot have different entries from leader eventually, so this property is ensured.

Leader Commits in Parallel. After leader sends an append entries request to followers, it waits for responses until majority acknowledgments are received. After this, leader can commit the entry and the transaction associated, then replies to client to inform the result of the transaction. The sequence restriction of Raft asks leader to commit entries in order, but in DP-Raft, entries can be committed in parallel. Section 3 has mentioned that parallel commit should obey the sequence of RAW dependency, so DP-Raft must ensure all the RAW dependent entries are committed, so that DP-Raft can commit the entry, as lines 3–8 in Algorithm 1 show. Checks on dependent entries enable DP-Raft to satisfy transaction commit semantics, which is different from ParallelRaft [5] that do not ensure the commit order between dependent entries. DP-Raft maintains a *commitFlag* to represent the status of entries in the sliding window, so leader will update *commitFlag* to set the bit of the commit entry to 1 (line 11) after committing it. To be mentioned, we use atomic *CAS* operation to modify *commitFlag*. Because the flag is a high contention resource when commit in parallel, so the latch-free *CAS* operation will reduce the conflict on the mutex which used to protect *commitFlag*. Next time leader appends entries or sends a heartbeat, *commitFlag* and *windowHead* will be attached to inform followers of which entry has been committed so that followers can apply it into their state machine.

Followers Apply in Parallel. After follower receives *commitFlag* and *windowHead*, it will check the status of log entries. For entries before *windowHead*, which indicates that they are committed, if some of them are still missing, follower has to notify leader to send the missing entries again so that log holes in follower can be filled. For entries in the sliding window, follower will update status based on *commitFlag*. If an entry is set to committed, it can be applied. Same as acknowledge and commit, apply also need to do in strict order for the correctness of state machine in Raft. As for DP-Raft, follower apply entries in parallel based on the dependency information in each entry. Lines 14–20 in Algorithm 1 show that, before applying an entry, the WAW dependent log entries need to be checked if they are applied. If so, DP-Raft can apply it; otherwise, it needs to wait for the dependent entries to apply.

Log consistency is ensured by *Log Matching* property, so if we want to make sure *State Machine Safety*, the order to apply logs also needs to be consistent. As described in Sect. 3, only the entries without WAW dependency can be applied in parallel, so *DP-Raft* can make sure the correctness of state machine by preserving dependency and applying write conflict entries according to the WAW dependency order.

4.4 Leader Election

Raft's leader is a strong leader who responsible for generating log entries and replicating them to followers, thus its log is the newest and must contain the complete committed log entries, which is called *Leader Completeness*. This property is easy to be ensured in Raft because with no log holes, the one who has the most up-to-date log will be elected as leader, which means the new leader has the complete logs. But the new leader in DP-Raft may lose some committed entries because of log holes. Therefore, an extra merge stage can make sure that the new leader will find all missing entries from other servers and become a complete leader. DP-Raft obeys the similar rules of Raft to vote, but when a candidate receives majority votes, it still unable to provide service until all holes in its log are filled. So DP-Raft introduces a new state called *pre-leader*, after pre-leader merges logs to fill its log holes, it can transform into a real leader.

Request Votes. When leader election is triggered, follower may transform into candidate and sends a voting request. In the request, the term and index of the lastest log entry will be included. Different from Raft, because the merge stage needs to merge the entries from the first uncommitted in pre-leader to the latest, so the first uncommitted log entry's index of candidate will also be attached to the voting request in DP-Raft. When the voting request is received, DP-Raft has two constraints to vote: (1) Follower first checks the term and index of the last log entry to see if the candidate's log is at least the same new as itself. If the candidate has a newer log, it grants vote; if candidate falls behind, it will refuse to vote; also if they are the same new, follower will use constraint (2) to see the complete status of candidate's log. (2) If the first uncommitted index of candidate is smaller than the local first uncommitted entry's index, it replies false; otherwise, it replies true to vote. By the first constraint, logs of the elected per-leader are at least as up-to-date as logs in majority servers, and only one pre-leader could be elected, which satisfies the *Election Safety* property. Besides this, the second constraint makes sure the number of log entries to be merged is as small as possible.

Merge Log Entries. According to the aforementioned, the pre-leader can transform into leader after the merge stage. In order to merge, followers need to send log entries between the first uncommitted index of pre-leader and the lastest

Is missing in pre-leader?	Is the entry committed?	Other conditions	Decision	Example in Fig.5.
No	Yes	(1) local entry committed	use the local entry	Entry 5
		(2) local entry uncommitted, but committed on other node	merge the committed entry to local log	
	No	(3) empty on all other nodes	use the local entry	Entry 7
		(4) different entry on other nodes	choose the entry with the highest term	Entry 9
Yes	Yes	(5) committed on other node	merge the committed entry to local log	Entry 4
	No	(6) empty on all other nodes	set the entry to NULL	Entry 6
		(7)different entry on other nodes	choose the entry with the highest term	Entry 8

Fig. 4. Seven conditions and their decisions during merging.

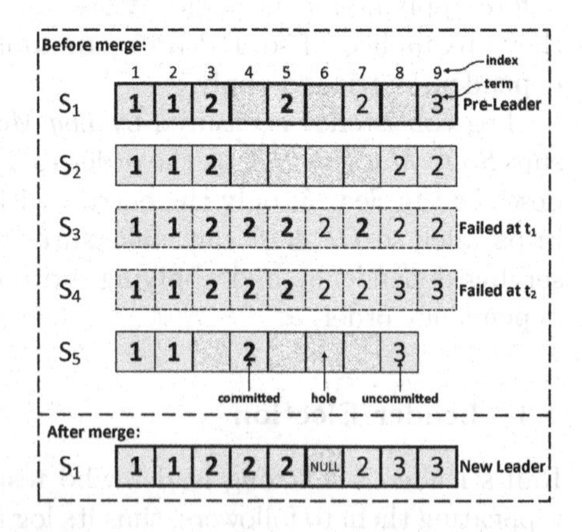

Fig. 5. An example of merging. S_3 and S_4 failed, S_1 is pre-leader now.

index in local to pre-leader. Imagining a follower was far behind leader but received the newest log entry before leader failed. If this follower starts an election, it is likely to be elected as the pre-leader because it has the newest entry. In this scenario, the number of entries to be sent for merging is too large, which wastes network bandwidth and increase the merging time. So the second constraint of voting ensures the range to be merged is small, so that the merging is vary fast.

After pre-leader receives log entries from majority servers, it starts to merge. There are several conditions of an entry as shown in Fig. 4. For a committed entry, it at least exists on one server if the system is still available. This means the committed entries can always be found on someone. In scenario (1), if an entry on pre-leader is not empty, and the entry is committed, it just use the local entry; but if a committed entry is uncommitted or missing on pre-leader (scenario (2) and (5)), the commited entry can be found at least on some other server, then pre-leader merges the committed entry to its own log. For an uncommitted entry, if pre-leader holds it, but it is missing on all other servers, pre-leader should preserve it. Because the entry may has been replicated to majority servers and committed, but has not notified the followers to commit, then some servers failed include leader. Now the pre-leader is the only one who has this entry and survived, so this entry cannot be abandoned, as shown in scenario (3). But if one entry is missing on both pre-leader and other servers, which means it cannot be committed, so we can ignore it and set the entry as NULL (scenario (6)). In the last two scenarios, (4) and (7), if an entry is uncommitted and is different among servers, pre-leader will choose the one with the highest term.

To understand how to merge more clearly, we give an example in Fig. 5. S_1-S_5 represent different servers, and entries with bold number and normal number represent committed and uncommitted entries respectively. At first S_3 was leader in term2, but it failed at t_1, then S_4 got votes from S_1 and S_5 to be the leader, and replicated entry 8 and 9 to others. At t_2, S_4 failed, S_1 started leader election and be voted by S_2 and S_5, so S_1 is the pre-leader now. After got entries between 4 and 9 from other servers, S_1 starts the merge stage. For entry 4, missing in its local log, but is committed on S_5, so S_1 merges S_5's entry 4 to its local log; entry 5 is stored and committed in local, it will be skipped; entry 6 is missing on all three servers, it is set to NULL and ignored; entry 7 is only stored in pre-leader, it will be preserved; for entry 8 and 9, different servers hold different entries, so two green entries of term 3 will be chosen. After the merge stage, logs in pre-leader is shown at the bottom of Fig. 5, then S_1 can provide service.

5 Evaluation

We implement DP-Raft in a database prototype with about 8300 lines Golang codes to verify the efficiency of our method. We adopt *Optimistic Concurrency Control* (OCC) as our concurrency control mechanism. The experimental setup and the benchmark used in the evaluation are given below.

Cluster Platform. We deploy the prototype on 7 machines, each of them is equipped with a *2-socket Intel Xeon Silver 4100 @ 2.10 GHz processor* (a total of 16 physical cores), and *192 GB DRAM*, connected by a *10 GB switch*.

Competitors. We compare DP-Raft to the standard Raft [14] and ParallelRaft [5]. Because ParallelRaft is designed for file system and is not suitable for transactional database systems, so we modify it by recording both the read and write key of a transaction in its log entry to replace the logical block address when used in file systems. When an entry enters the commit phase, it will check to see if the read and write set are overlapping with other entries, so that to support commit semantics of transaction when commit in parallel.

Benchmark. We adopt YCSB [6] to evaluate three methods. The schema contains a single table with each record is 100 bytes. The table is initialized to consist of 10 million records. To verify the performance of DP-Raft with RAW and WAW dependency, we modify the workload with one read and one write operation in each transaction, so that each log entry may have RAW and WAW dependencies with other entries.

5.1 Replication Performance

We first measure the throughput and latency under the YCSB workload. All methods have three replicas and we varying the client number to see the replication performance.

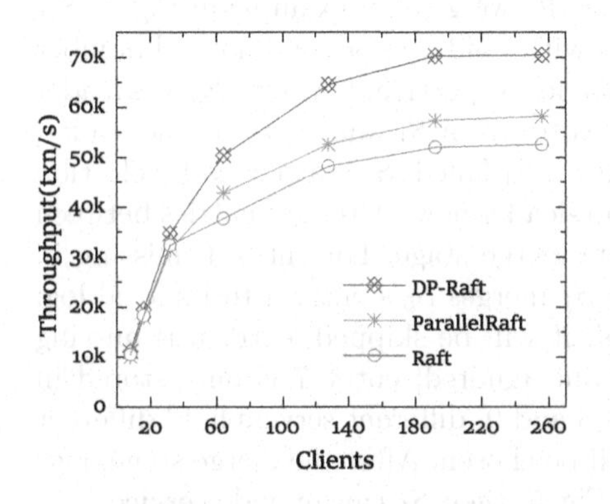

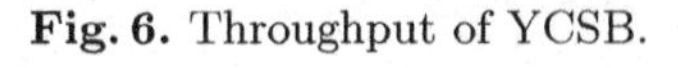

Fig. 6. Throughput of YCSB.

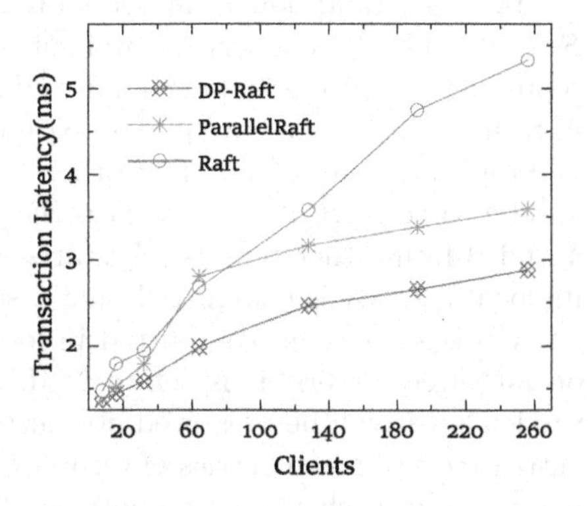

Fig. 7. Latency of YCSB.

The experiment results are shown in Fig. 6 and Fig. 7. In the beginning, all three methods grow quickly, but when the number of clients exceeds 32, the trend of Raft becomes slow because it must replicate logs one by one. Due to the parallel replication, DP-Raft and ParallelRaft will not wait for the result of previous entries and can utilize the bandwidth to replicate logs, therefore their throughput is higher than Raft. But because ParallelRaft needs to maintain the status of all entries from the first uncommitted, and need to compare both read and write set when committing an entry, which consumes a lot of CPU time and weighs on its throughput. On the contrary, DP-Raft only needs to check if the RAW dependent entries are committed and uses a *commitFlag* to record entries' status, so the low overhead brought by these inspires the throughput of DP-Raft, which is 1.34× than Raft and 1.21× than ParallelRaft. As for latency, Raft's serial replication causes a high latency because later entries have to wait. DP-Raft can replicate entries concurrently, so the latency gap between DP-Raft and Raft almost reaches to 50%. Although ParallelRaft can also replicate in parallel, due to the high overhead in commit checking and status maintenance, its latency is still 20% higher than DP-Raft.

5.2 Skewed Workload Results

Because DP-Raft's commit needs to obey the order of RAW dependencies, it is obvious that skewed data accesses will impact on the performance. So we vary the skewness θ of the Zipfian distribution to see how the performance changes. We experiment by using 256 clients and the same workload as in Sect. 5.1 under three replicas. Figure 8 gives the result that when $\theta < 0.6$, the performance is stable, but when $\theta > 0.7$, all three methods' throughput decrease sharply. This is because, with θ increasing, some records become "hot" and are frequently accessed, so the chance of a transaction conflicting with others is significantly increased. To be mentioned, the performance is influenced by both concurrency control mechanism and log replication, and because Raft does not check the conflict of log entries, so we display the normalized performance of DP-Raft and ParallelRaft against to Raft in Fig. 9 to see the influence of tracking dependency when commit. We can observe that before 0.8, DP-Raft achieves better performance compared to Raft, and when θ is 0.99, the throughput of DP-Raft is only half of Raft. The reason is that if θ is large, an entry is likely to have RAW dependencies with others which leads to a high probability of being blocked to wait for other entries to be committed. This is even worse for ParallelRaft

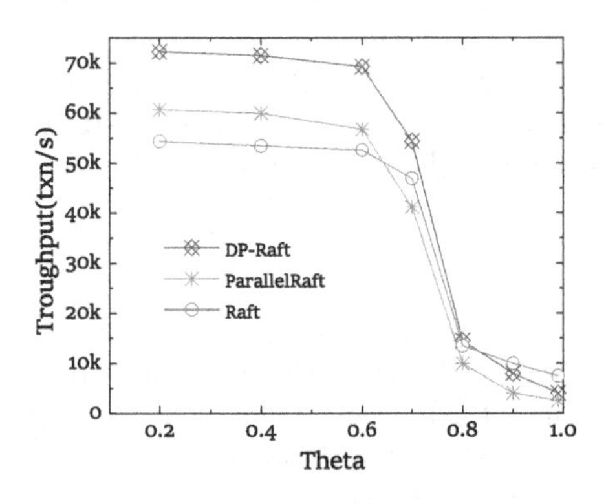

Fig. 8. Skewed workload.

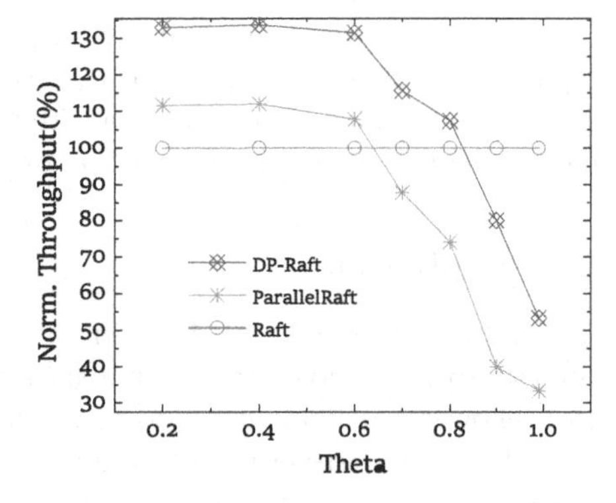

Fig. 9. Skewed workload (normalized).

because it should check if the read and write set are overlapping between entries, which means more kinds of dependencies should be checked, so it is more likely to be blocked. As a result, its throughput is only 30% of Raft.

5.3 The Number of Replicas

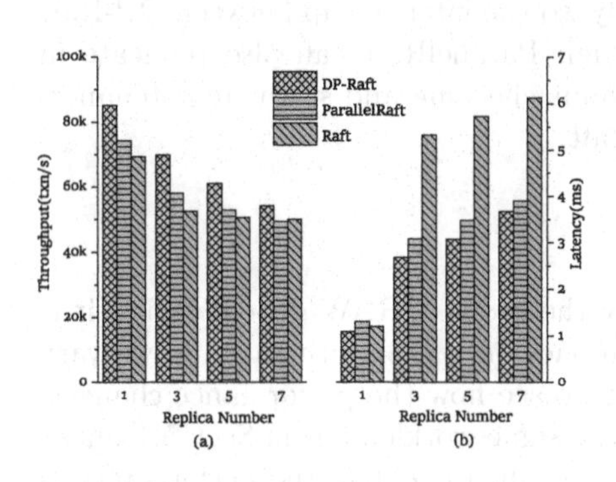

Fig. 10. Performance over replicas number.

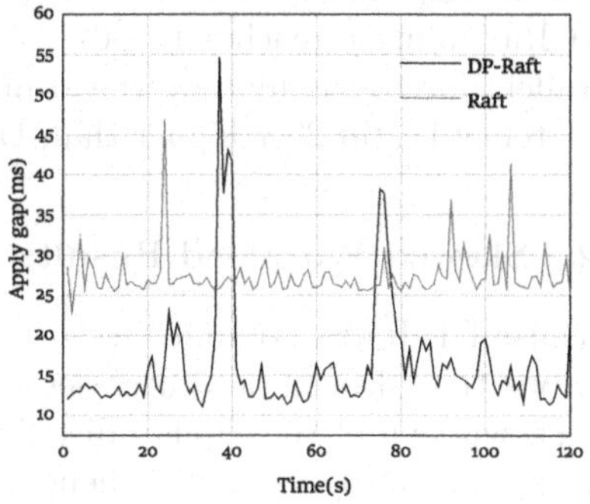

Fig. 11. Apply gap in 120 s.

We further explore the impact of the number of replicas on performance to verify the scalability of three methods. We measure the throughput and latency under different replica numbers with 256 clients and the θ is fixed to 0.5. Results in Fig. 10(a) show that the performance decreases significantly when the replica number changes from one to three. This is because transaction's commit under three-replica needs to go through the network, so the latency is obviously increased, as Fig. 10(b) shows. When the replica number changes to 5 and 7, Raft's performance remains stable, while DP-Raft and ParallelRaft have declined, but DP-Raft still outperforms Raft. The reason for the performance degradation is that with more replicas, DP-Raft and ParallelRaft need to replicate to more servers concurrently and process their results, so the efficiency of replication is reduced. In Fig. 10(b), we can see the latency of Raft is the highest. Therefore, with 3 or 5 replicas that are commonly adopted in practical, DP-Raft has the highest throughput and the lowest latency than ParallelRaft and Raft.

5.4 Apply Gap

Recall from Sect. 4.3, DP-Raft enables apply in parallel, so we use multiple threads to apply logs, which will narrow the gap between the state of leader and followers. We define t_l and t_f as the time of leader and follower apply the log entry of a transaction. Once a server applies an entry to its state machine, the results of the transaction are visible. So the gap between the same visible state of leader and follower is donated as $t_f - t_l$. We measure the apply gap in

120 s as shown in Fig. 11. The result shows that DP-Raft achieves a 2× lower apply gap than Raft, which means followers in DP-Raft can reach the up-to-date state in a short time after leader modified its state. We also notice that there are several jitters in DP-Raft's apply gap, by our analysis, it is because some hole entries on "hot" records block the later entries to apply. Once the hole entries are filled, the apply gap will return to a normal range. Therefore, apply in parallel can minimize the visible gap between leader and followers, which can benefit the read operations on followers to see the most up-to-date state as soon as possible.

5.5 Leader Election Time

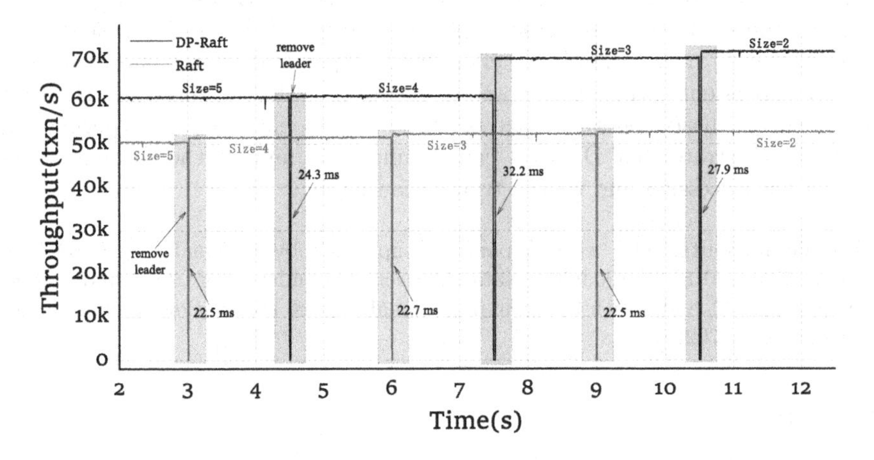

Fig. 12. Leader election in 12 s.

As mentioned in Sect. 4.4, to ensure the *Leader Completeness* property, pre-leader needs to merge followers' log and its own log to fill all the missing entries. An additional merge stage will increase the leader election time. To measure the time cost of the merge stage, we shut down leader three times in 12 s and record the time taken by each election, the result is shown in Fig. 12. Election time contains heartbeat timeout (20 ms by default) and the time to elect leader. There are five replicas initially. At 4.5 s, leader is shut down and trigger a leader election, the first election of DP-Raft takes 24.3 ms and merges 32 entries. Then leader is shut down again at 7.5 s, this time uses 32.2 ms to elect a new leader with 256 entries to be merged. And the third election at 10.5 s merges 128 entries and takes 27.9 ms. The reason for the difference in election time is that each election merges a different number of entries. Same as DP-Raft, we shut down leader three times for Raft at 3 s, 6 s, and 9 s. Each of the elections takes 22.5, 22.7 and 22.5 ms separately. Because Raft does not need to merge, so its election time is stable. We can observe even in the bad case as the second election of DP-Raft, it takes 10 ms more than Raft, which is acceptable because leader election is not frequent. What's more, we notice that for DP-Raft, when the size of consensus group is 5 and 4, the throughput is almost the same because both of them need to replicate to at least 3 servers. But when size is reduced to 3 at 7.5 s, the throughput rapidly increases because it only needs to replicate to 2 servers. As for Raft, the throughput remains stable regardless of the group size, which is the same as the result in Sect. 5.3.

6 Conclusion

In this paper, we target on how to break the strict serialization of Raft to replicate database logs concurrently. We propose a new variant of Raft protocol named DP-Raft for transactional database systems, which use log dependency to ensure the safety of out-of-order replication. In short, DP-Raft satisfies commit semantics of transaction by tracking RAW dependency and committing logs in RAW dependent entries' order, and to ensure the safety of the state, DP-Raft applies entries based on the order of WAW dependent entries. Experimental results demonstrate that DP-Raft can significantly improve the throughput of transaction processing and reduce transaction latency.

Acknowledgments. This work is partially supported by National Key R&D Program of China (2018YFB1003404), National Science Foundation of China under grant number 61672232, and Youth Program of National Science Foundation of China under grant number 61702189.

References

1. etcd. https://etcd.io/
2. Tidb. https://pingcap.com/
3. Arora, V., et al.: Leader or majority: Why have one when you can have both? improving read scalability in raft-like consensus protocols. In: HotCloud (2017)
4. Baker, J., Bond, C., Corbett, J.C., et al.: Megastore: providing scalable, highly available storage for interactive services. In: CIDR, pp. 223–234 (2011)
5. Cao, W., et al.: PolarFS: an ultra-low latency and failure resilient distributed file system for shared storage cloud database. PVLDB **11**(12), 1849–1862 (2018)
6. Cooper, B.F., Silberstein, A., Tam, E., et al.: Benchmarking cloud serving systems with YCSB. In: SoCC, pp. 143–154 (2010)
7. Corbett, J.C., Dean, J., Epstein, M., et al.: Spanner: Google's globally distributed database. ACM Trans. Comput. Syst. **31**(3), 8:1–8:22 (2013)
8. Guo, J., Chu, J., Cai, P., et al.: Low-overhead paxos replication. Data Sci. Eng. **2**(2), 169–177 (2017)
9. Hong, C., Zhou, D., Yang, M., et al.: KuaFu: closing the parallelism gap in database replication. In: ICDE, pp. 1186–1195 (2013)
10. Kemme, B., Alonso, G.: Don't be lazy, be consistent: Postgres-r, A new way to implement database replication. In: VLDB, pp. 134–143 (2000)
11. Lamport, L.: The part-time parliament. ACM Trans. Comput. Syst. **16**(2), 133–169 (1998)
12. Lamport, L., et al.: Paxos made simple. ACM SIGACT News **32**(4), 18–25 (2001)

13. Ongaro, D.: Consensus: Bridging theory and practice. Ph.D. thesis, Stanford University (2014)
14. Ongaro, D., Ousterhout, J.K.: In search of an understandable consensus algorithm. In: 2014 USENIX Annual Technical Conference, USENIX ATC, pp. 305–319 (2014)
15. Poke, M., Hoefler, T.: DARE: high-performance state machine replication on RDMA networks. In: HPDC, pp. 107–118 (2015)
16. Rao, J., Shekita, E.J., Tata, S.: Using paxos to build a scalable, consistent, and highly available datastore. PVLDB **4**(4), 243–254 (2011)
17. Stonebraker, M.: Concurrency control and consistency of multiple copies of data in distributed INGRES. IEEE Trans. Software Eng. **5**(3), 188–194 (1979)
18. Wiesmann, M., Schiper, A., Pedone, F., et al.: Database replication techniques: a three parameter classification. In: SRDS, pp. 206–215 (2000)
19. Zheng, J., Lin, Q., Xu, J., et al.: Paxosstore: high-availability storage made practical in WeChat. PVLDB **10**(12), 1730–1741 (2017)

Adaptive Method for Discovering Service Provider in Cloud Composite Services

Lei Yu[1(✉)] and Yifan Li[2]

[1] Department of Computer Science, Inner Mongolia University, Hohhot, China
yuleiimu@sohu.com
[2] Inner Mongolia Meteorological Bureau, Hohhot, China

Abstract. Application service providers implement Software-as-a-Service applications through a large number of Cloud Computing infrastructures. It is an increasingly challenging demand to discover trusted service providers based on services' outputs. However, the quality of service output may descend due to: (a) internal application logic of a Cloud service and (b) competition of resources in the sharing-based Cloud systems. Therefore, we propose an efficient method of trusted service provider discovery, called TSD (Trusted Service Discovery), to ensure that each service instance of composite services in Cloud systems is trustworthy. TSD treats all services as black boxes, and evaluates the outputs of service providers in service classes to obtain their equivalent or nonequivalent relationships. According to the equivalent or nonequivalent relationships, trusted service providers can be found easily. TSD improves accuracy of processing results as shown in experiments.

Keywords: Trusted service · Cloud composite service · Service discovery · Knowledge graph

1 Introduction

Cloud computing has become a cost-effective resource leasing model that eliminates the needs for users to maintain complex physical computing infrastructures on their own. Based on the concept of service-oriented architecture (SOA), cloud systems enable service providers to implement Software-as-a-Service (SaaS) applications through a large number of cloud computing infrastructures.

However, due to many factors, services as black boxes [1] can cause low trustiness. Taking an example in weather forecast domain, there are some global forecast centers, such as European Centre for Medium-Range Weather Forecasts, Moscow's meteorological center, Japan Meteorological Agency and National Meteorological Center of CMA. They use different forecast models to predict weather trend while they share the same historic weather information. Even though the format of service output is the same (REST and WSDL, Fahrenheit and centigrade, etc.), for the same query to weather forecast, some services may return inaccurate forecast data. Although the researchers have conducted extensive research on service trustiness issues, this trustiness problem in cloud composite services needs more investigation.

Y. Nah et al. (Eds.): DASFAA 2020, LNCS 12112, pp. 246–260, 2020.
https://doi.org/10.1007/978-3-030-59410-7_15

For cloud systems, traditional technology uses majority voting methods to detect untrusted service providers which will consume many resources. TSD (Trusted Service Providers Discovery) provides a robust and practical method for trusted service provider discovery. The traditional method based on majority voting assumes that trusted service providers in each service function account for the majority. However, in a large cloud system, multiple untrusted service providers may be the majority in certain service functions. To address this challenge, TSD comprehensively examines the equivalent relationships (the same outputs) for different SaaS providers across the cloud computing system, and effectively checks the equivalent relationships of each service class. These relationships can be obtained by analyzing two graphs: Regional Graph and Global Graph. The two graphs are constructed from service relations to find untrusted service providers. Therefore, TSD can find untrusted service providers and trusted service providers, even if the untrusted service providers are majority in certain service class. TSD not only finds untrusted services effectively, but also replaces the data generated by the untrusted service providers.

The contributions of this article are:

A robust and efficient service provider discovery method (TSD) in large-scale cloud computing infrastructures is proposed. The method can test service providers automatically.

Based on collected data, a knowledge graph is constructed for discovering an effective rule, which determines whether a service provider is trusted or untrusted. The knowledge graph can be expanded easily for more incoming service data, and our method is capable to process new data.

2 Related Works

Ontology Web Language for Service (OWL-S) is a semantic model in the present service discovery, and semantic elements are introduced into service discovery problem. Jiao et al. [2] put forward an OWL-S method from latent semantic analysis, and quality attributes of Web services are expressed in the tree-like structure. Ben Mahmoud et al. [3] define a learning semantic Web service for each learning object. This service is an extension of OWL-S that encompasses the description of the learning intention and the use of context that characterizes a learning object. They propose a discovery mechanism based on learning intention and context guided by the learner's intention and profile in order to offer a personalized learning path.

Semantic Web services discovery approach can mine the underlying semantic structures of interaction interface parameters to help users find and employ Web services, and can match interfaces with high precision when the parameters of those interfaces contain meaningful synonyms, abbreviations, and combinations of disordered fragments [4]. Zeshan et al. [5] present a service discovery framework for distributed embedded real-time system which uses context-aware ontology of embedded and real-time systems and a semantic matching algorithm to facilitate the discovery of device services. Chen [6] defines a semantic similarity measure combining functional similarity and process similarity, and presents a service discovery mechanism that utilizes semantic similarity measure for service matching. Ma et al. [7] proposes a business-rule-based service discovery approach based on a business rule annotation mechanism that includes condition rules, enumeration rules, and applied utility references. It retrieves suitable single services as well as service sets using the proposed two-point and incremental service query relaxation mechanism.

Social-based and graph-based service discovery are also investigated. Rodriguez [8] presents a theoretical analysis of graph-based service composition in terms of dependency with service discovery. Chen [9] propose linked social service-specific principles based on linked data principles for publishing services on the open Web as linked social services. They suggest a framework for constructing the social service network following linked social service-specific principles based on complex network theories. In some special domains, mobile social networks represent a convergence between mobile communications and service-oriented paradigms, which are supported by the large availability and heterogeneity of resources and services offered by recent mobile devices [10]. Liu [11] proposes a Socio-ecological Service Discovery model for advanced service discovery in Machine-to-Machine communication networks.

Hierarchical agglomerative clustering based approach is often used for service discovery. Two related models are proposed [12]: Output Similarity Model (OSM) and Total Similarity Model (TSM). OSM computes similarity between services using solely the outputs of services while clustering services. TSM computes similarity between services using both inputs and outputs of services. Chen [13] proposes a measure of semantic similarity, which enables more accurate service-request comparison by treating different conceptual relationships in ontologies such as is-a and has-a differently. Each service or request is represented by vectors of words that characterize both the interface signature and textual description. The overall semantic similarity is computed as a weighted aggregation of interface similarity and description similarity.

Machine learning is a promising area for service researchers to estimate equivalently. Machine learning system effectively "learns" how to estimate from training set of services [14]. Yin [15] proposes CloudScout, a non-intrusive approach that is capable of automatically discovering dependent service components. CloudScout analyzes the correlation among service components based on the time-series information from system monitoring logs. Cassar et al. [16] use probabilistic machine-learning techniques to extract latent factors from semantically enriched service descriptions. The latent factors are used to construct a model to represent different types of service descriptions in a vector form. With this transformation, heterogeneous service descriptions can be represented, discovered, and compared on the same homogeneous plane. Samper [17] describes a multi-agent platform for a traveler information system, allowing travelers to find the road traffic information web service that best fits their requirements. Theories and applications about data provenance in various domains are proposed for various data supply chains [18–24].

Above related works discuss service discovery in the perspective of semantics, social networks, machine learning and data provenance. However, they did not consider the influences from untrusted output values of services while the services meet other criteria, such as semantics, social networks and QoS, etc.

3 Trusted Cloud Service System

It is common that different service providers provide the same service function in large SaaS cloud. The service provider can create the same service instance to achieve fault tolerance and load balancing. In order to support the automatic combination of services, a portal can be deployed as a gateway for users to access cloud services. According to the user's requirements, the portal can aggregate different service instances to become a composite service. Unlike other distributed systems, SaaS cloud system has a unique set of features. In order to protect intellectual property, service providers often need to shield implementation details of software services. For privacy protection, only SaaS portal has detailed information about the service provider providing services through the SaaS cloud. Cloud users and other service providers have no detailed information about SaaS cloud. In addition, SaaS does not install hardware on the service provider, whether the hardware is from a cloud infrastructure provider or from a third-party service provider.

As a service-oriented architecture, SaaS allows service providers provide data stream processing service in cloud computing infrastructure, for example, Amazon and Google provide a set of enterprise applications for large data processing. Each service have one or more inputs. Each service provides specific data processing functions: filtering, sorting, association, and data mining. Figure 1 shows a running example of weather data stream processing service. A_1 is a collection of weather stations monitoring wind speed, humidity, and pressure, etc. A_2 is a collection of weather data processing sites analyzing weather treads. A_3 is a collection of multimedia studios producing weather forecasting services to public. All of them have API interfaces. In another example for weather forecast, each circle in Fig. 1 represents a data processing node (complex prediction function).

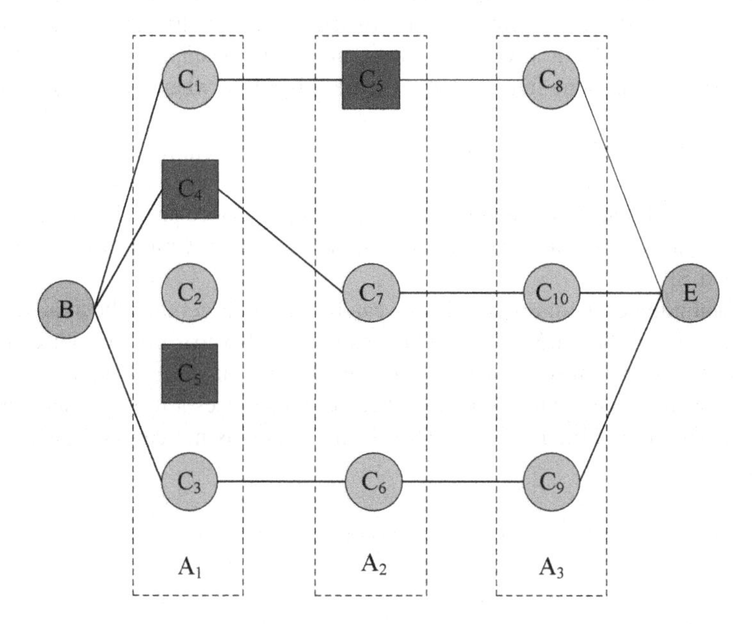

Fig. 1. A composite service for data stream processing

Following text descriptions are understandable enough for a reader to know the proposed method. Moreover, corresponding formalized definitions of text descriptions are also shown for reasoning:

Service Provider (SP): a company that provides software services to users, it is denoted by C in figures of this paper.

Service Instance (SI): SaaS software service that implements service function. A service provider in real world can provide multiple service instances with different functions.

Service Class (SC): a collection of service instances that implement the same service function. For instance, A_1 is a service class. Service classes are also known as service functions. A service provider can provide multiple service instances with different functions.

Composite Services (CS): a composite service formed by multiple service instances according to user needs. One composite service is a path which begins in node B and ends in node E in Fig. 1.

Trusted Service Provider (TSP): a set of service providers that produce the same output for an input in a service class. USP $\subset$ SP. It is represented by circular in figures. We assume the format of service output is the same. Similar to ensemble methods in machine learning, it is reasonable to believe that most services with the same output in a service class will be trusted by users in most cases. Furthermore, if a service provider is trusted in a service class, it has greater probability to be a trusted service provider in another service class. This phenomenon can be observed in many domains. Google has advantages in search engine, artificial intelligence, and other domains it involved.

Untrusted Service Provider (USP): a set of service providers that produce different output from trusted service providers for an input. USP $\subset$ SP. It is represented by square in figures. In our application domain, all services provided by untrusted service providers are untrusted. US means untrusted service instances. Likewise, all services provided by trusted service providers are trusted. TS means trusted service instances.

Regional Graph (RG): a RG shows equivalent relationships among service providers that provide the same functionality. Figure 2 is a RG, where each node represents a SP.

Global Graph (GG): a Global Graph shows nonequivalent relationships among the service providers in all functions. Figure 3 is a GG, where each node represents a SP.

Remaining Graph (ReG): a graph after removing a service provider and its adjacent service providers from the GG. Figure 4 is a Remaining Graph, where each node represents a SP.

The total number of untrusted services can be less or larger than the total number of trusted services in the SaaS system. Our solution is also feasible in both conditions. Based on the user-defined similarity function, the output data of services is deterministic. For the same input, we assume that trusted services always produce the same or similar output, and the nonequivalence of the results is not caused by hardware or software failures.

4 Mechanism and Algorithm

4.1 Trusted Service Discovery Scheme

To detect trusted services and find untrusted service providers, our algorithm needs to have an equivalent check of input data to obtain an equivalent or nonequivalent relationship among service providers. For example, to prove that the three service providers C_1, C_2 and C_3 provide the same output in function A_1, SaaS portal inputs the original data T to C_1, and it returns the result T_1. The portal then inputs copy of T (Tc) to C_2 and gets the result T_2. The portal finally compares T_1 and T_2 to see if C_1 and C_2 are equivalent. Each service provider shares a key and a sequence ID with the portal so that the portal uses different keys to encrypt the T to avoid message replay from faked providers.

If the two service providers have different processing results for the same input, at least one of them is untrusted. We do not send test data at the same time. On the contrary, when received the original data processing results, we resend to other service providers. In this way, the untrusted service will produce different results for the original data. However, this scheme is slow.

For all input data, if the two service providers make an equivalent output, then there is an equivalent relationship between them, otherwise there is a nonequivalent relationship between them. Two trusted service providers, however, may make similar but not exactly the same outputs. For instance, there may be small differences when getting the same city's weather forecast from different weather agencies. Therefore, users are allowed to set a distance function to determine the maximum acceptable difference.

A composite service consists of multiple service classes, each service class consists of a collection of service instances provided by service providers. For composite services, we propose a random probability test technique, i.e., random send test data. For an input data T, the portal performs a trustiness test with a certain probability. In Fig. 1, the portal first sends the data T to the predefined service paths C_1-C_5-C_8 according to the service functions A_1-A_2-A_3. After receiving the output of T, the portal resends T on another path C_3-C_6-C_9.

We can test the service providers with the same functionality, then obtain their equivalence. For the same input, when the output of the two services is the same, there is an equivalent link between the two service providers.

Then we build a RG for each function to get an equivalent relationship among the service providers that provide the same functionality. Figure 2 (assuming that the red service is untrusted) shows an equivalent relationship. If the relationship of two service providers is equivalent, there is a link between them. C_1 C_2 C_3 are equivalent, and C_4 C_5 are equivalent.

It is obvious to find out that the two service providers with red color in above figure are distinguished with others. They may be untrusted by intuition, and we just assume they are untrusted currently. However, only the RG is not enough, because it just groups SPs together and it is still hard to determine which groups of SP are untrusted. Therefore both the RG and the GG are used. The GG reflects the nonequivalence of service providers in an invocation chain. In Fig. 3, we use a GG to get all nonequivalence of service providers.

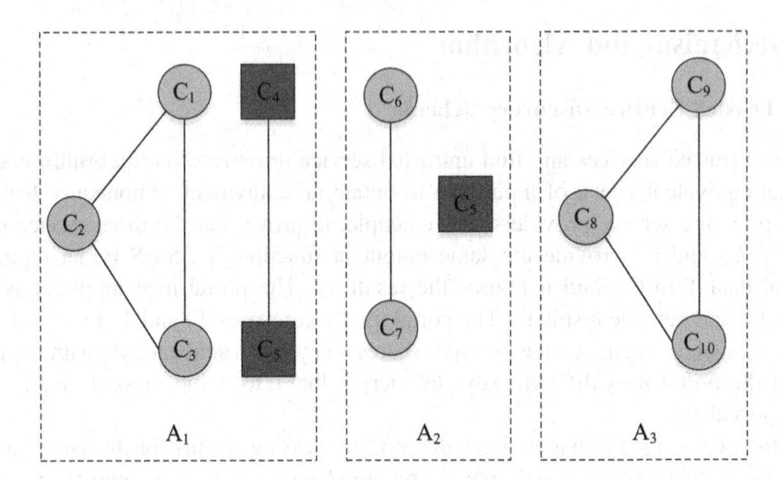

Fig. 2. RGs (SPs are divided by outputs to form groups)

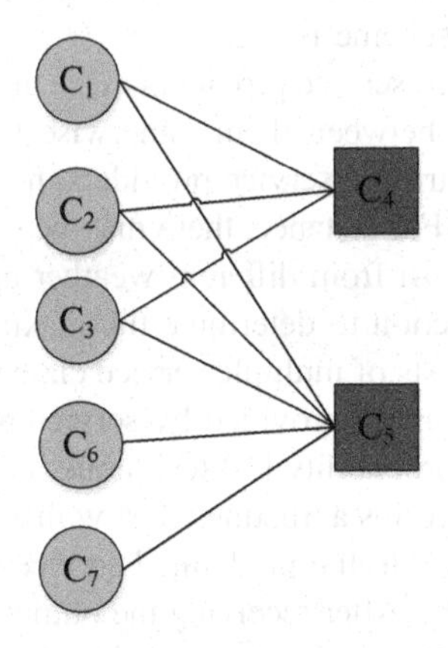

Fig. 3. GG (USP shows their unique character in GG)

For the same input, when the output of the two services is different, there is a nonequivalent link between the two services. As long as the two service providers are nonequivalent, they are linked in the GG. Therefore, we can get a more comprehensive relationship through the GG. Figure 3 shows a GG, its nodes include all service providers and its links are called nonequivalent links. Notice that service provider C_5 provides service function A_1 and A_2. The portal is responsible for building and maintaining the RG and the GG.

4.2 Trusted Service Discovery Method

The Trusted Service Discovery method is divided to following steps: Analysis of RG, Analysis of GG and Identified service check.

(1) Analysis of RG.

We first check the RGs to find any suspicious service provider. An equivalent link in the RG tells us which groups of service providers are equivalent with the specific service functions. For a service function, trusted service providers will be equivalent with each other and will form a group with equivalent links. For example, in Fig. 1, C_1, C_2, and C_3 are trusted service providers, and they form a group. We propose an algorithm to identify untrusted service providers and trusted service providers. In the same function, if the number of trusted service providers is greater than the number of untrusted service provider, then the number of trusted services will be greater than N/2, where N is the total number of service providers. Therefore, we can identify suspicious services by identifying groups with numbers fewer than N/2. For example, in Fig. 2, C_4 and C_5 are identified as suspicious nodes because their group is less than N/2. However, untrusted services producing the same output can be majority sometimes. Thus we need to integrate the RGs and the GG to achieve more robust results.

(2) Analysis of GG.

The GG contains a collection of trusted nodes and a collection of untrusted nodes. If the total number of untrusted service providers in the entire system does not exceed M (the number of untrusted service providers), we can find a part of the truly untrusted service providers. Through a nonequivalent link between the two service providers, we can deduce that at least one of them is untrusted because a pair of trusted service providers must be equivalent with each other. Therefore, by examining the minimum vertex coverage of the GG, we can get the minimum number of untrusted service providers. The smallest vertex covering of a graph is the smallest vertex set of the graph. For example, in Fig. 3, C_4 and C_5 form a minimum vertex coverage. A minimum vertex cover is a vertex cover with minimal cardinality, which is a NP problem, and it is solved by a greedy algorithm we provided in our website.

Given a GG, the number of untrusted service providers approximate the minimum vertex coverage of GG, because untrusted service providers have nonequivalent links with trusted service providers. However, the number of untrusted service providers is not equals to the minimum vertex coverage of GG in some cases.

According to statistical analysis, a fact is discovered: given M, when *Degrees* + *Coverage* > M, the service provider must be untrusted. *Degrees* is the number of degrees (neighbors) of the service providers in GG, and *Coverage* is the smallest vertex coverage of the remaining graph after removing the service provider and its neighbor service providers from GG. For instance, Fig. 4 shows the remaining graph after deleting service provider C_4.

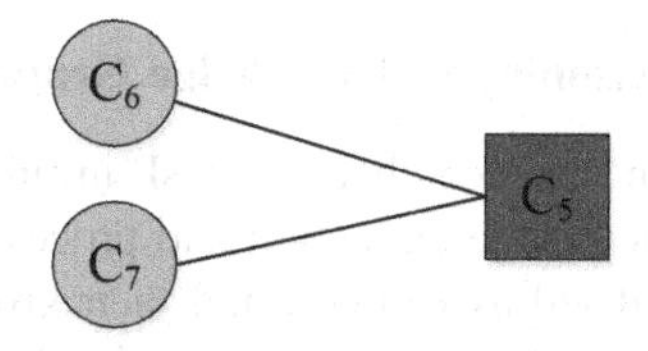

Fig. 4. ReG (USP will be separated according to ReG)

The expression above can be formalized as following axiom and will be further discussed in the experiment section:

$$\exists x\{\text{SP}(x) \wedge [\text{Degrees}(x) + \text{Coverage}(x) > \text{M}]\} \rightarrow \text{USP}(x) \quad (1)$$

Where SP (x) means $x \in$ SP, and USP (x) means $x \in$ USP. Axiom 1 determines which service provider is untrusted, while a trusted service provider can be discovered by reversing the operator:

$$\exists x\{\text{SP}(x) \wedge [\text{Degrees}(x) + \text{Coverage}(x) \leq \text{M}]\} \rightarrow \text{TSP}(x) \quad (2)$$

For instance, we suppose that the number of untrusted service providers does not exceed 2 in Fig. 3. At the beginning, it is not known that whether C_4 is trusted or untrusted. After removing C_4 and its neighbors C_1, C_2 and C_3 from the GG, the minimum vertex coverage is one. 3 + 1 > 2, thus C_4 is untrusted. If C_1 is checked at the beginning, it is not known that whether C_1 is trusted or untrusted. After removing C_1 and its neighbors C_4 and C_5, the remaining GG will become a graph without any connection, and its minimum vertex coverage is zero. Since C_1 has two adjacent services, $2 + 0 \leq 2$. Thus C_1 is a trusted service provider. If the trusted service provider and the untrusted service provider provide different functions, they will be separated in the GG.

(3) Identified service check

For each service class, separated connected sub-graphs are natural indicators of different kinds. For a service class *i*, Mi is a list of suspiciously untrusted services obtained by analyzing the RG, and V is a list of suspiciously untrusted services obtained by analyzing GG. Vi represents a subset of V for a service class *i*. If it is not empty while Vi intersects with Mi, the non-intersected services in Mi should be added to the identified untrusted service set.

5 Experiment

The real environment and simulation environment are used as the experimental basis, and a service generation simulation software was developed. Axiom verification is discussed and a knowledge graph is constructed for experiments. Through the experimental environment, we adjust the distance function, the number of trusted service providers, the number of untrusted service providers and other parameters. By calculating the success rate or identification rate, we verify the effectiveness of our system.

5.1 Construction and Reasoning of Knowledge Graphs

Axiom 1 is formalized as an instance checking task in this paper. Knowledge Graph [25–27] is a database for knowledge extraction and procession in our experiment. All service instances can be modeled as vertices, and their group, invocation, equivalent, nonequivalent relations can be modeled as different kinds of edges in the knowledge graph [28–31]. Neo4J is used for constructing the knowledge graph and reasoning for trusted service providers. Below is the knowledge graph for trusted service provider discovery, where many edges are not presented and group connect service providers for concise illustration (Fig. 5).

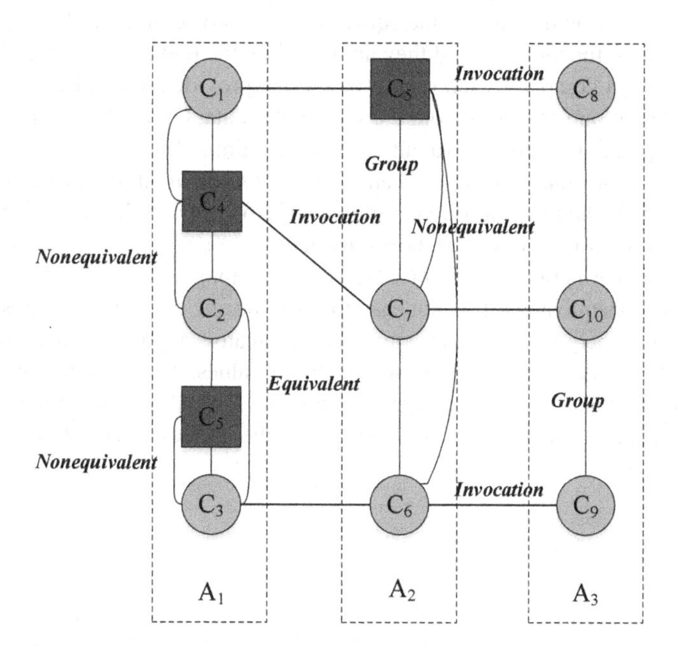

Fig. 5. Knowledge graph

Neo4J uses a language called ‘cypher’ to query and reasoning. Below is a cypher clause to count the number of *Degrees* of service provider ‘C2’ in axiom 1.

```
MATCH p = (Service_ Provider: SP) - [: Nonequivalent] - (: SP)
WHERE Service_ Provider. SP_ID = 'C2'
RETURN COUNT (p);
```

The value of count (p) should be stored as an attribute in the vertex of the service provider ‘C2’. Likewise, the smallest vertex coverage of the remaining graph after removing a service and its neighbor services from GG can be calculated in advance, and the calculated value is stored as an attribute in the vertex of the service provider ‘C2’. Once all *Degrees* and *Coverage* of service providers are computed, axiom 1 can be verified by below cypher clause.

```
MATCH (Service_ Provider: SP)
WHERE Service_ Provider. Degrees + Service_ Provider. Coverage > M
RETURN SP;
```

This cypher clause is a query that returns all untrusted service providers in the composite service.

5.2 Settings of Distance Function

The distance function determines the equivalence of two services. In Fig. 1, C_1 and C_3 provide the same function A_1, and they are tested by the portal. After receiving the test results, the portal compares each output of service providers C_1 and C_3. If C_1 and C_3 produce different output results from the same input data, C_1 and C_3 are nonequivalent, otherwise C_1 and C_3 are equivalent in this function. Assuming that the distance function is set as a value of no more than 20, for the same input, if C_1 output is 500 and C_3 output is 505, then C_1 and C_3 are equivalent. However, if C_1 output is 500 and C_3 output is 550, then C_1 and C_3 are nonequivalent.

If the distance between two services is large enough, the two services are considered to be nonequivalent. It is not easy to set a specific distance as a threshold value to distinguish nonequivalence and equivalence, because application domains are various which needs domain experts to set threshold values. To avoid differences among application domains, percentage representation is a suitable measurement. Figure 6 shows the effect of the output difference on the success rate of trusted service identification.

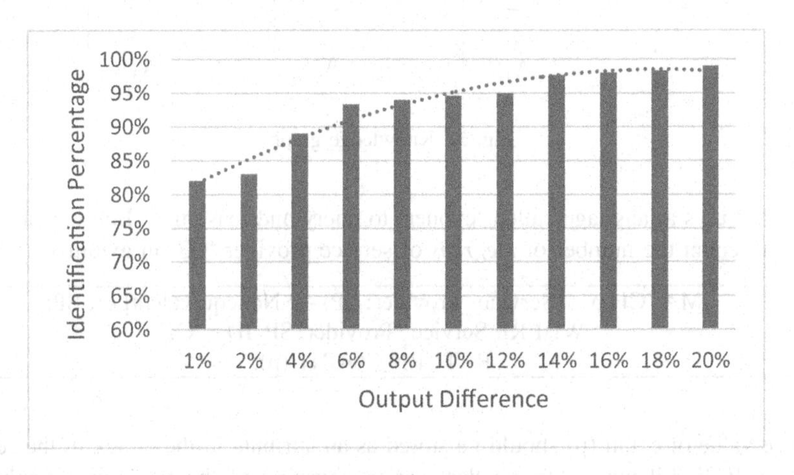

Fig. 6. Effect of output difference

The dotted line (trend line) in Fig. 6 shows that the success rate of trusted service identification increases with the increase of output difference. To a certain extent, strict difference settings (1% and 2%) also limits the success rate of trusted service identification, because some trusted service providers are mistakenly considered to be untrusted. However, slight more difference improves success rate. For instance, it is acceptable that weather forecasting services provide a small temperature difference.

5.3 Estimate the Number of Untrusted Service Providers

The settings of M affect the identification rate of trusted service, and their relationship should be discovered by further experiments. However, the third step of our approach tries to minimize the negative effect. For example, Fig. 2 and Fig. 3 show the RGs and the GG. If M is set to 4, the analysis of the GG will reveal the untrusted C_5, but will not find the untrusted C_4. The reason is that when M is 4, C_4 does not satisfy *Degrees* + *Coverage* > M (axiom 1), so C_4 will not be found. Since C_5 has 5 adjacent providers, the minimum vertex coverage of the remaining graph after removing C_5 and its five neighbors is zero. 5 + 0 > 4 satisfies axiom 1, so C_5 is untrusted. However, by checking the RG A_1, we find that $V_1 = \{C_5\}$ which overlaps with $M_1 = \{C_4, C_5\}$. According to the third step of our approach, C_4 is also untrusted. Finally C_4 and C_5 are discovered to be untrusted.

The settings of M affect the identification rate of trusted service. N represents the total number of providers in the cloud system. If we assume that M (the total number of untrusted providers) is lower than the total number of trusted providers, then the number of untrusted providers must not exceed N/2. The number of untrusted service providers must be no less than R (the minimum coverage of the GG). Therefore, the value of M is limited by its lower limit R and upper limit N/2.

Table 1. Identification rate

Number of N					
50		150		250	
M#	I.R.	M#	I.R.	M#	I.R.
4	81%	10	84%	20	89%
8	87%	30	91%	40	94%
15	94%	50	95%	70	95%
18	96%	60	95%	90	96%
20	95%	65	92%	100	95%
25	88%	75	90%	125	89%

Table 1 shows the effect of M on the Identification Rate (I.R.) of trusted services in different scenarios. In the experiment, we begin from the lower limit of M and an untrusted service set V, then we gradually increase M. For each value of M, we can get a set of untrusted services. In the case of a larger M, the number of services satisfying axiom 1 becomes smaller, which will make V decrease. When V is empty, we stop increasing M, because we no longer get more untrusted services by any larger M. Intuitively, when M becomes larger, fewer services will satisfy axiom 1. Therefore, we can only identify a small part of the untrusted services. On the contrary, when M becomes smaller, there will be more services that may satisfy axiom 1, which may erroneously treat the trusted services as an untrusted services. In order to avoid false positive, a large enough M must be chosen, so that our method can find a group of truly untrusted service providers.

5.4 Identification Rate vs. Network Size

A method of decentralized trustworthy service discovery (DTSD) is proposed by LI [32], and it is compared with our method w.r.t. identification rate and network size. Below is a figure showing that our method achieves better performance than DTSD. The dotted lines in the figure show linear prediction of identification rate, which indicates TSD has better scalability in large service networks (Fig. 7).

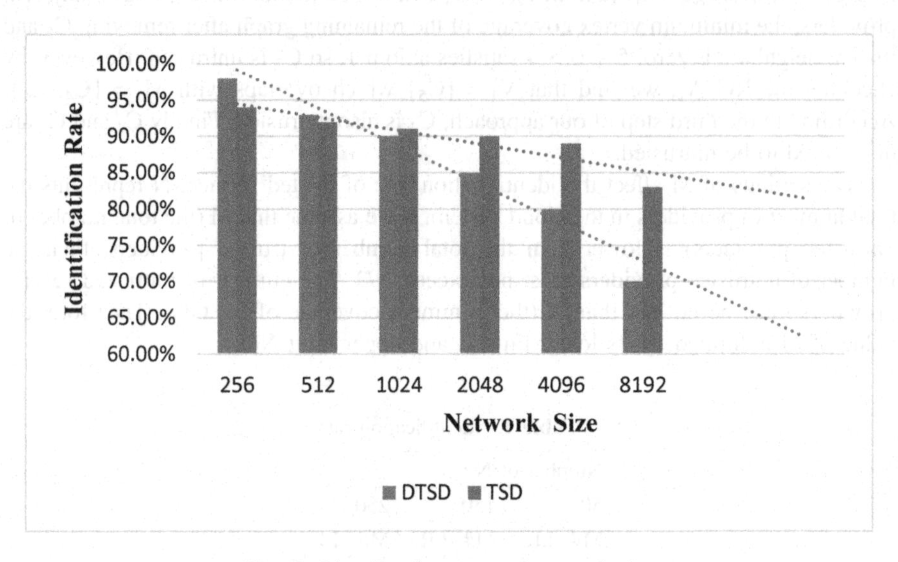

Fig. 7. Identification rate and network size

6 Conclusion

The goal of TSD is to help SaaS cloud systems discover trusted service providers. TSD treats all services as black boxes, and evaluates the outputs of service providers in service classes to obtain their equivalent or nonequivalent relationships. According to the equivalent or nonequivalent relationships, trusted service providers can be found easily. We test and validate outputs of service providers with a random probability at current implementation. In the future, we plan to concurrently test and validate outputs of services with the same functionalities for large cloud systems.

Acknowledgments. This work was supported by grants from NSFC under Grant (No. 619-62040).

References

1. Narock, T., Yoon, V., March, S.: A provenance-based approach to semantic web service description and discovery. Decis. Support Syst. **87**, 105–106 (2016)
2. Jiao, H., et al.: Research on cloud manufacturing service discovery based on latent semantic preference about OWL-S. Int. J. Comput. Integr. Manuf. **30**(4-5SI), 433–441 (2017)
3. Ben Mahmoud, C., et al.: Discovery mechanism for learning semantic web service. Int. J. Seman. Web inf. Syst. **12**(1), 23–43 (2016)
4. Cheng, B., et al.: A web services discovery approach based on mining underlying interface semantics. IEEE Trans. Knowl. Data Eng. **29**(5), 950–962 (2017)
5. Zeshan, F., et al.: Ontology-based service discovery framework for dynamic environments. IET Softw. **11**(2), 64–74 (2017)
6. Chen, F., et al.: Web service discovery among large service pools utilising semantic similarity and clustering. Enterp. Inf. Syst. **11**(3), 452–469 (2017)
7. Ma, S., et al.: QoS-aware query relaxation for service discovery with business rules. Future Gener. Comput. Syst. **60**, 1–12 (2016)
8. Rodriguez-Mier, P., et al.: An integrated semantic web service discovery and composition framework. IEEE Trans. Serv. Comput. **9**(4), 537–550 (2016)
9. Chen, W., Paik, I., Hung, P.C.K.: Constructing a global social service network for better quality of web service discovery. IEEE Trans. Serv. Comput. **8**(2), 284–298 (2015)
10. Girolami, M., Chessa, S., Caruso, A.: On service discovery in mobile social networks: survey and perspectives. Comput. Netw. **88**, 51–71 (2015)
11. Liu, L., et al.: A socioecological model for advanced service discovery in machine-to-machine communication networks. ACM Trans. Embed. Comput. Syst. **15**(382SI), 1–26 (2016)
12. Surianarayanan, C., Ganapathy, G.: An approach to computation of similarity, inter-cluster distance and selection of threshold for service discovery using clusters. IEEE Trans. Serv. Comput. **9**(4), 524–536 (2016)
13. Chen, F., et al.: A semantic similarity measure integrating multiple conceptual relationships for web service discovery. Expert Syst. Appl. **67**, 19–31 (2017)
14. Bhardwaj, K.C., Sharma, R.K.: Machine learning in efficient and effective web service discovery. J. Web Eng. **14**(3–4), 196–214 (2015)
15. Yin, J., et al.: CloudScout: a non-intrusive approach to service dependency discovery. IEEE Trans. Parallel Distrib. Syst. **28**(5), 1271–1284 (2017)
16. Cassar, G., Barnaghi, P., Moessner, K.: Probabilistic matchmaking methods for automated service discovery. IEEE Trans. Serv. Comput. **7**(4), 654–666 (2014)
17. Samper Zapater, J.J., et al.: Semantic web service discovery system for road traffic information services. Expert Syst. Appl. **42**(8), 3833–3842 (2015)
18. Neisse, R., Steri, G., Nai-Fovino, I.: A blockchain-based approach for data accountability and provenance tracking. In: ACM International Conference on Availability, Reliability and Security, p. 14 (2017)
19. Ramachandran, A., Kantarcioglu, M.: Using blockchain and smart contracts for secure data provenance management (2017)
20. Lu, Q., Xu, X.: Adaptable blockchain-based systems: a case study for product traceability. IEEE Softw. **34**(6), 21–27 (2017)
21. Li, P., Wu, T.Y., Li, X.M., et al.: Constructing data supply chain based on layered PROV. J. Supercomput. **73**(4), 1509–1531 (2016). https://doi.org/10.1007/s11227-016-1838-0
22. Curcin, V., Fairweather, E., Danger, R., et al.: Templates as a method for implementing data provenance in decision support systems. J. Biomed. Inf. **65**(C), 1–21 (2016)

23. Bart, A.C., Tibau, J., Tilevich, E., et al.: BlockPy: an open access data-science environment for introductory programmers. Computer **50**(5), 18–26 (2017)
24. Jiang, L., Yue, P., Kuhn, W., et al.: Advancing interoperability of geospatial data provenance on the web: gap analysis and strategies. Comput. Geosci. **117**, 21–31 (2018)
25. Bellomarini, L., Sallinger, E., Gottlob, G.: The vadalog system: datalog-based reasoning for knowledge graphs. In: 44th International Conference on Very Large Data Bases, vol. 11, no. 9, pp. 975–987 (2018)
26. Zhang, Q., Yao, Q.: Dynamic uncertain causality graph for knowledge representation and reasoning: utilization of statistical data and domain knowledge in complex cases. IEEE Trans. Neural Netw. Learn. Syst. **29**(5), 1637–1651 (2018)
27. Zhang, Y., Dai, H., Kozareva, Z., Smola, A.J., Song, L.: Variational reasoning for question answering with knowledge graph (2017)
28. Thost, V.: Attributed description logics: reasoning on knowledge graphs. In: Proceedings of the Twenty-Seventh International Joint Conference on Artificial Intelligence, pp. 5309–5313 (2018)
29. Wang, Z., Chen, T., Ren, J.S.J., Yu, W., Cheng, H., Lin, L.: Deep reasoning with knowledge graph for social relationship understanding. In: International Joint Conference on Artificial Intelligence, pp. 1021–1028 (2018)
30. Xiong, W., Hoang, T., Wang, W.Y.: DeepPath: a reinforcement learning method for knowledge graph reasoning. In: Conference on Empirical Methods in Natural Language Processing, pp. 564–573 (2017)
31. Trivedi, R., Dai, H., Wang, Y., Song, L.: Know-evolve: deep temporal reasoning for dynamic knowledge graphs. In: 34th International Conference on Machine Learning, pp. 3462–3471 (2017)
32. Li, J., Bai, Y., Zaman, N., et al.: A decentralized trustworthy context and QoS-aware service discovery framework for the internet of things. IEEE Access **5**(99), 19154–19166 (2017)

Auction-Based Order-Matching Mechanisms to Maximize Social Welfare in Real-Time Ride-Sharing

Bing Shi[1,2,3](✉), Yikai Luo[1], and Liquan Zhu[1]

[1] Wuhan University of Technology, Wuhan 430070, China
{bingshi,lyk,zlqlovecode}@whut.edu.cn
[2] Shenzhen Research Institute, Wuhan University of Technology, Shenzhen 518000, China
[3] State Key Laboratory for Novel Software Technology, Nanjing University, Nanjing 210023, China

Abstract. Ride-sharing has played an important role in reducing travel costs, traffic congestion and air pollution. However, existing works of order matching in the ride-sharing usually aim to minimize the total travel distances or maximize the profit of the platform running the ride-sharing service. Inefficient matching may result in loss for drivers, and they may not want to participate in the ride-sharing business. In this paper, we intend to solve the order matching issue by maximizing the social welfare of the platform and vehicle drivers. Specifically, we propose two truthful auction based order matching mechanisms, SWMOM-VCG and SWMOM-SASP, where vehicle drivers bid for the orders published by the platform to accomplish the orders and make profits. Compared with SWMOM-VCG, SWMOM-SASP can match a vehicle with multiple orders at each time slot and can do the order matching quicker with only a slight sacrifice of social welfare. We theoretically prove that both mechanisms satisfy the properties such as truthfulness, individual rationality, profitability and so on. We then evaluate the performance of both mechanisms in the real taxi order data in New York city and demonstrate that our mechanisms can achieve higher social welfare than the state-of-the-art approaches.

Keywords: Ride-sharing · Order matching · Mechanism design · Truthfulness

1 Introduction

Car-hailing service companies (e.g. Uber[1] and Didi Chuxing[2]) have promoted the ride-sharing business in recent years where a vehicle can accept a new riding order while serving the current riding orders. In such a system, the ride-sharing

[1] http://www.uber.com.
[2] http://www.didiglobal.com.

Y. Nah et al. (Eds.): DASFAA 2020, LNCS 12112, pp. 261–269, 2020.
https://doi.org/10.1007/978-3-030-59410-7_16

platform should be able to match the incoming orders with the available vehicles efficiently.

There exist various works including Crowdsourcing based solutions [1,7] were proposed to incentive vehicle drivers to "share rides" among passengers to utilize the empty seats in order to reduce social costs including travel costs, traffic congestion and global pollution. There also exists work discussing how to maintain the budget balance of the ride-sharing platform [10]. However, existing works usually try to maximize the profit of the ride-sharing platform, reduce the payments of passengers or minimize the traveling distance in the order matching, such as [8,11]. They ignore the fact that the vehicle drivers also try to make more profits. Actually, inappropriate order matching may result in loss for the drivers, and they may not have incentives to provide the riding service. In this paper, we propose auction based truthful mechanisms under which drivers bid for orders to maximize the social welfare of both the platform and the drivers, in order to ensure that both sides are willing to participate in this business.

In more detail, this paper advances the state of art in the following ways. Firstly we propose a VCG mechanism, SWMOM-VCG, to solve the order matching problem in real-time ride-sharing with the constraint that each vehicle can be matched with at most one order at each time slot. We further propose another mechanism, SWMOM-SASP, to overcome high computation time and one order constraint of SWMOM-VCG, which is based on a sequential second-price auction and can match the vehicle with multiple orders at each time slot. Both SWMOM-VCG and SWMOM-SASP can achieve four properties including truthfulness, individual rationality, profitability [12] and computational efficiency. Finally, we run experiments to evaluate the performance of our proposed mechanisms against the state-of-art approaches in terms of social welfare, social cost, ratio of served orders. The experiments show that our mechanisms can generate higher social welfare. We also show that the social welfare of SWMOM-SASP is just slightly lower than SWMOM-VCG, but can do the order matching quicker.

The structure of the paper is as follows. In Sect. 2, we give the problem formulation. In Sect. 3 and 4, we introduce the proposed mechanisms in detail, and in Sect. 5, we experimentally evaluate our proposed mechanisms. Finally, we conclude in Sect. 6.

2 Problem Formulation

The same as [11,12], we model the order matching process as an one-side reverse auction which runs in a set of time slot $\mathcal{T}$. Each time slot $t \in \mathcal{T}$ is a predefined time segment, e.g. one minute. At the beginning of time slot t, the platform collects the set of orders $\mathcal{O}_t = \{o_i\}$, including new incoming orders at this time slot and unserved orders at the previous time slot. In the below, we give the definition of an order.

Definition 1 (Order). *Order o_i can be represented by a tuple $\langle l_i^s, l_i^e, t_i^r, t_i^w, d_i, n_i, f_i \rangle$, where l_i^s and l_i^e represent the pick-up and drop-off locations of the order* *respectively,* t_i^r *and* t_i^w *represent the time when* o_i *is raised and the maximum waiting time respectively,* d_i *is the maximum detour ratio,* n_i *represents the number of passengers in* o_i *and* f_i *is the trip fare of the order.*

Now the platform will compute the trip fares of orders immediately, and passengers then decide whether accepting the fares or not. After collecting the prepaid orders, the platform will search the set of feasible vehicles $\mathcal{V}_t$ from available vehicles. We define the vehicle in the following.

Definition 2 (Vehicle). *Vehicle* v_j *can be represented by a tuple* $\langle l_j, s_j, c_j, n_j, N_j \rangle$*, where* l_j *is the current location of,* s_j *is the travel plan,* c_j *is the cost per kilometer of,* n_j *is the number of seats currently available, and* N_j *is the seat capacity.*

For each vehicle, its travel plan is a sequence comprised by both pick-up and drop-off locations of the uncompleted orders. In the following, sometimes we use $o_i \in s_j$ to represent $l_i^s, l_i^e \in s_j$ for convenience. For the vehicle, serving a new order may result in a detour for the orders in its original travel plan. We use $d(i,j)$ to represent the detour ratio of o_i in s_j and it is equal to $\frac{len_j(l_i^s,l_i^e)-len(l_i^s,l_i^e)}{len(l_i^s,l_i^e)}$, where $len(\cdot)$ is the shortest driving distance between two locations and $len_j(l_i^s, l_i^e)$ is the distance between l_i^s and l_i^e actually traveled in s_j. Based on the above definitions, we define **Feasible Travel Plan**.

Definition 3 (Feasible Travel Plan). *Given a travel plan* s_j *at time slot* t *and average speed* V_{avg} *of vehicles, we say* s_j *is feasible for* v_j *if the following conditions hold:*

1. $\sum_{o_i \in s_j} n_i \leq N_j$.
2. $\forall o_i \in s_j,\ d(i,j) \leq d_i$.
3. $\forall l_i^s \in s_j,\ V_{avg} \cdot (t_i^w + t_i^r - t) \geq len(l_j, l_i^s)$.

Condition *1* means that the number of passengers in the vehicle cannot exceed its' seat capacity at any time. Conditions *2* and *3* mean that the travel plan should not exceed order o_i's maximum detour ratio and maximum waiting time.

Definition 4 (Feasible Pair). *A feasible pair is a tuple* (o_i, v_j) *if the travel plan of* v_j *is still feasible after* o_i *inserted.*

After collecting orders from passengers, the platform calculates all feasible pairs at time slot t which constitutes the feasible pair set FP_t. Then each vehicle calculates the additional cost of serving each new order according to the feasible pair set and uses it as a bid submitted to the platform. Since route planning has been proved as an NP-Hard problem [11], in this paper, we assume that vehicles adopt the method in [3] to insert incoming order's pick-up, drop-off locations into the original travel plan. We use $cost(i, s_j)$ to denote the additional cost that v_j serves o_i, which is computed as follows:

$$cost(i, s_j) = c_j \cdot \Delta D(i, s_j) \tag{1}$$

where $\Delta D(i, s_j)$ represents the additional travel distance of v_j caused by serving o_i. We use $\widehat{cost}(i, s_j)$ to denote the additional cost which is bid by the driver. Note that $\widehat{cost}(i, s_j)$ may be not equal to $cost(i, s_j)$ if the driver bids untruthfully. All bids in time slot t constitute a set:

$$B_t = \{\widehat{cost}(i, s_j) | o_i \in O_t,\ v_j \in V_t,\ (o_i, v_j) \in FP_t\} \tag{2}$$

Finally, the platform acts as an auctioneer, which decides the matching according to the bids. We use $\sigma_t(\cdot)$ to represent the matching result. If $\sigma_t(j) = \{i\}$, it means that $\{o_i | i \in \sigma_t(j)\}$ is assigned to v_j at t. Here, we use $\mathcal{W}_t \subseteq \mathcal{V}_t$ to denote the set of winning vehicles which obtain the orders in the order matching. The platform also computes the payment received by the winning vehicle drivers who are matched with the orders, denoted by $\mathcal{P}_t = \{p_{j,i}\}$. Furthermore, we can compute the profit of vehicle v_j, u_j, and the profit of the platform, U_p, as follows:

$$u_j = \begin{cases} \sum_{i \in \sigma_t(j)} (p_{j,i} - cost(i, s_j)), & \text{if } v_j \in \mathcal{W}_t. \\ 0, & \text{otherwise.} \end{cases} \tag{3}$$

$$U_p = \sum_{v_j \in \mathcal{W}_t} \sum_{i \in \sigma_t(j)} (f_i - p_{j,i}) \tag{4}$$

Based on Eq. 3 and Eq. 4, the social welfare can be computed as follows:

$$SW_t = U_p + \sum_{v_j \in \mathcal{W}_t} u_j = \sum_{v_j \in \mathcal{W}_t} \sum_{i \in \sigma_t(j)} (f_i - cost(i, s_j)) \tag{5}$$

Now, we give the definition of the problem addressed in this paper.

Definition 5 (Social Welfare Maximization Order Matching). *Given the set of orders $\mathcal{O}_t$ and the set of vehicles $\mathcal{V}_t$ at time slot t, **Social Welfare Maximization Order Matching (SWMOM)** is to match orders with vehicles in order to maximize the social welfare (see Eq. 5) under the constraint of each vehicle's travel plan being **feasible** after order matching.*

3 SWMOM-VCG Mechanism

SWMOM-VCG consists of two parts: **Order Matching** and **Pricing**, which is described in Algorithm 1. Here, at each time slot, each vehicle can be only matched with at most one order, i.e. $|\sigma_t(j)| \le 1$. We model the order matching as a bipartite graph matching where the weight of each edge is the partial social welfare of each matching pair (o_i, v_j):

$$\widehat{sw}(i, j) = f_i - \widehat{cost}(i, s_j) \tag{6}$$

Then, we adopt the maximum weighted matching (MWM) algorithm [5] to maximize the social welfare of the platform and the vehicle drivers (line 3). We use SW^* to denote the optimal social welfare at time slot t. In **Pricing** we apply

Algorithm 1: SWMOM-VCG Mechanism

Input: $\mathcal{O}_t, \mathcal{V}_t, FP_t, B_t, t$
Output: $\mathcal{W}_t, \mathcal{P}_t, \mathcal{O}_t, \sigma_t$
1 Compute $\widehat{sw}(i,j)$ for each matching pair (o_i, v_j);
2 Use non-negative $\widehat{sw}(i,j)$ to create a bipartite graph G;
3 $\mathcal{W}_t, SW_t^*, \sigma_t = \mathbf{MWM}(G)$;
4 Set $\mathcal{P}_t = \emptyset$;
5 **for** $v_{j^*} \in \mathcal{W}_t$ **do**
6   Remove v_{j^*} and Convert G into $G_{-v_{j^*}}$;
7     $\mathcal{W}_{-v_{j^*},t}, SW^*_{-v_{j^*},t}, \sigma_{-v_{j^*},t} = \mathbf{MWM}(G_{-v_{j^*}})$;
8     $i^* = \sigma_t(j^*)$;
9     $p_{j^*,i^*} = \min(\widehat{cost}(i^*, s_{j^*}) + (SW_t^* - SW^*_{-v_{j^*},t}), f_{i^*})$;
10     $\mathcal{P}_t = \mathcal{P}_t \cup \{p_{j^*,i^*}\}$;
11 **end**
12 $\mathcal{O}_t = \mathcal{O}_t \setminus \{o_{\sigma_t(j^*)} | v_{j^*} \in \mathcal{W}_t\}$;
13 **Return:** $\mathcal{W}_t, \mathcal{P}_t, \mathcal{O}_t, \sigma_t$;

VCG[3] pricing method to compute the payment received by the winning vehicle driver. The payment received by the winning vehicle is:

$$p_{j^*,i^*} = \min(p^{vcg}_{j^*,i^*}, f_{i^*}) \tag{7}$$

where $p^{vcg}_{j^*,i^*} = \widehat{cost}(i^*, s_{j^*}) + (SW_t^* - SW^*_{-v_{j^*},t})$ is the VCG price. The driver will receive the payment of the minimal value between the VCG price and the upfront fare of the order (line 9). Finally, the platform takes the unmatched orders which do not exceed their maximum waiting time to the next time slot for the future matching (line 12).

3.1 Theoretical Analysis

In the following, we prove that SWMOM-VCG holds the properties of truthfulness, individual rationality, profitability, and computational efficiency. Note that truthfulness and individual rationality can be easily proved by VCG price. Also, probability can be guaranteed by Eq. 7.

Theorem 1. *SWMOM-VCG achieves computational efficiency.*

Proof. In the beginning of Algorithm 1, it takes $\mathcal{O}(nm)$ for creating a bipartite graph G, where $n = |\mathcal{O}_t|$ and $m = |\mathcal{V}_t|$. In the **Order Matching** stage, it takes $\mathcal{O}(\lambda^3)$ for maximum weight matching [5], where $\lambda = \max(n, m)$. In the **Pricing** stage, it takes $\mathcal{O}(\kappa\lambda^3)$ to compute the payments for all winning drivers, where $\kappa = \min(n, m)$. Therefore, SWMOM-VCG achieves computational efficiency.

[3] Vickrey-Clarke-Groves (VCG) [4,6,9] may be the most well-known efficient auction mechanism where the participants reveal their information truthfully.

4 SWMOM-SASP Mechanism

We propose an improved mechanism based on sequential second-price auction, SWMOM-SASP, to allow the vehicle to be matched with multiple orders at a time slot and to deal with larger number of vehicles and incoming orders while guaranteeing real-time performance. This mechanism is described in Algorithm 2, and consists of two parts: **Order Matching** and **Pricing**. In **Order matching**, the platform selects an order with the maximum fare and assigns it to the driver with the lowest bid. In **Pricing**, we adopt the second highest price as the payment to the winning driver (line 9). The platform also asks the winning driver to update his bid for other unmatched orders if the vehicle still has available seats (line 19).

4.1 Theoretical Analysis

In the following, we prove that SWMOM-SASP holds the properties of truthfulness, individual rationality, profitability and computational efficiency. Individual rationality can be easily proved by second highest price in line 9 of Algorithm 2, and profitability can be easily proved by line 10–13 of Algorithm 2.

Algorithm 2: SWMOM-SASP Mechanism

Input: $\mathcal{O}_t, \mathcal{V}_t, FP_t, B_t, t$
Output: $\mathcal{W}_t, \mathcal{P}_t, \mathcal{O}_t, \sigma_t$

1 **Set** $\mathcal{O}' = \emptyset$;
2 **while** $\mathcal{O}_t \neq \emptyset$ **do**
3 $\quad o_i = \arg\max_{o_{i^*} \in \mathcal{O}_t} f_i$;
4 $\quad$ **if** $\forall v_{j^*} \in \mathcal{V}_t (o_i, v_{j^*}) \notin FP_t$ **then**
5 $\quad\quad \mathcal{O}_t = \mathcal{O}_t \setminus \{o_i\}$, $\mathcal{O}' = \mathcal{O}' \cup \{o_i\}$;
6 $\quad\quad$ **continue**
7 $\quad$ **end**
8 $\quad v_j = \arg\min_{v_{j^*} \in \mathcal{V}_t, cost(i,s_{j^*}) \in B_t} cost(i, s_{j^*})$;
9 $\quad p_{j,i} = \arg\min_{v_{j^*} \in \mathcal{V}_t \setminus \{v_j\}, cost(i,s_{j^*}) \in B_t} cost(i, s_{j^*})$;
10 $\quad$ **if** $p_{j,i} > f_i$ **then**
11 $\quad\quad \mathcal{O}_t = \mathcal{O}_t \setminus \{o_i\}$, $\mathcal{O}' = \mathcal{O}' \cup \{o_i\}$;
12 $\quad\quad$ **continue**
13 $\quad$ **end**
14 $\quad \sigma_t(j) = \sigma_t(j) \cup \{i\}$;
15 $\quad$ **if** v_j *not in* $\mathcal{W}_t$ **then**
16 $\quad\quad \mathcal{W}_t = \mathcal{W}_t \cup \{v_j\}$
17 $\quad$ **end**
18 $\quad \mathcal{P}_t = \mathcal{P}_t \cup \{p_{j,i}\}$, $\mathcal{O}_t = \mathcal{O}_t \setminus \{o_i\}$;
19 $\quad$ Update $cost(i', j)$ with $i' \neq i, cost(i', s_j) \in B_t$;
20 **end**
21 $\mathcal{O}_t = \mathcal{O}_t \cup \mathcal{O}'$;
22 **Return:** $\mathcal{W}_t, \mathcal{P}_t, \mathcal{O}_t, \sigma_t$;

Theorem 2. *SWMOM-SASP achieves truthfulness.*

Proof. We will show that for any driver v_j, he cannot achieve a higher profits by bidding untruthfully. We prove it case by case.

Case 1: v_j always loses whether bidding truthfully nor not. From Eq. 3, $u_j = 0$ and v_j cannot improve u_j.

Case 2: v_j always wins whether bidding truthfully or not. From Algorithm (line 10), the payment is independent to the bid of v_j. So v_j cannot improve u_j.

Case 3: v_j wins by bidding untruthfully but loses bidding truthfully. When v_j bids untruthfully, we assume that o_{i^*} is assigned to v_j, instead of v_{j^*}. Since v_j, who bids truthfully, will lose in the auction, we have $\widehat{cost}(i^*, s_j) \leq cost(i^*, s_{j^*}) \leq cost(i^*, s_{j^*})$. We can get $u_j = p_{j,i^*} - cost(i^*, s_j) = cost(i^*, s_{j^*}) - cost(i^*, s_j) \leq 0$.

According to the above analysis, SWMOM-SASP achieves truthfulness.

Theorem 3. *SWMOM-SASP achieves computational efficiency.*

Proof. In Algorithm 2, it takes $\mathcal{O}(|\mathcal{O}_t|)$ in line 3, and takes $\mathcal{O}(|\mathcal{V}_t|)$ in each line of 8–9, and takes $\mathcal{O}(|\mathcal{O}_t|)$ in line 19, and the whole loop iterates $|\mathcal{O}_t|$ times. So the time complexity of SWMOM-SASP is $\mathcal{O}((2|\mathcal{O}_t| + 2|\mathcal{V}_t|) \cdot |\mathcal{O}_t|)$. Therefore SWMOM-SASP achieves computational efficiency.

5 Experimental Evaluation

In this section, we use the real taxi order data in New York city[4] to conduct experiments to evaluate the performance of our mechanisms. Initial parameters, such as maximum detour ratio, maximum waiting time, the number of vehicles, seat capacity, and average vehicle speed are shown in Table 1. We divide a day into 1440 time slots and one time slot is one minute. After the order matching, the driver with unaccomplished orders will go ahead according to the travel plan, while an idle vehicle will walk randomly to search for potential passengers. In the experiments, we evaluate our mechanisms against **Nearest-Matching** and **SPARP** [2] in terms of social welfare, social cost and ratio of served orders.

Table 1. Parameter Values

Parameter	Values
Maximum detour ratio (d_i)	0.25, 0.50, 0.75, 1.00
Maximum waiting time (t_i^w)	3, 4, 5, 6, 7, 8 (min)
Number of vehicles	500, 600, ..., 1900, 2000
Seat capacity (N_j)	2, 3, 4, 5, 6
Average vehicle speed (V_{avg})	12 (mph)

[4] https://www1.nyc.gov/site/tlc/about/tlc-trip-record-data.page.

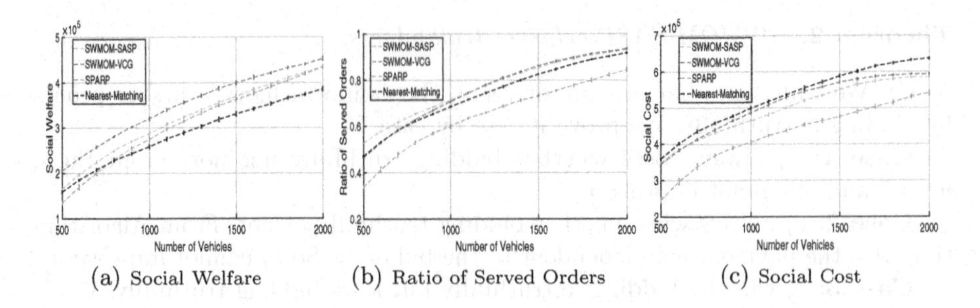

(a) Social Welfare (b) Ratio of Served Orders (c) Social Cost

Fig. 1. Results of one day

From Fig. 1(a), we find that as the number of vehicles increases, the social welfares of all approaches increase since the absolute amount of successfully served orders increases. We also find that both **SWMOM-VCG** and **SWMOM-SASP** have better social welfares than others. Furthermore, we can see in Fig. 1(b) that the ratio of served orders in **SWMOM-SASP** is higher than **SWMOM-VCG** when there are a small number of vehicles. However, as the number of vehicles entering the platform increases, the ratios of served orders in both **SWMOM-VCG** and **SWMOM-SASP** are approximately the same since the current feasible vehicle may satisfy current riding demands even though the vehicle is only matched with one order. Also, we can see that the social welfare in **SWMOM-SASP** is just slightly lower than that in **SWMOM-VCG**, but **SWMOM-SASP** takes a significantly less computation time than **SWMOM-VCG**. In Fig. 1(c), we can see that the social cost of **SPARP** is the lowest due to the fact that **SPARP** enumerates all possible route plans when inserting new orders, but it may adjust the original travel plan in this process.

6 Conclusion and Future Work

In this paper, we focus on maximizing social welfare in the ride-sharing system. We theoretically prove that our mechanisms hold several desirable properties. Experiments show that our mechanisms can achieve good performance in terms of social welfare. Furthermore, SWMOM-SASP can outperform SWMOM-VCG in terms of the computation time, with a slight sacrifice in the social welfare. In the future, we would like to extend our work to consider the dynamic changes of arriving orders in different time slots.

Acknowledgement. This paper was funded by the Humanity and Social Science Youth Research Foundation of Ministry of Education (Grant No. 19YJC790111), the Philosophy and Social Science Post-Foundation of Ministry of Education (Grant No. 18JHQ060), Shenzhen Basic Research Foundation (General Program, Grant No. JCYJ20190809175613332) and the Innovation Foundation for Industry, Education and Research in Universities of Science and Technology Development Center of Ministry of Education (Grant No. 2018A02030).

References

1. Asghari, M., Deng, D., Shahabi, C., Demiryurek, U., Li, Y.: Price-aware real-time ride-sharing at scale: an auction-based approach. In: Proceedings of the 24th ACM SIGSPATIAL International Conference on Advances in Geographic Information Systems, pp. 3:1–3:10 (2016)
2. Asghari, M., Shahabi, C.: An on-line truthful and individually rational pricing mechanism for ride-sharing. In: Proceedings of the 25th ACM SIGSPATIAL International Conference on Advances in Geographic Information Systems, pp. 7:1–7:10 (2017)
3. Cheng, P., Xin, H., Chen, L.: Utility-aware ridesharing on road networks. In: Proceedings of the 2017 ACM International Conference on Management of Data, pp. 1197–1210 (2017)
4. Clarke, E.H.: Multipart pricing of public goods. Public Choice **11**, 17–33 (1971). https://doi.org/10.1007/BF01726210
5. Galil, Z.: Efficient algorithms for finding maximum matching in graphs. ACM Comput. Surv. **18**(1), 23–38 (1986)
6. Groves, T.: Incentives in teams. Econometrica **41**(4), 617–631 (1973)
7. Huang, Y., Bastani, F., Jin, R., Wang, X.S.: Large scale real-time ridesharing with service guarantee on road networks. PVLDB **7**(14), 2017–2028 (2014)
8. Kleiner, A., Nebel, B., Ziparo, V.A.: A mechanism for dynamic ride sharing based on parallel auctions. In: Proceedings of the 22nd International Joint Conference on Artificial Intelligence, pp. 266–272 (2011)
9. Vickrey, W.: Counterspeculation, auctions, and competitive sealed tenders. J. Finance **16**(1), 8–37 (1961)
10. Zhao, D., Zhang, D., Gerding, E.H., Sakurai, Y., Yokoo, M.: Incentives in ridesharing with deficit control. In: International Conference on Autonomous Agents and Multi-Agent Systems, pp. 1021–1028 (2014)
11. Zheng, L., Chen, L., Ye, J.: Order dispatch in price-aware ridesharing. PVLDB **11**(8), 853–865 (2018)
12. Zheng, L., Cheng, P., Chen, L.: Auction-based order dispatch and pricing in ridesharing. In: 35th IEEE International Conference on Data Engineering, pp. 1034–1045 (2019)

GDS: General Distributed Strategy for Functional Dependency Discovery Algorithms

Peizhong Wu, Wei Yang(✉), Haichuan Wang, and Liusheng Huang

University of Science and Technology of China, Hefei, China
qubit@ustc.edu.cn

Abstract. Functional dependencies (FDs) are important metadata that describe relationships among columns of datasets and can be used in a number of tasks, such as schema normalization, data cleansing. In modern big data environments, data are partitioned, so that single-node FD discovery algorithms are inefficient without parallelization. However, existing parallel distributed algorithms bring huge communication costs and thus perform not well enough.

To solve this problem, we propose a general parallel discovery strategy, called GDS, to improve the performance of parallelization for single-node algorithms. GDS consists of two essential building blocks, namely FD-Combine algorithm and affine plane block design algorithm. The former can infer the final FDs from part-FD sets. The part-FD set is a FD set holding over part of the original dataset. The latter generates data blocks, making sure that part-FD sets of data blocks satisfy FD-Combine induction condition. With our strategy, each single-node FD discovery algorithm can be directly parallelized without modification in distributed environments. In the evaluation, with p threads, the speedups of FD discovery algorithm FastFDs exceed $\sqrt{p}$ in most cases and even exceed $p/2$ in some cases. In distributed environments, the best multi-threaded algorithm HYFD also gets a significant improvement with our strategy when the number of threads is large.

Keywords: FD discovery · Parallelization · FD-Combine · General distributed strategy

1 Introduction

Functional dependencies (FDs) are important metadata that describe relationships among columns of dataset, and many single-node algorithms have been proposed to do automatic FD discovery, e.g., TANE [1], FDep [2], Fun [4], Dep-Miner [3], FastFDs [5], Dfd [6], and HYFD [10]. However, although excellent techniques used in single-node algorithms make their performance close to the best in single-node environment, the exponential complexity of algorithms raise the need of parallelization and distribution.

Y. Nah et al. (Eds.): DASFAA 2020, LNCS 12112, pp. 270–278, 2020.
https://doi.org/10.1007/978-3-030-59410-7_17

A direct solution to do parallel and distributed discovery is: each compute node locally extracts FDs from local data, then takes some operations on local dependencies to get the final dependencies. But when the operation is simply intersection, it faces the problem that the FDs holding on the part of dataset may not still hold on all the dataset. For example, each part of dataset may have the FD $A \rightarrow B$ locally holding, but the $A \rightarrow B$ is not holding when record (a_1, b_1) in the first part of dataset and record (a_1, b_2) in second part of dataset.

In order to get the final FDs correctly with this solution, we need to solve two problems first.

One is how to infer the final FDs from part-FD sets of compute nodes. A part-FD set is FDs extracting by one compute node from local records it stored. For this problem, we construct B_L boolean algebra to formalize it and propose FD-Combine algorithm, which is developed from the deduction-induction techniques of data-driven algorithms [11] such as FDep, DepMiner. These deduction-induction techniques infer the final FDs from a set of non-FDs. Since FD-Combine is able to process non-FDs in arbitrary order. And the induction results of parts non-FDs are part-FD sets, therefore FD-Combine can infer the final FDs from part-FD sets.

The other is how to divide original dataset into multiple smaller datasets, making sure that part-FD sets from smaller datasets are sufficient to derive the final FDs. If data is stored partitioned without redundant records, some information between records will be lost. For this problem, we introduce affine plane block design to divide dataset. According to FD-Combine, any necessary non-FD should be contained in at least one smaller dataset. Affine plane block design divides original dataset into $k(k+1)$ smaller datasets, called *data blocks*, and ensures that every pair of records is contained in at least one data block, so that each non-FD is contained in at least one data block.

Combined with the above two building blocks, our general distributed strategy GDS allows us to parallelize the single-node algorithm directly, fully takes advantages of single-node algorithms, and reduce the communication costs in distributed environments. In summary, our contributions are as follows:

(1) **General Distributed Strategy.** We propose GDS, a general distributed strategy that is able to parallelize any single-node FD discovery algorithm without redesign. The strategy reduces the communication costs in distributed environment. With our strategy, the performances of multi-threaded HYFD have a significant improvement when the number of threads is large.
(2) **FD-Combine Algorithm and Block Design Algorithm.** To address the challenging problems in GDS, we put forward two building blocks. Proposed FD-Combine algorithm solves the how-to-infer problem, which can derive the final FDs from part-FD sets. Affine plane block design is introduced to solve the how-to-divide problem, which is used to divide dataset and ensures that each non-FD is contained in at least one data block.
(3) **Evaluation.** To evaluate the performance of GDS, we apply our strategy on single-node algorithms, include FDep, FastFDs, TANE, multi-threaded

HYFD. We conduct the experiments on real and synthetic big datasets with different block number strategies in distributed environments. Experimental results show that column-efficient algorithms perform better with GDS, and their speedups are usually nearly $\sqrt{p}$ or even exceed $\sqrt{p}$.

2 Related Works

There are some works on parallel and distributed dependency discovery, such as PARADE [7], DFDD [8], the analysis framework [11] with implementation [5,11] and multi-threaded HYFD [10].

PARADE [7] and DFDD [8] are parallelization of schema-driven algorithms. Each node will call part data or intermediate results frequently from other nodes during FD validation, which leads to huge communication cost.

In [11], Saxena et al. first aim at reducing intermediate results communication costs in distributed environments. However, in experiments of past works [10,11], single-node multi-threaded HYFD is shown to be most efficient algorithm, and outperforms other algorithms include distributed version of HYFD [11] with same number threads.

In GDS, there is only one round of communication on data blocks and part-FD sets without other data transferring. In our evaluation, the HYFD with GDS in distributed environments has a better performance than the original multi-threaded HYFD, especially when the number of threads is large.

3 Preliminaries

Let $R = \{A_1, A_2, ..., A_m\}$ be a database table schema and r be an instance of R. For $X \subseteq R$ and $t \in r$, $t[A]$ represents the projection of t to X.

Definition 1. ***Functional dependency:*** *A functional dependency (FD) written as $X \to A$ is a statement over a relational schema R, where $X \subset R$ and $A \in R$. The FD is satisfied by an instance r of R, iff for all pairs of tuples $t_1, t_2 \in r$, the following is true: if $t_1[X] = t_2[X]$ then $t_1[A] = t_2[A]$.*

Definition 2. ***Non-FD:*** *A non functional dependency(non-FD) written as $X \nrightarrow A$ is a statement over a relational schema R, where $X \subseteq R$ and $A \in R$. The non-FD is satisfied by an instance r of R, if exists a pair of tuples $t_1, t_2 \in r$, the following is true: $t_1[X] = t_2[X]$ and $t_1[A] \neq t_2[A]$.*

Definition 3. ***Maximal negative cover:*** *The negative cover NC of a dataset r is the set of all non-FDs that violates the r. The NC_{max} is maximal if it only contains maximal non-FDs. For an attribute A, the negative cover of A is written as NC(A), and NC(A) = {lhs(non-fd) | rhs(non-fd) = A and non-fd $\in$ NC}.*

Definition 4. ***Minimal positive cover:*** *The positive cover PC of a dataset r is the set of all valid FD satisfied by r. The PC_{min} is minimal if it only contains non-trivial minimal FDs. For an attribute A, the positive cover of A is written as PC(A), and* $PC(A) = \{lhs(fd) \mid rhs(fd) = A \text{ and } fd \in PC\}$.

Definition 5. ***Part-FD:*** *A part holding functional dependency(part-FD) written as $X \rightarrow A$ is a statement over a relational schema R, where $X \subseteq R$ and $A \in R$. $NC(A)$ is the negative of dataset r with attribute A. The part-FD is satisfied by dataset r if part-FD $\in INDU(NC'(A)) = PC'(A)$ where $NC'(A) \subset NC(A)$, INDU is a deduction-induction algorithm.*

4 GDS: FD-Combine and Affine Block Design

In this section, we introduce GDS and its building blocks: FD-Combine algorithm and affine plane block design. Section 4.1 constructs boolean algebra B_L to infer the FD-Combine algorithm. Section 4.2 introduces the affine plane block design algorithm. Section 4.3 introduces three phases strategy GDS and gives an example of GDS.

4.1 FD-Combine: A Distributed Deduction-Induction Algorithm

Boolean algebra system B_L is used to express positive cover and negative cover of an attribute, the elements of B_L are sets of attribute combinations. $[X]_S$ and $[X]_G$ are basic terms in B_L, which are used to express positive cover and negative cover. We define B_L and $[X]_S, [X]_G$ as follow.

Definition 6. ***B_L algebra system:*** *Let $R = \{A_1, A_2, ..., A_m\}$ be a database schema, for arbitrary $A \in R$, the LHS of a FD is expressed as a attribute set. Each possible valid LHS of attribute A is an element in set LS where $LS = P(L)$ and $L = R \setminus \{A\}$. The B_L is the algebra system defined as $<P(LS),\cup,\cap,-,\emptyset,LS>$ or $<P(LS),+,\cdot,-,\emptyset,LS>$. $\cup$ or $+$ is set union operation, $\cap$ or $\cdot$ is set intersection operation, - is complementation operation. LS containing all LHSs is maximal element in P(LS), and $\emptyset$ is minimal element. P(LS) is the power set of LS.*

Property 1. B_L is a boolean algebra, idempotent laws, commutative laws, De Morgan's laws and other boolean algebra properties are satisfied in B_L.

Definition 7. ***Basic term $[X]_S$ and $[X]_G$:*** *In boolean algebra B_L, we define elements $[X]_S$ and $[X]_G$ in B_L. For attribute combination $X \in LS$, $[X]_S = \{Y|Y \in LS \text{ and } X \subseteq Y\}$, $[X]_G = \{Y|Y \in LS \text{ and } Y \subseteq X\}$.*

Theorem 1. ***Additional Simplification Rules:*** *$[X]_S, [X]_G$ are defined as Definition 7, then they satisfy the following properties:*

(1) Rule 1*: For arbitrary attribute combination sets $X_1, X_2 \in LS$,*

$$[X_1]_G \cap [X_2]_G = [X_1 \cap X_2]_G, [X_1]_S \cap [X_2]_S = [X_1 \cup X_2]_S. \quad (1)$$

(2) Rule 2: *For $L \in LS$, and $\emptyset \in LS$,*

$$\overline{[\emptyset]_G} = \bigcup_{C_i \in L} [C_i]_S, \overline{[L]_G} = \emptyset, \overline{[\emptyset]_S} = \emptyset, \overline{[L]_S} = \bigcup_{C_i \in L} [L \setminus C_i]_G. \quad (2)$$

(3) Rule 3: *For attribute combination set $X \in LS$, $X \neq \emptyset$ and $X \neq L$:*

$$\overline{[X]_G} = \bigcup_{C_i \in (L \setminus X)} [C_i]_S, \overline{[X]_S} = \bigcup_{C_i \in X} [L \setminus \{C_i\}]_G. \quad (3)$$

Example 1. When $L = \{B, C, D\}$, $\overline{[BC]_S} = [BD]_G \cup [CD]_G$, $\overline{[BD]_G \cup [CD]_G} = \overline{[BD]_G} \bigcap \overline{[CD]_G} = [C]_S \cap [B]_S = [BC]_S$.

Theorem 2. ***Expressions of PC(A) and NC(A):*** *Let $PC(A)$ be the positive cover of attribute A, and $NC(A)$ be the negative cover of attribute A, then the following corollaries hold:*

(1) According to Definition 7, $PC(A)$ and $NC(A)$ can be express as elements in B_L. $PC(A) = \overline{NC(A)}$ and

$$PC(A) = \bigcup_{X \to A \text{ is FD}} [X]_S, NC(A) = \bigcup_{X \nrightarrow A \text{ is non-FD}} [X]_G. \quad (4)$$

(2) $PC_{min}(A)$ contains the minimum number of basic terms. Let $PC_{sim}(A)$ be the simplest form of any $PC(A)$ expression, then $PC_{min}(A) = PC_{sim}(A) = Simplify(\overline{NC(A)})$.

Theorem 3. ***FD-Combine:*** *In a dataset r, for attribute A, $NC(A) = \cup_{i=1}^{n}[X_i]_G$. Let $r_1, r_2, ..., r_p$ be samples of r. Let $NC_i(A)$ denote negative cover of attribute A on sample r_i. Let $PC_i(A)$ denote positive cover of attribute A on sample r_i, the $PC_i(A)$ is also part-FD set of dataset r. If $\cup_{i=1}^{p} NC_i(A) = NC(A)$, then $PC_{min}(A) = Simplify(\overline{NC(A)}) = Simplify(\overline{\cup_{j=1}^{p} NC_j(A)}) = Simplify(\cap_{j=1}^{p}(PC_j(A))$.*

4.2 Affine Plane Block Design

According to the FD-Combine, each non-FD should be contained into at least one block after dataset partition, therefore, every pair of records should be contained into at least one block. We introduce 2-$(m^2, m, 1)$ affine plane block design in combinatorics. This design generates $m(m+1)$ blocks, block size is m. Each pair of m^2 elements is contained in one block. We divide dataset into m^2 parts records to fit this block design.

In Algorithm 1, the m can be determined by the number of compute nodes. Function *datasplit* in line 3 could be any function splitting records into m^2 parts.

Algorithm 1. Affine plane block design

Require: $reocrds, m$
Ensure: $dataPoints, blocksIndex$
1: $dataPoints \leftarrow$ DATASPLIT$(records, m^2)$
2: Array $blocksIndex$ size $m(m+1)$
3: **for** $p = 0 \rightarrow m-1, q = 0 \rightarrow m-1$ **do**
4: $\quad blocksIndex[p * m + q] \leftarrow \{(i + mj) \bmod m \mid (i, j) \text{ in } y = px + q \bmod m\}$
5: **end for**
6: **for** $p = 0 \rightarrow m-1$ **do**
7: $\quad blocksIndex[m^2 + p] \leftarrow \{(i + mj) \bmod m \mid (i, j) \text{ in } x = p \bmod m\}$
8: **end for**
9: **return** $dataPoints, blocksIndex$

4.3 General Distributed Strategy

GDS consists of FD-Combine and affine plane block design. There are three phases in our strategy: In phase 1, the input dataset is divided into b data blocks using affine plane design algorithm. In phase 2, p compute nodes extract part-FD sets from b data blocks. In phase 3, the final FDs are derived from the part-FD sets by FD-Combine.

In the example, records are divided into 6 blocks (Fig. 1).

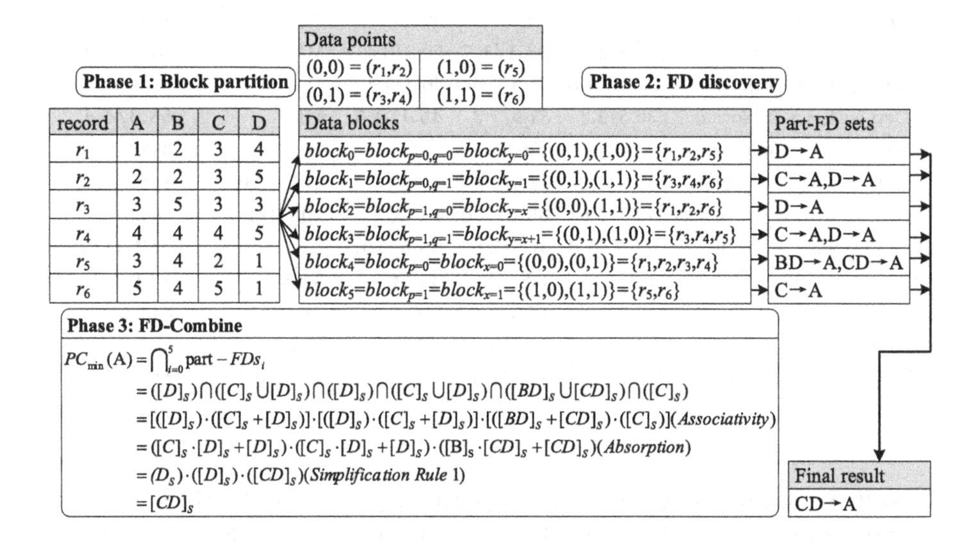

Fig. 1. *An example of GDS data flow*

5 Evaluation

Experimental Setup: We run all our experiments on a server with 56 Intel Xeon E5-2683 2.00 GHz CPUs with 62 GB RAM. The server runs on CentOS 7.6 and uses OpenJDK 64-Bit as Java environment. All single-node algorithms we used are from *Metanome* data profiling framework. *Spark-2.4.0* is used to manage threads and HDFS of *Hadoop-2.9.1* is used to store original dataset, data points and part-FD sets.

Evaluation with Smaller Datasets: For datasets in [9], Table 1 lists runtimes/speedups of FDep, FastFDs, and TANE with different parameter strategies. Algorithm with $m(m+1)$ blocks strategy is executed by $m(m+1)$ threads. TL indicates that runtime exceeds 2 h, and ML indicates memory usage exceeds 62 GB. Runtime unit is second.

Table 1. Runtimes in seconds and speedups for different datasets

Items	Chess	Nursery	Letter	Fd-reduced-30	Adult
Cols[#]	7	9	17	30	15
Rows[#]	28,056	12,960	20,000	250,000	48,842
Size[KB]	519	1,024	695	69,581	3,528
FDs[#]	1	1	61	89,571	78
FDep in single node	119.7	40.2	202.5	TL	298.7
FDep with 2×3 blocks	37.8/3.2	**23.1/1.7**	59.9/3.4	ML	133.5/2.2
FDep with 3×4 blocks	**33.8/3.5**	25.5/1.6	46.4/4.4	ML	81.9/3.6
FDep with 4×5 blocks	36.5/3.3	31.9/1.3	**45.1/4.5**	ML	**68.3/4.4**
FastFDs in single node	243.3	106.8	951.2	241.2	3830.4
FastFDs with 2×3 blocks	63.0/3.9	44.0/2.4	191.2/5.0	117.8/2.0	1119.6/3.4
FastFDs with 3×4 blocks	43.3/5.6	**35.0/3.0**	107.3/8.9	**109.9/2.2**	568.0/6.7
FastFDs with 4×5 blocks	**41.7/5.8**	37.7/2.8	**88.5/10.7**	130.7/1.8	**312.4/12.2**
TANE in single node	**1.7**	**1.8**	**349.6**	**34.8**	**77.9**
TANE with 2×3 blocks	18.7/0.08	17.6/0.1	ML	74.4/0.5	84.9/0.9
TANE with 3×4 blocks	23.2/0.07	24.4/0.07	ML	91.2/0.4	86.0/0.9
TANE with 4×5 blocks	32.7/0.05	32.8/0.05	ML	120.1/0.3	ML

Result Summary: Column-efficient algorithms like FDep, FastFDs will have a better performance with our strategy, and the speedups are usually nearly $\sqrt{p}$ or even exceed $\sqrt{p}$ closing to p. Because column-efficient algorithms are sensitive to the number of records. In contrast, row-efficient algorithm may benefit less on runtime from our strategy, since smaller block size doesn't reduce runtime.

Evaluation with Larger Datasets and Multi-threaded HYFD: We evaluate p-threaded HYFD and $\frac{p}{6}$-threaded HYFD with 2×3 blocks. Since memory is not enough, we evaluate the $\frac{p}{6}$-threaded HYFD with 6 block in a simulated way: For p threads, phase 1 and phase 3 are executed normally. In phase 2, we compute each block one by one with $\frac{p}{6}$ threads to simulate executing process of p threads on 6 blocks, then take the maximal runtime each block as runtime of phase 2.

Dataset *lineitem* [10] is 16 columns, 6 million rows with 3984 FDs. The size is 1051 MB. Dataset *ATOM_SITE_1350000r_31c* [10] is 31 columns,1.35 million rows with 7263 FDs. The size is 181 MB. Figure 2 and Fig. 3 shows the runtimes and speedups.

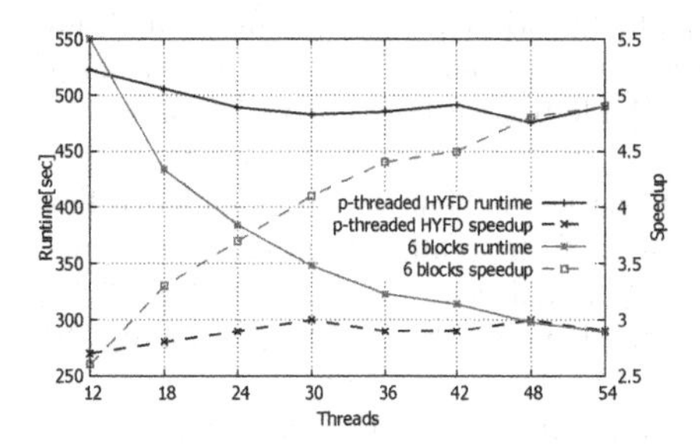

Fig. 2. *lineitem*

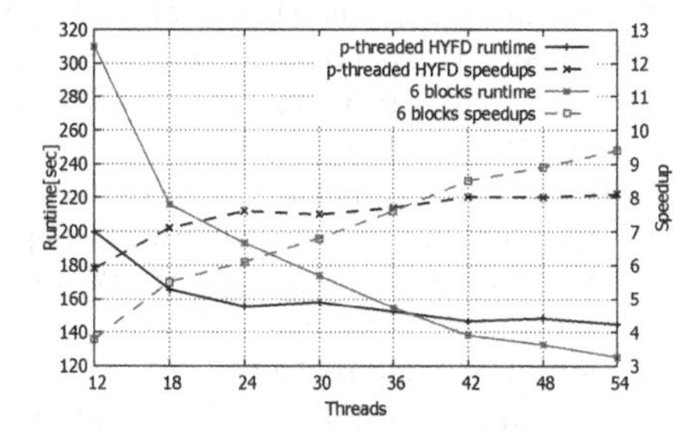

Fig. 3. *ATOM_SITE_1350000r_31c*

Result Summary: The performance of multi-threaded HYFD does not improve with the increasing the number of threads p, when p is large enough. But with our strategy, the performance could be improved again. Especially for dataset *lineitem* with more records, the improvement is significant.

6 Conclusion

In this paper, we proposed GDS, a strategy that parallelizes single-node FD discovery algorithms without any modification in distributed environments. In order to avoid large communication costs of intermediate results, GDS uses FD-Combine and affine plane block design, so that part-FD sets computed locally could be combined to get the final FDs holding over the whole dataset. In the experiments, column-efficient algorithms have a better performance with our strategy. Especially, the speedups of FastFDs are close to upper bound p with p threads in some cases. In most cases of past works, multi-threaded HYFD gets best results with the same threads. With our strategy, the multi-threaded HYFD gets a better performance when the number of threads is large.

Acknowledgments. This work was supported by the Anhui Initiative in Quantum Information Technologies (No. AHY150300).

References

1. Huhtala, Y., et al.: TANE: an efficient algorithm for discovering functional and approximate dependencies. Comput. J. **42**(2), 100–111 (1999)
2. Flach, P.A., Savnik, I.: Database dependency discovery: a machine learning approach. Ai Commun. **12**(3), 139–160 (1999)
3. Lopes, S., Petit, J.-M., Lakhal, L.: Efficient discovery of functional dependencies and armstrong relations. In: Zaniolo, C., Lockemann, P.C., Scholl, M.H., Grust, T. (eds.) EDBT 2000. LNCS, vol. 1777, pp. 350–364. Springer, Heidelberg (2000). https://doi.org/10.1007/3-540-46439-5_24
4. Novelli, N., Cicchetti, R.: FUN: an efficient algorithm for mining functional and embedded dependencies. In: Van den Bussche, J., Vianu, V. (eds.) ICDT 2001. LNCS, vol. 1973, pp. 189–203. Springer, Heidelberg (2001). https://doi.org/10.1007/3-540-44503-X_13
5. Wyss, C., Giannella, C., Robertson, E.: FastFDs: a heuristic-driven, depth-first algorithm for mining functional dependencies from relation instances extended abstract. In: Kambayashi, Y., Winiwarter, W., Arikawa, M. (eds.) DaWaK 2001. LNCS, vol. 2114, pp. 101–110. Springer, Heidelberg (2001). https://doi.org/10.1007/3-540-44801-2_11
6. Abedjan, Z., Schulze, P., Naumann, F., et al.: DFD: efficient functional dependency discovery. In: CIKM (2014)
7. Garnaud, E., et al.: Parallel mining of dependencies. In: HPCS (2014)
8. Li, W., Li, Z., Chen, Q., et al.: Discovering functional dependencies in vertically distributed big data. In: WISE (2015)
9. Papenbrock, T., et al.: Functional dependency discovery: an experimental evaluation of seven algorithms. In: VLDB (2015)
10. Papenbrock, T., Naumann, F.: A hybrid approach to functional dependency discovery. In: ICMD (2016)
11. Saxena, H., Golab, L., Ilyas, I.F., et al.: Distributed implementations of dependency discovery algorithms. In: Very Large Data Bases, vol. 12, no. 11, pp. 1624–1636 (2019)

Efficient Group Processing for Multiple Reverse Top-k Geo-Social Keyword Queries

Pengfei Jin[1], Yunjun Gao[1(✉)], Lu Chen[2], and Jingwen Zhao[3]

[1] College of Computer Science, Zhejiang University, Hangzhou, China
{jinpf,gaoyj}@zju.edu.cn
[2] Department of Computer Science, Aalborg University, Aalborg, Denmark
luchen@cs.aau.dk
[3] AIPD, Tencent, Shenzhen, China
jingwenzhao@tencent.com

Abstract. A Reverse Top-k Geo-Social Keyword Query (RkGSKQ) aims to find all the users who have a given geo-social object in their top-k geo-social keyword query results. This query is practical in detecting prospective customers for online business in social networks. Existing work on RkGSKQ only explored efficient approaches in answering a single query per time, which could not be efficient in processing multiple queries in a query batch. In many real-life applications, multiple RkGSKQs for multiple query objects can be issued at the same time. To this end, in this paper, we focus on the efficient batch processing algorithm for multiple RkGSKQs. To reduce the overall cost and find concurrently results of multiple queries, we present a group processing framework based on the current state-of-the-art indexing and group pruning strategies to answer multiple RkGSKQs by sharing common CPU and I/O costs. Extensive experiments on three data sets demonstrate the effectiveness and efficiency of our proposed methods.

Keywords: Reverse Top-k Geo-Social Keyword Query · Social network · Batch processing · Algorithm

1 Introduction

Volumes of location-based social network (LBSN) data trigger extensive studies on Geo-Social Keyword Queries (GSKQ) [1,13,14]. Among various GSKQ query types, the Reverse Top-k Geo-Social Keyword Query (RkGSKQ) is useful in detecting potential customers in LBSNs. Existing work on RkGSKQ only studied how to process a single query per time, which is not effective in processing multiple RkGSKQs at the same time. In some real-life applications, performing RkGSKQs for a group of objects simultaneously is needed. For example, managers of companies like KFC and Starbucks would prefer to require RkGSKQ results of all chain stores to better understand the overall market trend. Toward this, this paper studies the problem of Batch RkGSKQ (BRkGSKQ) to achieve more efficient multiple RkGSKQs retrieving.

To solve this problem, two challenges should be addressed in our solution. First, the similarity computation is much more complicated since both geo-textual and social

Y. Nah et al. (Eds.): DASFAA 2020, LNCS 12112, pp. 279–287, 2020.
https://doi.org/10.1007/978-3-030-59410-7_18

information are considered and users as well as objects are located in the road network. To this point, we utilize the current state-of-the-art index GIM-Tree [15] and devise group pruning strategies to effectively shrink the search space of several queries. Second, duplicate query costs exist when answering each query one by one. Thus, we design a group processing framework to answer multiple queries with reduced overall costs by sharing their common CPU and I/O costs. In summary, the key contributions of this paper are summarized as follows:

- We define and study the BRkGSKQ problem. To the best of our knowledge, this is the first attempt to handle batch processing on multiple RkGSKQs.
- We propose an efficient group processing framework to support BRkGSKQ by sharing common CPU and I/O costs among different RkGSKQs.
- We conduct extensive experiments to verify the effectiveness and performance of our presented methods.

The rest of the paper is organized as follows. Section 2 reviews the related work. Section 3 introduces the preliminaries of this work and gives a baseline solution. The BRkGSKQ processing framework is elaborated in Sect. 4, and experimental results and our findings are reported in Sect. 5. Finally, Sect. 6 concludes this paper.

2 Related Work

Reverse k Nearest Neighbor (RkNN) Query. As one of the popular GSKQ types [1,13,14], the RkGSKQ is basically extended from the traditional reverse k nearest neighbor (RkNN) search problem which was first proposed by Korn et al. [7] and further studied and optimized in latter researches [8,12]. More recently, the development of web services has driven the fusion of spatial and textual information, thus more interesting RkNN query variants are considered in advanced applications [4,6,9]. However, all of these efforts are limited on efficient single query processing, so the batch processing of multiple RkNN queries is still an open problem.

Batch Query Processing. Three work exist related to batch query processing on geo-textual or graph data. Wu et al. [11] first explore the problem of joint Top-k Boolean Spatial Keyword Query, which retrieves the results for multiple Top-k Boolean Spatial Keyword Queries jointly. Chen et al. [2] study the batch processing for Keyword Queries on Graph Data. Recently, Choudhury et al. [5] investigate batch processing for ranked top-k Spatial Keyword Queries. Nonetheless, these studies only studied batch processing for Top-k queries. While this paper focuses on batch processing for RkGSKQ, a variant of RkNN query, which totally differs to the aforementioned work.

3 Preliminaries

Problem Settings. Our problem is considered in a road network $\mathcal{G}_r$ and a social network $\mathcal{G}_s$ defined as a graph $\{\mathcal{V}, \mathcal{E}\}$. Here, $\mathcal{V}$ is a set of vertexes, and $\mathcal{E}$ is a set of edges. In addition, a geo-social object set $\mathcal{O}$ and a user set $\mathcal{U}$ are located on $\mathcal{G}_r$. Each object

$o = \{o.loc, o.key, o.\mathcal{CK}\}$ in $\mathcal{O}$ is a point of interests (POI), where $o.loc = \{N_i, N_j, dis\}$ (N_i and N_j are vertexes of the edge o located on, and dis is the minimum distance from o to vertexes) describes the spatial information, $o.key$ is the textual description, and $o.\mathcal{CK}$ is the set of check-ins on o. A user $u = \{u.loc, u.key, F(u)\}$ in $\mathcal{U}$ contains spatial and textual information similar as o, with $F(u)$ the set of his/her friends. Here, $\mathcal{U}$ forms the vertex set of $\mathcal{G}_s$ and links between $F(u)$ and $u \in \mathcal{U}$ form its edge set.

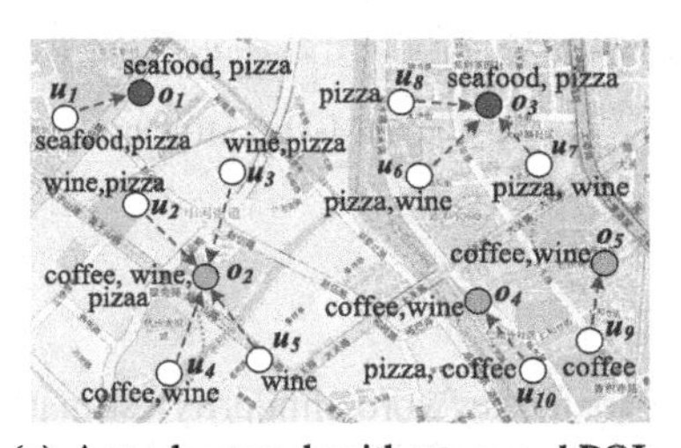

(a) A road network with users and POIs

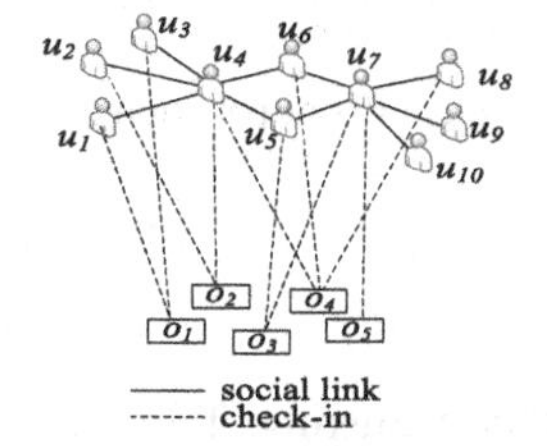

(b) A social network with check-ins

Fig. 1. Example of BRkGSKQ

Definition 1. Top-k Geo-Social Keyword Queries (TkGSKQ) [1]. *Given a set $\mathcal{O}$ of objects located in $\mathcal{G}_r$, a TkGSKQ is issued by a user u to retrieve a set $S_t(u)$ containing the k most relevant objects ranked by the following score function:*

$$F_{GSK}(o, u) = \frac{\alpha \times f_s(o, u) + (1 - \alpha) \times f_t(o, u)}{f_g(o, u)} \quad (1)$$

where $f_g(o, u)$ is the spatial proximity between o and u measured by their shortest path distance on the road network, $f_t(o, u)$ denotes their textual similarity computed by the traditional TF-IDF metric [3,5], and $f_s(o, u)$ is the social relevance and its detailed formulation can be found in work [14]. The parameter α in [0, 1] balances the importance of social and textual score. Note Equation 1 adopts the ratio function instead of the linear function [1, 14] to combine scores since it avoids expensive network distance normalization [10]. However, our approach supports both of these functions.

Definition 2. Reverse Top-k Geo-Social Keyword Queries (RkGSKQ) [14]. *Given a set $\mathcal{U}$ ($\mathcal{O}$) of users (objects) located on a road network $\mathcal{G}_r$, an RkGSKQ is issued for a specified object $o \in \mathcal{O}$, to find all users $u \in \mathcal{U}$ who have o in their TkGSKQ result set $S_t(u)$. Formally, the RkGSKQ result set is $S_r(o) = \{u \in \mathcal{U} | o \in S_t(u)\}$.*

Definition 3. Batch RkGSKQ (BRkGSKQ) retrieving. *Given a query set $\mathcal{Q} \subset \mathcal{O}$, BRkGSKQ is to concurrently retrieve all the RkGSKQ results for every $q \in \mathcal{Q}$. Formally, the result set of BRkGSKQ is $S_r(\mathcal{Q}) = \{S_r(q) | q \in \mathcal{Q}\}$.*

Example 1. An example of BRkGSKQ is illustrated in Fig. 1, assuming *KFC* has three chain stores o_2, o_4 and o_5. Given $k = 1$ and $\mathcal{Q} = \{o_2, o_4, o_5\}$, the answers of BR$k$GSKQ are $S_r(o_2) = \{u_2, u_3, u_4, u_5\}$, $S_r(o_4) = \{u_{10}\}$ and $S_r(o_5) = \{u_9\}$.

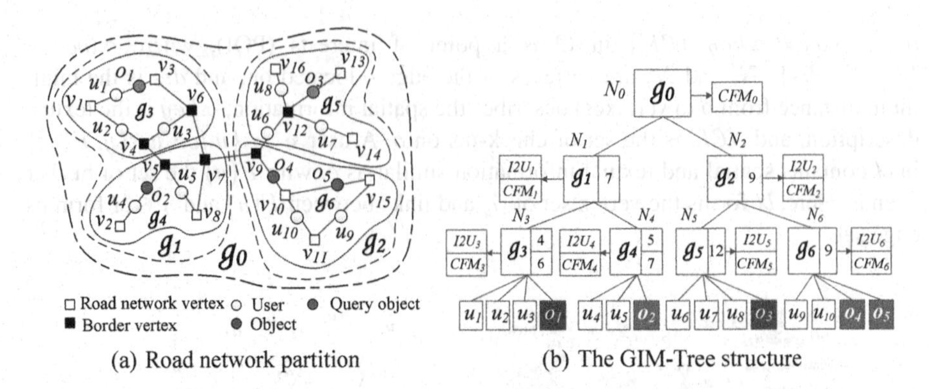

Fig. 2. Example of GIM-tree

The *GIM-Tree* Indexing. To accelerate geo-social keyword similarity estimation, the GIM-tree [14] is used in our algorithms. Basically, the GIM-tree is a G-Tree [15] integrated with an Intersection-Union (*I2U*) file and a Check-in&Friendship Matrix (*CFM*). It recursively partitions the road network into subnetworks as shown in Fig. 2(a), and pre-computes the shortest path distances between the borders of subnetworks, which can be employed to speed up the shortest path distance computation. The *I2U* file and the *CFM* describe the global textual and social information of objects located in a subgraph indexed by a GIM-tree node. An example of the GIM-tree is depicted in Fig. 2(b), due to space limitation we omit its detailed description which can be found in [14].

Baseline Approach. We first propose a baseline approach called BA to solve multiple RkGSKQs. BA clusters query objects into groups by clustering algorithms [11,13]. For objects in one cluster, BA combines their textual and social information, extracts the smallest subnetwork they co-located on, and constructs a super-object [4]. Such super-object is feeded into the RkGSKQ (RG) algorithm [14] which concurrently retrieves the results for query objects within the cluster. BA shares costs among queries within clusters, but the overheads on clustering could be large when we have many query objects. Moreover, BA will degrade to RG algorithm if too many query clusters are generated. Toward this, we present a framework for efficient BRkGSKQ processing.

4 Group Processing Framework for Batch RkGSKQ

BRkGSKQ is solved by the group processing framework which includes three phases, namely, query partitioning, group filtering, and group verification.

i) Query Partitioning. We first group query objects within the same GIM-tree leaf node into a query partition. Then for each partition in the leaf node level, we find their least common ancestor node (denoted as LCA) in the GIM-tree, from each leaf node of query partition to LCA(P) we traverse the GIM-Tree in a bottom-up manner. During the node traversing, all the visited nodes are extracted to generate a subtree. We couple each node of this subtree with combined textual and check-in information from query objects within this partition. Now the original query objects are organized by a query partition tree (denoted as $\mathcal{QT}$) that will be used in the next phase.

ii) Group Filtering. This phase is to eliminate unpromising users by group pruning strategies and accessing the GIM-Tree index only once. Before details, related definition and lemma in [14] are needed for us to better understand how it works.

Definition 4. *($\mathcal{U}_j.CL^l$ [14]). Given k and a user set $\mathcal{U}_j$, the lower bound counting list of $\mathcal{U}_j$(denoted as $\mathcal{U}_j.CL^l$) is a sequence of t ($1 \leq t \leq k$) tuples $\langle r_i,\ \mathcal{O}_i,\ s_i\rangle$, with the elements sorted in descending order of r_i value. Here, t is the minimum value satisfying $\sum_{i=1}^{t} \mathcal{U}_j.CL^l.s_i \geq k$, and values of s_i and r_i are correlated and determined as follows.*

$$s_i = \begin{cases} |\mathcal{O}_i| & if\ r_i = MinF_{GSK}(\mathcal{O}_i, \mathcal{U}_j) \\ 1 & if\ r_i = TightMinF_{GSK}(\mathcal{O}_i, \mathcal{U}_j) \end{cases}$$

Lemma 1 *[14]. if the t-th element in $\mathcal{U}_j.CL^l$ (i.e., $\mathcal{U}_j.CL^l.r_t$) is not less than the maximal similarity ($MaxF_{GSK}(o, \mathcal{U}_j)$) between an object o and the user set $\mathcal{U}_j$, o can not be RkGSKQ results of any user in $\mathcal{U}_j$, and thus, $\mathcal{U}_j$ can be safely pruned.*

Due to limited space, we ommit how the above geo-social keyword similarity bounds (i.e., $MaxF_{GSK}$, $MinF_{GSK}$, $TightMinF_{GSK}$ involved in Definition 4 and Lemma 1) are derived, interested readers can refer to [14] for details. Here, our focus is to extend Lemma 1 into our solutions and present group pruning strategies as follows.

Lemma 2. *Given a node $\mathcal{Q}_i$ of the query partition tree $\mathcal{QT}$ and a user set $\mathcal{U}_j$, if $\mathcal{U}_j.CL^l.r_t$ is no less than the maximal geo-social keyword similarity between $\mathcal{Q}_i$ and $\mathcal{U}_j$ (i.e., $MaxF_{GSK}(\mathcal{Q}_i, \mathcal{U}_j)$), all the users in $\mathcal{U}_j$ can not have any $o \in \mathcal{Q}_i$ in their TkGSKQ results.*

Definition 5 *($\mathcal{U}_j.\mathcal{AL}_q$). The associating query list of $\mathcal{U}_j$, denoted as $\mathcal{U}_j.\mathcal{AL}_q$, is a sequence of tuples in a form of $\langle l_i, u_i, \mathcal{Q}_i, \rangle$ where l_i is $MinF_{GSK}(\mathcal{Q}_i, \mathcal{U}_j)$ and u_i is $MaxF_{GSK}(\mathcal{Q}_i, \mathcal{U}_j)$, elements in $\mathcal{U}_j.\mathcal{AL}_q$ is sorted in a descending order by u_i value.*

Lemma 3. *If a query partition $\mathcal{Q}_i$ in $\mathcal{U}_j.\mathcal{AL}_q$ is found with $MaxF_{GSK}(\mathcal{O}_i, \mathcal{U}_j) \leq \mathcal{U}_j.CL^l.r_t$, then $\mathcal{Q}_i$ is removed from $\mathcal{U}_j.\mathcal{AL}_q$. When $\mathcal{U}_j.\mathcal{AL}_q$ becomes empty, users in $\mathcal{U}_j$ can not be in RkGSKQ results of any query object, and hence, it is pruned.*

Based on this, Group Filtering (GF) phase works as follows. Initially, all users are associated with all query objects, so the root of $\mathcal{QT}$ is added to $\mathcal{U}.\mathcal{AL}_q$. After this, GF conducts group pruning strategies iteratively. In each iteration, the lower bound counting list ($\mathcal{U}_i.CL^l$) of each user set $\mathcal{U}_i$ is computed and all the query partitions $\mathcal{Q}_j$ currently in $\mathcal{U}_i.\mathcal{AL}_q$ are evaluated and updated according to Lemma 2 and 3. If $\mathcal{U}_i.\mathcal{AL}_q$ remains some query partitions, all of them are replaced by their child nodes and added into the lower bound counting list of the child node of $\mathcal{U}_i$ for further evaluation in the next iteration. When $\mathcal{U}_i.\mathcal{AL}_q$ becomes empty, $\mathcal{U}_i$ is discarded. The GIM-tree is accessed from top to bottom in a single pass. Finally after all user sets are evaluated, unpruned user sets are stored in candidate user set $\mathcal{CU}$ which is returned for the next phase.

iii) Group Verification. This phase aims to determine which user in $\mathcal{CU}$ really has some query objects in their TkGSKQ results. Since retrieving TkGSKQ results for some users is inevitable, here the issue is how to avoid unnecessary TkGSKQ computation when $|\mathcal{CU}|$ is large and share common CPU and I/O costs during TkGSKQ retrieving of different users. Based on this, Group Verification (GV) works as follows. It first initializes a set $S_r(o)$ for $o \in \mathcal{Q}$, then verifies each user set $\mathcal{U}_i$ in $\mathcal{CU}$. For each $u \in \mathcal{U}_i$, the lower bound counting list of u is computed with the objects in $\mathcal{U}_i.CL^l$ to update the associated query partitions of u. If $u.\mathcal{AL}_q$ becomes empty, we do not need to retrieve $S_t(u)$ of u and u is removed from $\mathcal{U}_i$. When there are still users in $\mathcal{U}_i$, GV performs Group TkGSKQ (GRP_TkGSKQ) procedures to obtain their results jointly. After TkGSKQ results of each $u_k \in \mathcal{U}_i$ are obtained, GV examines whether the query object $q_o \in u_k.\mathcal{AL}_q$ is really included in $S_t(u_k)$ to update $S_r(q_o)$. Finally after all elements in $\mathcal{CU}$ are verified, we return $S_r(q)$ for each $q \in \mathcal{Q}$.

We continue to introduce how GRP_TkGSKQ works. Basically, GRP_TkGSKQ expands the road network, examines each visited object to find TkGSKQ results of $u_k \in \mathcal{U}_i$. It creates a max-heap for each $u_k \in \mathcal{U}_i$ to store their current TkGSKQ results and maintains a global min-priority queue Q to keep track of the objects $\mathcal{O}_j$ which are yet to be examined. Here, the value of shortest path distance serves as the key in Q, so objects with smaller network distance are examined first. This algorithm works via traversing the GIM-tree in a bottom-up manner, which is similar to the kNN algorithm on road networks [15]. Note that with Lemma 2, GRP_TkGSKQ does not need to examine each visited object set. Examined objects are accessed from the disk for $u \in \mathcal{U}_i$ only once such that their common CPU and I/O costs can be shared. GRP_TkGSKQ terminates when all users in $\mathcal{U}_i$ obtain their TkGSKQ and can not get better results.

Table 1. Statistics of the datasets used in our experiments

	LAS	BRI	GOW
# of road network vertices	26,201	175,813	1,070,376
# of road network edges	25,744	179,179	1,356,398
# of objects	12,773	196,591	1286,591
# of users	28,243	58,228	196,228
Average keywords per object	5.5	7.7	8.1
Average keywords per user	2.8	4.3	3.1
Average check-in counts per object	2.7	5.5	6.8
Average social links per user	9.1	8.7	7.4

Table 2. Parameter settings

Parameter	Range
k (# of most relevant object required)	10, 15, **20**, 25, 30
$\lvert\mathcal{Q}\rvert$ (# of query objects)	1, 10, 50, 100, **200**
α (balanceing parameter in ranking function)	0.0, 0.2, 0.4, **0.6**, 0.8

5 Experimental Evaluation

i) Experimental Settings. Experiments are conducted on three datasets, namely, LAS, BRI, and GOW. In particular, LAS is a real-life dataset from the Yelp[1], which contains POIs and users located in Las Vegas. The road network of LAS is obtained from openstreetmap[2], we map each POI and user to the nearest edge on the road network. BRI and GOW are synthetic datasets generated by combining the data from the DIMACS4[3] and Stanford Large Network Dataset Collection2[4]. The textual information in BRI and GOW are randomly attached following Zipf distribution [11]. Statistics of all datasets are summarized in Table 1.

For algorithms evaluated in the experiments, we test BA algorithm, the algorithm of Group Processing (GP) framework for BRkGSKQ, and the single RkGSKQ (RG) algorithm [14]. All the algorithms were implemented in C++ and conducted on a PC with an Intel Core i-7 3.6 GHz CPU and 32 GB RAM running Ubuntu 14.04 OS. The main memory of the Virtual Machine is 16G. We compare algorithms by their average runtime per query, the total runtime and the I/O cost (total number of accessed blocks). Each test is repeated 10 times and the average result is reported. The parameter settings are listed in Table 2 with values in bold the default settings.

ii) Effectiveness Study. We use the real-life dataset LAS to demonstrate the effectiveness of our methods. We compare BA, GP with the RG algorithm [14] that processes each query $q \in \mathcal{Q}$ one by one. The results are shown in Fig. 3. As depicted in Fig. 3, RG and BA are consistently outperformed by GP. The costs of RG and BA increase with $|\mathcal{Q}|$, while the costs in GP are relatively stable. This is because RG and BA repeatedly invoke RG which causes large redundant CPU and I/O costs. In contrast, GP benefits from I/O and CPU sharing, thus is more efficient. When $|\mathcal{Q}|$ is large (i.e., $|\mathcal{Q}| > 100$), the superiority of GP is more prominent since the ratio of common CPU and I/O sharing increases as $|\mathcal{Q}|$. It is of interest to find that when $|\mathcal{Q}| = 1$, GP still outperforms RG and BA. This is because the group verification phase in GP eliminates more unnecessary and redundant I/O and CPU costs during the candidate users verification.

iii) Performance Study

Effect of k. We explore the impact of different k values on the performance of proposed algorithms in this paper, the results are depicted in Fig. 4. It is observed that the efficiency of GP is better than BA by 1–2 orders of magnitude and the I/O costs of GP are about 10% of BA. This is because GP uses group processing framework to eliminate more redundant costs and also share more common costs. The query costs increase slightly when k grows, since more objects become relevant to users, which increases the costs of TkGSKQ.

---

[1] http://www.yelp.com/balancing_challenge/.
[2] https://www.openstreetmap.org/.
[3] http://www.dis.uniroma1.it/challenge9/.
[4] http://snap.stanford.edu/data/index.html.

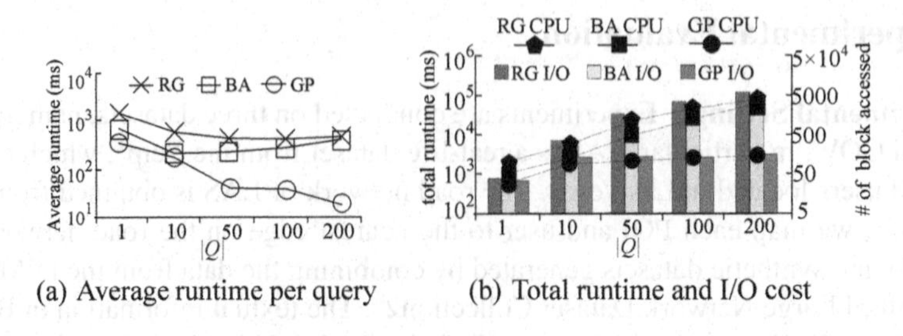

(a) Average runtime per query (b) Total runtime and I/O cost

Fig. 3. Comparison of different solutions in processing multiple RkGSKQs

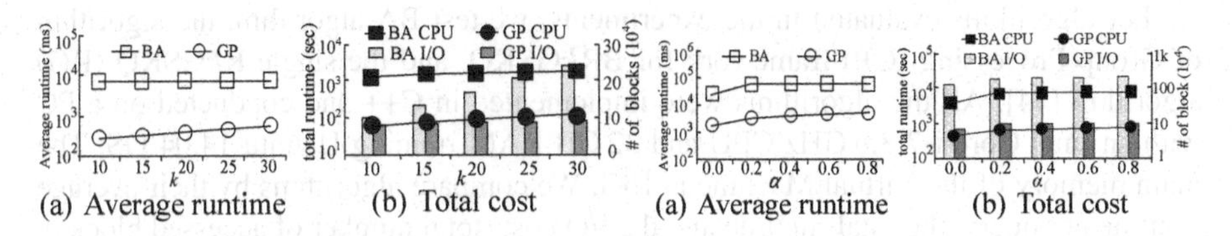

(a) Average runtime (b) Total cost (a) Average runtime (b) Total cost

Fig. 4. Performance vs. k on BRI **Fig. 5.** Performance vs. α on GOW

Effect of α. We investigate the impact of α on the processing performance, where α controls the importance of social relevance. As illustrated in Fig. 5, the average runtime and the total costs of all algorithms ascend when α varies from 0.0 to 0.2, but become stable when $\alpha \geq 0.2$. This is because, when $\alpha = 0.0$, social relevance has no affect in the ranking function making the similarity computation more efficient. When $\alpha \geq 0.2$, more users can be influenced by their friends but the effect of their textual similarity is weaken in turn, so the number of candidate users will not change very much.

6 Conclusions

This paper studies the problem of BRkGSKQ. Different from the existing work on RkGSKQ, we aim to develop efficient methods to accelerate multiple RkGSKQs processing in a query batch. We present a group processing framework to jointly retrieve the results of multiple RkGSKQs by sharing common CPU and I/O costs. Extensive experiments demonstrate the effectiveness and efficiency of our proposed methods.

Acknowledgments. This work was supported in part by the National Key R&D Program of China under Grant No. 2018YFB1004003, the NSFC under Grant No. 61972338, the NSFCZhejiang Joint Fund under Grant No. U1609217, and the ZJU-Hikvision Joint Project. Yunjun Gao is the corresponding author of the work.

References

1. Ahuja, R., Armenatzoglou, N., Papadias, D., Fakas, G.J.: Geo-social keyword search. In: Claramunt, C., et al. (eds.) SSTD 2015. LNCS, vol. 9239, pp. 431–450. Springer, Cham (2015). https://doi.org/10.1007/978-3-319-22363-6_23

2. Chen, L., Liu, C., Yang, X., Wang, B., Li, J., Zhou, R.: Efficient batch processing for multiple keyword queries on graph data. In: CIKM, pp. 1261–1270 (2016)
3. Choudhury, F.M., Culpepper, J.S., Bao, Z., Sellis, T.: Finding the optimal location and keywords in obstructed and unobstructed space. VLDB J. **27**(4), 445–470 (2018). https://doi.org/10.1007/s00778-018-0504-y
4. Choudhury, F., Culpepper, J., Sellis, T., Cao, X.: Maximizing bichromatic reverse spatial and textual k nearest neighbor queries. In: VLDB, pp. 1414–1417 (2016)
5. Choudhury, F.M., Culpepper, J.S., Bao, Z., Sellis, T.: Batch processing of top-k spatial-textual queries databases. ACM Trans. Spat. Algorithms Syst. **3**(e), 1–13 (2018)
6. Gao, Y., Qin, X., Zheng, B., Chen, G.: Efficient reverse top-k Boolean spatial keyword queries on road networks. TKDE **27**(5), 1205–1218 (2015)
7. Korn, F., Muthukrishnan, S.: Influence sets based on reverse nearest neighbor queries. In: SIGMOD, pp. 201–212 (2000)
8. Lee, K.C., Zheng, B., Lee, W.C.: Ranked reverse nearest neighbor search. TKDE **20**(7), 894–910 (2008)
9. Lu, J., Lu, Y., Cong, G.: Reverse spatial and textual k nearest neighbor search. In: SIGMOD, pp. 349–360 (2011)
10. Rocha-Junior, J., Norvag, K.: Top-k spatial keyword queries on road networks. In: EDBT, pp. 168–179 (2012)
11. Wu, D., Yiu, M., Cong, G., Jensen, C.S.: Joint top-k spatial keyword query processing. TKDE **24**(10), 1889–1903 (2012)
12. Yiu, M.L., Papadias, D., Mamoulis, N., Tao, Y.: Reverse nearest neighbors in large graphs. TKDE **18**(4), 540–553 (2006)
13. Zhao, J., Gao, Y., Chen, G., Chen, R.: Why-not questions on top-k geo-social keyword queries in road networks. In: ICDE, pp. 965–976 (2018)
14. Zhao, J., Gao, Y., Chen, G., Jensen, C.S., Chen, R., Cai, D.: Reverse top-k geo-social keyword queries in road networks. In: ICDE, pp. 387–398 (2017)
15. Zhong, R., Li, G., Tan, K.L., Zhou, L., Gong, Z.: G-tree: an efficient and scalable index for spatial search on road networks. TKDE **27**(8), 2175–2189 (2015)

C2TTE: Cross-city Transfer Based Model for Travel Time Estimation

Jiayu Song[1(✉)], Jiajie Xu[1,3], Xinghong Ling[1], Junhua Fang[1], Rui Zhou[2], and Chengfei Liu[2]

[1] School of Computer Science and Technology, Soochow University, Suzhou, China
20195227006@stu.suda.edu.cn, {xujj,lingxinghong,jhfang}@suda.edu.cn
[2] Swinburne University of Technology, Melbourne, Australia
{rzhou,cliu}@swin.edu.au
[3] State Key Laboratory of Software Architecture (Neusoft Corporation), Shenyang, China

Abstract. Travel time estimation (TTE) is of great importance to many traffic related applications, such as traffic monitoring, route planning and ridesharing. Existing approaches mainly utilize deep neural networks to achieve accurate travel time estimation. These models usually require large-scale trajectory data of the target city, while in reality, it is always difficult for the service providers to obtain sufficient trajectory data in the target city for training the model. To deal with this problem, we propose a cross-city knowledge transfer based travel time estimation solution, namely C2TTE, which is the first to address TTE task via transfer learning. The C2TTE models the travel time in spatial grid granularity first, using not only the sparse trajectory data in the target city, but also the knowledge of data-rich source city via transfer learning. After matching the spatial grid in target city to a suitable grid in source city, we fuse the knowledge between target city and source city to balance their importance adapting to the contextual conditions. In addition, we further use sequential model to represent the path in spatial grid granularity, so that the travel time of the path can be accurately estimated. Comprehensive experiments on two real trajectory datasets have been conducted to demonstrate the superior performance of C2TTE over existing approaches.

Keywords: TTE · Transfer learning · Sparse data

1 Introduction

The estimation of vehicle travel time is an important technique for many location based services, including route planning [3,20,21], departure time suggestion and vehicle dispatching. Existing TTE methods [14,22] mainly utilize deep neural networks for traffic prediction. Specially, [10] is first to employ the stacked auto encoder to predict traffic states of different nodes. However, these models heavily rely on large scale trajectory data of the target city, while it is not always available for some reasons, such as data privacy issues and difficulties of

Y. Nah et al. (Eds.): DASFAA 2020, LNCS 12112, pp. 288–295, 2020.
https://doi.org/10.1007/978-3-030-59410-7_19

data collection. In such cases, travel time estimation over sparse trajectory data becomes a useful but challenging problem.

To deal with data sparsity, a suitable solution can be designing TTE model in coarser spatial granularity (e.g. region of spatial grids), rather than on GPS coordinates or road segment granularity [1,4,9]. Following this method, a model called DEEPTRAVEL is proposed in [22] to model the travel time for every spatial grid first, and then derive the results by aggregating them. Though this method successfully alleviates the problem of data sparsity, its performance is still unsatisfactory when using a small public trajectory dataset for training.

In fact, we can easily obtain some large-scale public dataset of trajectories for cities like Chengdu. These data are potentially helpful for the TTE task of target city, like the knowledge of a data-rich city (i.e. Chengdu) to a data-sparse city (i.e. Beijing). To this end, transfer leaning [17] is an advanced technique for knowledge transfer, and it has been proven to be successful in applications like recommendation [2] and computer vision [19]. Therefore, towards the TTE problem, an inspiring method to deal with data sparsity can be designing a cross-city knowledge transfer based model.

However, we face several challenges when conducting more accurate TTE by cross-city transferred knowledge via transfer learning. Due to the skewed spatial distribution data, each grid in target city may face either data rich or data sparsity circumstances. For data rich regions, travel time within these regions can be well predicted using the trajectory in the target city. However, for data sparse regions, the key challenge is how to obtain an relatively accurate model using the suitable pattern of source city by knowledge transfer. Also, for the target path, it is necessary to accurately aggregate the travel time of different regions, using the knowledge learned from data in both of the target and source cities.

In this paper, we propose a cross-city knowledge transfer based Travel Time Estimation model (C2TTE model), which is the first to address TTE task via transfer learning. Specially, toward each spatial grid in target city, we first model the travel time in this grid region using not only the trajectory data in target city, but also the transfered knowledge from a data-rich source city. A matching function is designed to link one spatial grid in target city to another in source city, such that their grid-level TTE patterns tend to be similar, so that correct knowledge can be transferred after fine-tuning. At last, we properly aggregate the features of each region to the TTE of the path by a sequential model. Extensive experiments have been conducted on two real datasets: Beijing and Chengdu. The results show that C2TTE can improve the accuracy of the travel time prediction of a path.

2 Related Work

Travel Time Estimation. Estimating travel time has been studied extensively and it generally falls into three categories: path-based method, Origin-Destination based method and GPS coordinate-based method. The GPS coordinate-based method estimates the travel time based on the GPS points. [14] proposes a DeepTTE model to capture the relevance between the path and trajectories in GPS coordinates. But when the input is a path with sparse GPS points, its performance drops significantly. The path-based method takes the road network structure, including using map matched trajectories [8], into consideration and build the model based on sub-path, which is a supplement to the GPS method. [12] develops a non-parametric method for route travel time estimation using low-frequency floating car data (FeD). [18] finds the optimal concatenation of trajectories for an estimation through a dynamic programming solution. The origin-destination (OD) based method estimates the travel time only based on the origin and destination instead of path, which alleviates the problem of path data sparsity. [6] combines the travel distance prediction with the time of day to predict the travel time. [15] proposes to estimate travel time using neighboring trips from the large-scale historical data. These methods alleviate the problem of data sparsity, but their performance is still unsatisfactory when using a small number of public trajectories.

Transfer Learning. Transfer learning [11] is a machine learning approach to address the problem when labeled training data is scare. It has been adopted in various areas for various tasks. For recommendation area, [7] separates the city-specific topics of each city from the common topics shared by all cities and transfers real interests from the source cities to the target city by the medium of the common topics. [13] utilizes transfer learning from adults models to children models for children Automatic Speech Recognition. For visual recognition area, [23] introduces a weakly-supervised cross-domain dictionary learning method. For spatio-temporal field, [16] proposes the cross-city transfer learning framework for deep spatio-temporal prediction. As far as we know, C2TTE is the first to employ cross-domain transfer learning to address the trajectory sparsity problem to further predict travel time.

3 The C2TTE Model

Given the little trajectories in target city D and rich trajectories given in source city D^*, we aim to predict the travel time T^t of the target path P through the knowledge of each grid g_i from target city and source city (via transfer). In C2TTE, it not only adopts a courser granularity in grid level to model the travel time, but also transfers the knowledge from a data-rich city.

The C2TTE model is composed of four components: grid-level travel time estimation, cross-city grid matching, grid-level model fusion and path-level travel time estimation. The grid-level TTE aims to predict the travel time within a spatial grid region. The cross-city grid matching means to link the grid of a target city to a best-matched grid in (data rich) source city. The grid-level model fusion is to fine-tune the grid-level TTE (so that the grid-level pattern can be properly used) and fuse the knowledge from target and source grid. The path-based travel time estimation is designed to predict the total travel time by using the fused knowledge of target city in grid-level model fusion.

3.1 Grid-Level Travel Time Estimation

As shown in Fig. 1, we first extract feature of each grid g_i. Like [22], for spatial feature, V_{sp}, we adopt distributed representation to represent each grid, such as Word2Vec. For temporal feature, V_{tp}, we divide a day into several time-bins and use distributed representation to represent each time-bin. For driving state features, we divide the trajectory into three stages: the *starting stage*, the *middle stage* and the *ending stage*, and we use a vector V_{dir} to represent: (1) the stage of current grid in the trajectory, using three 0–1 bits; (2) the proportion of the trip that has traveled by a percentage. For short-term traffic features, we use historical traffic speed and traffic flow observations of this grid as the input of LSTM [5], and V_{short} is the last hidden state of the LSTM network. Then, we combine these features as a feature vector $V = \{V_{sp}, V_{tp}, V_{dir}, V_{short}\}$.

Last, we feed the feature vector V of each grid g_i into a CNN network. The network mainly includes convolution layer and fully-connected layer. In convolution layer, the hidden features $\mathbf{X}_{g_i}^{trep}$, which contains the hidden traffic information of each grid, can be written as $f_c(W_c * V + b_c)$, where $*$ denotes the convolution operator followed by an activation f_c, and W_c and b_c are the parameters. Then, we pass the hidden features $\mathbf{X}_{g_i}^{trep}$ to the fully-connected layer to predict travel time of each grid. After training the CNN network, we get the results of convolution layer as outputs, $\mathbf{X}_{g_i}^{trep}$, of the Grid-level Travel Time Estimation.

3.2 Cross-city Grid Matching

In this section, our purpose is to link the grid of a target city to a best-matched grid in (data rich) source city. Here we adopt the matching function proposed in [16] to calculate the correlation coefficient between target and source grid with the corresponding traffic data. Then we select the grid in source city with the highest coefficient as the best matching grid. The matching function $\mathcal{F}$ is as follows:

$$\rho_{g_i,g_j^*} = \frac{corr(s_{g_i}, s_{g_j^*})}{\sqrt{D(s_{g_i})}\sqrt{D(s_{g_j^*})}}, \qquad g_i \in \mathbb{C}_D, g_j^* \in \mathbb{C}_{D'} \tag{1}$$

$$\mathcal{F}_{g_i} = g_j^*, \qquad if\ \rho_{g_i,g_j^*} > \rho_{g_i,g_j^*}{}' \tag{2}$$

Here s_{g_i} is the data of traffic time in the i-th target spatial grid g_i, $s_{g_j^*}$ is the data of traffic time in the j-th source spatial grid g_j^*.

3.3 Grid-Level Model Fusion

The grid-level model fusion aims to utilize the knowledge of two cities in a reasonable way (target city and source city). As shown in Fig. 2, with the proposed grid-level TTE, we reference the parameter optimization method in [16] to fine-tune the network structure for source city. We use the network parameters θ_D

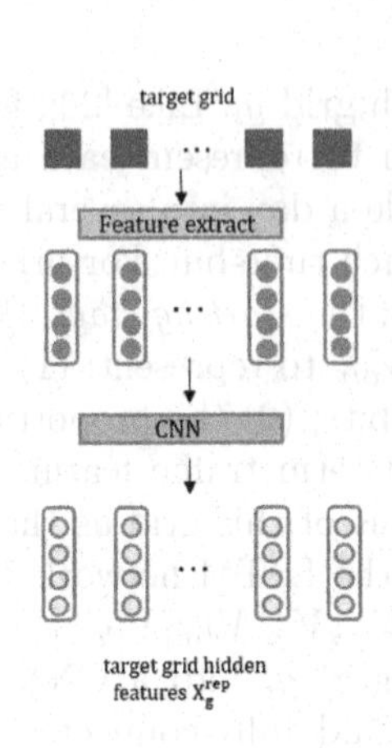

Fig. 1. Grid-level travel time estimation.

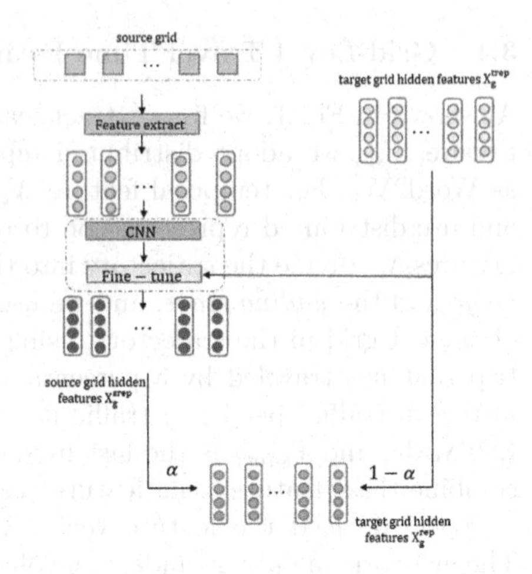

Fig. 2. Grid-level model fusion.

learned from target model (Sect. 3.1) as pre-trained network parameters to train the model in the source city D^* with its rich spatio-temporal data. During this training process, we also get hidden features $\mathbf{X}_{g_i^*}^{srep}$ of each source grid at the same time. Since our objective is to minimize hidden features divergence between matched grid pairs, the squared error is,

$$\min_{\theta_D} \sum_{g_i \in \mathbb{C}_D} \|\mathbf{X}_{g_i}^{trep} - \rho_{g_i,g_i^*} \cdot \mathbf{X}_{g_i^*}^{srep}\|^2, where\, g_i^* = \mathcal{F}(g_i) \tag{3}$$

Finally, the target grid hidden features learned from target and source city is the summation of the hidden features, i.e., $\mathbf{X}_{g_i}^{rep} = \alpha_1 \mathbf{X}_{g_i}^{trep} + \alpha_2 \mathbf{X}_{g_i^*}^{srep}$. Here α_1 and α_2 are weights, and $\alpha_1 + \alpha_2 = 1$.

3.4 Path-Level Travel Time Estimation

We feed the hidden features $\mathbf{X}_{g_i}^{rep}$ of each target grid g_i to BiLSTM to get the forward and backward hidden state $\overrightarrow{h_i}$ and $\overleftarrow{h_i}$, and then concatenate them to get the final i-th hidden state $h_i = [\overrightarrow{h_i}, \overleftarrow{h_i}]$.

Then, we adopt the dual interval loss method [22] to optimize the path-based model. This method divides the path travel time into two parts (the forward interval time and the backward interval time). The forward interval time is the time from the start point to each intermediate GPS point and the backward interval time is the time from each intermediate GPS point to the end point. Formally,

$$\tilde{T}^{t_f} = W_t^T[h_1, h_1 + h_2, ..., \sum_{i=1}^{n-1} h_i, \sum_{i=1}^{n} h_i] + b_t \quad (4)$$

$$\tilde{T}^{t_b} = [W_t^T[\sum_{i=2}^{n} h_i, \sum_{i=3}^{n} h_i, ..., h_{n-1}, h_n] + b_t, 0] \quad (5)$$

where $W_t^T h_i + b_t$ represents the time spent only on grid g_i, W_t and b_t are the shared weights.

Besides, $M \in \{0,1\}^n$ is to represent whether there is a point sampled in the spatial grid g_i, and use T^{t_f}, T^{t_b} to record the forward and backward interval ground truth of the path. Our objective is to minimize the estimation loss of the whole path. We adopt the equation of the whole path loss in [22]:

$$\mathcal{L} = \frac{M^T * ((\tilde{T}^{t_f} - T^{t_f})[/]T^{t_f})^{[2]} + M^T * ((\tilde{T}^{t_b} - T^{t_b})[/]T^{t_b})^{[2]}}{\mathbf{1}^T \cdot (M[*]2)} \quad (6)$$

4 Experiments

In this paper, we use two real trajectory dataset in different cities, Beijing (target city) and Chengdu (source city), to evaluate the performance of C2TTE and baseline methods.

We adopt *mean absolute error* (MAE), *mean absolute percentage error* (MAPE) and *root-mean-squared error* (RMSE) as the major performance metrics.

Table 1. Performance comparison of C2TTE and its competitors in Beijing.

Metrics	MAE (sec)	RMSE (sec)	MAPE
Grid-CNN	313.95	469.56	0.4432
Grid-based Transfer	260.32	400.22	0.3845
DEEPTRAVEL-NLTraf	205.53	321.13	0.3013
C2TTE	**191.31**	**300.87**	**0.2836**

4.1 Performance Comparison

For competitions, we implement several models as baselines.

- **Grid-CNN:** Grid-CNN uses the convolutional neural network (CNN) model to perform the estimation based on the data of target city. It is the same as our first part of our model, with two convolution layers, each having 64 3×3 filters with stride 1, and 2 max_pooling layers, each in the size of 2×2 and a fully-connected layer with 512 units and sigmoid activation.

- **Grid-based Transfer:** Grid-based Transfer uses the convolutional neural network (CNN) model to obtain the hidden features (like C2TTE). Then it adopts the method of fusion in C2TTE to fuse the hidden features both from target city spatial grid and its similar matched spatial grid in source city to predict the travel time of each target city spatial grid. Finally it adopts a fully-connected layer like Grid-CNN. This method is based on the spatial grid rather than the whole path.
- **DEEPTRAVEL-NLTraf:** DEEPTRAVEL-NLTraf uses the same method in [22] to estimate the travel time of the whole path. It directly uses the historical trajectories to estimate the travel time. It adopts a new loss function for auxiliary supervision to optimize the model and extracts multiple features that affect the travel time. However, as Beijing has only a few days of data, the features in DEEPTRAVEL-NLTraf do not include long-term traffic features.

As we can see in Table 1, Grid-based Transfer performs better than grid-CNN. This is because that the data of Beijing is sparse, but Grid-based Transfer can transfer the knowledge from the similar spatial grid in Chengdu to the spatial grid in Beijing to help estimate the travel time of the Beijing spatial grid. In this way, the spatial grid in Beijing can have more information reflecting the traffic condition to predict travel time. Note that, DEEPTRAVEL-NLTraf shows a good performance than these above methods. It reflects that the relationship between adjacent spatial grids is of great importance in the prediction of a whole path. However, our model further significantly outperforms DEEPTRAVEL-NLTraf. We combine those features both from Beijing (data sparse) and Chengdu (data rich) to predict travel time. That is to say, just using the sparse data of the target city is not sufficient and we need utilize the knowledge of another city which has rich data to help us estimate the travel time based on the whole path.

5 Conclusion

To address the data scarcity issue in a query path travel time prediction tasks, this paper proposes a cross-city knowledge transfer based travel time estimation model, called C2TTE. We design a matching function to link target city grid and source city grid, so that we can fine-tune the model based on target and source data to transfer knowledge correctly. Finally, we predict the travel time of the path with the fine-tuned knowledge. We conduct experiments on real GPS trajectory datasets to understand the importance of target city and source city in a query path TTE and to demonstrate the outperformance of C2TTE.

References

1. Asif, M.T., et al.: Spatiotemporal patterns in large-scale traffic speed prediction. IEEE Trans. **15**(2), 794–804 (2013)
2. Chen, L., Zheng, J., Gao, M., Zhou, A., Zeng, W., Chen, H.: Tlrec: transfer learning for cross-domain recommendation. In: ICBK, pp. 167–172. IEEE (2017)

3. Dai, J., Liu, C., Xu, J., Ding, Z.: On personalized and sequenced route planning. WWW **19**(4), 679–705 (2016). https://doi.org/10.1007/s11280-015-0352-2
4. De Fabritiis, C., Ragona, R., Valenti, G.: Traffic estimation and prediction based on real time floating car data. In: ITSC, pp. 197–203. IEEE (2008)
5. Hochreiter, S., Schmidhuber, J.: Long short-term memory. Neural Comput. **9**(8), 1735–1780 (1997)
6. Jindal, I., Chen, X., Nokleby, M., Ye, J., et al.: A unified neural network approach for estimating travel time and distance for a taxi trip (2017). arXiv preprint arXiv:1710.04350
7. Li, D., Gong, Z., Zhang, D.: A common topic transfer learning model for crossing city poi recommendations. IEEE Trans. Cybern. **49**(12), 4282–4295 (2019)
8. Li, Y., Liu, C., Liu, K., Xu, J., He, F., Ding, Z.: On efficient map-matching according to intersections you pass by. In: Decker, H., Lhotská, L., Link, S., Basl, J., Tjoa, A.M. (eds.) DEXA 2013. LNCS, vol. 8056, pp. 42–56. Springer, Heidelberg (2013). https://doi.org/10.1007/978-3-642-40173-2_6
9. Lv, Y., Duan, Y., Kang, W., Li, Z., Wang, F.Y.: Traffic flow prediction with big data: a deep learning approach. IEEE Trans. **16**(2), 865–873 (2014)
10. Lv, Z., Xu, J., Zheng, K., Yin, H., Zhao, P., Zhou, X.: Lc-rnn: A deep learning model for traffic speed prediction. In: IJCAI, pp. 3470–3476 (2018)
11. Pan, S.J., Yang, Q.: A survey on transfer learning. IEEE Trans. **22**(10), 1345–1359 (2009)
12. Rahmani, M., Jenelius, E., Koutsopoulos, H.N.: Route travel time estimation using low-frequency floating car data. In: ITSC, pp. 2292–2297. IEEE (2013)
13. Shivakumar, P.G., Georgiou, P.: Transfer learning from adult to children for speech recognition: Evaluation, analysis and recommendations (2018). arXiv preprint arXiv:1805.03322
14. Wang, D., Zhang, J., Cao, W., Li, J., Zheng, Y.: When will you arrive? estimating travel time based on deep neural networks. In: AAAI, pp. 2500–2507. AAAI Press (2018)
15. Wang, H., Tang, X., Kuo, Y.H., Kifer, D., Li, Z.: A simple baseline for travel time estimation using large-scale trip data. TIST **10**(2), 1–22 (2019)
16. Wang, L., Geng, X., Ma, X., Liu, F., Yang, Q.: Cross-city transfer learning for deep spatio-temporal prediction. In: IJCAI-19, pp. 1893–1899, July 2019
17. Wang, L., Geng, X., Ma, X., Zhang, D., Yang, Q.: Ridesharing car detection by transfer learning. AI **273**, 1–18 (2019)
18. Wang, Y., Zheng, Y., Xue, Y.: Travel time estimation of a path using sparse trajectories. In: Proceedings of the 20th ACM SIGKDD, pp. 25–34. ACM (2014)
19. Whatmough, P.N., Zhou, C., Hansen, P., Venkataramanaiah, S.K., Seo, J.s., Mattina, M.: Fixynn: Efficient hardware for mobile computer vision via transfer learning (2019). arXiv preprint arXiv:1902.11128
20. Xu, J., Chen, J., Zhou, R., Fang, J., Liu, C.: On workflow aware location-based service composition for personal trip planning. Future Gener. Comput. Syst. **98**, 274–285 (2019)
21. Xu, J., Gao, Y., Liu, C., Zhao, L., Ding, Z.: Efficient route search on hierarchical dynamic road networks. Distrib. Parallel Databases **33**(2), 227–252 (2015)
22. Zhang, H., Wu, H., Sun, W., Zheng, B.: Deeptravel: a neural network based travel time estimation model with auxiliary supervision. In: IJCAI, pp. 3655–3661 (2018). www.ijcai.org
23. Zhu, F., Shao, L.: Weakly-supervised cross-domain dictionary learning for visual recognition. Int. J. Comput. Vis. **109**(1–2), 42–59 (2014)

Migratable Paxos

Low Latency and High Throughput Consensus Under Geographically Shifting Workloads

Yanzhao Wang, Huiqi Hu(✉), Weining Qian, and Aoying Zhou

School of Data Science and Engineering,
East China Normal University, Shanghai, China
51175100009@stu.ecnu.edu.cn, {hqhu,wnqian}@dase.ecnu.edu.cn,
ayzhou@sei.ecnu.edu.cn

Abstract. Global web services or storage systems have to respond to changes in clients' access characteristics for lower latency and higher throughput. As access locality is very common, deploying more servers in datacenters close to clients or moving related data between datacenters is the common practice to respond to those changes. Now Paxos-based protocols are widely used to tolerate failures, but the Reconfiguration process (i.e., the process to changing working servers) of existing Paxos-based protocols is slow and costly, thus has a great impact on performance. In this paper, we propose *Migratable Paxos* (MPaxos), a new Paxos-based consensus protocol that minimizes the duration of reconfiguration and accelerates the migration of data between datacenters, without losing low commit latency and high throughput.

Keywords: Global web services · Paxos consensus protocol · Shifting workloads · Migration

1 Introduction

The access hot spots in global web services usually come from the same geographical region (e.g., a country, a state, or a time zone). And such hot spots may move frequently, as a function of local time. For example, a global electronic trading system will witness the moving of hot spots between countries because of different stock opening times. If such web services use servers in fixed locations, the latency will increase significantly as the hot spots migrate to a location far away from the working servers. Different applications have different tolerances for delays, but high latency generally leads to a decline in revenue.

Now Paxos-based protocols [1,2] are widely used in web services [3,9] and database systems [10] to tolerate server failures and provide high availability. But existing Paxos protocols are not adaptive for the frequent changing workloads, two problems lie ahead of these protocols:

1. Agreements are reached by a majority. Such that the total number of nodes (denote N) is related to tolerated failures (denote F), which satisfies $N = 2F + 1$. Then larger N results in a larger F. This property limits the number of nodes that can be deployed in a Paxos instance and makes it impossible to place a server near every possible client.

Y. Nah et al. (Eds.): DASFAA 2020, LNCS 12112, pp. 296–304, 2020.
https://doi.org/10.1007/978-3-030-59410-7_20

2. In data migration, usually only a specific part of data needs to be migrated. But existing reconfiguration mechanisms transferring the entire state machine, which adds extra cost.

In this paper, we proposed MPaxos to solve these problems. We proposed the *Working Cluster allocation* mechanism. A Working Cluster is the set of nodes explicitly specified for one object, a command can and only can be committed in a majority of the related Working Cluster. In this way, we decoupled quorum size $(F+1)$ from the number of nodes in cluster (N) by introducing W (the number of nodes in a Working Cluster), where $W <= N$ and $W = 2F+1$, thus solved problem 1. Also, the Working Clusters of different objects work independently, so the migration of one object will not affect the read and write of other objects, thus alleviate problem 2. Moreover, MPaxos provides a scheduling framework for automating Working Cluster selection and migration to maintain low latency.

We briefly introduce the relevant background of Paxos and other related works in Sect. 2. Then the design and implementation of MPaxos are present in Sect. 3. The scheduling framework is introduced in Sect. 4. Section 5 shows the performance evaluation. The paper concludes in Sect. 6 with a summary.

2 Related Work

Leader-based Paxos protocols use a master to determine the execution order of all the command, while the Leaderless Paxos protocols usually don't care the order of irrelevant commands and only establish order constraints between commands for the same object. *Egalitarian Paxos* [4] is an efficient leaderless Paxos protocol and it uses a totally decentralized approach to commit commands and handle interferes: in the process of choosing a command in a log entry, each participant attaches ordering constraints to that command, and the agreement is achieved when a majority agree with that constraints, thus irrelevant commands can be committed by different replicas. Therefore, EPaxos is born with high throughput.

The commit protocol of EPaxos is divided into 3 phases. When there are no interferes between commands from different command leaders, these commands can be committed on the Fast-Path (involves only Phase 1 and 3); otherwise interfered commands will be committed on the Slow-Path (involves Phase 1, 2, and 3). On the Slow-Path, each command proposed by command leader requires replies from a majority of replicas, while in the fast path, $F + \lfloor \frac{F+1}{2} \rfloor$ replies is needed to guarantee the correctness, where $F = \lfloor \frac{N}{2} \rfloor$ and N is the number of replicas.

There are also numerous works proposed for the migration of services and workload burst handling. Some early works use the idea of VM live migration, in which the VMs containing the services are migrated between datacenters. This method has been widely used in practice [7,8]. These works adopt the shared disk technology for faster migration but face long latency in accessing a shared disk image. Then another VM based protocol Supercloud [6] proposed a Data Propagation method that implements the storage layer much like a state machine, modified blocks are transferred between datacenters to maintain a consistent storage view.

3 The Design of MPaxos

MPaxos is proposed to achieve lower latency and higher throughput than existing Paxos based protocols in wide-area deployment. High throughput is reached by implementing a decentralized command committing method similar to EPaxos, as the read and write requests can be distributed across different nodes. But a Fast-Path in EPaxos consisting of roughly $\frac{3}{4}$ of nodes makes the commit latency even higher. So MPaxos made the following changes to EPaxos: 1. Introduce the concept of *Working Cluster*. 2. Introduce the concept of *Reorganization* that changes the Working Cluster. 3. Let each object has its own Working Cluster.

Below we describe the components in MPaxos in more detail.

3.1 Working Cluster

The ***Working Cluster*** is a subset of the overall cluster. A replica inside the Working Cluster is called ***Working Replica***. The committing of the commands need to be performed by a Working Replica. The replica outside of the Working Cluster on receiving a command from the client should redirect it to a Working Replica.

Through the Working Cluster mechanism, we can reduce the Fast-quorum size to $F_W + \lfloor \frac{F_W+1}{2} \rfloor$, slow-quorum size to $F_W + 1$ (where $F_W = \lfloor W/2 \rfloor$, so F_W is the actual number of tolerated failures under Working Cluster with size W). Thus we can deploy more idle machines in the cluster, while keeping a small quorum size.

3.2 Reconfiguration and Reorganization Algorithm

The process of changing the Working Cluster (i.e. ***Reorganization***) is essentially a reconfigure process for the Working Cluster, except that the migration does not involve the startup and shut down of replicas. The process of reconfiguration can be divided into three steps: 1. Stop the old state machine, 2. Transfer the state, 3. Start the new state machine. Below we present a detailed description of steps 1 and 2. As the third step is simple and trivial, we don't discuss it here.

The general way to stop the old state machine is to submit a stop command (it is usually done by committing a RECONFIG command), and there can be no other valid commands in old state machines after the stop command (only NOP commands are permitted) [5]. Due to the multileader style of MPaxos, it is possible to have multiple RECONFIG commands committed at the same time, and their contents may be different. One solution is to modify the commit protocol to refusing the old RECONFIG and using another round of communication to confirm this RECONFIG. But this could cause a livelock (different replicas alternately send new RECONFIG and no RECONFIG command can be confirmed successfully). Thus we chose another way: allow multiple potentially different reconfig commands to be committed, but only the earlier one will take effect. To do this, two settings need to be introduced:

Definition 1. *RECONFIG commands conflict with each other.*

Definition 2. *RECONFIG commands conflict with read/write commands.*

With Definition 1, the concurrent RECONFIG commands will establish an execution order. Definition 2 establishes an execution order between the read/write commands and the RECONFIG commands. The read/write commands have to be set to NOP when there is a RECONFIG command in its dependencies, this guarantees no valid command after RECONFIG.

We abstract the reconfiguration process of MPaxos into three states: NORMAL, RECONFIGURING, and TRANSFERING. RECONFIGURING implies that some replica has sent a RECONFIG command, and the command is not yet committed; TRANSFERRING state means that the RECONFIG command has been committed and the transfer of states is in progress. To know when transfer finishes, *TRANSFER-FINISH* command is defined and commit it in a majority of the new config. A receiver confirms this log after its transfer process is completed. Upon this command is committed successfully, the transfer state ends. The replica cannot submit normal commands in the RECONFIGURING and TRANSFERING states, while the RECONFIG command can be submitted at any time.

The reorganization process inherits the 3 steps and the 3 replica state from reconfiguration, and introduce 1 extra log type *REORGANIZE*. But reorganization conceptually just alter the roles some set of replicas plays. Hence it has less impact on the performance. Figure 1 shows the pseudocode of the protocol for choosing commands in MPaxos, and Fig. 2 shows the Execution logic of the REORGANIZE command.

4 Scheduling Framework

In this chapter, we present the scheduling framework of MPaxos which is responsible for making migration decisions. Suppose there are N replicas deployed in MPaxos: $\{d_1, d_2, ..., d_n\}$. A Working Cluster placement plan for some object θ with k nodes is denoted as $P = \{p_1, p_2, ...p_k\}$, where d_{p_i} is a replica in the Working Cluster. Periodically, MPaxos measures end-to-end latency between different replicas and stores the results in matrix L, where $L(i, j)$ is the round-trip-time (RTT) from d_i to d_j. The workload statistics is denoted as $S = \{(r_1, w_1), (r_2, w_2), ..., (r_n, w_n)\}$, (r_i, w_i) is the read and write workload on replica d_i.

To evaluate a placement plan, we provide a function $f(P, S, L)$ that evaluates a placement plan under a certain workload:

$$f(P, S, L) = -\frac{\sum_{i=0}^{n}(\alpha \cdot C(P, i) \cdot r_i + \beta \cdot C(P, i) \cdot w_i)}{n}$$

where $C(P, i)$ is replica d_i's commit latency under the placement plan P. α and β are weights indicated the importance of read and write latency.

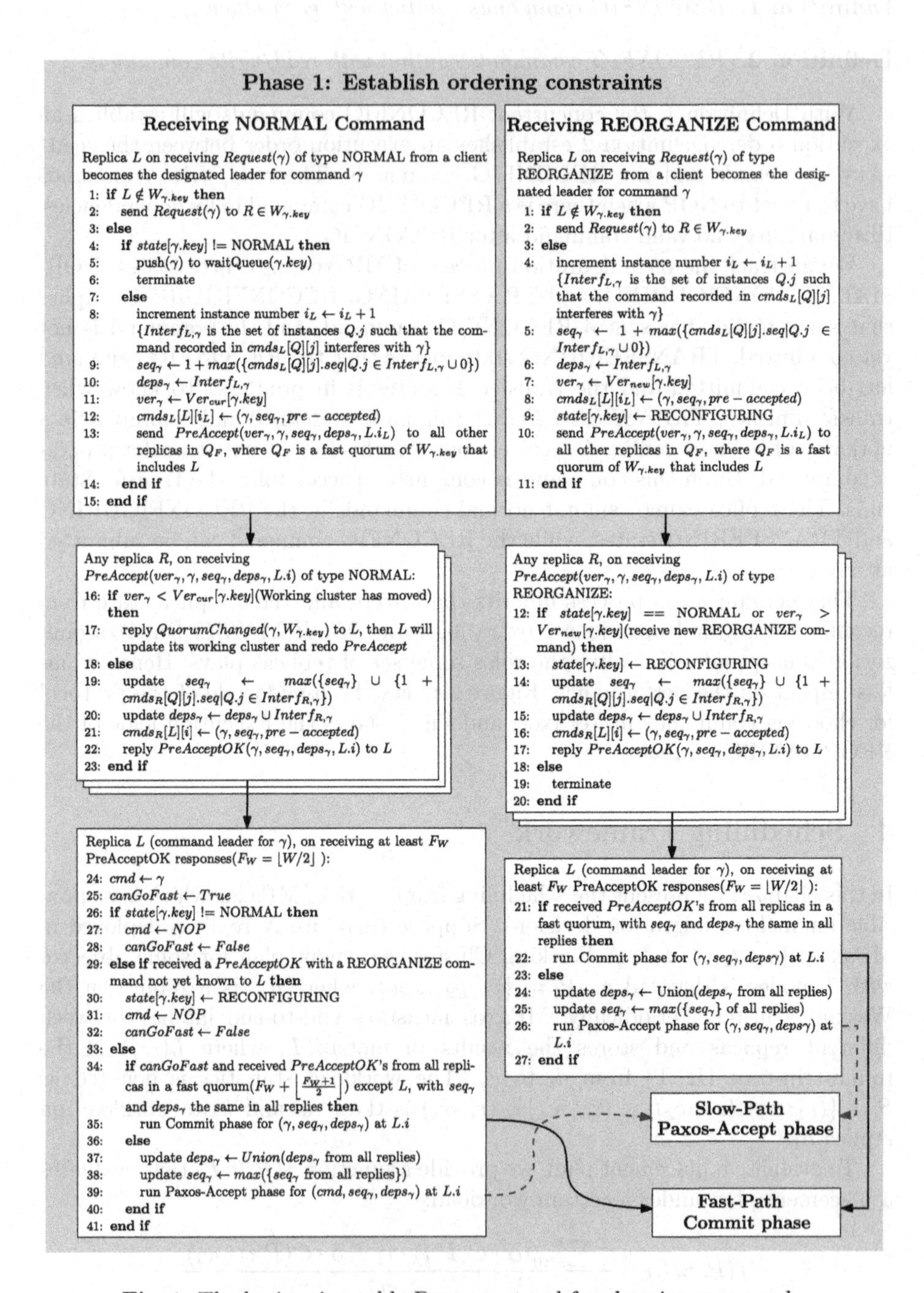

Fig. 1. The basic migratable Paxos protocol for choosing commands

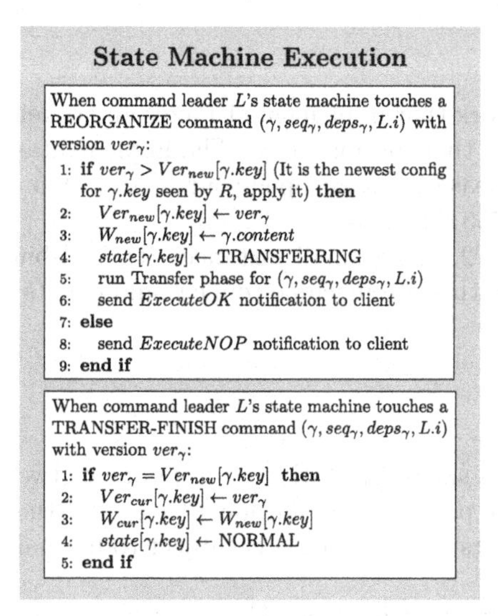

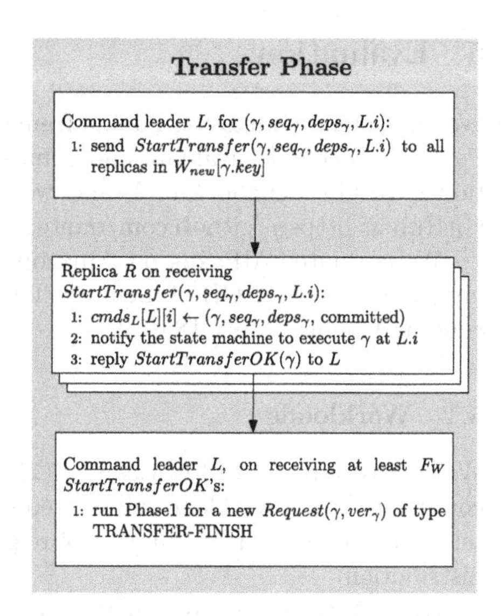

Fig. 2. The execution and transfer phase of migratable paxos

The evaluation of $C(P, i)$ is split into 2 steps:

1. Send a command from d_i to the closest replica in P (denoted p_L), the latency is:

$$C_1 = \begin{cases} L(i, p_L), & \text{if } i \text{ not in } P \\ 0, & \text{if } i \text{ in } P \end{cases}$$

2. Commit latency. The command could be committed in the Fast-Path or the Slow-Path, we specify a parameter e to represent the possibility of going through Fast-Path, then the latency is:

$$C_2 = e \cdot Fast(P, p_L) + (1 - e) \cdot Slow(P, p_L)$$

The Fast-Path only involves one round-trip between a Fast-quorum of working replicas. As a Fast-quorum contains $F_W + \lfloor \frac{F_W+1}{2} \rfloor$ replicas ($F_W = \lfloor W/2 \rfloor$, with W indicates the cardinality of P), the network latency of commit in Fast-Path is roughly the same as the third quartile of latencies between p_l and other replicas in P, that is:

$$Fast(P, pL) = 3rd\ Quartile(\{L(p_l, p_i) \mid p_i\ in\ P\})$$

The Slow-Path involves two round-trips between a Slow-Quorum (i.e. a majority). So the network latency is 2-times the median of the latencies from p_l to replicas in P.

$$Slow(pl) = 2 \cdot Median(\{L(p_l, p_i) \mid p_i\ in\ P\})$$

5 Evaluation

We implement MPaxos on Paxi, a framework that implements EPaxos and other Paxos protocols, to evaluate and compare their performance. The implementation is in Go, version 1.11.2, and we release it as an open-source project on GitHub at https://github.com/dante159753/MPaxos.

We evaluated MPaxos on Amazon EC2, using small instances (two 64-bit virtual cores with 2 GB of memory) for both state machine replicas and clients, running Ubuntu Linux 18.04.2.

5.1 Workloads

We specify two types of workloads: (1) hot spots are static and from one or two continents; (2) a workload with a request peak at 9:00 am local time, and the relationship between the number of requests and local time is subject to normal distribution.

Our tests also capture conflicts, an important workload characteristic – conflict is a situation when potentially interfering commands reach replicas in different orders. Conflicts affect EPaxos and MPaxos. As write requests usually occupy no more than 2% of all requests, we believe that 0% and 2% command interference rates are the most realistic [4].

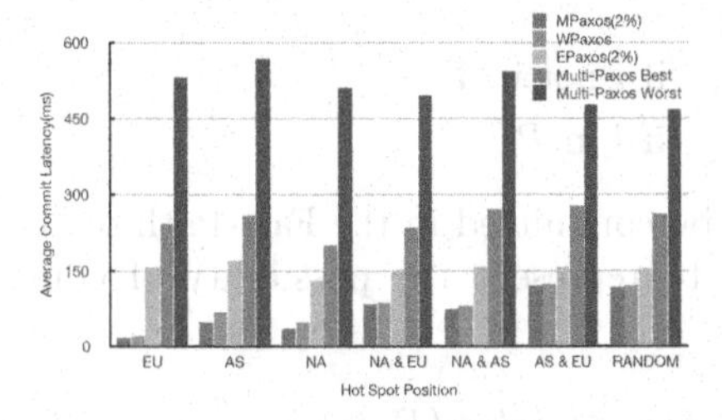

Fig. 3. Commit latency under static workloads

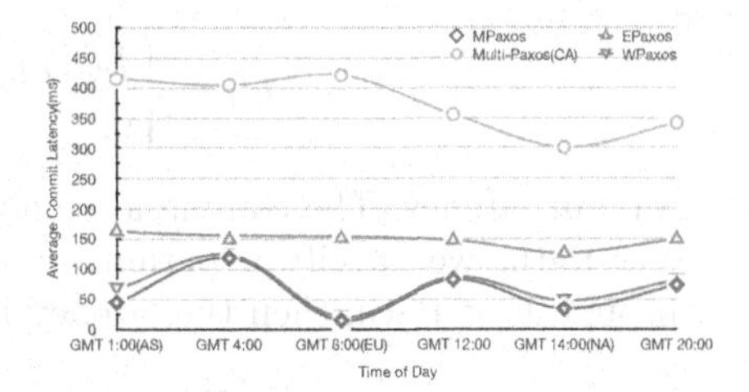

Fig. 4. Latency under regularly shifting workloads

5.2 Latency in Wide Area

We evaluate MPaxos with nine replicas (tolerating one failure, so the minimal size of Working Cluster is 3). The replicas are located in Amazon EC2 datacenters in California (CA), Virginia (VA) , Oregon (OR) , Japan (JP), Korea (KR), Singapore (SG), London (LON), Paris (PAR), and Sweden (SE). We set $F_W = 1$ for MPaxos and $F = 0, f = 1$ for WPaxos [11].

Figure 3 shows the average client latency for MPaxos, EPaxos, Multi-Paxos and WPaxos under static workloads, where WPaxos is a recent leader-based

Paxos protocol optimized for migration scenario. The X-Axis indicates the positions of clients. With nine replicas, protocols with static clusters – such as EPaxos and Multi-Paxos with static clusters – produce high latency, while protocols with migratable clusters – such as MPaxos and WPaxos – achieve lower latency, and MPaxos outperforms WPaxos because of the leaderless command committing fashion. And the last group of the test (RANDOM), in which requests come from a random client of the world, shows that MPaxos also works well under irregular workloads.

Figure 4 shows the average client latency for these protocols under the shifting workload. MPaxos also achieves the lowest commit latency by timely responding to the shifting workload. WPaxos performs very close to MPaxos. Nevertheless, as shown in Fig. 5, MPaxos outperforms WPaxos in migration cost. Although WPaxos only need 1 round of communication during migration instead of 2 round in MPaxos, it suffers from a larger Phase-1 quorum size (at least 6 replicas), where MPaxos only need to communicate with 2 replicas twice.

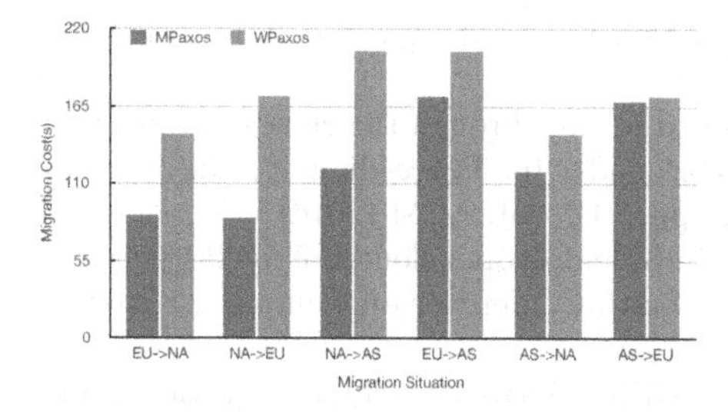

Fig. 5. Migration cost between different continents

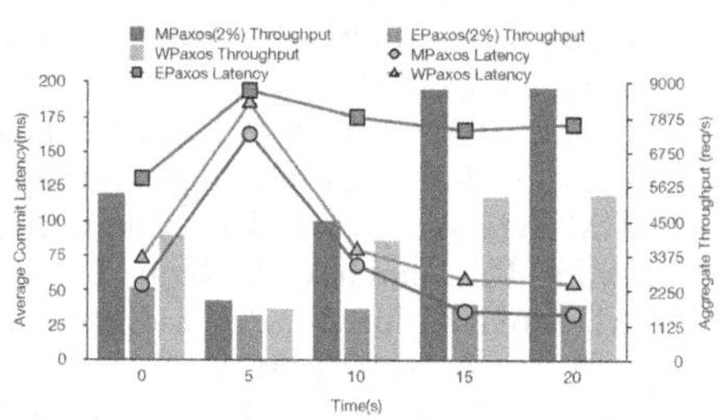

Fig. 6. Latency and workload when an emergency occurred in the 3rd second.

To evaluate how MPaxos and other protocols perform under emergency, we initialize a workload by deploy 300 clients in NA, then simulate the emergency by shutting down all clients in NA and starting 300 new clients in AS in the 3rd second. Figure 6 shows how latency and throughput change. It shows that MPaxos consistently retain the lowest latency and highest throughput by timely responding to the changes in client characteristics. The latency of WPaxos is roughly as good as MPaxos, but the single-leader-per-object style limits the throughput.

6 Conclusion

In this paper, we propose MPaxos, a Paxos-based protocol for shifting workloads. We show that designing specifically for the client characteristics yields significant performance rewards. MPaxos includes two main proposals: (1) use Working Clusters to make Replication quorums small and close to users, (2) Reorganization that enables Working Clusters timely responding to workload shifting with low migration cost. These proposals improve performance significantly as we show on a real deployment across 9 datacenters.

Acknowledgement. This work is supported by National Key R&D Program of China (2018YFB 1003404), National Science Foundation of China under grant number 61672232 and Youth Program of National Science Foundation of China under grant number 61702189.

References

1. Lamport, L.: The part-time parliament. ACM Trans. Comput. Syst. (TOCS) **16**(2), 133–169 (1998)
2. Lamport, L.: Paxos made simple. ACM SIGACT News **32**(4), 18–25 (2001)
3. Hunt, P., Konar, M., Junqueira, F.P., et al.: ZooKeeper: Wait-free coordination for internet-scale systems. In: USENIX Annual Technical Conference, vol. 8(9) (2010)
4. Moraru, I., Andersen, D.G., Kaminsky, M.: Egalitarian paxos. In: ACM Symposium on Operating Systems Principles (2012)
5. Lamport, L., Malkhi, D., Zhou, L.: Reconfiguring a state machine. SIGACT News **41**(1), 63–73 (2010)
6. Shen, Z., Jia, Q., Sela, G.E., et al.: Follow the sun through the clouds: application migration for geographically shifting workloads. In: Proceedings of the Seventh ACM Symposium on Cloud Computing, pp. 141–154. ACM (2016)
7. Bryant, R., Tumanov, A., Irzak, O., et al.: Kaleidoscope: cloud micro-elasticity via VM state coloring. In: Proceedings of the Sixth Conference on Computer Systems, pp. 273–286. ACM (2011)
8. Hines, M.R., Gopalan, K.: Post-copy based live virtual machine migration using adaptive pre-paging and dynamic self-ballooning. In: Proceedings of the 2009 ACM SIGPLAN/SIGOPS International Conference on Virtual Execution Environments, pp. 51–60. ACM (2009)
9. Burrows, M.: The chubby lock service for loosely-coupled distributed systems. In: Proceedings of the 7th Symposium on Operating Systems Design and Implementation, pp. 335–350 (2006)
10. Guo, J., Chu, J., Cai, P., Zhou, M., Zhou, A.: Low-overhead Paxos replication. Data Sci. Eng. **2**(2), 169–177 (2017). https://doi.org/10.1007/s41019-017-0039-z
11. Ailijiang, A., Charapko, A., Demirbas, M., et al.: WPaxos: wide area network flexible consensus. IEEE Trans. Parallel Distrib. Syst. **31**(1), 211–223 (2019)

GDPC: A GPU-Accelerated Density Peaks Clustering Algorithm

Yuxuan Su, Yanfeng Zhang(✉), Changyi Wan, and Ge Yu

Northeastern University, Shenyang, China
{yuxuansu,wanchangyi}@stumail.neu.edu.cn, {zhangyf,yuge}@mail.neu.edu.cn

Abstract. Density Peaks Clustering (DPC) is a recently proposed clustering algorithm that has distinctive advantages over existing clustering algorithms. However, DPC requires computing the distance between every pair of input points, therefore incurring quadratic computation overhead, which is prohibitive for large data sets. To address the efficiency problem of DPC, we propose to use GPU to accelerate DPC. We exploit a spatial index structure VP-Tree to help efficiently maintain the data points. We first propose a vectorized GPU-friendly VP-Tree structure, based on which we propose GDPC algorithm, where the density ρ and the dependent distance δ can be efficiently computed by using GPU. Our results show that GDPC can achieve over 5.3–78.8× acceleration compared to the state-of-the-art DPC implementations.

1 Introduction

Density Peaks Clustering (DPC) [6] is a novel clustering algorithm proposed recently. Given a set of points, DPC computes two metrics for every point p: (i) the local density ρ and (ii) the dependent distance δ. The *local density* ρ_i of data point p_i is the number of points whose distance to p_i is smaller than d_c.

$$\rho_i = |\{p_j | d_{ij} < d_c\}| \tag{1}$$

where d_{ij} is the distance from point p_i to point p_j, and d_c is called the cutoff distance. The *dependent distance* δ_i of point p_i is computed as

$$\delta_i = \min_{j:\rho_j > \rho_i} (d_{ij}) \tag{2}$$

It is the minimum distance from point p_i to any other point whose local density is higher than that of point p_i. Suppose point p_j is point p_i's the nearest neighbor with higher density, i.e., $p_j = argmin_{j:\rho_j > \rho_i}(d_{ij})$. We say that point p_i is *dependent* on point p_j and name point p_j as the *dependent point* of point p_i.

Figure 1 illustrates the process of DPC through a concrete example. Figure 1a shows the distribution of a set of 2-D data points. Each point p_i is depicted on a *decision graph* by using (ρ_i, δ_i) as its x-y coordinate as shown in Fig. 1b. By observing the decision graph, the *density peaks* can be identified in the top right region since they are with relatively large ρ_i and large δ_i. Since each point

Y. Nah et al. (Eds.): DASFAA 2020, LNCS 12112, pp. 305–313, 2020.
https://doi.org/10.1007/978-3-030-59410-7_21

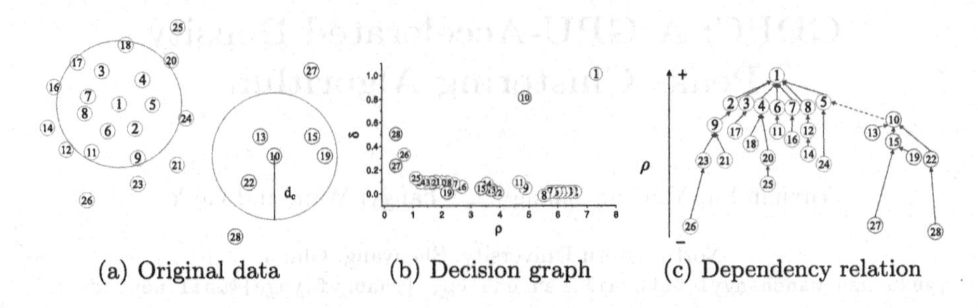

(a) Original data (b) Decision graph (c) Dependency relation

Fig. 1. An illustrative example of density peaks clustering.

is only dependent on a single point, we can obtain a dependent tree [2] rooted by the absolute density peak as shown in Fig. 1c. The height of each point implies the density. The length of each link implies the dependent distance. For each point there is a dependent chain ending at a density peak. Then each remaining point is assigned to the same cluster as its dependent point.

Compared with previous clustering algorithms [5], DPC has many advantages. 1) Unlike Kmeans, DPC does not require a pre-specified number of clusters. 2) DPC does not assume the clusters to be "balls" in space and supports arbitrarily shaped clusters. 3) DPC is more deterministic, since the clustering results have been shown to be robust against the initial choice of algorithm parameters. 4) The extraction of (ρ, δ) provides a two dimensional representation of the input data, which can be in very high dimensions.

While DPC is attractive for its effectiveness and its simplicity, the application of DPC is limited by its computational cost. In order to obtain the density values ρ, DPC computes the distance between every pair of points. That is, given N points in the input data set, its computational cost is $O(N^2)$. Moreover, in order to obtain the dependent distance values δ, a global sort operation on all the points based on their density values (with computational cost $O(Nlog(N))$) and $\frac{N(N-1)}{2}$ comparison operations are required. As a result, it can be very time consuming to perform DPC for large data sets. The recent advance of GPU technology is offering great prospects in parallel computation. There exist several related works [1,4] having been devoted to accelerate DPC using GPU's parallelization ability. However, these methods only consider utilizing GPU's hardware features to accelerate DPC without paying attention to parallelizable index structures that can maximize GPU performance.

In this paper, we exploit a spatial index structure vantage point tree (VP-Tree) [7] to help efficiently maintain clustering data. With VP-Tree, data points are partitioned into "hypershells" with decreasing radius. Comparing with other spatial index structures, VP-Tree is more appropriate in DPC algorithm, because the decreasing-radius hypershell structure can well support the point density computation (that obtains a point's nearby points within a pre-defined radius) and the dependent distance computation (that obtains the distance to a nearest neighbor with higher density). More importantly, the construction and the search of VP-Tree can be well parallelized to adapt to GPU's structure. Based on the GPU-based VP-Tree, we propose *GDPC* algorithm, where the density ρ and the dependent distance δ can be efficiently calculated. Our results show that GDPC can achieve over 5.3–78.8× acceleration compared to the state-of-the-art DPC implementations.

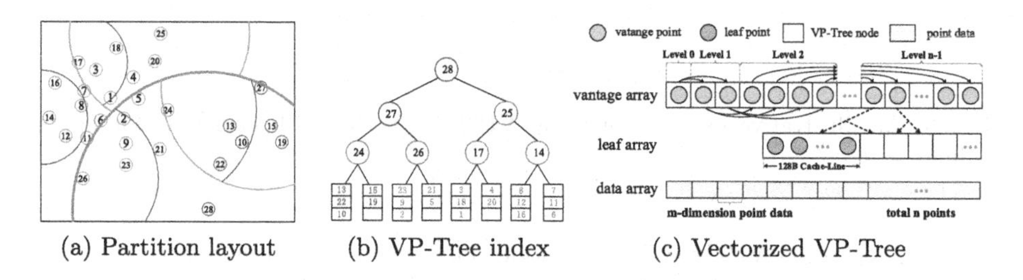

(a) Partition layout (b) VP-Tree index (c) Vectorized VP-Tree

Fig. 2. VP-Tree

2 Vectorized VP-Tree Layout

In DP clustering, the calculations of the density value ρ and the dependence value δ for each data point are the two key steps, which take up most of the computation time. According to Eq. (1), the computation of the density values requires a huge amount of nearest neighbors (NN) search operations, especially for big data clustering. According to Eq. (2), the computation of a point's dependence value also requires to access the point's NNs since the point's dependent point is likely to be close. A common approach for speeding up NN search is to exploit spatial index.

Based on our observation and analysis, Vantage Point Tree (VP-Tree) [7] is the best spatial index candidate. Each node of the tree contains one of the data points, and a radius. Under the left child are all points which are closer to the node's point than the radius. The other child contains all of the points which are farther away. The construction of VP-Tree can be explained with an illustrative example. As shown in Fig. 2, point 28 is firstly chosen as the *vantage point* (vp) as it is far away from other points. Point 28 is also picked as the level-0 vp (root node) of the VP-Tree as shown in Fig. 2b. We then draw a ball centered at point 28 with carefully computed radius r such that half of the points are in the ball while half are outside. All the points in the ball are placed in the root node's left subtree, while all the points outside are placed in the right subtree. The process is recursively applied for the inside-ball points and outside-ball points respectively. Finally, we will obtain such a VP-Tree as shown in Fig. 2b. The tree only requires a distance function that satisfies the properties of a metric space [3]. It does not need to find bounding shapes (hyperplanes or hyperspheres) or find points midway between them. Furthermore, the construction and the search of VP-Tree can be efficiently parallelized with CUDA since only a few data dependencies are required to handle.

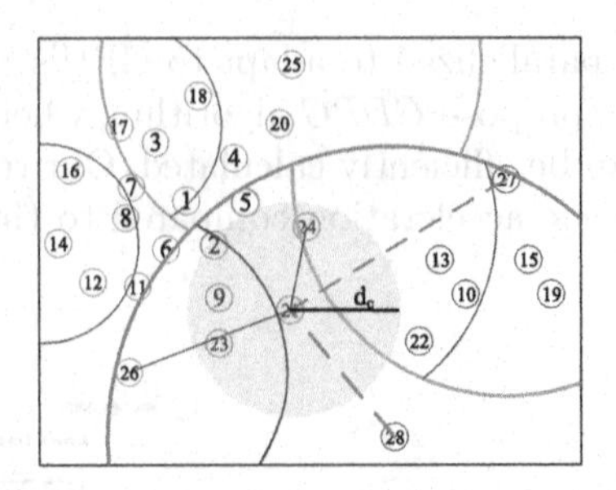

Fig. 3. An illustrative example of using VP-Tree (better with color) (Color figure online)

In the original VP-Tree, a child node reference is a pointer referring to the location of next level child. Since the memory locations of these tree nodes are randomly spread out in memory space, it is difficult to utilize the GPU memory hierarchy to explore the data locality and could result in memory divergence. Therefore, a vectorized GPU-friendly VP-Tree structure is desired instead of the pointer-based tree structure. In our approach as shown in Fig. 2c, the VP-Tree nodes are arranged in a breadth-first fashion in a one dimensional array (or vector) instead of pointers. The root node is stored at position 0 in the array. Suppose a node's position is i, we can obtain its left child position as $2i + 1$ and its right child position as $2i + 2$. Since a node's child position is known, there is no need to store pointers. This design requires less memory and provides higher search throughput due to coalesced memory access.

3 GDPC Based on VPTree

3.1 Computing Density Values ρ

Our basic idea is to utilize the VP-Tree index to avoid unnecessary distance measurements. We illustrate the use of existing VP-Tree through an illustrative example. When computing a point's density value, it is required to access the points in point 21's d_c range. Let us compute point 21's density value (i.e., count the number of points within the grey circle) based on an existing VP-Tree's space partition result as shown in Fig. 3. We first evaluate the distance from point 21 to the level-0 vantage point 28. Since the grey circle with radius d_c is totally inside the level-0 ball (with green arc line), i.e., $|p21, p28| + d_c \leq vantage[0].radius$ where $|\cdot|$ is distance measurement, it is enough to search the left child, where the vantage point is point 27. Vantage point 27's ball (with orange arc line) intersects with the grey circle, i.e., $|p21, p27| - d_c \leq vantage[1].radius$ (the grey circle has a part inside the orange ball) and $|p21, p27| + d_c \geq vantage[1].radius$ (the grey circle has a part outside the orange ball), so we need to search both the left child (with vantage point 24) and the right child (with vantage point 26). Similarly, we find the grey circle is totally inside vantage point 24's ball

Algorithm 1: GDPC Algorithm based on VP-Tree

Input: cut-off distance dc, data array $data[]$, vantage array $vantage[]$, vatange array length n, and leaf array $leaf[]$

Output: density $\rho[]$, dependent distance $\delta[]$, point-cluster assignment $cluster[]$

```
1  foreach point pid parallel do
2      Stack S.push(0) ;                          // push root node id into stack
3      while S is not empty do
4          i ← S.pop();
5          if i ≥ n then
6              cover_leafs.append(i − n) ;        // this is a covered leaf node
7          if |data[vantage[i].id], data[pid]| − d_c ≤ vantage[i].radius then
8              S.push(2i + 1) ;                   // search left child node
9          if |data[vantage[i].id], data[pid]| + d_c ≥ vantage[i].radius then
10             S.push(2i + 2) ;                   // search right child node
11     ρ[pid] ← count the number of points in all leaf[l] ∈ cover_leafs whose
         distance to pid is less than d_c
12 peak_candidates ← ∅;        // initialize the density peak candidates set
13 foreach point pid parallel do
14     dep[pid], δ[pid] ← find the nearest neighbor point that has higher density
         than ρ[pid] in all leaf[l] ∈ cover_leafs and compute its distance ;
15     if point pid has the highest density among the covered leaf nodes then
16         add pid into peak_candidates;
17 foreach point pid ∈ peak_candidates parallel do
18     dep[pid], δ[pid] ← find the nearest neighbor point that has higher density
         than ρ[pid] among all points and compute its distance ;
19 peak[] ← determine the density peaks which have both larger ρ and larger δ;
20 cluster[] ← assign points to clusters based on peak[] and dep[];
```

but intersecting with vantage point 26's ball, so we can locate the covered leaf nodes, i.e., vantage point 24's left leaf node (containing point 24, 13, 10, 22), vantage point 26's left leaf node (containing point 26, 23, 9, 2), and vantage point 26's right leaf node (containing point 5, 21, 28). These points are the candidate points for further distance calculations. We describe the details more formally in Algorithm 1. Line 1–11 depicts the ρ computation process.

We design a GPU-friendly search algorithm on the vectorized VP-Tree to achieve high parallelism and coalesced memory access. Specifically, we use several parallel optimizations: **1) Arrange calculation order** (Line 1). During tree traversal, if multiple threads in a warp execute random queries, it is difficult to achieve a coalesced memory access because they might traverse the tree along different paths. If multiple queries share the same traversal path, the memory accesses can be coalesced when they are processed in a warp. We design our warp parallelism in terms of VP-Tree properties. Because the points assigned to the same leaf node share the same traversal path, we assign the threads in the same warp to process the points in the same leaf node. That is, we execute warp-parallelism between leaf nodes and execute thread-parallelism within each leaf node. By this way, we can mitigate warp divergence. **2) Ballot-Counting optimization** (Line 11). CUDA's *ballot* function takes a boolean expression and returns a 32 bit integer, where the bit at every position i is the boolean value of thread i within the current thread's warp. The intrinsic operation can enable an efficient implement of the per-block scan. By combining the `__ballot()` and `__popc()` intrinsics, we can efficiently count the number of points within d_c. **3) Fully contained leaf nodes**. In original VP-Tree, the points in vantange node might be not contained in leaf nodes. That means, we have to check internal vantage nodes in order not to miss d_c range points. In our implementation, the vantage points also have their copies in leaf nodes to avoid memory divergence.

3.2 Computing Dependent Distances δ

Given the computed density values, we can calculate the dependent distance values as shown in Line 12–18 in Algorithm 1. Recall that the dependent distance of a point is its distance to the nearest neighbor with higher density as shown in Eq. (2). We can again leverage the VP-Tree index. Since a point's nearest neighbor with higher density is highly likely to be in its leaf nodes, we first locally search among its leaf nodes (Line 14). If such a point does not exist (when the point has the highest density among its leaf nodes), we sort the candidate points in the descending order of their density values and then search globally by checking all the other points with higher density (Line 18).

3.3 Assigning Points to Clusters

After picking a set of density peaks (Line 19), we should assign each point to a certain cluster. We perform this by tracing the assignment chain till meeting a certain density peak. The assignment dependency relationship (as shown in Fig. 1c) is recorded when computing δ (Line 14, 18). To adapt to GPU's parallelization ability, we need to build a reverse index of the assignment chain. Then, the point assignment is similar to a label propagation process starting from a number of density peaks in a top-down manner (Line 20), where the label is a certain density peak's id and the reversed dependencies can be regarded as the underlying graph edges. This label propagation process can be easily parallelized since the propagations on different sub-trees are totally independent.

4 Experiments

Preparation. We conduct all experiments on an 8-core server with an NVIDIA RTX2080Ti GPU. Our implementation is compiled by CUDA 10 along with nvcc optimization flag -O3. Table 1 lists the data sets used in our experiments. These include two small sized 2D data sets, and four real world large high-dimensional data sets, all of which are available online. To obtain clustering

Table 1. Data sets

Data set	No. instances	No. dimensions
Aggregation	788	2
S2	5,000	2
Facial	27,936	300
KDD	145,751	74
3Dspatial	434,874	3
BigCross500K	500,000	57

result in a reasonable time period, we construct a smaller BigCross500K data set by randomly sampling 500,000 points from the original data set.

Performance Comparison with State-of-the-Art DPC Methods. We first compare with the state-of-the-art GPU-based DPC algorithm CUDA-DP [1] on smaller datasets, which returns out-of-memory error on larger datasets 3Dspatial, KDD, and BigCross500K. CUDA-DP just optimizes the distance calculations without leveraging spatial index structure. In Fig. 4, we can see that our GDPC can achieve a 5.3×–17.8× speedup attributed to our vectorized VP-Tree design and GPU-friendly parallel algorithm. In addition, in order to evaluate our algorithm on larger datasets, we compare with a state-of-the-art distributed DP clustering algorithm LSH-DDP [8], which is implemented based on Hadoop

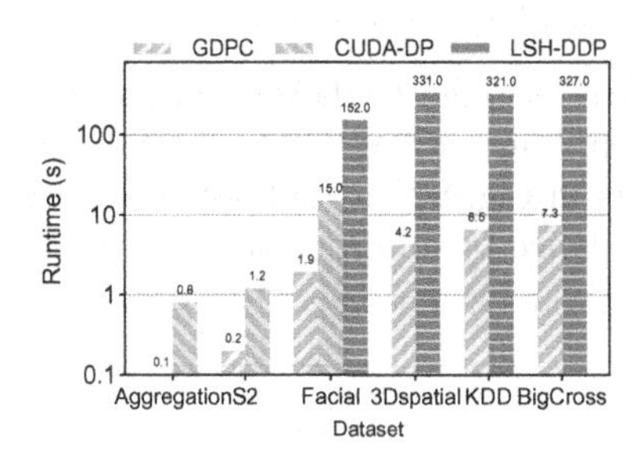

Fig. 4. Runtime comparison

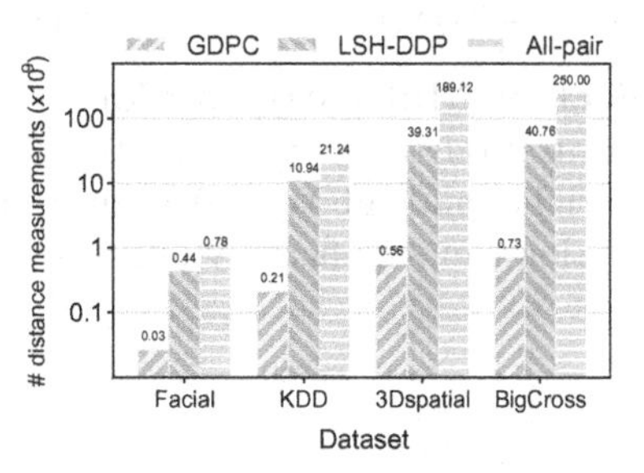

Fig. 5. Computational cost

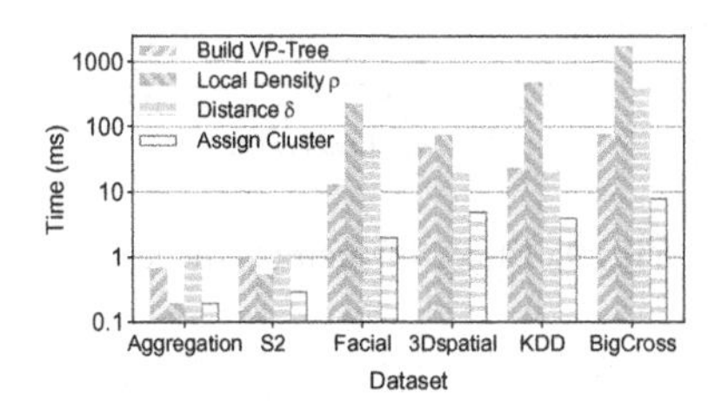

Fig. 6. Runtime breakdown

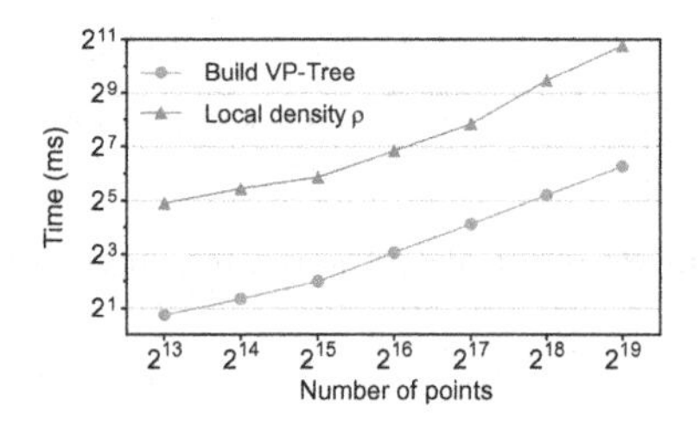

Fig. 7. Scaling performance

MapReduce and utilizes locality-sensitive hashing index to improve the ρ and δ calculations. The distributed LSH-DDP experiments are performed on a cluster with 5 machines, each equipped with an Intel I5-4690 3.3G 4-core CPU, 4 GB memory. We can see that our GDPC can achieve 44.8–78.8× speedup.

Computational Cost Analysis. The distance calculations for computing ρ and δ is the most expensive part, especially for large and high-dimensional data. GDPC utilizes VP-Tree to avoid large number of unnecessary distance calculations due to its excellent support for nearest neighbors search. Similarly, LSH-DDP also leverages LSH index to avoid unnecessary distance calculations. We evaluate the computational cost of naive all-pair computation, LSH-DDP, and GDPC by comparing their number of distance calculations during the clustering process and show the results in Fig. 5. Our GDPC requires significantly fewer exact distance calculations than prior works, say only 1.4–6.8% of LSH-DDP and 0.3–3.8% of all-pair calculations.

Runtime Breakdown Analysis. There are four main steps to obtain the final clustering result, which is VP-Tree construction, density ρ calculation, dependent distance δ calculation, and point-to-cluster assignment. In Fig. 6, we can see that for larger dataset, the ρ computation is always the most expensive part since it requires large number of distance measurements.

Scaling Performance. We randomly choose 2^{13}–2^{19} number of points from the BigCroos500K dataset to generate multiple datasets with different sizes. The runtime in Fig. 7 exhibits linearly growth when increasing the data size, while the all-pair distance calculations will exhibit quadratic growth. This experiment shows our GDPC algorithm can achieve great scaling performance.

Effect of Multi-stream Construction. We leverage CUDA's multi-stream optimization to improve the parallelism when constructing left child sub-tree and right child sub-tree. We can see the results in Fig. 8 that multi-stream optimization can significantly improve the performance by an order of magnitude.

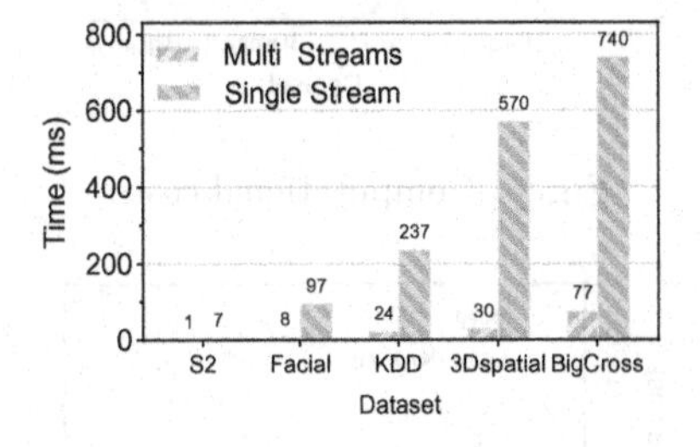

Fig. 8. Effect of multi-stream processing

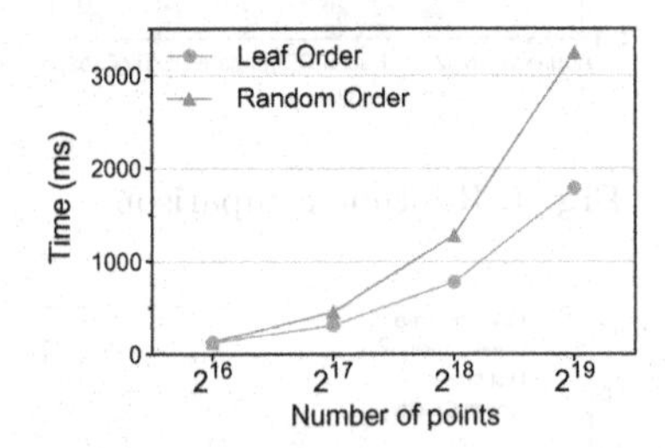

Fig. 9. Effect of memory access

Effect of Coalesced Memory Access. To understand the performance improvement by our optimization, we use the random processing order to see the advantages by using the leaf-based processing order. In Fig. 9, our approach shows significant better performance especially when increasing the data size.

5 Conclusion

In this paper, we propose a parallel density peaks algorithm named GDPC, which can fully utilize the powerful computation resources of GPU. It leverages a GPU-friendly spatial index VP-Tree to reduce the unnecessary distance calculations. The VP-Tree construction process and the DP clustering process are greatly improved by utilizing GPU's parallel optimizations. Our result show that GDPC can achieve 5.3–78.8× speedup over the state-of-the-art DPC implementations.

Acknowledgements. This work was partially supported by National Key R&D Program of China (2018YFB1003404), National Natural Science Foundation of China (61672141), and Fundamental Research Funds for the Central Universities (N181605017, N181604016).

References

1. Ge, K., Su, H., Li, D., Lu, X.: Efficient parallel implementation of a density peaks clustering algorithm on graphics processing unit. Front. Inf. Technol. Electron. Eng. **18**(7), 915–927 (2017). https://doi.org/10.1631/FITEE.1601786
2. Gong, S., Zhang, Y., Yu, G.: Clustering stream data by exploring the evolution of density mountain. Proc. VLDB Endow. **11**(4), 393–405 (2017)
3. Kramosil, I., Michálek, J.: Fuzzy metrics and statistical metric spaces. Kybernetika **11**(5), 336–344 (1975)
4. Li, M., Huang, J., Wang, J.: Paralleled fast search and find of density peaks clustering algorithm on gpus with CUDA. In: SNPD '2016. pp. 313–318 (2016)
5. Patil, C., Baidari, I.: Estimating the optimal number of clusters k in a dataset using data depth. Data Sci. Eng. **4**(2), 132–140 (2019)
6. Rodriguez, A., Laio, A.: Clustering by fast search and find of density peaks. Science **344**(6191), 1492–1496 (2014)
7. Yianilos, P.N.: Data structures and algorithms for nearest neighbor search in general metric spaces. In: SODA '93. pp. 311–321 (1993)
8. Zhang, Y., Chen, S., Yu, G.: Efficient distributed density peaks for clustering large data sets in mapreduce. IEEE Trans. on Knowl. Data Eng. **28**(12), 3218–3230 (2016)

RS-store: A SkipList-Based Key-Value Store with Remote Direct Memory Access

Chenchen Huang, Huiqi Hu(✉), Xuecheng Qi, Xuan Zhou, and Aoying Zhou

School of Data Science and Engineering,
East China Normal University, Shanghai, China
{cchuang,xcqi}@stu.ecnu.edu.cn, {hqhu,xzhou,ayzhou}@dase.ecnu.edu.cn

Abstract. Many key-value stores use RDMA to optimize the messaging and data transmission between application layer and storage layer, most of which only provide point-wise operations. Skiplist-based store can support both point operations and range queries, but its CPU-intensive access operations combined with the high-speed network will easily lead to the storage layer reaches CPU bottlenecks. In this paper, we present RS-store, a skiplist-based key-value store with RDMA, which can overcome the cpu handle of the storage layer by enabling two access modes: local access and remote access. In RS-store, we redesign a novel data structure *R-skiplist* to save the communication cost in remote access, and implement a latch-free concurrency control mechanism to ensure all the concurrency during two access modes. At last, our evaluation on a RDMA-capable cluster shows that the performance of RS-store over *R-skiplist* is 0.6 ×–1 × higher than the existing skiplist, and it supports application layer's high scalability.

Keywords: Skiplist · Key-value store · RDMA

1 Introduction

Key-value store is a critical cornerstone for Internet services to deliver low-latency access to large-scale data sets. In recent years, RDMA (Remote Direct Memory Access) - a necessary network technology for high performance data centers, has been applied in many key-value stores to further reduce the latency of messaging and data transmission, which include some commodity stores such as RAMCloud [13] and Memcached [9], and some research stores like Pliaf [12], HERD [10] and HydraDB [17]. However, most of these RDMA-enabled key-value stores only provide point-wise operations.

Skiplist is widely-used in key-value stores, as it supports range query and is more parallelizable than other search data structures because of the fine-grained data access and relaxed structure hierarchy. The typical representatives of the key-value stores are LevelDB [1] and RocksDB [2]. Thus, we can combine RDMA and skiplist to build a key-value store that enables high performance range query. However, traditional skiplist designs are node-based, providing little data locality and high irregular access patterns, both of which are significant drawbacks

Y. Nah et al. (Eds.): DASFAA 2020, LNCS 12112, pp. 314–323, 2020.
https://doi.org/10.1007/978-3-030-59410-7_22

for CPU performance. If using the traditional request processing method, the storage layer is mainly responsible for performing operations, and the application layer simply sends RPC requests and receives replies from the storage layer, then the CPU-intensive access of skiplist combined with high-speed network is easy to make the storage layer's CPU reach the bottleneck.

In a RDMA-enabled key-value store, the common solution to the CPU bottlenecks of storage layer is offloading some operations into the application layer and using RDMA to perform remote access, which can bypass the CPU of storage layer. However, this method is only used in the hash table-based store. There is a problem when it used in the skiplist-based store: in the traditional skiplist, a node is accessed from the head node. As one RDMA round trip can only obtain one node, then traversing skiplist on the application layer by RDMA requires many rounds of network communications.

In this paper, a skiplist-based key-value store with RDMA (RS-store) is designed, it not only supports local access skiplist at storage layer, but also supports remote access by RDMA on the application layer, which makes performance can continue to improve after the storage layer's CPU becomes a bottleneck. To save the communication cost in remote access, we redesign a novel skiplist *R-skiplist* by packing many nodes into a block. Thus a block can contain multiple nodes, and one RDMA communication always retrieves a block instead of a node. Meanwhile, concurrency control is the key to high scalability. In RS-store, besides the concurrency control between local access operations, concurrency control is also required between local access and remote access operations. So we implement a latch-free concurrency control strategy to enable the concurrency among all operations in RS-store.

2 Background

RDMA. Remote Direct Memory Access (RDMA) enables low-latency network access by directly accessing memory from remote machines. There are two types of commands in RDMA for remote memory access: two-sided verbs and one-sided verbs. In this paper, we mainly use one-sided verbs to send message and transmit data, because it bypasses the CPU of remote machines and provides relatively higher bandwidth and lower latency than two-sided verbs.

RDMA Optimizations in Key-Value Stores. RDMA has been adopted in several key-value stores to improve performance [10,12,17]. Pilaf [12] optimizes the get operation using multiple RDMA read commands at the client side, which offloads hash calculation burden from remote servers to clients. HERD [10] implements both get and put operations using the combination of RDMA write and UD send. HydraDB [17] is a versatile key-value middle ware that achieves data replication to guarantee fault-tolerance and awareness for NUMA architecture, and adds client-side cache to accelerate the get operation. Most of these systems are hash table-based and only provide point-wise operations. In addition, they exclusively use local access or remote access, which can lead to unilateral bottlenecks in CPU or network.

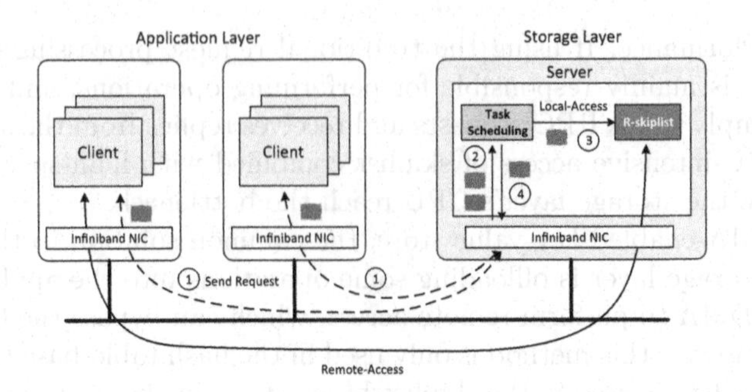

Fig. 1. The architecture of RS-store.

SkipList. Pugh introduced the skiplist data structure in [14], and described highly concurrent implementation of skiplists using locks in [15]. Fraser in [6] discussed various practical non-blocking programming abstractions, including the skiplist. Herlihy et al. [7] presented blocking skiplist implementations that are based on optimistic synchronization. Herlihy et al. [8] also provided a simple and effective approach to design nonblocking implementation of skiplists. However, all these concurrency control methods only supports the concurrency between local access operations, but not the concurrency between remote access and local access operations. In RS-store, we design a latch-free concurrency control mechanism to support the concurrency among all operations.

3 RS-store Design

In this section, we first give the architecture of RS-store and then discuss its main component: the data structure of *R-skiplist*.

3.1 Architecture

Figure 1 shows the architecture of RDMA-enabled skiplist-based store (RS-store), which consists of two parts: application layer and storage layer. For convenience, we respectively call the nodes in application layer and storage layer as client and server in the following. In the architecture, data are stored on server named *R-skiplist*, and clients and server are communicated with RDMA.

RS-store supports two access modes: local-access and remote-access.

- **Local-Access** performs the read (*get*, *scan*) and write (*insert*, *delete*) operations on the server;
- **Remote-Access** is that clients directly access the data by one-side RDMA read, it only supports read (*get*, *scan*) operations.

When a request need to be processed in RS-store, 1) the client first sends it to server using self-identified RPC [11], 2) then the server determines an access mode for the request by detecting the current CPU resource utilization.

3) If the determined result is local-access, a local thread will perform the request and return the result to client. 4) If the determined result is remote-access, this request will be immediately returned to the client and then execute remote-access through one-side RDMA read.

Note that we do not provide write operations for remote-access with two reasons: 1) RS-store aims to serve real-world workloads, which are mostly read-heavy [3]; 2) performing write operations on the client leads to complex and fragile designs, such as remote memory management and remote locking, which introduces the risk of clients failing.

3.2 Data Structure of *R-skiplist*

To facilitate remote-access, we redesign the data structure of *R-skiplist* based on a conventional skiplist (*skiplist*) [14]. Figure 2(a) gives an instance of *skiplist* ($p = 1/3$). The bottommost layer is a sorted linked list of data nodes, and the up layers are called index layers, each of which is composed of the keys randomly selected with a fixed probability p from the next lower layer. Traversing *skiplist* on the client by RDMA requires many rounds of network communications since one round trip can only obtain one node. To save the cost, *R-skiplist* takes the straightforward way by packing many nodes into a block, each block is constructed on a continuous address with fixed memories. Thus one RDMA communication can retrieves a block instead of a node.

Figure 2(b) gives an instance of *R-skiplist* based on Fig. 2(a). As shown in the figure, each layer is split into many blocks, and each block contains a sorted linked list of multiple nodes. At the same layer, one block links with the next block using a *next* pointer. For two adjacent layers, layer $i + 1$ links with layer i by a *down* pointer in the index node, which records the address of block that contains the node with the same key at next layer.

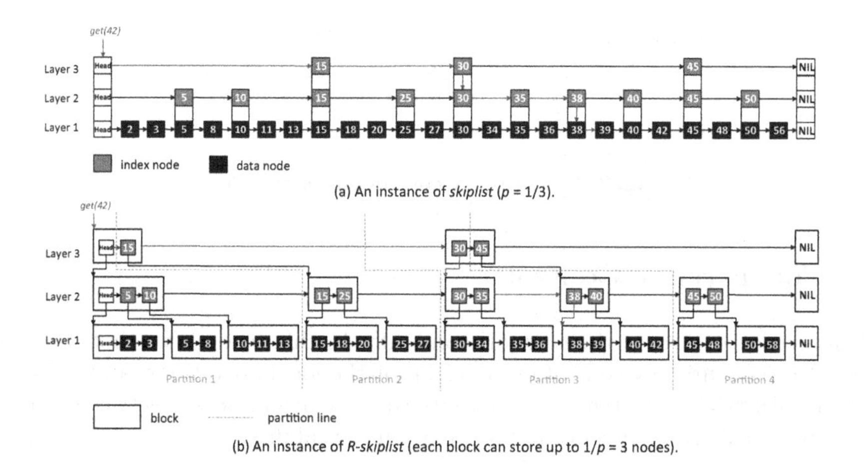

(a) An instance of *skiplist* ($p = 1/3$).

(b) An instance of *R-skiplist* (each block can store up to $1/p = 3$ nodes).

Fig. 2. *skiplist* and *R-skiplist*.

To construct the *R-skiplist*, we support inserting records. When inserting a data (*key*, *value*) into *R-skiplist*, we first find each layer's target block where the new data will be inserted and generate a random height h with probability p, then the new data is inserted into the target blocks from layer 1 to layer h. During the insert process, the target blocks will split in two cases: (i) since the block size is fixed, the target block will split equally when its memory capacity is not enough. (ii) the target block at layer i needs to be split into two blocks from the inserted *key* if it requires to be continuously inserted into layer $i+1$. Due to the second splitting rule, a key in layer $i+1$ must link to the fist node of a block in layer i with its down pointer. To delete data, we first mark all the target nodes and then delete them from top to bottom just as the *skiplist*.

R-skiplist is traversed from the first block at the top layer. At each layer, we find the maximum node smaller than (or equal with) the target by fetching one block at a time, then we use its down pointer to visit the next layer. The process continues until it reaches the bottom layer.

In comparison of performance, *R-skiplist* introduces block structures, which increase a bit of overhead for the insertion because of its memory copy cost of splitting nodes. However, the block structure is inherently friendly to CPU cache, which makes *R-skiplist* much more CPU efficient than *skiplist* when traveling the index. As most of operations are read operations and insert operations can also benefit from traversing the index, *R-skiplist* achieves better performance (also see experiment in Sect. 5.2).

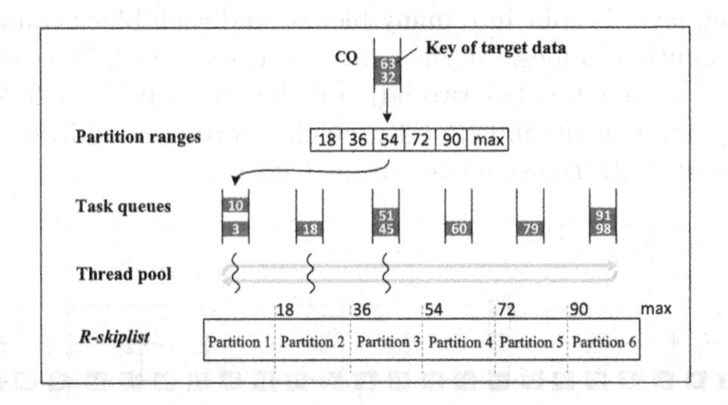

Fig. 3. Concurrency control on server side.

4 Concurrency Control

Concurrency control is the crux of scalability. In this section, we first design a latch-free concurrency control mechanism for the local operations, and then use a verification-based technique to support the concurrent access between remote read and local write. Both of them satisfy linearizability.

4.1 Server Side Concurrency Control

On the server side, latch contention on the data structure severely constrains the multi-core scalability [4]. To eliminate latches, we adopt the exclusive access strategy in concurrency control, which divides *R-skiplist* and tasks (i.e. operation requests) into different partitions and each partition is exclusively accessed by one thread. Figure 3 gives an instance of the overall process, the detailed implementation is as follows.

Exclusive Access. We first partition *R-skiplist* according to the keys in its top-layer nodes (see Fig. 2(b)), where a number of keys in the top-layer will be picked and divides data into different ranges. After that, we partition the tasks using the same way. As shown in Fig. 3, there are the same number of task queues as *R-skiplist*'s partitions, where one partition corresponds to one task queue. When the server successfully receives a task (when using RDMA, a task is popped from the Complete Queue (CQ)), it first push the task into the corresponding task queue according to the key, then each thread exclusively run tasks popped from a task queue.

As the tasks are partitioned by range of data, there may appear skewness. To this end, we take two countermeasures: (*i*) We use a thread pool, and all threads in the pool access task queues in a round-robin manner. The number of partitions is set larger than that of threads. Frequently switch task queues for threads causes context switching, thus we use an epoch-based method, where one thread binds to a queue with an epoch of time instead of just popping a task. If a task queue lacks tasks, we also release the thread from the task queue and make it serve other queues. (*ii*) We also adjust the partition ranges according to the skewness of task queues. To adjust the partition, we initialize another group of task queues. All subsequent tasks are pushed into the new task queues. Threads will not access the new task queues until all the old tasks are executed.

The partition method can also facilitate the traversal of *R-skiplist*. After *R-skiplist* is partitioned, each partition can be treated as an independent sub-list. Thus, we can perform traversal from the head node of each partition instead of from the head node of the complete *R-skiplist*, which can reduce the block accessed of the top-layer. Due to the structure of *R-skiplist*, the access method also ensures that two threads never access the same block below the top-layer, which eliminates most access conflicts. However, if the result set of a range query across multiple partitions, there will be conflicts when accessing the target data. For this case, we can use a verification-based technique to ensure the atomicity of block reading in other partitions, which will be introduced in the next subsection.

4.2 Client-Server Concurrency Control

To support concurrency between client and server, we guarantee the remote read atomicity using a verification-based technique, which can ensure the remote read and local write are linearizable.

For each block, we verify weather it is being modified by other threads on the client side. We add a flag *is_update* on each block. To update a block on

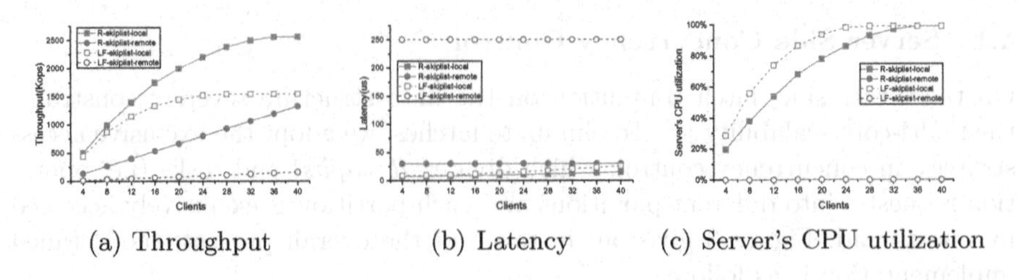

(a) Throughput (b) Latency (c) Server's CPU utilization

Fig. 4. Search performance of remote access and local access with the increase of clients

the server, the thread first sets *is_update* to true, performs the update, then resets *is_update* to false. When a remote read get a block from the server, it first verifies the block is being updated by checking the value of *is_update*: if false, it can directly access the block; if true, we retry to read the block. Note that, when trying to access the next block based a current block, the next block on the server may be in a split, which will cause the next block we read to contain only part of nodes. However, this would not affect the correctness of reading, because it can still continue to read the rest nodes using the next pointer of the next block (which has been split) like B-link tree [16].

5 Experiment

We run our experiments on a cluster with 8 machines, equipped with a Mellanox MT 27710 ConnectX-3 Pro 56 Gbps InfiniBand NIC, two 8-core Intel Xeon Silver 4208 processors and 128 GB of DRAM. In the cluster, a server is ran on one machine to store the data, and the clients are distributed among the remaining machines. Yahoo! Cloud Serving Benchmark (YCSB) [5] is chose to generate the workloads. By default, we run the experiment on an uniform workload and the data set has 20M records, there are 36 clients sending request in parallel, the server runs 8 threads to process local access, *R-skiplist* is divided into 16 partitions and its p value equals 0.03125. We also implement a Lock-Free Skiplist (*LF-skiplist*) [8] to make a compare with *R-skiplist*, which achieves lock-free using CAS instruction.

5.1 Remote Access vs. Local Access

Search Performance. Figure 4 gives the search throughput, latency and server's CPU utilization of remote access and local access over two indexes with the increase of clients. Firstly, whether *R-skiplist* or *LF-skiplist*, the throughput of local access is higher than that of remote access, because remote access requires network communications. However, with the increase of clients, the throughput of local access increases more and more slowly and stops increasing when the number of clients increased to a value, but the throughput of remote access increases almost linearly. Figure 4(c) shows that the throughput of local access

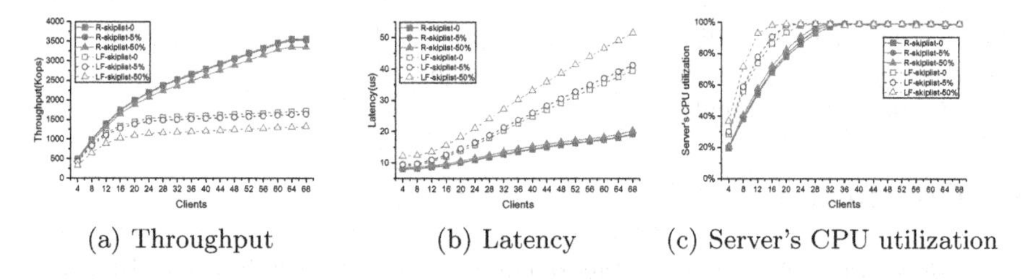

(a) Throughput (b) Latency (c) Server's CPU utilization

Fig. 5. Point operation performance of RS-store with the increase of clients

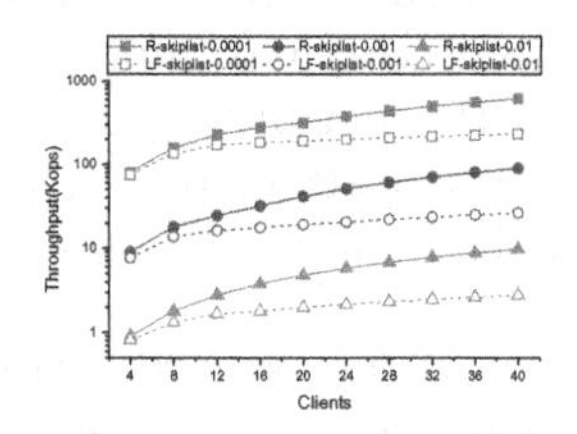

Fig. 6. Range query performance of RS-store with the increase of clients

stops increasing is because the server's CPU reaches a bottleneck, but the remote access does not consume the server's CPU at all. Secondly, the throughputs of remote access and local access over *R-skiplist* are always better than that over *LF-skiplist*, their latency speedup about 7 $\times$ and 0.6 $\times$ respectively, which shows that the block-based *R-skiplist* is very effective.

5.2 RS-store

Client-Side Scalability. Figure 5 gives the point operation throughput, latency and server's CPU utilization of RS-store over two indexes under the workload with different write ratios and clients. As the increase of clients, the throughputs over two indexes are increasing almost all the time. This is because when the server's CPU becomes a bottleneck, RS-store will call remote access to continue improving the performance. What's more, the throughput of *R-skiplist* under the workload with write operations is still higher than that of *LF-skiplist*, which demonstrates that the load caused by block split is relatively small compared with the performance improvement by our optimization. In addition, as the number of clients increases, the throughput of *R-skiplist* increases faster than that of *LF-skiplist*, especially when the write ratio is higher. That's because write operations will cause conflicts, *LF-skiplist* needs take time to deal with the conflicts, but *R-skiplist* avoids most conflicts by task partitioning, so the scalability of *R-skiplist* is better than *LF-skiplist*.

Figure 6 shows the range query throughput of RS-store over two indexes with different selectivity and clients. In the figure, the throughput of *R-skiplist* is still higher than that of *LF-skiplist*. In addition, with the increase of selectivity, the performance improvement of *R-skiplist* is more and more. This is because its block structure is friendly to the CPU cache in local access and reduces the communication cost in remote access, the higher the selectivity, the greater the performance improvement.

6 Conclusion

This paper presents a skiplist-based key-value store with RDMA (RS-store), which can avoid the client-side scalability from being blocked by the server's CPU bottleneck. In RS-store, we redesign a novel data structure *R-skiplist* to save the communication cost of remote access, and implement a latch-free concurrency control mechanism to support the concurrency of all operations. Experimental result shows that the performance of RS-store over *R-skiplist* is better than that over the existing lock-free skiplist, and it supports high client-side scalability.

Acknowledgements. This work is supported by National Key R&D Program of China (2018YFB1003303), National Science Foundation of China under grant number 61772202, Youth Program of National Science Foundation of China under grant number 61702189, Youth Science and Technology - Yang Fan Program of Shanghai under Grant Number 17YF1427800. Thanks to the corresponding author Huiqi Hu.

References

1. Leveldb.: In: http://code.google.com/p/Leveldb/
2. Rocksdb.: In: https://github.com/facebook/rocksdb/
3. Atikoglu, B., Xu, Y., Frachtenberg, E.: Workload analysis of a large-scale key-value store. In: SIGMETRICS. pp. 53–64 (2012)
4. Cha, S.K., Hwang, S., Kim, K., Kwon, K.: Cache-conscious concurrency control of main-memory indexes on shared-memory multiprocessor systems. In: 27th International Conference on Very Large Data Bases. pp. 181–190 (2001)
5. Cooper, B.F., Silberstein, A., Tam, E.: Benchmarking cloud serving systems with YCSB. In: 1st ACM Symposium on Cloud Computing. pp. 143–154 (2010)
6. Fraser, K.: Practical lock-freedom. Ph.D. thesis, University of Cambridge (2004)
7. Herlihy, M., Lev, Y., Luchangco, V., Shavit, N.: A simple optimistic skiplist algorithm. In: SIROCCO. pp. 124–138 (2007)
8. Herlihy, M., Lev, Y., Shavit, N.: Concurrent lock-free skiplist with wait-free contains operator. In: US Patent 7,937,378 (2011)
9. Jose, J., Subramoni, H., Luo, M.: Memcached design on high performance RDMA capable interconnects. In: ICPP. pp. 743–752 (2011)
10. Kalia, A., Kaminsky, M., Andersen, D.G.: Using RDMA efficiently for key-value services. In: ACM SIGCOMM 2014 Conference. pp. 295–306 (2014)
11. Lu, Y., Shu, J., Chen, Y.: Octopus: an rdma-enabled distributed persistent memory file system. In: 2017 USENIX Annual Technical Conference. pp. 773–785 (2017)
12. Mitchell, C., Geng, Y., Li, J.: Using one-sided RDMA reads to build a fast, cpu-efficient key-value store. In: Annual Technical Conference. pp. 103–114 (2013)
13. Ousterhout, J.K., Gopalan, A., Gupta, A.: The ramcloud storage system. ACM Trans. Comput. Syst. **33**(3), 7:1–7:55 (2015)
14. Pugh, W.: Skip lists: a probabilistic alternative to balanced trees. In: Workshop on Algorithms & Data Structures. pp. 668–676 (1990)

15. Pugh, W.: Concurrent maintenance of skip lists (1998)
16. Shasha, D.E.: Review - efficient locking for concurrent operations on b-trees. ACM SIGMOD Digital Review **1** (1999)
17. Wang, Y., Zhang, L., Tan, J.: Hydradb: a resilient rdma-driven key-value middleware for in-memory cluster computing. In: SC. pp. 22:1–22:11 (2015)

Balancing Exploration and Exploitation in the Memetic Algorithm via a Switching Mechanism for the Large-Scale VRPTW

Ying Zhang(✉), Dandan Zhang, Longfei Wang, Zhu He, and Haoyuan Hu

Zhejiang Cainiao Supply Chain Management Co., Ltd., Hangzhou, China
{youzhu.zy,danae.zdd,shouchu.wlf,zhuhe.hz,haoyuan.huhy}@cainiao.com

Abstract. This paper presents an effective memetic algorithm for the large-scale vehicle routing problem with time windows (VRPTW). Memetic algorithms consist of an evolutionary algorithm for the global exploration and a local search algorithm for the exploitation. In this paper, a switching mechanism is introduced to balance quantitatively between exploration and exploitation, to improve the convergent performance. Specifically, a similarity measure and a sigmoid function is defined to guide the crossover. Experimental results on Gehring and Homberger's benchmark show that this algorithm outperforms previous approaches and improves 34 best-known solutions out of 180 large-scale instances. Although this paper focuses on the VRPTW, the proposed switching mechanism can be applied to accelerate more general genetic algorithms.

Keywords: Vehicle routing · Memetic algorithm · Switching mechanism · Exploration and exploitation · Best-known solutions

1 Introduction

The vehicle routing problem with time windows (VRPTW) has been one of the most important and widely investigated NP-hard optimization problems in transportation, supply chain management, and logistics. It is an important problem that is faced by every delivery company when they have to divide up their packages to a set of trucks. The VRPTW involves finding a set of routes, starting and ending at a depot, that together cover a set of customers. Each customer should be visited exactly once by a single vehicle within a given time interval, and no vehicle can service more customers than its capacity permits.

The VRPTW is defined on a complete directed graph $G = (V, E)$, where $V = \{0, 1, ..., N\}$ is the set of vertices and $E = \{(i, j)|i \neq j \in V\}$ is the set of edges. Vertex 0 represents the depot and the other vertices represent the customers. Associated with each vertex i is a non-negative demand q_i and a time window $[e_i, l_i]$, where e_i and l_i represent the earliest and latest time to visit vertex i. Associated with each edge (i, j) is a travel cost d_{ij} and a non-negative travel time t_{ij}, where t_{ij} includes the service time at vertex i.

Y. Nah et al. (Eds.): DASFAA 2020, LNCS 12112, pp. 324–331, 2020.
https://doi.org/10.1007/978-3-030-59410-7_23

A feasible solution to the VRPTW is a set of m routes in graph G such that the following conditions hold: (1) each route starts and ends at the depot, (2) each customer i belongs to exactly one route, (3) the total demand of the visited customers does not exceed the vehicle capacity Q, and (4) the service at each customer i begins between e_i and l_i. The cost of one route is equal to the sum of the distance of the edges traveled. The problem often has a hierarchical objective: 1) Minimize number of vehicles 2) Minimize total distance.

2 Related Work

Some exact algorithms have been proposed for the VRPTW, e.g. [1,3], and work well for small-scale problems. By now, most of the instances in Solomon's benchmarks (100 customers) [14] have been solved to optimality. However, due to its NP-hardness, the computation time of exact methods for large-scale problems is not acceptable. Therefore, recent studies mainly focus on various heuristic algorithms, which can find acceptable, but not necessarily optimal solutions quickly. These heuristics consist of tabu search [4], particle swarm optimization [7], ant colony optimization [16] and memetic algorithm (MA)[2,10,12].

Among all these heuristics, the MA proves to be extremely effective, and our method belongs to this category. MA is a population based heuristic that combines evolutionary algorithm for the global search (exploration), with local search algorithm for the intensive search (exploitation) [10]. An edge assembly crossover (EAX) operator is usually applied to generate new children from parents. [9,12] demonstrate the potential of EAX in routing problems. In each generation, individuals are *randomly ordered* to conduct global exploration.

Recently, several studies reveal the importance of route diversity and similarity in evolutionary algorithms. The parent selection techniques used in the simple genetic algorithm (GA) only take into account the individual fitness, regardless of the parenthood or likeness. This process is random mating. Nevertheless, [5] and [13] separately propose a similarity measure based on Jaccard's similarity coefficient [8] and incorporate it into evolutionary algorithms. Parents are recombined under both fitness and similarity, which avoid early convergence and get an appropriate balance between the exploration and exploitation.

3 Our Focus and Contributions

The classical MA includes a random paired EAX operation. We find that the convergence of the algorithm could be slow, especially for large-scale problems. The local search operator contributes little to the acceleration of convergence. Based on the distance measurement, we focus on guiding the "ordering procedure" quantitatively, instead of random ordering.

The main idea is to control the order of parents been paired, so that the EAX operation can realize both global exploration and local exploitation. Specifically, a *similarity measure* is defined to quantify the likeness between two individual solutions. Low similarity means that two solutions are far away from each other in the solution space, and vice versa. Thus, if solutions are paired with *low* similarity, the EAX operation is more like conducting global exploration; analogously, *high* similarity means local exploitation. A steady step counter C records the non-improvement iterations. As C increases, the EAX operation will switch from exploitation to exploration gradually. Once the objective value gets improved, reset $C = 0$, a local exploitation is then performed.

Our method is tested on the well-known benchmark problems of [6]. Experimental results show that this method can significantly speed up the solving procedure. Moreover, this switching mechanism can also be applied to other memetic and genetic algorithms. To the best of our knowledge, this method is the first to quantitatively control exploration and exploitation regarding crossover operation in MA.

4 Our Method

4.1 An Overview of the MASM

Follows the classical MA (refer to [10] for more details), proposed MASM is also a two-stage approach where the number of routes and the total travel distance are independently minimized. The significant difference is in updating counter C to record the non-improvement iterations.

In each generation, the individuals with size N_{pop} are arranged in a tour for the crossover, where the probability for each individual solution σ_i to be selected as the k_{th} one in the tour is determined by $p_k(\sigma_i)$. For each pair of adjacent parents, the EAX operation generates N_{ch} offspring solutions. Each offspring σ_{ch} is then fixed by the repair operator to eliminate the infeasibility, and further improved by the local search methods. At the end of each generation, if the best solution improves, C is reset to 0, otherwise incremented by 1. How the counter C and the probability $p_k(\sigma_i)$ control the EAX will be presented below.

4.2 A Similarity Measure

To quantify the similarity between two individual solutions, we refer to the Jaccard Index [8], which is widely used to compare the similarity of sample sets. Let σ_A and σ_B be two solutions in one generation, and E_A and E_B be the sets of directed edges consisting of σ_A and σ_B, respectively. The Jaccard's similarity between σ_A and σ_B is defined as $s(\sigma_A, \sigma_B) = \frac{|E_A \cap E_B|}{|E_A \cup E_B|}$, where $|E_A \cap E_B|$ and $|E_A \cup E_B|$ are the number of edges in the intersection and union of E_A and E_B, respectively.

Considering that the variance of similarity between solution pairs may change along with the evolution, we normalize $s(\sigma_A, \sigma_B)$ by

$$S(\sigma_A, \sigma_B) = \frac{s(\sigma_A, \sigma_B) - \min\limits_{\sigma_A \neq \sigma_B} s(\sigma_A, \sigma_B)}{\max\limits_{\sigma_A \neq \sigma_B} s(\sigma_A, \sigma_B) - \min\limits_{\sigma_A \neq \sigma_B} s(\sigma_A, \sigma_B)}. \tag{1}$$

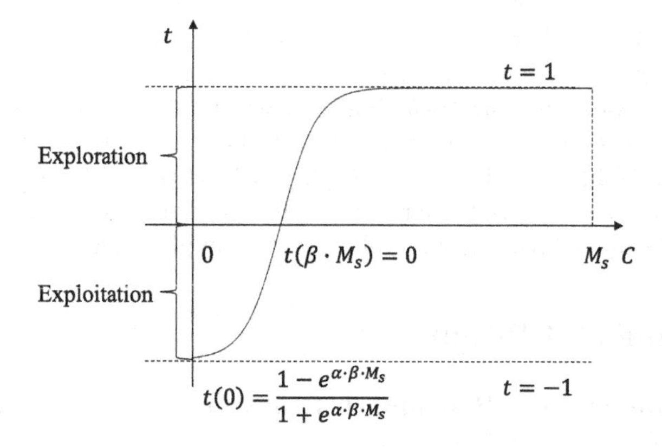

Fig. 1. The indicator function for switching between exploration and exploitation

Note that the calculation of similarity measures and the normalization are repeated in each generation. If the offspring is generated from two parents with low (high) similarity, the EAX indicates an exploration (exploitation). Thus, we could embed both exploration and exploitation in the process of EAX by controlling the parents paired.

4.3 The Switching Mechanism

The process starts with exploitation and repeats until the value of C keeps increasing. We define an indicator t to control the switching from exploitation to exploration according to C, which is defined by the following sigmoid function

$$t = \frac{1 - e^{\alpha(\beta M_s - C)}}{1 + e^{\alpha(\beta M_s - C)}}, \tag{2}$$

where M_s is the maximum steady steps, α controls the shape of the function and β controls the frequency of switching.

In each generation, all the individual solutions are arranged in a tour, and for each pair of adjacent solutions, the EAX operator will generate an offspring solution. In our implementation, the first solution $\sigma_{r(1)}$ is randomly selected. Suppose that if $k-1$ individuals have been put into the tour, the probability for each of the remaining individuals to be selected as the k_{th} one is determined by

$$p_{r(k)}(\sigma_i) = \frac{0.5 - t[S(\sigma_{r(k-1)}, \sigma_i) - 0.5]}{\sum\limits_{\sigma_i \in \Delta_k} \{0.5 - t[S(\sigma_{r(k-1)}, \sigma_i) - 0.5]\}}. \tag{3}$$

Specially,

$$p_{r(k)}(\sigma_i) = \begin{cases} \frac{S(\sigma_{r(k-1)}, \sigma_i)}{\sum\limits_{\sigma_i \in \Delta_k} S(\sigma_{r(k-1)}, \sigma_i)}, & t = -1 \\ \frac{[1 - S(\sigma_{r(k-1)}, \sigma_i)]}{\sum\limits_{\sigma_i \in \Delta_k} [1 - S(\sigma_{r(k-1)}, \sigma_i)]}, & t = 1 \end{cases}, \tag{4}$$

so $t = -1$ indicates that the probability for each individual to be selected is *positively* correlated with its similarity with the last individual in the tour. In this way, individual solutions with *high* similarity will be paired to generate offspring for *exploitation* (see Fig. 1 when $-1 < t < 0$). On the contrary, $t = 1$ indicates that individual solutions with *low* similarity will be paired for *exploration* (see Fig. 1 when $0 < t < 1$). As C increases, t changes from -1 to 1, and the EAX operator gradually switches from exploitation to exploration.

5 Experimental Results

5.1 Environment and Benchmark Problems

The algorithm is implemented in C++ under CUDA V8.0 environment and run on Nvidia P100 GPU (3584 CUDA cores, 16 GB memory). It is tested on the well-known Gehring and Homberger (GH) benchmark [6], which consists of five sets of 200–1000 customers. Each set is divided into six groups: C1, C2, R1, R2, RC1, and RC2, each containing 10 instances. The customers in C1 and C2 classes are located in clusters, while the customers in R1 and R2 are at random positions. The RC1 and RC2 classes contain a mix of clustered and random customers. The best-known solutions are collected from the Sintef website[1].

5.2 Experimental Setting

Two groups of experiments are conducted. In the first group, we compare the traditional MA and the proposed MASM on all the instances with 400–1000 customers from the GH benchmarks. The population size N_{pop} is set to 100. The maximal number of generations M_{gens} is 2000. In the second group, we only apply the MASM on those instances with 600–1000 customers (resulting 180 instances in total). N_{pop} and M_{gens} are respectively set to 1000 and 5000.

Other parameters remain the same in all experiments. In detail, the maximum number of steady generation M_s is 1000, the number of children in each reproduction step N_{ch} is 20. The proposed indicator (2) is a transformation of

$$f(x) = \frac{1 - e^x}{1 + e^x} \sim \begin{cases} -1, & x \leq -5 \\ 1, & x \geq 5 \end{cases} \tag{5}$$

Parameters α and β are determined by grid search using small scale instances (100 customers). By making $\alpha\beta M_s$ close to 5, we find it work well with $\alpha = \frac{1}{20}$ and $\beta = \frac{1}{10}$. Results show that the performance of the proposed MASM is relative robust with respect to these values.

[1] Sintef website: https://www.sintef.no/projectweb/top/vrptw/.

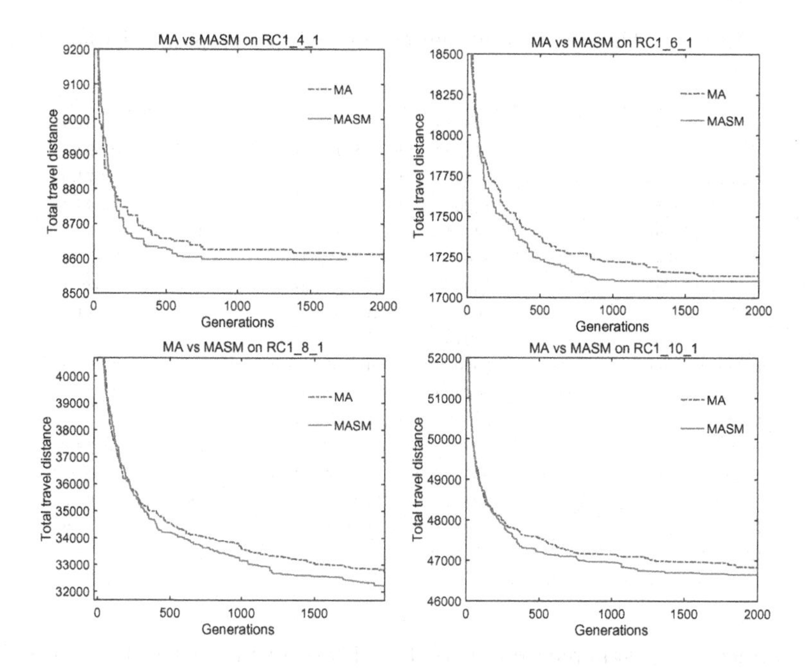

Fig. 2. The performance of MA with random EAX operation and the MASM

5.3 Computational Results

The results of the first experiments are illustrated in Fig. 2. Results demonstrate that the MASM outperforms the previous MA with random EAX operation among most cases, in terms of both convergence speed and the solution quality. Due to space limited, we only show the convergence curves of the first one from each category (i.e., RC1_4_1, RC1_6_1, RC1_8_1 and RC1_10_1). This category covers both the random and cluster customers. For instance RC1_4_1, the MASM terminates early as the condition of maximum steady generations is satisfied. Note that to calculate the similarity, we only need to compare the neighborhood for each customer, and the computational complexity is $O(N)$.

The results of the second experiments are shown in Table 1. We compare our results with those published on the Sintef website and report the new best-known solutions that we have found. The GH benchmark is a standard set for evaluating the performance of emerging techniques. A lot of teams have kept updating the world-best solutions [2,11,15]. In total, 34 best-known solutions out of 180 instances are improved. The gap between the "bad results" (total distance) and corresponding world-best solutions is smaller than 1%, we ignore the details due to space limit.

Most new best solutions are obtained within 10 h, which is close to those reported in the literature (e.g., 11 h in [11] and 6 h in [15]). Finally, we need to make a note about the computing platform. According to the recent literature, when solving the GH benchmarks, servers are usually used. For example, the

Table 1. The best-known solutions updated by proposed MASM

Instance	Customers	Vehicles	Distance	Instance	Customers	Vehicles	Distance
C2_6_2	600	17	8258.20	RC2_8_2	800	16	18151.95
C2_6_9	600	17	7911.61	RC2_8_4	800	15	10999.03
R1_6_1	600	59	21394.95	RC2_8_5	800	15	19074.02
R2_6_1	600	11	18205.58	RC2_8_6	800	15	18143.04
R2_6_2	600	11	14754.13	RC2_8_8	800	15	15759.14
R2_6_4	600	11	8008.14	C1_10_2	1000	90	42222.96
RC1_6_2	600	55	15914.70	C2_10_3	1000	28	16829.47
RC2_6_4	600	11	7057.94	R1_10_10	1000	91	47414.38
RC2_6_6	600	11	11913.11	R2_10_1	1000	19	42182.57
RC2_6_8	600	11	9990.40	R2_10_4	1000	19	17851.96
RC2_6_9	600	11	9574.99	R2_10_9	1000	19	32995.71
C1_8_2	800	72	26540.53	RC1_10_1	1000	90	45830.62
R1_8_2	800	72	32322.85	RC2_10_1	1000	20	30276.27
R1_8_5	800	72	33529.73	RC2_10_3	1000	18	19913.46
R2_8_2	800	15	22795.79	RC2_10_4	1000	18	15693.28
R2_8_6	800	15	20412.02	RC2_10_6	1000	18	26741.27
R2_8_10	800	15	20358.61	RC2_10_7	1000	18	25017.97

performance of the supercomputer used in [11] is 5 times better than ours. Parallel computing is a promising way to reduce the computing time, and GPU is suitable to run thousands of tasks concurrently. The records are updated by different teams around the world monthly. Our algorithm is running and continuing to find better solutions.

6 Conclusion and Future Work

The paper describes a memetic algorithm to solve large scale VRPTWs. The proposed algorithm incorporates a mechanism to switch between exploration and exploitation. The central aspects of the proposed mechanism are a similarity measure and a sigmoid function to control the paired solutions for crossover. Specifically, the contribution is a modification to EAX so that instead of on solutions ordered at random, a similarity measure is used to pair individuals for crossover. The rationale is that such guidance will accelerate the convergence of the memetic algorithm on large instances. Based on the GH benchmark, proposed MASM is shown to have faster convergence speed then the classical MA, and improves 34 best-known solutions out of 180 large-scale instances.

Although this paper focuses on the VRPTW, the proposed switching mechanism can be applied to more general genetic algorithms to accelerate the search. By controlling the parents been paired for crossover rather than conducting this operation randomly, the algorithm can find better solutions much more quickly. There are several potential directions for future research. One is to automatically determine the parameters used in the switching mechanism for different problems. Another one is to further analyze the mechanism of this algorithm, and investigate the conditions under which the algorithm performs best.

References

1. Bettinelli, A., Ceselli, A., Righini, G.: A branch-and-cut-and-price algorithm for the multi-depot heterogeneous vehicle routing problem with time windows. Transp. Res. Part C Emerg. Technol. **19**(5), 723–740 (2011)
2. Blocho, M., Czech, Z.J.: A parallel memetic algorithm for the vehicle routing problem with time windows. In: Eighth International Conference on P2P, Parallel, Grid, Cloud and Internet Computing, pp. 144–151. IEEE (2013)
3. Chabrier, A.: Vehicle routing problem with elementary shortest path based column generation. Comput. Oper. Res. **33**(10), 2972–2990 (2006)
4. Cordeau, J.F., Laporte, G., Mercier, A.: A unified tabu search heuristic for vehicle routing problems with time windows. J. Oper. Res. Soc. **52**(8), 928–936 (2001). https://doi.org/10.1057/palgrave.jors.2601163
5. Garcia-Najera, A., Bullinaria, J.A.: Bi-objective optimization for the vehicle routing problem with time windows: using route similarity to enhance performance. In: Ehrgott, M., Fonseca, C.M., Gandibleux, X., Hao, J.-K., Sevaux, M. (eds.) EMO 2009. LNCS, vol. 5467, pp. 275–289. Springer, Heidelberg (2009). https://doi.org/10.1007/978-3-642-01020-0_24
6. Gehring, H., Homberger, J.: A parallel hybrid evolutionary metaheuristic for the vehicle routing problem with time windows. In: Proceedings of EUROGEN99, vol. 2, pp. 57–64. Citeseer (1999)
7. Gong, Y.J., Zhang, J., Liu, O., Huang, R.Z., Chung, H.S.H., Shi, Y.H.: Optimizing the vehicle routing problem with time windows: a discrete particle swarm optimization approach. IEEE Trans. Syst. Man Cybern. Part C (Appl. Rev.) **42**(2), 254–267 (2012)
8. Jaccard, P.: Nouvelles recherches sur la distribution florale. Bull. Soc. Vaud. Sci. Nat. **44**, 223–270 (1908)
9. Nagata, Y.: Edge assembly crossover for the capacitated vehicle routing problem. In: Cotta, C., van Hemert, J. (eds.) EvoCOP 2007. LNCS, vol. 4446, pp. 142–153. Springer, Heidelberg (2007). https://doi.org/10.1007/978-3-540-71615-0_13
10. Nagata, Y., Bräysy, O., Dullaert, W.: A penalty-based edge assembly memetic algorithm for the vehicle routing problem with time windows. Comput. Oper. Res. **37**(4), 724–737 (2010)
11. Nalepa, J., Blocho, M.: Co-operation in the parallel memetic algorithm. Int. J. Parallel Prog. **43**(5), 812–839 (2015). https://doi.org/10.1007/s10766-014-0343-4
12. Nalepa, J., Blocho, M.: Adaptive memetic algorithm for minimizing distance in the vehicle routing problem with time windows. Soft Comput. **20**(6), 2309–2327 (2015). https://doi.org/10.1007/s00500-015-1642-4
13. Shunmugapriya, P., Kanmani, S., Fredieric, P.J., Vignesh, U., Justin, J.R., Vivek, K.: Effects of introducing similarity measures into artificial bee colony approach for optimization of vehicle routing problem. Int. J. Comput. Inf. Eng. **10**(3), 651–658 (2016)
14. Solomon, M.M.: Algorithms for the vehicle routing and scheduling problems with time window constraints. Oper. Res. **35**(2), 254–265 (1987)
15. Vidal, T., Crainic, T.G., Gendreau, M., Prins, C.: A hybrid genetic algorithm with adaptive diversity management for a large class of vehicle routing problems with time-windows. Comput. Oper. Res. **40**(1), 475–489 (2013)
16. Yu, B., Yang, Z.Z., Yao, B.: An improved ant colony optimization for vehicle routing problem. Eur. J. Oper. Res. **196**(1), 171–176 (2009)

Machine Learning

Game Recommendation Based on Dynamic Graph Convolutional Network

Wenwen Ye[1(✉)], Zheng Qin[1], Zhuoye Ding[2], and Dawei Yin[2]

[1] School of Software Engineering, Tsinghua University, Beijing, China
{yeww14,qingzh}@mails.tsinghua.edu.cn
[2] Data Science Lab, JD.com, Beijing, China
dingzuoye@jd.com, yindawei@outlook.com

Abstract. Recent years have witnessed the popularity of game recommendation. Different from the other recommendation scenarios, the user and item properties in game recommendation usually exhibit highly dynamic properties, and may influence each other in the user-item interaction process. For taming such characters, so as to design a high quality recommender system tailored for game recommendation, in this paper, we design a dynamic graph convolutional network to highlight the user/item evolutionary features. More specifically, the graph neighbors in our model are not static, they will be adaptively changed at different time. The recently interacted users or items are gradually involved into the aggregation process, which ensures that the user/item embeddings can evolve as the time goes on. In addition, to timely match the changed neighbors, we also update the convolutional weights in a RNN-manner. By these customized strategies, our model is expected to learn more accurate user behavior patterns in the field of game recommendation. We conduct extensive experiments on real-world datasets to demonstrate the superiority of our model.

Keywords: Recommendation · Graph convolutional network · Dynamic graph · Game

1 Introduction

On-line game, as an important entertainment method, is becoming more and more popular among young people. For providing users with their potentially favorite games, game recommendation has attracted a lot of attentions.

Comparing with traditional recommendation scenarios, game recommendation is unique in two aspects: On one hand, **both of the users and items can be dynamic.** Dynamic User preference has been noticed and verified by quite a lot of previous work. Similar to these research, user features in game recommendation can also be changeable. For example, people may like leisure games in the beginning, and shift their preference towards shooting games later on. In traditional recommendation settings, the item properties are static, and

Y. Nah et al. (Eds.): DASFAA 2020, LNCS 12112, pp. 335–351, 2020.
https://doi.org/10.1007/978-3-030-59410-7_24

the time effects are usually reflected by the user's dynamic modeling. However, in game recommendation, the item properties are no longer unchangeable. The game developer may always alter the game contents and styles to match users' up-to-date preferences. How to effectively capture user-item relations under such dual-dynamic settings bring more challenges for the game recommendation. On the other hand, **the interacted users and items can mutually influence each other.** (1) **item** → **user**: different from the other items, game usually requires users' deeper engagement. In the playing process, users are more likely to shift their preferences after experience some attractable and surprising episodes. (2) **user** → **item**: in order to keep competitiveness, the game developers need to continually change their products according to the users' preferences. For a real example, *Don't Starve* was initially released single-player, open-world, action-adventure game. However, in December 2014, after numerous requests and feed-back form reviews, the developer finally decided add multiplayer gameplay for the game.

The above characters are important for understanding user preference or item properties, and should be taken seriously when building the game recommendation systems [1,16,21]. While there are already many recommender systems tailored for the game domain, most of them fail to systematically consider the dynamic nature of this field. To bridge this gap, in this paper, we propose to build a dyn**A**mic g**R**aph convolu**T**ional mod**E**l **F**or g**A**me re**C**ommenda**T**ion (called **ARTEFACT** for short). Our model is built upon the graph structure. Each node corresponds a user or an item. The node embedding is derived by aggregating information from itself as well as its neighbors. For modeling the user/item evolving characters, we design a dynamic random sampling strategy, where the recent interacted items/users are more likely to be involved in the aggregation process. By the final BPR-like optimization function, our model is expected to accurately capture users' implicit feedback in a dynamic manner. Our contributions can be concluded as follows:

- We analyse the characters of the game recommendation, highlighting its dynamic natures in the real applications.
- We design a dynamic graph convolutional network, where the neighbors are gradually involved in the aggregation process for capturing the dynamic properties.
- We conduct extensive experiments to demonstrate the superiorities of our framework. Comparing with traditional methods, our model can obtain significant improvements on real-world datasets.

In the following, we first conduct some data analysis to verify our basic assumptions, based on which we formally define the task of game recommendation. Then we detail our model architectures, and analysis why we make such designs for game recommendation. In the next, we conduct extensive experiments for demonstrating the effectiveness of our model. And the related work and conclusion of this paper come at last.

2 Preliminaries

In this section, we first conduct primary analysis based on a real-world dataset to verify our assumption: 1) games' features are always changeable in real scenarios. 2) users and items can influence each other in the interaction process. Then we formally define the task of game recommendation.

2.1 Data Analysis

We base our analysis on the Steam dataset[1], and to alleviate the influence of product diversity, we focus our analysis in the game category.

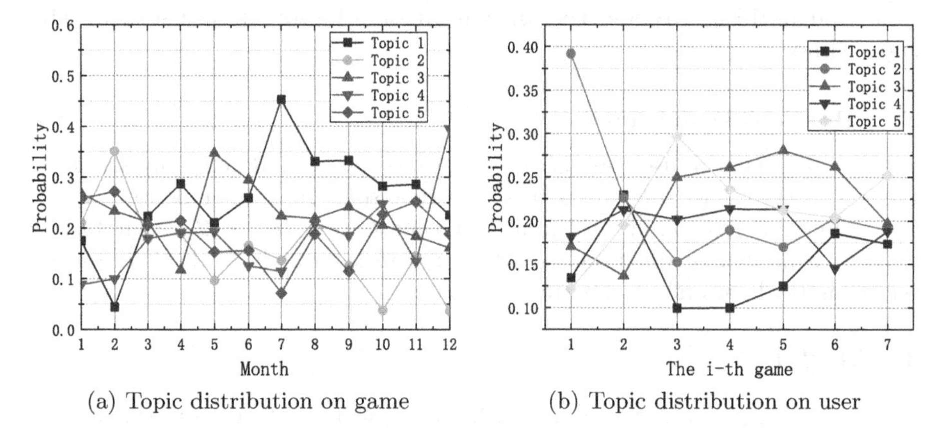

(a) Topic distribution on game (b) Topic distribution on user

Fig. 1. The topic distribution on game (steam id: 578080) and user (steam id: Zodux) across 2018. The number of topics in LDA is set as 5. The x coordinate in (b) represents the order of the games he has purchased.

The Dynamic Features of the Game. On many game platforms, users are allowed to express their opinions in the form of textual reviews. These reviews can usually reflect extensive item properties. In order to observe whether the item features vary at different time, we leverage user review as a proxy to analysis whether the review contents will change along the time axis. More specifically, for each item, we merged all the reviews in the same month into a document, and project the documents for different months into a unified topic space based on Latent Dirichlet Allocation (LDA) [2]. The results are presented in Fig. 1(a), we can see: (1) The game features are changing along the time axis. For example, at the beginning of the year, the topic distribution is $[0.18, 0.21, 0.27, 0.08, 0.25]$. While eight months later, the distribution became $[0.45.130.22.110.07]$. (2) The per-topic importances are always changing. In February, the least mentioned

[1] We collected the dataset from www.steampowered.com, which is released at https://github.com/wenye199100/SteamDataset.

topic in the reviews is second topic. (with the probability of 0.04), while in July, this topic became the most cared topic (with the probability of 0.45).

The Influence Between the User and Item. The above analysis manifest that the game features are changeable. Intuitively, people usually talk about their cared games' aspects in the review information, so we also use LDA to project the his reviews on all the games into the topic distribution (as shown in Fig. 1(b)). We can see that after each purchase of a new game, the user's topic distribution will change appropriately. And this indicates that users' cared games features are affect by the games he purchased. These results manifest that games' features are always affected by users and changeable, and user's preference is also affected the features of the game he purchased which motivates us to provide adaptive approaches to satisfy the co-evolution between user and game and **making proper recommendation of games to users at the right time.**

2.2 Problem Definition

Suppose we have a user set $\mathcal{U} = \{u_1, u_2, \cdots, u_m\}$ and an item set $\mathcal{I} = \{i_1, i_2, \cdots, i_n\}$. The user behaviors are represented as a tuple set: $\mathcal{O} = \{(u, i, t)|\text{user } u \text{ interacted with item } i \text{ at time } t, u \in \mathcal{U}, i \in \mathcal{I}\}$. Our task is to learn a predictive function, such that it can accurately rank all the items according to the users' up-to-date preferences.

3 Model

Our model is presented in Fig. 2, which mainly includes three components: (1) The first module is a graph convolutional operation, which allows us to capture multi-hop connectivities for neighbor aggregation. (2) The second component is a recurrent mechanism to update the self-evolution and the co-evolution parameters to capturing the dynamism of embeddings over time. (3) At last, we use BPR-like objective function to optimize our model parameters.

3.1 Time-Aware Convolutional Operation

As mentioned before, the user preference and item properties will continuously change as the time goes on. A user's current preference is usually determined by her previous characters as well as the games she recently interacted with. Similar intuitions also apply for the games. In our model, we consider these characters in the user/item embedding process. Specifically, each entity embedding[2] is allowed to evolve along the time axis. The current embedding is derived from its previous embedding (self-evolution) as well as the neighbors it interacted with recently (co-evolution). In our model, the user-item relations are formulated by a graph. Each node corresponds an entity, and the edges are leveraged to depict the interaction behaviors. In the following, we begin with the self-evolution mechanism, and then extend to the co-evolution strategy.

[2] Here, we use "entity" as an umbrella work to represent a user or an item.

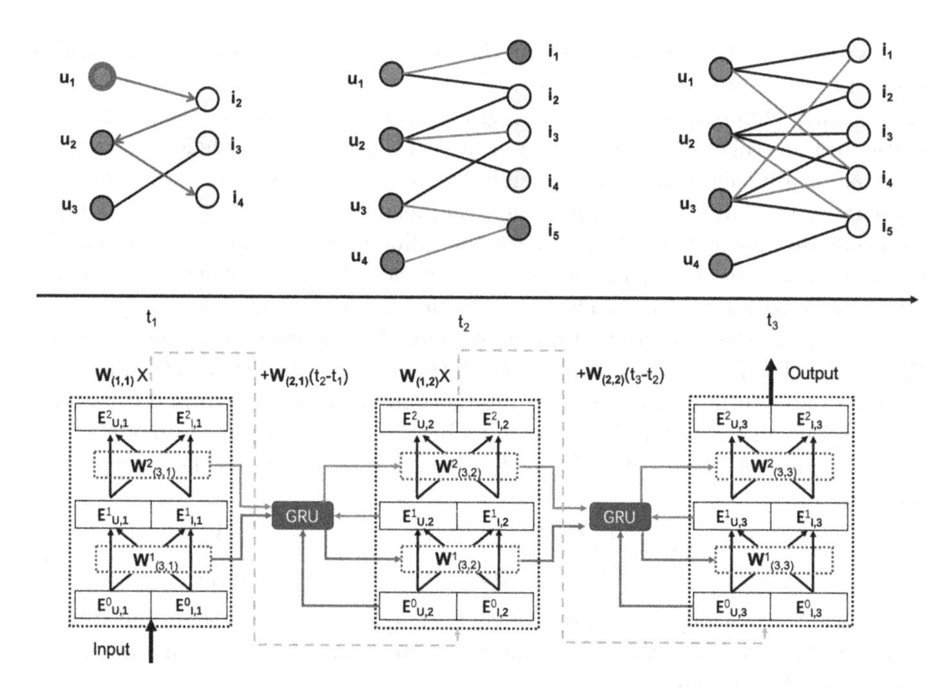

Fig. 2. The overall framework. Red arrows in user-item interaction graph represent the propagation of collaborative information. Blue points represent user node and yellow points represent item node. Green points are new user/item nodes and green lines are new interaction compared with previous interaction graph. (Color figure online)

Self-evolution. For simplicity, we uniformly segment the time period into Z parts (*e.g.*, $[0, t_1, t_2, \cdots, t_Z]$), and for each time step t_k, we have a separate user/item embedding, that is, $\boldsymbol{e}_{(u,t_k)}$ for user u and $\boldsymbol{e}_{(i,t_k)}$ for item i, respectively. For involving entity self-information, we compute $\boldsymbol{e}_{(u\leftarrow u,t_k)} = \boldsymbol{W}_{(1,t_k)}\boldsymbol{e}_{(u,t_k)}$ and $\boldsymbol{e}_{(i\leftarrow i,t_k)} = \boldsymbol{W}_{(1,t_k)}\boldsymbol{e}_{(i,t_k)}$, where $\boldsymbol{W}_{(1,t_k)} \in \mathbb{R}^{d\times d}$ is a trainable weight matrices, and $\boldsymbol{e}_{u\leftarrow u}$ and $\boldsymbol{e}_{i\leftarrow i}$ are the propagated self-information, which will be used to derive the final user/item embeddings. Careful readers may find that $\boldsymbol{e}_{(u\leftarrow u)}$ and $\boldsymbol{e}_{(i\leftarrow i,t_k)}$ are still not related with the time interval information, which can be useful in modeling users' dynamic preferences (*e.g.*, long time interval may degrade the influence from the former action to the current behavior). To capture this information, we further change $\boldsymbol{e}_{(u\leftarrow u,t_k)}$ and $\boldsymbol{e}_{(i\leftarrow i,t_k)}$ by:

$$\boldsymbol{e}_{(u\leftarrow u,t_k)} = \boldsymbol{W}_{(1,t_k)}\boldsymbol{e}_{(u,t_k)} + \boldsymbol{w}_{(2,t_k)}(t_k - t_{k-1}) \tag{1}$$

$$\boldsymbol{e}_{(i\leftarrow i,t_k)} = \boldsymbol{W}_{(1,t_k)}\boldsymbol{e}_{(i,t_k)} + \boldsymbol{w}_{(2,t_k)}(t_k - t_{k-1}) \tag{2}$$

where $\boldsymbol{w}_{(2,t_k)} \in \mathbb{R}^d$ is a scaling parameter casting time information on different user embedding dimensions.

Co-evolution. As mentioned before, the user/item embedding is not only influenced by itself, but also can be affected by its interacted entities (neighbors on the graph). Different from previous graph methods, where all the edges are static and leveraged to profile the user/item properties as a whole, we propose to dynamically aggregate entity neighbors to capture user/item evolutionary properties. In traditional graph model, the neighbors of an entity are assumed to be equally important, which ignores the influence of the time information. In our framework, the neighbors of an entity are not always the same, the recently interacted entities have more chances to be leveraged for aggregation. As can be seen in Fig. 2, the interaction graph in our model is changeable at different time, the left graph represents is composed of all the interactions between 0 and t_1, the middle graph summarizes user-item interactions in the period from 0 to t_2, and the right graph further adds interactions between t_2 and t_3 in the model. Suppose $\boldsymbol{e}_{u\leftarrow i}$ and $\boldsymbol{e}_{i\leftarrow u}$ are the propagated information between different entities. Then we have:

$$\begin{aligned} \boldsymbol{e}_{(u\leftarrow i,t_k)} &= \text{AGGREGATE}(\{\boldsymbol{e}_{(i,t_k)}, \forall i \in \mathcal{N}_{(u,t_k)}\}) \\ \boldsymbol{e}_{(i\leftarrow u,t_k)} &= \text{AGGREGATE}(\{\boldsymbol{e}_{(u,t_k)}, \forall u \in \mathcal{N}_{(i,t_k)}\}) \end{aligned} \tag{3}$$

where $\mathcal{N}_{(u,t_k)}$ and $\mathcal{N}_{(i,t_k)}$ are the entity neighbor sets of user u and item i, respectively. AGGREGATE($\cdot$) is a merging function, which is empirically set as average pooling in our model.

Neighbor Selection. We follow previous methods to select neighbors (*i.e.*, $\mathcal{N}_{(u,t_k)}$ and $\mathcal{N}_{(i,t_k)}$) based on the random sampling method. However, in order to select reasonable neighbor sets $\mathcal{N}_{(u,t_k)}$ and $\mathcal{N}_{(i,t_k)}$, we revise the random sampling method by taking time information into consideration. The recently interacted entities are more likely to be selected. In specific, suppose we have Z time segments, then the edges added in the kth segment are selected with the probability of $\frac{k}{\sum_{i=1}^{Z} i}$, which ensures that the later involved entities are more likely to be picked up.

By combining both self- and co-evolutionary operation together, the final entity embedding in our model can be computed as:

$$\begin{aligned} \boldsymbol{e}'_{(u,t_k)} &= \sigma(\boldsymbol{W}_{(3,t_k)} \cdot \text{CONCAT}(\boldsymbol{e}_{(u\leftarrow u,t_k)}, \boldsymbol{e}_{(u\leftarrow i,t_k)})) \\ \boldsymbol{e}'_{(i,t_k)} &= \sigma(\boldsymbol{W}_{(3,t_k)} \cdot \text{CONCAT}(\boldsymbol{e}_{(i\leftarrow i,t_k)}, \boldsymbol{e}_{(i\leftarrow u,t_k)})) \end{aligned} \tag{4}$$

where $\boldsymbol{W}_{(3,t_k)}$ is a weighting parameter, and $\sigma(\cdot)$ is the sigmoid activation function. And CONCAT($\cdot$) is a concatenate function, which concatenates two vector over the axis.

Multi-hop Relation Modeling. Multi-hop modeling can help to aggregate information from a longer distance.

In this section, we stack more convolutional layers to model multi-hop relations. The higher layer embedding can be derived by the lower ones as:

$$\begin{aligned} \boldsymbol{e}^{l}_{(x\leftarrow y,t_k)} &= \text{AGGREGATE}_l(\{\boldsymbol{e}^{l-1}_{(y,t_k)}, \forall y \in \mathcal{N}(x,t_k)\}) \\ \boldsymbol{e}^{l}_{(x,t_k)} &= \sigma(\boldsymbol{W}^{l}_{(3,t_k)} \cdot \text{CONCAT}(\boldsymbol{e}^{l-1}_{(x\leftarrow x,t_k)}, \boldsymbol{e}^{l}_{(x\leftarrow y,t_k)})) \end{aligned} \tag{5}$$

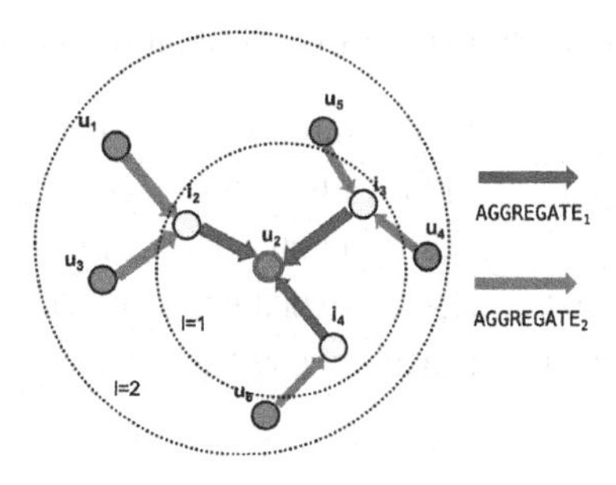

Fig. 3. Visual illustration of the multi-hop aggregation.

where $\boldsymbol{W}_{(3,t_k)}$ is a weighting parameter, and this equation applies for both $x = u, y = i$ and $x = i, y = u$. And Fig. 3 is the visual illustration of aggregation.

In this formula, the neighbors are aggregated towards the target node, and by recursively execute this operation, more nodes will be involved in the propagation process. In practice, too many propagation layers may result in over-smoothing effect, while too little layers may not enough for improving the representation power. We should select a moderate value for the number of layers according to the real datasets.

3.2 Temporal Evolution

For sufficiently capture the dynamic properties, we further design a weights evolution mechanism. In specific, the weights in different parts of Fig. 2 are modeled by a RNN model. We use the following formula to update the weighting parameters:

$$\begin{aligned} \boldsymbol{W}_{(3,t_k)}^{(l)} &= \mathrm{GRU}\big(\mathrm{Pool}(\boldsymbol{E}_{t_k}^{(l)}, d), \boldsymbol{W}_{(3,t_{k-1})}^{(l)}\big) \\ \boldsymbol{E}' &= \mathrm{Pool}(\boldsymbol{E}_{t_k}^{(l)}, d) = \Big(\mathrm{topK}\big(\boldsymbol{E}_{t_k}^{(l)} \circ \tanh\big(\frac{\boldsymbol{E}_{t_k}^{(l)}\boldsymbol{p}}{||\boldsymbol{p}||}\big)\big)\Big) \end{aligned} \tag{6}$$

where $\boldsymbol{E}_{t_k}^{(l)} = [\boldsymbol{e}_{(u_1,t_k)}^{(l)}, \boldsymbol{e}_{(u_2,t_k)}^{(l)}, \cdots, \boldsymbol{e}_{(i_1,t_k)}^{(l)}, \boldsymbol{e}_{(i_2,t_k)}^{(l)}, \cdots]$, $\boldsymbol{p}$ is a learnable vector and $||\cdot||$ is the L_2 norm. topK$(\cdot)$ function denotes to select the top-k indices from the given input vector and Pool$(\boldsymbol{E}_{t_k}^{(l)}, d)$ means merge $\boldsymbol{E}_{t_k}^{(l)}$'s $m+n$ rows into d rows. $\circ$ is the Hadamard (element-wise) product. This operation requires only a pointwise projection operation and slicing into the original feature and adjacency matrices, and therefore trivially retains sparsity. The input of the GRU unit is the concatenated embeddings of users and items and the weight matrix $\boldsymbol{W}_{(3,t_k)}^{(l)}$ is the GRU hidden state respectively. Therefore, we can bring the self-evolution and co-evolution information at this moment into the next embedded propagation. And the collaborative signals of different periods are passed along the path $[\boldsymbol{E}_{t_{k-1}}^{(l)}, \boldsymbol{W}_{(3,t_{k-1})}^{(l)}] \rightarrow \boldsymbol{W}_{(3,t_k)}^{(l)} \rightarrow \boldsymbol{E}_{t_k}^{(l)}$ through this weight evolution (Table 1).

Table 1. Statistics of the evaluation datasets.

Dataset	Users	Items	Interactions	Items/user	Users/item
Steam-1M	60,600	4,143	1,012,000	16.70	244.26
Steam-5M	454,542	19,013	4,999,970	11.05	262.97
Steam-10M	967,194	34,853	10,003,060	10.34	287.01
Steam-All	5,172,897	39,312	13,652,588	2.639	347.29

3.3 Prediction and Loss

We use BPR-like object function to optimize our parameter, that is,

$$l_{bpr} = - \sum_{\substack{(u,i,t) \in \mathcal{O} \\ (u,j,t) \notin \mathcal{O}}} \ln \text{sigmoid}(\hat{y}^{u,i} - \hat{y}^{u,j}) + \lambda \|\Theta\|_2^2 \tag{7}$$

where $\hat{y}^{u,i}$ and $\hat{y}^{u,j}$ represent the positive and negative user-item interactions, respectively. Θ is the parameter set need to be regularized.

4 Experiments

In this section, we evaluate our model by comparing it with several state-of-the-art models and focus on three research questions, that is:
RQ 1: Whether our model can enhance the performance of game recommendation as compared with state-of-the-art methods?
RQ 2: What's the effects of different components in our model?
RQ 3: How the convolutional operation impact our model performance?
RQ 4: What's the effect of the co-evaluation operation?

We first introducing the experimental setup, and then report and analyze the experimental results.

4.1 Experiments Setup

Datasets: Our dataset is crawled from a famous on-line game platform–Steam. It contains 5,172,897 users, 39,312 games and 13,652,588 English only reviews from October 2010 to March 2019. The dataset also includes meta-information for reviews, details of game (tags, developer, gameplay, price and etc.) and users' recent activities (total playtime and last playtime for each game). For promoting our research direction, we released our dataset and the code used for crawling the data at https://github.com/wenye199100/SteamDataset/Crawler.

Baselines: In our experiments, the following representative models are selected as the baselines:

- **MF-BPR:** [19] is a classic method for learning personalized rankings from implicit feedback. Biased matrix factorization is used as the underlying recommender.

- **NGCF:** [22] is a graph-based method which injects the user-item graph structure into the embedding learning process and leverages high-order connectivities in the user-item integration graph as collaborative signal.
- **T-POP:** For a user u_i at time t_j, this baseline recommends the most popular items for the current time t_j.
- **TimeSVD++:** [14] Extension of SVD++, which further takes temporal dynamics into consideration.
- **DeepCoevolve:** [10] This baseline model the nonlinear co-evolution nature of users' and items' embeddings. And it uses the user-item one-hop neighbors to captures the user and item's evolving and coevolving process.

Implementation Details. For each user behavior sequence, the last and second last interactions are used for testing and validation, while the other interactions are left for training. In our model, the batch size as well as the learning rate are determined in the range of {50, 100, 150} and {0.001, 0.01, 0.1, 1}, respectively. The user/item embedding size K is tuned in the range of {8, 16, 32, 64, 128}, and we will discuss its influence on the model performance in the following sections. The baselines designed for Top-N recommendation are revised to optimize the F1 and NDCG score.

4.2 RQ1: Top-N Recommendation

The overall comparison between our model and the baselines are presented in Table 2, we can see:

Table 2. The results of comparing our model with the baselines. BPR, POP and DeepC are short for MF-BPR, T-POP and DeepCoevolve, respectively.

Datasets	M@10(%)	(a) BPR	(b) NGCF	(c) POP	(d) SVD	(e) DeepC	(f) ARTEFACT	imp (%)
Steam-1M	F1	2.111	2.834	3.581	3.676	4.136	**4.498**	8.752
	NDCG	4.322	4.672	7.835	9.011	9.612	**10.78**	12.15
Steam-5M	F1	2.152	2.902	3.841	3.830	4.381	**4.793**	9.404
	NDCG	4.339	4.744	9.328	9.923	11.23	**12.62**	12.38
Steam-10M	F1	2.178	2.951	3.882	3.902	4.478	**4.986**	11.34
	NDCG	4.503	4.831	9.625	10.45	11.68	**13.17**	12.76

- By integrating temporal information, T-POP, TimeSVD++, DeepCoevolve and ARTEFACT obtained better performance than BPR and NGCF in most cases, which verifies the effectiveness of temporal information for the task of Top-N recommendation.

- Surprisingly, T-POP has achieved quite good results. Its performance is even better than TimeSVD++ on the dataset of Steam-5M. As can be seen from the statistics, the number of games is much smaller than the number of users. The development cycle of a Triple-A[3] game often takes several years, so it is highly regarded by players. When it releases, many players are willing to follow the trend to buy it.
- DeepCoevolve achieved better performance than the other baselines in most cases. This is not surprising, because DeepCoevolve considers the co-evaluation between user and item for user/item profiling.
- Encouragingly, we find that our ARTEFACT model was better than DeepCoevolve across different datasets in most cases. This result ascertains the effectiveness of our proposed model and positively answers RQ1. As mentioned before, collaborative signal is important for recommendation. However, DeepCoevolve only learns the one-hop neighbors as co-evaluation information. In contrast, our model utilizes recursively gcn to focus on multi-hop neighborhoods, which helps to better capture user-item's co-evolution information and eventually improve the recommendation performance.

4.3 RQ2: Ablation Study

Our model is composed of many components, it should be helpful to understand our method by analyzing their contributions for the final results. In specific, we respectively remove the self-evolution, co-evolution and temporal evolution from our final framework, and we name these variants as ARTEFACT(-self), ARTEFACT(-co) and ARTEFACT(-tem), respectively. The parameters follow the above settings, and we present their comparison in Table 3.

Table 3. F1@10(%) of variants

Ablations	Steam-1M	Steam-5M	Steam-10M
ARTEFACT(-self)	3.659	3.809	3.955
ARTEFACT(-co)	3.665	3.824	3.962
ARTEFACT(-temp)	3.214	3.425	3.658
ARTEFACT	4.502	4.805	4.993

- From the results, we can see, all the three variants perform worse than ARTEFACT which demonstrates that self-evolution, co-evolution and temporal evolution are all useful in our model. And this observation also verifies the importance of self-evolution, co-evolution and temporal evolution for game recommendation.

[3] AAA is a classification term used for games with the highest development budgets and levels of promotion. A title considered to be AAA is therefore expected to be a high quality game or to be among the year's bestsellers.

- In addition, the performance of removing self- and co-evolution are almost the same. Removing temporal evolution perform worst among these methods, which manifest that temporal weighting mechanism is more important than the former two components. By combining all the components, our final model can obtain the best performance.

4.4 RQ3: Study on the Convolutional Operation

Influence of the Stack Number L. We use L times embedding propagation layer to achieve the purpose of capturing multi-hop neighbors' collaborative signals between user-item interactions. We vary the model depth L to investigate whether ARTEFACT can benefit from multi-hop embedding propagation. We set $L = \{1, 2, 3, 4\}$ and results shown in Table 4. We can see from the result:

Table 4. F1@10(%) of embedding stack numbers (L)

L	Steam-1M	Steam-5M	Steam-10M
1	4.251	4.449	4.568
2	4.386	4.592	4.861
3	4.488	4.805	4.993
4	4.502	4.801	4.990

1) Increasing the number of L to a certain number can improve the recommendation performance. When $L = 1$, our model considers the first-order neighbors only. And we can see that the ablations($L > 1$) achieve consistent improvement over the ablation ($L = 1$) across all the datasets, which means that collaborative signals carried by the multi-hop neighbors play an important role in co-evolution.
2) ARTEFACT achieves the best when $L = 4$ on Steam-1M dataset while achieves the best when $L = 3$ on the other two datasets. Steam-5M and Steam-10M have much more interactions than Steam-1M which lead the over-fitting on big datasets.

The Effectiveness of the Dynamic Sampling Method. In order to demonstrate the effectiveness of our dynamic sampling method, we compare it with totally uniform sampling, where all the nodes at different time can be selected with equal probabilities. The results can be see in Table 5.

1) We can see, by deploying dynamic sampling method, our model's performance can be greatly enhanced. This verifies the effectiveness of our design, and manifests that in the game recommendation field, dynamic nature is an important factor, which should be explicitly modeled in the framework.

Table 5. F1@10(%) of sampling neighbors, dynamic sampling method is described at Sect. 3.1 *Neighbor selection* and random represents random sampling.

Datasets	Steam-1M		Steam-5M		Steam-10M	
Sample Number	Dynamic	Random	Dynamic	Random	Dynamic	Random
5	4.141	4.122	4.412	4.397	4.718	4.678
10	4.301	4.276	4.604	4.577	4.755	4.713
15	4.383	4.335	4.666	4.620	4.852	4.824
20	4.401	4.362	4.720	4.688	4.903	4.881
25	4.478	4.401	4.761	4.711	4.948	4.923
30	**4.502**	**4.433**	**4.805**	**4.782**	**4.993**	**4.961**
40	4.439	4.353	4.758	4.677	4.886	4.832

2) ARTEFACT achieves the best when sampling number is 30 on all dataset. However, when sampling number becomes larger, the performance of ARTEFACT and ARTEFACT (random sampling) become worse.

Influence of the Time Steps Number Z**.** To investigate whether different time steps Z can affect the performance of our model. In particular, we search the time steps numbers in the range of $\{1, 2, 3, 4, 5, 6, 7, 8, 9, 10\}$. We have the following observations (Fig. 4):

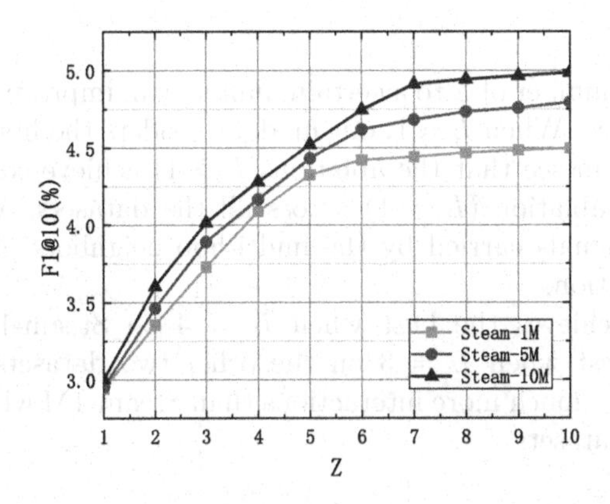

Fig. 4. F1@10(%) of time steps Z.

1) When $Z = 1$, it means we build all the interactions in the dataset in a single "static" graph and there is no temporal evolution (both time drift and weight evolution). This is not surprising that this "static" method performs the worst. This also shows that the dynamics in our model are necessary for the recommendation.

2) The performance of the model continues to increase as Z increases. This also verifies that more temporal evolution can improve the performance of the model.
3) Based on 2), another observation is that the growth rate when $Z < 6$ is greater than the $Z < 6$. During the training of the model, we need to calculate Z times normalize adjacency matrix for user-item interactions graph and $Z - 1$ times temporal evolution, which is very time consuming. So our maximum time step is set to 10. But when $Z > 6$, the improvement of the model's performance becomes slower, so we can infer that even if we continue to increase Z, the final result will not be much better than $Z = 10$.

In order to evaluate our recommendations from both qualitative and quantitative perspectives, we use the review information to analyze whether the recommendation results are reasonable and compare the prediction between ARTEFACT and DeepCoevolve.

4.5 RQ4: Qualitative and Quantitative Analysis on the Co-evaluation Operation

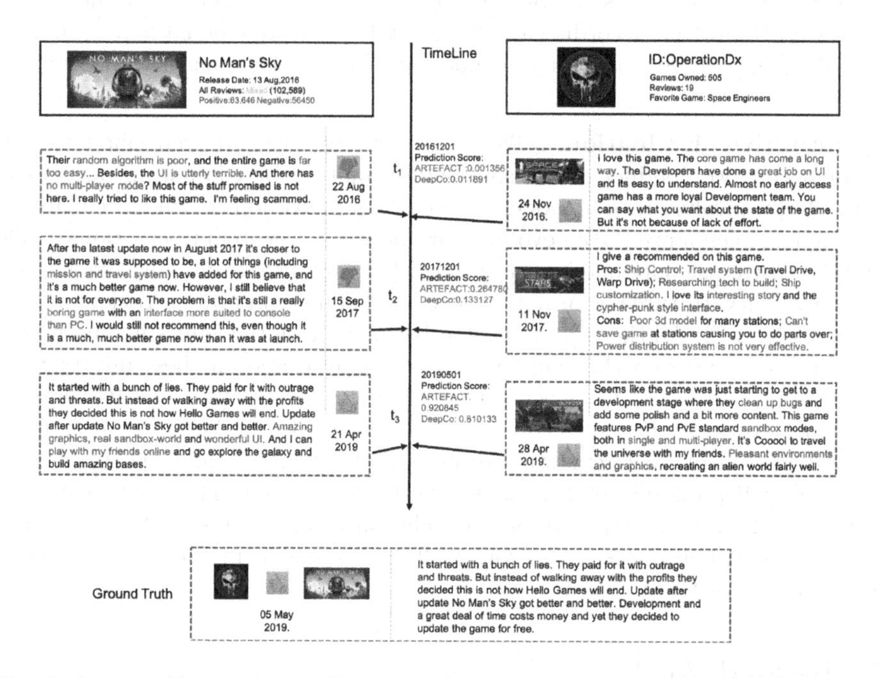

Fig. 5. A case of leveraging qualitative evaluation for recommendation. (Color figure online)

Qualitative Evaluation. To provide intuitive analysis about our model's performance on co-evaluation, we present an example in Fig. 5. In specific, we calculate the prediction score between the same pair of game and user at three different time. We use the reviews to shown the game's features and user's preference. We use red to highlight the negative features and blue to the positive features. We can see:

(1) Before t_1 (20161201), the game has pretty poor content and terrible UI and the user likes the space background game with rich content and good interface, Therefore, at t_1, both the prediction of ARTEFACT and DeepCoevolve is not high; At t_2 (20171201), the developer added the mission and travel system in game according the users' feedback. However it's still a boring game with poor interface. Meanwhile the user was attracted by another game's rich travel and ship system and interesting story. The prediction scores of ARTEFACT and DeepCoevolve are still not high. (2) Then after about a year and a half of development, the game became a good quality game with amazing graphics and sandbox-world at t_3 (20190501). The developers also added the multiplayer gameplay for this game. On the other hand, the user has just purchased a multiplayer game which can explore the universe in it and enjoy playing with friends. At this time, the features of this game are completely in line with the user's current preferences, the score of ARTEFACT 0.920845 is very high and higher than DeepCoevolve. According to the ground truth, user bought this game at May 05. This can demonstrate our model performs better than DeepCoevolve.

Quantitative Evaluation. To quantitatively evaluate our model's co-evolutionary, we conduct crowd-sourcing evaluation by comparing our model with DeepCoevolve. In our model, we consider multi-hop neighbors collaborative signals to enhance the co-evolutionary between user-item interactions, while DeepCoevolve use only one-hop neighbors. For each user-item pair, the workers are provided with three information sources: (1) The user's real review for the current game. (2) All the reviews posted by the user on the games he purchased, and (3) All reviews on this game at various time. Based on these information, the workers are asked to select one from three options (i.e. A:ARTEFACT, B:DeepCoevolve, C:Tie) to answer: **Q:** Which model's prediction score is more accurate based on real reviews? We employed 3 workers, and one result was valid only when more than 2 workers give the same result. The result is A:63%, B:25%, C:12%. And we can draw a conclusion from this result: the multi-hop neighbors collaborative signals work well than one-hop in the co-evolution of user-item interactions.

5 Related Work

Graph-Based Recommendation. Nowadays we have witnessed the great success of the recommendation system in many fields, such as e-commerce [24,25] and media-service platform [5–7]. In recent years, several convolutional neural network architectures [3,9,11,13,18] for learning over graphs have been proposed. [12] introduced a method for computing node representations in an inductive manner.

The existing recommendation methods also successfully uses the graph convolution [26,28]. [28] proposed a spectral convolution operation to discover all possible connectivity between the user and the item in the spectrum domain. But the feature decomposition leads to high computational complexity, which is very time consuming and difficult to support large scale recommendation scenarios.

[22] devised a new framework NGCF, which achieves the target by leveraging high-order connectivities in the user-item integration graph by GCN [13].

[23] alleviates this problem by combining a graphics-based approach with an embedded-based approach. It performs a random walk to enrich the user's interaction with multi-hop interactions to build a recommender.

Dynamic Graphs. Dynamic graphs based methods are often extensions of those for a static one, with an additional focus on the temporal information and update schemes. Recent approaches which focus on this are combinations of GNNs and recurrent network. There are two combinations: the first one uses a GCN to obtain node embeddings, then feds them into the LSTM to learns the dynamism [15,17,20]; and the second is to modify LSTM that takes node features as input but replaces the fully connected layers therein by graph convolutions [4].

6 Conclusion

In this paper, we propose to model the evolutionary feature of user and item and co-evolutionary feature of their multi-hop collaborative signals over continuous and seamlessly apply our approach to a novel recommendation scenario: game recommendation. Based on our designed model, we can improve the performance of user-item rating prediction as compare with several state-of-the-art methods. We conduct extensive experiments to demonstrate the superiority of our model. As for future work, we will investigate the potential advantages of explainable recommendation [8,27] to explicitly describe the user dynamic preference and item dynamic feature with context. Furthermore, we will also study how to integrate game images and game video into our model for more comprehensive recommendation explanations.

References

1. Anwar, S.M., Shahzad, T., Sattar, Z., Khan, R., Majid, M.: A game recommender system using collaborative filtering (gambit). In: 14th International Bhurban Conference on Applied Sciences and Technology (IBCAST), pp. 328–332. IEEE (2017)
2. Blei, D.M., Ng, A.Y., Jordan, M.I.: Latent Dirichlet allocation. J. Mach. Learn. Res. **3**, 993–1022 (2003)
3. Bruna, J., Zaremba, W., Szlam, A., LeCun, Y.: Spectral networks and locally connected networks on graphs. arXiv preprint arXiv:1312.6203 (2013)

4. Cangea, C., Veličković, P., Jovanović, N., Kipf, T., Liò, P.: Towards sparse hierarchical graph classifiers. arXiv preprint arXiv:1811.01287 (2018)
5. Chen, X., et al.: Personalized fashion recommendation with visual explanations based on multimodal attention network: towards visually explainable recommendation. In: Proceedings of the 42nd International ACM SIGIR Conference on Research and Development in Information Retrieval, pp. 765–774 (2019)
6. Chen, X., et al.: Sequential recommendation with user memory networks. In: Proceedings of the Eleventh ACM International Conference on Web Search and Data Mining, pp. 108–116. ACM (2018)
7. Chen, X., Zhang, Y., Ai, Q., Xu, H., Yan, J., Qin, Z.: Personalized key frame recommendation. In: Proceedings of the 40th International ACM SIGIR Conference on Research and Development in Information Retrieval, pp. 315–324 (2017)
8. Chen, X., Zhang, Y., Qin, Z.: Dynamic explainable recommendation based on neural attentive models. In: Proceedings of the AAAI Conference on Artificial Intelligence, vol. 33, pp. 53–60 (2019)
9. Dai, H., Dai, B., Song, L.: Discriminative embeddings of latent variable models for structured data. In: International Conference on Machine Learning, pp. 2702–2711 (2016)
10. Dai, H., Wang, Y., Trivedi, R., Song, L.: Deep coevolutionary network: embedding user and item features for recommendation. arXiv preprint arXiv:1609.03675 (2016)
11. Defferrard, M., Bresson, X., Vandergheynst, P.: Convolutional neural networks on graphs with fast localized spectral filtering. In: Advances in Neural Information Processing Systems, pp. 3844–3852 (2016)
12. Hamilton, W., Ying, Z., Leskovec, J.: Inductive representation learning on large graphs. In: Advances in Neural Information Processing Systems, pp. 1024–1034 (2017)
13. Kipf, T.N., Welling, M.: Semi-supervised classification with graph convolutional networks. arXiv preprint arXiv:1609.02907 (2016)
14. Koren, Y.: Collaborative filtering with temporal dynamics. In: Proceedings of the 15th ACM SIGKDD International Conference on Knowledge Discovery and Data Mining, pp. 447–456. ACM (2009)
15. Manessi, F., Rozza, A., Manzo, M.: Dynamic graph convolutional networks. Pattern Recogni. **97**, 107000 (2019)
16. Meidl, M., Lytinen, S.L., Raison, K.: Using game reviews to recommend games. In: Tenth Artificial Intelligence and Interactive Digital Entertainment Conference (2014)
17. Narayan, A., Roe, P.H.: Learning graph dynamics using deep neural networks. IFAC-PapersOnLine **51**(2), 433–438 (2018)
18. Niepert, M., Ahmed, M., Kutzkov, K.: Learning convolutional neural networks for graphs. In: International Conference on Machine Learning, pp. 2014–2023 (2016)
19. Rendle, S., Freudenthaler, C., Gantner, Z., Schmidt-Thieme, L.: BPR: Bayesian personalized ranking from implicit feedback. In: Proceedings of the Twenty-fifth Conference on Uncertainty in Artificial Intelligence, pp. 452–461. AUAI Press (2009)
20. Seo, Y., Defferrard, M., Vandergheynst, P., Bresson, X.: Structured sequence modeling with graph convolutional recurrent networks. In: Cheng, L., Leung, A.C.S., Ozawa, S. (eds.) ICONIP 2018. LNCS, vol. 11301, pp. 362–373. Springer, Cham (2018). https://doi.org/10.1007/978-3-030-04167-0_33
21. Sifa, R., Bauckhage, C., Drachen, A.: Archetypal game recommender systems. In: LWA, pp. 45–56 (2014)

22. Wang, X., He, X., Wang, M., Feng, F., Chua, T.S.: Neural graph collaborative filtering. arXiv preprint arXiv:1905.08108 (2019)
23. Yang, J.H., Chen, C.M., Wang, C.J., Tsai, M.F.: Hop-rec: high-order proximity for implicit recommendation. In: Proceedings of the 12th ACM Conference on Recommender Systems, pp. 140–144. ACM (2018)
24. Ye, W., Qin, Z., Li, X.: Deep tag recommendation based on discrete tensor factorization. In: Cheng, L., Leung, A.C.S., Ozawa, S. (eds.) ICONIP 2018. LNCS, vol. 11301, pp. 70–82. Springer, Cham (2018). https://doi.org/10.1007/978-3-030-04167-0_7
25. Ye, W., Zhang, Y., Zhao, W.X., Chen, X., Qin, Z.: A collaborative neural model for rating prediction by leveraging user reviews and product images. In: Sung, W.K., et al. (eds.) AIRS 2017. Lecture Notes in Computer Science, vol. 10648, pp. 99–111. Springer, Cham (2017)
26. Ying, R., He, R., Chen, K., Eksombatchai, P., Hamilton, W.L., Leskovec, J.: Graph convolutional neural networks for web-scale recommender systems. In: Proceedings of the 24th ACM SIGKDD International Conference on Knowledge Discovery & Data Mining, pp. 974–983. ACM (2018)
27. Zhang, Y.: Explainable recommendation: theory and applications. arXiv preprint arXiv:1708.06409 (2017)
28. Zheng, L., Lu, C.T., Jiang, F., Zhang, J., Yu, P.S.: Spectral collaborative filtering. In: Proceedings of the 12th ACM Conference on Recommender Systems, pp. 311–319. ACM (2018)

From Code to Natural Language: Type-Aware Sketch-Based Seq2Seq Learning

Yuhang Deng[1], Hao Huang[1], Xu Chen[1(✉)], Zuopeng Liu[2], Sai Wu[3], Jifeng Xuan[1], and Zongpeng Li[1]

[1] School of Computer Science, Wuhan University, Wuhan, China
{dengyh,haohuang,xuchen,jxuan,zongpeng}@whu.edu.cn

[2] Xiaomi Technology Co., Ltd., Beijing, China
liuzuopeng@xiaomi.com

[3] College of Computer Science and Technology, Zhejiang University, Hangzhou, China
wusai@zju.edu.cn

Abstract. Code comment generation aims to translate existing source code into natural language explanations. It provides an easy-to-understand description for developers who are unfamiliar with the functionality of source code. Existing approaches to code comment generation focus on summarizing multiple lines of code with a short text, but often cannot effectively explain a single line of code. In this paper, we propose an asynchronous learning model, which learns the code semantics and generates a fine-grained natural language explanation for each line of code. Different from a coarse-grained code comment generation, this fine-grained explanation can help developers better understand the functionality line-by-line. The proposed model adopts a type-aware sketch-based sequence-to-sequence learning method to generate natural language explanations for source code. This method incorporates the type of source code and the mask mechanism with the Long Short Term Memory (LSTM) network via encoding and decoding phases. We empirically compare the proposed model with state-of-the-art approaches on real data sets of source code and description in Python. Experimental results demonstrate that our model can outperform existing approaches on commonly used metrics for neural machine translation.

Keywords: Code comment generation · Sketch · Attention mechanism

1 Introduction

Program comprehension is a time-consuming activity in software development and maintenance [12,13]. To improve program comprehension, code comment generation is proposed to automatically generate natural language explanations

Y. Nah et al. (Eds.): DASFAA 2020, LNCS 12112, pp. 352–368, 2020.
https://doi.org/10.1007/978-3-030-59410-7_25

for existing source code [6,8]. For developers who are unfamiliar with the functionality of source code, it provides an easy-to-understand description. In addition, since automatic code comment generation can add comprehensible comments to source code without the manual effort by human developers, it also addresses a widely-existed issue of daily software development, i.e., the lack of code comments.

Code comment generation can be viewed as the translation from source code to natural language explanations. Existing approaches to this problem can be mainly categorized into two groups: rule-based models and deep learning-based models. The rule-based models chase the goal of exactly matching the functionality via heuristics or manually-written templates, i.e., rules [9,10]. These models can provide accurate comments in particular cases but fail in unexpected cases due to the inflexibility of rules. The deep learning-based models shape the rules of code summarization with training data. Most of the deep learning-based models follow an encoder-decoder framework of neural machine translation that often contains the attention mechanism [1,18]. In an encoder-decoder framework, a source code snippet is parsed into an Abstract Syntax Tree (AST) and encoded into a sequence-to-sequence [4,24] (abbreviated as seq2seq) model [12,25]. However, most of these models focus on generating short textual descriptions for code snippets. Furthermore, these models simultaneously learn the textual descriptions of identifiers (including variable names and strings) and the semantics of source code. In this way, the interaction between textual learning and semantic learning may degrade the convergence performance of the model training.

In this paper, we propose a simple yet effective asynchronous learning model, which first learns the code semantics and then generates a fine-grained natural language explanation for each line of source code. Compared with a coarse-grained summarization for code snippets, this fine-grained explanation can help a developer, who is unfamiliar with the source code, better understand the role in the functionality of each line of code.

The proposed model adopt a type-aware sketch-based seq2seq learning method to generate natural language explanations for source code. This method incorporates types of source code and the mask mechanism with the LSTM [11] Recurrent Neural Network (RNN). Before the model training, we prepare natural language sketch from training data by removing low-frequency tokens and replacing tokens with their types. In the encoding phase, we mask variable names and values in source code to reduce the volume of sketch vocabulary and improve the training accuracy and efficiency. Then, we employ a Bidirectional LSTM network (BiLSTM) [23] to encode the source code with type and position information. In the decoding phase, we first generate a natural language sketch corresponding to a source code structure, and then use an LSTM decoder with the copying mechanism [7], which decides whether to select existing words from the natural language vocabulary or copy from source code based on the sketch and code context, to generate natural language explanations for source code under the guidance of the sketch. We empirically compare our model with state-of-the-art approaches on real data sets of Python code. Experimental results demonstrate that our model can outperform these approaches on commonly used metrics for neural machine translation.

Our key contribution is two-fold. (1) We propose an asynchronous learning framework to model the semantics of source code and the textual description of identifiers. (2) We design a new method to generate a fine-grained explanation for each line of source code, and it outperforms state-of-the-art approaches on real data sets of Python code.

The remainder of the paper is organized as follows. We first review the related work. Then, we present a problem formulation and elaborate our type-aware sketch-based seq2seq learning method, followed by reporting experimental results and our findings before concluding the paper.

2 Related Work

The existing approaches to code comment generation can be categorized into two main groups: (1) rule-based methods, and (2) deep learning-based methods.

2.1 Rule-Based Methods

Traditional code comment generation techniques usually first extract words from identifiers in source code and then stitching these words into natural language sentences base on rules and templates [9,10]. Moreno et al. create a rule-based model using information selected by stereotype information with predefined heuristics to generate text description for Java classes. McBurney and McMillan generate English descriptions for Java methods by combining keywords and contextual information, including the dependencies of the method and other methods relying on method outputs. To improve the rule-based models which generally rely on templates to get the natural language sentences, Oda et al. propose a method adopting the statistical machine translation framework based on the extracted relationship and statistical models [15]. They use the phrase-based and tree-to-string machine translation to convert source code into pseudo-code.

Although the rule-based models often achieve promising performance on code comment generation, they may not understand the semantics of code as their results mainly rely on the extracted words from identifiers in source code.

2.2 Deep Learning-Based Methods

With the successful development of deep learning, some studies have tried introducing neural networks to code comment generation. CODE-NN [14] is the first neural network model, which adopts LSTM with the attention mechanism to produce text descriptions for C# code snippets and SQL queries. Code-GRU [16] uses a new recursive neural network called Code-RNN to encode the parse tree of source code into one vector and then feed the vector into decoder to generate text description of the source code. To improve the qualities of code comment generation, some methods utilize Abstract Syntax Trees (AST) to capture the structures and semantics of source code [12,25].

Nevertheless, different from existing deep learning-based methods that produce coarse-grained code comments, we aim to generate a fine-grained natural language explanation for each line of source code.

3 Problem Formulation

The goal of our work is to generate a syntactically correct and semantically fluent natural language explanation for each line of source code. Let $X = \{x_1, ..., x_N\}$ denote a line of source code with length N, and $Y = \{y_1, ..., y_M\}$ denote a natural language sentence with length M, respectively. Given input X and Y, our model learns $p(Y|X)$, i.e., the probability of natural language expression Y conditioned on source code X, to generate appreciate natural language expressions based on given source code. To make the Y syntactically correct and semantically fluent, we utilize a sketch to help standardise the final Y and learn the sketch based on X. Then, the probability $p(Y|X)$ can be decomposed as follows.

$$p(Y|X) = p(Y|X, a)p(a|X) \tag{1}$$

where $a = \{a_1, ..., a_M\}$ is a natural language sketch that represents the natural language explanations of keywords in source code X.

Given the above equation, our model training consists of the following two stages:

First, the model learns the probability $p(a|X)$ of sketch a conditioned on X as follows.

$$p(a|X) = \prod_{t=1}^{N} p(a_t|a_{<t}, X) \tag{2}$$

where $a_{<t} = a_1, ..., a_{t-1}$, and t refers to t-th time step of the decoding phase in this stage.

Second, the model utilize the BiLSTM to encode the sketch into vectors which in turn guide and constrain the decoding of Y. The generation probability is factorized as follows.

$$p(Y|X, a) = \prod_{t=1}^{M} p(y_t|y_{<t}, X, a) \tag{3}$$

where $y_{<t} = y_1, ..., y_{t-1}$.

In what follows, we present how to learn the above probabilities to generate sketch and natural language explanations in detail.

4 Methodology

We employ an encoder-decoder framework [3,4] to build our model for code comment generation. Figure 1 illustrates the architecture of the model, which mainly consists of three parts, namely (1) an input encoder, (2) a sketch decoder, and (3) a natural language explanation decoder. In the section, we elaborate how each of the three parts works, and conclude the section by introducing how to train the model.

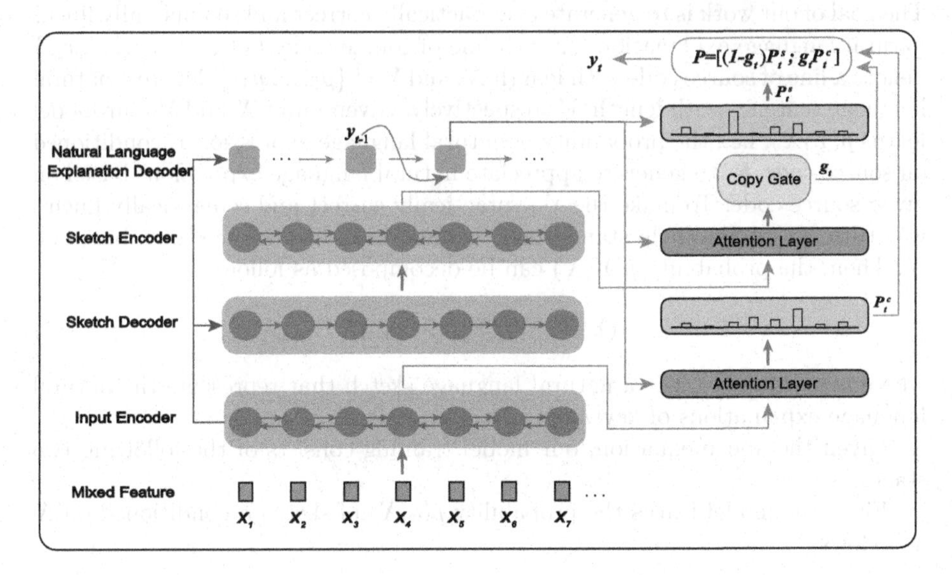

Fig. 1. The proposed model architecture. It consists of two decoders with an attention mechanism and copying mechanism. The first decoder generates natural language sketch, which guides the second decoder to generate natural language explanations.

4.1 Input Encoder

The input encoder is used to encode the embedding vector of all the input data into a latent vector space, and learns the latent features from the input. It applies BiLSTM to encode the input embedding. A BiLSTM is composed of two LSTM networks. The forward LSTM reads in a left-to-right direction, obtaining hidden states $(\overrightarrow{h_1}, ..., \overrightarrow{h_N})$. The backward LSTM reads reversely and gets $(\overleftarrow{h_1}, ..., \overleftarrow{h_N})$. Then, the final representation for the code is $(h_1, ..., h_N)$, where $h_i = [\overrightarrow{h_i}, \overleftarrow{h_i}]$. To enhance the feature extraction capabilities of the encoder, we adopt the following two strategies.

Type Information. In addition to the word embedding, we also utilize additional information including position embedding and type embedding. The position information is obtained directly from position of the token in the list of the tokenized source code and helps the encoder to extract structure features of different variable names. We use the built-in lexical scanner to tokenize the code and obtain token types, and this information is important for the decoder to get the attributes of variable names. For example, for a Python code "source[:1].higher() = 'tcp'", the token type is "NAME OP OP NUMBER OP OP NAME OP OP OP STRING". We concatenate the embedding of words and their corresponding position and types for the encoder.

Mask Mechanism. Furthermore, to deal with the variable names that are usually low-frequency words, we use a masking strategy to set embeddings of variable names in code vector to 0, so that the information of variable names is represented only by the type and position features. The reason is two-fold. First, the type and position features of identifiers in code are more important than their exact meanings in the natural language sketch generation process. Second, the mask mechanism substantially reduces the number of parameters that need to be learned, and thus reduces the model complexity.

After learning the latent features from the input, we feed the feature vectors into a sketch decoder to generate a most possible sketch for current source code.

4.2 Sketch Decoder

The sketch decoder adopts LSTM with an attention mechanism. The first hidden state of the decoder LSTM is initialized by concatenating the encoding vector $d_0 = [\overrightarrow{h_N}, \overleftarrow{h_1}]$. At each time step t, the hidden state vector is computed by $d_t = f_{LSTM}(d_{t-1}, a_{t-1})$, where a_{t-1} is the embedding of a previously predicted token. After we get d_t from LSTM, we use an attention mechanism to learn soft alignments and predict token to obtain a_t based on the following equation,

$$p(a_t|a_{<t}, X) = \text{softmax}(W_0 d_t^{att} + b_0) \tag{4}$$

where d_t^{att} refers to a multiplicative attention [18], which can be calculated as follows.

$$d_t^{att} = \tanh(W_1 d_t + W_2 c_t) \tag{5}$$

where

$$c_t = \sum_{k=1}^{N} s_{t,k} h_k \tag{6}$$

$$s_{t,k} = \frac{\exp(d_t \cdot h_k)}{\sum_{j=1}^{N} \exp(d_t \cdot h_j)} \tag{7}$$

Note that in modern programming language, a few keywords are enough to express the main programming logics, indicating that a limited number of natural language words are also sufficient to explain the purpose of a specific code. Therefore, we choose some high-frequency words as the candidate words for sketch, forming a sketch vocabulary and it is about 2000 words in this dateset. The sketch decoder predicts tokens from the sketch vocabulary, in which words usually do not have a high correlation with the exact meaning of identifiers (i.e., variable names and strings). Then, the sketch decoder may focus on generating words that correspond to the keywords in source code. For example, "def" corresponds to "define", and "+" corresponds to "sum". In the sketch, we use the type words such as "NAME", "NUMBER" or "STRING" to replace the rare words which may be variable names, and the replaced words are either copied or generated by another decoder. For example, the sketch decoder would predict a sketch "NAME and NAME are integer NUMBER" for "pos = last = 0", and we can see that the "NAME" in sketch refers to variable name "pos" or "last". The type information in sketch can be used to constrain generation in the second decoder phase. For example, the sketch token "NUMBER" specifies that number type words should be emitted.

After learning sketch from the feature vectors of source code, we combine the sketch and the feature vectors to generate a natural language explanation for the source code.

4.3 Natural Language Explanation Decoder

The natural language explanation decoder is based on recurrent neural networks with an attention mechanism and a copying mechanism. With the help of sketch, the generated natural language explanations could be more syntactically correct [5]. To make the explanations more semantically fluent, we propose to use the sketch in a soft-embedded way. To be specific, at each time step t, we use BiLSTM to encode the sketch as sketch hidden states, and then an attentive vector y_{t-1}^{att} is generated through the attention mechanism by taking the sketch hidden states and encoding target word y_{t-1} as input. With the attentive vector y_{t-1}^{att} and the hidden state e_{t-1} at last time step $t-1$, the hidden state e_t of current time step t can be updated by LSTM. Formally,

$$e_t = f_{LSTM}(e_{t-1}, y_{t-1}^{att}) \tag{8}$$

where the hidden state e_0 at time step 0 can be initialized as $e_0 = [\overrightarrow{h_N}, \overleftarrow{h_1}]$. In this way, the attention mechanism applies the sketch to constrain the predicted words in a soft way. The attention weights indicate how the model distracts attention to the different encoder hidden states during the decoding process. If the sketch token has a stronger relationship with the predicted word y_{t-1}, the corresponding sketch hidden state will have a larger attention weight, so the decoder will focus on generating words related to the sketch hidden state.

There are two strategies to generate words for natural language explanations, namely, selecting words from the natural language vocabulary and copying words from the source code X. The later one enable us to generate a few Out-Of-Vocabulary (OOV) words, such as user-defined variable names, to describe the source code more accurately. The selection probabilities of words in the vocabulary form a vector P_t^s, which can be calculated by feeding the hidden state e_t into a linear layer with dropouts, i.e.,

$$P_t^s = \text{softmax}(W_3 e_t + b_3) \tag{9}$$

The copy probabilities of words in source code X form a vector P_t^c, which can be calculated as follows.

$$P_t^c = u_t X \tag{10}$$

where u_t is an attention score vector of y_{t-1} over $(h_1, \ldots, h_N)$. The product of the two parameters indicate which words to copy from source code X. To decide which strategy the model will use to generate a specific word in final natural language explanation, the model first checks the type of the word in source code. If it is "NAME", "NUMBER", or "STRING", then the model will use the copying strategy; otherwise, the model will use the selection strategy.

By concatenating selection probabilities and copy probabilities, the generation probability for the natural language explanation of given source code X is calculated as follows.

$$P(y_t | y_{<t}, X, a) = [(1 - g_t) P_t^s, g_t P_t^c]) \tag{11}$$

where

$$g_t = \text{sigmoid}(W_4 y_{t-1}^{att} + b_4) \tag{12}$$

and the g_t works as a copy gate. A higher value of g_t indicates a higher probability that the copying strategy will be applied.

4.4 Model Training

The training objective of the model is to maximize the log likelihood of the generated sketch and natural language explanation for a given source code X, i.e.,

$$\max \sum_{(X,a,Y) \in D} \log P(Y|X, a) + \log P(a|X) \tag{13}$$

where D refers to data set of training pairs, each of which consists of the expression X of a line of source code, and the corresponding real sketch a^* and code comment Y^*.

In the model training, $P(Y|X, a)$ is estimated by Eq. (11), and $P(a|X)$ is estimated by Eq. (4). Then, the training objective is to find an optimal parameter set (which including parameters W_0, b_0, W_1, W_2, W_3, b_3, W_4, b_4, etc.) that maximizes Eq. (13). We utilize random gradient descent to update parameters, using the Cross-Entropy as loss function for both the sketch decoder and natural language explanation decoder.

5 Experimental Evaluation

In this section, we first introduce the experimental setup, and then verify the effectiveness of our proposed model on real data sets of Python code. To this end, we compare the performance of our model with that of state-of-the-art models on a set of commonly used metrics, and then investigate the effect of different components and the effect of code length on the performance of our model. We conclude this section with a case study to provide some insights on our model's efficacy and minor flaws. All experiments in this paper are conducted on a workstation with a 3.5 GHz Intel Core i7 CPU, 32 GB 2400 MHz DDR4 RAM, and a GTX 1080Ti with 11 Gb memory, running Ubuntu 18.04. The source code of our model and all data sets are available at https://github.com/code2nl/code2nl.

5.1 Experimental Setup

Data Sets. We evaluate the performance of our model with a data set of Python code [19], which was build upon the Django library. The data set contains 18805 pairs of Python statements and corresponding natural language expressions. For cross-validation, we shuffle the data set and split it into 3 parts with 16000 for training, 1000 for validation and 1805 for testing. The natural language sketch is generated from the training data set. For a pair of code and comment in the training data set, the sketch is generated by replacing some tokens with their types in the comment. These replaced tokens are identifiers and numbers appearing in the code and comment. We adopt a lexical scanner of Python to tokenize the code and obtain token types. Table 1 illustrates statistics for the length of the code and corresponding comment on the testing data. We can observe that about 93% code statements have no more than 15 tokens and about 95% comments have no more than 30 words.

Performance Metrics. We evaluate the performance of our model on three metrics that are widely-used in neural machine translation, including BLEU [20], METEOR [2], and ROUGE-L [17]. BLEU measures the translating precision by

Table 1. Statistics for code lengths and comments lengths

Code lengths					
Avg	Mode	Median	Max	<10	<15
8.11	6	6	527	76.12%	93.44%
Comments lengths					
Avg	Mode	Median	Max	<20	<30
13.03	5	10	70	78.84%	95.67%

how much the words in translation sentences appear in reference sentences at the corpus level, with a penalty for short sentences. METEOR is recall-oriented and measures the extent to which a model captures content from references in the output. ROUGE-L takes into account sentence level structure similarity and identifies the longest co-occurring in sequence n-grams automatically.

Baselines. We compare our proposed model with the following four state-of-the-art approaches. We download the source codes of these approaches from open source websites and run them on the Python code data set.

- **Reduced-T2SMT** [19] is a high performance framework of statistical machine translation to generate natural language expressions for Python source codes.
- **Seq2Seq** [24] is a classical deep learning model widely used in translation tasks. It encodes a source sentence into a hidden space, and then decodes it into target sentences. In the experiment, its encoder and decoder are implemented based on LSTM.
- **CODE-NN** [14] is the first deep learning-based model which uses LSTM with attention mechanism to produce summarization of code snippets.
- **Code-GRU** [16] adopts a new recursive neural network called Code-RNN to extract features from source code and embed them into a vector, and then utilizes a new recurrent neural network to generated natural language sentences based on the vector.

Furthermore, to better understand the importance of different components used in our model, we also evaluate the performance of the following four versions of our model.

- **Code2NL (no copying strategy)** does not use the copying strategy in the natural language explanation decoder. It generates words only based on the established vocabulary, and removes each rare word with a frequency below 3.
- **Code2NL (no type information)** uses only the word embeddings and position embeddings in the code input without any type information, and generate each sketch by replacing the words that appear in both code and comment with token "$< unk >$".
- **Code2NL (no sketch decoder)** has no sketch decoder. It calculates the copy probability by attention mechanism using decoder hidden state and input context vector as input, and generates natural language expression directly with an encoder-decoder architecture.
- **Code2NL** refers to the complete version of our model.

Training Details. Our method is implemented with PyTorch [21]. The vocabulary is collected from the training data set and we keep the top 2000 frequent words for the sketch data set. The embedding dimension of the sketch word vector and comment word vector is set to 100 and initialized by Glove [22]. The Python code word vectors are initialized randomly. The type feature is embedded to 10-dimensional vectors and the position feature is also embedded to 10-dimensional

Table 2. Evaluation results of different models on Python code data set

Methods	BLEU-4	METEOR	ROUGE-L
Reduced-T2SMT [19]	0.5476	0.3907	0.7986
Seq2Seq [24]	0.2826	0.2490	0.6549
CODE-NN [14]	0.4051	0.3071	0.7047
Code-GRU [16]	0.5081	0.3767	0.7864
Code2NL (no copying strategy)	0.4864	0.3699	0.7738
Code2NL (no type information)	0.4339	0.3324	0.6872
Code2NL (no sketch decoder)	0.3494	0.3161	0.5884
Code2NL	**0.5654**	**0.4146**	**0.8117**

vectors. All embeddings in the model are trainable. We employ a single layer BiLSTM with hidden size 250 for all encoders, and a single layer LSTM with hidden size 250 for the two decoders. The dropout rate is set to 0.3. We perform gradient descent by Adam optimizer [26] to optimize the Cross-Entropy loss function. The initial learning rate is set to 0.005 and it decays from the first epoch with the decay rate 0.985. The model is trained for up to 10 epochs and the mini-batch size for each update is set to 64.

5.2 Performance Comparison

Table 2 compares the performance of our Code2NL model with the four existing models on Python code data set. We can observe that Code2NL achieves the best performance on all metrics. The reasons behind are as follows. First, we incorporate a copying strategy to copy the words from source code to the generated sentences. Therefore, a few OOV words may be generated to help better describe the source code. Second, we use the type information to labeled the source code. This strategy helps our model solve the problem that some variable names cannot be well encoded and recognized by the model. Third, we adopt an asynchronous learning framework, which first learns natural language sketch for the source code, and then uses the sketch in a soft-embedded way to guide the generation of natural language explanations for the source code. This strategy make the generated natural language explanations to be more syntactically correct and semantically fluent. By jointly using the above three strategies, our Code2NL model can achieve a reasonably better performance for code comment generation, compared against the four existing models.

5.3 Effect of Components

To study the effect of different components used in our model, we compare the performance of Code2NL with that of its three different versions, in which the copying strategy, type information, and sketch decoder are removed respectively.

Table 2 reported the comparison results on the Python code data set, from which we can observe that (1) the copying strategy slightly improves the performance of our model. As an alternative of the selection strategy which selects words from the natural language vocabulary, the copying strategy can generate a few OOV words which helps better describe the source code. (2) The type information reasonably improves the performance of our model. This is because the type information helps to accurately define the attributes of the identifiers, and thus improves the accuracy of predicting low frequency words. On the other hand, without type information, the sketch will lack effective semantics to guild the natural language generation. For example, for source code "for pos, elt in enumerate (source) :" and its corresponding comment "for every pos and elt in enumerate iterable source", if we have type information like "OP, NAME, OP, NAME, OP, OP, OP, NAME, OP, OP", then we can built a more accurate sketch by replacing variable names in code comment, like "for every NAME and NAME in enumerate iterable NAME"; otherwise, we can only built a sketch by replacing words that appearing in both code and comment, like "unk every unk and unk in unk iterable unk". (3) The sketch decoder significantly improves the performance of our model. Without the sketch decoder, our model degenerates into a simple seq2seq model with attention and copying strategy, and its performance will be worse than Code-NN which also is a seq2seq model with attention mechanism. This is because without the guidance and constraints of sketch, the copying strategy in our model will copy a lot of repeated OOV words which may be not high-quality descriptions of the source code.

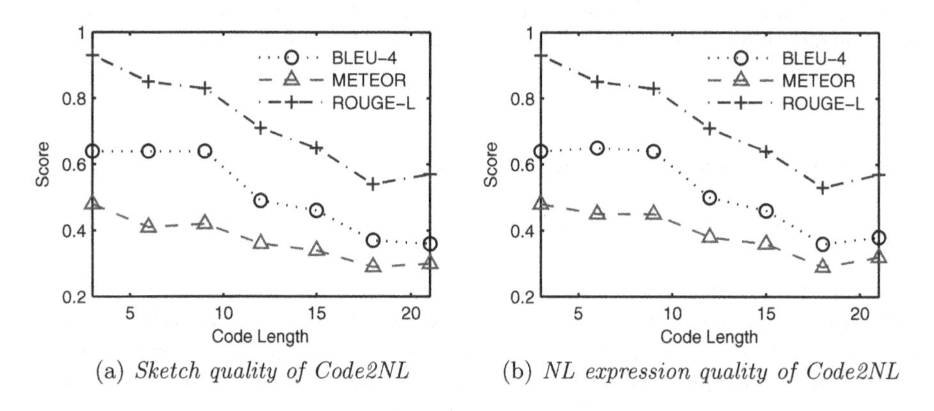

(a) *Sketch quality of Code2NL* (b) *NL expression quality of Code2NL*

Fig. 2. Effect of code length on qualities of sketch and natural language expression.

5.4 Effect of Code Length

To study the effect of code length on the performance of our model, we evaluate our model with source codes that have different code length (varying from 3 to 21 with an interval of 3).

Figure 2(a) and (b) illustrate the scores of sketch and final natural language expression of our Code2NL model on BLEU-4, METEOR, and ROUGE-L metrics, and Fig. 3 compares the BLEU-4 scores of natural language expressions generated by each tested model. From Fig. 2(a) and (b), we can observe that the score curve trends of sketch and final natural language expression of our Code2NL model are almost the same. This indicates that the performance of our model is dominated by the performance of our sketch decoder. From Fig. 3, we can observe that for all the evaluated models, although the qualities of natural language expression often become lower with growth of source code length, our Code2NL model always generates a relatively better natural language expression. Similar results can also be observed on the scores of the other metrics.

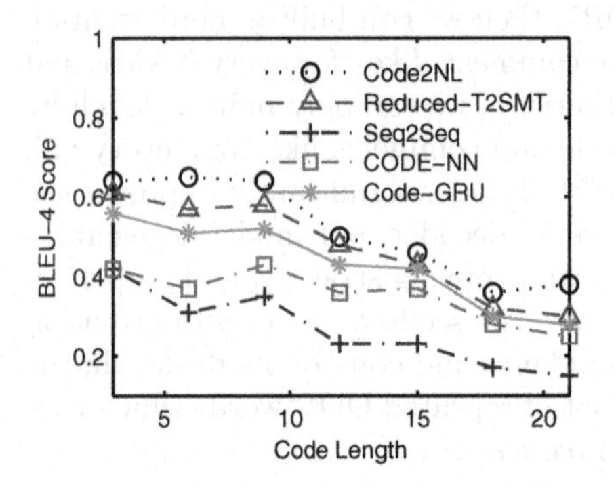

Fig. 3. Comparison of different models performance under different code lengths

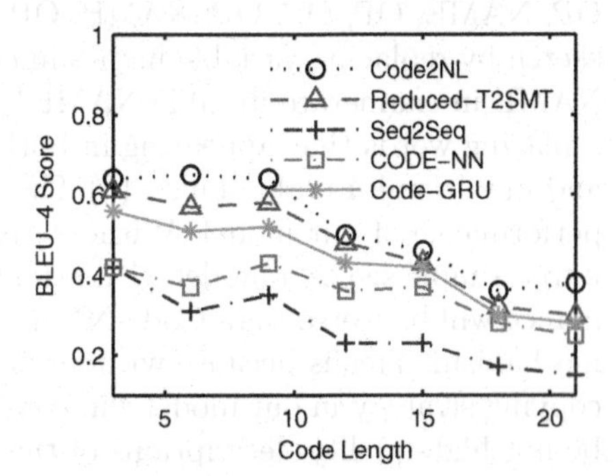

Fig. 4. Effect of sketch vocabulary size on qualities natural language expression.

5.5 Effect of Sketch Vocabulary Size

The sketch data set is generated by replacing some words with types in comments. The sketch plays an important role in our model and it guides the natural language generation. As the sketch vocabulary size increases, the ability of the sketch to guide natural language generation may weaken. To study the effect of the sketch vocabulary size in our model, we compare the performance of our model under different vocabulary size. The sketch vocabulary preferentially remove low frequency words in training data set.

From Fig. 4, we can observe that as we gradually reduce the sketch vocabulary size, the model performance is slowly improved. This is because the low frequency words are removed from the training data set. However, with the further reduction of the sketch vocabulary size, the performance is significantly reduced. This is due to the deletion of some commonly-used words, which often causes the decline of the expression ability of the model.

5.6 Case Study

To qualitative analyse the performance of our proposed model, we carry out a case study which is illustrated in Table 3. This table provides three examples of Python source code, and the corresponding real sketch and code comments written by human. Below each real sketch and real comment, we list the sketch and comments generated by our Code2NL model.

Table 3. Case study on the performance of Code2NL

Python	last = pos + 1
Real Sketch	increment NAME by one, substitute the result for NAME
Sketch	sum NAME and integer NUMBER, substitute the result for NAME
Real Comment	increment pos by one, substitute the result for last
Comment	sum pos and integer 1, substitute the result for last

Python	params = [source[1]]
Real Sketch	NAME is a list with second element of NAME as an element
Sketch	NAME is an list containing element of NAME
Real Comment	params is a list with second element of source as an element
Comment	params is an list containing element of source

Python	t = super(SafeBytes, self)._add_(rhs)
Real Sketch	call the NAME method with an arguments NAME from the base class of the class NAME, substitute the result for NAME
Sketch	call the method NAME from the base class of the class NAME, substitute the result for NAME
Real Comment	call the _add_ method with an arguments rhs from the base class of the class SafeBytes, substitute the result for t
Comment	call the _add_ method from the base class of the class SafeBytes, substitute the result for t

From Table 3, we can observe that the sketch often captures relatively accurate semantic and type information, and the natural language explanation decoder is able to copy the right words from source code and generate correct comments. Nonetheless, there are still some minor flaws in each example, which may indicate the direction of possible improvements for our proposed model.

In the first example, although the generated comments hold relevant semantics, they may gain low scores on performance metrics. This is because although the real and generated comments basically describe similar functionalities, but they use different words or word orders. An enhancement with synonyms may reduce this difference.

In the second example, the generated sketch ignores the location information of the element in the list, although the semantics is correct. This indicates that our model cannot capture the exact meaning of the number. At the same time, it also lacks common sense about Python, such as the position of the elements in the Python list is counted from zero.

In the third example, the source code is a little more complicated than the previous two. Compared with the real comments, the comment generated by our model lacks some details about methods parameters. The data set is not complicated, but Table 3 shows that our work can accurately translate the meaning of a line of code.

6 Conclusion

We present a type-aware sketch-based sequence-to-sequence learning model to translate each line of source code into a natural language explanation. Our model has an important and unique two-stage text generation process. The first generation stage focuses on learning the code semantic and the second stage focuses on learning identifiers and numbers from source code. In this model, we parse the source code and generate the sketch to guide natural language generation; meanwhile, we use an LSTM decoder to make a decision between selecting words from natural language vocabulary and copying from source code. Empirical evaluation on a Python data set reveals that our model outperforms state-of-the-art approaches on commonly used metrics for neural machine translation.

Acknowledgments. This work was supported in part by the National Key R&D Program of China (2018YFB1003901), the NSFC Grants (61976163, 61872315 and 61872273), the Advance Research Projects of Civil Aerospace Technology ("Intelligent Distribution Technology for Domestic Satellite Information"), the Technological Innovation Major Projects of Hubei Province (2017AAA125), the Natural Science Foundation of Hubei Province (ZRMS2020000714), the High-End Industry Development Fund of Beijing City ("Smart Home-Oriented AI Open Innovation Platform"), the Science and Technology Program of Wuhan City (2018010401011288), the Open Funding of CETC Key Laboratory of Aerospace Information Applications, and the Xiaomi-WHU AI Lab.

References

1. Bahdanau, D., Cho, K., Bengio, Y.: Neural machine translation by jointly learning to align and translate. arXiv preprint arXiv:1409.0473 (2014)
2. Banerjee, S., Lavie, A.: METEOR: an automatic metric for MT evaluation with improved correlation with human judgments. In: Proceedings of the ACL Workshop on Intrinsic and Extrinsic Evaluation Measures for Machine Translation and/or Summarization, pp. 65–72 (2005)

3. Cho, K., Van Merriënboer, B., Bahdanau, D., Bengio, Y.: On the properties of neural machine translation: Encoder-decoder approaches. arXiv preprint arXiv:1409.1259 (2014)
4. Cho, K., et al.: Learning phrase representations using RNN encoder-decoder for statistical machine translation. arXiv preprint arXiv:1406.1078 (2014)
5. Dong, L., Lapata, M.: Coarse-to-fine decoding for neural semantic parsing. arXiv preprint arXiv:1805.04793 (2018)
6. Eddy, B.P., Robinson, J.A., Kraft, N.A., Carver, J.C.: Evaluating source code summarization techniques: Replication and expansion. In: Proceedings of the 21st IEEE International Conference on Program Comprehension, pp. 13–22 (2013)
7. Gu, J., Lu, Z., Li, H., Li, V.O.: Incorporating copying mechanism in sequence-to-sequence learning. arXiv preprint arXiv:1603.06393 (2016)
8. Haiduc, S., Aponte, J., Marcus, A.: Supporting program comprehension with source code summarization. In: Proceedings of the 32nd ACM/IEEE International Conference on Software Engineering-Volume 2, pp. 223–226 (2010)
9. Haiduc, S., Aponte, J., Moreno, L., Marcus, A.: On the use of automated text summarization techniques for summarizing source code. In: Proceedings of the 17th Working Conference on Reverse Engineering, pp. 35–44 (2010)
10. Hill, E., Pollock, L., Vijay-Shanker, K.: Automatically capturing source code context of NL-queries for software maintenance and reuse. In: Proceedings of the 31st International Conference on Software Engineering, pp. 232–242 (2009)
11. Hochreiter, S., Schmidhuber, J.: Long short-term memory. Neural Comput. **9**(8), 1735–1780 (1997)
12. Hu, X., Li, G., Xia, X., Lo, D., Jin, Z.: Deep code comment generation. In: Proceedings of the 26th Conference on Program Comprehension, pp. 200–210 (2018)
13. Hu, X., Li, G., Xia, X., Lo, D., Lu, S., Jin, Z.: Summarizing source code with transferred API knowledge (2018)
14. Iyer, S., Konstas, I., Cheung, A., Zettlemoyer, L.: Summarizing source code using a neural attention model. In: Proceedings of the 54th Annual Meeting of the Association for Computational Linguistics (Volume 1: Long Papers), pp. 2073–2083 (2016)
15. Koehn, P.: Statistical Machine Translation. Cambridge University Press, New York (2009)
16. Liang, Y., Zhu, K.Q.: Automatic generation of text descriptive comments for code blocks. In: Proceedings of the 32nd AAAI Conference on Artificial Intelligence, pp. 5229–5236 (2018)
17. Lin, C.Y.: ROUGE: a package for automatic evaluation of summaries. In: Proceedings of the Workshop on Text Summarization Branches Out, pp. 74–81 (2004)
18. Luong, M.T., Pham, H., Manning, C.D.: Effective approaches to attention-based neural machine translation. arXiv preprint arXiv:1508.04025 (2015)
19. Oda, Y., et al.: Learning to generate pseudo-code from source code using statistical machine translation (t). In: Proceedings of the 2015 30th IEEE/ACM International Conference on Automated Software Engineering, pp. 574–584 (2015)
20. Papineni, K., Roukos, S., Ward, T., Zhu, W.J.: BLEU: a method for automatic evaluation of machine translation. In: Proceedings of the 40th Annual Meeting on Association for Computational Linguistics, pp. 311–318 (2002)
21. Paszke, A., et al.: Automatic differentiation in PyTorch (2017)
22. Pennington, J., Socher, R., Manning, C.: GloVe: global vectors for word representation. In: Proceedings of the 2014 Conference on Empirical Methods in Natural Language Processing, pp. 1532–1543 (2014)

23. Schuster, M., Paliwal, K.K.: Bidirectional recurrent neural networks. IEEE Trans. Signal Process. **45**(11), 2673–2681 (1997)
24. Sutskever, I., Vinyals, O., Le, Q.V.: Sequence to sequence learning with neural networks. In: Advances in Neural Information Processing Systems, pp. 3104–3112 (2014)
25. Wan, Y., et al.: Improving automatic source code summarization via deep reinforcement learning. In: Proceedings of the 33rd ACM/IEEE International Conference on Automated Software Engineering, pp. 397–407 (2018)
26. Kingma, D.P., Ba, J.: Adam: a method for stochastic optimization. arXiv preprint arXiv:1412.6980 (2014)

Coupled Graph Convolutional Neural Networks for Text-Oriented Clinical Diagnosis Inference

Ning Liu[1], Wei Zhang[2], Xiuxing Li[1], Haitao Yuan[1], and Jianyong Wang[1(✉)]

[1] Tsinghua University, Beijing, China
victorliucs@gmail.com, jianyong@mail.tsinghua.edu.cn
{lixx16,yht16}@mails.tsinghua.edu.cn
[2] East China Normal University, Shanghai, China
zhangwei.thu2011@gmail.com

Abstract. Text-oriented clinical diagnosis inference is to predict a set of diagnoses for a specific patient given its medical notes. Due to the great potential of automatic diagnosis inference, machine learning methods have began to be applied to this domain. However, existing approaches focus on performing either labor-intensive feature engineering or sequential modeling of each medical note separately, without considering the information sharing among similar patients, which is essential for evidence-based medicine, an emerging new diagnosis process. Motivated by this issue and the recently proposed graph convolutional network (GCN) for text classification, we propose to apply GCN for the text-oriented clinical diagnosis inference task. To encode the comorbidity of diagnoses into the GCN model and allow information sharing between patients, we devise a coupled graph convolutional neural networks (CGCN), where a note-dependent graph and a label-dependent graph are learned collaboratively with hyperplane projection to ensure they are in the same semantic space. The comprehensive results on two real datasets show that our method outperforms the state-of-art methods in text-oriented diagnosis inference.

Keywords: Medical data mining · Graph neural network · Diagnosis inference

1 Introduction

The computer aided auto-diagnosis system can reduce the burden of clinical diagnosis inference by providing the most probable diagnostic options. Among the different types of sources, medical notes can give more descriptive information about the patients such as the past disease information, cure measurements, etc. The experts make accurate diagnoses from two aspects: one is from the patient's medical records and the other is from the disease information.

Y. Nah et al. (Eds.): DASFAA 2020, LNCS 12112, pp. 369–385, 2020.
https://doi.org/10.1007/978-3-030-59410-7_26

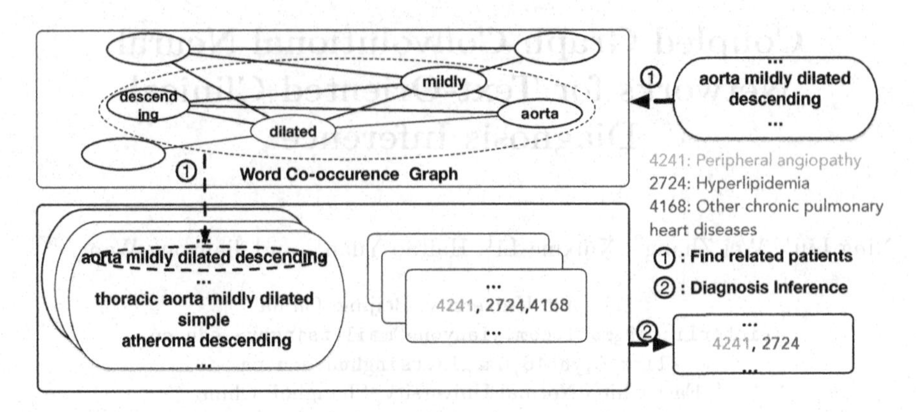

Fig. 1. The process of evidence-based medicine(left bottom part) and information sharing via word co-occurrence graph (left upper part). When diagnosing a patient, the experts firstly find the related patients in the patient corpus and give the diagnoses upon their knowledge and the existing evidences from the related patients. Relations between the two patients can be described by the common words(aorta,..,descending).

However, automatic text-oriented clinical diagnosis inference can be challenging. Firstly, learning a better representation of patients via their medical documents is difficult. With the rapid development of the medicine, evidence-based medicine (EBM) [26] has been widely used in clinical decision-making process where the diagnoses of a patients are not only judged by the experts' knowledge but also influenced by the shared information with related patients who have similar symptoms. The simulated procedure of evidence-based medicine is shown in Fig. 1. Compared with the traditional decision-making process, EBM encourages the evidences shared among similar patients to improve the quality of the healthcare [8]. Since the information sharing is eagerly stressed in the process of evidence-based medicine, the embedding of a patient should consider not only the text information describing the symptoms of the patient but also the evidences obtained from other related patients for making accurate diagnoses. Secondly, compared to traditional multilabel classification, learning a good embedding of diagnoses considering diagnosis information is quite important. Among the disease information, disease comorbidity is an essential factor and the disease comorbidity is quite common in many diseases such as Epilepsy [13], Rheumatoid Arthritis [25], Cluster headache [19] and so on.

Recently, deep learning methods have been applied to the document classification such as the convolutional neural networks [14] and recurrent neural networks [2,22] to capture the word sequence information hidden in the texts. And LEAM [29] uses attention mechanism to jointly learn both word embeddings and label embeddings. These methods can capture semantic and syntactic information in local consecutive word sequences well. And a condensed memory neural network is proposed for clinical diagnosis inference which uses outer knowledge such as Wikipedia [24]. However, they represent text as a vector and ignore the information sharing among the patient corpus. In the previous work [34], the texts are represented as a graph to learn a text graph neural networks which shows great improvements on multi-class text classifications due to the information transfer among the text corpus. Yet it is not applied to the clinical diagnosis inference problem. Therefore, in the previous works, those methods either not explicitly model the disease embeddings with the disease information or unable to combine the information from other related patients' medical documents.

To address the above challenges, we propose a novel Coupled Graph Convolutional Network (CGCN) and model the text-oriented clinical diagnosis inference in two aspects: one is from the aspects of patients and the other is from the aspects of diagnoses. In order to allow information sharing from the related patients' medical documents, we build a large heterogeneous two-level text corpus graph where the upper level node is the medical document node and the bottom level node is the word node where the relations between the medical documents can be linked via the common words in the global word co-occurrence graph (See left upper part of Fig. 1). And then, we formulate the process of patient's embedding learning as the learning process of document node embedding learning in a graph and apply Graph Convolution Neural Network on the text corpus graph to gather high order neighbor information. Similarly, we construct diagnosis comorbidity graph and consider diagnoses as the nodes and apply Graph Convolutional Neural Network to update the embeddings of diagnoses via the comorbidity information. After that, we use a hyperplane project method to project the embeddings of diagnoses to the related hyperplane of the patients' embedding space and compute the probability of each diagnosis to each patients and make the final predictions.

To summarize, our contributions are listed as the followings:

- We propose a novel Coupled Graph Neural Network (CGCN) in clinical diagnosis inference. To the best of our knowledge, we are the first to apply graph convolutional neural networks in the clinical diagnosis inference problem.
- The CGCN uses two separate Graph Conventional Networks to jointly learn embeddings of patients and diagnoses considering the relations between the patients and the relations between the diagnoses.
- In order to compute the match score in the same hyperplane, we use hyperplane projection to project the diagnosis embeddings into the related hyperplane of the patient embeddings.
- Experiments on two medical datasets show the CGCN outperforms the state-of-art methods. Besides, we develop an inductive learning framework for CGCN and Text GCN and external experiments show that our model still outperforms the state-of-art methods on the inductive settings.

2 Related Work

2.1 Clinical Diagnosis Inference

The problem of clinical diagnosis inference has been studied for over twenty five years. Binaghi et al. use an artificial neural network in the diagnosis of acute coronary occlusion [3]. And rule based methods are performed on clinical datasets [4]. In the previous studies, deep learning technologies show great strengths over traditional methods and have been widely used in clinical diagnosis inference. Recurrent neural networks are used for modeling the time dependency in medical data [6,17] and hierarchical attention neural networks are used to model long medical texts [2]. Besides, medical knowledge is further used to improve the performance of deep learning models. For example, C-MemNN [24] uses Wikipedia as external knowledge in predicting patients' diagnosis. However, these methods have several drawbacks. Firstly, these methods are either based on structured features or sequential representation of notes which cannot meet the requirements of evidence-based medicine with the information sharing among the patients. Secondly, the relations between diagnoses are not modelled.

2.2 Text Classification

Text Classification studies mainly focus on obtaining better text representations and can be categorized into two groups. One group focus on feature engineering. Manevitz et al.[20] uses different features, including term frequency representation and term frequency-inverse document frequency (TF-IDF) to represent texts. In [31], support vector machines and navies Bayes are used with word-level features as well. Some other works focus on indirect features learned from other models. For example, Ghassemi et al. [9] uses topic distributions learned from LDA to make mortality predictions.

In the past few years, deep learning has been widely used in document classification and achieved remarkable success. Convolutional neural networks are used for sentence classification [14]. Wang et al. [30] used convolutional neural networks which combine knowledge and character level features for classifying short text. Despite convolutional networks, recurrent neural networks [2,33] and recursive neural networks [11] also find their applications in text classification. Zhang et al. [35] learned label embeddings via multitask framework while LEAM [29] uses attention mechanism in learning the word and label embeddings. Further, Liu et al. [18] uses convolutional neural networks for the text based mortality prediction. These deep learning based methods consider the text as a sequential representation and cannot capture advanced text representations. In order to capture the global word co-occurrence information, Yao et al. [34] exploited graph convolutional neural networks in text based multi-class classification and achieve the state-of-art performance. Xue et al. [32] developed adversarial mutual learning to address the text classification under the unsupervised domain adaptation setting. However, they are not used in the clinical diagnosis inference problem and the diagnosis comorbidity is not modelled.

2.3 Graph Neural Networks

Graph neural networks are designed to handle graph structure data. While the traditional neural networks are capable of handling with structured data such as text sequences or image, they cannot handle the semi-structured data such as graphs, trees and so on, which drives the studies of graph neural networks [28,36]. Traditional convolutional neural networks with the properties of local connection, shared weights have gained success in various applications such as image classification, text classification and motivate the research on convolutions on the arbitrarily graphs [28]. The spectral-based graph convolutional networks [5,10,16] are based on spectral theory in the graph signal processing and use the orthonormal basis formed by eigenvectors of normalized graph Laplacian to transformer the input signals and have been applied into many domains due to their information sharing among the neighbors. Recently, GCNs are widely used to encode advanced graph structures in various tasks such as computational drug development and discovery [28], relation classification [27], machine translation [1], multi-class text classification [7,16,23,34].

3 Coupled Graph Neural Networks

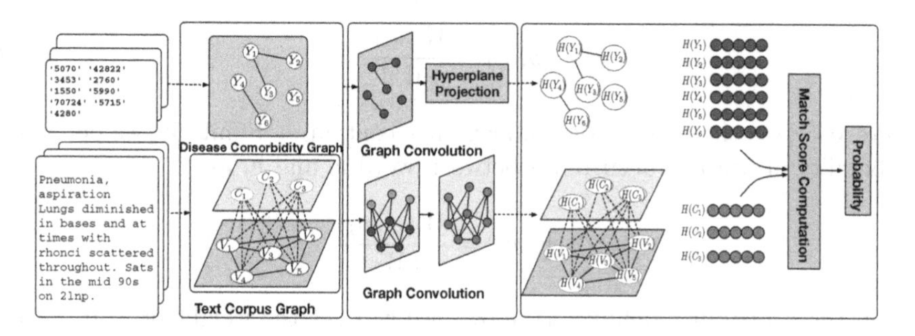

Fig. 2. The overall structure of Coupled Graph Convolutional Neural Network. Y_i indicates the diagnosis, C_i indicates the medical notes indexed by i and $H(.)$ indicates the representations of words, medical documents or diagnoses.

In this section, we will give a detailed information of Coupled Graph Convolutional Networks(CGCN). To begin with, we formulate the problem. And then, we introduce the process of constructing text corpus graph and diagnosis comorbidity graph. After that, we give a brief introduction of graph convolutional neural networks. Finally, we give a detailed overview of our model structure.

3.1 Problem Formulation

Formally, the basic elements can be defined as $(\mathcal{V}, \mathcal{C}, \mathcal{T}, \mathcal{Y}, \mathcal{G}_\mathcal{C}, \mathcal{G}_\mathcal{T})$ where $\mathcal{V}$ is the vocabulary set, $\mathcal{C}$ is the text corpus set, $\mathcal{T}$ is the diagnosis set, Y_i is the diagnoses of the document i, $\mathcal{G}_C$ is the text corpus graph and $\mathcal{G}_T$ is the diagnosis comorbidity graph. Therefore, the goal of our method is to learn the function $\mathcal{F}$ to predict the set of diagnoses given the text $c \in \mathcal{C}$, $\mathcal{G}_C$ and $\mathcal{G}_T$:

$$\mathcal{F}(c, \mathcal{G}_C, \mathcal{G}_T) \rightarrow Y_i \tag{1}$$

3.2 Graph Construction

In this section, we will give the detailed information of constructing text corpus graph $\mathcal{G}_C$ and diagnosis comorbidity graph $\mathcal{G}_T$.

For text corpus graph construction, we build a large and heterogeneous text graph which consider texts and words as nodes of the graph. The graph is built in two levels: one is from the bottom word co-occurrence information and the other is built from word to text relations.

For word graph construction, we use sliding window based methods to compute the co-occurrence of words. To get the global word co-occurrence information, we use a fix size of sliding window over the text corpus $\mathcal{C}$ and get a window set defined as W and then the word co-occurrence graph between the words are computed based on the window set. We use the point-wise mutual information to define the relations between words. For each word $v \in \mathcal{V}$, we define the probability of occurrence using the sliding window based methods, that as:

$$p(v_i) = \frac{W(v_i)}{\mid W \mid} \tag{2}$$

where $W(v_i)$ is the number of windows that contain v_i. Similarly, we define the probability of word co-occurrence between word v_i and v_j as:

$$p(v_i, v_j) = \frac{W(v_i, v_j)}{\mid W \mid} \tag{3}$$

Given the probability definitions above, we can get the weight of word v_i and word v_j :

$$R(v_i, v_j) = log \frac{p(v_i, v_j)}{p(v_i)p(v_j)} \tag{4}$$

Follow the above constructions, we can get the word graph as $G_w = (V, E_w)$, where

$$E_w = \{(i, j, R(i, j)) \mid i, j \in \mathcal{V}, R(i, j) > 0\} \tag{5}$$

For word-document graph construction, some methods such as bag of words(BOW), term frequency-inverse document frequency(TF-IDF) can be used to describe the relations between words and corresponding texts. In our method, we use TF-IDF to model the relations between the words and medical documents because TF-IDF will assign a higher weight to those more descriptive words to ensure the correct information sharing between words and medical documents. Therefore, the word to document graph G_d can be defined as $G_d = (N, E_d)$ where $N = \mathcal{C} \cup \mathcal{V}$ and E_d is defined as:

$$E_d = \{(i, j, TFIDF(i, j)) \mid i \in \mathcal{V}, j \in \mathcal{C} \mid\mid i \in \mathcal{C}, j \in \mathcal{V}\} \quad (6)$$

Follow the definitions of G_w and G_d, we get the text corpus graph $\mathcal{G}_\mathcal{C} = (N, E_C)$ where $N = \mathcal{C} \cup \mathcal{V}$ and $E_C = E_d \cup E_w \cup \{(i, i, 1) \mid i \in N\}$

For the diagnosis comorbidity graph construction, as the number of diagnoses is further less than the number of documents and words, we compute the frequency of any two diagnoses in the diagnosis set and construct the diagnosis comorbidity graph. And the comorbidity score between diagnosis i and j is defined as:

$$s(i, j) = \frac{No(i, j)}{\mid \mathcal{C} \mid} \quad (7)$$

where $No(i, j)$ indicates the number of patients are diagnosed with disease i and j. Therefore, the diagnosis comorbidity graph can be defined as $\mathcal{G}_\mathcal{T} = (\mathcal{T}, E_T)$ where

$$E_T = \{s(i, j) \mid i, j \in \mathcal{T}\} \quad (8)$$

From the above definitions, we can get text corpus graph $\mathcal{G}_\mathcal{C}$ and diagnosis comorbidity graph $\mathcal{G}_\mathcal{T}$. Although there are some knowledge graph built for diagnoses, such as the relations in ICD-10[1], we find that the disease hierarchical relation matrix is very sparse so that we use the statistics from the dataset directly.

3.3 Graph Convolutional Neural Network

In formal, the graph can be defined as $G = (V, E)$ where V is the set of nodes and E is the set of edges and the node features are defined as a matrix defined as $X \in R^{|V| \times d}$ where d is the number of node features. For computation efficiency, the graph G is represented in an adjacency matrix $A \in R^{|V| \times |V|}$ and we define D as the degree matrix of A where D is a diagonal matrix and $D_{ii} = \sum_j A_{ij}$. Graph Convolutional Neural Networks(GCN) take the node features X and graph structure A as its input and update the embeddings of the nodes using the information from the neighbors. In the spectral graph convolution theory, the graph convolution of a filter g_θ and graph G in is defined as:

$$X *_G g_\theta = U g_\theta U^T X \quad (9)$$

where $L = I_n - D^{-\frac{1}{2}} A D^{-\frac{1}{2}} = UMU^T$ is a normalized graph Laplacian matrix and U is the eigenvectors ordered by eigenvalues. Following the above definitions, Kipf and Welling et al. [16] provide a simple approximation for graph convolution computation. And the convolution over graph is defined as the following:

$$x *_G g_\theta \approx \theta(I_n + D^{-\frac{1}{2}} A D^{-\frac{1}{2}} x) \quad (10)$$

And then, Eq. 10 is further modified to a compositional layer which is defined as :

$$H = f(\tilde{A} X \Theta)) \quad (11)$$

[1] https://en.wikipedia.org/wiki/ICD-10.

where $\tilde{A} = D^{-\frac{1}{2}} A D^{-\frac{1}{2}} + I_n$ and $\Theta \in R^{|V| \times m}$ is a weight matrix and f is the activation function. In order to reduce the numerical instability, GCN introduces a normalised trick and the graph convolution can be defined as the followings:

$$H = f(\hat{A} X \Theta)) \tag{12}$$

where $\hat{A} = \hat{D}^{-\frac{1}{2}} (A + I_n) \hat{D}^{-\frac{1}{2}}$ with $\hat{D} = \sum_j (A + I_n)_{ij}$.

Following the above definitions, we can define one layer graph convolution as the following:

$$H_0 = f(\hat{A} X \Theta_0)) \tag{13}$$

And the stacked multi-layer graph convolution which can obtain higher order information in graphs can be defined in a recursive way:

$$H_j = f(\hat{A} H_{j-1} \Theta_j) \tag{14}$$

3.4 Overall Structure

The overall structure of our proposed model is illustrated in Fig. 2.

Embedding Learning. With the text corpus graph $\mathcal{G}_\mathcal{C}$ and diagnosis comorbidity graph $\mathcal{G}_\mathcal{T}$ constructed, we build a novel neural network upon the graph data which takes the related information of medical text corpus and diagnosis comorbidity into account. We explicitly model the word embeddings, document embeddings and diagnosis embeddings in a coupled way. One is for document and word embedding learning and the other is for diagnosis embedding learning.

For document and word embedding learning, we use a two graph convolution layers to model the text corpus graph. Though the text corpus graph does not model the document relations, the two layer graph convolution allow the message passing between documents. We define the adjacency matrix of the text corpus graph $\mathcal{G}_\mathcal{C}$ as $A^\mathcal{C}$. Then the word and document embeddings of one layer graph convolution can be obtained by:

$$H_0^\mathcal{C} = f(\hat{A}^\mathcal{C} X^\mathcal{C} \Theta_0^\mathcal{C}) \tag{15}$$

where $\hat{A}^\mathcal{C} = \hat{D}^{-\frac{1}{2}} A^\mathcal{C} \hat{D}^{-\frac{1}{2}}$ is the normalised matrix. Then the embeddings of words and documents after two graph convolution layers can be obtained by:

$$H_1^\mathcal{C} = f(\hat{A}^\mathcal{C} H_0^\mathcal{C} \Theta_1^\mathcal{C}) \tag{16}$$

where $H_1^\mathcal{C} \in R^{(|\mathcal{V}|+|\mathcal{C}|) \times k}$ and k is the hidden size.

For diagnosis embedding learning, we use a one layer graph convolution to get information from the neighbors. In this part, we can get the hidden representation of diagnoses by:

$$H_0^\mathcal{T} = f(\hat{A}^\mathcal{T} X^\mathcal{T} \Theta_0^\mathcal{T}) \tag{17}$$

where $\hat{A}^{\mathcal{T}}$ is the normalized matrix of $A^{\mathcal{T}}$ and $A^{\mathcal{T}}$ is the adjacency matrix of diagnosis comorbidity graph $\mathcal{G}_{\mathcal{T}}$.

In our model, we use Tanh function as the activation function and get the hidden representations of words, documents and diagnosis which hold the information from the graph structures to compute the similarities. The Tanh function can be defined as the followings:

$$f = \frac{e^x - e^{-x}}{e^x + e^{-x}} \tag{18}$$

Match Score Computation. From the above learning process, we get the hidden representations of words, documents and diagnoses. In this section, we will describe how to compute the similarities between the document representations and diagnosis representations.

It is clear that we can apply matrix dot to compute the scores. However, this can lead to inconsistency of the vector spaces between the embeddings of documents and diagnoses. In order to solve the above issue, we simply apply a hyperplane projection over the diagnosis space by using an orthogonal matrix and transfer the diagnosis representations into the related hyperplane of the document embeddings. In formal, the projection process can be defined as the followings:

$$\hat{H}^{\mathcal{T}} = H_0^{\mathcal{T}} - w_p^T H_0^{\mathcal{T}} w_p \tag{19}$$

where w_p is a matrix related to the relavance hyperplane of the document embeddings. Then, the hidden representations $\hat{H}^{\mathcal{T}}$ contain the diagnosis information and the hidden state $H^{\mathcal{C}}$ contain the embeddings of words and documents. Therefore, we use a matrix dot operation to compute the score between the document $i \in \mathcal{C}$ and the diagnosis $t \in \mathcal{T}$:

$$s(t, i) = H_i^{\mathcal{C}} \cdot (\hat{H}_t^{\mathcal{T}})^T \tag{20}$$

where $H_i^{\mathcal{C}}$ indicates the embeddings of the document i and $\hat{H}_t^{\mathcal{T}}$ is the embedding of diagnosis t.

Then, we can compute the probability of document $i \in \mathcal{C}$ is diagnosed with the diagnosis $j \in \mathcal{T}$ given the disease comorbidity graph $\mathcal{G}_{\mathcal{T}}$ and text corpus graph $\mathcal{G}_{\mathcal{C}}$ via a sigmoid function.

$$p(1 \mid i, j, \mathcal{G}_{\mathcal{T}}, \mathcal{G}_{\mathcal{C}}) = \frac{1}{1 + e^{-s(j,i)}} \tag{21}$$

where $s(j \mid i)$ is defined in Eq. 20.

Loss Function. With the probabilities of diagnoses given the documents learned, we use a multilabel binary cross-entropy loss [22].

$$\begin{aligned} \mathcal{L} = &- \frac{1}{N} \sum_{i,j} y_{ij} log(p(1 \mid i, j, \mathcal{G}_{\mathcal{T}}, \mathcal{G}_{\mathcal{C}})) \\ &- \frac{1}{N} \sum_{i,j} (1 - y_{ij}) log(p(0 \mid i, j, \mathcal{G}_{\mathcal{T}}, \mathcal{G}_{\mathcal{C}})) \end{aligned} \tag{22}$$

where $p(1 \mid i, j, \mathcal{G}_T, \mathcal{G}_C)$ is defined in Eq. 21 and y_i is an indicator vector which records the real diagnoses of the patient i.

4 Experiments

4.1 Dataset

In this paper, we use two datasets in our final experiments. As illustrated in [24], the distribution of diagnoses in the dataset has a very long tail and some diagnoses only lie in few documents. Therefore, we use the 50 most common labels in our final datasets.

We extract the medical notes with the most recent records of patients from the MIMIC III dataset [12] and use two subcategories of the medical documents in the dataset: one is from the discharge summary and the other is from the nursing. And we only use the most recent records of the patients and we remove the documents whose diagnoses is not among the most 50 common diagnoses to get the final datasets. Table 1 gives the basic information of the datasets used in the paper.

Table 1. Basic description of the nursing dataset

Dataset	No. samples	Average length
Nursing#50	6900	131.7
Dischargesummary# 50	38238	763.7

Data Processing. For data prepossessing, we apply the traditional text data processing procedure. Vocabularies for each medical document are generated by first tokenizing the free text and lemmatizing the words in the texts and then removing stop words. In order to find the most representative words in the text corpus, we further analyze the tokenized text corpus and restrict the vocabulary size to 10000 for the Dischargesummary dataset and 5000 for the Nursing dataset.

Model Configuration. We use Adam [15] to train our CGCN model with the learning rate set to 0.02. We randomly split the dataset into the training dataset, validation dataset and test dataset with the ratio 8:1:1. We find the best score on the validation dataset and report the performance on the test dataset. Our model is trained on a Ubuntu Server (16.04) with a Titan XP GPU (12 G) and we use pytorch to implement our model.

4.2 Models

In this section, we introduce the methods we used for comparison as the followings:

- **MLP.** We use the average of word embeddings as the document representation and feed them into a two layer linear layer with Tanh as the activation. The embeddings of words are learned during the training.
- **CNN.** Convolutional Neural Network. We use 50 filters with kernel size set to 2 and 3 to capture bigram information and trigram information of text sequences and then the feature map is fed into a max pooling layer to extract the most important information of a document. In our results, CNN_2 is the Convolutional Neural Networks with the kernel size set to 2 and CNN_3 is the Convolutional Neural Networks whose kernel size is 3. For the multilabel settings, we use the sigmoid as the final activation function.
- **HA_GRU.** Hierarchical attention GRU. It is first introduced in [33] which first encodes sentence to a sentence representation and then use sentence representations to learn a document representation with a hierarchical attention mechanism.
- **KV.** Key Value Memory Network. The model is proposed in [21] and keeps a long term memory which can be retrieved in multiple hops. In our experiments, we use 3-hop Key Value Memory(KV_3) and 5-hop Key Value Memory Networks(KV_5).
- **C_MM.** Condensed Memory Networks. The model is proposed in [24] and uses a condensed way for compressing the information learned from the long term memory. And in our experiments, we use 3-hop Condensed Memory Network(C_MM_3) and 5-hop Condensed Memory Network(C_MM_5).
- **LEAM.** The model is first introduced in [29] and use label-wise attention to jointly learn the word embeddings and label embeddings in the multiclass text classification and the label descriptions are used in the classification task. For the clinical diagnosis inference problem, we change the last layer of activation function to the sigmoid function.
- **Text GCN.** The model is first addressed in [34] and explores power of graph neural networks in the multiclass text classification. We use the same settings as in [34] and change the activation of the predictive layer to the sigmoid function.
- **CGCN-base.** The model is a simple version of our CGCN where we remove the hyperplane projection part of the proposed CGCN.

4.3 Test Performance

Metrics. In the experimental part, we report auc(macro), precision@5(pre@5), recall@5(rec@5) and f1_score@5(f1@5) in our final results. For real applications, precision@5 indicates the ratio of relevance labels in the top 5 predictions and the recall@5 indicates the ratio of relevance labels of real diagnoses. Then the f1_score@5 can be computed by:

$$f1@5 = \frac{2 * pre@5 * rec@5}{pre@5 + rec@5} \tag{23}$$

Table 2. Results on nursing #50

Methods	auc(macro)	rec@5	pre@5	f1@5
MLP	0.6734	0.3580	0.3380	0.3477
CNN_2	0.6729	0.3587	0.3467	0.3526
CNN_3	0.6642	0.3617	0.3455	0.3534
HA_GRU	0.6250	0.3365	0.3244	0.3303
KV_3	0.6729	0.3587	0.3528	0.3557
KV_5	0.6731	0.3594	0.3536	0.3568
C_MM_3	0.6590	0.3599	0.3412	0.3503
C_MM_5	0.6598	0.3504	0.3397	0.3450
LEAM	0.6788	0.3544	0.3344	0.3441
Text GCN	0.6842	0.3790	0.3579	0.3681
CGCN-base	0.6875	0.3731	0.3513	0.3619
CGCN	**0.6982**	**0.3909**	**0.3684**	**0.3793**

Results. In both of the two datasets, our proposed CGCN outperforms the baselines proposed in Table 2 and Table 3. The Text GCN performs the best among the baselines while our CGCN outperforms the Text GCN about 1.5% higher on Nursing#50 dataset and 1% higher than the Text CGN on Discharge-summary#50 dataset. And for f_score@5, our CGCN still outperforms the Text GCN about 1% higher on both datasets.

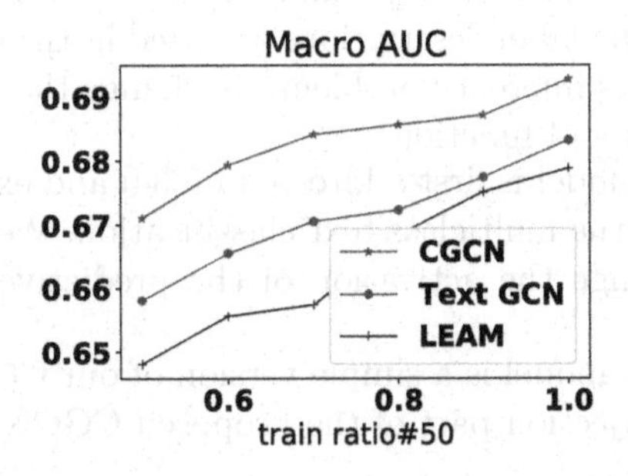

Fig. 3. Performance on varying ratio of training data

In order to illustrate the predictive power of our model with less train data, we randomly sample different ratios of the train data on the Nursing#50 dataset for training and test on the whole test samples. From Fig. 3, among the baselines, Text GCN outperforms the other baselines. What is more, even with the 60% of the training data for training, the performance of our CGCN is still competitive.

Table 3. Results on dischargesummary #50

Methods	auc(macro)	rec@5	pre@5	f1@5
MLP	0.8263	0.5394	0.4219	0.4729
CNN_2	0.8074	0.5362	0.4204	0.4713
CNN_3	0.8142	0.5446	0.4287	0.4798
HA_GRU	0.8154	0.5299	0.4212	0.4693
KV_3	0.8171	0.5126	0.4034	0.4515
KV_5	0.8164	0.5123	0.4025	0.4508
C_MM_3	0.8259	0.5359	0.4184	0.4693
C_MM_5	0.8258	0.5330	0.4191	0.4692
LEAM	0.8377	0.5806	0.4517	0.5081
Text GCN	0.8634	0.5980	0.4639	0.5225
CGCN-base	0.8718	0.6000	0.4696	0.5267
CGCN	**0.8730**	**0.6023**	**0.4726**	**0.5296**

4.4 Effect of Graph Convolution Layers

In this section, in order to figure out the effect of graph convolution layers, we do ablation study on our model by varying the number of the graph convolution operation both on the disease level and the text corpus level. Since there are coupled graph in our proposed model, we vary the number of graph convolution layers on one graph while keeping the same structure as our CGCN on the other graph. And then, we conduct experiments on the Nursing dataset and report the auc(macro) as the performance in Fig. 4.

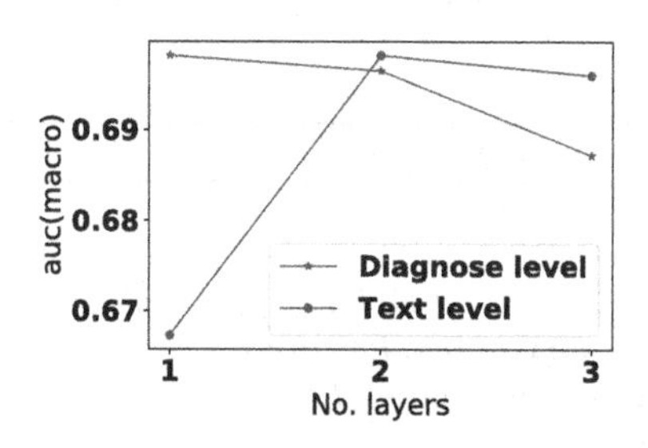

Fig. 4. Performance by varying the number of graph convolutions.

In Fig. 4, we can conclude that the performance(auc(macro)) dose not benefit when adding the number of graph convolutions. As more graph convolution layers are appended on the diagnosis level, the embeddings of diagnoses may not be distinctive, causing the performance down. Therefore, we only use one layer of graph convolution on the diagnose level. Besides, we vary the number of graph convolutions on the text level while setting the number of the graph convolution on the diagnose level to 1. From the Fig. 4, with the growing number of graph convolution layers on the text level, the performance achieve the best when the number of graph convolutions is set to 2 and adding more layers dose not gain any benefits. Since the text level graph is a heterogeneous graph where the upper node is the document node and the bottom node is the word node, one layer of graph convolution cannot transfer the information among the documents. With more layers of graph convolution, more information from high order neighbors are collected to the center document node.

4.5 Discussion

In this section, we discuss the performance of CGCN on the inductive setting. This is especially useful when we need to perform classification on a new patient. To this end, we develop the inductive learning algorithm for CGCN described in Algorithm 2.

In the training phase, we construct the training graph containing the training dataset, validation dataset and train the CGCN in the transductive mode. In the test phase, for each new patient, we generate a graph generated by Algorithm 1 and feed the graph to the CGCN without re-training. We use the TF-IDF metric as the document node features and the one-hot representations as the vocabulary node features. The results are listed in Table 4:

Cost Analysis. The cost of applying new patients to the CGCN can be divided into two parts. One comes from the graph augmentation, and the other comes from the procedure of forward stage. With the given feature of a new patient (e.g. TF-IDF metric), we do not need to re-compute the word graph so that the cost of graph augmentation depends on the procedure of adding a new test node

---

Algorithm 1: GraphAugmentation

1: **procedure** GRAPHAUGMENTATION(d,X_G,$\mathcal{G}$)
2: $V \leftarrow \mathcal{G}.V, E \leftarrow \mathcal{G}.E$;
3: Compute set $m = \{(i, j, TFIDF(i, j))|i \in \mathcal{V}, j = d||i = d, j \in \mathcal{V}\}$;
4: Compute $\mathcal{G}_{new} = (V \cup \{d\}, E \cup m)$;
5: Append the representation of d to X_G;
6: **return** $X_G, \mathcal{G}_{new}$
7: **end procedure**

Algorithm 2: Inductive learning For CGCN

1: Construct the text corpus $\mathcal{G}_{train}$ on training and validation dataset;
2: Train the CGCN with node features X_{train} and text corpus graph $\mathcal{G}_{train}$;
3: For a new patient d, $X_{test}, \mathcal{G}_{test} \leftarrow GraphAugmentation(d, X_{train}, \mathcal{G}_{train})$;
4: Feed the $X_{test}, \mathcal{G}_{test}$ to the CGCN

---

and linking the node to the current word graph (see Algorithm 1(3–4)). The process of appending a new node depends on the number of words in the test node. In the graph propagation stage, we use the same parameters learned in the training stage. As such, we can reduce the cost of dealing with a new patient.

Table 4. Performance on inductive settings

Methods	Nursing #50	Dischargesummary #50
Text GCN	0.6747	0.8402
CGCN	0.6961	0.8562

From Table 4, we can conclude that, under the inductive setting, our model still outperforms the baselines (e.g., LEAM with 0.6788 on Nursing#50 dataset and Text GCN with 0.8402 on Dischargesummary#50 dataset) and suggests that the new patient can learn some knowledge form the existing patients in the training dataset.

5 Conclusion

In this paper, we propose a novel coupled graph neural network (CGCN) in the clinical diagnosis inference. We emphasize the challenges of the text based clinical diagnosis inference and take the information sharing and disease comorbidity into the consideration and use a hyperplane projection method to project the diagnosis embeddings into the related hyperplane of the patient embeddings. The CGCN outperforms the baselines on several datasets. We have shown that the information sharing and disease comorbidity are essential in the clinical diagnosis inference and need to be further studied in the computer aided diagnosis inference systems.

Acknowledgement. This work was supported in part by National Natural Science Foundation of China under Grant No. 61532010 and 61521002, and Beijing Academy of Artificial Intelligence (BAAI).

References

1. Bastings, J., Titov, I., Aziz, W., Marcheggiani, D., Sima'an, K.: Graph convolutional encoders for syntax-aware neural machine translation. arXiv (2017)
2. Baumel, T., Nassour-Kassis, J., Cohen, R., Elhadad, M., Elhadad, N.: Multi-label classification of patient notes: case study on ICD code assignment. In: AAAI (2018)
3. Baxt, W.G.: Use of an artificial neural network for data analysis in clinical decision-making: the diagnosis of acute coronary occlusion. Neural Comput. **2**(4), 480–489 (1990)
4. Binaghi, E.: A fuzzy logic inference model for a rule-based system in medical diagnosis. Expert Syst. **7**(3), 134–141 (1990)

5. Bruna, J., Zaremba, W., Szlam, A., Lecun, Y.: Spectral networks and locally connected networks on graphs (2014)
6. Choi, E., Bahadori, M.T., Schuetz, A., Stewart, W.F., Sun, J.: Doctor AI: predicting clinical events via recurrent neural networks. In: MLHC, pp. 301–318 (2016)
7. Defferrard, M., Bresson, X., Vandergheynst, P.: Convolutional neural networks on graphs with fast localized spectral filtering. In: NIPS, pp. 3844–3852 (2016)
8. Djulbegovic, B., Guyatt, G.H.: Progress in evidence-based medicine: a quarter century on. Lancet **390**(10092), 415–423 (2017)
9. Ghassemi, M., et al.: Unfolding physiological state: Mortality modelling in intensive care units. In: SIGKDD, pp. 75–84. ACM (2014)
10. Henaff, M., Bruna, J., Lecun, Y.: Deep convolutional networks on graph-structured data. arXiv (2015)
11. Iyyer, M., Enns, P., Boyd-Graber, J., Resnik, P.: Political ideology detection using recursive neural networks. In: ACL, vol. 1, pp. 1113–1122 (2014)
12. Johnson, A.E., et al.: MIMIC-III, a freely accessible critical care database. Sci. Data **3**, 160035 (2016)
13. Keezer, M.R., Sisodiya, S.M., Sander, J.W.: Comorbidities of epilepsy: current concepts and future perspectives. Lancet Neurol. **15**(1), 106–115 (2016)
14. Kim, Y.: Convolutional neural networks for sentence classification. arXiv (2014)
15. Kingma, D.P., Ba, J.: Adam: a method for stochastic optimization. arXiv (2014)
16. Kipf, T., Welling, M.: Semi-supervised classification with graph convolutional networks (2017)
17. Lipton, Z.C., Kale, D.C., Elkan, C., Wetzel, R.: Learning to diagnose with LSTM recurrent neural networks. arXiv (2015)
18. Liu, N., Lu, P., Zhang, W., Wang, J.: Knowledge-aware deep dual networks for text-based mortality prediction. In: ICDE, pp. 1406–1417. IEEE (2019)
19. Lund, N., Petersen, A., Snoer, A., Jensen, R.H., Barloese, M.: Cluster headache is associated with unhealthy lifestyle and lifestyle-related comorbid diseases: results from the Danish cluster headache survey. Cephalalgia **39**(2), 254–263 (2019)
20. Manevitz, L.M., Yousef, M.: One-class SVMS for document classification. JMLR **2**, 139–154 (2001)
21. Miller, A., Fisch, A., Dodge, J., Karimi, A.H., Bordes, A., Weston, J.: Key-value memory networks for directly reading documents. arXiv (2016)
22. Nam, J., Kim, J., Loza Mencía, E., Gurevych, I., Fürnkranz, J.: Large-scale Multi-label Text Classification - Revisiting Neural Networks. arXiv (2013)
23. Peng, H., et al.: Large-scale hierarchical text classification with recursively regularized deep graph-CNN. In: WWW, pp. 1063–1072 (2018)
24. Prakash, A., et al.: Condensed memory networks for clinical diagnostic inferencing. In: AAAI (2017)
25. Ramos, A.L., Redeker, I., Hoffmann, F., Callhoff, J., Zink, A., Albrecht, K.: Comorbidities in patients with rheumatoid arthritis and their association with patient-reported outcomes: results of claims data linked to questionnaire survey. J. Rheumatol. **46**(6), 564–571 (2019)
26. Sackett, D.L., Rosenberg, W.M., Gray, J.M., Haynes, R.B., Richardson, W.S.: Evidence based medicine: what it is and what it isn't (1996)
27. Schlichtkrull, M., Kipf, T.N., Bloem, P., van den Berg, R., Titov, I., Welling, M.: Modeling relational data with graph convolutional networks. In: Gangemi, A., et al. (eds.) ESWC 2018. LNCS, vol. 10843, pp. 593–607. Springer, Cham (2018). https://doi.org/10.1007/978-3-319-93417-4_38

28. Sun, M., Zhao, S., Gilvary, C., Elemento, O., Zhou, J., Wang, F.: Graph convolutional networks for computational drug development and discovery. Brief. Bioinform. **21**(3), 919–935 (2019)
29. Wang, G., et al.: Joint embedding of words and labels for text classification. arXiv (2018)
30. Wang, J., Wang, Z., Zhang, D., Yan, J.: Combining knowledge with deep convolutional neural networks for short text classification. In: IJCAI, pp. 2915–2921. AAAI Press (2017)
31. Wang, S., Manning, C.D.: Baselines and bigrams: simple, good sentiment and topic classification. In: ACL, pp. 90–94. ACL (2012)
32. Xue, Q., Zhang, W., Zha, H.: Multi-label classification of patient notes: case study on ICD code assignment. In: AAAI (2020)
33. Yang, Z., Yang, D., Dyer, C., He, X., Smola, A., Hovy, E.: Hierarchical attention networks for document classification. In: NAACL, pp. 1480–1489 (2016)
34. Yao, L., Mao, C., Luo, Y.: Graph convolutional networks for text classification. In: AAAI, vol. 33, pp. 7370–7377 (2019)
35. Zhang, H., Xiao, L., Chen, W., Wang, Y., Jin, Y.: Multi-task label embedding for text classification. arXiv (2017)
36. Zhou, J., Cui, G., Zhang, Z., Yang, C., Liu, Z., Sun, M.: Graph neural networks: a review of methods and applications. arXiv (2018)

Vector-Level and Bit-Level Feature Adjusted Factorization Machine for Sparse Prediction

Yanghong Wu[1], Pengpeng Zhao[1(✉)], Yanchi Liu[2], Victor S. Sheng[3], Junhua Fang[1], and Fuzhen Zhuang[4,5]

[1] Institute of Artificial Intelligence, School of Computer Science and Technology, Soochow University, Suzhou, China
yhwu1@stu.suda.edu.cn, {ppzhao,jhfang}@suda.edu.cn
[2] NEC Labs America, Princeton, USA
yanchi@nec-labs.com
[3] Department of Computer Science, Texas Tech University, Lubbock, USA
victor.sheng@ttu.edu
[4] Key Lab of IIP of CAS, Institute of Computing Technology, Beijing, China
zhuangfuzhen@ict.ac.cn
[5] University of Chinese Academy of Sciences, Beijing 100049, China

Abstract. Factorization Machines (FMs) are a series of effective solutions for sparse data prediction by considering the interactions among users, items, and auxiliary information. However, the feature representations in most state-of-the-art FMs are fixed, which reduces the prediction performance as the same feature may have unequal predictabilities under different input instances. In this paper, we propose a novel Feature-adjusted Factorization Machine (FaFM) model by adaptively adjusting the feature vector representations from both vector-level and bit-level. Specifically, we adopt a fully connected layer to adaptively learn the weight of vector-level feature adjustment. And a user-item specific gate is designed to refine the vector in bit-level and to filter noises caused by over-adaptation of the input instance. Extensive experiments on two real-world datasets demonstrate the effectiveness of FaFM. Empirical results indicate that FaFM significantly outperforms the traditional FM with a 10.89% relative improvement in terms of Root Mean Square Error (RMSE) and consistently exceeds four state-of-the-art deep learning based models.

Keywords: Sparse prediction · Factorization machines · Feature adjustment

1 Introduction

Prediction analysis plays a vital role in many information retrievals (IR) and data mining (DM) tasks, including recommender systems [7,12,17,22,23] and

Y. Nah et al. (Eds.): DASFAA 2020, LNCS 12112, pp. 386–402, 2020.
https://doi.org/10.1007/978-3-030-59410-7_27

targeted advertising [4,9,11,16]. The goal is to predict user behaviors, i.e., gaining the probability that a user clicks on an item under certain conditions. In literature, Factorization Machines (FMs) are a class of universal and effective solutions, which have been widely used for sparse data prediction by considering the interactions among users, items, and auxiliary information.

Existing efforts on FMs can be divided into two categories, traditional factorization models and deep learning based models. Factorization Machines [14] is a typical representative of traditional models, which embeds features into a latent space and auto-cross features via vector inner product. Field-aware Factorization Machines [9] also adopts the idea of matrix factorization to reduce model parameters when constructing feature interactions significantly. Compared to FM, FFM introduces the concept of the field and associates the same properties to one field. Despite their widespread usage, its performance is greatly limited due to capturing second-order interactions only. Although there were attempts for high-level interactions, the high computational cost is required when conducting high-order interaction via the above methods. Even though Higher-Order Factorization Machines [2] presents efficient algorithms for training arbitrary-order feature interactions, it still relies on a large number of parameters. In general, as the order increases, the number of model parameters continues to grow, possibly resulting in noise due to invalid interactions.

With the wide application of deep learning, many deep learning based models have emerged. Most of them optimize FM from the perspective of improving interactions. A Factorization-Machine based Neural Network [4] optimizes FM by combining low-order and high-order feature interactions. Neural Factorization Machines [6] extends FM to higher-order interactions through the b-interaction layer and the MLP structure. Attentional Factorization Machines [20] assigns different weights to combine second-order interactions through the attention mechanism. However, these methods implicitly assume features have invariant representation capabilities in different input instances. Fixed feature representation greatly limits the prediction accuracy of the models and may even introduce noise, which adversely affects models. In practical applications, features do not always have equal predictability and effectiveness. For example, the feature $female$ in the input instance $\{userID, coffee, female, latte\}$ and $\{userID, shoes, female, red\}$ has a completely different responsibility. For the former one, $female$ is insignificant but is vital for the later one. Intuitively, gender has more influence on whether to buy shoes than coffee. Therefore, An input-aware factorization machine [21] improves FM from the input perspective instead of interactive standing, which exploits fully connected layers to learn input-aware factor for refining weights and feature vectors to adapt each input instance. However, IFM only refines feature embedding in vector-level, which is not sufficient and may even introduce noise caused by over-adaptation of the input instance.

In this paper, we focus on enhancing FM to solve the problems mentioned above. We propose a novel model for prediction under a sparse setting, named Feature-adjusted Factorization Machine (FaFM), which enhances FM via adaptively refining feature representation in both vector-level and bit-level. Inspired by IFM [21], we refer to a similar method to adjust feature embedding in vector-level via a neural network, which learns a specific weight value for one feature in a non-linear way. For bit-level, we adopt a novel gating structure for the adjustment, which utilizes a user and item interaction-aware structure. The main contributions of this paper are summarized as follows:

- To the best of our knowledge, we are the first to adjust feature vectors in both vector-level and bit-level and integrate the two modules in a natural order.
- We develop a novel model to strengthen FM via an adaptive vector-level adjustment structure and an interaction-aware bit-level adjustment structure for the feature adaptation.
- We conduct extensive experiments on two real-world datasets, and the results demonstrate the superiority of our proposed method.

The rest of the paper is organized as follows. We first introduce the preliminaries about FM in Sect. 2. Then we present the general framework and model details of FaFM in Sect. 3. After showing our experiment results in Sect. 4, we briefly summarize the work of this paper in the last section.

2 Preliminaries

Factorization Machines is proposed for sparse data prediction, which combines the advantages of Support Vector Machines (SVMs). Given a dataset $D = \{(\mathbf{x}_1, y_1), (\mathbf{x}_2, y_2), ..., (\mathbf{x}_N, y_N)\}$ with N samples, where $\mathbf{x}$ is a feature vector including a user, an item, and relative contexts. $y \in \{-1, 1\}$ is the label denoting whether the user responds to the item or not, where $y = 1$ represents a positive response. FM estimates the target by modeling all feature interactions via factorized interaction parameters:

$$\hat{y}_{FM}(\mathbf{x}) = w_0 + \sum_{i=1}^{n} w_i x_i + \sum_{i=1}^{n} \sum_{j=i+1}^{n} \langle \mathbf{v}_i, \mathbf{v}_j \rangle x_i x_j \tag{1}$$

where $\mathbf{x}$ is an input instance consisting of n features, and w_0 is the global bias. w_i captures the relationship weight of the i-th feature to the target. The term $\langle \mathbf{v}_i, \mathbf{v}_j \rangle$ models the interaction between the i-th and j-th feature, where $\langle \cdot, \cdot \rangle$ is the dot product. $\mathbf{v}_i \in \mathbb{R}^d$, and d is the embedding size of the vector. Note that due to the influence of coefficient $x_i x_j$, only the interactions between non-zero features are taken into account. According to FM [13], pairwise interactions can be reformulated:

$$\sum_{i=1}^{n} \sum_{j=i+1}^{n} \langle \mathbf{v}_i, \mathbf{v}_j \rangle x_i x_j = \frac{1}{2} \sum_{q=1}^{d} \left[\left(\sum_{j=1}^{n} v_{j,q} x_j \right)^2 - \sum_{j=1}^{n} v_{j,q}^2 x_j^2 \right] \tag{2}$$

where $v_{j,q}$ denotes the element of the q-th dimension of $\mathbf{v}_j$.

Although FM is widely used in academia and industry, its fixed feature representation limits the accuracy of the model and may bring noise, which adversely affects the predictive power of the model. Moreover, its linear property makes it difficult to model complex real-world scenarios.

3 Feature-Adjusted Factorization Machine

In this section, we first introduce the framework of our proposed model FaFM. Then we discuss the relationship between FM, AFM, IFM, and FaFM.

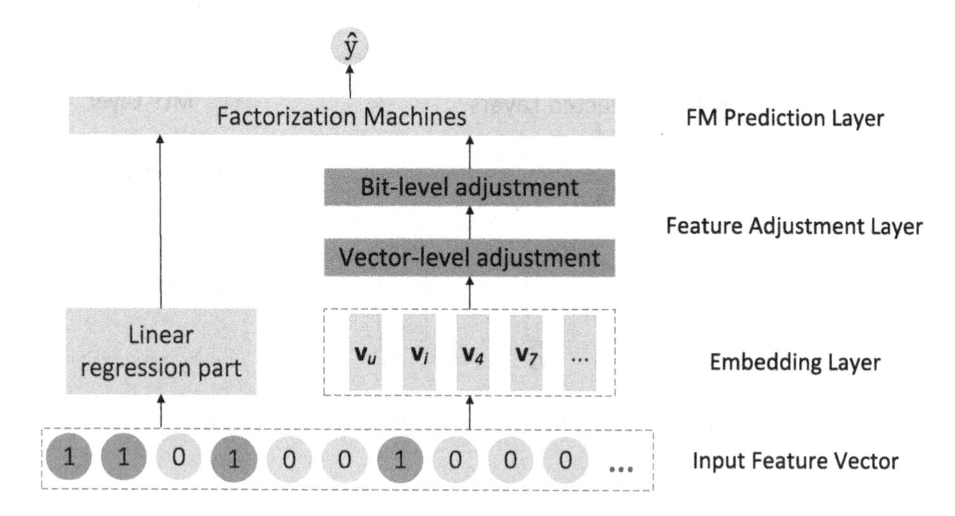

Fig. 1. The global framework of Feature-adjustment Factorization Machines.

3.1 The FaFM Model

We optimize FM from vector-level and bit-level, and adaptively adjust the feature representation according to different input instances. Compared to FM, we add two feature adjustment modules. Given an input feature vector $\mathbf{x} \in \mathbb{R}^n$, the target of FaFM is formulated as:

$$\hat{y}_{FaFM}(\mathbf{x}) = w_0 + \sum_{i=1}^{n} w_i x_i + \sum_{i=1}^{n} \sum_{j=i+1}^{n} \langle f(\mathbf{v}_i), f(\mathbf{v}_j) \rangle x_i x_j \tag{3}$$

where w_0 and w_i represent the global bias and the weight of the i-th feature. The term $f(\cdot)$ is feature adjustment function, which is the core component of FaFM for modeling the transformation process of vector-level adjustment and bit-level adjustment, where $\mathbf{v}_i$ denotes the original feature embedding, which is obtained by an embedding table lookup. We represent the network structure in Fig. 1, where the blue blocks are extra parts compared to the FM model.

Input and Embedding Layer. The input feature is a sparse vector $\mathbf{x}$, including userID, itemID, and related attributes. Then each feature is mapped to a dense embedding vector $\mathbf{v}_i \in \mathbb{R}^d$, where d is the embedding dimension. x_i represents the value of the i-th feature. Owing to the sparsity of $\mathbf{x}$, we only include the non-zero feature vectors. Formally, let $\mathcal{V}_\mathbf{x}$ denote the set of non-zero feature embeddings. $\mathcal{V}_\mathbf{x} = \{x_i \mathbf{v}_i\}$, where $x_i \neq 0$.

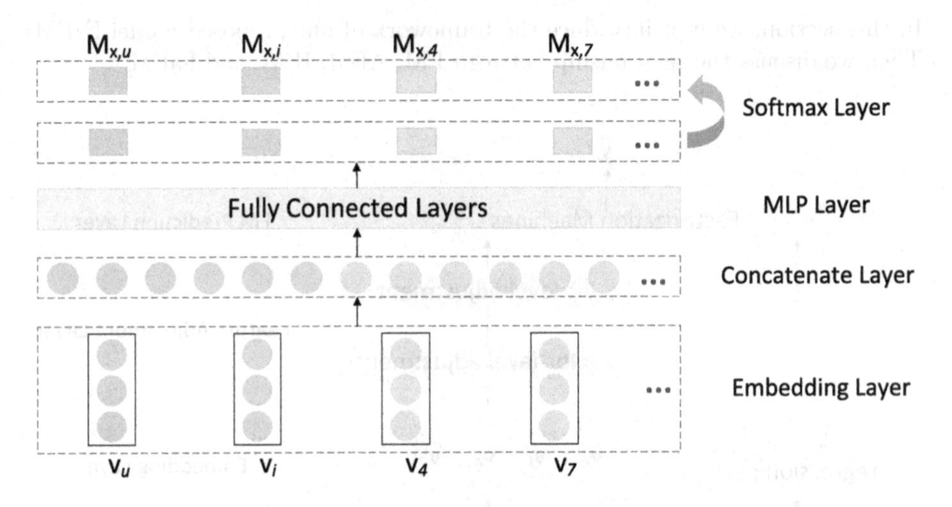

Fig. 2. The neural network of vector-level adjustment.

Vector-Level Adjustment. First, we adjust features from the vector-level. To make the identical feature own different representations in different inputs, we will consider the effect of the input when conducting vector adjustment, i.e., input-aware adjustment. We adopt the method in IFM to learn vector-level adjustment by introducing an input-aware factor which allows each feature to have different contributions in different input. The input-aware factor is calculated by fully connected layers, which assigns different weights to each feature. Figure 2 illustrates the structure, where we only show the non-zero features.

We concatenate the non-zero feature vectors in $\mathcal{V}_\mathbf{x}$ as a vector $\mathbf{e}_\mathbf{x}$, i.e., the concatenate layer in Fig. 2. After concatenation, we utilize the fully connected layers to gain mixed input representation $\mathbf{r}_\mathbf{x}$, which is equal to the output of the fully connected layer:

$$\begin{aligned} \mathbf{z}_1 &= \sigma_1(\mathbf{W}_1 \mathbf{e}_\mathbf{x} + \mathbf{b}_1), \\ \mathbf{z}_2 &= \sigma_2(\mathbf{W}_2 \mathbf{z}_1 + \mathbf{b}_2), \\ &\cdots\cdots \\ \mathbf{r}_\mathbf{x} = \mathbf{z}_L &= \sigma_L(\mathbf{W}_L \mathbf{z}_{L-1} + \mathbf{b}_L) \end{aligned} \tag{4}$$

where $\mathbf{e_x}$ is a concatenate representation of feature vectors. L denotes the number of hidden layers. $\mathbf{W}_l$ and $\mathbf{b}_l$ represent the weight and the bias of the l-th layer, respectively. σ_l denotes activation function, and we adopt ReLu [3] as the activation function in this paper. The output vector of the fully connected layers $\mathbf{r_x}$ is used to learn an input-aware weight $m_{\mathbf{x},i}$ by softmax:

$$\mathbf{m}'_{\mathbf{x}} = \mathbf{r_x}\mathbf{P}, \mathbf{P} \in \mathbb{R}^{t \times h},$$
$$m_{\mathbf{x},i} = h \times \frac{\exp(m'_{\mathbf{x},i})}{\sum_{j=1}^{h} \exp(m'_{\mathbf{x},j})} \tag{5}$$

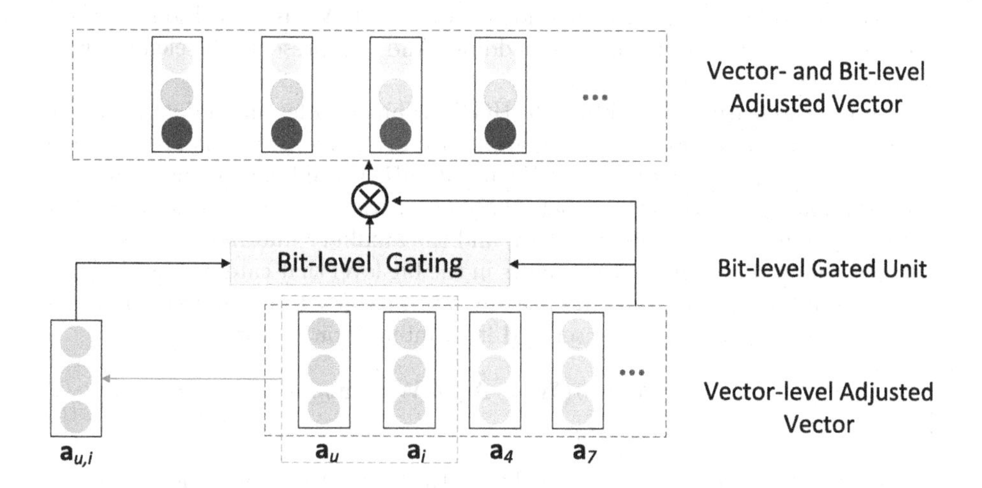

Fig. 3. The framework of bit-level adjustment.

where $\mathbf{r_x} \in \mathbb{R}^t$, and t represents the number of neurons in the last hidden layer. $\mathbf{P}$ is a weight matrix, which transforms $\mathbf{r_x}$ into a h-dimension vector $\mathbf{m}'_{\mathbf{x}}$, where h is equal to the number of non-zero vectors. The element $m'_{\mathbf{x},i}$, where $i \in [1, h]$ in $\mathbf{m}'_{\mathbf{x}}$, is normalized through a softmax function, which multiplies h to make the sum of all elements equals to h. The input-aware weight $m_{\mathbf{x},i}$ is used to refine the i-th feature vector, which is calculated in a non-linear way. Note that, unlike IFM, once the input-aware weight $\mathbf{m_x}$ is obtained, we only adjust the embedding vector $\mathbf{v}_i$, instead of refining both the feature weight w_i and embedding vector. In experiments, we find that just refining the embedding vector is sufficient to achieve input perception. Now we can adjust the features in the vector-level by multiplying the input perception weights:

$$\mathbf{a}_i = m_{\mathbf{x},i}\mathbf{v}_i \tag{6}$$

where $\mathbf{a}_i$ is the adjusted embedding vector of the i-th feature about the current input. Through the adjustment of the vector-level layer, the feature representation incorporates the influence of specific input instance and realizes non-linear prediction of the model.

Bit-Level Adjustment. Next, we adjust features from the bit-level, as shown in Fig. 3. We learn from the HGN framework [10], which is for sequential recommendation via hierarchical gating network selecting what information deserves to be passed to downstream layers. HGN exploits a personalized feature gate to select important bit dimensions. To model item features tailored to users' preferences, it adopts a user-specific gate to filter bit-level information. The feature gate in HGN is shown as follows:

$$\mathbf{S}^F = \mathbf{S} \otimes \sigma(\mathbf{W}_1 \cdot \mathbf{S} + \mathbf{W}_2 \cdot \mathbf{u} + \mathbf{b}) \tag{7}$$

where $\mathbf{S}$ and $\mathbf{S}^F$ are the input and output embedding. $\mathbf{W}_1$, $\mathbf{W}_2$, $\mathbf{b}$ are learnable parameters, and σ is a sigmoid function. The term of $\mathbf{W}_2 \cdot \mathbf{u}$ is used to filter user-specific features, where $\mathbf{u}$ is user embedding. And $\otimes$ represents the element-wise product.

We use the feature gate similar to HGN for bit-level feature adjustment. We argue that not all features enjoy equal importance while using the factorization machines to make predictions. UserID and itemID should have higher contributions than auxiliary features. In actual scenario, what we care about is whether the user purchases (or clicks) the item, and the auxiliary information is only the relevant condition. To refine the vectors in the bit-level that cater to the user and item interaction, we improve the feature gate in HGN to an interaction-aware gate structure by introducing user and item interaction vector $\mathbf{a}_{u,i}$:

$$\begin{aligned} \mathbf{O}_{\mathbf{x}} &= \mathbf{A}_{\mathbf{x}} \otimes \sigma(\mathbf{W}_{g1} \cdot \mathbf{A}_{\mathbf{x}} + \mathbf{W}_{g2} \cdot \mathbf{a}_{u,i} + \mathbf{b}), \\ \mathbf{a}_{u,i} &= \mathbf{a}_u \otimes \mathbf{a}_i \end{aligned} \tag{8}$$

Let $\mathbf{A}_{\mathbf{x}} = \{\mathbf{a}_i\}$.$\mathbf{A}_{\mathbf{x}} \in \mathbb{R}^{d \times h}$ is obtained from the vector-level adjustment network. $\mathbf{W}_{g1}, \mathbf{W}_{g2} \in \mathbb{R}^{d \times d}$ and $\mathbf{b} \in \mathbb{R}^d$ are learnable weights and bias, respectively. $\otimes$ represents the element-wise product. The term $\mathbf{a}_{u,i}$ represents the interaction between user and item, which is used to filter feature information that is really relevant to the user and item interaction. $\mathbf{a}_{u,i}$ is calculated by the element-product of userID and itemID embeddings. $\mathbf{O}_{\mathbf{x}}$ is the combination matrix of feature embeddings after two adjustments. $\mathbf{o}_i \in \mathbf{O}_{\mathbf{x}}$ is the representation of the input feature $\mathbf{v}_i$ after two adjustments. And the vector-level and the bit-level adjustment function $f(\cdot)$ is defined as follows:

$$f(\mathbf{v}_i) = \mathbf{o}_i \tag{9}$$

Prediction Layer. At last, the final refined feature vector $f(\mathbf{v}_i)$ is fed into Factorization Machines for prediction as Eq. 3. Similarly, the Equation can be reformulated like FM to reduce the running time of this layer to linear time complexity $O(kn)$:

$$\sum_{i=1}^{n}\sum_{j=i+1}^{n} \langle f(\mathbf{v}_i), f(\mathbf{v}_j)\rangle x_i x_j = \frac{1}{2}\sum_{q=1}^{d}\left[\left(\sum_{j=1}^{n} f(v_{j,q})x_j\right)^2 - \sum_{j=1}^{n} f(v_{j,q})^2 x_j^2\right] \tag{10}$$

To summarize, $\theta = \{w_0, \{w_i, \mathbf{v}_i\}, \{\mathbf{W}_l, \mathbf{b}_l, \mathbf{P}\}, \mathbf{W}_{g1}, \mathbf{W}_{g2}, \mathbf{b}\}$ concludes all model parameters of the model. Comparing with FM, the additional model parameters of FaFM mainly are $\{\mathbf{W}_l, \mathbf{b}_l, \mathbf{P}\}$ and $\{\mathbf{W}_{g1}, \mathbf{W}_{g2}\}$, which are learned for the vector-level adjustment and the bit-level adjustment, respectively.

The Relationship with FM, AFM and IFM. FM is a linear model, which can be seen as a special case of FaFM. By redefining the feature adjustment function $f(\mathbf{v}_i) = \mathbf{v}_i$ in Eq. 3, we can precisely recover the FM model.

We define $s(\cdot)$ and $g(\cdot)$ as the vector-level and the bit-level adjustment function, respectively. We set $g(\mathbf{v}_i) = \mathbf{v}_i$, meanwhile we replace a_{ij} and $\mathbf{P}^{\mathbf{T}}$ with $m_{\mathbf{x},i}m_{\mathbf{x},j}$ and constant $[1, \cdots, 1]$, respectively. The model of AFM [20] can be roughly recovered by our model as follows:

$$\begin{aligned}
\hat{y}_{AFM}(\mathbf{x}) &= w_0 + \sum_{i=1}^{n} w_i x_i + \mathbf{P}^{\mathbf{T}} \sum_{i=1}^{n} \sum_{j=i+1}^{n} a_{ij}(\mathbf{v}_i \odot \mathbf{v}_j) x_i x_j \\
&\approx w_0 + \sum_{i=1}^{n} w_i x_i + \sum_{i=1}^{n} \sum_{j=i+1}^{n} m_{\mathbf{x},i} m_{\mathbf{x},j} \langle \mathbf{v}_i, \mathbf{v}_j \rangle x_i x_j \\
&= w_0 + \sum_{i=1}^{n} w_i x_i + \sum_{i=1}^{n} \sum_{j=i+1}^{n} \langle g(s(\mathbf{v}_i)), g(s(\mathbf{v}_j)) \rangle x_i x_j \\
&= w_0 + \sum_{i=1}^{n} w_i x_i + \sum_{i=1}^{n} \sum_{j=i+1}^{n} \langle f(\mathbf{v}_i), f(\mathbf{v}_j) \rangle x_i x_j
\end{aligned} \tag{11}$$

At last, for IFM, it is easy to recover via setting the bit-level adjustment function $g(\mathbf{v}_i) = \mathbf{v}_i$ and fixing the feature weights.

3.2 Model Learning

FaFM can be applied in many prediction tasks like regression, classification, and ranking, as expanding FM from a data perspective. Specifically, for regression, a common practice is to use the squared loss as the objective function:

$$L_{reg} = \sum_{\mathbf{x} \in \mathcal{X}} (\hat{y}(\mathbf{x}) - y(\mathbf{x}))^2 \tag{12}$$

where $\mathcal{X}$ is the training data. $\hat{y}(\mathbf{x})$ and $y(\mathbf{x})$ denote the predict label and the target label w.r.t. input $\mathbf{x}$, respectively. For the classification task, hinge loss or log loss [7] can be used as the loss function. And pairwise personalized ranking loss [19] is an optional loss for ranking tasks. In this work, we mainly focus on the regression task, so we adopt a square loss to optimize Eq. 3. The regularization term is optional, and we omit it in Eq. 12, mainly including the $\mathbf{W}$ and $\mathbf{P}$ matrices of the vector-level and the bit-level adjustment layer. Pre-trained feature weights w_i and embedding vectors $\mathbf{v}_i$ are also supported. To be fair, we do not use model pre-training while evaluating the model.

To optimize the objective function, we employ a universal solver for neural network training—Adaptive Gradient descent (AdaGrad). The key of it is to obtain the derivative of each parameter w.r.t. squared loss L_{reg} via the adaptive learning rate.

When a complex feed-forward neural network is trained on a small dataset, it is easy to cause overfitting [8]. In machine learning models, if the model has many parameters and few training samples, the trained model is prone to overfitting. Dropout [18] is a regularization technique to avoid overfitting problems. The key idea of dropout is to drop some neurons during training randomly. In FaFM, we adopt dropout on the fully connected layer of the vector-level adjustment to mitigate overfitting.

Table 1. Statistics of the evaluation datasets.

Dataset	Record	Feature	User	Item
Frappe	96,193	5,382	957	4,082
MovieLens	668,953	90,445	17,045	23,743

4 Experiments

In this section, we conduct experiments to answer the following questions:

- **Q1:** How does FaFM perform compared to the state-of-the-art methods for sparse prediction?
- **Q2:** Can the vector-level and the bit-level adjustment layer effectively refine the feature embedding for the given input instance? How much do the vector-level and the bit-level contribute to the adjustment?
- **Q3:** How do the critical hyper-parameters of FaFM (e.g., the dropout rate and the depth of hidden layers) impact its performance?

4.1 Experimental Settings

Datasets. To demonstrate the performance, we conduct experiments on two public datasets: Frappe[1] and MovieLens[2].

Frappe. The Frappe [1] dataset has been used for context-aware recommendation, which contains 96,193 app usage log with 957 users and 4,082 apps. Each log includes user ID, app ID, and 8 context variables (e.g., weather, city, daytime, etc.). The target value 1 indicates that the user used the app under the current content. We randomly sampled two unused apps for one log in the content to guarantee the generalization of the model.

[1] http://baltrunas.info/research-menu/frappe.
[2] https://grouplens.org/datasets/movielens/latest/.

MovieLens. MovieLens is a collection of movie ratings, which is built by GroupLens [5]. Each record includes user ID, item ID, and tag information. MovieLens consists of 668,953 records with 17,045 users and 23,743 items. Similarly, the target value 1 denotes the user and the item have an interaction under the current content, and we sample two negative examples for each record. The negative target value is specified as -1. And we randomly split the MovieLens dataset into train, validation and test set for 7:2:1. The statistics of the datasets are summarized in Table 1.

Evaluation Metrics. To accurately evaluate the performance of models, we adopt MAE (Mean Absolute Error) and RMSE (Root Mean Square Error) as metrics, where a smaller value indicates a better performance.

Baselines. We compared our method with the following competitive models, which are specifically designed for sparse data prediction, including one traditional method and four deep learning based models:

- **LibFM** [14]**:** A traditional method has been proven with a strong performance for personalized tag recommendation and context-aware prediction [15] by constructing second-order feature interactions.
- **DeepFM** [4]**:** This model mixes the original FM and an MLP to generate recommendations.
- **AFM** [20]**:** This model assigns different weights to each second-order interaction through the attention network.
- **NFM** [6]**:** This model converts all features into a vector through a pool operation, and then sends it to an MLP layer to learn high-order interactions in a non-linear way.
- **IFM** [21]**:** This model refines feature weights and embedding vectors via a Factor Estimating Network for different input instance.

Parameter Settings. For a fair comparison, all methods are optimized by the squared loss, seeing Eq. 12. Besides, all models adopt the early stop strategy, where if the root mean square error of the validation set continues to increase for four consecutive epochs, we stop the model training. For the optimizer, all methods use Adagrad with a learning rate 0.01 except NFM, which adopts 0.05 as the learning rate. The batch size of Frappe and MovieLens are set to 2048 and 4096, respectively. In addition, for all datasets and all models, the embedding dimension is set as 256. For DeepFM, IFM and FaFM, the hidden layer is set as 3 layers, and the number of neurons is [200, 200, 200], [512, 256, 128], and [512, 256, 128]. NFM only sets one hidden layer, and the layer size is 256, as He et al. [6] showed a hidden layer that could achieve the best results in the NFM model. For AFM, We set the attention factor to 16 and the L_2 regularization of attention net to 2. To reduce the inevitable experimental errors, we run ten times for each experiment and average the results.

4.2 Performance Comparison (Q1)

In this section, we compare our model FaFM with five state-of-the-art methods, and Table 2 presents experimental comparison results on the Frappe and the MovieLens datasets. We have the following observations:

- First, we can observe that FaFM achieves better results than all baselines. In particular, FaFM is better than LibFM with 11.79% and 9.99% relative improvement in terms of RMSE on Frappe and MovieLens, respectively. And FaFM outperforms the suboptimal model with 3.06% and 4.68%. This proves the effectiveness of FaFM, adjusting feature vectors from two levels based on the input instances.
- Second, the bit-level adjustment module adopts a user and item specific gate, which can enhance the accuracy of feature representations, and filter noise caused by over-adaptation to the input data. FaFM has improved IFM with 4.78% and 6.51% in terms of RMSE on two datasets, which illustrates the effectiveness of bit-level adjustments. Note that unlike IFM, we do not refine feature weights as we find adjusting the feature weights in our model has little effect and may even weaken the model. The performance of IFM on the MovieLens dataset is not as good as that on Frappe. We argue that the features appearing in the testing set all exist in the training set of Frappe, but 0.2% of feature vectors in MovieLens' testing set do not appear in the training set, involving 217 records. These features are over-adapted in IFM, which adversely affects the learning of input-aware factor. Besides, each record in the MovieLens dataset has only three features, which will also affect the representation of IFM.
- Finally, both FaFM and IFM optimize FM from the perspective of data, which only considers the second-order interaction of features. FaFM's performance exceeds that of NFM and DeepFM on the Frappe dataset in terms of Mean Absolute Error (MAE), which considers higher-order interactions. Besides, FaFM is also superior to the model considering higher-order interactions on MovieLens. We can conclude that it is not the best to consider high-level interactions, and a properly designed model can achieve excellent performance.

Table 2. Comparison of model prediction capabilities on Frappe and MovieLens.

Method	Frappe		MovieLens	
	MAE	RMSE	MAE	RMSE
LibFM	0.1639	0.3300	0.2848	0.4753
DeepFM	0.1090	0.3177	0.2662	0.4718
AFM	0.1577	0.3225	0.2829	0.4694
NFM	0.0982	0.3003	0.1728	0.4488
IFM	0.0682	0.3057	0.2239	0.4576
FaFM	**0.0676**	**0.2911**	**0.1534**	**0.4278**

4.3 Impact of Feature Adjustment Layer (Q2)

We now focus on two feature adjustment modules in this section. First, we compare FaFM and FM as a whole to prove that the combination of the two works. Subsequently, we conduct experiments on the vector-level module and the bit-level module separately to discuss the improvement of each module relative to FM and their respective contributions to FaFM.

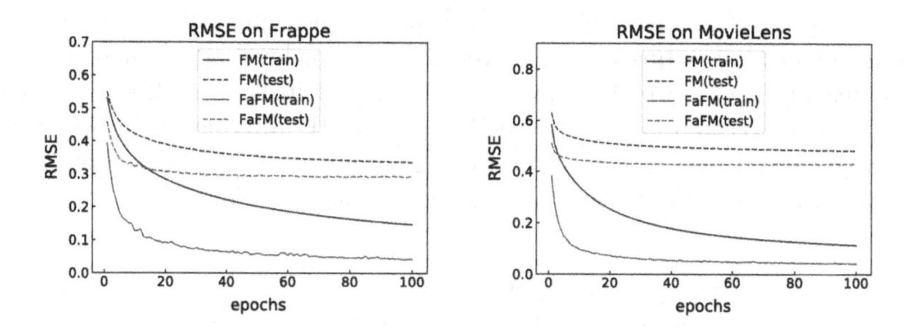

Fig. 4. The convergence behaviors of FM and FaFM.

Our proposed model FaFM optimizes FM by adjusting the representation of the feature vectors. Figure 4 illustrates that FaFM can converge faster and outperforms FM on both the training and validation sets, which proves the validity of FaFM. Besides, this indicates that the fixed feature representation limits the predictive ability of the model, and adjusting the feature representation according to input data can strengthen the model's expressive ability.

Table 3. The improvement rate of two adjustment modules in Frappe and MovieLens.

Method	Module	Frappe	MovieLens
		IR	IR
FM	–	–	–
v-FaFM	Vector-level adjustment	7.95%	7.22%
b-FaFM	Bit-level adjustment	7.94%	7.22%
FaFM	Vector- and bit-level adjustment	11.79%	9.99%

Mechanism Analysis. To better analyze the impact of the vector-level and the bit-level vector adjustment on the model, we conduct detailed experiments separately. Table 3 shows the experimental results, where v-FaFM denotes that only vector-level adjustments are applied, and b-FaFM denotes only bit-level feature adjustments are applied. Note that v-FaFM is different from IFM. The former one only adjusts the vector-level feature vectors based on the input instance, while IFM adjusts not only feature embeddings but also feature weights. We adopt the improvement rate (IR) to represent the improvement rate of RMSE relative to FM. We can observe that the model can be optimized by only the vector-level adjustment or only the bit-level adjustment, which proves that refining the feature representation according to the input data in the vector level and the bit level are all effective strategies. In particular, for RMSE, the two modules improve almost the same. That is, the contribution of the two modules is nearly 1:1. Besides, FaFM improves 11.79% and 9.99% over FM on the Frappe and MovieLens dataset, respectively. This indicates that adjusting features from two levels is better than adjusting one level. The vector-level adjustment can learn specific feature vector weights based on each input instance, while the bit-level mainly adjusts vectors based on userID and itemID interaction information to filter noise caused by over-adaptation to the input instance. Therefore, it is not sufficient to perform the feature adjustment at only one level.

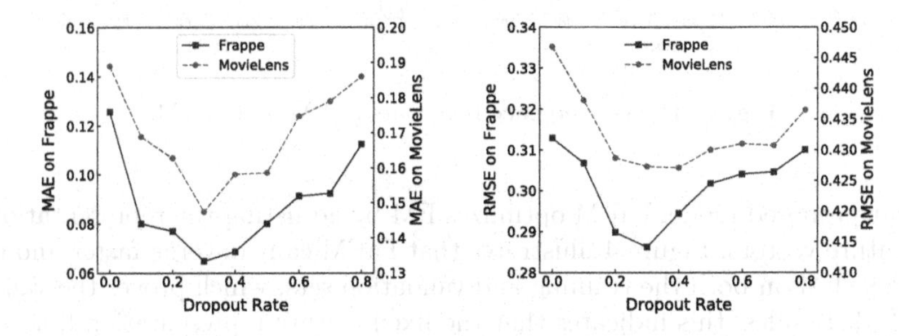

Fig. 5. The impact of the dropout rate.

4.4 Hyper-Parameter Study (Q3)

In this section, we explore the effect of hyper-parameters, including the dropout rate, the depth of the hidden layer, the activation function and embedding size. We construct experiments on the validation set to learn the effects of parameters.

Dropout Rate. In the vector-level adjustment layer, the dropout regularization technology is used in the fully connected layer for preventing overfitting. We set the dropout rate in [0,0.1,0.2,...,0.8] in each hidden layer. Figure 5 shows the impact of dropout on FaFM in terms of MAE and RMSE on the two datasets. No dropout or large dropout rates can lead to poor performance. This proves that proper tuning can improve model capabilities. The dropout rate 0 leads to poor results. This demonstrates the effectiveness of dropout. However, high dropout rates will inhibit the expression of the model. Specifically, on the Frappe dataset, using dropout rates 0.3 results in the lowest validation in terms of MAE and RMSE. And the optimal dropout rate for MovieLens is 0.3 and 0.4. Appropriate dropout rates not only can improve the model expression ability but also can have a positive impact on the generalization ability of FaFM.

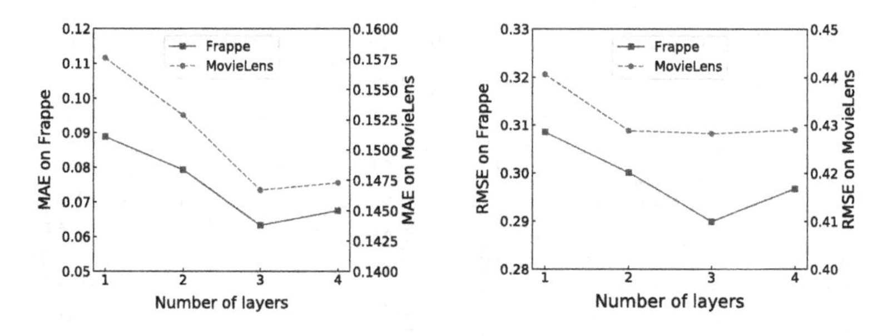

Fig. 6. The impact of hidden layers.

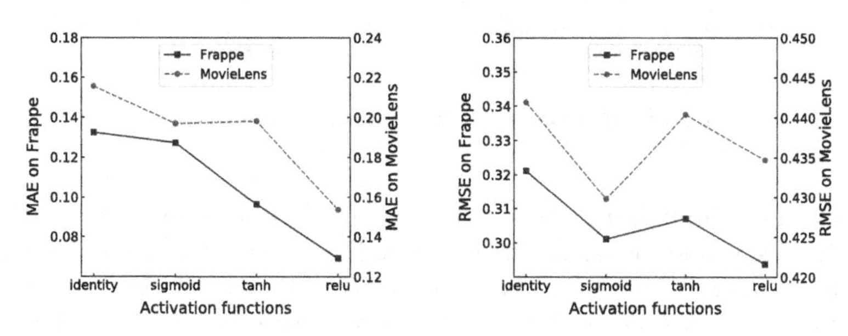

Fig. 7. The impact of activation functions.

Depth of Hidden Layer. The hidden layers of the vector-level adjustment layer play a vital role in capturing the internal connections of certain input instances. To illustrate the effect of the number of hidden layers, we set the layers from 1 to 4 for experiments. To be fair, the number of neurons in each layer is set the same, which is equal to 256. Figure 6 shows the validation error of FaFM. We can observe that as the number of layers increases, the loss decreases and then increases. In general, FaFM performs the best when the number of layers is 3. Specifically, for the MovieLens dataset, the depth of hidden layers adopts both two and three are all excellent choices. As the depth of the network increases, the learning ability of the model becomes stronger. However, when the number of layers is large, it is easy to overfit.

Activation Functions. For deep learning based models, a common strategy is to employ activation functions to hidden neurons, which provides the neurons with the non-linear characteristics and is necessary to simulate complex non-linear datasets. Figure 7 discusses the effect of different activation functions for FaFM, including identity, sigmoid, tanh, and relu. From the figure, we can notice that the relu activation function is the most suitable for our model on Frappe. For MovieLens, both sigmoid and relu activation functions are proper options. We can observe that the sigmoid activation function has a better adaptability in terms of RMSE. We guess that RMSE is too sensitive to outliers, and the sigmoid function has a large signal gain for the central region and a small signal gain for the two regions, which can complement each other.

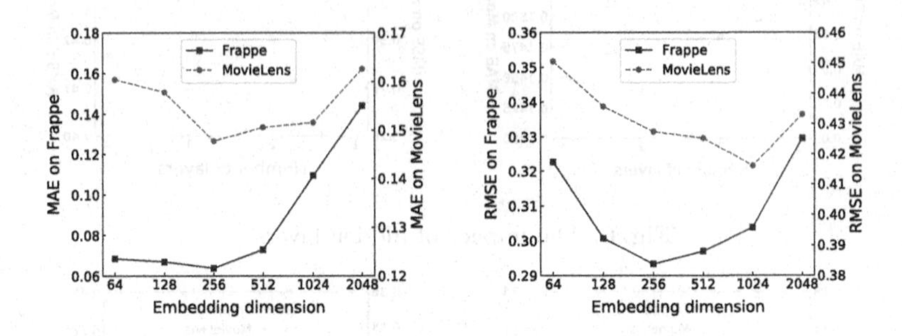

Fig. 8. The impact of the embedding dimension.

Embedding Dimension. We set the embedding dimension in [64, 128, 256, 512, 1024, 2048], and the results are shown in Fig. 8. We can observe that embedding size has a great impact on the performance of the model. As the embedding size increases, the loss decreases first and then increases. Specifically, in terms of MAE, the minimum is achieved at the embedding size = 256 on two datasets. In terms of RMSE, 256 is still the best choice for Frappe. For MovieLens, the embedding size is optimal when equal to 1024. Small embedding size weakens learning ability, while the large size makes it easy to overfit. MovieLens is relatively large, and larger embedding sizes can enhance performance.

5 Conclusion

In this paper, we proposed a vector-level and bit-level Feature-adjusted Factorization Machine (FaFM) model to optimize FM from the data perspective, which generates different feature representations for diverse input instances. The fixed representation is not enough to model the dynamic variability and flexibility of the feature for a specific input instance, which significantly limits the predictive ability of the model. To tackle this problem, we used a fully connected layer to learn the internal effects of input instances, which refines feature embedding from the vector-level. Besides, we adopted a gate structure to refine vectors in the bit level, which can also select useful information related to user and item interactions. To the best of our knowledge, FaFM is the first model to adjust feature representation from both vector-level and bit-level. We conducted extensive experiments on two real-world datasets, and the experiment results demonstrated that FaFM could consistently outperform the state-of-the-art models by a significant margin.

Acknowledgments. This research was partially supported by NSFC (No. 6187 6117, 61876217, 61872258, 61728205), Suzhou Science and Technology Development Program(SYG201803), Open Program of Key Lab of IIP of CAS (No. IIP2019-1) and PAPD.

References

1. Baltrunas, L., Church, K., Karatzoglou, A., Oliver, N.: Frappe: understanding the usage and perception of mobile app recommendations in-the-wild. CoRR abs/1505.03014 (2015)
2. Blondel, M., Fujino, A., Ueda, N., Ishihata, M.: Higher-order factorization machines. In: NIPS, pp. 3351–3359 (2016)
3. Glorot, X., Bordes, A., Bengio, Y.: Deep sparse rectifier neural networks. In: AISTATS, pp. 315–323 (2011)
4. Guo, H., Tang, R., Ye, Y., Li, Z., He, X.: DeepFM: a factorization-machine based neural network for CTR prediction. In: IJCAI, pp. 1725–1731 (2017)
5. Harper, F.M., Konstan, J.A.: The MovieLens datasets: history and context. ACM Trans. Interact. Intell. Syst. (TiiS) **5**(4), 19 (2016)
6. He, X., Chua, T.S.: Neural factorization machines for sparse predictive analytics. In: SIGIR, pp. 355–364. ACM (2017)
7. He, X., Liao, L., Zhang, H., Nie, L., Hu, X., Chua, T.S.: Neural collaborative filtering. In: WWW, International World Wide Web Conferences Steering Committee, pp. 173–182 (2017)
8. Hinton, G.E., Srivastava, N., Krizhevsky, A., Sutskever, I., Salakhutdinov, R.R.: Improving neural networks by preventing co-adaptation of feature detectors. CoRR abs/1207.0580 (2012)
9. Juan, Y., Zhuang, Y., Chin, W.S., Lin, C.J.: Field-aware factorization machines for CTR prediction. In: RecSys, pp. 43–50. ACM (2016)
10. Ma, C., Kang, P., Liu, X.: Hierarchical gating networks for sequential recommendation. In: KDD, pp. 825–833 (2019)
11. McMahan, H.B., et al.: Ad click prediction: a view from the trenches. In: KDD, pp. 1222–1230. ACM (2013)
12. Qian, Y., et al.: Interaction graph neural network for news recommendation. In: Cheng, R., Mamoulis, N., Sun, Y., Huang, X. (eds.) WISE 2020. LNCS, vol. 11881, pp. 599–614. Springer, Cham (2019). https://doi.org/10.1007/978-3-030-34223-4_38
13. Rendle, S.: Factorization machines. In: ICDM, pp. 995–1000. IEEE (2010)
14. Rendle, S.: Factorization machines with libFM. TIST **3**(3), 57 (2012)
15. Rendle, S., Gantner, Z., Freudenthaler, C., Schmidt-Thieme, L.: Fast context-aware recommendations with factorization machines. In: SIGIR, pp. 635–644 (2011)
16. Richardson, M., Dominowska, E., Ragno, R.: Predicting clicks: estimating the click-through rate for new ads. In: WWW, pp. 521–530. ACM (2007)
17. Shan, Y., Hoens, T.R., Jiao, J., Wang, H., Yu, D., Mao, J.: Deep crossing: web-scale modeling without manually crafted combinatorial features. In: KDD, pp. 255–262. ACM (2016)
18. Srivastava, N., Hinton, G., Krizhevsky, A., Sutskever, I., Salakhutdinov, R.: Dropout: a simple way to prevent neural networks from overfitting. J. Mach. Learn. Res. **15**(1), 1929–1958 (2014)
19. Wang, X., He, X., Nie, L., Chua, T.S.: Item silk road: recommending items from information domains to social users. In: SIGIR, pp. 185–194. ACM (2017)

20. Xiao, J., Ye, H., He, X., Zhang, H., Wu, F., Chua, T.S.: Attentional factorization machines: learning the weight of feature interactions via attention networks. In: IJCAI, pp. 3119–3125 (2017)
21. Yu, Y., Wang, Z., Yuan, B.: An input-aware factorization machine for sparse prediction. In: IJCAI, pp. 1466–1472. AAAI Press (2019)
22. Zhang, T., et al.: Feature-level deeper self-attention network for sequential recommendation. In: IJCAI, pp. 4320–4326 (2019)
23. Zhao, P., et al.: Where to go next: a spatio-temporal gated network for next POI recommendation. AAAI **33**, 5877–5884 (2019)

Cross-Lingual Transfer Learning for Medical Named Entity Recognition

Pengjie Ding[1,2], Lei Wang[2(✉)], Yaobo Liang[3], Wei Lu[1], Linfeng Li[2,4], Chun Wang[6], Buzhou Tang[5], and Jun Yan[2]

[1] School of Information and DEKE, Renmin University of China, Beijing, China
{pengjie,lu-wei}@ruc.edu.cn
[2] Yidu Cloud (Beijing) Technology Co., Ltd., Beijing, China
{lei.wang01,jun.yan}@yiducloud.cn
[3] Microsoft Research Asia, Beijing, China
[4] Institute of Information Science, Beijing Jiaotong University, Beijing, China
[5] Harbin Institute of Technology, Shenzhen, China
[6] Department of Cardiac Surgery, The First Hospital of China Medical University, Shenyang, China

Abstract. Extensive technologies have been employed to explore a best way for cross-lingual transfer learning. In medical domain, Named Entity Recognition is pivotal for many downstream tasks, such as medical entity linking and clinical decision support systems. Nevertheless, the lack of annotation limits the applicability in many languages without enough labeled data. To alleviate this issue and make use of languages with sufficient annotated data, we find a new way to obtain medical parallel corpus from medical terminology systems and knowledge bases and propose a methodology which combines cross-lingual language model pretraining and bilingual word embedding alignment with the help of the parallel corpus. Moreover, our combined architecture which maintains the framework of pretrained model can not only be used for NER task but also other downstream NLP tasks. Experiments demonstrated that incorporating Chinese and English medical data can effectively improve the performance for an English medical NER dataset (i2b2).

Keywords: Transfer learning · Cross-lingual pretraining · Word embedding alignment · Medical terminology systems · Medical NER

1 Introduction

Cross-lingual transfer learning have spurred renewed interest in recent years owing to its ability of making use of annotated data from other languages, which can compensate the deficiency of annotated data in target language.

A large body of methods have been proposed for cross-lingual transfer learning. State-of-the-art methods for learning cross-lingual word embeddings have relied on bilingual dictionaries or parallel corpora. While in recent years, various algorithms have been employed for semi-supervised and unsupervised learning for the lack of annotated data, such as [18,38]. Bilingual word representation

Y. Nah et al. (Eds.): DASFAA 2020, LNCS 12112, pp. 403–418, 2020.
https://doi.org/10.1007/978-3-030-59410-7_28

requires parallel corpus in preliminary stage of related research, but unsupervised method such as adversarial training have been proposed to get rid of the dependence for it [2]. And its performance either is on par, or outperforms supervised state-of-the-art methods.

But considering the particularity and expertise in medical field, it is more efficient for utilizing medical bilingual dictionaries or medical terminology systems, for their advantages that containing substantial medical knowledge and linguistic information. In addition, medical field contains a large amount of unsupervised medical data such as electronic medical record (EMR) and terminology systems which can be utilized as semi-supervised data. Whereas, recent work didn't fully develop latent potentials of existing international standard of medical terminology such as ICD, causing a massive waste for natural language processing task.

Moreover, the amazing performance of Bert [5] enables rapid development of pretraining techniques, which brings another way to deal with data hunger problem for low-resource tasks. An increasingly popular approach to alleviate this issue is to first learn general language representations on unlabeled data, which are then integrated in task-specific downstream systems. These kind of approach are first proposed by [23,26], but the representation for language is static, and has recently been superseded by sentence-level representations such as [4,5,27]. However, all these works learn a language model for a single language and unable be applied to different languages. Thus, Facebook have proposed their cross-lingual language modeling XLM [17] to apply pretraining techniques to cross-lingual transfer learning, which achieve state-of-the-art results on cross-lingual classification, unsupervised and supervised machine translation. XLM introduces a new unsupervised method for learning cross-lingual representations and a new supervised learning objective that improves cross-lingual pretraining when parallel data is available.

In this paper, we first put forward a feasible way to make fully use of various kinds of medical data, such as parallel corpora from a series of international standard of medical terminologies and medical corpora from electronic medical record. We proposed a hybrid architecture combines cross-lingual language model pretraining and bilingual word embedding alignment. Inspired by pretraining of language model, which have spurred a tremendous response recently, we pretrain a Chinese-English language model with medical unsupervised corpora and parallel corpora we extracted above. Notably, we proposed a new Mixture Language Modeling objective for our cross-lingual pretrained model.

Moreover, we utilize English-Chinese word representation alignment to enhance performance of transfer in the level of word embedding with the additional advantage that we also make use of the parallel corpus in pretraining procedure. That is, we utilize parallel corpus as a seed dictionary to align bilingual word embedding owing to our parallel corpus contains abundant medical knowledge and information, which means a good alignment may cause a desirable improvement for cross-lingual task.

At last we combine the cross-lingual pretrained model and word embedding alignment module as a joint feature extractor. Then we attach a BiLSTM-CRF architecture behind to deal with NER task. After fine-tuning with a source language annotated dataset, the target language can also benefit in our proposed framework due to its trait of having aligned two language's vector spaces.

In summary, our contributions are as follows.

- We introduce a new idea about how to make use of medical terminology systems and other knowledge bases to leverage the paucity of lacking in parallel corpus in cross-lingual transfer learning scenario.
- We pretrain a bilingual language model with unsupervised medical corpus and parallel corpora which extracted via previous step, improving performance of cross-lingual transfer learning.
- We propose a new Mixture Language Modeling objective for pretrained language model to make a better alignment for bilingual language representation, and it brings a benefit in cross-lingual downstream task.
- We train a transformation matrix with the purpose of aligning bilingual word embedding spaces to further augment performance of cross-lingual transfer learning.
- We significantly outperform the present baseline model on medical cross-lingual NER task.

The remainder of this paper is organized as follows. We review the related work in Sect. 2, and elaborate our proposed method in Sect. 3, following which we report experimental results and our analyze in Sect. 4. At last, we conclude the paper in Sect. 5.

2 Related Work

2.1 Cross-Lingual Transfer Learning

Traditional cross lingual transfer learning usually concentrate on annotation projection approaches and direct model transfer. Annotation projection approaches create weakly labeled training data in target language. The basic way to create weakly labeled data is through aligned parallel corpus or translations between source language and target language, such as [7,36,39]. For direct model transfer, a basic idea is to train a model in source language domain with language-invariant features, which makes the transfer model can also work on target language utilizing the universal features, such as [33]. Another method is to learn cross-lingual embedding mappings. [22] proposed to exploit the similarity of bilingual word embeddings by learning a linear mapping from a source to a target embedding space. Since then, plenty of works [1,8,30] following this idea have been done to improve performance, yet still rely on bilingual word lexicons. Moreover, [24] have proposed a hybrid method that combine annotation projection approaches and direct model transfer with requirement of annotated data in source language and parallel corpus.

While traditional cross lingual transfer learning rely on parallel corpus or seed dictionary, adversarial training has recently produced satisfying results in fully unsupervised scenarios [18,38]. On another strand of work, [2] showed that an iterative self-learning method is able to bootstrap a high quality mapping from very small seed dictionaries. However, their analysis reveals that the self-learning method gets stuck in poor local optima when the initial solution is not good enough, thus failing for smaller training dictionaries.

Generative pretraining of sentence encoders [5,14] are ubiquitously used on numerous natural language processing tasks. Recent developments in learning and evaluating cross-lingual sentence representations in many languages aim at mitigating the English-centric bias and suggest that it is possible to build universal cross-lingual encoders that can encode any sentence into a shared embedding space, such as [3,17].

2.2 Medical Named Entity Recognition

Medical NER aims to recognize entities from medical texts. It is a flagship for building intelligent medical systems, and has aroused general interest in recent years. Traditional methods for medical NER concentrate on manually designed rules [9,12,13], as well as recent open source ones such as cTAKES [28] and HiTEX [37], which rely on existing biomedical vocabularies to identify clinical entities.

Nevertheless, digging suitable rules to extract medical named entities requires enormous in-domain knowledge, which makes rule-based medical NER systems are expensive to build and usually have low flexibility. Thus, both rule-based [6,31] and machine learning based methods [19,25], as well as hybrid methods [20] have been developed to extract medication entities.

Recently a large body of neural network based model have been probed to handle medical NER task. [11,35] utilized a LSTM-CRF architecture while [10] add a extra LSTM layer. Nonetheless, there exist a large amount of in-domain knowledge for medical entity, which can not be covered in limited scale of training dataset, making supervised training can not get a desirable performance. Consequently, enormous methods are proposed to incorporate medical dictionary knowledge for medical NER. Such as [34], they proposed a method about extracting dictionary features via entity matching and then concatenate with the character embeddings or hidden representations in the LSTM-CRF architecture for Chinese medical NER.

3 Proposed Method

In this section, we present our methods for bilingual transfer learning throughout this work, which consists of three subsections: 3.1 introduce the acquisition of bilingual parallel text used for subsequent steps. 3.2 demonstrate procedure of cross-lingual language model pretraining with the utilization of unsupervised medical corpus and parallel corpus obtained in step 1. While 3.3 shows bilingual word embeddings alignment step that further improves the performance of downstream task.

3.1 Acquisition of Bilingual Parallel Text

For the sake of restricting the scale of parallel text, it is difficult to find an accurate **relationship** between two languages, for unsupervised method can not get a desirable result in domain-specific scenario. To alleviate this issue, we extract parallel corpora and bilingual dictionary from medical terminology systems. For our work, we use International Classification of Diseases 11th Revision (ICD) as the medical terminology system.

International Classification of Diseases 11th Revision. ICD is the foundation for the identification of health trends and statistics globally, and the international standard for reporting diseases and health conditions. It is the diagnostic classification standard for all clinical and research purposes. ICD defines the universe of diseases, disorders, injuries and other related health conditions, listed in a comprehensive, hierarchical fashion. And a new version called ICD-11 was released on 18 June 2018 to allow Member States to prepare for implementation. We collected English version and Chinese version of ICD-11 dataset and combine them after striking out some unaligned data. Then we extract qualified parallel corpora from the corpus preprocessed as our medical parallel corpus. Some samples of extracted parallel corpora are illustrated in Table 1.

Table 1. Some samples of parallel corpora

English	Chinese
Intestinal infection due to other Vibrio	其他弧菌的肠道感染
Foodborne staphylococcal intoxication	葡萄球菌食物中毒
Foodborne Clostridium perfringens intoxication	产气荚膜梭状芽孢杆菌食物中毒
Symptomatic late neurosyphilis	有症状性晚期神经梅毒
Tuberculosis of the respiratory system	呼吸系统结核病

3.2 Pre-training with Mixture Language Model

Based on the work of Facebook XLM [17], we propose a new Mixture Language Modeling objective with the help of parallel corpus obtained in Sect. 3.1. The pretraining methodology will be explained as follows.

Firstly, the unsupervised dataset used in this paper consist of the original XLM corpus, a public English electronic medical record dataset (MIMIC III) and our own Chinese electronic medical record corpus. MIMIC is an openly available dataset developed by the MIT Lab for Computational Physiology, comprising unidentified health data associated with 60,000 intensive care unit admissions. It includes demographics, vital signs, laboratory tests, medications, and more. MIMIC supports a diverse range of analytic studies spanning epidemiology, clinical decision-rule improvement, and electronic tool development. And our Chinese medical corpus is collected from three different hospitals, containing around 4000 deidentified real-world electronic medical records.

Considering the way for word embedding is not char-level but BPE [29] and some specific word units can represent strong semantic meaning in medical domain, which is not covered in original BPE vocabulary of XLM. We re-build a BPE vocabulary after tokenizing all our English and Chinese corpus and re-train a new bilingual language model from scratch. BPE is based on the intuition that various word classes are translatable via smaller units than words, with the purpose of alleviating the issue of OOV and scaling down the size of vocabulary.

The original cross lingual XLM includes three objectives: CLM, MLM and TLM. Causal language modeling (CLM) task consists of a Transformer language model trained to model the probability of a word given the previous words in a sentence. Masked language modeling (MLM), also known as Cloze task [32], is a common language modeling task in recent days. Following the original method, XLM sample randomly 15% of the BPE tokens from the text streams, replace them by a [MASK] token 80% of the time, by a random token 10% of the time, and keep them unchanged 10% of the time. Whereas CLM and MLM cannot be utilized for leveraging parallel data, which is the main reason that translation language modeling (TLM) objective have been proposed. TLM objective is an extension of MLM, which instead of considering monolingual text streams, concatenating parallel sentences. TLM randomly mask words in both the source and target language sentences. Take an English-Chinese sentence pair for example, to predict a word masked in an English sentence, the model can either attend to surrounding English words or to the Chinese translation, encouraging the model to align the English and Chinese representations.

Besides the training objectives used in XLM: Causal Language Modeling (CLM), Masked Language Modeling (MLM) and Translation Language Modeling (TLM), we proposed a new Mixture Language Modeling objective which can make use of parallel data at word level (we will explain how to get word level parallel data in next section). See Fig. 1 for a schematic, our new Mixture Language Modeling objective is a variant of TLM, we supersede some word in source language with their parallel word in target language to get English-Chinese-mixed corpus. Then we randomly concatenate these sentence as our input sequences (unlike TLM, the sentences are concatenated randomly, not parallel). The strategy for masking is making the superseded words to be masked in a higher probability (80% for default), in contrast with TLM's randomly masking. Then the subsequent training steps are almost the same as MLM. The Mixture Language Modeling objective is reasonable because there is less ambiguity in medical domain and it forces our model to learn the word alignment information. Besides Mixture Language Modeling, we also deploy original MLM and TLM in default setting. That is, in our proposed pretraining language model, we have one monolingual MLM objective and two bilingual objectives TLM and MixLM.

Bilingual language model pretraining obtain a better initialization of sentence encoders for few-shot cross-lingual downstream tasks. A Chinese annotated dataset is used to fine-tuning the pretraining model and then we conduct a few-shot learning via a small scale of target language dataset.

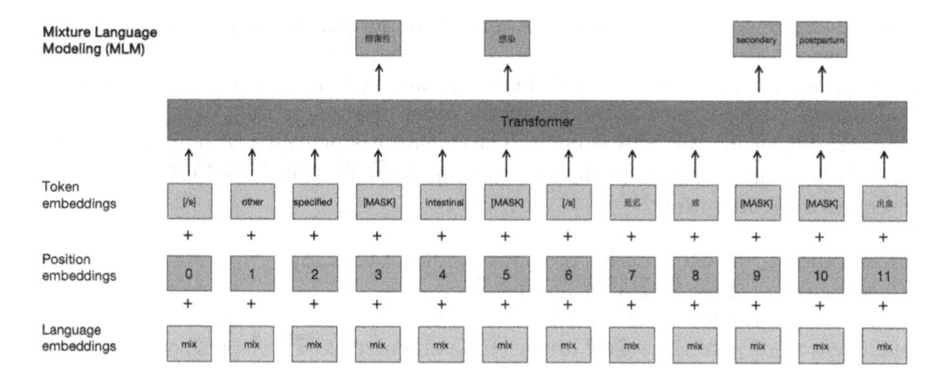

Fig. 1. Mixture Language Modeling. For Mixture Language Modeling objective, we replace some words in source language sentence with their correspond words in parallel corpus. For example, we replace bacteria with its Chinese word "细菌性". The subsequent step of masking are same as TLM. Namely, we mask some words in source language and predict their correspond words in target language to establish a robust connection between two language modeling vector space. The acquisition of word-level parallel data will be introduced in 3.3.

3.3 Bilingual Word Embeddings Alignment

Various technologies have been employed for cross-lingual word embeddings. While in medical domain, different with other domains, exists many terminology systems, which can be regarded as a supplementary for medical corpus. And owing to their hallmark of having multi-lingual version, we can extract a large amount of multi-lingual sentence pair from it, which can be utilized as parallel corpora. This incurs an apparent improvement to those unsupervised methods. We speculate that this method of relying on parallel corpora for achieving word alignment would be effective. Inspired by [22], learning a linear projection between vector spaces that represent languages can augment the performance for multilingual transfer learning. Meanwhile, medical corpus contains much more terms, whose semantic meaning is precise and fixed, inducing less semasiological variation. We optimize this procedure with a data selection schema which will be introduced subsequently.

Extraction of bilingual dictionary. To learn a linear projection between two language vector spaces requires monolingual word vectors and a seed dictionary. From our medical parallel corpus we can extract a set of bilingual dictionary as our seed for aligning two vector spaces. The extraction is conducted via Moses[1] which uses GIZA++[2], a statistical machine translation toolkit for word alignment. The result of alignment is illustrated in Table 2. The accuracy of our linear projection and the target-language NER model, depend on the quality of seed dictionary extracted by Moses. To alleviate this drawback we design a **data**

[1] http://www.statmt.org/moses/.

[2] http://www.statmt.org/moses/?n=FactoredTraining.AlignWords/.

Table 2. En-sentence and Zh-sentence come from our parallel corpus such as ICD-11. We use word align tool to get a coarse-grained alignment result showed in line 3. The words in bilingual sentence are split into sequence by blank space so the numbers in final alignment words such as a-b means the order of words in their own sequence. For instance, 4-0 means the fifth word in 'Gastroenteritis or colitis of infectious origin', which is 'infectious', equals to the first word in '感染 性 胃肠炎 或 结肠炎', for index of sequence begins with 0.

En-sentence	Gastroenteritis or colitis of infectious origin
Zh-sentence	感染 性 胃肠炎 或 结肠炎
alignment result	4-0 5-1 0-2 1-3 2-4
alignment word	infectious-感染 origin-性 Gastroenteritis-胃肠炎 or-或 colitis-结肠炎

selection schema to choose good-quality bilingual word-pairs as seed dictionary. We build a frequent table F which stores all the words in the parallel corpus and number of their occurrences. For a sentence **y** consists of n words $(x_1, x_2, ..., x_n)$, we calculate a score of this sentence by averaging each word's occurrences.

$$q(y) = \frac{\sum_{i=1}^{n} F(x_i)}{n} \quad (1)$$

where n denote the number of words in this sentence. So we choose the bilingual words extracted from sentences whose score can satisfy:

$$q(y) > q \quad (2)$$

where q is a threshold and we can change this parameter to get a desirable set of bilingual dictionary.

Word embedding alignment. Next, we build two monolingual models for Chinese and English using larger amount of unsupervised medical text corpus. The representations of languages are learned utilizing distributed Continuous Bag of Words (Cbow) or Skip-gram proposed by [21]. For the sake of restricting the scope of this paper, we don't introduce it in detail and concentrate on the subsequent application. The proposed method is based on the assumption of vector representations of similar words in different languages were related by a linear transformation. Thus, if we know some bilingual word pairs from Chinese to English, we can learn the transformation matrix bases on it. For our bilingual dictionary $((E_1, Z_1), (E_2, Z_2), ..., (E_n, Z_n))$, E_1 is the first word in English and Z_1 is the respond Chinese word. Suppose we have a set of word pairs sized n and their vector representations (Word2vec) e_i, z_i, e_i denote distributed representation of word i in English, while z_i is the vector representation of its translation in Chinese. Then the word-pairs are utilized for training a transformation matrix W so We_i approximates z_i. The objective for training this matrix is to minimize

$$\sum_{i=1}^{n} \|Wx_i - z_i\|^2 \quad (3)$$

which we can solve with stochastic gradient descent.

3.4 Transfer Learning for NER

Combine pretrain language model with word embeddings alignment module for NER task. See Fig. 2 for a schematic, the basic model we deployed is a BiLSTM-CRF architecture which contains a feature concatenation layer to combine embedding vectors generated by pretrain language model and word embeddings alignment. To be specific, after obtaining pretrain language model and word embedding linear transformation matrix mentioned above, we utilize them to encode input sentence separately and concatenation them as jointly features for input sequence.

For a large amount of Chinese NER annotated dataset, we use it to fine-tuning the model for purpose of learning knowledge from Chinese medical domain. The encoding of each input sentence include embedding from pretrain language model and Chinese word2vec which we trained to obtain transformation matrix. And for limited amount of English NER dataset, we encode each input sentence by pretrain language model and English word2vec, but, different with Chinese dataset, we use the transformation matrix to map the vector representation of word2vec to Chinese word embedding space.

Besides the embedding layer and feature concatenation layer, Bi-directional LSTM and CRF layer [16], as a baseline neural approach for Namely Entity Recognition task, are utilized to decode label jointly from the input feature sequence.

Long short term memory network (LSTM) is desirable to capture long-distance dependency, and has been used in a large body of work. Moreover, since the contexts in both directions are useful for representation learning, we used a Bi-LSTM layer to learn the hidden character representations by capturing global contexts. The input of this layer is the sequence of concatenated features and the output is a sequence of the hidden states of LSTM in both directions, which is denoted as $[r_1, r_2, ..., r_M]$. CRF layer was used to capture the dependency between labels and decode label jointly in our approach. Usually, labels of neighbor characters have relatedness. For example, "I-Problem" can not be the next tagging of "B-Test". After the CRF layer we can get final NER results with viterbi algorithm.

4 Experiments

In this section, we first introduce the experimental setup in terms of dataset and experimental setting, and the design of experiments. Then, we empirically demonstrate and interpret the experimental results, based on which we elucidate a summarization about the impact of our two proposed methods.

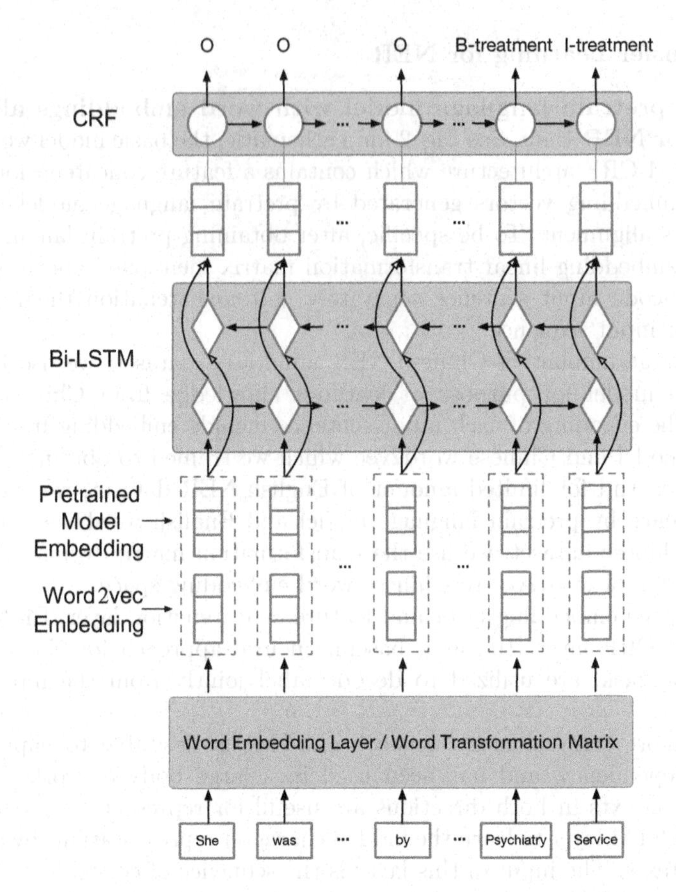

Fig. 2. Model architecture. We utilize a concatenated feature layer which contains cross-lingual language model embedding and word2vec embedding as a joint input. Special for target language domain, which is i2b2 task in our work, the transformation matrix we introduced in 3.3 will be used to align bilingual word embedding spaces. The rest of model are standard bidirectional LSTM-CRF architecture.

4.1 Dataset

The dataset we used can be divided into different branches: monolingual unsupervised corpus, parallel corpus for pretraining a bilingual language model and bilingual word embedding transformation matrix. Moreover, we need annotated dataset for downstream NER task.

Monolingual corpus. The monolingual unsupervised dataset for pretraining bilingual language model are MIMIC dataset [15] and our own Chinese electronic medical records corpus. The latest version of MIMIC comprises 61,532 intensive care unit stays: 53,432 stays for adult patients and 8,100 for neonatal patients. The data spans June 2001 - October 2012. And our Chinese medial corpus contains more than 4000 deidentified real-world electronic medical records.

In addition, we add our own Chinese wiki corpus as Bert and XLM via WikiExtractor[3].

Parallel corpus. The acquisition of parallel corpus for training language model and transformation matrix of bilingual word embedding have been introduced in 3.1. The medical terminology system and knowledge base we choose is ICD-11. We obtain the English version and Chinese version of ICD-11, for the quality of data can be guaranteed and they can be used as parallel data.

Annotated NER dataset. And for downstream NER task, the Chinese medical NER dataset we choose is from CCKS 2019 task 1. This task is a continuation of CCKS's series of evaluations on Chinese electronic medical records. It is extended and expanded from CCKS 2017 and 2018 medical named entity recognition tasks. The predefined categories of entities in it include diagnosis of disease, medical checking, treatment method, lab test, medicine and body site of disease. While for final English NER task, we utilize i2b2 2010 NER dataset. The 2010 i2b2/VA Workshop on Natural Language Processing Challenges for Clinical Records presented three tasks, one among them is concept extraction task focused on the extraction of medical concepts from patient reports, and the predefined categories of concepts include problem, test and treatment.

4.2 Experiments Design

Firstly we use a Bi-directional LSTM and CRF architecture to conduct a supervised training in i2b2 dataset as our baseline model. i2b2 dataset consist of total of 394 training reports and 477 test reports. We construct three different scale of i2b2 dataset for testing the effectiveness of our bilingual transfer learning method. The whole 394 training reports and 477 test reports is the **full version** of i2b2 dataset, while the scale of **medium version** and **tiny version** are 1/2 and 1/4 of the full version respectively. And during process of training, we split training dataset into training sets and validation sets with 70% of sentences for training and the remaining for validation.

Then we use the pretrained language model and transformation matrix which is described in Sect. 3 separately to observe the improvement on same training dataset and validation dataset.

Pretrained language model. Specifically, CCKS 2019 dataset and i2b2 training dataset are utilized to fine-tuning pretrained language model with a Bi-directional LSTM and CRF architecture attaching behind. Then we record performance of 3 different scale i2b2 dataset in terms of this method.

Word embedding alignment. And for transformation matrix trained above, we use a totally same Bi-LSTM and CRF architecture but random initialization for word embedding layer is superseded by word2vec. Namely, we use Chinese medical NER dataset (CCKS 2019 dataset) to conduct a supervised training procedure with pretrained word2vec as initialization of word embeddings first,

[3] https://github.com/attardi/wikiextractor.

then we use English medical NER dataset to fine-tuning the model, with the additional advantage that utilizing transformation matrix to map English word embedding to Chinese word embedding space.

An extension of our work is combining pretrained language model and word embedding alignment to encoding input sentence jointly as joint features of a given input sentences.

4.3 Training Details

Pretraining details. For pretraining procedure, we use a Transformer architecture with 768 hidden units, 6 heads, GELU activations akin to XLM training. We set the dropout rate to 0.3 and train the model with Adam optimizer and learning rates varying from 5×10^{-4} to 1×10^{-4}. For MLM (mask language model) objective, we use streams of 256 tokens and set batch size to 256. Averaged perplexity over languages are chosen as a stopping criterion for training.

Word embedding alignment details. We use English version and Chinese version of ICD-11 coding as our parallel corpus. The sum of parallel sentence is more than 30,000. We preprocess the data for filtering out parallel sentence that only contains numeric and punctuation. As for threshold of data selection schema mentioned in 3.3, we choose $q = 3$. When training transformation matrix we use a linear regression to converge the alignment of two language word embedding spaces.

Finetuning details. For finetuning procedure, we use a one layer Bi-LSTM with 1024 embedding units, 200 hidden units. The maximum length of input sequence is set to 256 and dropout rate to 0.3. In addition, we finetuning with Adam optimizer and learning rate set to 1×10^{-4}.

4.4 Results Analysis

We display our result to measure the contribution of each component of our method. We compare our proposed method with a baseline BiLSTM-CRF architecture in target domain (English). It should be noted that we conducted three independent experiments to evaluate the effectiveness of our method for three scale of i2b2 dataset, which is full version, medium version and tiny version mentioned before.

For result of full version i2b2 dataset demonstrated in Table 3, we can noticed that XLM pretraining language model which only uses MLM and TLM objectives, can obtain a 79.24 F1 score. Noted that this XLM (MLM) pretrained model comes from Facebook's official repository, which is trained with open-domain corpora for 15 languages. After adding extra medical corpus and our parallel corpus extracted via 3.1, an increase of 3.16 F1 score is obtained for our three-objective pretrained language model. Remarkably, our new pretrained language model only can be used for Chinese and English, and its vocabulary is not the same as 15 languages XLM model's for the difference in training corpora.

Then we evaluate the effectiveness of word embedding alignment procedure, which brings a 1.09 boosting for F1 score.

After combining pretrained language model and bilingual word embedding alignment, we get the highest F1 score of 82.23. But the improvement towards baseline architecture (BiLSTM-CRF) is not equal to addition of improvement of single XLM (MLM+TLM+MixLM) and Word Embedding Alignment, which we deduce that pretrained language model have learned most word embedding alignment knowledge and cover the effectiveness of latter.

And for medium version presented in Table 4 and tiny version in Table 5, our combined method can also bring a huge boost of performance, especially for tiny scale of finetuning dataset. From two result tables, we can noticed that both pretraining method and word embedding alignment are effective with different scale of dataset. Even the combined method (pretraining and word embedding

Table 3. Transfer result for **full** version of i2b2 dataset

Method	Precision	Recall	F1
Target Domain Supervised Training	80.39	76.15	78.21
XLM (MLM+TLM:15 languages)	80.50	78.02	79.24
XLM (MLM+TLM+MixLM)	82.47	80.29	81.37
Word Embedding Alignment	80.92	77.75	79.30
XLM (MLM+TLM+MixLM) + Word Embed Align	**82.55**	**81.91**	**82.23**

Table 4. Transfer result for **medium** version of i2b2 dataset

Method	Precision	Recall	F1
Target Domain Supervised Training	67.11	69.82	68.43
XLM (MLM+TLM:15 languages)	68.78	68.95	68.86
XLM (MLM+TLM+MixLM)	71.46	**71.31**	71.38
Word Embedding Alignment	69.38	68.82	69.10
XLM (MLM+TLM+MixLM) + Word Embed Align	**72.83**	71.29	**72.05**

Table 5. Transfer result for **tiny** version of i2b2 dataset

Method	Precision	Recall	F1
Target Domain Supervised Training	53.59	51.48	52.51
XLM (MLM+TLM:15 languages)	54.36	54.57	54.46
XLM (MLM+TLM+MixLM)	**57.71**	56.68	57.19
Word Embedding Alignment	53.31	54.01	53.66
XLM (MLM+TLM+MixLM) + Word Embed Align	57.68	**58.19**	**57.93**

alignment) do not get the best result for precision and recall, the gap is small. In addition, the total improvement of F1 score for tiny version of i2b2 dataset is more than 5.5, comparing to 4.02 of full version and 3.62 of medium version.

5 Conclusion

In this work, we experimentally demonstrated the strong impact of bilingual transfer learning in medical domain with exploiting two methods of utilizing medical parallel corpora. And most importantly, we explore a innovative way of extracting parallel corpora from medical terminology systems or medical knowledge bases. The experiments validate the effectiveness of pretrained language model with the help of medical parallel corpora and bilingual word embedding alignment whose improvement depend on accuracy of bilingual dictionary. We plan to explore more medical data to bring an enforcement to our task and probe some unsupervied methods in medical domain in future work.

References

1. Artetxe, M., Labaka, G., Agirre, E.: Learning principled bilingual mappings of word embeddings while preserving monolingual invariance, pp. 2289–2294 (2016)
2. Artetxe, M., Labaka, G., Agirre, E.: Learning bilingual word embeddings with (almost) no bilingual data, vol. 1, pp. 451–462 (2017)
3. Artetxe, M., Schwenk, H.: Massively multilingual sentence embeddings for zero-shot cross-lingual transfer and beyond (2018). arXiv Computation and Language
4. Conneau, A., Kiela, D., Schwenk, H., Barrault, L., Bordes, A.: Supervised learning of universal sentence representations from natural language inference data (2017). arXiv Computation and Language
5. Devlin, J., Chang, M.W., Lee, K., Toutanova, K.: Bert: pre-training of deep bidirectional transformers for language understanding (2018). arXiv preprint arXiv:1810.04805
6. Doan, S., Bastarache, L., Klimkowski, S., Denny, J.C., Xu, H.: Integrating existing natural language processing tools for medication extraction from discharge summaries. J. Am. Med. Inf. Assoc. **17**(5), 528–531 (2010)
7. Ehrmann, M., Turchi, M., Steinberger, R.: Building a multilingual named entity-annotated corpus using annotation projection, pp. 118–124 (2011)
8. Faruqui, M., Dyer, C.: Improving vector space word representations using multilingual correlation, pp. 462–471 (2014)
9. Friedman, C., Alderson, P.O., Austin, J.H.M., Cimino, J.J., Johnson, S.B.: A general natural-language text processor for clinical radiology. J. Am. Med. Inf. Assoc. **1**(2), 161–174 (1994)
10. Gridach, M.: Character-level neural network for biomedical named entity recognition. J. Biomed. Inf. **70**, 85–91 (2017)
11. Habibi, M., Weber, L., Neves, M.L., Wiegandt, D.L., Leser, U.: Deep learning with word embeddings improves biomedical named entity recognition. Bioinformatics **33**(14), i37–i48 (2017)
12. Haug, P.J., Christensen, L.M., Gundersen, M.L., Clemons, B., Koehler, S.B., Bauer, K.: A natural language parsing system for encoding admitting diagnoses, pp. 814–818 (1997)

13. Haug, P.J., Koehler, S., Lau, L.M., Wang, P., Rocha, R.A., Huff, S.M.: Experience with a mixed semantic/syntactic parser, pp. 284–288 (1995)
14. Howard, J., Ruder, S.: Universal language model fine-tuning for text classification, vol. 1, pp. 328–339 (2018)
15. Johnson, A.E.W., et al.: Mimic-iii, a freely accessible critical care database. Sci. Data **3**(1), 160035 (2016)
16. Lample, G., Ballesteros, M., Subramanian, S., Kawakami, K., Dyer, C.: Neural architectures for named entity recognition, pp. 260–270 (2016)
17. Lample, G., Conneau, A.: Cross-lingual language model pretraining (2019). arXiv preprint arXiv:1901.07291
18. Lample, G., Conneau, A., Ranzato, M., Denoyer, L., Jegou, H.: Word translation without parallel data (2018)
19. Li, Z., Liu, F., Antieau, L.D., Cao, Y., Yu, H.: Lancet: a high precision medication event extraction system for clinical text. J. Am. Med. Inf. Assoc. **17**(5), 563–567 (2010)
20. Meystre, S.M., Thibault, J., Shen, S., Hurdle, J.F., South, B.R.: Textractor: a hybrid system for medications and reason for their prescription extraction from clinical text documents. J. Am. Med. Inf. Assoc. **17**(5), 559–562 (2010)
21. Mikolov, T., Chen, K., Corrado, G., Dean, J.: Efficient estimation of word representations in vector space (2013)
22. Mikolov, T., Le, Q.V., Sutskever, I.: Exploiting similarities among languages for machine translation (2013). arXiv Computation and Language
23. Mikolov, T., Sutskever, I., Chen, K., Corrado, G., Dean, J.: Distributed representations of words and phrases and their compositionality, pp. 3111–3119 (2013)
24. Ni, J., Dinu, G., Florian, R.: Weakly supervised cross-lingual named entity recognition via effective annotation and representation projection, vol. 1, pp. 1470–1480 (2017)
25. Patrick, J., Li, M.: High accuracy information extraction of medication information from clinical notes: 2009 i2b2 medication extraction challenge. J. Am. Med. Inf. Assoc. **17**(5), 524–527 (2010)
26. Pennington, J., Socher, R., Manning, C.D.: Glove: Global vectors for word representation, pp. 1532–1543 (2014)
27. Peters, M.E., et al.: Deep contextualized word representations (2018). arXiv Computation and Language
28. Savova, G., et al.: Mayo clinical text analysis and knowledge extraction system (ctakes): architecture, component evaluation and applications. J. Am. Med. Inf. Assoc. **17**(5), 507–513 (2010)
29. Sennrich, R., Haddow, B., Birch, A.: Neural machine translation of rare words with subword units (2015). arXiv Computation and Language
30. Smith, S.L., Turban, D.H.P., Hamblin, S., Hammerla, N.Y.: Offline bilingual word vectors, orthogonal transformations and the inverted softmax (2017). arXiv Computation and Language
31. Spasic, I., Sarafraz, F., Keane, J.A., Nenadic, G.: Medication information extraction with linguistic pattern matching and semantic rules. J. Am. Med. Inf. Assoc. **17**(5), 532–535 (2010)
32. Taylor, W.L.: "cloze procedure": A new tool for measuring readability. Journalism Bull. **30**(30), 415–433 (1953)
33. Tsai, C., Roth, D.: Cross-lingual wikification using multilingual embeddings, pp. 589–598 (2016)

34. Wang, Q., Xia, Y., Zhou, Y., Ruan, T., Gao, D., He, P.: Incorporating dictionaries into deep neural networks for the chinese clinical named entity recognition (2018). arXiv Computation and Language
35. Xu, K., Zhou, Z., Gong, T., Hao, T., Liu, W.: Sblc: a hybrid model for disease named entity recognition based on semantic bidirectional lstms and conditional random fields. BMC Med. Inf. Decis. Making **18**(5), 114 (2018)
36. Yarowsky, D., Ngai, G., Wicentowski, R.: Inducing multilingual text analysis tools via robust projection across aligned corpora, pp. 1–8 (2001)
37. Zeng, Q.T., Goryachev, S., Weiss, S.T., Sordo, M., Murphy, S.N., Lazarus, R.: Extracting principal diagnosis, co-morbidity and smoking status for asthma research: evaluation of a natural language processing system. BMC Med. Inf. Decis. Making **6**(1), 30–30 (2006)
38. Zhang, M., Liu, Y., Luan, H., Sun, M.: Adversarial training for unsupervised bilingual lexicon induction, vol. 1, pp. 1959–1970 (2017)
39. Zitouni, I., Florian, R.: Mention detection crossing the language barrier, pp. 600–609 (2008)

A Unified Adversarial Learning Framework for Semi-supervised Multi-target Domain Adaptation

Xinle Wu[1,2], Lei Wang[2(✉)], Shuo Wang[1], Xiaofeng Meng[1], Linfeng Li[2,3], Haitao Huang[4], Xiaohong Zhang[5], and Jun Yan[2]

[1] School of Information, Renmin University of China, Beijing, China
{xinle.wu,shuowang,xfmeng}@ruc.edu.cn
[2] Yidu Cloud (Beijing) Technology Co., Ltd., Beijing, China
{lei.wang01,Linfeng.Li,jun.yan}@yiducloud.cn
[3] Institute of Information Science, Beijing Jiaotong University, Beijing, China
[4] The second Department of Neurology, Liaoning People's Hospital, Shenyang, China
[5] The Fourth Affiliated Hospital, China Medical University, Taichung, Taiwan, R.O.C.

Abstract. Machine learning algorithms have been criticized as difficult to apply to new tasks or datasets without sufficient annotations. Domain adaptation is expected to tackle this problem by establishing knowledge transfer from a labeled source domain to an unlabeled or sparsely labeled target domain. Most existing domain adaptation models focus on the single-source-single-target scenario. However, the pairwise domain adaptation approaches may lead to suboptimal performance when there are multiple target domains available, because the information from other related target domains is not being utilized. In this work, we propose a unified semi-supervised multi-target domain adaptation framework to implement knowledge transfer among multiple domains (a single source domain and multiple target domains). Specifically, we aim to learn an embedded space and minimize the marginal probability distribution differences among all domains in the space. Meanwhile, we introduce Prototypical Networks to perform classification, and extend it to semi-supervised settings. On this basis, we further align the conditional probability distributions among the domains by generating pseudo-labels for the unlabeled target data and training the model with bootstrapping method. Extensive sentiment analysis experiments show that our approach significantly outperforms several state-of-the-art methods.

Keywords: Domain adaptation · Adversarial learning · Semi-supervised · Prototypical networks · Self-training · Sentiment analysis

1 Introduction

Supervised learning algorithms have achieved great success in many fields with the availability of large quantities of labeled data. However, it is costly and time-consuming to annotate such large-scale training data for new tasks or datasets.

Y. Nah et al. (Eds.): DASFAA 2020, LNCS 12112, pp. 419–434, 2020.
https://doi.org/10.1007/978-3-030-59410-7_29

A naive idea is directly applying the model trained on a labeled source domain to the related and sparsely labeled target domain. Unfortunately, the model usually fails to perform well in the target domain due to domain shifts [24]. Domain adaptation (DA) is proposed to address this problem by transferring knowledge from a labeled source domain to a sparsely labeled target domain.

Existing DA methods can be divided into: supervised DA (SDA) [14,20,26], semi-supervised DA (SSDA) [11,21,23,31], and unsupervised DA (UDA) [4,5,9, 15]. SDA methods assume that there are some labeled data in the target domain, and perform DA algorithms only use the labeled data. Conversely, UDA methods do not need any target data labels, but they require large amounts of unlabeled target data to align the distributions between domains. Considering that it is cheap to annotate a small number of samples and a few labeled data often leads to significant performance improvements, we focus on SSDA which exploits both labeled and unlabeled data in target domains.

Typical DA methods are designed to embed the data from the source and target domains into a common embedding space, and align the marginal probability distributions between the two domains. There are two approaches to achieve this, adversarial training [4,10,20,27] and directly minimizing the distance between the two distributions [15,18,28,35]. Both of the methods can generate domain-invariant feature representations for input data, and the representations from the source domain are used to train a classifier, which is then generalized to the target domain. However, only aligning the marginal distributions is not sufficient to ensure the success of DA [4,5,12,33], because the conditional probability distributions between the source and target domains may be different.

Most DA algorithms focus on the single-source-single-target setting. However, in many practical applications, there are multiple sparsely labeled target domains. For example, in the sentiment analysis task of product reviews, we can take the reviews of Books, DVDs, Electronics and Kitchen appliances as different domains. If we only have access to sufficient labeled data of Book reviews (source domain), and hope to transfer knowledge to the other domains, then each of the other domains can be seen as a target domain. In this case, pairwise adaptation approaches may be suboptimal, especially when there are shared features between the source and multiple target domains or the source and the target domain are associated through another target domain [9]. This is due to that these methods fail to leverage the knowledge from other relevant target domains. In addition, considering the distribution differences among multiple target domains, simply merging multiple target domains into a single one may not be the optimal solution.

To address these problems, we propose semi-supervised multi-target domain adaptation networks (MTDAN). Specifically, we use a shared encoder to extract the common features shared by all domains, and a private encoder to extract the domain-specific features of each domain. For feature representations generated by the two encoders, we train a domain discriminator to distinguish which domain they come from. To ensure that the shared representation is domain-invariant, the shared encoder is encouraged to generate the representation cannot be correctly distinguished by the domain discriminator. Given that there are only a few labeled data in each target domain, we introduce Prototypical Networks to perform classification, which is more superior than deep classifiers in few-shot scenarios [25]. We further leverage unlabeled data to refine prototypes, and extend Prototypical Networks to semi-supervised scenarios. Moreover, we utilize the self-training algorithm to exploit unlabeled target data, and we show that it can also align the class-conditional probability distributions among multiple domains.

Contributions. Our contributions are: a) We propose a unified adversarial learning framework for semi-supervised multi-target DA. b) We show that the prototype-based classifier can achieve better performance than the deep classifier when target domains have only a few labeled data and large amounts of unlabeled data. c) We show that the self-training algorithm can effectively align the class-conditional probability distributions among multiple domains. d) Our method outperforms several state-of-the-art DA approaches on sentiment analysis dataset.

2 Related Work

Domain Adaptation. Numerous domain adaptation approaches have been proposed to solve domain shift [29]. Most of them seek to learn a shared embedded space, in which the representations of source domain and target domain cannot be distinguished [27]. Based on that, the classifier trained with labeled source data can be generalized to the target domain. There are two typical ways to learn cross-domain representations: directly minimizing the distance between two distributions [15,17,18] and adversarial learning [6,7,26,27].

For the first method, several distance metrics have been proposed to measure the distance between source and target distributions. One common distance metric is the Maximum Mean Discrepancy (MMD) [2], which computes the norm of the difference between two domain means in the reproducing Kernel Hilbert Space (RKHS). Specifically, the DDC method [28] used both MMD and regular classification loss on the source to learn representations that are discriminative and domain invariant. The Deep Adaptation Network (DAN) [15] applied MMD to the last full connected layers to match higher order statistics of the two distributions. Most recently, [18] proposed to reduce domain shift in joint distributions of the network activation of multiple task-specific layers. Besides, Zellinger et al. proposed Center Moment Discrepancy (CMD) [32] to diminish the domain shift by aligning the central moment of each order across domains.

The other method is to optimize the source and target mappings using adversarial training. The idea is to train a domain discriminator to distinguish whether input features come from the source or target, whereas the feature encoder is trained to deceive the domain discriminator by generating representations that cannot be distinguished. [6] proposed the gradient reversal algorithm (ReverseGrad), which directly maximizes the loss of the domain discriminator by reversing its gradients. DRCN in [8] takes a similar approach in addition to learning to reconstruct target domain images. [3] enforced these adversarial losses in a shared feature space, while learned a private feature space for each domain to avoid the contamination of shared representations.

[4,27,33] argued that only aligning the marginal probability distributions between the source and target is not enough to guarantee successful domain adaptation. [16] proposed to align the marginal distributions and conditional distributions between the source and target simultaneously. [20] extended the domain discriminator to predict the domain and category of the embedded representation at the same time to align the joint probability distributions of input and output, and achieved a leading effect in the supervised domain adaptive scene. [5] proposed to align the class-conditional probability distributions between the source and target.

Recently, Zhao et al. [34] introduced an adversarial framework called MDAN, which is used for multi-source-single-target domain adaption. They utilized a multi-class domain discriminator to align the distributions between multiple source and a target domain. [9] proposed an information theoretic approach to solve unsupervised multi-target domain adaptation problem, which maximizes the mutual information between the domain labels and domain-specific features, while minimizes the mutual information between the the domain labels and the domain-invariant features. Unlike their approach, we base our method on self-training rather than entropy regularization. Moreover, we introduce prototypical networks to perform classification, which is more effective than deep classifiers in SSDA scenarios.

Semi-supervised Learning. Recently, some works treat domain adaptation as a semi-supervised learning task. [11] proposed a Domain Adaptive Semi-supervised learning framework (DAS) to jointly perform feature adaptation and semi-supervised learning. [21] applied a co-training framework for semi-supervised domain adaptation, in which the shared classifier and the private classifier boost each other to achieve better performance. [22] re-evaluated classic general-purpose bootstrapping approaches under domain shift, and proved that the classic bootstrapping algorithms make strong baselines on domain adaptation tasks.

3 Preliminaries

In this section, we introduce the notations and definitions related to single-source-multi-target DA.

Notations. We use $\mathcal{D}$ to denote a domain, which consists of an m-dimensional feature space $\mathcal{X}$ and a marginal probability distribution $P(\mathrm{x})$, i.e., $\mathcal{D} = \{\mathcal{X}, P(\mathrm{x})\}$, where $\mathrm{x} \in \mathcal{X}$. We use $\mathcal{T}$ to denote a task which consists of a C-cardinality label set $\mathcal{Y}$ and a conditional probability distribution $P(y|\mathrm{x})$, i.e., $\mathcal{T} = \{\mathcal{Y}, P(y|\mathrm{x})\}$, where $y \in \mathcal{Y}$.

Problem Formulation (Single-Source-Multi-target Domain Adaptation). Let $\mathcal{D}_s = \{(\mathrm{x}_l^s, y_l^s)\}_{l=1}^{n_s}$ be a labeled source domain where n_s is the

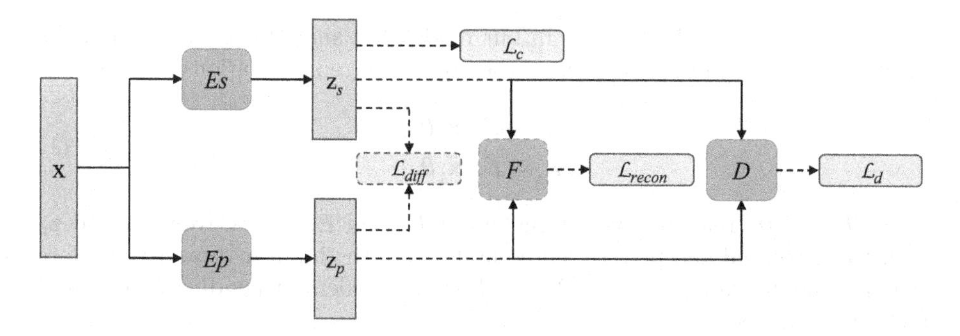

Fig. 1. The network structure of the proposed framework. The shared encoder E_s captures the common features shared among domains, while the private encoder E_p captures the domain-specific features. The shared decoder F reconstruct the input samples by using both the shared and private representations. The domain classifier D learns to distinguish which domain the input representations come from. The orthogonality constraint loss $\mathcal{L}_{diff}$ encourages E_s and E_p to encode different aspects of the inputs. The prototype-based classifier is computed on-the-fly, and the classification loss $\mathcal{L}_c$ is only used to optimize E_s.

number of labeled samples and let $\mathcal{D}_t = \{\mathcal{D}_{t_i}\}_{i=1}^{K}$ be multiple sparsely labeled target domains where $\mathcal{D}_{t_i} = \{(\mathrm{x}_l^{t_i}, y_l^{t_i})\}_{l=1}^{n_{l_i}} \bigcup \{\mathrm{x}_u^{t_i}\}_{u=1}^{n_{u_i}}$, K is the number of target domains, and $n_{l_i}(n_{l_i} \ll n_s)$ and $n_{u_i}(n_{u_i} \gg n_{l_i})$ refer to the number of labeled and unlabeled samples of i-th target domain respectively. We assume that all domains share the same feature space $\mathcal{X}$ and label space $\mathcal{Y}$, but the marginal probability distributions and the conditional probability distributions of source domain and multiple target domains are different from each other. The goal is to learn a classifier using the labeled source data and a few labeled target data, that generalizes well to the target domain.

4 Methodology

In this section, we describe each component and the corresponding loss function of the proposed framework in detail.

4.1 Proposed Approach

Our model consists of four components as shown in Fig. 1. A shared encoder E_s is trained to learn cross-domain representations, a private encoder E_p is trained to learn domain-specific representations, a shared decoder F is trained to reconstruct the input sample, and a discriminator D is trained to distinguish which domain the input sample comes from. Task classification is performed by calculating the distance from the domain-invariant representations to prototype representations of each label class.

Domain-Invariant and Domain-Specific Representations. We seek to extract domain-invariant (shared) and domain-specific (private) representations for each input $\mathbf{x}$ simultaneously. In our model, the shared encoder E_s and the private encoder E_p learn to generate the above two representations respectively:

$$\begin{aligned} \mathbf{z}_s &= E_s(\mathbf{x}, \boldsymbol{\theta}_s) \\ \mathbf{z}_p &= E_p(\mathbf{x}, \boldsymbol{\theta}_p) \end{aligned} \tag{1}$$

Here, $\boldsymbol{\theta}_s$ and $\boldsymbol{\theta}_p$ refer to the parameters of E_s and E_p respectively, $\mathbf{z}_s$ and $\mathbf{z}_p$ refer to the shared and private representations of the input $\mathbf{x}$ respectively. Note that E_s and E_p can be MLP, CNN or LSTM encoders, depending on different tasks and datasets.

Reconstruction. In order to avoid information loss during the encoding, we reconstruct input samples with both shared and private representations. We use $\hat{\mathbf{x}}$ to denote the reconstruction of the input $\mathbf{x}$, which is generated by decoder F:

$$\hat{\mathbf{x}} = F(\mathbf{z}_s + \mathbf{z}_p, \boldsymbol{\theta}_f), \tag{2}$$

where $\boldsymbol{\theta}_f$ are the parameters of F. We use mean square error to define the reconstruction loss $\mathcal{L}_{Recon}$, which is applied to all domains:

$$\mathcal{L}_{Recon} = \frac{\lambda_r}{N} \sum_{i=1}^{N} \frac{1}{C} \|\mathbf{x}_i - \hat{\mathbf{x}}_i\|_2^2, \tag{3}$$

where C is the dimension of the input $\mathbf{x}$, N is the total number of samples in all domains, $\mathbf{x}_i$ refers to the i-th sample, λ_r is the hyper-parameter controlling the weight of the loss function, and $\|\cdot\|_2^2$ is the squared L_2-norm.

Orthogonality Constraints. To minimize the redundancy between shared and private representations, we introduce orthogonality constraints to encourage the shared and private encoders to encode different aspects of inputs. Specifically, we use $\mathbf{H}_s$ to denote a matrix, each row of which corresponds to the shared representation of each input $\mathbf{x}$. Similarly, let $\mathbf{H}_p$ be a matrix, each row of which corresponds to the private representation of each input $\mathbf{x}$. The corresponding loss function is:

$$\mathcal{L}_{Diff} = \lambda_{diff} \|\mathbf{H}_s^\top \mathbf{H}_p\|_F^2, \tag{4}$$

where λ_{diff} is the scale factor, $\|\cdot\|_F^2$ is the squared Frobenius norm.

Adversarial Training. The goal of adversarial training is to regularize the learning of the shared encoder E_s, so as to minimize the distance of distributions among source and multiple target domains. After that, we can apply the source classification model directly to the target representations. Therefore, we first train a domain discriminator D with the domain labels of the shared and private representations (since we know which domain each sample comes from, it is obvious that we can generate a domain label for each sample). The discriminator D is a multi-class classifier designed to distinguish which domain the

Algorithm 1. MTDAN Algorithm

Input: labeled source domain examples L_s, labeled multi-target domain examples $L_t = \{L_{t_i}\}_{i=1}^{K}$, unlabeled multi-target domain examples $U_t = \{U_{t_i}\}_{i=1}^{K}$
Hyper-parameters: coefficients for different losses: $\lambda_r, \lambda_d, \lambda_c, \lambda_{diff}$, mini-batch size b, learning rate η

1: initialize $\boldsymbol{\theta}_s, \boldsymbol{\theta}_p, \boldsymbol{\theta}_f, \boldsymbol{\theta}_d$
2: **repeat**
3: **repeat**
4: Sample a mini-batch from $\{L_s, L_t\}$
5: Train F by minimizing $\mathcal{L}_{Recon}$
6: Train D by minimizing $\mathcal{L}_D$
7: Train E_p by minimizing $\mathcal{L}_P$
8: Train E_s by minimizing $\mathcal{L}_S$
9: **until** Convergence
10: Apply Eq.(9) to label U_t
11: Select the most confident p positive and n negative predicted examples U_t^l from U_t
12: Remove U_t^l from U_t
13: Add examples U_t^l and their corresponding labels to L_t
14: **until** obtain best performance on the developing dataset

input representation comes from. Thus, D is optimized according to a standard supervised loss, defined below:

$$\mathcal{L}_D = \mathcal{L}_{D_p} + \mathcal{L}_{D_s}, \tag{5}$$

$$\mathcal{L}_{D_p} = -\frac{\lambda_d}{N}\sum_{i=1}^{N} \mathbf{d}_i^\top \log D(E_p(\mathbf{x}_i, \boldsymbol{\theta}_p), \boldsymbol{\theta}_d), \tag{6}$$

$$\mathcal{L}_{D_s} = -\frac{\lambda_d}{N}\sum_{i=1}^{N} \mathbf{d}_i^\top \log D(E_s(\mathbf{x}_i, \boldsymbol{\theta}_s), \boldsymbol{\theta}_d), \tag{7}$$

where $\mathbf{d}_i$ is the one-hot encoding of the i-th sample's domain label, $\boldsymbol{\theta}_d$ is the parameter of D, and λ_d is the scale factor.

Second, we train the shared encoder E_s to fool the discriminator D by generating cross-domain representations. We guarantee this by adding $-\mathcal{L}_{D_s}$ to the loss function of the shared encoder E_s. On the other hand, we hope the private encoder only extracts domain-specific features. Thus, we add $\mathcal{L}_{D_p}$ to the loss function of E_p to generate representations that can be distinguished by D.

Prototypical Networks for Task Classification. The simplest way to classify the target samples is to train a deep classifier, however, it may only achieve suboptimal performance as we can see in Table 1. The reason is that there are only a few labeled samples in each target domain, which is not enough to fine-tune a deep classifier with many parameters, so that the classifier is easy to over fit the source labeled data. Although we could generate pseudo-labeled data for target domains, the correctness of the pseudo-labels can not be guaranteed due to the poor performance of the deep classifier.

To efficiently utilize the labeled samples in target domains, we refer to the idea of prototypical networks [25]. Prototypical networks assume that there is a prototype in the latent space for each class, and the projections of samples belonging to this class cluster around the prototype. The classification is then performed by computing the distances to prototype representations of each class in the latent space. By reducing parameters of the model, the prototype-based classifier can achieve better performance than the deep classifier when labeled samples are insufficient. Note that we refine prototypes during self-training by allowing unlabeled samples with pseudo-labels to update the prototypes. Specifically, we compute the average of shared representations belonging to each class in a batch as prototypes:

$$\mathbf{c}_k = \frac{1}{n_k}\sum_{i=1}^{n_k} E_s(\mathbf{x}_i, \boldsymbol{\theta}_s), \tag{8}$$

where n_k is the number of samples belonging to class k in a batch. Then we calculate a distribution by applying softmax function to distances between a shared representation with a prototype:

$$p(y = k|\mathbf{x}) = \frac{exp(-d(\mathbf{z}_s, \mathbf{c}_k))}{\sum_{k'} exp(-d(\mathbf{z}_s, \mathbf{c}'_k))}, \tag{9}$$

where $d(\cdot)$ is a distance measure function. We use the squared Euclidean distance in this work. The classification loss is defined as:

$$\mathcal{L}_C = -\lambda_c \log p(y = k|\mathbf{x}), \tag{10}$$

where λ_c is the scale factor.

Self-training for Conditional Distribution Adaptation. As described in [5,19], only aligning the marginal probability distributions between source and target is not enough to guarantee successful domain adaptation. Because this only enforces alignment of the global domain statistics with no class specific transfer. Formally, we can achieve $P_s(E_s(\mathrm{x}_s)) \approx P_{t_i}(E_s(\mathrm{x}_{t_i}))$ by introducing adversarial training, but $P_s(y_s|E_s(\mathrm{x}_s)) \neq P_{t_i}(y_{t_i}|E_s(\mathrm{x}_{t_i}))$ may still hold, where $P_s(y_s|E_s(\mathrm{x}_s))$ can be regarded as the classifier trained with source data.

Here, we tackle this problem by further reducing the difference of conditional probability distributions among source domain and target domains. In practice, we replace conditional probability distributions with class-conditional probability distributions, because the posterior probability is quite involved [16].

However, it is nontrivial to adapt class-conditional distributions, as most of the target samples are unlabeled. We address this problem by producing pseudo-labels for unlabeled target samples, and train the whole model in a bootstrapping way. As we perform more learning iterations, the number of target samples with correct pseudo-labels grows and progressively enforces distributions to align class-conditionally.

To be specific, we first train our model on labeled source and target samples. Then, we use the model to generate a probability distribution over classes for each unlabeled target sample. If the probability of a sample on a certain class is higher than a predetermined threshold τ, the sample would be added to the training set with the class as its pseudo-label.

Loss Function and Model Training. We alternately optimize the four modules of our model.

For E_p, the goal of training is to minimize the following loss:

$$\mathcal{L}_P = \mathcal{L}_{Recon} + \mathcal{L}_{Diff} + \mathcal{L}_{D_p} \tag{11}$$

For E_s, the goal of training is to minimize the following loss:

$$\mathcal{L}_S = \mathcal{L}_{Recon} + \mathcal{L}_{Diff} - \mathcal{L}_{D_s} + \mathcal{L}_C \tag{12}$$

For F and D, the losses are $\mathcal{L}_{Recon}$ and $\mathcal{L}_D$, respectively. The detailed training process is shown in algorithm 1.

5 Experiments

5.1 Dataset

We evaluate our proposed model on the Amazon benchmark dataset [1]. It is a sentiment classification dataset[1], which contains Amazon product reviews from four different domains: Books (B), DVD (D), Electronics (E), and Kitchen appliances (K). We remove reviews with neutral labels and encode the remaining reviews into 5000 dimensional feature vectors of unigrams and bigrams with binary labels indicating sentiment.

We pick two product as the source domain and the target domain in turn, and the other two domains as the auxiliary target domains, so that we construct 12 single-source-three-target domain adaptation tasks. For each task, the source domain contains 2,000 labeled examples, and each target domain contains 50 labeled examples and 2,000 unlabeled examples. To fine-tune the hyper-parameters, we randomly select 500 labeled examples from the target domain as the developing dataset.

---

[1] https://www.cs.jhu.edu/mdredze/datasets/sentiment/.

5.2 Compared Method

We compare MTDAN with the following baselines:

(1) **ST:** The basic neural network classifier without any domain adaptation trained on the labeled data of the source domain and the target domain.
(2) **CoCMD:** This is the state-of-the-art pairwise SSDA method on the Amazon benchmark dataset [21]. The shared encoder, private encoder and reconstruction decoder used in this model are the same as ours.
(3) **MTDA-ITA:** This is the state-of-the-art single-source-multi-target UDA method on three benchmark datasets for image classification [9]. We implemented the framework and extend it to semi-supervised DA method. The shared encoder, private encoder, reconstruction decoder and domain classifier used in this model are the same as ours.
(4) **c-MTDAN:** We combine all the target domains into a single one, and train it using MTDAN. Similarly, we also report the performance of c-CoCMD and c-MTDA-ITA.
(5) **s-MTDAN:** We do not use any auxiliary target domains, and train MTDAN on each source-target pair.

5.3 Implementation Details

Considering that each input sample in the dataset is a tf-idf feature vector without word ordering information, we use a multilayer perceptron (MLP) with an input layer (5000 units) and one hidden layer (50 units) and sigmoid activation functions to implement both shared and private encoders. The reconstruction decoder consists of one dense hidden layer (2525 units), tanh activation functions, and relu output functions. The domain discriminator is composed of a softmax layer with n-dimensional outputs, where n is the number of the source and target domains. For MTDA-ITA, we follow the framework proposed by [9], and use the above modules to replace the original modules in the framework. Besides, the task classifier for MTDA-ITA is a fully connected layer with softmax activation functions.

The network is trained with Adam optimizer [13] and with learning rate 10^{-4}. The mini-batch size is 50. The hyper-parameters $\lambda_r, \lambda_d, \lambda_c$ and λ_{diff} are empirically set to $1.0, 0.5, 0.1$ and 1.0 respectively. The threshold τ for producing pseudo-labels is set to 0.8. Following previous studies, we use classification accuracy metric to evaluate the performances of all approaches.

5.4 Results

The performances of the proposed model and other state-of-the-art methods are shown in Table 1. Key observations are summarized as follows. (1) The proposed model MTDAN achieves the best results in almost all tasks, which proves the effectiveness of our approach. (2) c-CoCMD has worse performance in all tasks compared with CoCMD, although c-CoCMD exploits labeled and unlabeled data from auxiliary target domains for training. Similar observation can also be observed by comparing MTDA-ITA with c-MTDA-ITA and MTDAN with c-MTDAN. This demonstrates that simply combine all target domains into a single one is not an effective method to solve the multi-target DA problem. (3) Our model outperforms CoCMD by an average of nearly 2.0%, which indicates that our model can effectively leverage the labeled and unlabeled data from multiple target domains. Similarly, our model performs better than its variant, s-MTDAN, which does not leverage the data from auxiliary target domains. This also shows that it is helpful to mine knowledge from auxiliary target domains. (4) Although MTDA-ITA is also a multi-target domain adaptation method, its performance is worse than that of MTDAN. This can be due to (i) self-training is a superior method than entropy regularization to exploit unlabeled target data, (ii) the prototype-based classifier is more efficient than the deep classifier in semi-supervised scenarios, (iii) we introduce orthogonality constraints to further reduce the redundancy between shared and private representations. (5) In the K→E task, MTDAN performs slightly worse than s-MTDAN. This can be explained that domain K is closer to domain E than the other domains as shown in Fig. 2 (a), and MTDAN leads to negative transfer when using relevant target domains to help domain adaptation. (6) s-MTDAN outperforms CoCMD in 9 of the 12 tasks, note that both of them do not use the auxiliary domains. This indicates that our model is more effective than CoCMD in pairwise domain adaptation task. (7) All models achieve better performance than the basic ST model, which demonstrates that domain adaptation methods are crucial when there exist a domain gap between the source domain and the target domain.

Table 1. Average classification accuracy with 5 runs on target domain testing dataset. The best is shown in bold. c-X: combining all target domains into a single one and performing pairwise domain adaptation with model X. s-X: performing pairwise domain adaptation between the original source and target domains with model X

Method	B→D	B→E	B→K	D→B	D→E	D→K	E→B	E→D	E→K	K→B	K→D	K→E
ST	81.6	75.8	78.2	80.0	77.0	80.4	74.7	75.4	85.7	73.8	76.6	85.3
CoCMD	83.1	83.0	85.3	81.8	83.4	85.5	76.9	78.3	87.3	77.2	79.6	87.2
c-CoCMD	82.7	82.2	84.5	80.6	83.0	84.8	76.3	77.6	87.1	75.9	79.4	86.1
MTDA-ITA	83.8	83.2	83.7	81.8	83.6	85.4	76.6	78.9	87.7	77.0	78.8	86.8
c-MTDA-ITA	83.3	82.3	83.2	81.4	83.0	85.0	76.0	79.3	87.6	76.7	78.5	87.0
s-MTDAN	83.3	83.9	84.7	81.6	83.7	84.7	78.0	80.2	87.9	78.6	79.9	**87.8**
c-MTDAN	84.0	84.0	85.5	81.7	84.3	85.9	80.2	80.7	88.1	79.8	80.5	87.0
MTDAN	**84.5**	**84.3**	**86.0**	**82.3**	**85.3**	**87.2**	**80.5**	**81.2**	**88.9**	**80.0**	**80.9**	87.4

5.5 Ablation Studies

We performed ablation experiments to verify the importance of each component of our proposed model. We report the results of removing orthogonality constraints loss (set λ_{diff}=0), self-training process, the prototype-based classifier (replaced by the deep classifier) respectively.

As we can see from Table 2, removing each of the above components causes performance degradation. To be specific, disabling self-training degrades the performance to the greatest extent, with an average decrease of 5.1%, which shows the importance of mining information from the unlabeled data of target domains. Similarly, replacing prototype-based classifiers with deep classifiers also leads to performance degradation, with an average decrease of 1.4%, which shows that the prototype-based classifiers is more effective than deep classifiers in semi-supervised scenarios. Besides, disabling the orthogonality constraints loss leads to a performance degradation of 0.7%, which indicates that encouraging the disjoint of shared and private representations can make the shared feature space more common among all domains.

We did not test the performance degradation caused by disabling reconstruction loss and multi-class adversarial training loss, because they have been proved in previous work [3,9]. To summarize, each of the proposed components helps improve classification performance, and using all of them brings the best performance.

Table 2. Ablations. Performance of the proposed model when one component is removed or replaced. woDiff means without orthogonality constraints loss, woSelf means without self-training procedures, woProto means replace the prototype-based classifier with the deep classifier.

Method	B→D	B→E	B→K	D→B	D→E	D→K	E→B	E→D	E→K	K→B	K→D	K→E
MTDAN-*woDiff*	84.1	83.6	85.9	81.5	85.0	86.7	79.7	80.5	88.3	79.6	80.0	87.0
MTDAN-*woSelf*	82.8	77.6	80.0	81.6	78.8	81.5	74.8	75.6	86.7	74.5	77.3	86.7
MTDAN-*woProto*	83.3	83.8	84.1	81.8	83.8	86.9	79.2	78.9	87.9	78.0	80.3	86.4
MTDAN	**84.5**	**84.3**	**86.0**	**82.3**	**85.3**	**87.2**	**80.5**	**81.2**	**88.9**	**80.0**	**80.9**	**87.4**

5.6 Feature Visualization

In order to understand the behavior of the proposed model intuitively, we project the shared and private encoder outputs into two-dimensional space with principle component analysis (PCA) [30] and visualize them. For comparison, we also show the visualization result of the basic ST model. Due to space constraints, we only show the visualization results of MTDAN with B as the source domain, E as the target domain, D and K as the auxiliary target domains. The results are shown in Fig. 2.

Figure 2 (a) shows the encoder output distribution of the ST model. As we can see, the distributions of domain B and domain D (called group 1) are similar and the distributions of domain E and domain D (called group 2) are similar, while the distributions of cross-group domains are relatively different. That's why the ST model gets worse classification performance when the source domain

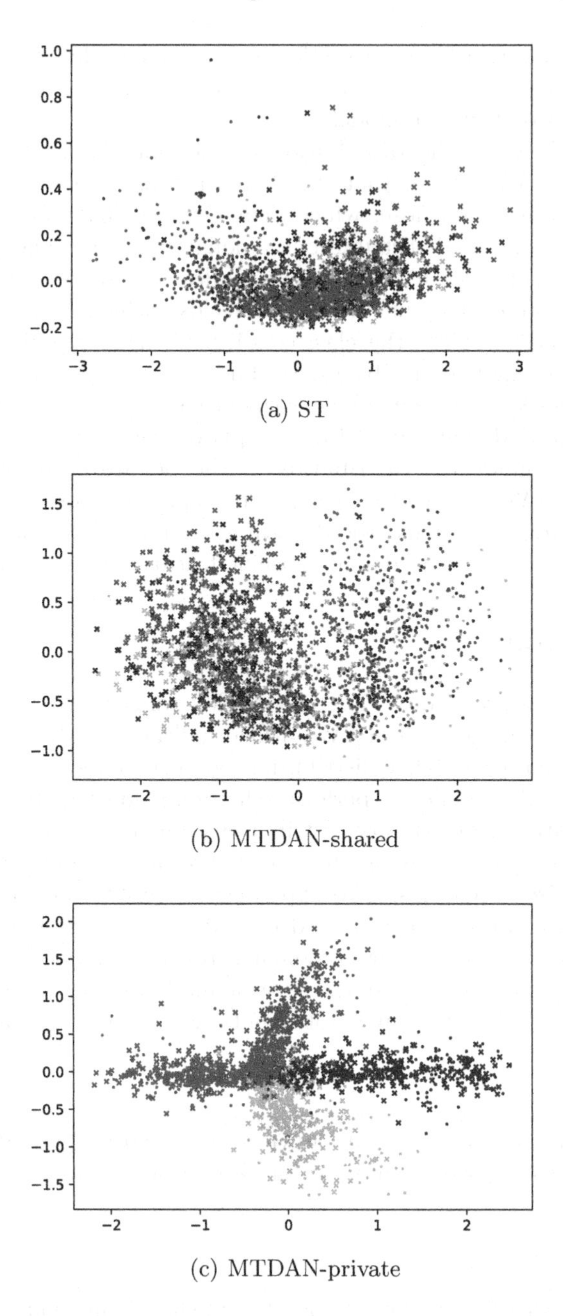

(a) ST

(b) MTDAN-shared

(c) MTDAN-private

Fig. 2. Feature visualization for the embedding of source and target data. The red, blue, yellow and green symbols denote the samples from B, D, E and K respectively. The symbol 'x' is used for positive samples and '.' is for negative samples. (a) the distribution of the encoder output of ST, (b) the distribution of shared representations of MTDAN, (c) the distribution of private representations of MTDAN. For ST and MTDAN, we take B as the source domain and D, E and K as the target domains.

and the target domain belong to different groups. Besides, there is no obvious boundary between positive and negative samples, which is consistent with the poor performance of the ST model.

Figure 2 (b) shows the distribution of the shared encoder output of the MTDAN model. We can see that the shared representations of the source and target domains are very close, which demonstrates that our model can effectively align the marginal distributions among the source and multiple target domains. Meanwhile, for each class of samples, the shared representations of the source and target domains are also very close, which demonstrates that our model can effectively align the class-conditional distributions among multiple domains. Comparing (a) and (b), we can find that the boundary of positive and negative samples in (b) is more obvious than that in (a), which means the shared representations of MTDAN model have superior class separability.

Figure 2 (c) shows the distribution of the private encoder output of the MTDAN model. We can see that the private representations have good domain separability, partially because the domain discriminator D encourages the private encoder E_p to generate domain-specific feature representations.

6 Conclusion

In this paper, we propose MTDAN, a unified framework for semi-supervised multi-target domain adaptation. We utilize multi-class adversarial training to align the marginal probability distributions among source domain and multiple target domains. Meanwhile, we perform self-training on target unlabeled data to align the conditional probability distributions among the domains. We further introduce Prototypical Networks to replace the deep classifiers, and extend it to semi-supervised scenarios. The experimental results on sentiment analysis dataset demonstrate that our method can effectively leverage the labeled and unlabeled data of multiple target domains to help the source model achieve generalization, and is superior to the existing methods. The proposed framework could be used for other domain adaptation tasks, and we leave this as our future work.

Acknowledgment. This work was supported by National Natural Science Foundation of China (Grant No: 91646203, 91846204, 61532010, 61941121, 61532016 and 61762082). The corresponding author is Xiaofeng Meng.

References

1. Blitzer, J., Dredze, M., Pereira, F.: Biographies, bollywood, boom-boxes and blenders: Domain adaptation for sentiment classification. In: Proceedings of the 45th Annual Meeting of the Association of Computational Linguistics, pp. 440–447 (2007)
2. Borgwardt, K.M., Gretton, A., Rasch, M.J., Kriegel, H.P., Schölkopf, B., Smola, A.J.: Integrating structured biological data by kernel maximum mean discrepancy. Bioinformatics **22**(14), e49–e57 (2006)

3. Bousmalis, K., Trigeorgis, G., Silberman, N., Krishnan, D., Erhan, D.: Domain separation networks. In: Advances in Neural Information Processing Systems, pp. 343–351 (2016)
4. Cicek, S., Soatto, S.: Unsupervised domain adaptation via regularized conditional alignment. arXiv preprint arXiv:1905.10885 (2019)
5. Gabourie, A.J., Rostami, M., Pope, P.E., Kolouri, S., Kim, K.: Learning a domain-invariant embedding for unsupervised domain adaptation using class-conditioned distribution alignment. In: 2019 57th Annual Allerton Conference on Communication, Control, and Computing (Allerton), pp. 352–359. IEEE (2019)
6. Ganin, Y., Lempitsky, V.S.: Unsupervised domain adaptation by backpropagation. In: Proceedings of the 32nd International Conference on Machine Learning, ICML 2015, Lille, France, 6–11 July 2015, pp. 1180–1189 (2015)
7. Ganin, Y., et al.: Domain-adversarial training of neural networks. J. Mach. Learn. Res. **17**(1), 2030–2096 (2016)
8. Ghifary, M., Kleijn, W.B., Zhang, M., Balduzzi, D., Li, W.: Deep reconstruction-classification networks for unsupervised domain adaptation. In: Leibe, B., Matas, J., Sebe, N., Welling, M. (eds.) ECCV 2016. LNCS, vol. 9908, pp. 597–613. Springer, Cham (2016). https://doi.org/10.1007/978-3-319-46493-0_36
9. Gholami, B., Sahu, P., Rudovic, O., Bousmalis, K., Pavlovic, V.: Unsupervised multi-target domain adaptation: An information theoretic approach. arXiv preprint arXiv:1810.11547 (2018)
10. Guo, J., Shah, D.J., Barzilay, R.: Multi-source domain adaptation with mixture of experts. In: Proceedings of the 2018 Conference on Empirical Methods in Natural Language Processing, Brussels, Belgium, 31 October–4 November 2018, pp. 4694–4703 (2018)
11. He, R., Lee, W.S., Ng, H.T., Dahlmeier, D.: Adaptive semi-supervised learning for cross-domain sentiment classification. In: Proceedings of the 2018 Conference on Empirical Methods in Natural Language Processing, Brussels, Belgium, 31 October–4 November 2018, pp. 3467–3476 (2018)
12. Hosseini-Asl, E., Zhou, Y., Xiong, C., Socher, R.: Augmented cyclic adversarial learning for low resource domain adaptation. In: 7th International Conference on Learning Representations, ICLR 2019, New Orleans, LA, USA, 6–9 May 2019 (2019)
13. Kingma, D.P., Ba, J.: Adam: a method for stochastic optimization. In: 3rd International Conference on Learning Representations, ICLR 2015, San Diego, CA, USA, 7–9 May 2015, Conference Track Proceedings (2015)
14. Koniusz, P., Tas, Y., Porikli, F.: Domain adaptation by mixture of alignments of second-or higher-order scatter tensors. In: Proceedings of the IEEE Conference on Computer Vision and Pattern Recognition, pp. 4478–4487 (2017)
15. Long, M., Cao, Y., Wang, J., Jordan, M.I.: Learning transferable features with deep adaptation networks. In: Proceedings of the 32nd International Conference on Machine Learning, ICML 2015, Lille, France, 6–11 July 2015, pp. 97–105 (2015)
16. Long, M., Wang, J., Ding, G., Sun, J., Yu, P.S.: Transfer feature learning with joint distribution adaptation. In: Proceedings of the IEEE International Conference on Computer Vision, pp. 2200–2207 (2013)
17. Long, M., Zhu, H., Wang, J., Jordan, M.I.: Unsupervised domain adaptation with residual transfer networks. In: Advances in Neural Information Processing Systems, pp. 136–144 (2016)
18. Long, M., Zhu, H., Wang, J., Jordan, M.I.: Deep transfer learning with joint adaptation networks. In: Proceedings of the 34th International Conference on Machine Learning, vol. 70, pp. 2208–2217 (2017). JMLR. org

19. Luo, Z., Zou, Y., Hoffman, J., Fei-Fei, L.F.: Label efficient learning of transferable representations across domains and tasks. In: Advances in Neural Information Processing Systems, pp. 165–177 (2017)
20. Motiian, S., Jones, Q., Iranmanesh, S., Doretto, G.: Few-shot adversarial domain adaptation. In: Advances in Neural Information Processing Systems, pp. 6670–6680 (2017)
21. Peng, M., Zhang, Q., Jiang, Y.g., Huang, X.J.: Cross-domain sentiment classification with target domain specific information. In: Proceedings of the 56th Annual Meeting of the Association for Computational Linguistics (Volume 1: Long Papers), pp. 2505–2513 (2018)
22. Ruder, S., Plank, B.: Strong baselines for neural semi-supervised learning under domain shift. In: Proceedings of the 56th Annual Meeting of the Association for Computational Linguistics, ACL 2018, Melbourne, Australia, 15–20 July 2018, Volume 1: Long Papers, pp. 1044–1054 (2018)
23. Saito, K., Kim, D., Sclaroff, S., Darrell, T., Saenko, K.: Semi-supervised domain adaptation via minimax entropy. arXiv preprint arXiv:1904.06487 (2019)
24. Shimodaira, H.: Improving predictive inference under covariate shift by weighting the log-likelihood function. J. Stat. Plann. Inference **90**(2), 227–244 (2000)
25. Snell, J., Swersky, K., Zemel, R.: Prototypical networks for few-shot learning. In: Advances in Neural Information Processing Systems, pp. 4077–4087 (2017)
26. Tzeng, E., Hoffman, J., Darrell, T., Saenko, K.: Simultaneous deep transfer across domains and tasks. In: Proceedings of the IEEE International Conference on Computer Vision, pp. 4068–4076 (2015)
27. Tzeng, E., Hoffman, J., Saenko, K., Darrell, T.: Adversarial discriminative domain adaptation. In: Proceedings of the IEEE Conference on Computer Vision and Pattern Recognition, pp. 7167–7176 (2017)
28. Tzeng, E., Hoffman, J., Zhang, N., Saenko, K., Darrell, T.: Deep domain confusion: maximizing for domain invariance. arXiv preprint arXiv:1412.3474 (2014)
29. Wang, M., Deng, W.: Deep visual domain adaptation: a survey. Neurocomputing **312**, 135–153 (2018)
30. Wold, S., Esbensen, K., Geladi, P.: Principal component analysis. Chemometr. Intell. Lab. Syst. **2**(1–3), 37–52 (1987)
31. Yao, T., Pan, Y., Ngo, C.W., Li, H., Mei, T.: Semi-supervised domain adaptation with subspace learning for visual recognition. In: Proceedings of the IEEE Conference on Computer Vision and Pattern Recognition, pp. 2142–2150 (2015)
32. Zellinger, W., Grubinger, T., Lughofer, E., Natschläger, T., Saminger-Platz, S.: Central moment discrepancy (CMD) for domain-invariant representation learning. In: 5th International Conference on Learning Representations, ICLR 2017, Toulon, France, 24–26 April 2017, Conference Track Proceedings (2017)
33. Zhao, H., des Combes, R.T., Zhang, K., Gordon, G.J.: On learning invariant representations for domain adaptation. In: Proceedings of the 36th International Conference on Machine Learning, ICML 2019, 9–15 June 2019, Long Beach, California, USA, pp. 7523–7532 (2019)
34. Zhao, H., Zhang, S., Wu, G., Moura, J.M., Costeira, J.P., Gordon, G.J.: Adversarial multiple source domain adaptation. In: Advances in Neural Information Processing Systems, pp. 8559–8570 (2018)
35. Zhuang, F., Cheng, X., Luo, P., Pan, S.J., He, Q.: Supervised representation learning: Transfer learning with deep autoencoders. In: Twenty-Fourth International Joint Conference on Artificial Intelligence (2015)

MTGCN: A Multitask Deep Learning Model for Traffic Flow Prediction

Fucheng Wang[1], Jiajie Xu[1,2(✉)], Chengfei Liu[3], Rui Zhou[3], and Pengpeng Zhao[1]

[1] Institute of Artificial Intelligence, A School of Computer Science and Technology, Soochow University, Suzhou, China
wfcandni@163.com, {xujj,ppzhao}@suda.edu.cn
[2] State Key Laboratory of Software Architecture (Neusoft Corporation), Suzhou, China
[3] Swinburne University of Technology, Melbourne, Australia
{cliu,rzhou}@swin.edu.au

Abstract. The prediction of traffic flow is of great importance to urban planning and intelligent transportation systems. Recently, deep learning models have been applied to study this problem. However, there still exist two main limitations: (1) They do not effectively model dynamic traffic patterns in irregular regions; (2) The traffic flow of a region is strongly correlated to the transition-flow between different regions, while this issue is largely ignored by existing approaches. To address these issues, we propose a multitask deep learning model called MTGCN for a more accurate traffic flow prediction. First, to process the input traffic network data, we propose using graph convolution in place of traditional grid-based convolution to model spatial dependencies between irregular regions. Second, as original graph convolution can not well respond to traffic dynamics, we design a novel attention mechanism to capture dynamic traffic patterns. At last, to obtain a more accurate prediction result, we integrate two correlated tasks which respectively predict two types of traffic flows (region-flow and transition-flow) as a whole, by combining the representations learned from each task in a rational way. We conduct extensive experiments on two real-world datasets and the results show that our proposed method achieves better performance compared with other baseline models.

Keywords: Traffic flow prediction · Graph convolutional networks · Multitask learning · Spatio-temporal

1 Introduction

In recent years, trajectory data have been largely accumulated in enterprises. These data contain abundant traffic information and can help us understand traffic dynamics in greater detail [8,24]. Traffic flow prediction is recognized as an important problem in the construction of Intelligent Transportation Systems

Y. Nah et al. (Eds.): DASFAA 2020, LNCS 12112, pp. 435–451, 2020.
https://doi.org/10.1007/978-3-030-59410-7_30

(ITS) [2,17,23]. It aims at predicting traffic flows in a region at a given time interval. Accurate prediction is of great significance as it benefits many applications such as traffic control, risk assessment and public safety [21].

Most existing methods focus on predicting traffic flows in regular grid-like regions, using convolutional neural networks (CNN) to model spatial dependencies [7,12,21]. However, the regions in a city are actually separated by road networks, which naturally segment urban areas into sub-regions with varying sizes and shapes [19]. Compared with regular grid-like regions, irregular regions tend to have more semantic meanings such as the functions (e.g. residential areas and entertainment areas) in a city. Therefore, predicting traffic flows in such irregular regions becomes more meaningful. However, contrary to traditional grid-based data, irregular regions are actually described in graph-based structure. Hence, CNN based models which require input to be grid-like data are not applicable in this scenario. To tackle this problem, Graph Convolutional Networks have been introduced by researchers [4,14,18].

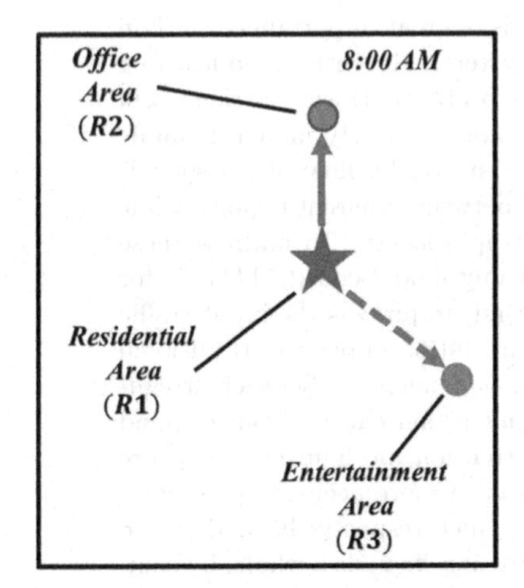

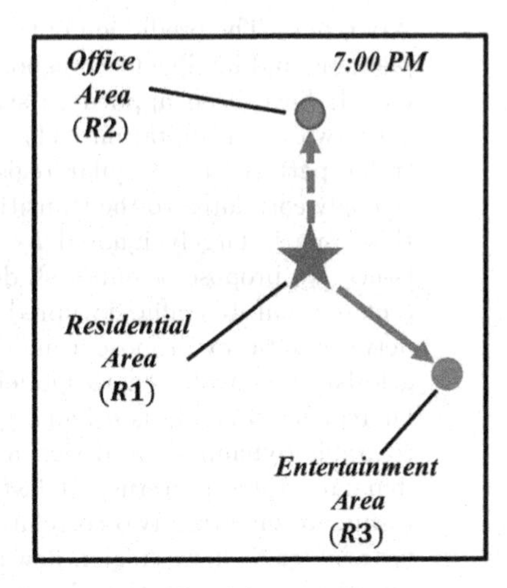

Fig. 1. Dynamic spatial dependencies at different times

However, traffic flows change over time, showing great dynamics and uncertainties, so do the spatial dependencies at different times. As Fig. 1 shows, at 8 a.m. in the morning, traffic flows tend to transfer from residential area (R1) to office area (R2) as most people go to work by vehicles (assuming that it is a workday). When it becomes 7 p.m. in the evening, there will be a great many people flocking to entertainment area (R3) for recreation. Based on this observation, we can find that the outflow of R1 is more related to the inflow of R2 than that of R3 at 8 a.m. While at 7 a.m., it is exactly the opposite situation. However, contrary to the above observation, most existing GCN models assume that the contributing weights of R3 and R2 w.r.t the outflow of R1 are fixed and do not change with time, which inevitably leads to inaccurate prediction.

Furthermore, we find that the traffic flow of a region is strongly correlated to the transition-flow between different regions. As Fig. 1 shows, R1 has two out-transitions which are respectively directed to R2 and R3. To be more specific, the outflow of R1 is the aggregated result of its out-transitions to R2 and R3, while the out-transitions to R2 and R3 depict the individual correlations between pair-wise regions. Most studies merely focus on the region-flow other than the transition-flow, which results in incomprehensive prediction. In addition, although there exist so many combinations of transitions, some of them may not occur in reality and some combinations do not happen for many times, which makes the transition-flow highly sparse and poses a challenge for training deep learning models.

To tackle the above challenges, we propose a multitask deep learning model called MTGCN for a more accurate traffic flow prediction. First, to make it accessible to predict traffic flows over irregular regions, we adopt graph convolution to process the graph-structured data. As traditional convolution requires input to be standard grid-based data, graph convolution can process both grid-like and graph-based inputs, thus becoming more general than other models. Second, to obtain a more accurate prediction, we consider integrating two correlated tasks which respectively predict two types of traffic flows (region-flow and transition-flow) as a whole. Since the representation learned from each task merely depicts traffic flows from one perspective, we combine these two tasks in a rational way for a more general representation of latent traffic patterns. Specifically, we design three components in each prediction task to model three-granularity temporal properties: neighbouring, daily-periodicity and weekly-periodicity. Furthermore, we add an embedding layer to deal with the transition sparsity in the task of transition-flow prediction. We summarize our contributions as follows:

- We formulate the traffic flow prediction problem in irregular regions and adopt graph convolution to model spatial dependencies.
- We incorporate attention mechanism into graph convolution, by which our model effectively captures traffic dynamics in traffic flow.
- By integrating two tasks which respectively predict region-flow and transition-flow as a whole, we propose a multitask deep learning model called MTGCN for traffic flow prediction.
- We evaluate our model on two real-world datasets. The results show that our proposed method outperforms several baseline approaches.

The remaining parts are organized as follows: First, we review the related work about traffic flow prediction and graph convolution in Sect. 2. Then, we give several definitions and formulate the problem in Sect. 3. In Sect. 4, we introduce our solution and related techniques in detail. Finally, we show the extensive experimental results on two real-world datasets in Sect. 5 and conclude our work in Sect. 6.

2 Related Work

In this section, we review the related work about traffic flow prediction and graph convolution.

2.1 Traffic Flow Prediction

Being an important problem in urban planning, traffic flow prediction has been studied for many years. A number of researchers viewed it as a time series problem, using ARIMA and H-ARIMA [11] to predict future flows. Some supervised learning models, such as LR, VAR and SVR [16] considered it to be a regression problem. Although these methods modeled temporal dependencies very well, they ignored the spatial correlations in the surrounding areas, thus making inaccurate predictions.

Recently, deep learning models have been widely used in traffic flow prediction. Convolutional neural networks (CNN) can effectively extract spatial correlation based on grid-like data, thus many researchers designed CNN based models to predict traffic flows. DeepST [22] viewed traffic flow information as heat maps, using convolution operations to model spatial dependencies. In addition, it summarized temporal properties into three categories: temporal closeness, period and trend. Furthermore, ST-ResNet [21] proposed a residual framework to model long-range spatial dependencies throughout a city. Meanwhile, it also considered external factors such as weather, holiday and events for a more accurate prediction. Based on ST-ResNet, [20] proposed a multitask deep learning framework to simultaneously predict inflow/outflow and transitions between regions. DeepSTN+ [7] proposed a ConvPlus operation in place of ordinary convolution to directly capture long-range spatial dependencies. Furthermore, it introduced PoI distributions to depict the attributes of different regions. Apart from above CNN based models, the combination of recurrent models with CNN such as ConvLSTM [13] and Periodic-CRN [25] were also proposed, but the training of recurrent models always consumed a lot of time.

However, CNN based models require input to be grid-like data. When input becomes graph-structured data, they may not be suitable any more. Particularly, regions in a city are actually irregular and are of varying shapes and sizes. The above models ignore the real conditions and are probably not perfect to be applied to real-world applications.

2.2 Graph Convolution

Recently, generalizing CNN to work on graph-structured data has been studied by many researchers. There are mainly two types of methods to implement graph convolution: spatial methods and spectral methods. The spatial methods focus on directly performing convolution operations on each node of a graph. [10] proposes a heuristic linear method to select the neighbours of each node. Contrary to spatial methods, spectral methods carry out convolution operations in spectral domains [1,6]. However, this method requires Laplacian eigenvalue decomposition, which is computationally expensive. To solve this problem, [3] proposed a fast approximation method by a multi-order chebyshev polynomial (ChebNet). [15] proposed using self-attention mechanism to address the shortcomings of graph convolution. Later on, researchers began to utilize GCN to predict traffic flow. STGCN [18] was proposed to use graph convolutional networks to model spatio-temporal dependencies based on graph-structured data. Then, MVGCN [14] and ASTGCN [4] were further proposed to effectively integrate spatial modeling and temporal modeling as a whole.

Based on our observations, the region-flow is strongly correlated to the transition-flow between different regions, as these two types of traffic flows mutually influence each other. The existing methods overlooked this important relation, which could potentially increase the prediction accuracy. In our MTGCN, we integrate two correlated tasks which respectively predict two types of traffic flows (region-flow and transition-flow) as a whole, enabling our model to obtain a more comprehensive understanding of traffic patterns, thus achieving better performance than other models.

3 Problem Definition

In this section, we first give several definitions related to traffic flow prediction and then formulate the problem.

Definition 1 (Urban graph). An urban graph is denoted as $G = (V, E, A)$, where V represents the set of regions and E represents the edges between regions (if two regions are geographically connected, there will be an edge between them). A is a binary unweighted adjacency matrix. Figure 2(a) shows an example of urban graph, where $V = \{R_1, R_2, R_3, R_4\}$, E consists of five edges and the adjacency matrix A can be obtained by V and E.

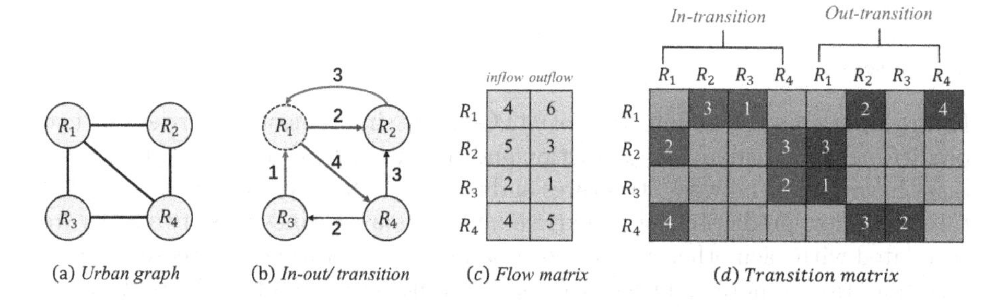

(a) *Urban graph* (b) *In-out/ transition* (c) *Flow matrix* (d) *Transition matrix*

Fig. 2. Region-flow and transition-flow in different regions

Definition 2 (Inflow/Outflow). In this paper, we mainly study two types of region-flows: inflow and outflow. At a given time interval, inflow is the total number of traffic flows entering a region while outflow is the total number of traffic flows leaving a region for other places. At time interval t, we use $X_t \in R^{N\times C}$ to denote the region-flow of N nodes. Specifically, $C = 2$. $X_t[i,0]$ and $X_t[i,1]$ respectively denote inflow and outflow of node i at time interval t. Then we can obtain a flow matrix like Fig. 2(c) from Fig. 2(b) which shows the traffic flow transitions between different regions. Taking R_1 as an example, the transition from R_1 to R_2 and that from R_1 to R_4 constitute the outflow of R_1, while the transition from R_2 to R_1 and that from R_3 to R_1 comprise the inflow of R_1.

Definition 3 (Transition-flow). We mainly study two types of transition-flows: in-transition and out-transition. At time interval t, we use $Z_t \in R^{N\times 2N}$ to denote the transition-flow of N nodes. Specifically, $Z_t[i,k]$ $(k = 0,1,...,N-1)$ denotes the in-transitions from all N nodes to node i and $Z_t[i,k]$ $(k = N, N+1,...,2N-1)$ denotes the out-transitions to other nodes from node i. Then we can obtain a transition matrix like Fig. 2(d) from Fig. 2(b). Taking R_1 as an example, it has up to 8 possibilities of transition-flows. The in-transition of R_1 includes the transitions from R_2 and R_3 while the out-transition is composed of the transitions from R_1 to R_2 and R_1 to R_4.

Problem Formulation (Traffic flow prediciton). Given a graph $G = (V, E, A)$ and historical observations of region-flow and transition-flow $\{X_t, Z_t | \text{t} = 0,...,\text{T}-1\}$, predict region-flow $X = \{X_j | \text{j}=\text{T},...,\text{T}+z-1\}$ and transition-flow $Z = \{Z_j | \text{j}=\text{T},...,\text{T}+z-1\}$, where z is the number of time intervals to be predicted.

4 MTGCN: Multi-task Graph Convolutional Network

In this section, we introduce our multitask framework in detail. As it seamlessly integrates two tasks of predicting region-flow and transition-flow as a whole, our proposed MTGCN is able to obtain a comprehensive prediction result.

4.1 Overview

Figure 3 presents the architecture of MTGCN. It simultaneously carries out two prediction tasks: region-flow prediction and transition-flow prediction. These two tasks have similar network structures and the only difference is that the model of transition-flow prediction has an extra embedding layer. Since these two tasks are correlated with each other, we aim to combine representations learned from each task to obtain a more comprehensive understanding of latent traffic patterns.

To be specific, our MTGCN mainly consists of the following components: an embedding layer, a spatio-temporal modeling module and a fusion layer. First, the embedding layer is used to deal with the transition sparsity problem. It encodes the sparse and high-dimensional transition-flow into a low-dimensional vector. Second, in each prediction task, we model spatio-temporal dependencies in traffic data. Specifically, for temporal modeling, we summarize temporal properties from three-granularity: neighbouring, daily-periodicity and weekly-periodicity. As for spatial modeling, we design a novel AGC unit which is composed of a graph convolution layer integrated with attention mechanism. Since one AGC unit only captures local correlation, we stack L such units to capture long-range spatial correlation. Finally, the representations learned from each task are combined by a fusion mechanism. After fusion, we obtain a more comprehensive representation of latent traffic patterns and utilize it to predict both region-flow and transition-flow.

4.2 Embedding Layer

According to our observations, the transition-flow among regions reflects the dynamic region-to-region correlation. To obtain a more accurate prediction, it is necessary to take this transition-flow feature into account.

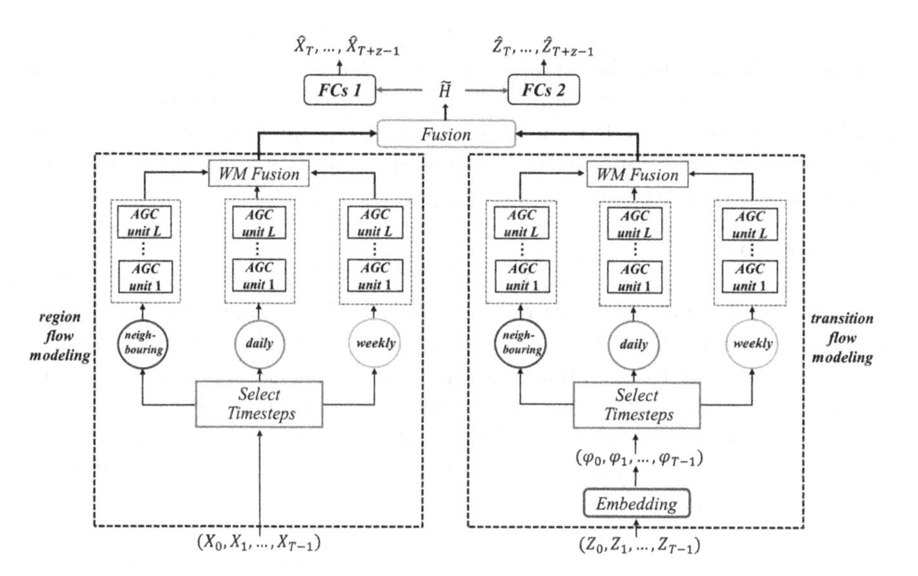

Fig. 3. Architecture of MTGCN. AGC unit: attentional graph convolution unit; FCs: Fully-connected layers; WM Fusion: Weight-matrix based fusion

Furthermore, we find that the transition-flow that really occurs between regions only accounts for a small portion. Hence, the transition-flow becomes extraordinarily sparse and high-dimensional (because of N nodes). This data sparsity really poses a challenge for training deep learning models. Therefore, we aim to adopt embedding strategy to solve this problem.

Transition Embedding. The embedding operation is aimed to encode the individual transition-flow into a low-dimensional feature vector, which can be used to reveal the region-to-region correlation in traffic flow. Specifically, the embedding process can be written as follows:

$$\varphi_t = \sigma(Z_t W_z + b_z) \tag{1}$$

where $W_z \in R^{2N \times k}$ and $b_m \in R^k$ are learnable parameters, $Z_t \in R^{N \times 2N}$ is the transition-flow matrix at time interval t, $\sigma(\cdot)$ is a non-linear function (e.g. ReLU). In our experiment, we adopt a two-layer feedforward neural network to implement embedding operation. After embedding, we obtain a k-dimension feature vector of transition-flow.

4.3 Spatio-Temporal Modeling

In this section, we aim to model complex spatio-temporal dependencies in traffic flows. To be specific, we model traffic patterns mainly from two perspectives: temporal modeling and spatial modeling. Moreover, to dynamically adjust the correlation strength between different regions, we incorporate attention mechanism into our model.

Temporal Modeling. We model temporal dependencies mainly from three granularities: neighbouring, daily-periodicity and weekly-periodicity. Each temporal property corresponds to an independent component and all of them share the same structure. Supposing the current time interval is T−1 and we aim to predict next z timesteps (i.e. T, T + 1, ..., T + z − 1).

1) neighbouring:
   $\Psi_n = (X_{T-l_n}, X_{T-(l_n-1)}, ..., X_{T-1}) \in R^{l_n \times N \times C}$. Ψ_n is composed of neighbouring timesteps which are relevant to z predicted timesteps and l_n is the length of neighbouring timesteps. Intuitively, the traffic flow at current time is more relevant to neighbouring timesteps than that of distant timesteps. For instance, the traffic congestion occurring at highway road will inevitably affect the subsequent traffic conditions. Based on this observation, we design neighbouring component to capture neighbouring correlation in traffic data.
2) daily-periodicity:
   $\Psi_d = (X_{T-l_d \cdot d}, ..., X_{T-l_d \cdot d+z-1}, ..., X_{T-d}, ..., X_{T-d+z-1}) \in R^{z \cdot l_d \times N \times C}$. Ψ_d consists of daily timesteps w.r.t. z predicted timesteps, l_d is the length of daily-periodicity and d is the daily timespan. We find that daily routines always follow repeated patterns, e.g.. peak hours in a day are similar during a week. Thus, we utilize the daily-periodicity component to capture day-level temporal correlation in traffic flow.
3) weekly-periodicity:
   $\Psi_w = (X_{T-l_w \cdot w}, ..., X_{T-l_w \cdot w+z-1}, ..., X_{T-w}, ..., X_{T-w+z-1}) \in R^{z \cdot l_w \times N \times C}$. Ψ_w comprises weekly timesteps w.r.t. z predicted timesteps, l_w is the length of weekly-periodicity and w is the weekly timespan. Generally, the traffic pattern of one day shows great similarity during continuous weeks. Therefore, we design this weekly-periodicity component to model week-level temporal dependencies in traffic flow data.

Spatial Modeling. Due to the geographical connection among urban areas, the traffic flow of a region is naturally affected by that of nearby regions. Furthermore, owing to the rapid development of urban transportation systems, people can reach distant areas just within a short time. Thus, we need to consider both near and distant correlations in spatial modeling. Particularly, to model spatial dependencies between irregular regions, we propose using graph convolutional networks in spatial modeling.

Contrary to traditional grid-based convolution on image-like data, researchers define graph convolution in the spectral domain [1], which is denoted as $*_G$. By this definition, the graph signal $\Phi_i \in R^{N \times F}$ at an arbitary time interval i (F is the dimension of features) is accompanied with the filter $g_w = diag(w)$ which is parameterized by $w \in R^N$. Then the convolution in the spectral domain can be written as follows:

$$g_w *_G \Phi_i = g_w(L)\Phi_i = g_w(U \Lambda U^T)\Phi_i \tag{2}$$

where $U \in R^{N \times N}$ is Fourier basis, $\Lambda \in R^{N \times N}$ is the diagonal matrix composed of eigenvalues of the normalized Laplacian matrix $L = I_N - D^{\frac{1}{2}} A D^{-\frac{1}{2}} = U \Lambda U^T \in R^{N \times N}$. I_N is the identity matrix, $D \in R^{N \times N}$ is a diagonal matrix and $D_{ii} = \sum_j A_{ij}$. Taking graph signal at time interval i as an example, the graph Fourier transformation is defined as $\tilde{\Phi}_i = U^T \Phi_i$ while the inverse Fourier transformation is $\Phi_i = U \tilde{\Phi}_i$. Based on the above formulae, we can understand graph convolution as follows: Φ_i is transformed into the spectral domain, operated with g_w and then be transformed back. However, this convolution operation needs to perform eigenvalue decomposition, which is computationally expensive. Therefore, Chebyshev polynomials are proposed for a fast approximation:

$$g_w *_G \Phi_i = g_w(L)\Phi_i = \sum_{k=0}^{K-1} w_k T_k(\tilde{L})\Phi_i \tag{3}$$

where $T_k(\tilde{L})$ is the k-th order Chebyshev polynomial of scaled Laplacian matrix $\tilde{L} = \frac{2}{\lambda_{max}} L - I_N$, λ_{max} is the maximum eigenvalue of L and $w_k \in R^K$ is a vector of Chebyshev coefficients. The recursive definition of the Chebyshev polynomial is $T_k(\tilde{L}) = 2\tilde{L}T_{k-1}(\tilde{L}) - T_{k-2}(\tilde{L})$, $T_0(\tilde{L}) = I_N$ and $T_1(\tilde{L}) = \tilde{L}$. For each node, the K-order chebyshev polynomials aim to extract features of 0 to $(K-1)$-hop neighbours. Intuitively, we can stack several graph convolution layers to capture spatial correlation between distant areas. In our experiment, we further adopt a residual structure [5] to speed up the training process of our network. The residual structure is shown in Fig. 4. Assuming that the input of l-th layer is H^l, the output H^{l+1} is computed by

$$H^{l+1} = f(H^l) + H^l \tag{4}$$

where f is the non-linear transformation in AGC unit.

Graph Based Attention. The K-order chebyshev polynomial based graph convolution can effectively extract information from surrounding $(K-1)$-hop

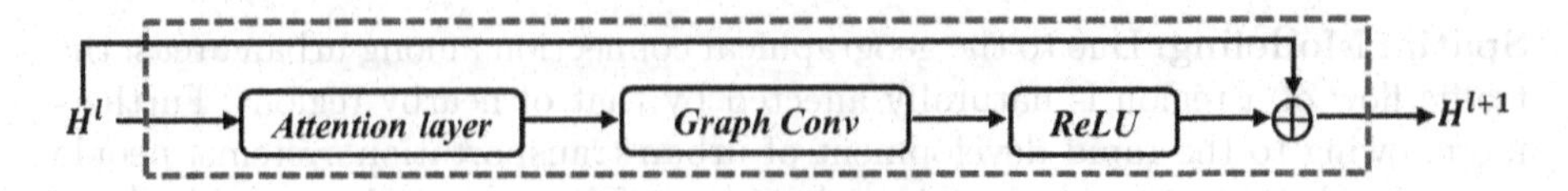

Fig. 4. AGC unit with residual structure

neighbours. However, traffic conditions change over time, so do the spatial dependencies among different regions. The Laplacian matrix used in graph convolution assigns fixed contributing weights to the neighbours of a target node, which can not depict the differences among regions. Therefore, we adopt the attention mechanism to adaptively adjust the spatial correlations between regions.

Supposing that the l-th graph convolution layer transforms input F features into high-level F' features. Here, we take node i and node j as an example, $h_i \in R^F$ and $h_j \in R^F$ are corresponding feature vectors, then the operations in the attention layer can be written as follows:

$$\gamma_{i,j} = a(Wh_i, Wh_j) \tag{5}$$

$$\gamma'_{i,j} = softmax_j(\gamma_{i,j}) = \frac{exp(\gamma_{i,j})}{\sum_{j=1}^{N} exp(\gamma_{i,j})} \tag{6}$$

where $W \in R^{F' \times F}$, a is a single-layer feedforward network which is parameterized by a $2F'$-dimension weight vector. $\gamma_{i,j}$ denotes the correlation strength between region i and region j. To make coefficients easily comparable across different nodes, we normalize them across all j using the softmax function.

After obtaining the attention matrix, we fuse it with initial Laplacian matrix to dynamically adjust correlation strength between regions. To be specific, $\gamma' \odot T_k(\tilde{L})$ will replace initial $T_k(\tilde{L})$ when implementing graph convolution.

WM Fusion. As mentioned above, the traffic flow of a region is affected by three-granularity temporal properties, but at varying time intervals, the degree of influence may be quite different. Hence, we propose a weight-matrix based fusion strategy to assign different weights to these three temporal properties. Taking the outputs of transition-flow modeling as an example, the result of WM fusion is computed by

$$H_{trans} = W_n \odot O_n + W_d \odot O_d + W_w \odot O_w \tag{7}$$

Where W_n, W_d and W_w are learnable parameters which reflect weights of different temporal properties, O_n, O_d and O_w are respectively the outputs of the above three components. Similarly, we can obtain the result of region-flow modeling H_{region} by same operations.

4.4 Representation Fusion

After spatio-temporal modeling, we obtain two representations: $H_{region} \in R^{N \times C}$ and $H_{trans} \in R^{N \times D}$, which respectively denote the representations learned from region-flow modeling and transition-flow modeling. Then we fuse them for a more comprehensive understanding of traffic patterns. In this paper, we mainly consider the following fusion strategies:

Sum Fusion. The sum fusion mechanism directly sums up H_{region} and H_{trans}, the output is as follows:

$$\tilde{H} = H_{region} + H_{trans} \tag{8}$$

Apparently, these two inputs H_{region} and H_{trans} must have the same shape, i.e. C must be equal to D.

Concatenation Fusion. The concatenation fusion has no constraint that two inputs must have the same shape. To be specific, it concatenates two inputs in the feature dimension. The process can be written as

$$\tilde{H}[r, : C] = H_{region}[r, : C] \quad r = 0, 1, ..., N \tag{9}$$

$$\tilde{H}[r, C : C + D] = H_{trans}[r, : D] \quad r = 0, 1, ..., N \tag{10}$$

It is obvious that the concatenation fusion extends the dimension of above feature vectors. It also reserves initial information from region-flow and transition-flow.

Gating Fusion. Furthermore, we propose a gating fusion method as follows:

$$\tilde{H} = \lambda \odot H_{region} + (1 - \lambda) \odot H_{trans} \tag{11}$$

$$\lambda = \sigma(H_{region}W_{region} + H_{trans}W_{trans} + b_{\lambda}) \tag{12}$$

where W_{region}, W_{trans} and b_{λ} are learnable parameters, $\odot$ represents the element-wise product, $\sigma(\cdot)$ denotes the sigmoid activation function. The gating fusion mechanism adaptively controls the influence of these two signals at each time interval. Meanwhile, it also requires two inputs to have the same shape.

After fusion, we obtain a more comprehensive representation of traffic flow: $\tilde{H} \in R^{N \times F}$, where N equals to the region amount and F is the feature dimension after fusion. Then, we adopt fully-connected networks to adapt the feature dimension of this representation to each prediction task. To be specific, the feature dimension F is mapped to 2 (region-flow) and $2N$ (transition-flow).

4.5 Multi-task Loss

In this paper, we have two prediction tasks: region-flow prediction and transition-flow prediction. Let $\hat{X}_T$ and $\hat{Z}_T$ be the predicted results of region-flow and transition-flow, X_T and Z_T are corresponding true values. We use mean squared error as our loss function, which is defined as follows:

$$L(\Theta) = \lambda_{region}\|X_T - \hat{X}_T\|_2^2 + \lambda_{trans}\|Z_T - \hat{Z}_T\|_2^2 \tag{13}$$

where Θ are all learnable parameters in our MTGCN, λ_{region} and λ_{trans} are respectively weight coefficients of two tasks.

5 Experiments

In this section, we mainly conduct experiments based on taxi trajectories from Beijing and Chengdu to evaluate the effectiveness of MTGCN.

5.1 Datasets

We use two real-world datasets from Beijing and Chengdu[1] to evaluate our model. The details are as follows:

- **Beijing:** The trajectory data was collected from 1st June to 31st July in 2016. There are about 2 million trajectories covering the road network every day. In our experiment, we mainly extracted the area within the fourth ring road of Beijing and got 256 regions. 80% of the data are used to train our model and the remaining 20% are the test set.
- **Chengdu:** It is the trajectory data collected from 1st Nov to 30th Nov in 2016, which is published by Didi. There are more than 20 thousand trajectoris obtained every day and we got 64 regions within the central area of whole city. We use the last 6 days as test set and the others are used to train our model.

5.2 Settings

We implement our MTGCN model based on PyTorch framework. Traffic data are aggregated into every 10-min interval from raw data. We train our network with the following hyper-parameter settings: The number of terms of Chebyshev polynomial K is 3. The number of AGC unit is set to be 2. Mini-batch size is 64 and the learning rate is 0.0001. $l_n = 3, l_d = 1, l_w = 1$. We set $\lambda_{region} = \lambda_{trans} = 1$. For the transition fusion strategy, sum fusion achieves the best performance in all experiments. We use Adam optimizer to optimize the parameters in our network. In addition, data input are normalized by *Z-Score* method.

5.3 Baselines

We compare our model with the following baseline approaches:

- **HA:** Historical Average, which uses the average values of previous timesteps to predict the future values.
- **ARIMA:** Auto-Regressive Integrated Moving Average (ARIMA) is a well-known model for time series analysis and predicting future values.
- **SVR**[16]: Support Vector Regression (SVR) is a great regression method of powerful generalization ability.
- **LSTM** [9]: As a variant of RNN, LSTM can effectively capture long-range temporal dependencies among traffic flow data.
- **MVGCN** [14]: Multi-View Graph Convolutional Networks. We design a simplified version of MVGCN without considering external factors.
- **ASTGCN** [4]: A Spatio-Temporal Graph Convolutional Network which incorporates attention mechanism into traffic flow prediction.

[1] https://gaia.didichuxing.com.

5.4 Evaluation Metrics

In this paper, we measure the performance of different approaches by Root Mean-Squared Error (RMSE) and Mean Absolute Error (MAE). As most existing methods only predict region-flow, we merely compare the performance of region-flow prediction between different approaches. RMSE and MAE are computed as follows:

$$RMSE = \sqrt{\frac{1}{M}\sum_{i}(x_i - \hat{x}_i)^2} \tag{14}$$

$$MAE = \frac{1}{M}\sum_{i}|x_i - \hat{x}_i| \tag{15}$$

where M is the total number of all predicted values, x_i is the true value and $\hat{x}_i$ is the predicted value.

5.5 Performance Comparison

Table 1 gives a comprehensive comparison between MTGCN and several baseline models. The result differences between two datasets are owing to different total traffic flow volumes. Specifically, the average values of traffic flow volume are 270.58 (Beijing) and 23.88 (Chengdu) respectively. We can find that our MTGCN achieves the best performance on two datasets, showing 7.30%–11.28% improvement than the state-of-the-art model ASTGCN. We can also observe that the performance of HA method is not bad on two datasets, which proves that periodical patterns in traffic flow really exist. Traditional time series methods such as SVR and ARIMA are not greatly better than HA, because they assume that the change of traffic flow is stable and linear, which is totally different in real-world conditions. The well-known sequence model LSTM can effectively capture long-range temporal dependencies, but it ignores the spatial correlation between regions, thus showing worse performance than GCN based models.

Table 1. Performance comparison between different methods

Dataset	*Beijing*		*Chengdu*	
Metrics	RMSE	MAE	RMSE	MAE
HA	38.83	27.19	14.96	9.47
SVR	35.70	25.85	12.17	7.66
ARIMA	33.20	23.15	10.12	7.97
LSTM	31.58	21.27	10.98	8.73
MVGCN	29.37	19.79	5.68	3.91
ASTGCN	28.51	19.06	5.34	3.63
MTGCN	**25.38**	**16.91**	**4.95**	**3.35**

Among three GCN based models, the simplified MVGCN takes spatio-temporal dependencies into account, yet ignoring the dynamics in spatial correlation and the transition-flow which is correlated with region-flow, thus performing worst among them. ASTGCN proposes a spatio-temporal attention mechanism to model dynamic spatial dependencies, yet neglecting the correlated task of transition-flow prediction. Our proposed MTGCN comprehensively considers traffic dynamics and seamlessly integrates two models which respectively predict two types of traffic flows, thus showing the best performance on two datasets.

5.6 Influence of Attention and Multi-task Framework

To prove the effectiveness of proposed attention mechanism and multitask framework, we also design another two GCN models: one is called TSGCN (Three-input-Stream GCN), which gets rid of attention mechanism and multitask framework; the other is called AGCN (Attentional GCN), which considers attention mechanism but only considers one prediction task.

Table 2. RMSE comparison between different GCN models

Dataset	TSGCN	AGCN	**MTGCN**
Beijing	28.97	27.42	**25.38**
Chengdu	5.71	5.28	**4.95**

As Table 2 shows, TSGCN neither considers the traffic dynamics in modeling spatial dependencies nor utilizes the correlated task of predicting transition-flow, thus performing worst among three models. Moreover, by incorporating attention mechanism into spatio-temporal modeling, AGCN successfully captures traffic dynamics in spatial correlation and obtains about 5.4%–7.5% performance improvement than TSGCN. Furthermore, our proposed MTGCN not only considers traffic dynamics in spatial dependencies, but also combines two correlated prediction tasks to form a multitask framework. By this multitask framework, MTGCN obtains a more comprehensive understanding of latent traffic patterns than other two models, achieving 6.3%–7.4% improvement than AGCN.

5.7 Results on Multi-step Prediction

The performance comparison of multi-step prediction are as follows:

Our proposed MTGCN can be applied to make both short-term and long-term predictions. In our experiment, we find that with the increase of time intervals to be predicted, the above GCN based models show relatively stable performance, while the performance of other baseline models fluctuates a lot. Therefore, to prove the stability and effectiveness of our proposed MTGCN on multi-step prediction, we mainly compare the performance of the above GCN

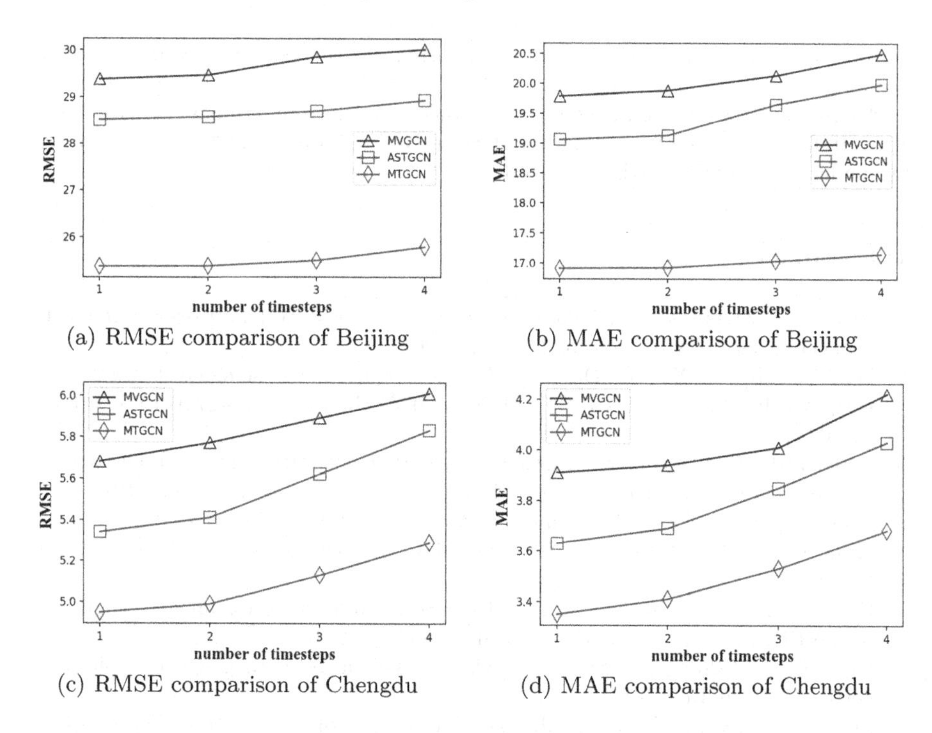

(a) RMSE comparison of Beijing

(b) MAE comparison of Beijing

(c) RMSE comparison of Chengdu

(d) MAE comparison of Chengdu

Fig. 5. Multi-step prediction

based models. Particularly, the initial time interval is 10-min, we predict traffic flow of future 40 min at most.

As is shown in Fig. 5, with the increase of timesteps to be predicted, the performance of all GCN based models become worse. Obviously, this is mainly caused by the decrease of correlation between time intervals being predicted and current moment. Nevertheless, our MTGCN always achieves the best performance compared with simplified MVGCN and ASTGCN on Beijing and Chengdu dataset. Furthermore, our proposed model shows highly stability regardless of increase of the timesteps to be predicted.

6 Conclusion

In this paper, we propose a novel multi-task based graph convolutional network called MTGCN for traffic flow prediction. As traditional convolution is not applicable to graph-structured data, we use graph convolution to model spatial dependencies in irregular regions. Moreover, to obtain a more accurate prediction result, we adopt attention mechanism and propose a multitask framework which integrates the prediction of region-flow and transition-flow as a whole. Finally, we conduct extensive experiments on two real-world datasets and the results show that our proposed method is superior to several baseline models.

Acknowledgements. This work was supported by the National Natural Science Foundation of China under Grant Nos. 61872258, 61772356, 61876117, and 61802273, the Australian Research Council discovery projects under grant numbers DP170104747, DP180100212, and the Open Program of State Key Laboratory of Software Architecture under item number SKLSAOP1801.

References

1. Bruna, J., Zaremba, W., Szlam, A., LeCun, Y.: Spectral networks and locally connected networks on graphs. In: ICLR (2014)
2. Dai, J., Liu, C., Xu, J., Ding, Z.: On personalized and sequenced route planning. World Wide Web **19**(4), 679–705 (2015). https://doi.org/10.1007/s11280-015-0352-2
3. Defferrard, M., Bresson, X., Vandergheynst, P.: Convolutional neural networks on graphs with fast localized spectral filtering. In: NIPS, pp. 3837–3845 (2016)
4. Guo, S., Lin, Y., Feng, N., Song, C., Wan, H.: Attention based spatial-temporal graph convolutional networks for traffic flow forecasting. In: AAAI, pp. 922–929 (2019)
5. He, K., Zhang, X., Ren, S., Sun, J.: Deep residual learning for image recognition. In: CVPR, pp. 770–778. IEEE Computer Society (2016)
6. Kipf, T.N., Welling, M.: Semi-supervised classification with graph convolutional networks. In: ICLR (2017)
7. Lin, Z., Feng, J., Lu, Z., Li, Y., Jin, D.: Deepstn+: Context-aware spatial-temporal neural network for crowd flow prediction in metropolis. In: AAAI, pp. 1020–1027 (2019)
8. Lv, Z., Xu, J., Zheng, K., Yin, H., Zhao, P., Zhou, X.: LC-RNN: a deep learning model for traffic speed prediction. In: IJCAI, pp. 3470–3476 (2018)
9. Ma, X., Tao, Z., Wang, Y., Yu, H., Wang, Y.: Long short-term memory neural network for traffic speed prediction using remote microwave sensor data. Transp. Res. Part C Emerg. Technol. **54**, 187–197 (2015). https://doi.org/10.1016/j.trc.2015.03.014
10. Niepert, M., Ahmed, M., Kutzkov, K.: Learning convolutional neural networks for graphs. In: ICML, pp. 2014–2023 (2016)
11. Pan, B., Demiryurek, U., Shahabi, C.: Utilizing real-world transportation data for accurate traffic prediction. In: ICDM, pp. 595–604. IEEE Computer Society (2012)
12. Peng, S., Shen, Y., Zhu, Y., Chen, Y.: A frequency-aware spatio-temporal network for traffic flow prediction. In: Li, G., Yang, J., Gama, J., Natwichai, J., Tong, Y. (eds.) DASFAA 2019. LNCS, vol. 11447, pp. 697–712. Springer, Cham (2019). https://doi.org/10.1007/978-3-030-18579-4_41
13. Shi, X., Chen, Z., Wang, H., Yeung, D., Wong, W., Woo, W.: Convolutional LSTM network: A machine learning approach for precipitation nowcasting. In: NIPS, pp. 802–810 (2015)
14. Sun, J., Zhang, J., Li, Q., Yi, X., Zheng, Y.: Predicting citywide crowd flows in irregular regions using multi-view graph convolutional networks. CoRR abs/1903.07789 (2019). http://arxiv.org/abs/1903.07789
15. Velickovic, P., Cucurull, G., Casanova, A., Romero, A., Liò, P., Bengio, Y.: Graph attention networks. In: ICLR (2018)
16. Wu, C., Ho, J., Lee, D.: Travel-time prediction with support vector regression. IEEE Trans. Intell. Transp. Syst. **5**(4), 276–281 (2004)

17. Xu, J., Chen, J., Zhou, R., Fang, J., Liu, C.: On workflow aware location-based service composition for personal trip planning. Future Gener. Comput. Syst. **98**, 274–285 (2019)
18. Yu, B., Yin, H., Zhu, Z.: Spatio-temporal graph convolutional networks: a deep learning framework for traffic forecasting. In: IJCAI, pp. 3634–3640 (2018)
19. Yuan, J., Zheng, Y., Xie, X.: Discovering regions of different functions in a city using human mobility and POIs. In: SIGKDD, pp. 186–194. ACM (2012)
20. Zhang, J., Zheng, Y., Sun, J., Qi, D.: Flow prediction in spatio-temporal networks based on multitask deep learning. In: TKDE, pp. 1 (2019). https://doi.org/10.1109/TKDE.2019.2891537
21. Zhang, J., Zheng, Y., Qi, D.: Deep spatio-temporal residual networks for citywide crowd flows prediction. In: AAAI, pp. 1655–1661 (2017)
22. Zhang, J., Zheng, Y., Qi, D., Li, R., Yi, X.: DNN-based prediction model for spatio-temporal data. In: SIGSPATIAL, pp. 92:1–92:4. ACM (2016)
23. Zhang, J., Wang, F., Wang, K., Lin, W., Xu, X., Chen, C.: Data-driven intelligent transportation systems: a survey. IEEE Trans. Intell. Transp. Syst. **12**(4), 1624–1639 (2011)
24. Zhao, J., Xu, J., Zhou, R., Zhao, P., Liu, C., Zhu, F.: On prediction of user destination by sub-trajectory understanding: a deep learning based approach. In: CIKM, pp. 1413–1422 (2018)
25. Zonoozi, A., Kim, J., Li, X., Cong, G.: Periodic-CRN: a convolutional recurrent model for crowd density prediction with recurring periodic patterns. In: IJCAI, pp. 3732–3738 (2018)

SAEA: Self-Attentive Heterogeneous Sequence Learning Model for Entity Alignment

Jia Chen[1], Binbin Gu[2], Zhixu Li[1,3](✉), Pengpeng Zhao[1], An Liu[1], and Lei Zhao[1]

[1] School of Computer Science and Technology, Soochow University, Suzhou, China
jchen0812@stu.suda.edu.cn, {zhixuli,ppzhao,anliu,zhaol}@suda.edu.cn
[2] University of California, Santa Cruz, USA
gu.binbin@hotmail.com
[3] IFLYTEK Research, Suzhou, China

Abstract. We consider the problem of entity alignment in knowledge graphs. Previous works mainly focus on two aspects: One is to improve the TransE-based models which mostly only consider triple-level structural information i.e. relation triples or to make use of graph convolutional networks holding the assumption that equivalent entities are usually neighbored by some other equivalent entities. The other is to incorporate external features, such as attributes types, attribute values, entity names and descriptions to enhance the original relational model. However, the long-term structural dependencies between entities have not been exploited well enough and sometimes external resources are incomplete and unavailable. These will impair the accuracy and robustness of combinational models that use relations and other types of information, especially when iteration is performed. To better explore structural information between entities, we novelly propose a Self-Attentive heterogeneous sequence learning model for Entity Alignment (SAEA) that allows us to capture long-term structural dependencies within entities. Furthermore, considering low-degree entities and relations appear much less in sequences prodeced by traditional random walk methods, we design a degree-aware random walk to generate heterogeneous sequential data for self-attentive learning. To evaluate our proposed model, we conduct extensive experiments on real-world datasets. The experimental results show that our method outperforms various state-of-the-art entity alignment models using relation triples only.

Keywords: Knowledge graph · Entity alignment · Degree-aware random walk

1 Introduction

Knowledge graphs (KGs) are playing an increasingly important role in many basic applications of artificial intelligence, like question answering and

Y. Nah et al. (Eds.): DASFAA 2020, LNCS 12112, pp. 452–467, 2020.
https://doi.org/10.1007/978-3-030-59410-7_31

recommendation systems. Many KGs including the most popular ones such as DBpedia [10], YAGO [16] and Freebase [2] are created by different methods in different languages, which makes them inevitably heterogeneous. Therefore, Entity alignment is proposed aiming to find entities in different KGs referring to the same real-world object.

Conventional methods [12,15,21] for entity alignment mainly rely on string similarities, such as entity names and attribute values. Nevertheless, the computation cost increases exponentially with the increase in the number of entities in KGs, and when it comes to a cross-lingual scenario, symbolic features are hard to be extracted. Recently, many methods leverage KG embedding techniques to address the above problems. Their key idea is to encode entities and relations into semantic space and find alignment between entities according to their embedding similarities in this space. TransE [3] is the most popular one in the KG embedding area which makes each relation triple (h, r, t) meet the requirement of $\mathbf{h}+\mathbf{r}\approx\mathbf{t}$ in vector space. For instance, MTransE [5] utilizes TransE to learn vector representations of entities in different KGs in separated embedding space and then maps each embedding vector of seed alignments to its corresponding counterpart via different transitions. However, the number of seed alignments is limited. SEA [13] is then proposed, leverages both seed alignments and plenty of unlabeled entities in a semi-supervised way and alleviates the effect of entity's degree differences in TransE-based KG embedding methods. The above two alignment models can be noted as TransE-based methods which are known as triple-level learning. It is obvious that there exists low expressiveness and inefficient information propagation in triple-level learning as it learns entity embeddings from triples and disseminates alignment information employing limited seed entities.

Many other models use iteration and extra resources to improve alignment results. IPTransE [26] and BootEA [18] design elaborate iteration strategies to acquire more high-confidence aligned entities for training. JAPE [17], AttrE [19], GCN-Align [22] and MultiKE [25] jointly model structure and external resources like attributes and names to refine entity embeddings. However, the above methods are still TransE-based methods where the potential of structural semantics has not been exploited well enough, not to mention that sometimes external information is incomplete and unreliable or unavailable, making models that combining external resources not applicable in some scenarios.

To better explore structural information between entities, RSNs [7] is proposed focusing on learning from relational paths which are composed of entities and relations alternately and have achieved state-of-the-art performance using relation triples only. However, its RNN-based sequence learning model is **difficult to learn dependencies between distant positions** as the output at time step t is determined by the current input and a mix of information of all the previous inputs. Thus, it cannot capture long-term dependencies between entities effectively and efficiently. Beyond that, when making use of random walk to sample relational paths, it ignores that random walk is **biased to entities of high degree** which results in extremely uneven information collection between long-tail entities and norm entities.

Considering limitations of the above method, we believe it is of critical importance to develop a model that can not only capture long-term dependencies between entities in an efficient manner but also lay more emphasis on low-degree entities. Towards this end, inspired by the new sequential model Transformer [20], which has achieved better performance than traditional recurrent models in machine translation tasks but have not been explored in KGs for entity alignment, we propose a brand-new **S**elf-**A**ttentive heterogeneous sequence learning model for **E**ntity **A**lignment (**SAEA**). The key idea of SAEA is to model dependencies without regard to the distance between items in the sequence by solely using self-attention mechanisms. However, it is necessary to customize due to the heterogeneity in sequences. Furthermore, unlike RNN-based models assuming that the next element in a relational path depends on the current input and hidden state which is inappropriate for paths in KGs, we adapt the original residual connection in Transformer to a special crossed residual module.

In addition, to generate ideal relational paths for sequence learning, we design a **degree-aware random walk**, which pays more attention to low-degree entities. It differs from previous random walks in the network embedding area in two points. One is that the sampled path is made up of two kinds of nodes—entities and relations. The other is we control the bias according to the frequency of relations and degree of entities. The output of the degree-aware random walk is then sent to our heterogeneous sequence learning model to learn embeddings. After sequence learning, we get alignment results via calculating embedding similarities.

Our main contributions in this paper are summarized as follows:

- We propose a self-attentive model for entity alignment. To the best of our knowledge, we are the first to manage to apply self-attention mechanisms to heterogeneous sequences in KGs for alignment.
- We also propose to generate heterogeneous sequences in KGs with a designed degree-aware random walk.
- We conduct extensive experiments on four real-world datasets and the result demonstrates our model outperforms TransE-based methods significantly and achieves state-of-the-art performance.

Roadmap. The rest of the paper is organized as follows: We discuss the related works in Sect. 2. After stating the problem definition in Sect. 3, our proposed method is presented in Sect. 4 and followed by reporting our empirical study in Sect. 5. We finally conclude the whole paper in Sect. 6.

2 Related Work

Entity alignment is a subtask of ontology alignment consisting of schema and instance matching. In this section, we first discuss the existing methods for entity alignment from two aspects and then introduce sequence learning which is at the core of our model.

2.1 Entity Alignment

According to the number of resources used, the existing entity alignment models fall into two categories.

Single-Resource-Based. In most cases, entity alignment can achieve high accuracy by simply comparing string similarities between entity names. However, as the number of entities increases, the computation cost is very high, not to mention that sometimes names are unavailable and difficult to compare in cross-lingual scenarios. In the past few years, much work has been done on the problem of KG embedding. As the entities can be represented by vectors, entity alignment can be implemented by comparing the distance between vectors. MTransE [5] encodes relation triples by utilizing TransE method and provides transitions for entity embeddings in one KG to map into the counterpart entity in the other KG. The loss function of MTransE is the weighted sum of TransE and alignment model. GCN-Align [22] employs graph convolutional networks (GCNs) to embed entities based on adjacent entities linked by different relations. To train the above two alignment models, a set of aligned entities between KGs are needed, leaving the abundant unaligned entities without consideration. SEA [13] is targeted at designing a semi-supervised entity alignment model learning from both aligned and unaligned entities rather than IPTransE [26] and BootEA [18] focus on acquiring more aligned entities using iteration strategy. These models use single-resource—structural information between entities.

Multi-Resource-Based. In addition to using structural information, many works leverage extra resources, such as OWL properties [8], entity descriptions [24] and attribute information of entities. JAPE [17] jointly embeds the structures of two KGs into a unified vector space and further refines it by leveraging attribute correlations in entities. KDcoE [4] iteratively trains two components embedding models on multilingual KG structures and entity descriptions respectively using co-training strategy. AttrE [19] generates attribute character embeddings for large numbers of attribute triples to shift the entity embeddings from two KGs into the same space by computing the similarity between entities based on their attributes. MultiKE [25] uses three different strategies to combine three representative views—name, relation and attribute features each of which uses an appropriate model to learn entity embeddings. Such methods are usually limited by the availability of the extra information in KGs. Therefore, we still aim at learning from structural information between entities as it is the intrinsic feature and stable. However, different from previous single-resource-based models which are hard to propagate alignment information effectively, our approach is capable of learning embeddings through relational paths that containing richer information than triples and neighbors.

2.2 Sequence Learning

In recent years, numerous works try to apply sequence models to graph structures, as sequence learning has been widely used in many fields, for instance, machine translation, speech recognition and music generation, and achieved excellent performance on many tasks. However, in the network and KG embedding area, inputs are naturally represented as graphs rather than sequence, thus existing Seq2Seq models [23] face a significant challenge in achieving appropriate sequence from graphs. DeepWalk [14] is the pioneer of generating sequences in graph. It introduces uniform random walk to sample paths while node2vec [6] uses biased random walk to explore nodes in a breadth-first-search and depth-first-search fashion. In this paper, we design a degree-aware random walk in KGs concentrating on low-degree entities and relations of low frequency in case the vast majority of sampled paths point to a small set of high-frequent entities. Besides, although the dominant sequence models are based on combining complicated recurrent or convolutional networks with attention mechanisms, their inherent sequential nature precludes parallelization. Therefore, stimulated by the Transformer [20], a purely self-attention based sequence model achieving the start-of-the-art performance and efficiency, we seek to build a sequential alignment model based upon it. The model needs to be specially designed due to the differences between entity alignment and traditional sequence tasks.

3 Problem Definition

KGs represent knowledge about the real-world entities as triples. It is a directed multi-relational graph where nodes denote entities and edges have directions and labels indicate that there exist a specific relation from one entity to the other.

Formally, a KG can be noted as $KG = \{E, R, T\}$, where E is the set of entities, R is the set of relations, and T is the set of triples, each of which is a triple (h, r, t), including the head entity h, the relation r and the tail entity t. Each triple can be presented as $(\mathbf{h}, \mathbf{r}, \mathbf{t})$ by KG embedding techniques in which $\mathbf{h}$, $\mathbf{r}$, $\mathbf{t}$ represent the embeddings of head entity h, relation r and tail entity t, respectively.

Given two KGs: $KG_1 = (E_1, R_1, T_1)$, $KG_2 = (E_2, R_2, T_2)$ and a set of pre-aligned entities $AS = \{(e_1^i, e_2^j)|e_1^i \in E_1^a, e_2^j \in E_2^a\}$ between KG_1 and KG_2 where E_1^a and E_2^a are subsets of E_1 and E_2 respectively, entity alignment is a task to find the remaining semantically same entity set $US = \{(e_1^i, e_2^j)|e_1^i \in E_1^u, e_2^j \in E_2^u\}$ where $E_1^u = E_1 \setminus E_1^a$ and $E_2^u = E_2 \setminus E_2^a$.

4 Our Approach

The framework of our proposed method is shown in Fig. 1: The whole process can be divided into two phases where the sequence generating phase is the prerequisite of the sequence learning phase. In phase one, the input layer is composed of two KGs and a set of known aligned entity pairs between them. The basic idea of our approach is to utilize an appropriate random walk method to generate heterogeneous sequences from KGs so that sequence learning model is allowed to be applied in our situation. After the well-designed phase two, entity alignment is predicted by measuring the cosine similarity in the embedding space.

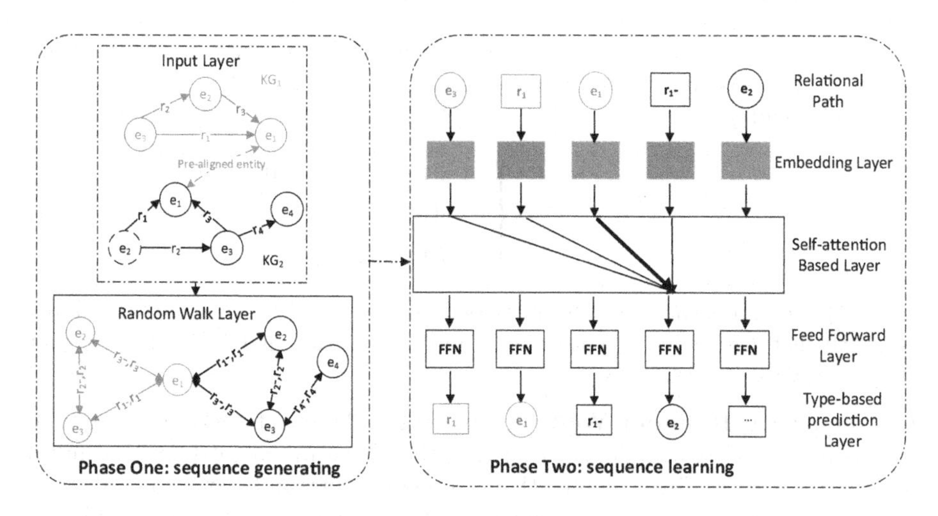

Fig. 1. Framework of our approach

In the following of this section, we start with degree-aware random walk for sequence generating in Sect. 4.1. Then, in Sect. 4.2. We describe self-attention based sequence learning for entity alignment in detail.

4.1 Sequence Generating with Random Walk

Random walk is the most basic way to learn node embeddings in the network embedding area by generating sequences. Their core innovation is to optimize the node embeddings so that nodes have similar embeddings if they tend to co-occur on short random walk paths over the graph. Towards KG embeddings, the edges between nodes have labels and directions, making it different from traditional network embeddings and, for entity alignment, there are two KGs for sampling paths in which the information may be complementary. Inspired by node2vec able to trade off between embeddings that emphasize local structural roles and embeddings that emphasize community structures, we propose a degree-aware random walk between KGs, which allows for deep relational paths and is sensitive to the degree of nodes at the same time.

Traditional Random Walk. We first clarify that the sampled path in KGs is a cross composition of entities and relations which means it has two types of nodes. When applying traditional random walk node2vec to KGs, it follows the equation below to calculate the probability of passing the next entity:

$$P(e_{i+1}|e_i) = \begin{cases} \alpha_{pq}(e_i, e_{i+1}) \cdot w_r & \exists r \in R, (e_i, r, e_{i+1}) \in T \\ 0 & otherwise \end{cases} \tag{1}$$

where e_i is the ith entity in this walk. The walk now needs to decide on the next entity so it evaluates the transition probability from e_i to e_{i+1} if there is a relation r between them. w_r is the edge weight between entity e_i and e_{i+1} and $\alpha_{pq}(e_i, e_{i+1})$ is computed as follows by introducing two random walk hyperparameters p and q:

$$\alpha_{pq}(e_i, e_{i+1}) = \begin{cases} \frac{1}{p} & \text{if } d_{e_{i-1} \to e_{i+1}} = 0 \\ 1 & \text{if } d_{e_{i-1} \to e_{i+1}} = 1 \\ \frac{1}{q} & \text{if } d_{e_{i-1} \to e_{i+1}} = 2 \end{cases} \quad (2)$$

where e_{i-1} denotes the previous entity of e_i and $d_{e_{i-1} \to e_{i+1}}$ represents the shortest path distance between entity e_{i-1} and e_{i+1}. Note that it is obvious that $\{0, 1, 2\}$ covers all situations.

Degree-aware Random Walk. In the entity alignment task, there are two graphs to sample paths. Considering that they have pre-aligned entities linked with each other, we first combine them as one joint graph by swapping aligned-entities in their triples and add reverse relations between entities to enhance connectivity. Second, entities and relations in KGs have different degrees, to solve the problem of path imbalance caused by this, we introduce degree-aware bias to favor long-tail entities and relations. Formally, the degree-aware bias between e_i and e_{i+1}, denoted by $w_{e_i \to e_{i+1}}$, is defined as follows:

$$w_{e_i \to e_{i+1}} = \begin{cases} \frac{1}{d_{e_{i+1}} + f_r} & \exists r \in R, (e_i, r, e_{i+1}) \in T \\ 0 & \textit{otherwise} \end{cases} \quad (3)$$

where $d_{e_{i+1}}$ is the degree of entity e_{i+1} which indicates the number of entities connected with e_{i+1} and f_r refers to the frequency of relation r that connecting entity e_i and e_{i+1}.

Besides, to prevent the path from walking backward, we leverage the idea of node2vec to control the behavior of random walk to generate deep and acyclic paths. We define the following function to reach our goal:

$$\alpha_q = \begin{cases} q & \text{if } d_{e_{i-1} \to e_{i+1}} = 2 \\ 1 - q & \text{if } d_{e_{i-1} \to e_{i+1}} \neq 2 \end{cases} \quad (4)$$

The overall transition probability distribution of next entity can be achieved by replacing $\alpha_{pq}(e_i, e_{i+1})$ and w_r in Eq. (1) with α_q and $w_{e_i \to e_{i+1}}$ respectively. The complete degree-aware random walk process is shown in Algorithm 1.

4.2 Sequence Learning with Self-attention Mechanism

In the setting of sequence learning for entity alignment, we are given a heterogeneous sequence $p = (e_1, r_1, e_2, r_2, ..., e_l)$, and seek to predict the next item based on the previous items. As depicted in phase two in Fig. 1, it is easy to see that the model's output is a shifted version of the input sequence. In this subsection, we demonstrate how the sequential model operates to capture the long-term dependencies between entities via three components.

Algorithm 1: Degree-aware Random Walk

Input : $KG_1 = (E_1, R_1, T_1)$, $KG_2 = (E_2, R_2, T_2)$ and a set of pre-aligned entities $AS = \{(e_1^i, e_2^j)|e_1^i \in E_1^a, e_2^j \in E_2^a\}$, bias q, path length l, sampling times t.

Output: Paths containing entities and relations.

1. Combine two KGs as one $KG = (E, R, T)$;
2. Calculate transition probability distribution $P(e_{i+1}|e_i)$ for every couple of r and e_{i+1} if $\exists r \in R, (e_i, r, e_{i+1}) \in T$;
3. **for** $i = 1$ *to* t **do**
    4. **for** *each* $(e_i, r, e_{i+1}) \in T$ **do**
        5. $p = e_i \rightarrow r \rightarrow e_{i+1}$;
        6. **while** $length(p) < l$ **do**
            7. find the most likely couple of r' and e' in T followed by the last entity in p according to the probability;
            8. $p = p \rightarrow r' \rightarrow e'$;
        end
    end
end
return p;

Embedding Layer. As there are two types of nodes in the path, we create an entity embedding matrix $\boldsymbol{E}$ and a relation embedding matrix $\boldsymbol{R}$. A learnable position embedding $\boldsymbol{P}$ also needs adding to the input embeddings since the self-attention mechanism is not aware of the positions. Hence, the embedding layer can be denoted as $\hat{E}$, as follows:

$$\hat{E} = \begin{bmatrix} \boldsymbol{E}_1 + \boldsymbol{P}_1 \\ \boldsymbol{R}_1 + \boldsymbol{P}_2 \\ \boldsymbol{E}_2 + \boldsymbol{P}_3 \\ \dots \\ \boldsymbol{E}_l + \boldsymbol{P}_{2l-1} \end{bmatrix} \tag{5}$$

Self-attention Block. This module is made up of self-attention based layer and feed forward layer, and can be stacked to learn more complex features.

We start with the self-attention based layer by introducing the most commonly used scaled dot-product attention, defined as:

$$Attention(\boldsymbol{Q}, \boldsymbol{K}, \boldsymbol{V}) = softmax(\frac{\boldsymbol{Q}\boldsymbol{K}^T}{\sqrt{d}})\boldsymbol{V} \tag{6}$$

where $\boldsymbol{Q}$, $\boldsymbol{K}$ and $\boldsymbol{V}$ represent queries, keys and values respectively. The weighted sum computed by queries and keys is then assigned to values as the final output. To counteract the extremely small gradients caused by large values of dimension, $\frac{1}{\sqrt{d}}$ is implemented to scale the dot products. The vast majority of attention in sequence learning are used with $\boldsymbol{K} = \boldsymbol{V} \neq \boldsymbol{Q}$.

However, the self-attention mechanism makes queries, keys and values the same. In our scenario, take embedding $\hat{E}$ as input, we find it is beneficial to concatenate h self-attention functions whose input is projected versions of queries, keys and values. The equation is defined as:

$$\begin{aligned} S = SA(\hat{E}) = MultiHead(\boldsymbol{Q}, \boldsymbol{K}, \boldsymbol{V}) = Concat(head_1, ..., head_h) \\ where\, head_i = Attention(\hat{E}W_i^Q, \hat{E}W_i^K, \hat{E}W_i^V) \end{aligned} \tag{7}$$

where W_i^Q, W_i^K and W_i^V are projection matrices. This multi-head attention allows the model to learn representations independently and is capable of preventing overfitting by ensembling different attentions.

Owing to the uniqueness of paths in KGs that sampled in units of triples and the entanglements between the embedding of last visited item and all previous items after several self-attention blocks, instead of adding common residual connections to propagate the last item's features, we hold the view that when the next item is entity, the last entity is the most important feature and use a crossed residual connection (C) to realize it. This operation is simply omitted in the figure, but its visual display is the role of the thick arrow in the center of phase two. It is calculated with the following equation:

$$C_i = \begin{cases} S_i & i = odd \\ S_i + \hat{E}_{\frac{i}{2}} & i = even \end{cases} \tag{8}$$

where i denotes the ith item in the relational path.

Although through the self-attention and crossed residual function, the model is capable of aggregating features of all previous items and paying more attention to some specific items, it is unable to interact with different latent dimensions because it is essentially a linear model. Hence, a two-layer feed-forward network is applied to each position separately and identically:

$$F_i = FFN(C_i) = ReLU(C_i W_1 + b_1)W_2 + b_2 \tag{9}$$

where W_1 and W_2 are weight matrices, b_1 and b_2 are bias vectors and $ReLU$ is activation function.

The above three steps form one self-attention block. Though its output is an aggregation of all items which means it can capture long-term dependencies effectively, it is encouraging to stack more blocks to learn more complex item transitions and boost the performance. The bth block is defined as:

$$\hat{E}^{(b)} = SB(\hat{E}^{(b-1)}) \tag{10}$$

where the initial input is $\hat{E}$ and the output of the first block is $SB(\hat{E})$, a.k.a., $\hat{E}^{(1)}$.

However, simply stacking more blocks suffers from overfitting and unstable training process due to more parameters. Inspired by [20], we adopt layer normalization and dropout strategy in the self-attention layer and feed-forward layer to stabilize and accelerate the model. The entire formulas for them are shown as follows:

$$\begin{aligned} SA(x) &= Dropout(SA(LayerNorm(x))) \\ FFN(x) &= x + Dropout(FFN(LayerNorm(x))) \end{aligned} \tag{11}$$

Table 1. Statistics of datasets

Datasets	Source KGs	#Entity	#RelationN	#TripleN	#RelationD	#TripleD
DBP-WD	DBPdeia (English)	15,000	253	38,421	220	68,598
	Wikidata (English)	15,000	144	40,159	135	75,465
DBP-YG	DBPdeia (English)	15,000	219	33,571	206	71,257
	YAGO3 (English)	15,000	30	34,660	30	97,131
EN-FR	DBPdeia (English)	15,000	221	36,508	217	71,929
	DBPdeia (French)	15,000	177	33,532	174	66,760
EN-DE	DBPdeia (English)	15,000	225	38,281	207	56,983
	DBPdeia (German)	15,000	118	37,069	117	59,848

Type-based Prediction Layer. After stacking b self-attention blocks, we predict the next item based on previous items. As we can see, different from common sequence prediction tasks, we have two kinds of elements in our relational path, so it is simple and convenient to separate them and that is why we call our last layer type-based prediction layer. Besides, KGs usually have large amounts of entities and relations making every prediction time-consuming. To deal with the difficulty of having too many output vectors that need to be updated every epoch, we use negative sampling [11] to update a sample of them.

It is obvious that the output item should be kept in our sample, and we need to sample a few entities or relations as negative samples according to the predicted item is an entity or a relation. A specific probabilistic distribution is needed for the sampling process to generate suitable negative samples. In our case, for each positive entity, we generate entities whose frequency is in the first three-quarters of the overall frequency distribution as its negative samples, and the positive relations are handled in the same way. The overall loss is defined as follows:

$$\begin{aligned} L = &\sum_{i_e=1}^{p_e} \Big(-\log \sigma(F_{i_e} \cdot y_{i_e}) - \sum_{j_e=1}^{n_e} \big(\log \sigma(-F_{i_e} \cdot y_{j_e})\big)\Big) + \\ &\sum_{i_r=1}^{p_r} \Big(-\log \sigma(F_{i_r} \cdot y_{i_r}) - \sum_{j_r=1}^{n_r} \big(\log \sigma(-F_{i_r} \cdot y_{j_r})\big)\Big) \end{aligned} \tag{12}$$

where σ is the sigmoid activation function, F_{i_e} is the embedding of output entity, y_{i_e} is its label, p_e and n_e represent the number of positive and negative samples respectively. When the subscript is r, the prediction process of r is explained.

5 Experiments

In this section, we conduct experiments on four couples of real-world datasets with different entity distributions and evaluate our proposed model on them with several metrics.

5.1 Datasets

We use DBPedia, Wikidata and YAGO3 datasets in the experiment, which were built by [7]. The datasets contain English (EN), French (FR) and German (DE) knowledge graphs. To comprehensively evaluate the effectiveness of entity alignment models, the distribution of entities in those sampled datasets are guaranteed to follow the original KGs. Table 1 outlines the detail information of the datasets. Each dataset has a norm and a dense version. RelationN and TripleN represent the number of relation and triple in norm datasets while RelationD and TripleD represent that in dense datasets. However, whether the dataset is norm or dense, all KGs contain 15,000 entities.

5.2 Experiment Settings

As for the evaluation metrics, we adopt Hit@k and MRR to assess the performance of all the approaches. Hit@k measures the proportion of correctly aligned entities ranked in the top k candidates. MRR (Mean Reciprocal Rank) calculates the average ranking scores of the aligned entities whose score is the reciprocal of its rank. Both metrics are preferred to be higher to represent better performance.

For all the compared approaches, we follow the previous works using 30% of pre-aligned entities for training and the rest 70% for testing. The split and dimension of knowledge graph embeddings are the same for all methods.

For the parameters of our approach, we set dimension = 256, learning rate = 0.003, dropout = 0.5, q = 0.9, l = 15 and t = 2. The optimizer is the Adam optimizer, the number of self-attention block and head is 2 and 8 respectively.

5.3 Comparative Methods

To show the effectiveness of our model, we picked up two types of state-of-the-art entity alignment models for comparison.

The first type is single-resource-based models which only consider structural information without extra resources.

- **MTransE**: This is a simple baseline that finds alignment via a unidirectional transition matrix between the embeddings of KGs which is based on TransE.
- **SEA**: A novel semi-supervised model for alignment. SEA solves the problem existing in the present embedding methods which is caused by the degree difference of entities in different KGs. A cycled consistent loss is used where bidirectional transition matrices are trained to incorporate unlabeled entities.

- **RSNs**: A state-of-the-art entity alignment method that leverages recurrent neural networks to model relational paths instead of triples to capture long-term dependencies between entities and relations.
- **IPTranE**: An iterative version of the TransE model which jointly encodes both entities and relations of various KGs into a unified low-dimension semantic space.
- **BootEA**: A bootstrapping approach to embedding-based entity alignment. It uses a truncated uniform negative sampling method in the embedding phase and employs an alignment editing method to reduce error accumulation during iterations.

Table 2. Results on norm and monolingual datasets

Datasets	DBP-WD			DBP-YG		
Metrics	Hits@1	Hits@10	MRR	Hits@1	Hits@10	MRR
MTransE	22.3	50.1	0.32	24.6	54.0	0.34
SEA	31.0	51.9	0.38	30.3	45.9	0.36
RSNs	38.8	65.7	0.49	40.0	67.5	0.50
IPTransE	23.1	51.7	0.33	22.7	50.0	0.32
BootEA	32.3	63.1	0.42	31.3	62.5	0.42
JAPE	21.9	50.1	0.31	23.3	52.7	0.33
GCN-Align	17.7	37.8	0.25	19.3	41.5	0.27
SAEA w/o DB	38.8	69.2	0.50	42.4	70.3	0.52
SAEA	**44.1**	**70.7**	**0.53**	**44.5**	**71.0**	**0.53**

Table 3. Results on norm and cross-lingual datasets

Datasets	EN-FR			EN-DE		
Metrics	Hits@1	Hits@10	MRR	Hits@1	Hits@10	MRR
MTransE	25.1	55.1	0.35	31.2	58.6	0.40
SEA	25.8	40.0	0.31	42.5	59.6	0.49
RSNs	34.7	63.1	0.44	48.7	72.0	0.57
IPTransE	25.5	55.7	0.36	31.3	59.2	0.41
BootEA	31.3	62.9	0.42	44.2	70.1	0.53
JAPE	25.6	56.2	0.36	32.0	59.9	0.41
GCN-Align	15.5	34.5	0.22	25.3	46.4	0.33
SAEA w/o DB	37.5	64.8	0.46	50.1	72.8	0.58
SAEA	**38.3**	**65.8**	**0.47**	**51.3**	**73.4**	**0.59**

The second type is multi-resource-based approaches, which not only consider relation triples but also attribute triples to some extent.

Table 4. Results on dense and monolingual datasets

Datasets	DBP-WD			DBP-YG		
Metrics	Hits@1	Hits@10	MRR	Hits@1	Hits@10	MRR
MTransE	38.9	68.7	0.49	22.8	51.3	0.32
SEA	67.2	85.2	0.74	68.1	84.1	0.74
RSNs	76.3	92.4	0.83	82.6	95.8	0.87
IPTransE	43.5	74.5	0.54	23.6	51.3	0.33
BootEA	67.8	91.2	0.76	68.2	89.8	0.76
JAPE	39.3	70.5	0.50	26.8	57.3	0.37
GCN-Align	43.1	71.3	0.53	31.3	57.5	0.40
SAEA w/o DB	78.1	93.3	0.84	85.1	96.1	0.88
SAEA	**79.9**	**93.7**	**0.85**	**85.6**	**96.5**	**0.89**

Table 5. Results on dense and cross-lingual datasets

Datasets	EN-FR			EN-DE		
Metrics	Hits@1	Hits@10	MRR	Hits@1	Hits@10	MRR
MTransE	37.7	70.0	0.49	34.7	62.0	0.44
SEA	62.3	85.7	0.71	65.2	79.4	0.70
RSNs	75.6	92.5	0.82	73.9	89.0	0.79
IPTransE	42.9	78.3	0.55	34.0	63.2	0.44
BootEA	64.8	91.9	0.74	66.5	87.1	0.73
JAPE	40.7	72.7	0.52	37.5	66.1	0.47
GCN-Align	37.3	70.9	0.49	32.1	55.2	0.40
SAEA w/o DB	80.2	93.7	0.84	75.7	89.3	0.81
SAEA	**80.6**	**94.2**	**0.85**	**77.6**	**91.2**	**0.82**

- **JAPE**: This is the first model to learn embeddings of cross-lingual KGs while preserving their attribute information. By leveraging the attribute triples of KGs with attribute embedding, the original structural embeddings of entities can be refined.
- **GCN-Align**: A GCN-based approach for cross-lingual KG alignment. It generates node-level embeddings by encoding information about the nodes' neighborhoods. Both the relation and attribute triples in KGs are employed to discover alignments.

Since other entity alignment models make use of more resources that are difficult to achieve, we omit comparisons against them.

5.4 Results

The evaluation results are presented in Tables 2, 3, 4 and 5. SAEA w/o DB and SAEA represent our method. The latter one means that at the stage of random walk, degree-aware bias is implemented while the former one not. The best results are shown in bold among the group of methods. From the results, we have the following findings:

(1) **Our proposed SAS consistently outperforms the state-of-the-art approaches on all datasets under different evaluation metrics.** This observation verifies that SAEA can effectively model relational paths in KGs for improving the accuracy of entity alignment and is capable of capturing long-term dependencies between entities effectively. In particular, compared with the second-best baseline RSN, our method achieves more than 3% improvement when matching the correct entity in the first position. This improvement is more obvious on monolingual datasets as the heterogeneity among them is more severe than one KG with different languages. In addition, it is worth mentioning that SAEA showed larger superiority over iterative and multi-resource-based approaches which manifests that our model fully explores the inherent structural features of KGs and can be enhanced by employing existing iteration strategies or combing extra resources. All the results show the advantage of our method.

(2) **Our degree-aware random walk produces better entity representations for alignment.** This comparison is drawn from the comparison of improvement made by SAEA w/o DB and SAEA. Especially on norm monolingual datasets, SAEA has more than 2% improvement on Hits@1 while on dense datasets, the improvement is slight. This indicates that degree-aware random walk plays a more important role in sampling paths in KGs where richer relational triples are scarce. Degree bias makes long-tail entities and relations more likely to be sampled so that the information between norm entities and long-tail entities is balanced. How much this biased random walk can help is limited by the original entity distribution in KGs, but the improvement made from SAEA w/o DB to SAEA is justifiable.

6 Conclusions and Future Work

This paper presents a novel self-attentive heterogeneous sequence learning model for entity alignment. It discovers alignments based on learning embeddings from relational paths. In order not to ignore the unique triple structure of KGs, we design a crossed residual connection to focus on triples. Also, we introduce a degree-aware random walk for sampling paths in a balanced way. Our experiments on four real-world datasets demonstrated the effectiveness of our framework.

In future work, we plan to explore more advanced sequential models and graph neural networks for KG alignment task, such as MARINE [9] and MixHop [1]. Besides, how to combine extra resources in our approach is another interesting direction.

Acknowledgments. This research is partially supported by Natural Science Foundation of Jiangsu Province (No. BK20191420), National Natural Science Foundation of China (Grant No. 61632016, 61572336, 61572335, 61772356), Natural Science Research Project of Jiangsu Higher Education Institution (No. 17KJA520003, 18KJA520010), and the Open Program of Neusoft Corporation (No. SKLSAOP1801).

References

1. Abu-El-Haija, S., et al.: Mixhop: higher-order graph convolution architectures via sparsified neighborhood mixing. arXiv preprint arXiv:1905.00067 (2019)
2. Bollacker, K., Evans, C., Paritosh, P., Sturge, T., Taylor, J.: Freebase: a collaboratively created graph database for structuring human knowledge. In: Proceedings of the 2008 ACM SIGMOD International Conference on Management of Data, pp. 1247–1250. ACM (2008)
3. Bordes, A., Usunier, N., Garcia-Duran, A., Weston, J., Yakhnenko, O.: Translating embeddings for modeling multi-relational data. In: Advances in Neural Information Processing Systems, pp. 2787–2795 (2013)
4. Chen, M., Tian, Y., Chang, K.W., Skiena, S., Zaniolo, C.: Co-training embeddings of knowledge graphs and entity descriptions for cross-lingual entity alignment. arXiv preprint arXiv:1806.06478 (2018)
5. Chen, M., Tian, Y., Yang, M., Zaniolo, C.: Multilingual knowledge graph embeddings for cross-lingual knowledge alignment. arXiv preprint arXiv:1611.03954 (2016)
6. Grover, A., Leskovec, J.: node2vec: scalable feature learning for networks. In: Proceedings of the 22nd ACM SIGKDD International Conference on Knowledge Discovery and Data Mining, pp. 855–864. ACM (2016)
7. Guo, L., Sun, Z., Hu, W.: Learning to exploit long-term relational dependencies in knowledge graphs. arXiv preprint arXiv:1905.04914 (2019)
8. Hu, W., Chen, J., Qu, Y.: A self-training approach for resolving object coreference on the semantic web. In: Proceedings of the 20th International Conference on World Wide Web, pp. 87–96. ACM (2011)
9. Kawamae, N.: Marine: multi-relational network embeddings with relational proximity and node attributes. In: The World Wide Web Conference, pp. 470–479. ACM (2019)
10. Lehmann, J., et al.: DBpedia-a large-scale, multilingual knowledge base extracted from Wikipedia. Semantic Web **6**(2), 167–195 (2015)
11. Mikolov, T., Sutskever, I., Chen, K., Corrado, G.S., Dean, J.: Distributed representations of words and phrases and their compositionality. In: Advances in Neural Information Processing Systems, pp. 3111–3119 (2013)

12. Ngomo, A.C.N., Auer, S.: Limes-a time-efficient approach for large-scale link discovery on the web of data. In: Twenty-Second International Joint Conference on Artificial Intelligence (2011)
13. Pei, S., Yu, L., Hoehndorf, R., Zhang, X.: Semi-supervised entity alignment via knowledge graph embedding with awareness of degree difference. In: The World Wide Web Conference, pp. 3130–3136. ACM (2019)
14. Perozzi, B., Al-Rfou, R., Skiena, S.: Deepwalk: online learning of social representations. In: Proceedings of the 20th ACM SIGKDD International Conference on Knowledge Discovery and Data Mining, pp. 701–710. ACM (2014)
15. Raimond, Y., Sutton, C., Sandler, M.B.: Automatic interlinking of music datasets on the semantic web. In: LDOW, vol. 369 (2008)
16. Suchanek, F.M., Kasneci, G., Weikum, G.: YAGO: a core of semantic knowledge. In: Proceedings of the 16th International Conference on World Wide Web, pp. 697–706. ACM (2007)
17. Sun, Z., Hu, W., Li, C.: Cross-lingual entity alignment via joint attribute-preserving embedding. In: d'Amato, C., et al. (eds.) ISWC 2017. LNCS, vol. 10587, pp. 628–644. Springer, Cham (2017). https://doi.org/10.1007/978-3-319-68288-4_37
18. Sun, Z., Hu, W., Zhang, Q., Qu, Y.: Bootstrapping entity alignment with knowledge graph embedding. In: IJCAI, pp. 4396–4402 (2018)
19. Trisedya, B.D., Qi, J., Zhang, R.: Entity alignment between knowledge graphs using attribute embeddings. In: Proceedings of the AAAI Conference on Artificial Intelligence, vol. 33, pp. 297–304 (2019)
20. Vaswani, A., et al.: Attention is all you need. In: Advances in Neural Information Processing Systems, pp. 5998–6008 (2017)
21. Volz, J., Bizer, C., Gaedke, M., Kobilarov, G.: Discovering and maintaining links on the web of data. In: Bernstein, A., et al. (eds.) ISWC 2009. LNCS, vol. 5823, pp. 650–665. Springer, Heidelberg (2009). https://doi.org/10.1007/978-3-642-04930-9_41
22. Wang, Z., Lv, Q., Lan, X., Zhang, Y.: Cross-lingual knowledge graph alignment via graph convolutional networks. In: Proceedings of the 2018 Conference on Empirical Methods in Natural Language Processing, pp. 349–357 (2018)
23. Xu, K., Wu, L., Wang, Z., Feng, Y., Witbrock, M., Sheinin, V.: Graph2seq: graph to sequence learning with attention-based neural networks. arXiv preprint arXiv:1804.00823 (2018)
24. Yang, Y., Sun, Y., Tang, J., Ma, B., Li, J.: Entity matching across heterogeneous sources. In: Proceedings of the 21th ACM SIGKDD International Conference on Knowledge Discovery and Data Mining, pp. 1395–1404. ACM (2015)
25. Zhang, Q., Sun, Z., Hu, W., Chen, M., Guo, L., Qu, Y.: Multi-view knowledge graph embedding for entity alignment. arXiv preprint arXiv:1906.02390 (2019)
26. Zhu, H., Xie, R., Liu, Z., Sun, M.: Iterative entity alignment via joint knowledge embeddings. In: IJCAI, pp. 4258–4264 (2017)

TADNM: A Transportation-Mode Aware Deep Neural Model for Travel Time Estimation

Saijun Xu[1], Jiajie Xu[1,3](✉), Rui Zhou[2], Chengfei Liu[2], Zhixu Li[1], and An Liu[1]

[1] Institute of Artificial Intelligence, School of Computer Science and Technology, Soochow University, Suzhou, China
sjxu@stu.suda.edu.cn, {xujj,zhixuli,anliu}@suda.edu.cn

[2] Swinburne University of Technology, Melbourne, Australia
{rzhou,cliu}@swin.edu.au

[3] State Key Laboratory of Software Architecture (Neusoft Corporation), Shenyang, China

Abstract. Travel time estimation (TTE) has been recognized as an important problem in location-based services. Existing approaches mainly estimate travel time by learning from large-scale trajectories, they normally assume a path is in a single transportation mode (e.g., driving, biking), and could not provide accurate TTE for mixed-mode paths, which are indeed common in daily life. In this paper, we propose a transportation-mode aware deep neural model called TADNM, which considers both spatio-temporal characteristics and the heterogeneity of underlying transportation modes to achieve more accurate travel time estimation. Specifically, we estimate travel time using the knowledge from (sub-)trajectories not only roughly following the target path, but also being consistent with segments of the target path in terms of transportation mode. To this end, a well-designed neural network model is proposed to integrate the rich information extracted from trajectories first, and then to learn effective representations for capturing the spatial correlations, temporal dependencies and transportation mode effects from the trajectory data. Besides, the proposed model fully considers the transition time of switching transportation mode in the path, and a transportation-mode aware attention mechanism is used to better reflect the impact of transportation mode to the required travel time. Extensive experiments on real trajectory datasets demonstrate the effectiveness of our proposed model.

Keywords: Travel time estimation · Trajectory data mining · Deep learning

1 Introduction

The estimation of vehicle travel time of a planned target path is essential for many location-based applications, such as departure time suggestion [9], route

Y. Nah et al. (Eds.): DASFAA 2020, LNCS 12112, pp. 468–484, 2020.
https://doi.org/10.1007/978-3-030-59410-7_32

planning [2,15,21,22], vehicle dispatching [24], etc. Meanwhile, vehicle trajectory data has been accumulated to an extremely huge volume in online navigation and ridesharing enterprises (e.g. Uber and Google Map).

Large-scale historical trajectories provide us great opportunities to understand complex spatial dynamics of city crowd, which contributes to improving the performance of travel time estimation (TTE). However, it is difficult to achieve an accurate time estimation, because, on one hand, the travel time is affected by many dynamic factors such as variable traffic conditions on the road, different weather conditions at the departure time and different driving behaviors; on the other hand, effective information in the trajectory data such as spatial correlations, temporal dependencies has not been fully utilized.

In the existing literature, most travel time estimation solutions [6,14] are trajectory data driven, since travel time can be inferred from those trajectories roughly following the target path in history. More recently, TTE models [17,20,25] take advantage of trajectory embedding, which learns a suitable representation of raw trajectories, such that their relevance to the target path in 2D space can be accurately captured. In this way, they can easily identify the (sub-)trajectories relevant to the target path using their embeddings, and then estimate the travel time by proper aggregation of these (sub-)trajectories. However, all these solutions assume that trajectories are generated by users in a single transportation mode (e.g., driving, biking, walking). While in daily life, travel paths required by users may be composed of multiple transportation modes.

In reality, the travel time of a moving object is heavily affected by its corresponding transportation mode. For example, for two equal-length segments of a trajectory, the trajectory segment in driving mode tends to consume less travel time than the other in riding mode, however, they are equally treated by existing TTE models [17,20,25] for the estimation of travel time. Naturally, we can imagine that it is more effective to utilize the knowledge from (sub-)trajectories not only roughly following the target path, but also being consistent with segments of the target path in terms of transportation mode, especially those under similar contexts (e.g. weather and holidays). Unfortunately, this assumption cannot be well supported by existing TTE models.

To this end, this paper aims to investigate more accurate travel time estimation by considering the heterogeneity of trajectory transportation mode. This leads to several challenges: first, it calls for a trajectory representation that can fully capture not only spatio-temporal characteristics, but also the heterogeneity of the underlying transportation mode; second, it is challenging to learn the varying travel time pattern of different transportation modes by a uniform deep learning model; in addition, it may encounter data sparsity problem, which has an impact on the estimation accuracy, since there may be few or even no trajectories annotated to the required transportation mode in the specific region.

To address the above challenges, this paper proposes a transportation-mode aware deep neural model, namely TADNM, which considers the heterogeneity in trajectories with multiple transportation modes to achieve more accurate travel time estimation. First, the model divides a trajectory into segments of different transportation modes to distinguish (sub-)trajectories according to their transportation modes, aiming to mine effective information from the segments with specific transportation modes; and then the model maps each segment into a region which is divided into multiple grids to portray trajectories in geographical space. In this way, some sampled error in the trajectory points can be alleviated and it can also deal with spatial data sparsity issue. Second, the model adopts convolutional operation and GRU network to capture potential spatial patterns and temporal correlations in the trajectory sequence respectively. Last but not least, the model takes an attention mechanism to trade off different impact of multiple transportation modes on travel time and consider the transition time of switching transportation mode in the path. To sum up, the contributions of the paper are as follows:

- We propose a transportation-mode aware deep neural model, namely TADNM, for travel time estimation. We estimate the travel time by utilizing (sub-)trajectories which are not only roughly following the target path, but also being consistent with segments of the target path in terms of transportation mode.
- We integrate useful information including spatio-temporal features, trajectory information, transportation modes and other external factors (such as holidays, weekdays or weekends); and harness deep learning to design a neural model to learn effective representations, so that spatial correlations, temporal dependencies and transportation mode effects are captured from trajectories.
- We consider the transition time of switching transportation mode in the path and a transportation-mode aware attention mechanism is utilized to trade off the impact of different transportation modes on travel time.
- We conduct extensive experiments on the real dataset. The results demonstrate the advantages of our approach compared with baselines.

2 Related Work

Existing approaches of estimating travel time on the path could be classified into two categories, segment-based and path-based approaches. The former splits a path into a sequence of segments firstly. Then it estimates the travel time on each segment individually and combines them to estimate the entire travel time in a path [23]. These approaches adopt multiple techniques to estimate time on sub-segments such as Bayesian network, Hidden Markov Model, support vector regression, etc. However, segment-based approaches ignore the correlations between the road segments and the time that spends on the traffic lights. To deal with these drawbacks in segment-based methods, path-based approaches extract and aggregate similar sub-paths from historical trajectories for more accurate travel time estimation [19]. The path-based approaches decrease obvious errors in the segment-based approaches via taking the path continuity and delayed time on the road into consideration. [14] develops a non-parametric method and finds sub-paths of target path from historical trajectories to estimate the travel time of the path after incorporating a list of bias corrections. However, these path-based approaches suffer from data sparsity issue because there are many sub-paths which cannot be found in historical trajectory data. In general, neither segment-based nor path-based approaches could achieve satisfactory performance in travel time estimation.

On the other hand, sufficiently large datasets and enough computational power enable the success of deep learning these years. Some recent studies [17,20,25] aim to take advantage of deep learning for more effective TTE processing. WDR [20] first extracts a large set of effective features from the map-matched trajectory data which is mapped into the road network by using map-matching [12], and then trains wide linear models, deep neural networks and recurrent neural networks together to predict the travel time. It balances the memorization, generalization and representation abilities. Deeptravel proposed in [25] considers spatial and temporal embeddings, driving state features and traffic features in a granularity of spatial grids, and then uses Bi-LSTMs to predict the travel time. A trajectory embedding based model DeepTTE is proposed in [17], which utilizes geo-convolution in 2D space to find trajectories roughly following the target path for aggregation. It then stacks recurrent units to capture the temporal dependencies in the given path. However, none of those solutions mentioned above consider the impact of different transportation modes on travel time. Therefore, what we target to do is incorporating the heterogeneity of trajectory transportation mode into travel time estimation. Furthermore, to address the drawbacks in the existing solutions mentioned above, an ingenious deep-learning model which extracts rich features, captures spatial correlation, temporal memorization and transportation mode effects is sought after to deal with this estimation task.

In addition, except approaches of estimating travel time based on paths [14,17,19,20,23,25], travel time estimation also includes other aspects, such as origin-destination (OD) approaches, which aim to estimate the travel time without the actual travel path. It could be applied to the scenario where path information is unavailable or sampled trajectory data is very sparse. For instance, ST-NN [7] uses the origin and destination binned GPS coordinates and time information as inputs to MLP module for estimating travel time; TEMP [18] is a neighbor-based approach which estimates the travel time by using similar trips with a nearby origin and destination. MURAT [11] leverages the underlying road network and the spatio-temporal prior knowledge for origin-destination travel time estimation.

3 Problem Formulation

This section introduces some preliminaries and gives a formal statement of the problem studied in this paper.

Definition 1 (Trajectory). A trajectory is a sequence of GPS points. Each GPS point contains a location (longitude, latitude) and a timestamp.

Definition 2 (Trip). A travel trip $X^{(i)}$ is derived from a trajectory, which additionally contains transportation modes, it is defined as a tuple with five components, $X^{(i)} = (S^{(i)}, M^{(i)}, T^{(i)}, \tau^{(i)}, \delta^{(i)})$ where $S^{(i)} = (s_1^{(i)}, s_2^{(i)}, ..., s_L^{(i)})$ denotes a sequence of trajectory segments, $M^{(i)} = (m_1^{(i)}, m_2^{(i)}, ..., m_L^{(i)})$ represents the corresponding transportation modes on $S^{(i)}$, $T^{(i)} = (t_1^{(i)}, t_2^{(i)}, ..., t_L^{(i)})$ denotes the travel time on $S^{(i)}$, L is the number of transportation modes in $X^{(i)}$ and is 1 if $X^{(i)}$ contains only one transportation mode. Each trajectory segment $s_j^{(i)} (1 \leq j \leq L)$ consists of a sequence of GPS point locations and is labeled with its specific transportation mode $m_j^{(i)}$ by the user, the travel time of segment $s_j^{(i)}$ is $t_j^{(i)}$. $\tau^{(i)}$ denotes the departure time and $\delta^{(i)}$ is the entire travel time of the trip which is not equal to the summation of time $t_j^{(i)}$ on $T^{(i)}$ when $L > 1$ because $\delta^{(i)}$ contains the transition time when users change their transportation modes between two consecutive trajectory segments.

The historical trip dataset is accordingly represented by a set of trips which we denote as $D = \{X^{(i)} | i = 1, 2, 3, .., N\}$.

Definition 3 (Path). A path P consists of a sequence of sampled GPS point locations, it is divided into a sequence of trajectory segments S and corresponding transportation modes M are chosen to travel on S.

Problem Statement (Travel Time Estimation). Given a departure time τ and a query path P, our goal is to estimate the travel time δ of P by learning a deep neural network that considers the heterogeneity of trajectory transportation modes and captures spatial correlations as well as temporal dependencies from historical trip dataset D.

4 Model Architecture

Figure 1 presents the architecture of the TADNM model, which consists of four layers, spatial representation layer, temporal embedding layer, transportation-mode aware attention layer and time prediction layer. Firstly, the spatial representation layer is used to describe and capture the spatial distribution and shape information of (sub-)trajectories with specific transportation modes. Secondly, its output is fed to the temporal embedding layer so as to learn the temporal dependencies in the spatial sequence. Afterwards, the transportation-mode aware attention layer considers the impact of different transportation modes on the travel time of the path. Finally, the time prediction layer estimates the travel time of the given path and predicts the time on each trajectory segment simultaneously to make full use of supervised labels for better performance.

4.1 Spatial Representation Layer

The spatial representation layer mainly captures the spatial moving patterns of (sub-)trajectories which contain one transportation mode by finding a suitable

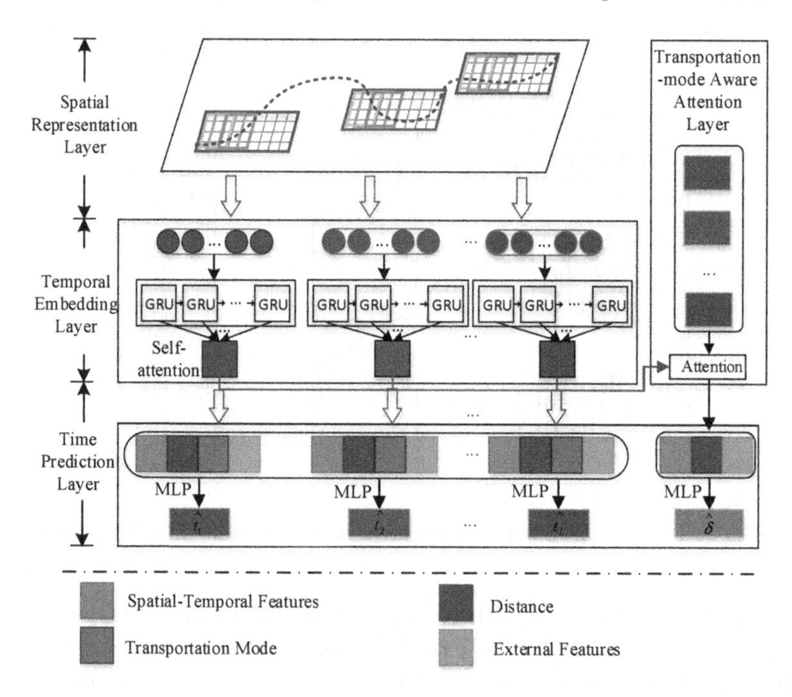

Fig. 1. The architecture of TADNM.

representation of the input raw trajectory data. It is important to portray the spatial shapes and distributions of trajectories generated by different transportation modes, the reason is that topographic ranges of trajectories differ especially depending on transportation modes, for example, the ranges of walking trajectories are often narrow while the ranges of trajectories travelling by train are broad. However, if we directly use the raw trajectory data as inputs to model the spatial patterns, on one hand, this process may encounter uncertainties due to different sampling intervals and sampling methods in raw trajectory point sequences; on the other hand, the sampling error in trajectory point sequences may be accumulated in this process; in addition, it is difficult to find an appropriate strategy to capture spatial features from raw trajectory data.

To deal with these issues, first, we map the trajectory segment into a region which is divided by grids, then raw trajectories can be represented as 2D image-like data called trajectory images [3]. Specifically, given a travel path P which consists of a sequence of trajectory segments $S = (s_1, s_2, ..., s_L)$ with their corresponding transportation modes $M = (m_1, m_2, ..., m_L)$, for each segment $s_i (1 \leq i \leq L)$, we need to portray the spatial information of s_i under specific transportation modes m_i. As shown in Fig. 2, we define a region G_i clipped by ranges of s_i's latitudes and longitudes and divide this region into $\omega \times \omega$ grids, accordingly, each grid in the area represents a pixel in the image, the value of pixels can be calculated in the following manner: since s_i includes a sequence

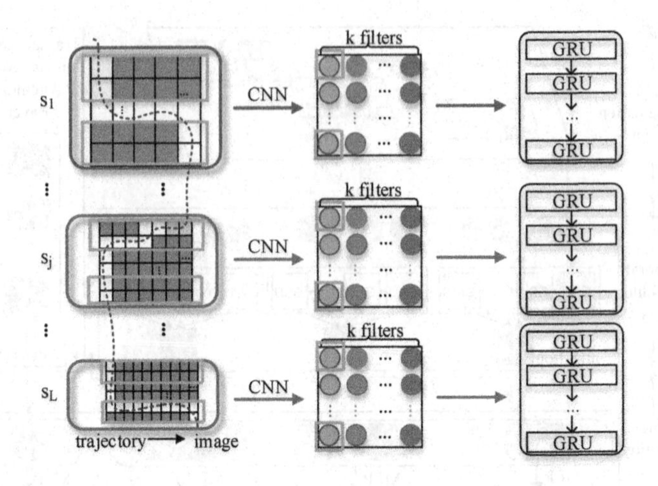

Fig. 2. The spatial representation layer and temporal embedding layer

of GPS points $U_{s_i} = (u_j)|_{j=1}^{N_{s_i}}$, where N_{s_i} denotes the number of GPS points in segment s_i and u_j represents the j-th GPS point location, if u_j is mapped into grid $G_i(x, y)$, we add a constant $c = 1$ to the pixel value of $G_i(x, y)$, therefore, the pixel value of each pixel is the number of GPS points. Then the generated trajectory image $G_i \in \mathbb{R}^{w \times w}$ can be treated as one-channel value per pixel like a grayscale image, the more GPS points in the grid, the larger the pixel value and the brighter the image, as shown in Fig. 2. In this way, the spatial information of the i-th trajectory segment s_i can be portrayed as the trajectory image G_i, the travel path P is transformed into a sequence of trajectory image sequences $I = (G_1, G_2, ..., G_L)$. Afterwards, considering the convolutional neural network (CNN) works well in spatial correlation understanding in the image recognition [10], we employ convolutional operations on each trajectory image G_i of the image sequence I to capture the spatial moving patterns of trajectory images of different transportation modes. As shown in Fig. 2, after generating trajectory image sequences I, we use k convolutional filters with the kernel size h to convolve each $G_i \in I (1 \le i \le L)$, $G_i \in \mathbb{R}^{\omega \times \omega}$. The matrix convolved by the j-th filter can be formulated as,

$$spt_{i,j,a} = ReLU(W_{conv_j} * G_{i,e\times(a-1)+1:e\times(a-1)+h} + b_j) \tag{1}$$

where $G_{i,e\times(a-1)+1:e\times(a-1)+h} \in \mathbb{R}^{h \times \omega}$ is the a-th sub-image from index $e \times (a - 1) + 1$ to $e \times (a - 1) + h$ of G_i divided in height dimension and e is the stride of convolution that controls the stride for the cross-correlation, $W_{conv_j} \in \mathbb{R}^{h \times \omega}$ denotes the parameter weights of the j-th filter and $*$ represents the convolutional operation, b_j is the bias. Here we use rectifier linear unit ($ReLU$) [10] as activation function and $ReLU(x) = max(0, x)$. Then we concatenate $spt_{i,j,a}$ convolved by k filters and obtain the spatial representations of the a-th sub-image in G_i as follows,

$$\mathcal{E}_{i,a} = \{spt_{i,1,a}, spt_{i,2,a}, ..., spt_{i,k,a}\} \tag{2}$$

Where $\mathcal{E}_{i,a} \in \mathbb{R}^{1\times k}$. Afterwards we scan sub-images of G_i in the chronological order and in a specified stride to obtain the spatial representations of G_i after the calculation of k filters, which is denoted as $\mathcal{V}_i \in \mathbb{R}^{N\times k}$, i.e.,

$$\mathcal{V}_i = \{\mathcal{E}_{i,1}, \mathcal{E}_{i,2}, ..., \mathcal{E}_{i,N}\} \tag{3}$$

where N is the number of sub-images to be scanned by filters whose value depends on the kernel size h and the stride e. Accordingly, the spatial representations of I which consists of a sequence of trajectory images G_i can be represented as,

$$\mathcal{SR} = \{\mathcal{V}_1, \mathcal{V}_2, ..., \mathcal{V}_L\} \tag{4}$$

where L is the number of trajectory segments of the path. In the end, the output of this layer is $\mathcal{SR}$ which portrays the spatial information of trajectories.

4.2 Temporal Embedding Layer

The previous layer learns the spatial representations $\mathcal{V}_i$ from the i-th trajectory image G_i which is generated from raw trajectory segment data. However, the temporal continuity in the trajectory segment has not been represented, therefore, in this section, we aim to capture the temporal dependencies in the trajectory segment.

We take advantage of sequential models and we adopt the gated recurrent unit(GRU) network [1] in our model. The inputs of the GRU network is the spatial representations $\mathcal{SR} = \{\mathcal{V}_1, \mathcal{V}_2, ..., \mathcal{V}_L\}$ which are the outputs of the previous spatial representation layer, specifically, $\mathcal{E}_{i,t} \in \mathcal{V}_i$ is the spatial representations of the t-th sub-image of G_i. Each GRU has a reset gate $r_{i,t}$ and an update gate $z_{i,t}$, $h_{i,t}$ is the t-th hidden state after we process the output $\mathcal{E}_{i,t}$ through GRU networks and its computation is in the following,

$$\begin{aligned}
h_{i,t} &= (1 - z_{i,t})h_{i,t-1} + z_{i,t}\tilde{h}_{i,t} \\
z_{i,t} &= \sigma(W_z\mathcal{E}_{i,t} + U_z h_{i,t-1}) \\
\tilde{h}_{i,t} &= tanh(W\mathcal{E}_{i,t} + U(r_{i,t} \odot h_{i,t-1})) \\
r_{i,t} &= \sigma(W_r\mathcal{E}_{i,t} + U_r h_{i,t-1})
\end{aligned} \tag{5}$$

where $h_{i,t-1}$ and $\tilde{h}_{i,t}$ respectively correspond to the previous memory content and new candidate memory content. The update gate $z_{i,t}$ decides how much the unit updates its content, it is computed based on the previous hidden states $h_{i,t-1}$ and the current input $\mathcal{E}_{i,t}$. W_z and U_z are the weight matrix, σ is a logistic sigmoid function. $r_{i,t}$ is the reset gate, W and U denote the weight matrix and $\odot$ is an element-wise multiplication. The reset gate $r_{i,t}$ allows a GRU to ignore the hidden states whenever it is deemed necessary considering the previous hidden states and the current input. W_r and U_r denote the weight matrix and the update mechanism helps GRU to capture long term dependencies by detecting the previous important memory content and allowing the model to reset if the previous content is unnecessary.

After using GRU networks on $\mathcal{V}_i$, we get a sequence of hidden states $\{h_{i,1},h_{i,2}, ..., h_{i,N}\}$ which capture the spatial and temporal dependencies in the i-th trajectory segment. Then we need to encode the hidden unit sequence into a representation which can highly summarize the contextual information. Self-attention [16] has been used successfully in a variety of tasks including abstract summarization, reading comprehension that relates different positions of a sequence to compute a representation in the intermediate process of the model. Therefore, we apply self-attention mechanism to $\{h_{i,1}, h_{i,2}, ..., h_{i,N}\}$ and calculate H_i as the comprehensive representations of the hidden states sequence. Finally, we use the method in L trajectory segments and get the embeddings $\mathcal{ST} = \{H_1, H_2, ..., H_L\}$ which capture both the spatial and temporal features from trajectories to prepare for the following time estimation.

4.3 Transportation-Mode Aware Attention Layer

The length of output $\mathcal{ST}$ in the previous layer is variable which prevents it from being applied directly to predict travel time. To solve this issue, we need to transform the sequence $\mathcal{ST}$ into a fixed-length vector. We can directly use the mean of all the H_i in $\mathcal{ST}$ as the final vector $H_{mean} = \frac{1}{L}\sum_{i=1}^{L} H_i$ to predict total travel time in path P, however, this transformation treats hidden units H_i of trajectory segments with different transportation modes equally. Actually, on one hand, different transportation modes at different segments of the path have different impacts on the travel time of the entire path; on the other hand, the transition time of switching transportation mode in the path is easy to be neglected when a user labels the transportation mode of his/her GPS data. Therefore, we design an attention mechanism that incorporates transportation modes m_i on the segment s_i to make a weighted summation of the sequence of hidden units $\{H_1, H_2, ..., H_L\}$, which can be expressed as,

$$H^* = \sum_{i=1}^{L} \alpha_i H_i \tag{6}$$

where α_i is the weights which are learned by incorporating transportation mode m_i on s_i, $\sum_{i=1}^{L} \alpha_i = 1$. $M_i = \{m_1, m_2, ..., m_L\}$ is a sequence of transportation modes in the path. Each categorical value m_i should be first encoded into value $\tilde{m}_i$ in a real space $\mathbb{R}^{1\times E}$ by using the embedding mechanism [13], which not only reduces the dimensionality of categorical values but also represents categories in the transformed space. We then calculate attention scores,

$$\theta_i = H_i W_a (\tilde{m}_i)^T \tag{7}$$

$$\alpha_i = \frac{e^{\theta_i}}{\sum_{j=1}^{L} e^{\theta_j}} \tag{8}$$

where $H_i \in \mathbb{R}^{1\times D}, \tilde{m}_i \in \mathbb{R}^{1\times E}$, $W_a \in \mathbb{R}^{D\times E}$ is the weight matrix that map H_i to a vector with the same dimension of $\tilde{m}_i$. Then the score θ_i is normalized by the softmax function to obtain the attention weight distribution α_i. By combining Eq. (6), Eq. (7) and Eq. (8), we compute H^* as the output of the transportation-mode aware attention layer.

4.4 Time Prediction Layer

In the time prediction layer, to fully leverage the supervised labels in historical trips and achieve more accurate results, we introduce two types of estimation tasks in the training phase: estimating travel time of queried path P and estimating time on each segment $s_i \in S$ simultaneously. The purpose of introducing the task of estimating time on each segment is to make full use of additional ground truth in the historical trips, i.e., travel time on each segment, to further train and adjust parameters of our TTE model, in this way, the model performance can be further improved. In particular, in the testing phase, we only need to estimate the travel time of the entire path. Specifically, for a trajectory segment sequence S, after the previous spatial representation layer and temporal embedding layer, we calculate the spatio-temporal features $\mathcal{ST} = \{H_1, H_2, ..., H_L\}$ for the following travel time estimation on segment sequence S; furthermore, we get H^* to represent spatio-temporal features of the entire trajectory after feeding $\mathcal{ST}$ to the transportation mode-aware attention layer for travel time estimation of the path.

Apart from spatio-temporal features mined from trajectories, we also incorporate other information which affects the travel time, including the distance d_i of each segment s_i which is replaced by the distance $\mathcal{D}$ of the entire path during the entire travel time estimation of the path; the transportation mode $\tilde{m}_i$ which is embedded into values in real space as described in Sect. 4.3; other external features $\mathcal{F}$ such as userID, the day of a week, holidays, etc. All these information can be concatenated as $V_{en} = \{H^*, \mathcal{D}, \mathcal{F}\}$ for entire time estimation and $V_{part,i} = \{H_i, d_i, \tilde{m}_i, \mathcal{F}\}$ for trajectory segment time estimation. Then we use multi-layer perceptron (MLP) [5] which is often applied in regression analysis in two estimation tasks. We feed $V_{part,i}$ into MLP to output the final estimated time on s_i, $\{\hat{t}_1, \hat{t}_2, ..., \hat{t}_L\}$ and feed V_{en} into another MLP to output the final estimated travel time $\hat{\delta}$ of the queried path.

Finally, we introduce the loss function in our model. This is an end-to-end model and the training goal is to minimize the loss function. We use the mean absolute percentage error (MAPE) as the objective function in the training phase because it is a relative error which can estimate time of both the short paths and long paths. In addition, we use multiple metrics to evaluate the estimation accuracy in the testing phase. We define the loss of travel time estimation on each trajectory segment as,

$$Loss_{part} = \frac{1}{L}\sum_{i=1}^{L}\left|\frac{t_i - \hat{t}_i}{t_i}\right| \tag{9}$$

Here L denotes the number of segments of the path, t_i is the ground truth travel time on segment s_i and $\hat{t}_i$ is the estimated value. At the same time, the loss of time estimation of the entire path is in the following,

$$Loss_{en} = \left| \frac{\delta - \hat{\delta}}{\delta} \right| \tag{10}$$

where $\hat{\delta}$ is the estimated travel time of the entire path and δ is the actual time. Finally we combine the loss function as the following,

$$Loss = \lambda Loss_{part} + (1 - \lambda) Loss_{en} \tag{11}$$

where λ is hyper-parameters to tune the impacts of $Loss_{part}$ and $Loss_{en}$. The goal of our model is trained to minimize *Loss*. In particular, during the testing phase, $\hat{\delta}$ is the estimated travel time of the queried path.

5 Experiments

In this section, we conduct experiments on GeoLife data to evaluate the performance of TADNM against several baselines for travel time estimation and compare our model under different parameter settings.

5.1 Datasets

- *Geolife-Pickup:* We use GeoLife dataset [26–28] published by Microsoft Research. The trajectory points were sampled at an interval of 1–5 s and 73 users have annotated labels on their trajectories with transportation mode over a period of six months. Each annotation contains a transportation mode with start and end time. The trajectories may be mixed-mode that contain more than one type of transportation modes or contain only one mode. We remove trajectories which users' annotation labels cannot be mapped into, because the start and end time in the annotation was not in the time period of these collected trajectories. We use the trajectory data of 58 users with 8 transportation modes (walk, bike, motorcycle, train, bus, subway, taxi, car) after the filtering operations above.
- *Geolife-Small:* In order to fully demonstrate the advantages of our method on mixed-mode trajectories which contain multiple transportation modes, we extract a subset of *Geolife-Pickup* called *Geolife-Small*, which contains all the mixed-mode trajectories in *Geolife-Pickup*.

5.2 Implementation Details

The architecture of our model TADNM is shown in Fig. 1, parameters of the model and experimental settings are described as follows:

- In the spatial representation layer, each trajectory segment is mapped into a region with $\omega \times \omega$ grids, we evaluate our model with different values of ω and finally set $\omega = 100$ on *Geolife-Pickup* and $\omega = 150$ on *Geolife-Small*; convolutional filter size h is 2 and we set the stride e of convolution which controls the stride for the cross-correlation as the same value as the filter size, the number of filters k is 32.
- In the temporal embedding layer, we use two GRU layers and fix the hidden units in the GRU as 128.
- In the transportation-mode aware attention layer, the transportation mode is embedded to $\mathbb{R}^{32}$.
- In the time prediction layer, userID, the day of the week, and holidays in the external features $\mathcal{F}$ are embedded to $\mathbb{R}^{16}$, $\mathbb{R}^{8}$, $\mathbb{R}^{8}$ respectively. For two MLPs which are used to predict partial and overall travel time, we use two hidden layers with ReLU activation in the MLPs.

We set the batch size of the experiment as 16, use Adam optimizer [8] as the optimization function, the learning rate is 0.0001. Since the number of segments of the path in a batch is different, we pad the number of segments of each path to the same length which equals to the maximum number of segments of the path in this batch. We fix the coefficient λ in the loss function as 0.3 on *Geolife-Pickup* and 0.2 on *Geolife-Small* respectively after evaluating different values from 0 to 1.0 on two datasets. We randomly split a dataset into three folds, 80% data as training set, 10% as validation set and 10% as test set. The experiment is implemented by PyTorch. We train and evaluate the model on the server with one NVIDIA GTX1080 GPU and 24 CPU cores.

5.3 Baselines

We compare our model with the following baselines in travel time estimation.

- **GBDT:** Gradient boosting decision tree (GBDT) [4] is a widely-used machine learning algorithm which is an ensemble model of decision trees. In our problem, the input of GBDT is the same as the input of TADNM, including the raw GPS sequence, the corresponding transportation modes, trajectory distance and other external factors (userID, the day of the week, holidays).
- **MlpTTE:** We use multi-layer perceptron (MLP) with ReLU activation to estimate the travel time. The input of MlpTTE is also the same as the input of TADNM. The categorical values (transportation modes and external factors) are embedded into the same dimensions as TADNM.
- **WDR:** WDR [20] is a model based on deep learning which extracts manually-craft features by feature engineering first and then feeds them into wide linear models, deep neural models and recurrent neural networks to jointly train.
- **Deeptravel:** Deeptravel [25] is a model based on deep learning which extracts spatial and temporal embeddings, driving state features and traffic features in a granularity of spatial grids and then uses Bi-LSTMs to estimate the travel time.

- **DeepTTE:** DeepTTE [17] is a model based on deep learning which transforms the raw GPS sequence to a series of feature maps to capture local spatial correlations, then uses LSTM to learn the temporal dependencies and estimate the travel time.

5.4 Performance Comparison

The comparison results of all the evaluated approaches are shown in Table 1. We use three metrics to evaluate the performance of our method, including the mean absolute percentage error (MAPE), the rooted mean squared error (RMSE) and the mean absolute error (MAE). From Table 1, we can find that, first, despite using the same inputs as TADNM, MlpTTE shows worse performance than the other among all the deep learning methods, which indicates that compared with introducing more appropriate features, it is more important to use effective learning model to extract the underlying information from data. Second, DeepTravel, WDR, DeepTTE achieve good performance in TTE because all those methods capture local spatial correlations and temporal patterns by finding suitable representations in the trajectory sequence. Furthermore, our model outperforms all the previous methods in terms of all the evaluated metrics, which demonstrates considering the heterogeneity of trajectory transportation mode has great influence on improving accuracy. Last but not least, we can also observe that the overall performance on *Geolife-Pickup* is better than that on *Geolife-Small*, this may due to there is larger number of trajectories on *Geolife-Pickup* compared with *Geolife-Small* and thus the model parameters can be fully optimized in the training phase. Furthermore, the improvement of MAE, RMSE, MAPE are 7.09, 5.54, 4.16% and 17.9, 33.93, 4.41% respectively on *Geolife-Pickup* and *Geolife-Small* compared with the best performance of the existing methods, which demonstrates our method is more effective in mixed-mode trajectories.

Table 1. Performance comparison of evaluated approaches in metrics MAE, RMSE and MAPE

Dataset	*Geolife-pickup*			*Geolife-small*		
Metrics	MAE	RMSE	MAPE	MAE	RMSE	MAPE
MlpTTE	49.92	106.63	20.58%	93.76	159.64	28.65%
GBDT [4]	50.72	95.34	22.61%	96.55	155.93	30.52%
DeepTravel [25]	43.06	89.38	18.51%	82.64	146.85	26.31%
WDR [20]	37.24	78.75	17.12%	78.73	133.28	25.09%
DeepTTE [17]	35.01	70.37	16.92%	73.32	127.09	24.57%
TADNM(Ours)	**27.92**	**64.83**	**12.76%**	**55.42**	**93.16**	**20.16%**

5.5 Model Analysis

Effect of Attention Mechanism. In this part, we evaluate the performance of transportation-mode aware attention mechanism comparing against without attention. The details are as follows:

- MeanModel: We remove the transportation-mode aware layer and use the mean of the outputs of temporal embedding layer $\{H_1, H_2, ..., H_L\}$, i.e., $H_{mean} = \frac{1}{L}\sum_{i=1}^{L} H_i$ and feed H_{mean} into the following time prediction layer to predict the travel time of the entire path.
- AttenModel: The outputs of temporal embedding layer are weighted into a hidden unit after attention mechanism which analyzes the impact of different transportation modes on the travel time, i.e., $H^* = \sum_{i=1}^{L} \alpha_i H_i$, and feed H^* into the time prediction layer to compute travel time.

Table 2. Effect of attention mechanism in metrics MAE, RMSE and MAPE

Dataset	*Geolife-pickup*			*Geolife-small*		
Metrics	MAE	RMSE	MAPE	MAE	RMSE	MAPE
MeanModel	32.89	68.71	15.84%	68.90	119.61	21.95%
AttenModel	**27.92**	**64.83**	**12.76%**	**55.42**	**93.16**	**20.16%**

The performance comparison is shown in Table 2. We can observe that the employment of transportation-mode aware attention boosts the accuracy of travel time estimation on two datasets, which verifies the effectiveness of considering the heterogeneity of trajectory transportation mode in travel time estimation.

Different Grid Size Comparison. In the spatial representation layer, the grid size $\omega \times \omega$ of the trajectory image affects the modeling of spatial moving patterns in trajectories. If ω is too small, it cannot obtain detailed movement and the sparse trajectory image degrades the generalization capability in the following temporal embedding layer; if ω is too large, it generates too long trajectory image sequences, which may encounter overfitting in the training phase. We evaluate our model with different grid size $\omega \times \omega$ from $\{50 \times 50, 100 \times 100, 150 \times 150, 200 \times 200, 250 \times 250\}$, the estimation results are shown in Fig. 3(a). We finally set $\omega = 100$ on *Geolife-Pickup* and $\omega = 150$ on *Geolife-Small* according to evaluation results.

The Impact of Segment-Based TTE Task. To investigate the effect of the task of TTE on each trajectory segment in the time prediction layer and find the most suitable parameter λ in Eq. 11. We test different values of λ from 0 to 1.0 on two datasets and the results are shown in Fig. 3(b). We can find that when we set $\lambda = 0.3$ on *Geolife-Pickup* and $\lambda = 0.2$ *Geolife-Small* respectively, the model achieves the best performance in TTE.

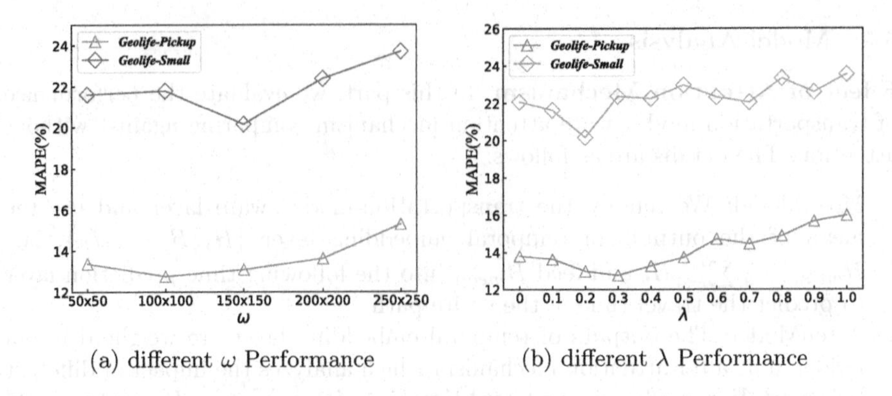

(a) different ω Performance (b) different λ Performance

Fig. 3. Performance with different parameters in metric MAPE

6 Conclusion

In this paper, we study the problem of travel time estimation. Existing approaches normally assume a path is in a single transportation mode and could not provide accurate TTE for mixed-mode paths, which are indeed common in daily life. To deal with this issue, we consider the heterogeneity of trajectory transportation mode and propose a transportation-mode aware deep neural model, namely TADNM, for more accurate travel time estimation. We first capture the spatial moving patterns of trajectory segments, which are divided by the trajectory according to their transportation modes, by finding a suitable representation in the spatial representation layer. Then GRU network is adopted in temporal embedding layer to mine the temporal dependencies in the spatial representation sequence. Furthermore, we consider the transition time of switching transportation mode in the path and take a transportation-mode aware attention layer to reflect the impact of different transportation modes on travel time. Extensive experimental results on real datasets demonstrate the effectiveness of our proposed method.

Acknowledgements. This work was supported by the National Natural Science Foundation of China under Grant Nos. 61872258, 61772356, 61876117, and 61802273, the Australian Research Council discovery projects under grant numbers DP170104747, DP180100212, the Open Program of State Key Laboratory of Software Architecture under item number SKLSAOP1801 and Blockshine corporation.

References

1. Cho, K., van Merrienboer, B., Bahdanau, D., Bengio, Y.: On the properties of neural machine translation: Encoder-decoder approaches. CoRR abs/1409.1259 (2014). http://arxiv.org/abs/1409.1259
2. Dai, J., Liu, C., Xu, J., Ding, Z.: On personalized and sequenced route planning. World Wide Web **19**(4), 679–705 (2015). https://doi.org/10.1007/s11280-015-0352-2

3. Endo, Y., Toda, H., Nishida, K., Kawanobe, A.: Deep feature extraction from trajectories for transportation mode estimation. In: Bailey, J., Khan, L., Washio, T., Dobbie, G., Huang, J.Z., Wang, R. (eds.) PAKDD 2016. LNCS (LNAI), vol. 9652, pp. 54–66. Springer, Cham (2016). https://doi.org/10.1007/978-3-319-31750-2_5
4. Friedman, J.H.: Greedy function approximation: a gradient boosting machine. Ann. Stat. **29**, 1189–1232 (2001)
5. Hornik, K., Stinchcombe, M.B., White, H.: Multilayer feedforward networks are universal approximators. Neural Networks **2**(5), 359–366 (1989)
6. Jenelius, E., Koutsopoulos, H.N.: Travel time estimation for urban road networks using low frequency probe vehicle data. Transp. Res. Part B Methodol. **53**, 64–81 (2013)
7. Jindal, I., Qin, T., Chen, X., Nokleby, M.S., Ye, J.: A unified neural network approach for estimating travel time and distance for a taxi trip. CoRR abs/1710.04350 (2017)
8. Kingma, D.P., Ba, J.: Adam: a method for stochastic optimization. In: ICLR (2015)
9. Kisialiou, Y., Gribkovskaia, I., Laporte, G.: The periodic supply vessel planning problem with flexible departure times and coupled vessels. Comput. Oper. Res. **94**, 52–64 (2018)
10. Krizhevsky, A., Sutskever, I., Hinton, G.E.: ImageNet classification with deep convolutional neural networks. In: NIPS, pp. 1106–1114 (2012)
11. Li, Y., Fu, K., Wang, Z., Shahabi, C., Ye, J., Liu, Y.: Multi-task representation learning for travel time estimation. In: KDD, pp. 1695–1704. ACM (2018)
12. Li, Y., Liu, C., Liu, K., Xu, J., He, F., Ding, Z.: On efficient map-matching according to intersections you pass by. In: Decker, H., Lhotská, L., Link, S., Basl, J., Tjoa, A.M. (eds.) DEXA 2013. LNCS, vol. 8056, pp. 42–56. Springer, Heidelberg (2013). https://doi.org/10.1007/978-3-642-40173-2_6
13. Mikolov, T., Chen, K., Corrado, G., Dean, J.: Efficient estimation of word representations in vector space. In: ICLR (2013)
14. Rahmani, M., Jenelius, E., Koutsopoulos, H.N.: Route travel time estimation using low-frequency floating car data. In: ITSC, pp. 2292–2297. IEEE (2013)
15. Shang, S., Liu, J., Zheng, K., Lu, H., Pedersen, T.B., Wen, J.-R.: Planning unobstructed paths in traffic-aware spatial networks. GeoInformatica **19**(4), 723–746 (2015). https://doi.org/10.1007/s10707-015-0227-9
16. Vaswani, A., et al.: Attention is all you need. In: NIPS, pp. 5998–6008 (2017)
17. Wang, D., Zhang, J., Cao, W., Li, J., Zheng, Y.: When will you arrive? estimating travel time based on deep neural networks. In: AAAI, pp. 2500–2507. AAAI Press (2018)
18. Wang, H., Kuo, Y., Kifer, D., Li, Z.: A simple baseline for travel time estimation using large-scale trip data. In: SIGSPATIALGIS, pp. 61:1–61:4. ACM (2016)
19. Wang, Y., Zheng, Y., Xue, Y.: Travel time estimation of a path using sparse trajectories. In: KDD, pp. 25–34. ACM (2014)
20. Wang, Z., Fu, K., Ye, J.: Learning to estimate the travel time. In: KDD, pp. 858–866 (2018)
21. Xu, J., Chen, J., Zhou, R., Fang, J., Liu, C.: On workflow aware location-based service composition for personal trip planning. Future Gener. Comput. Syst. **98**, 274–285 (2019)
22. Xu, J., Gao, Y., Liu, C., Zhao, L., Ding, Z.: Efficient route search on hierarchical dynamic road networks. Distrib. Parallel Databases **33**(2), 227–252 (2014). https://doi.org/10.1007/s10619-014-7146-x

23. Yang, B., Guo, C., Jensen, C.S.: Travel cost inference from sparse, spatio-temporally correlated time series using Markov models. PVLDB **6**(9), 769–780 (2013)
24. Yuan, N.J., Zheng, Y., Zhang, L., Xie, X.: T-finder: a recommender system for finding passengers and vacant taxis. IEEE Trans. Knowl. Data Eng. **25**(10), 2390–2403 (2013)
25. Zhang, H., Wu, H., Sun, W., Zheng, B.: Deeptravel: a neural network based travel time estimation model with auxiliary supervision. In: IJCAI, pp. 3655–3661 (2018)
26. Zheng, Y., Li, Q., Chen, Y., Xie, X., Ma, W.: Understanding mobility based on GPS data. In: UbiComp. ACM International Conference Proceeding Series, vol. 344, pp. 312–321. ACM (2008)
27. Zheng, Y., Xie, X., Ma, W.: Geolife: a collaborative social networking service among user, location and trajectory. IEEE Data Eng. Bull. **33**(2), 32–39 (2010)
28. Zheng, Y., Zhang, L., Xie, X., Ma, W.: Mining interesting locations and travel sequences from GPS trajectories. In: WWW, pp. 791–800. ACM (2009)

FedSel: Federated SGD Under Local Differential Privacy with Top-k Dimension Selection

Ruixuan Liu[1], Yang Cao[2], Masatoshi Yoshikawa[2], and Hong Chen[1](✉)

[1] Renmin University of China, Beijing, China
{ruixuan.liu,chong}@ruc.edu.cn
[2] Kyoto University, Kyoto, Japan
{yang,yoshikawa}@i.kyoto-u.ac.jp

Abstract. As massive data are produced from small gadgets, federated learning on mobile devices has become an emerging trend. In the federated setting, Stochastic Gradient Descent (SGD) has been widely used in federated learning for various machine learning models. To prevent privacy leakages from gradients that are calculated on users' sensitive data, local differential privacy (LDP) has been considered as a privacy guarantee in federated SGD recently. However, the existing solutions have a dimension dependency problem: the injected noise is substantially proportional to the dimension d. In this work, we propose a two-stage framework *FedSel* for federated SGD under LDP to relieve this problem. Our key idea is that not all dimensions are equally important so that we privately select Top-k dimensions according to their contributions in each iteration of federated SGD. Specifically, we propose three private dimension selection mechanisms and adapt the gradient accumulation technique to stabilize the learning process with noisy updates. We also theoretically analyze privacy, accuracy and time complexity of *FedSel*, which outperforms the state-of-the-art solutions. Experiments on real-world and synthetic datasets verify the effectiveness and efficiency of our framework.

Keywords: Local differential privacy · Federated learning

1 Introduction

Nowadays, massive peripheral gadgets, such as mobile phones and wearable devices, produce an enormous volume of personal data, which boosts the process of Federated Learning (FL). In the cardinal FL setting, Stochastic Gradient Descent (SGD) has been widely used [1–4] for various machine learning models, such as Logistic Regression, Support Vector Machine and Neural Networks. The server iteratively updates a global model for E epochs based on gradients of the objective function, which are collected from batches of m clients' d-dimensional local updates. Nevertheless, users' data are still under threats of privacy attacks

Y. Nah et al. (Eds.): DASFAA 2020, LNCS 12112, pp. 485–501, 2020.
https://doi.org/10.1007/978-3-030-59410-7_33

Table 1. Overview comparison (the number of epochs E, the dimension of a gradient vector d, the compressed dimension q, the privacy budget for training ϵ, the amount of participants in one aggregation m. See detailed analyses in Sects. 3.1 and 4.2).

LDP-SGD solutions	Upper bound of noise	Lower bound of batch size
Flat solution [12–14]	$O(\frac{E\sqrt{d \log d}}{\epsilon\sqrt{m}})$	$\Omega(\frac{E^2 d \log d}{\epsilon^2})$
Compressed solution [9]	$O(\frac{E\sqrt{q \log q}}{\epsilon\sqrt{m}})$	$\Omega(\frac{E^2 q \log q}{\epsilon^2})$
Our solution	$O(\frac{E\sqrt{\log d}}{\epsilon\sqrt{md}})$	$\Omega(\frac{E^2 \log d}{d\epsilon^2})$

[5–8] if the raw gradients are transmitted to an untrusted server. Local differential privacy (LDP) provides a rigorous guarantee to perturb users' sensitive data before sending to an untrusted server. Many LDP mechanisms have been proposed for different computational tasks or data types, such as matrix factorization [9], key-valued data [10,11] and multidimensional data [12–16].

However, applying LDP to federated SGD faces a nontrivial challenge when the dimension d is large. First, although some LDP mechanisms [12–14] proposed for multidimensional data are shown to be applicable to SGD (i.e., flat solution in Table 1), the injected noise is substantially proportional to the dimension d. Besides, in order to obtain an acceptable accuracy, the required batch size of clients (i.e., m) linearly depends on d. As the clients in federated learning have the full autonomy for their local data and can decide when and how to join the training process, a large required batch size impedes the model's applicability in practice. Second, a recent work [9] (i.e., compressed solution in Table 1) attempts to solve this problem by reducing the dimension from d to q with random projection [17], which is a well-studied dimension reduction technique. However, as this method randomly discards some dimensions, it introduces a large recovery error which may damage the learning performance.

Our idea is that not all the dimensions are equally "important" so that we privately select Top-k dimensions according to their contributions (i.e., the absolute values of gradients) in each iteration of SGD. A simple method for private Top-1 selection is to employ exponential mechanism [18] that returns a private dimension with a probability proportional to its absolute value of gradient. However, the challenge is how to design Top-k selection mechanisms under LDP for federated SGD with better selection strategies. Besides the absence of such private Top-k selection mechanisms, another challenge is that discarding delayed gradients causes the convergence issues, especially in our private setting where extra noises are injected. Although we can accumulate delayed gradients with momentum [19], it still requires additional design to stabilize the learning process with noisy updates.

Contributions: In this work, we take the first attempt to mitigate the dimension dependency problem in federated learning with LDP. We design, implement and evaluate a two-stage ϵ-LDP framework for federated SGD. As shown in Table 1, comparing to the state-of-the-art techniques, our framework relieves the effect of the number of dimensions d on the injected noise and the required batch size (i.e., size of clients in each iteration; the lower, the better). Our contributions are summarized below.

- First, we propose a two-stage framework *FedSel* with *Dimension Selection* and *Value Perturbation*. With our privacy analysis in Sect. 3.2, this framework satisfies ϵ-LDP for any client's local vector. Our theoretical utility analysis in Sect. 3.3 shows that it significantly reduces the dimensional dependency on LDP estimation error from $O(\sqrt{d})$ to $O(1/\sqrt{d})$. We also enhance the framework to avoid the loss of accuracy by accumulating delayed gradients. Intuitively, delayed gradients can improve the empirical performance and fix the convergence issues. In order to further stabilize the learning in the private setting with noises injected, we modify an existing accumulation[19]. Our analysis and experiments validate that this modification reduces the variance of noisy updates (Sect. 3).
- Second, we instantiate the *Dimension Selection* stage with three mechanisms, which are general and independent of the second stage of value perturbation. The privacy guarantee for the selection is provided. Besides, we show the advance of utility and computation cost by extending Top-1 to Top-k case with analysis and experiments (Sect. 4).
- Finally, we perform extensive experiments on synthetic and real-world datasets to evaluate the proposed framework and the private Top-k dimension selection mechanisms. We also implement a *hyper-parameters-free* strategy to automatically allocate the privacy budgets between the two stages with better utility. Significant improvements are shown in test accuracy comparing with the existing solutions [9,12–14] (Sect. 5).

The remainder of this paper is organized as follows. Section 2 presents the technical background. Section 3 illustrates the two-stage privatized framework with analyses. Section 4 proposes the Exponential Mechanism (EXP) for Top-1 selection and extends to Top-k case with Perturbed Encoding Mechanism (PE) and Perturbed Sampling Mechanism (PS). Section 5 provides results on both synthetic and real-world datasets and a *hyper-parameters-free* strategy. Section 6 gives an overview of related works. Section 7 concludes the paper. Due to the limited space, we put the complete pseudocodes and proofs in a full version of this paper.

2 Preliminaries

2.1 Federated SGD

Suppose a learning task defines the objective loss function $L(w; x, y)$ on example (x, y) with parameters $w \in \mathbb{R}^d$. The goal of learning is to construct an empirical minimization as $w^* = \arg\min_w \frac{1}{N} \sum_{i=1}^{N} L(w; x_i, y_i)$ over N clients' data. For a single iteration, a batch of m clients updates local models in parallel with the distributed global parameters. Then they transmit local model updates to the server for an average mean gradient to update the global model as: $w^t \leftarrow w^{t-1} - \alpha \frac{1}{m} \sum_{i=1}^{m} \nabla L(w^{t-1}; x_i, y_i)$. Without loss of generality, we describe our framework with the classic setting with one local update for each round.

2.2 Local Differential Privacy

Local differential privacy (LDP) is proposed for collecting sensitive data through local perturbations without any assumption on a trusted server. $\mathcal{M}$ is a randomized algorithm that takes a vector v as input and outputs a perturbed vector v^*. ϵ-LDP is defined on $\mathcal{M}$ with a privacy budget ϵ as follows.

Definition 1 (Local Differential Privacy [20]**).** *A randomized algorithm* $\mathcal{M}$ *satisfies* ϵ*-LDP if and only if the following is true for any two possible inputs* $v, v' \in \mathcal{V}$ *and output* v^**:* $Pr[\mathcal{M}(v) = v^*] \leq e^{\epsilon} \cdot Pr[\mathcal{M}(v') = v^*]$.

2.3 Problem Definition

This paper studies the problem of federated SGD with LDP. Note that in the practical non-private setting, the original gradient $g^t \leftarrow \nabla L(w^{t-1}; x, y)$ can be sparsified [21] or quantized [22] before a transmission. For generality, we do not limit the form of local gradient and use v_i to denote the local gradient calculated from client u_i's record (x_i, y_i) and the global parameters w^{t-1}.

Suppose the global model iterates for E epochs with a learning rate α and a total privacy budget ϵ for each client. For a single epoch, clients are partitioned into batches with size m. Then the privatized mechanism $\mathcal{M}$ privatizes m local updates before they are aggregated by the untrusted server for one iteration. The global model is updated as: $w^t \leftarrow w^{t-1} - \alpha \frac{1}{m} \sum_{i=1}^{m} \mathcal{M}(v_i)$. We aim to propose an ϵ-LDP framework with a specialized mechanism $\mathcal{M}$ for private federated training against the untrusted server. Moreover, we attempt to mitigate the dimension dependency problem for a higher accuracy.

3 Two-Stage LDP Framework: FedSel

In this section, we propose a two-stage framework *FedSel* with dimension selection and value perturbation as shown in Fig. 1. The framework and differences from existing works are presented in Sect. 3.1. We prove the privacy guarantee in Sect. 3.2 and analyze the stability improvement in Sect. 3.3.

3.1 Overview

We now illustrate the proposed framework in Algorithm 1 and compare it with the flat solutions [12–14] and the compressed solution [9]. In our framework, the server first initiates the ratio $\mu \in [0, 1]$ for privacy budget allocation and starts the iteration. For the procedure on a local device: (i) Current gradient g^t is accumulated with previously delayed gradients r^{t-1} (line 15). (ii) An important dimension index is privately selected by **Dimension Selection** (line 16). (iii) The value of the selected dimension plus its momentum (line 17) is perturbed by **Value Perturbation** (line 18). The accumulation in step(i) and the momentum in step(iii) derive from the work of Sun et al. [19] to compress local gradient with little loss of accuracy and memory cost. We adapt the accumulation in our framework to stabilize the iteration with noises. The step(ii) is designed to alleviate the dimensional bottleneck and we analyze the improvement of accuracy in Sect. 4.2.

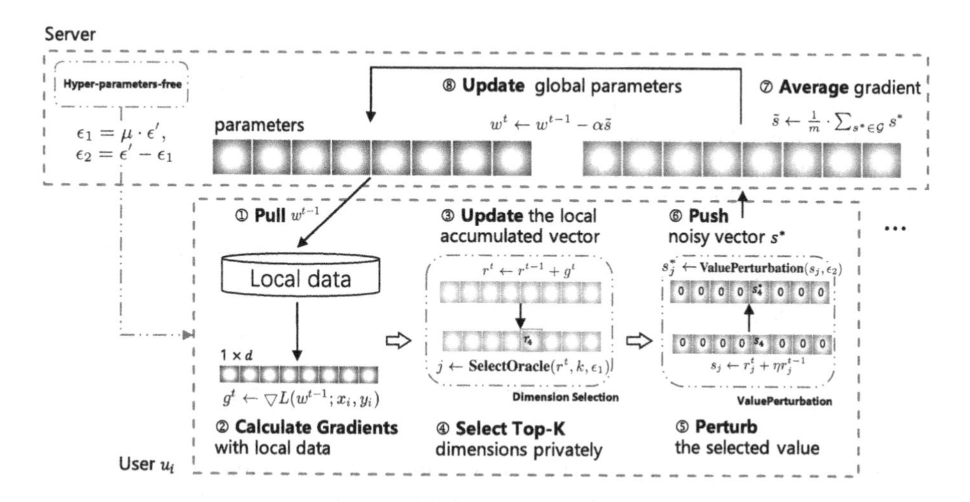

Fig. 1. Two-stage LDP framework.

Comparison with Existing Works. The most significant difference from existing works is the way of deciding which dimension to upload. In the flat solution, each client randomly samples and perturbs c dimensions from d. the perturbed value is enlarged by d/c for an unbiased mean estimation. Thus, the injected noise is also amplified. For the compressed solution, d dimensions of the gradient g is reduced to q dimensions by multiplying the vector g with a public random matrix $\Phi_{d\times q}$ drawn from the Gaussian distribution with mean 0 and variance $1/q$. Then an index is randomly sampled from q dimensions with its value perturbed. The estimated mean of compressed vector is then approximately recovered with the pseudo-inverse of Φ. Even if the random projection has the strict isometry property, it ignores the meaning of the gradient magnitude and brings recovery error. Our framework differs the compressed solution because we utilize the magnitude property for gradient value.

3.2 Privacy Guarantee

In Algorithm 1, both the local accumulated vector r and the vector with momentum s are true local vectors that are directly calculated from the private data.

Algorithm 1. Two-stage LDP framework of federated learning

$E, N, m, \mu, \alpha, \epsilon$ for the server; $\epsilon_1, \epsilon_2, \eta, k$ for N clients;

GlobalUpdate:

1: initialize $t = 1, w^0, r_i^0 \leftarrow \{0\}^d$ for $i \in [N]$
2: $\mu \leftarrow$ **HyperParametersFree**(m, ϵ, d)
3: initialize $\epsilon' = \epsilon/E, \epsilon_1 = \mu \cdot \epsilon', \epsilon_2 = \epsilon' - \epsilon_1$
4: **for** each epoch $1, \cdots, E$ **do**
5:     **for** each sample batch with size m **do**
6:         initialize $\mathcal{G}$ as an empty set
7:         **for** each client **do**
8:             $s^* \leftarrow$ LocalUpdate$(w^{t-1}, \epsilon_1, \epsilon_2, \eta, k)$
9:             add s^* to $\mathcal{G}$
10:        $\tilde{s} \leftarrow \frac{1}{m} \cdot \sum_{s^* \in \mathcal{G}} s^*$ # aggregation
11:        $w^t \leftarrow w^{t-1} - \alpha\tilde{s}, t = t + 1$
12: **return** final global model w

LocalUpdate: $(w^{t-1}, \epsilon_1, \epsilon_2, \eta, k)$

13: initialize $s^* \leftarrow \{0\}^d$
14: $g^t \leftarrow \nabla L(w^{t-1}; x, y)$
15: $r^t \leftarrow r^{t-1} + g^t$ # adapted accumulation
16: $j \leftarrow$ **SelectOracle**(r^t, k, ϵ_1)
17: $s_j \leftarrow r_j^t + \eta r_j^{t-1}$
18: $s_j^* \leftarrow$ **ValuePerturbation**$(s_j, \epsilon_2), r_j^t \leftarrow 0$
19: **return** s^*

So we abuse $v \in \mathbb{R}^d$ to denote them in the following statement. As previously defined by Shokri et al. [23], there are two sources of information that we intend to preserve for the local vector: (i) how a dimension is selected and (ii) the value of the selected dimension. Let $z = \{0,1\}^d$ indicate the ground-truth Top-k status. We decompose the protection goal into following privacy definitions for dimension selection and value perturbation. When combining the two stages together, we can provide an ϵ-LDP guarantee with Theorem 1.

Definition 2 (LDP Dimension Selection). *A randomized dimension selection algorithm $\mathcal{M}_1$ satisfies ϵ_1−LDP if and only if for any two status vectors $z, z' \in \{0,1\}^d$ and any output $j \in [d]$: $Pr[\mathcal{M}_1(z) = j] \leq e^{\epsilon_1} \cdot Pr[\mathcal{M}_1(z') = j]$.*

Definition 3 (LDP Value Perturbation). *A randomized value perturbation algorithm $\mathcal{M}_2$ satisfies ϵ_2−LDP if and only if for any two numeric values v_j, v_j' and any output v_j^*: $Pr[\mathcal{M}_2(v_j) = v_j^*] \leq e^{\epsilon_2} \cdot Pr[\mathcal{M}_2(v_j') = v_j^*]$.*

Theorem 1. *For a true local vector v, if the two-stage mechanism $\mathcal{M}$ first selects a dimension index j with $\mathcal{M}_1$ and then perturbs v_j with $\mathcal{M}_2$ under ϵ_1-LDP and ϵ_2-LDP respectively, $\mathcal{M}$ satisfies $(\epsilon_1 + \epsilon_2)$-LDP.*

Proof. Theorem 1 stands only when for any two possible local vectors v, v', the conditional probabilities for $\mathcal{M}$ to give the same output v^* satisfy the following condition: $Pr[v^*|v] \leq e^{\epsilon_1+\epsilon_2} Pr[v^*|v']$. Let z, z' denote selection status vectors of

v, v'. As we are considering the case where v, v' have the same output v^*, we end the proof with:

$$\frac{Pr[v^*|v]}{Pr[v^*|v']} = \frac{Pr[z|v]Pr[j|z]Pr[v_j^*|v_j]}{Pr[z'|v']Pr[j|z']Pr[v_j^*|v_j']} \leq \frac{Pr[z|v]}{Pr[z'|v']} \cdot e^{\epsilon_1} e^{\epsilon_2} = e^{(\epsilon_1+\epsilon_2)}.$$

3.3 Variance Analysis for Accumulation

In the existing non-private gradients accumulation [19,21,24,25], local vectors are accumulated as $r^t = r^{t-1} + \alpha g^t$. We adapt it to $r^t = r^{t-1} + g^t$ and scale it with the learning rate in line 11. This aims to reduce the variance of noisy local updates and stabilize the iteration. Suppose the jth dimension of a gradient is selected after T rounds of delay, and $s_{i,j}$ denotes the gradient value. In each iteration of Algorithm 1, the update for the global parameter w_j from user u_i is denoted as $\triangle_T w_{i,j} = \frac{\alpha}{m} s_{i,j}^*$. If αg is accumulated [19], the update is $\triangle_T' w_{i,j} = \frac{1}{m}(\alpha s_{i,j})^*$.

For LDP mechanisms of mean estimation, the input range is $[-1, 1]$. Thus all inputs for a perturbation mechanism should be clipped to the defined input range. When the value for the selected dimension is in the input range, we can easily have: $Var[\triangle_T w_{i,j}] = Var[\triangle_T' w_{i,j}]$. It should be noted that our selected gradient value has significantly larger magnitude. When initial values of both cases are clipped to the input domain, we define the clipped input as ξ. Then we have: $Var[\triangle_T w_{i,j}] = \frac{\alpha^2}{m^2} Var[\xi^*], Var[\triangle_T' w_{i,j}] = \frac{1}{m^2} Var[\xi^*]$. As the learning rate α is typically smaller than 1, such as 0.1, we then have Theorem 2. Therefore, the slight adaption of scaling procedure can reduce the variance of single client's noisy update. The advance of a smaller variance will lead to a more accurate learning performance as validated in the experimental part, Fig. 3(a).

Theorem 2. *Accumulating g instead of αg for the dimension j of user u_i has an update on the global model's parameter with a less variance:*

$$Var[\triangle_T w_{i,j}] \leq Var[\triangle_T' w_{i,j}], \text{ where } \alpha \in (0, 1).$$

4 Private Dimension Selection

To instantiate the ϵ_1-LDP dimension selection in Algorithm 1 line 16, we design LDP mechanisms from Top-1 to Top-k case. Then, we analyze the accuracy and time complexity of proposed LDP mechanisms.

4.1 Selection Mechanisms

Exponential Mechanism (EXP): Exponential mechanism [18] is a natural building block for selecting private non-numeric data in the centralized differential privacy. We modify this classic method to meet the selection requirement. A client's accumulated vector r is first sorted in ascending order of its absolute value. As a special case for Top-1 selection, the status vector in EXP is defined as: $z = \{1, \cdots, d\}^d$ instead of a binary vector. Intuitively, the dimension with the largest magnitude of its absolute value should be output with the highest probability. For the jth dimension, we assign the selection status as its ranking z_j. Thus, the index $j \in [d]$ is sampled unevenly with the probability $\frac{\exp(\frac{\epsilon_1 z_j}{d-1})}{\sum_{i=1}^{d} \exp(\frac{\epsilon_1 z_i}{d-1})}$. The privacy guarantee is shown Lemma 1.

Lemma 1. *EXP selection is* ϵ_1-*locally differentially private.*

Proof. Given any two possible ranking vectors as $z, z' \in \{1, \cdots, d\}^d$. j denotes any output index of EXP. The following conditional probability ends the proof:

$$\frac{Pr[j|z]}{Pr[j|z']} = \frac{\exp(\frac{\epsilon_1 z_j}{d-1})}{\sum_{i=1}^{d} \exp(\frac{\epsilon_1 z_i}{d-1})} / \frac{\exp(\frac{\epsilon_1 z'_j}{d-1})}{\sum_{i=1}^{d} \exp(\frac{\epsilon_1 z'_i}{d-1})} \leq \frac{\exp(\frac{\epsilon_1 \cdot d}{d-1})}{\exp(\frac{\epsilon_1 \cdot 1}{d-1})} = e^{\frac{\epsilon_1 (d-1)}{d-1}} = e^{\epsilon_1}.$$

In order to fit various learning tasks, k should be tunable. Thus, we propose two private Top-k methods to better control the selection.

Perturbed Encoding Mechanism (PE): The sorting step for the vector of absolute values $|r|$ in PE is the same as in EXP. Besides, a binary Top-k status vector z is derived. Then we perturb the vector z with the randomized response. Specifically, each status has a large probability p to retain its value and a small probability $1 - p$ to flip. For the privacy guarantee, $p = \frac{e^{\epsilon_1}}{e^{\epsilon_1}+1}$. Let $\acute{z}$ denote the privatized status vector. Since indices of non-zero elements in $\acute{z}$ are more likely to be Top-k dimensions, we gather these elements as the sample set $\mathbb{S}$. If $\mathbb{S}$ is empty, the client uploads $\perp$ and the server regards it as receiving a zero vector $s^* = \{0\}^d$. Elsewise the client randomly samples one dimension index from $\mathbb{S}$. The privacy guarantee is shown in Lemma 2.

Lemma 2. *PE selection is* ϵ_1-*locally differentially private.*

Proof. $\acute{z}$ denotes the perturbed status vector. The expected sparsity of $\acute{z}$ is:

$$\begin{aligned} l =& \mathbb{E}[||\acute{z}||_0] = \sum_{j=1}^{d} Pr[\acute{z}_j = 1|z_j] = \sum_{j=1}^{k} Pr[\acute{z}_j = 1|z_j = 1] + \sum_{j=k+1}^{d} Pr[\acute{z}_j = 1|z_j = 0] \\ =& k \cdot p + (d-k) \cdot (1-p), \text{ where } p = \frac{e^{\epsilon_1}}{e^{\epsilon_1}+1}. \end{aligned}$$

Given any two possible selection status vectors as $z, z' \in \{0, 1\}^d$ with k non-zero elements, there are two cases for the output j:(i) If the sample set $\mathbb{S}$ is not empty, $j \in \mathbb{S}$. (ii) If the sample set $\mathbb{S}$ is empty, $j = \perp$. For the first case, we have:

$$\frac{Pr[j|z]}{Pr[j|z']} = \frac{\frac{1}{l} Pr[\acute{z}_j = 1|z]}{\frac{1}{l} Pr[\acute{z}'_j = 1|z']} \leq \frac{\frac{1}{l} Pr[\acute{z}_j = 1|z_j = 1]}{\frac{1}{l} Pr[\acute{z}'_j = 1|z'_j = 0]} = \frac{e^{\epsilon_1}}{e^{\epsilon_1}+1} / \frac{1}{e^{\epsilon_1}+1} = e^{\epsilon_1}.$$

For the second case, we end the proof with the conditional probability:

$$\frac{Pr[j=\perp|z]}{Pr[j=\perp|z']}=\frac{(1-p)^k\cdot p^{d-k}}{(1-p)^k\cdot p^{d-k}}=1\leq e^{\epsilon_1}\text{ (for }\epsilon_1\geq 0)$$

Perturbed Sampling Mechanism (PS): PS selection has the same criterion as PE that regards Top-k as important dimensions. Intuitively, we define a higher probability p to sample an index j from the Top-k indices set $\{j\in[d]|z_j=1\}$ and elsewise, sample an index j from non-Top-k dimensions $\{j\in[d]|z_j=0\}$ with a smaller probability $1-p$. With the privacy guarantee in Lemma 3, we define $p=\frac{e^{\epsilon_1}\cdot k}{d-k+e^{\epsilon_1}\cdot k}$.

Lemma 3. *PS selection is ϵ_1-locally differentially private.*

Proof. Given any two possible Top-k status vector z, z' and the output index $j\in\{1,\cdots,d\}$, the following conditional probability ends the proof:

$$\frac{Pr[j|z]}{Pr[j|z']}\leq\frac{Pr[j|z_j=1]}{Pr[j|z'_j=0]}=\frac{p\frac{1}{k}}{(1-p)\frac{1}{d-k}}=e^{\epsilon_1},\text{where } p=\frac{e^{\epsilon_1}k}{d-k+e^{\epsilon_1}\cdot k}.$$

4.2 Analyses of Accuracy and Time Complexity

We analyze the accuracy improvement of the proposed two-stage framework by evaluating the error bound in Theorem 3 which stands independently of value perturbation algorithms in the second stage. The amount of noise in the average vector is $O(\frac{\sqrt{\log d}}{\epsilon_2\sqrt{md}})$ and the acceptable batch is $|m|=\Omega(\frac{\log d}{d\epsilon_2^2})$ which does not increase linearly with d. Since $\epsilon_2=\epsilon'(1-\mu)$ and $\epsilon'=\epsilon/E$, it is evident that $\Omega(E^2\log d/d\epsilon^2)<\Omega(E^2 d\log d/\epsilon^2)$ for $\mu<0.5$. Hence, we can improve the accuracy while keeping the same private guarantee. It also reminds us to allocate a small portion of privacy budget to the dimension selection.

Theorem 3. *For any $j\in[d]$, let $\tilde{s}=\frac{1}{m}\sum_{s^*\in\mathcal{G}}s^*$. $X=\frac{1}{m}\sum_{s^*\in\mathcal{G}}s$ denotes the mean of true sparse vectors without perturbations. With $1-\beta$ probability,*

$$\max_{j\in[1,d]}|\tilde{s}_j-X_j|=O(\frac{\sqrt{\log d/\beta}}{\epsilon_2\sqrt{md}}).$$

Compared with the non-private setting, LDP brings extra computation costs for local devices. For EXP, different utility scores and the summation can be initialized offline. Sorting a $d-$dimensional vector consumes $O(d\log d)$. Mapping all d dimensions to according utility values consumes $O(d^2)$ and sampling requires $O(d)$. Thus, each local device has extra time cost $O(d\log d+d^2+d)=O(d^2)$ for EXP. With a similar analysis, the extra time cost for PE is $O(d\log d+d+l)=O(d\log d)$ which is less than the time complexity $O(d^2)$ of EXP. Since PS avoids the perturbation for each dimension, it has a slightly less computation cost than PE with the magnitude of $O(d\log d)$. We validate this conclusion in experiments.

5 Experiments

In this section, we assess the performance of our proposed framework on real-world and synthetic datasets. We first evaluate our selection methods without the second stage of value perturbation. To evaluate the improvement of reducing injected noises, we compare the learning performance with the state-of-the-art works [9,13,14] and validate theoretical conclusions in Sects. 3.3 and 4.2. Moreover, we implement a *hyper-parameters-free* strategy that automatically initiates the budget allocation ratio μ to fit scenarios with dynamic population sizes.

5.1 Experimental Setup

Datasets and Benchmarks. For the convenience to control data sparsity, the synthetic data is generated with the existing procedure [26] with two parameters $C_1 = \{0.01, 0.1\}, C_2 = \{0.6, 0.9\}$ and dimensions {100(syn-L), 300(syn-H)} over {60,000, 100,000} records. The over real-world benchmark datasets includes BANK, ADULT, KDD99 which have {32, 123, 114} dimensions over {45,211, 48,842, 70,000} records. We follow a typical pre-process procedure in machine learning with one-hot-encoding every categorical attribute. We test on $l2$-regularized Logistic Regression and Support Vector Machine.

Table 2. Frameworks and variants for comparisons.

Solution	Abbreviation	Sparsification	Perturbation	Budget
Non-private	NP	Full/random/topk	-	∞
Flat [13,14]	PM/HM/Duchi	Random sampling	ϵ'	ϵ'
Compressed [9]	-RP	Random projection	ϵ'	ϵ'
Two-stage	EXP/PE/PS-	$\epsilon_1 = \mu \cdot \epsilon'$	$\epsilon_2 = \epsilon' - \epsilon_1$	ϵ'

Choices of Parameters. Since we observe that models on the above datasets can converge within 100 rounds of iterations, we set the batch size for one global model's iteration as $m = 0.01 \cdot N$. We report the average accuracy or misclassification rates of 10 times 5-folds cross-validations for one epoch unless otherwise stated. The discounting factor η and learning rate α are same in each case for a fair comparison. We set $k = 0.1d$, $\mu = 0.1$, $\lambda = 0.0001$ by default.

Comparisons with Competitors. The proposed *FedSel* framework is prefixed with selection mechanisms EXP/PE/PS. We compare it with non-private baselines (NP) of three different transmitting methods (-/RS/K): full gradient, random sampled dimension, Top-k($k = 1$) selection. We also compare with the flat solution [13,14] and the compressed solution [9] with random sampling and random projection respectively before perturbing the value. Due to limited space and variant baselines, we mainly demonstrate comparisons with the optimal competitor PM and show comparisons with other competitors in Fig. 2(i) and Fig. 2(h). Abbreviations of different variants are listed in Table 2.

5.2 Evaluation of the Dimension Selection

Convergence and Accuracy. We compare EXP/PE/PS with non-private baselines of NP, NP-RS and NP-K by visualizing the misclassification rate and accuracy of test set in Fig. 2(a) to Fig. 2(d). Note that this comparison only focuses on selection without the second stage of value perturbation. For NP-RS, we enlarge the value randomly sampled by each user for an unbiased estimation. This follows the same principle in the flat or compressed solution.

The advance of NP-K compared with NP-RS shows our essential motivation that Top-k is a more effective and accurate way to reduce the transmitted dimension. On 100-dimensional dataset, NP-K even approaches the full-gradient-uploading baseline NP in Fig. 2(a). Besides, EXP/PE/PS converge more stable and faster than NP-RS in Fig. 2(a) and Fig. 2(b) with $\epsilon_1 = 4$. With a larger privacy budget, there is a trend for EXP/PE/PS to approach the same accuracy performance as the NP-K($k = 1$) in Fig. 2(c) and Fig. 2(d). Moreover, even a small budget in dimension selection helps to increase the learning accuracy. We

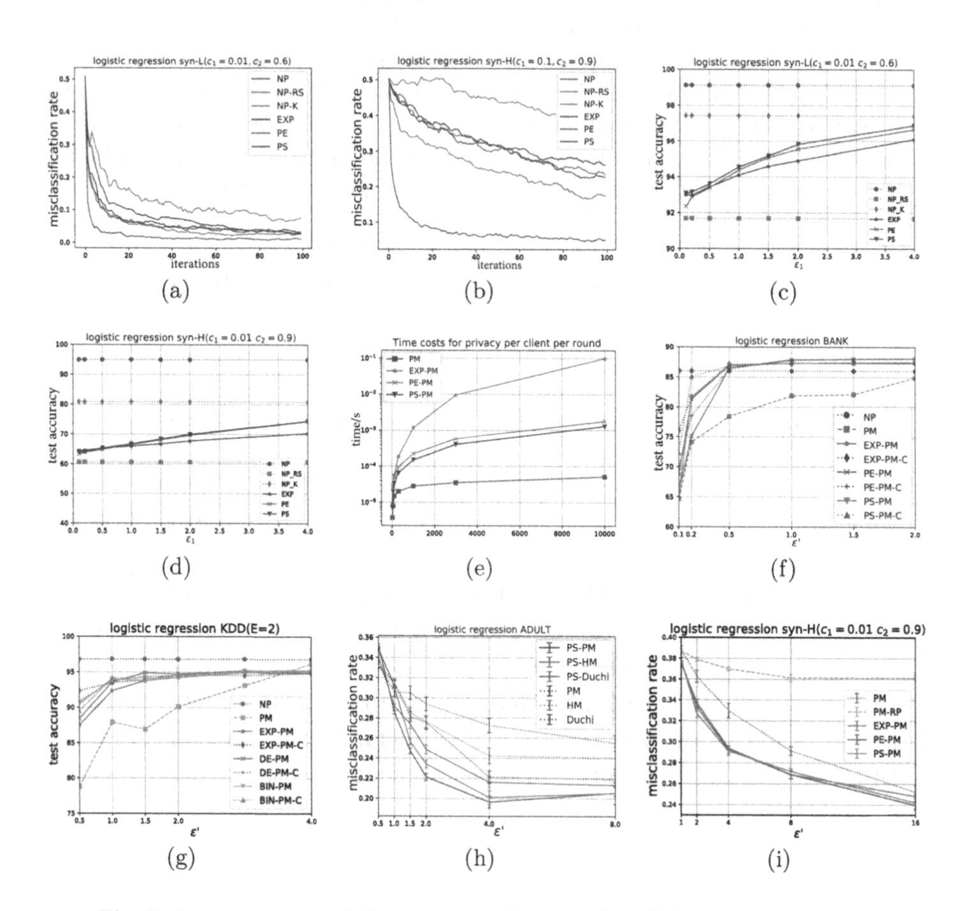

Fig. 2. Improvements of the two-stage framework and dimension selection.

can also observe that PE and PS methods which intuitively intend to select from the Top-k list have a better performance than EXP. Thus, our extension from Top-1 to Top-k is necessary.

Validation of Complexity Analysis. Here we analyze the time consumption for each client per transmission in Fig. 2(e). The time is counted by iterating over synthetic datasets with variant dimensions from 10 to 10,000. We observe that the selection stage indeed incurs extra computation cost. Consistent with previous analysis in Sect. 4.2, PS has the lowest computation cost.

5.3 Effectiveness of the Two-Stage Framework

Comparison with Existing Solutions. We compare the learning performance for EXP/PE/PS-PM with the flat solution PM/HM/Duchi [13,14] and the compressed solution [9]. Remark that we set control groups with postfix "-C" in Fig. 2(f) and Fig. 2(g) by allocating ϵ_1 for dimension selection and ϵ' for value perturbation. To further elucidate the trade-off between what we gain and what we lose, we qualify the benefit of private selection with the gap between the accuracy as the following, which is shown as Table 3. It is evident that what we gain is much larger than what we lose. This is because when we have enough privacy budget for value perturbation, increasing budget for value perturbation is not comparable to allocating surplus budget to privacy selection.

$$\text{gain} = \text{acc(EXP/PE/PS-PM-C)} - \text{acc(PM)},$$
$$\text{loss} = \text{acc(EXP/PE/PS-PM-C)} - \text{acc(EXP/PE/PS-PM)}.$$

Table 3. Gains and losses on accuracy for private selection with $\epsilon = 2$ (%).

Dataset	Model	EXP-gain	EXP-loss	PE-gain	PE-loss	PS-gain	PS-loss
syn-L-0.01-0.9	Logistic	**8.6074**	0.3517	**5.410**	1.192	**5.975**	0.4970
syn-L-0.01-0.9	SVM	**7.1950**	2.1593	**3.7704**	0.8533	**5.065**	2.0816
BANK	Logistic	**2.4197**	–0.157	**3.2338**	0.0464	**2.5525**	0.1463
BANK	SVM	**4.3823**	0.4436	**3.4369**	0.2530	**4.0244**	0.0164
KDD	Logistic	**2.0471**	0.5091	**2.5148**	0.2322	**2.0171**	0.3428
KDD	SVM	**1.85629**	–0.1625	**2.2168**	0.2288	**1.8291**	0.4465
ADULT	Logistic	**5.5745**	0.2935	**5.6445**	1.3096	**6.0535**	0.8091
ADULT	SVM	**5.5361**	0.1949	**5.6057**	0.9550	**5.1442**	0.3852

From Fig. 2(f) and Fig. 2(g), we observe that proposed two-stage solutions have higher test accuracy than the optimal private baseline PM on both models and all datasets. Given enough privacy budget on relatively low-dimensional datasets, proposed solutions even outperform the non-private baseline in Fig. 2(f). The key to this success is the inherent randomness in SGD. The slightly introduced stochasticity for privacy-preserving prevents the overfitting problem. In Fig. 2(g) with results of two epochs, our adapted local accumulation with momentum helps to reduce the impact of noisy gradients compared with the private baseline PM, especially when ϵ' is small.

From Fig. 2(h), we show a comparison with flat solutions of other perturbation methods which have the same optimal error bound as PM. Note that the value perturbation algorithms for each pair of comparison in Fig. 2(h) is the same and only differ in the selection stage. It is evident that our framework has a lower misclassification rate and standard deviation over 50 times tests. Therefore, we can conclude that this improvement is independent of value perturbation methods. We omit the comparison of EXP and PE for the same conclusion.

In Fig. 2(i), we compare with the compressed solution [9] with the same comparison ratio 0.1. The originally apply random projection in gradients of the Matrix Factorization and use another value perturbation method [12]. Since the dimension reduction idea is independent of value perturbation, for fairness comparison, we adopt the random projection idea and use the same method PM [13] to perturb value when implementing the competitor PM-RP. Our result shows that, even the error bound is reduced by random projection, the recovery error ruins the accuracy while our dimension selection still works.

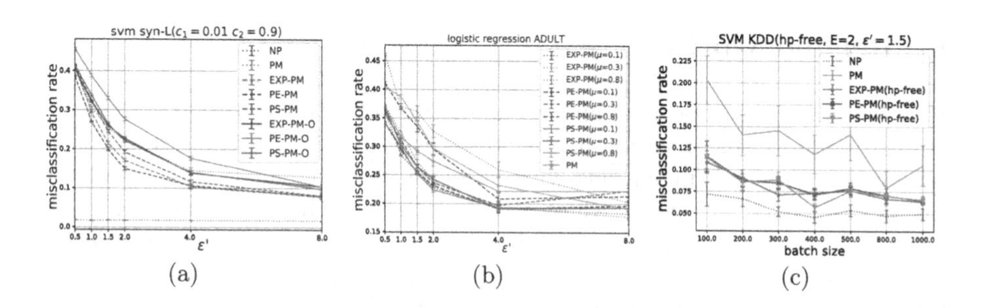

Fig. 3. Improvement of the adapted accumulation and impacts of μ

Effectiveness of the Adapted Accumulation. We validate the improvement of stability consistent with our analysis in Sect. 3.3 in Fig. 3(a). We set the variant of accumulating αg as the competitor (EXP/PE/PS-PM-O). The result shows accumulating αg directly will not improve the learning performance because what we gain by selection is offset by the larger turbulence. Thus, our adaptation is necessary for better compatibility with the private context.

Impacts of μ. As $\mu \in [0, 1]$ controls the privacy budget allocation in our two-stage framework, we evaluate μ in Fig. 3(b) with ADULT dataset while the same trends are found in other datasets. From Fig. 3(b), we observe that our framework with a small μ works no worse than the flat competitor, even when the total budget is small. In addition, $\mu = 0.1$ gives an optimal learning accuracy, and $\mu = 0.8$ leads to a worse performance with a higher misclassification rate and a significant standard deviation as expected. Thus, μ is an essential parameter that controls the trade-off between what we gain and what we lose.

The divergence of a large μ reminds us that a safe maximum threshold θ is required to guarantee this trade-off always benefits the model's accuracy. It is much easier to tune θ than μ as θ can be tested on synthetic datasets independently of total privacy budget and batch size. At the beginning of training, given the privacy budget per epoch ϵ', the model's dimension d, the available batch size m, our principle is to first allocate at least $\epsilon_2 = \Omega(\sqrt{d \log d/m})$ to the second stage. Then extra privacy budget can be allocated for dimension selection to improve the accuracy as a bonus.

In Fig. 3(c), we set a safe threshold as $\theta = 0.2$ empirically and validate the effectiveness of the proposed hyper-parameters-free strategy. Different batch sizes shown in the x-axis simulate the dynamic amount of participants when initiating a distributed learning task in practice. For the fairness to compare the test set accuracy with different batch sizes, we stop the learning process within the same number of iterations. We observe that the proposed solution with three dimension selection methods under this strategy significantly improve the model's accuracy. Besides, the proposed *hyper-parameters-free* strategy works steadily for dynamic batch sizes as the deviations among all 50 times tests are smaller than the private baseline.

6 Related Works

How we select the dimension and accumulate gradients are based on the well-studied gradient sparsification in the non-private setting. Strom et al. [27] propose only to upload dimensions with absolute values larger than a threshold. Instead of a fixed threshold, Aji et al. [25] introduce Gradient Dropping (GD) which sorts absolute values first and dropping a fixed portion of gradients. Wang et al. [26] drop gradients by trading off between the sparsity and variance. Alistarh et al. [21] show the theoretical convergence for Top-k selection. However, even if the gradient update is compressed, there still exist privacy risks because it is calculated directly with local data.

If local gradients are transmitted in clear, the untrusted server threatens clients' privacy. Nasr et al. [6] present the membership inference by only observing uploads or controlling the view of each participant. Wang et al. [8] propose a reconstruction attack in which the server can recover a specific user's local data with Generative adversarial nets (GANs). Secure attack [28] in FL is also an important topic but we focus on private issues in this paper.

Cryptography technologies face a bottleneck of heavy communication and computation costs. Bonawits et al. [29] present an implementation of Secure Aggregation, which entails four rounds interacts per iteration and several costs grow quadratically with the number of users. As for differential privacy(DP) [18] in distributed SGD, Shokri et al. [23] propose a asynchronous cooperative learning with privately selective SGD by sparse vector technique. It may lose accuracy as it drops delayed gradients instead of accumulating as our works. Agarwal et al. [30] combine gradient quantization and differential private mechanisms in synchronous setting, but it requires a higher communication cost for d-dimensional vector. It should be noticed that the privacy definition in the above works differentiate from LDP as it provides the plausible deniability for only single gradient value while LDP guarantees the whole gradient vector to be indistinguishable.

Many LDP techniques are proposed for categorical or numeric values. Randomized response (RR) method [31] is the classic method to perturb binary variables. Kairouz et al. [32] introduce a family of extremal privatization mechanisms k-RR to categorical attributes. With LDP mechanisms for mean estimation, Duchi et al. [14] suggest that LDP leads to an effective degradation in batch size. Wang et al. [13] show that compared with Duchi et al.'s work, their mean estimation mechanisms with lower worst-case variance lead to a lower misclassification rate when applied in SGD. Considering the required batch size is linearly dependent on the dimension, Bhowmick et al. [33] design LDP mechanisms for reconstruction attack with a large magnitude of privacy budget to get rid of the utility limitation of a normal locally differentially private learning.

7 Conclusions

This paper proposes a two-stage LDP privatization framework *FedSel* for federated SGD. The key idea takes the first attempt to mitigate the dimension problem in injected noises by delaying unimportant gradients. We further stabilize the global iteration by modifying the accumulation with a smaller variance on the noisy update. The improvement of proposed methods is theoretically analyzed and validated in experiments. The framework with *hyper-parameters-free* also outperforms baselines over variant batch sizes. In future work, we plan to formalize the optimal trade-off for utility and accuracy and extend *FedSel* to a more general case.

Acknowledgements. This work is supported by the National Key Research and Development Program of China (No. 2018YFB1004401), National Natural Science Foundation of China (No. 61532021, 61772537, 61772536, 61702522), JSPS KAKENHI Grant No. 17H06099, 18H04093, 19K20269, and Microsoft Research Asia (CORE16).

References

1. McMahan, B., Moore, E., Ramage, D., Hampson, S., Arcas, B.A.: Communication-efficient learning of deep networks from decentralized data. In: Artificial Intelligence and Statistics, pp. 1273–1282 (2017)
2. Bonawitz, K., et al.: Towards federated learning at scale: System design. arXiv preprint arXiv:1902.01046 (2019)

3. Yang, Q., Liu, Y., Chen, T., Tong, Y.: Federated machine learning: concept and applications. ACM Trans. Intell. Syst. Technol. (TIST) **10**(2), 1–19 (2019)
4. McMahan, H.B., Moore, E., Ramage, D., Arcas, B.A.: Federated learning of deep networks using model averaging. CoRR abs/1602.05629. arXiv preprint arXiv:1602.05629 (2016)
5. Zhu, L., Liu, Z., Han, S.: Deep leakage from gradients. In: NeurIPS, pp. 14747–14756 (2019)
6. Nasr, M., Shokri, R., Houmansadr, A.: Comprehensive privacy analysis of deep learning. In: IEEE SP (2019)
7. Fredrikson, M., Jha, S., Ristenpart, T.: Model inversion attacks that exploit confidence information and basic countermeasures. In: SIGSAC CCS, pp. 1322–1333 (2015)
8. Wang, Z., Song, M., Zhang, Z., Song, Y., Wang, Q., Qi, H.: Beyond inferring class representatives: user-level privacy leakage from federated learning. In IEEE INFOCOM, pp. 2512–2520 (2019)
9. Shin, H., Kim, S., Shin, J., Xiao, X.: Privacy enhanced matrix factorization for recommendation with local differential privacy. IEEE TKDE **30**(9), 1770–1782 (2018)
10. Gu, X., Li, M., Cheng, Y., Xiong, L., Cao, Y.: PCKV: locally differentially private correlated key-value data collection with optimized utility. In: USENIX Security Symposium (2020)
11. Ye, Q., Hu, H., Meng, X., Zheng, H.: PrivKV: key-value data collection with local differential privacy. In: IEEE SP, pp. 317–331 (2019)
12. Nguyên, T.T., Xiao, X., Yang, Y., Hui, S.C., Shin, H., Shin, J.: Collecting and analyzing data from smart device users with local differential privacy. arXiv preprint arXiv:1606.05053 (2016)
13. Wang, N., et al.: Collecting and analyzing multidimensional data with local differential privacy. In: IEEE ICDE, pp. 638–649 (2019)
14. Duchi, J.C., Jordan, M.I., Wainwright, M.J.: Minimax optimal procedures for locally private estimation. J. Am. Stat. Assoc. **113**(521), 182–201 (2018)
15. Gu, X., Li, M., Cao, Y., Xiong, L.: Supporting both range queries and frequency estimation with local differential privacy. In: IEEE Conference on Communications and Network Security (CNS), pp. 124–132 (2019)
16. Gu, X., Li, M., Xiong, L., Cao, Y.: Providing input-discriminative protection for local differential privacy. In: IEEE ICDE (2020)
17. Johnson, W.B., Lindenstrauss, J.: Extensions of Lipschitz mappings into a Hilbert space. Contempor. Math. **26**(189–206), 1 (1984)
18. Dwork, C., Roth, A.: The algorithmic foundations of differential privacy. Found. Trends Theor. Comput. Sci. **9**(3–4), 211–407 (2014)
19. Sun, H., et al.: Sparse gradient compression for distributed SGD. In: Li, G., Yang, J., Gama, J., Natwichai, J., Tong, Y. (eds.) DASFAA 2019. LNCS, vol. 11447, pp. 139–155. Springer, Cham (2019). https://doi.org/10.1007/978-3-030-18579-4_9
20. Duchi, J.C., Jordan, M.I., Wainwright, M.J.: Local privacy and statistical minimax rates. In: Annual Symposium on Foundations of Computer Science, pp. 429–438. IEEE (2013)
21. Alistarh, D., Hoefler, T., Johansson, M., Konstantinov, N., Khirirat, S., Renggli, C.: The convergence of sparsified gradient methods. In: NeurIPS, pp. 5973–5983 (2018)
22. Alistarh, D., Hoefler, T., Johansson, M., Konstantinov, N., Khirirat, S., Renggli, C.: The convergence of sparsified gradient methods. In NeurIPS, pp. 5973–5983 (2018)

23. Shokri, R., Shmatikov, V.: Privacy-preserving deep learning. In: SIGSAC CCS, pp. 1310–1321. ACM (2015)
24. Lin, Y., Han, S., Mao, H., Wang, Y., Dally, W.J.: Deep gradient compression: reducing the communication bandwidth for distributed training. In: ICLR (2018)
25. Aji, A.F., Heafield, K.: Sparse communication for distributed gradient descent. In: EMNLP, pp. 440–445 (2017)
26. Wangni, J., Wang, J., Liu, J., Zhang, T.: Gradient sparsification for communication-efficient distributed optimization. In: NeurIPS, pp. 1299–1309 (2018)
27. Strom, N.: Scalable distributed DNN training using commodity GPU cloud computing. In: INTERSPEECH (2015)
28. Fang, M., Cao, X., Jia, J., Gong, N. Z.: Local model poisoning attacks to Byzantine-robust federated learning. In: USENIX Security Symposium (2020)
29. Bonawitz, K., et al.: In: SIGSAC CCS, pp. 1175–1191, ACM (2017)
30. Agarwal, N., Suresh, A.T., Yu, F.X.X., Kumar, S., McMahan, B.: cpSGD: communication-efficient and differentially-private distributed SGD. In: NeurIPS, pp. 7564–7575 (2018)
31. Warner, S.L.: Randomized response: a survey technique for eliminating evasive answer bias. J. Am. Stat. Assoc. **60**(309), 63–69 (1965)
32. Kairouz, P., Oh, S., Viswanath, P.: Extremal mechanisms for local differential privacy. In: NeurIPS, pp. 2879–2887 (2014)
33. Bhowmick, A., Duchi, J., Freudiger, J., Kapoor, G., Rogers, R.: Protection against reconstruction and its applications in private federated learning. arXiv preprint arXiv:1812.00984 (2018)

BiGCNN: Bidirectional Gated Convolutional Neural Network for Chinese Named Entity Recognition

Tianyang Zhao[1], Haoyan Liu[1], Qianhui Wu[2], Changzhi Sun[3], Dongdong Zhan[3], and Zhoujun Li[1](✉)

[1] State Key Lab of Software Development Environment, Beihang University, Beijing, China
{tyzhao,haoyan.liu,lizj}@buaa.edu.cn

[2] Tsinghua University, Beijing, China
wu-qh16@mails.tsinghua.edu.cn

[3] East China Normal University, Shanghai, China
czsun.cs@gmail.com, ahnuzdd@gmail.com

Abstract. Recent advances on Chinese named entity recognition (NER) are mostly based on the recurrent neural network (RNN). Since RNNs are limited in parallel processing, some works apply the convolutional neural network (CNN) to perform NER. However, existing CNN-based models fail to explicitly distinguish the preceding and subsequent contexts, so they are difficult to handle cases that are sensitive to the location of the contexts. Moreover, they pay equal attention to the context within a convolution kernel, while not all the information is useful for semantic understanding. In this paper, we propose a novel CNN-based model, **Bi**directional **G**ated **C**onvolutional **N**eural **N**etwork (BiGCNN), to differentiate the entity-related information between preceding and subsequent contexts and filter out the convolution information adaptively. By incorporating automatic segmentation and glyph information, BiGCNN outperforms state-of-the-art models on four Chinese NER datasets. Additionally, benefiting from the parallelism processing, the proposed method enjoys higher training and testing efficiency, e.g., 12.04 times faster than RNN-based models, while with better performance.

1 Introduction

Named entity recognition (NER) is a task to identify text spans of named entities from text, and to classify them into predefined types like location (LOC), person (PER), organization (ORG), etc. It is an essential component in a variety of NLP applications such as event extraction [4], coreference resolution [9] and relation extraction [33]. And there is increasing interest in the field.

Most existing NER systems are based on recurrent neural network (RNN), especially long-short-term-memory (LSTM) [18,19,21]. But in RNNs, the outputs in each step rely on the previous step which hinders the parallel processing

Y. Nah et al. (Eds.): DASFAA 2020, LNCS 12112, pp. 502–518, 2020.
https://doi.org/10.1007/978-3-030-59410-7_34

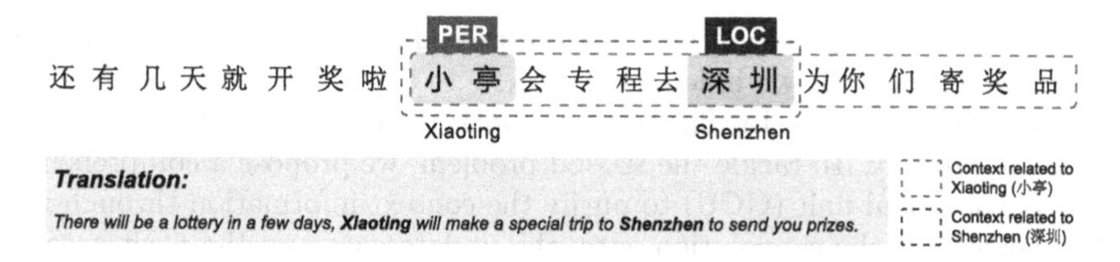

Fig. 1. An example from Weibo dataset where labels of the entities (小亭 $_{PER}$ and 深圳 $_{LOC}$) only depend on either preceding or subsequent context.

over an input sequence, so its computational speed is inevitably constrained. The convolutional neural network (CNN) operates all inputs simultaneously and thus allows parallelization over sequential inputs, which leads to higher efficiency. Therefore, some recent works proposed CNN-based approaches as an alternative to better capture the sequential information. For example, GRN [3] used CNNs with a gated relation structure and gained considerable improvement on English NER. The lastest method LR-CNN [11] adopted CNNs with a lexicon rethinking mechanism and achieved state-of-the-art performance on Chinese NER.

Although recent works adopting CNNs for NER task have achieved great success, these methods still face two limitations. **First**, existing CNN-based models use one convolution to capture both preceding and subsequent context features simultaneously and then combined them together through pooling operations, meaning that the unique information of these two parts cannot be explicitly distinguished. In many cases, the label of an named entity only depends on the preceding or the subsequent context. For example, as shown in the Chinese sentence in Fig. 1, 小亭 (*Xiaoting*) is a person, whose meaning can only be inferred from the the subsequent context "会专程去 ··· (*will make a special trip to*···)", and it is irrelevant to the previous "还有几天 ··· (*There will be*···)". Similarly, the label of 深圳 (*Shenzhen*), LOC, only depends on the preceding context "小亭 会专程去 ···(*Xiaoting will make a special trip to*···)" and has little relation with the subsequent "为你们寄奖品 (*to send you prizes*)". Therefore, to better capture the direction-sensitive cases with CNNs, it is necessary to take the bidirectionality into account for CNN structures. **Second**, current CNN-based methods lack an effective mechanism to control the context information. In traditional CNN models, all information within a certain kernel size will be propagated to the next computation stage. However, not all information is truly useful for semantic understanding. So it is difficult to select the meaningful feature based on the plain context. Furthermore, information could easily vanish through transformation. Existing gated linear unit (GLU) [6] only considers the output information to handle this issue but neglects to control the input and other fined-grained features. Hence, a more comprehensive mechanism to filter out context information becomes a pressing need for the CNN structure.

In this paper, we propose a **Bi**directional **G**ated **C**onvolutional **N**eural **N**etwork (BiGCNN), for Chinese NER task. To address the first problem, a novel and effective bidirectional CNN structure is introduced to better differentiate the entity-related information between the preceding and subsequent contexts. Specifically, It employs two independent CNNs to explicitly capture specific features from the two parts. In this manner, contexts of different directions can be modeled separately. To tackle the second problem, we propose a comprehensive gated convolutional unit (GCU) to purify the context information through multiple schemes, i.e.,the convolution gate, the update gate and the reset gate. As a result, the irrelevant information after the convolution update and the useful inputs are integrated to better represent context features. By incorporating the proposed bidirectional and gating mechanism, the CNN structure is capable of extracting the different information of the preceding and subsequent contexts adaptively and flexibly. Additionally, BiGCNN is completely based on CNN structure and thus it is much more efficient than RNN based models, which is quite beneficial to practical applications.

Specifically, we conduct extensive experiments on Chinese NER task. Compared with other languages, Chinese has the following substantial properties. Firstly, there are no explicit word boundaries for Chinese text. To avoid segmentation errors, many Chinese NER systems are based on characters rather than perform word-level NER [20,31]. Recent works further assigned words information to characters to fully use the word-level semantics [11,32]. In particular, the proposed model is character-based and further integrates characters with the automatic segmentation to take advantage of word-level features. Secondly, Chinese characters are logographic-based and the logographs always convey rich semantic meanings. For example, 河 (*river*), 湖 (*lake*), and 洋 (*ocean*) all include the radical 氵 (*water*). Hereby, Dong et al. [8] proposed to use radical sequences to model Chinese characters. However, this method ignores the case that completely different characters may share the same radical sequence. For instance, three characters 困 (*tired*), 呆 (*nerd*), and 杏 (*apricot*) can be only split into two radical sequences, i.e., "口, 木" and "木, 口". Recently, Glyce [29] collected various historical scripts and writing styles of characters and use the ensembles to encode Chinese characters. Accordingly, we also use the glyph information. But different from Glyce which uses the glyph to pre-train character representations, we treat it as an additional feature to enhance Chinese structural property. Moreover, without any manual work, we use an automatic and simple way to model the glyph feature effectively.

To summarize, the main contributions of this work are:

- We propose a novel and effective bidirectional CNN structure to distinguish the entity-related information between the preceding and subsequent contexts, which is beneficial to direction-sensitive NER cases.
- We introduce a comprehensive gating mechanism for CNN structure through multiple control schemes, to better filter out irrelevant information and retain the useful ones.
- Enhanced with segmentation and glyph features, BiGCNN achieves state-of-the-art performance on four Chinese NER datasets, and accelerates up to 12.04 times over RNN-based models because of the parallelism processing.

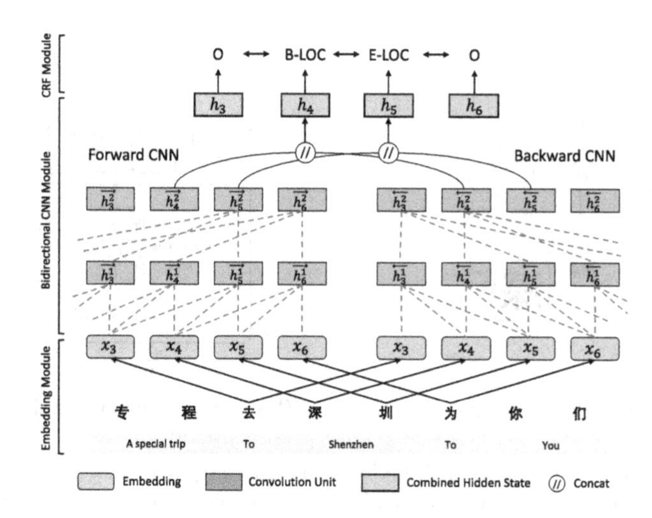

Fig. 2. Framework of BiGCNN. We take the processing of characters "深" and "圳" as an example to illustrate the architecture.

2 Model

In this section, we introduce the overall architecture of BiGCNN in detail. Formally, for an input sentence with T characters $\boldsymbol{s} = c_1, c_2, \ldots, c_T$, where c_i is the i-th character. The NER task is to predict a label sequence $\boldsymbol{y} = y_1, y_2, \ldots, y_T$, where y_i is the entity type of c_i. Suppose a sequence of characters $c_i^j = c_i, \ldots, c_j$ forms a word w, its corresponding segmentation sequence can be formulated as $w_i^j = w_i, \ldots, w_j$ where $w_i, \ldots, w_j = w, \ldots, w$. Take the sequence in Fig. 2 as an example, the characters c_4 = "深" and c_5 = "圳" compose a word "深 圳 (Shenzhen)", and then it comes to w_4 = "深圳" and w_5 = "深圳"..

As illustrated in Fig. 2, BiGCNN consists of three modules: the embedding module, the bidirectional CNN module, and the CRF module. We will elaborate on each of them in the following subsections.

2.1 Embedding Module

The embedding module focuses on mapping discrete characters into distributed semantic representations. As shown in Fig. 3a, for an input character c_i, the output of the embedding module x_i, is the concatenation of the character embedding x_i^c, the segmentation embedding x_i^w, and the glyph embedding x_i^g as:

$$x_i = [x_i^c; x_i^w; x_i^g], \tag{1}$$

where $x_i \in \mathbb{R}^{d_e}$ with d_e being the concatenated embedding size, and ; denotes the concatenation operation.

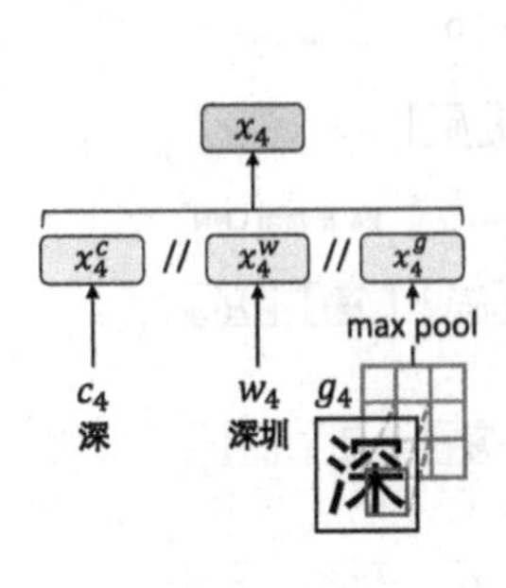

(a) Embedding Module.

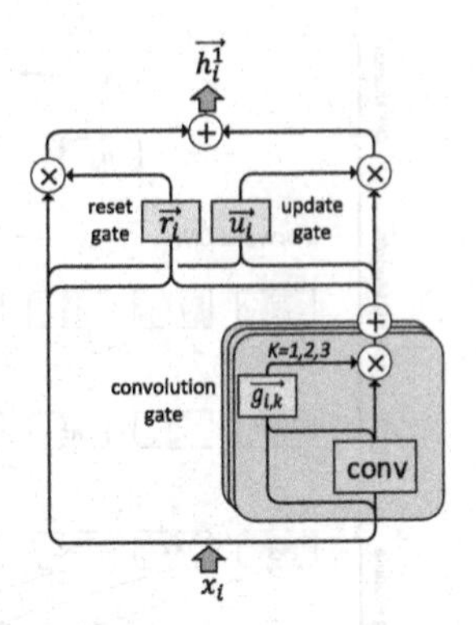

(b) Gated Convolutional Unit.

Fig. 3. Architectural elements in BiGCNN. (a) The output of the embedding module is the concatenation of the character embedding, the segmentation embedding and the glyph embedding. (b) The gated convolutional unit includes the convolution gate, the update gate and the reset gate.

Character Embedding. Given a character c_i, the character embedding x_i^c is defined as:

$$x_i^c = E^c(c_i), \tag{2}$$

where E^c is the random-initialized dictionary and be fine-tuned during training.

Segmentation Embedding. Simply using the character embedding ignores the inherent word information in the sentence. Therefore, we integrate the word segmentation into the character-level feature. For a word w_i corresponding to the character c_i, its segmentation embedding x_i^w is extracted as:

$$x_i^w = E^w(w_i), \tag{3}$$

where E^w is initialized with pre-trained word embedding and be fine-tuned during training.

Glyph Embedding. Character and word embedding can model most of the semantic information. But the embeddings of rare and unknown characters or words are less reliable. Inspired by [29], we propose a simplified glyph representation to better encode Chinese structural property. We generate a 24×24 pixel-sized image[1] for every character according to its glyph morphology. Characters are rendered to binary images via API calls to *pygame* library[2]. Defining I_i as the glyph image of the character c_i, we use a 2D convolution with a 3×3

[1] Empirically, smaller sizes than 24×24 lead to blurry glyph images, while larger sizes are unnecessary.

[2] https://www.pygame.org/docs/.

kernel to extract the glyph-based feature $\hat{x}_i^g$, and then perform a max pooling operation to aggregate the convolution result into the glyph embedding x_i^g as:

$$\begin{aligned} \hat{x}_i^g &= \text{conv}(I_i), \\ x_i^g &= \text{max\_pooling}(\hat{x}_i^g). \end{aligned} \tag{4}$$

Compared with the radical embedding in [8], the glyph embedding can uniquely encode each character based on its logographic property. Our experiments further verify its effectiveness and superiority over the radical embedding.

2.2 Bidirectional CNN Module

The bidirectional CNN module takes concatenated character-level feature as input and separately model preceding and subsequent contexts to high-level features for entity type prediction. We describe each component in detail as follows.

Bidirectional Structure. We propose a bidirectional structure typically for the convolutional operation, which leverages a forward CNN and a backward CNN to exploit the preceding and subsequent contexts, respectively.

For a T-character sentence with embeddings $\boldsymbol{x} = x_1, x_2, \ldots, x_T$, which are derived by Eq. 1, the output of the forward CNN at position i only relies on the i-th character and its preceding ones, i.e., $[x_{i-(k-1)d}, x_{i-(k-2)d}, ..., x_i]$, as:

$$\overrightarrow{h_{i,k}} = \text{conv}_k([x_{i-(k-1)d}, x_{i-(k-2)d}, ..., x_i]). \tag{5}$$

Here, $\overrightarrow{h_{i,k}} \in \mathbb{R}^{d_h}$ with d_h being the hidden size, k is the kernel size and d is the dilation factor which means each character in the sliding window skips over d characters in the input sentence. Considering that the semantic dependency between characters is not limited by a fixed distance, as revealed in [3], we utilize three different convolutions, each with a kernel size of k = 1, 2, and 3, respectively, to extract multi-scaled features.

The backward CNN performs in a similar way, except that its inputs are in the reverse order, i.e., $[x_i, x_{i+d}, ..., x_{i+(k-1)d}]$. For simplicity, we take the forward CNN as an example in the following procedures to elaborate on the module.

Gated Convolutional Unit. As shown in Fig. 3b, the gated convolutional unit (GCU) contains three kinds of control schemes, including the convolution gate, the update gate and the reset gate.

- **The convolution gate**: Generally, the contexts with different length contribute differently to understanding semantics of a given entity. Hence, it is necessary to clarify the importance of the convolutional output with a certain kernel size. For the input embedding x_i, the convolution gate corresponding to kernel size k is calculated as:

$$\overrightarrow{g}_{i,k} = \sigma\left(\boldsymbol{W}_g^k \overrightarrow{h_{i,k}} + \boldsymbol{V}_g^k x_i + \boldsymbol{b}_g^k\right), \tag{6}$$

where $\boldsymbol{W}_g^k \in \mathbb{R}^{d_h \times d_h}$, $\boldsymbol{V}_g^k \in \mathbb{R}^{d_e \times d_h}$, and $\boldsymbol{b}_g^k \in \mathbb{R}^{d_h}$ are learned parameters. $\sigma(\cdot)$ is the sigmoid function, we define its output as the convolution gate. The filtered representation of the input x_i is the gated combination of multi-kernel convolutions as:

$$\overrightarrow{h_i}' = \sum_{k=1,2,3} \overrightarrow{h_{i,k}} \otimes \overrightarrow{g_{i,k}}, \tag{7}$$

where $\otimes$ denotes the element-wise multiplication. The $\overrightarrow{h_i}'$ is the final combined convolutional hidden output, and is used to calculate the following update and reset gates.

- **The update gate**: The update gate focuses on filtering out the irrelevant information implied in the combined output above, which is defined as:

$$\overrightarrow{u_i} = \sigma\left(\boldsymbol{W}_u \overrightarrow{h_i}' + \boldsymbol{V}_u x_i + \boldsymbol{b}_u\right), \tag{8}$$

where $\boldsymbol{W}_u \in \mathbb{R}^{d_h \times d_h}$, $\boldsymbol{V}_u \in \mathbb{R}^{d_e \times d_h}$ and $\boldsymbol{b}_u \in \mathbb{R}^{d_h}$ are trainable parameters.

- **The reset gate**: As indicted in [5], the input embedding x_i is essential for prevent information decay. Therefore, we defined the reset gate to regulate the flow of input semantic information as:

$$\overrightarrow{r_i} = \sigma\left(\boldsymbol{W}_r \overrightarrow{h_i}' + \boldsymbol{V}_r x_i + \boldsymbol{b}_r\right), \tag{9}$$

which is parameterized by $\boldsymbol{W}_r \in \mathbb{R}^{d_h \times d_h}$, $\boldsymbol{V}_r \in \mathbb{R}^{e \times d_h}$ and $\boldsymbol{b}_r \in \mathbb{R}^{d_h}$.

The final hidden output of the gated convolutional unit is calculated by:

$$\overrightarrow{h_i} = \overrightarrow{h_i}' \otimes \overrightarrow{u_i} + x_i \otimes \overrightarrow{r_i}. \tag{10}$$

Similarly, we obtain $\overleftarrow{h_i}$ as the output of the backward CNN for the i-th input character x_i. Practically, we stack two convolutional layers of each direction to enlarge the receptive fields[3]. Denote $\overrightarrow{h_i^L}$ and $\overleftarrow{h_i^L}$ as the last hidden output of forward and backward CNN, the final result of the bidirectional CNN module is the concatenation as:

$$h_i = [\overrightarrow{h_i^L}; \overleftarrow{h_i^L}], \tag{11}$$

which will be mapped into entity type distributions and fed into the CRF module.

2.3 CRF Module

Conditional Random Field (CRF) is a probabilistic method that jointly models interactions between entity labels, which is incorporated in nearly all state-of-the-art NER models. Similarly, we utilize a CRF module over the bidirectional CNN module to calculate loss and perform label decoding.

[3] We also evaluated more convolution layers, but found the results comparable while the computational cost is higher.

Table 1. Statistics of Datasets.

Dataset	Type	Train	Dev	Test
OntoNotes4	Char	491.9 k	200.5 k	208.1 k
	Sentence	15.7 k	4.3 k	4.3 k
MSRA	Char	2169.9 k	–	172.6 k
	Sentence	46.4 k	–	4.4 k
Weibo	Char	73.8 k	14.5 k	14.8 k
	Sentence	1.4 k	0.27 k	0.27 k
Resume	Char	124.1 k	13.9 k	15.1 k
	Sentence	3.8 k	0.46 k	0.48 k

Loss Function. Suppose that the final output of the bidirectional CNN module forms a sequence $\boldsymbol{h} = h_1, h_2, \ldots, h_T$, where h_i corresponds to the hidden state of the i-th character derived from Eq. 11. Given a label sequence $\boldsymbol{y} = y_1, y_2, \ldots, y_T$, $p(\boldsymbol{y}|\boldsymbol{h})$ is defined as the probability of using $\boldsymbol{y}$ as the prediction sequence for the sentence as follows:

$$p(\boldsymbol{y}|\boldsymbol{h}) = \frac{\prod_{i=1}^{N} \phi_i(y_{i-1}, y_i, \boldsymbol{h})}{\sum_{y' \in \mathcal{Y}(\boldsymbol{h})} \prod_{i=1}^{N} \phi_i(y'_{i-1}, y'_i, \boldsymbol{h})}. \tag{12}$$

Here, $\mathcal{Y}(\boldsymbol{h})$ denotes the set of all possible label sequences. $\phi_i(y_{i-1}, y_i, \boldsymbol{h}) = \exp(\boldsymbol{W}_{\text{CRF}}^{y_i} h_i + \boldsymbol{b}_{\text{CRF}}^{y_{i-1} \rightarrow y_i})$, where $\boldsymbol{W}_{\text{CRF}} \in \mathbb{R}^{d_h \times d_l}$ and $\boldsymbol{b}_{\text{CRF}} \in \mathbb{R}^{d_l \times d_l}$ with d_l being the label vocabulary size. $\boldsymbol{W}_{\text{CRF}}^{y_i}$ is the column corresponding to label y_i, and $\boldsymbol{b}_{\text{CRF}}^{y_{i-1} \rightarrow y_i}$ is the transition probability from label y_{i-1} to y_i.

During training, the loss function $\mathcal{L}$ is defined as the negative log-likelihood:

$$\mathcal{L} = -\sum_h \log p(\boldsymbol{y}|\boldsymbol{h}). \tag{13}$$

Label Decoding. During inference, we predict the label sequence y^* with the maximal likelihood which can be efficiently settled by the Viterbi algorithm:

$$y^* = \arg\max_{y \in \mathcal{Y}(\boldsymbol{h})} p(\boldsymbol{y}|\boldsymbol{h}). \tag{14}$$

3 Experiments

To demonstrate the effectiveness of the bidirectional gated CNN structure, we evaluate BiGCNN over four widely-used Chinese NER datasets, and compare it with existing state-of-the-art methods.

3.1 Datasets

We conduct extensive experiments on four datasets including OntoNotes4 [28], MSRA [16], Weibo [22] and Resume [32]. Table 1 shows the statistics of the four datasets. Gold-standard segmentation is provided for OntoNotes4 and MSRA training set. Since the golden segmentation is not available for Weibo and Resume datasets as well as the test set of MSRA, we adopt the automatic neural segmentor as [32] to construct the word segmentation for each sentence. The data splits and tagging scheme(i.e., BIOES) follow those in [32].

3.2 Settings

For evaluation, standard precision (P), recall (R) and F1-score (F1) are used as metrics in our experiments. Other settings of our model are described as follows:

Embeddings. The dimension of character embedding is 80 and the embedding matrix is randomly initialized with kaiming uniform [13]. We use the word segmentation embedding following paper [34], which is trained on Chinese Baidu encyclopedia [17]. For the out-of-vocabulary words, we initialize them with a uniform distribution as in [21]. The dimension of glyph embeddings is set to 20.

Weight Initialization. We initialize all weights of convolution operations (Eq. 4 and Eq. 5) in the same way as in [10], while other weights are initialized with kaiming uniform [13] and bias with zero.

Network Structure. The output channel of bidirectional convolution (Eq. 5) is set as 400. The first layer of the bidirectional CNN module uses a dilation factor $d = 1$ and the second layer uses $d = 2$.

Training. We use stochastic gradient descent (SGD) with momentum as the optimizer, where the batch size is 10 and the momentum is 0.9. The learning rate is initialized as $\eta_0 = 0.02$. At the end of each epoch, we update the learning rate with $\eta_t = \frac{\eta_0}{1+\rho t}$, where $\rho = 0.05$ is the decay rate and t refers to the epoch index. Dropout layers are added upon both the inputs and outputs of the bidirectional CNN module, with a dropout rate of 0.5. We train our model for 100 epochs and report the average P/R/F1 results of 5 runs for each experiment.

3.3 Experimental Results

Table 2 shows the performance of BiGCNN. The first block of sub-tables lists the recent advances for Chinese NER. Among them, LR-CNN [11] is the latest Chinese NER model based on CNN. All others employ RNN structure as backbones. Note that WC-LSTM [19] adopts four strategies to encode word information, we listed their best results for better comparison. The second block lists three baselines with the same embedding module as BiGCNN (Eq. 1). Specifically, BiLSTM [14] is the base of most RNN-based varieties. GRN [3] is the recent CNN-based model for English NER. Transformer [27] constructs character representations through attention mechanism and enjoys the parallelism property similar to CNN. All of these models adopt CRF to perform label prediction.

Table 2. Performance comparisons on the four datasets. * denotes models based on RNN structure. † denotes models based on CNN structure. The "NE" and "NM" denote F1-scores for named entities and nominal entities, following [22].

Model	**OntoNotes4**			**Resume**		
	P(%)	R(%)	F1(%)	P(%)	R(%)	F1(%)
CAN-NER [34]*	73.63	70.82	72.20	95.05	94.82	94.94
LatticeLSTM [32]*	76.35	71.56	73.88	94.81	94.11	94.46
WC-LSTM [19]*	76.09	72.85	74.43	95.27	95.15	95.21
LR-CNN [11]†	76.40	72.60	74.45	95.37	94.84	95.11
BiLSTM*	75.94	70.42	73.03	93.69	94.76	94.23
GRN [3]†	75.81	70.97	73.33	93.92	94.89	94.40
Transformer [27]	73.25	71.39	72.31	93.61	92.95	93.27
BiGCNN w/o glyph	**76.32**	**73.66**	**74.97**	94.81	**95.50**	95.15
BiGCNN	**76.86**	**74.10**	**75.46**	94.84	**95.62**	**95.23**
Model	MSRA			Weibo		
	P(%)	R(%)	F1(%)	NE(%)	NM(%)	F1(%)
Cao et al. [2]*	91.73	89.58	90.64	54.34	57.35	58.70
CAN-NER [34]*	93.53	92.42	92.97	55.38	62.98	59.31
LatticeLSTM [32]*	93.57	92.79	93.18	53.04	62.25	58.79
WC-LSTM [19]*	94.58	92.91	93.74	52.55	67.41	59.84
LR-CNN [11]†	94.50	92.93	93.71	57.14	66.67	59.92
BiLSTM*	93.63	92.26	92.94	52.56	63.44	58.25
GRN [3]†	93.37	92.00	92.68	53.08	63.53	58.90
Transformer [27]	92.75	91.18	91.96	52.35	62.56	57.20
BiGCNN w/o glyph	**94.66**	**92.93**	**93.78**	**57.42**	66.82	**60.27**
BiGCNN	**94.63**	**93.14**	**93.88**	**57.60**	**67.56**	**61.54**

Comparisons with RNN-based Models. As shown in Table 2, without external labeled features, BiGCNN achieves the state-of-the-art performance on four datasets compared with all the RNN-based models. Even removing the glyph embedding, BiGCNN still outperforms the baselines on most of the datasets and is strongly competitive with the best result on Resume, despite the performance drop slightly. This well verifies the feasibility and effectiveness of replacing RNN with our well-designed CNN on Chinese NER task. Meanwhile, the extended glyph embedding is helpful for further improvement.

Comparisons with CNN and Transformer Based Models. BiGCNN outperforms the best CNN-based model LR-CNN [11] on four datasets. Especially on OntoNotes4 and Weibo, BiGCNN consistently increasing the F1 score by 1.01% and 1.62%. Without the glyph embedding, the improvement is also

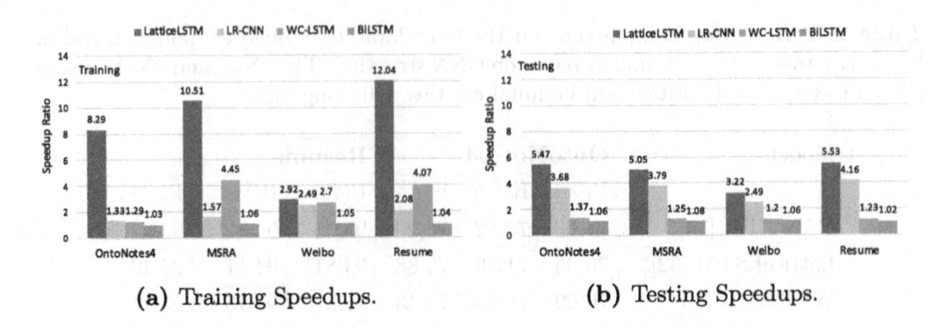

(a) Training Speedups. **(b)** Testing Speedups.

Fig. 4. Training and testing speedups gained by BiGCNN over baseline models.

considerable. Besides, compared with the latest CNN-based model for English NER, GRN [3], BiGCNN has an overall significant performance boost. Both LR-CNN and GRN are based on traditional convolutions. Hence, such improvements well demonstrate that the bidirectionality is beneficial for common CNN structures.

Additionally, the Transformer [27] also performs inferior to BiGCNN on the four datasets. We consider that the Transformer focuses on learning distant dependencies within a sequence. While for NER, the long-term dependency is not absolutely necessary [1]. Therefore, with limited data, extracting local features using CNNs is more helpful to improve NER performance than the Transformer.

3.4 Efficiency

In this section, we further evaluate the speedups gained by BiGCNN over latest state-of-the-art Chinese NER models and the basic BiLSTM model [21]. All models are trained for 10 epochs in total with the same batch size on the same physical machine and use CRF for label decoding. After each training epoch, we evaluate the learned model on the test set. We log the training and testing time costs for each epoch and calculate the average. Accordingly, the speedups are obtained for efficiency comparisons.

Speedups over State-of-the-art Chinese NER Models. LatticeLSTM [32], WC-LSTM [19] and LR-CNN [11] are the recent advanced models for Chinese NER. We run their publicly available implementations. As shown in Fig. 4, BiGCNN is much faster than the three baselines both in training and testing. Particularly on Resume, the training-time speedup is 12.04 times over LatticeLSTM. Despite that LR-CNN is also CNN-based, BiGCNN still gains noticeable speedups (an average of 1.86 and 3.53 times faster than LR-CNN for training and test). The reason may be that LR-CNN needs to stack multiple layers to extract multi-gram lexicon features, and readjust the weight of each layer after the convolutional operation, which inevitably limits efficiency. More importantly, BiGCNN does NOT sacrifice performance for speedup, meaning that it is more effective and efficient than the compared models.

Table 3. Experimental comparisons on the four datasets for enhancing BiGCNN with fine-tuned BERT. The results of BERT are listed following [29].

Model	OntoNotes4			Resume		
	P(%)	R(%)	F1(%)	P(%)	R(%)	F1(%)
BERT	78.01	80.35	79.16	96.12	95.45	95.78
Glyce	82.06	68.74	74.81	95.72	95.63	95.67
BiGCNN+BERT	**79.63**	**80.41**	**80.02**	**95.42**	**96.76**	**96.08**
Model	MSRA			Weibo		
	P(%)	R(%)	F1(%)	NE(%)	NM(%)	F1(%)
BERT	94.97	94.62	94.80	67.12	66.88	67.33
Glyce	93.86	93.92	93.89	53.69	55.30	54.32
BiGCNN+BERT	**95.51**	**95.80**	**95.65**	**67.50**	**73.25**	**68.91**

Table 4. Ablation study results (F1 score) on the four datasets.

Model	OntoNotes4	MSRA	Weibo	Resume
BiGCNN	**75.46**	**93.88**	**61.54**	**95.23**
BiGCNN w/o direction	73.85	92.70	59.02	94.06
BiGCNN w/o GCU	74.21	93.06	59.41	94.39
BiGCNN replace GCN w/GLU	74.83	93.36	60.34	95.02
BiGCNN w/o glyph	74.97	93.78	60.27	95.15
BiGCNN replace glyph w/radical	74.40	93.23	60.12	94.90

Speedups over BiLSTM Model. As mentioned before, BiLSTM is the base of many RNN-based models with more complicated structures. So taking BiLSTM by itself into comparison, which reflects the lower-bound of speedups that BiGCNN can achieve. Specifically, we replace the bidirectional CNN module of BiGCNN with a BiLSTM module, and keep other modules and hyper-parameters the same. As shown in Fig. 4, BiGCNN gains speedups of 1.06 and 1.08 on the largest dataset MSRA during training and testing, respectively. Since the experiment is derived in an end-to-end manner, the time costs for calculating the embedding and CRF modules are also non-negligible. Therefore, such speedups are still noticeable. Moreover, the efficiency is more meaningful for large-scale NER applications, which can substantially improve overall system throughput.

3.5 Enhancing BiGCNN with Pre-trained Language Model

In this section, we verify the pre-trained language model (LM) can enhance the performance of BiGCNN. BERT [7] is taken as a representative as it is a powerful and most influential pre-trained LM currently. Glyce [29] is a pre-trained Chinese character representations and is used as a strong baseline.

We concatenate the Chinese BERT feature with our character embeddings and feed them into the bidirectional CNN module. The parameters of BERT are fine-tuned during training. As shown in Table 3, the BERT has already achieved a remarkable performance on the four datasets. When combining BiGCNN with BERT, the P/R/F1 scores on the four datasets improve significantly, and outperform Glyce. Especially, the performance increases obviously by 14.59% on Weibo dataset. All these results demonstrate that BERT can enhance the performance of BiGCNN and further verify the effectiveness of BiGCNN.

3.6 Detailed Analysis

Ablation Study. We consider the following variant models:

1. *BiGCNN w/o direction*, which adopts traditional convolution layers with unidirectional kernels. To guarantee the traditional CNN to have the same coverage as the bidirectional ones, we apply three different convolutions with kernel sizes as 1, 3, 5, respectively;
2. *BiGCNN w/o GCU*, which removes the convolution gate, the update gate, and the reset gate (Eq. 6-9);
3. *BiGCNN replace GCN w/ GLU.* The gated convolutional unit (GCU) is replaced with the gated linear unit (GLU) in [6]. For a fair comparison, we use multiple convolutions with kernel sizes the same as BiGCNN for GLU;
4. *BiGCNN replace glyph w/ radical.* The glyph embedding is replaced with the radical embedding [8] to verify its superiority. We split character c_i into a radical sequence[4] and feed it to an LSTM layer, where the last hidden state is referred as the radical embedding x_i^r. Then x_i^g is replaced with x_i^r in Eq. 1.

Except for the above changes in structure, other modules and experimental settings are kept the same as BiGCNN. The CRF module is also included in all these variants. Table 4 presents the comparison results, which suggests that:

- Bidirectionality plays a crucial role in modeling context information. The model without direction only uses a single CNN to extract both preceding and subsequent context simultaneously, and the degradation of performance is substantial on the four datasets (drops 1.61% on OntoNotes4). The result verifies that distinguishing contexts can be helpful for better performance.
- The proposed GCU is beneficial to the CNN-based structure. Regarding the model without the GCU, its performance drops significantly on the four datasets, which indicates that filtering out irrelevant information and retaining useful inputs is important to improve performance. Meanwhile, the result of *BiGCNN replace GCU w/ GLU* is lower than BiGCNN, indicating the superiority of GCU over GLU and the necessity of controlling the convolution input for the gating mechanism.
- Glyph embedding can further lead to a performance boost. As mentioned in Sect. 3.3, without the glyph embedding, the performance of BiGCNN drops

[4] https://github.com/kfcd/chaizi.

Table 5. An example in OntoNotes4 test set. Characters with blue and red text highlight the correct and incorrect labeled entities, respectively.

Sentence	这 一 计 划 得 到 了 英 国 威 康 公 司 的 大 力 支 持 The plan was strongly supported by Wellcome UK.
Segment	这/一/计划/得到/了/英国/威康/公司/的/大力/支持
BiGCNN	这 一 计 划 得 到 了 [英 **B-ORG**] [国 **I-ORG**] [威 I-ORG] [康 I-ORG] [公 I-ORG] [司 E-ORG] 的 大 力 支 持 (✓)
- direction	这 一 计 划 得 到 了 [英 **B-GPE**] [国 **E-GPE**] [威 B-ORG] [康 I-ORG] [公 I-ORG] [司 I-ORG] 的 大 力 支 持 (×)
- gate	这 一 计 划 得 到 了 [英 **B-GPE**] [国 **E-GPE**] [威 O] [康 O] [公 O] [司 O] 的 大 力 支 持 (×)

slightly. In addition, when replacing the glyph feature with the radical feature, there is still a performance gap with BiGCNN. We attribute it to that, splitting characters into radical sequences can not only lead to different characters sharing the same sequence (Sect. 1) but also the loss of structural information. Therefore, it may not be as effective as the glyph embedding.

Case Study. Table 5 shows an example in OntoNotes4 test set. In this case, the correct label of "英国 (*the UK*)", i.e., `ORG`, should be inferred from the subsequent context "威康公司 (*Wellcome*)". Otherwise, it could easily be mislabeled as Geo-Political Entities (`GPE`). Benefiting from the bidirectionality, BiGCNN can concentrate on the subsequent context, and predict the label correctly. The model without directional structures fuses the preceding/subsequent context and may somehow confuse the important information, and thus result in incorrect predictions. Additionally, the model without gating mechanism fails to distill the useful context clues, which also affects the recognition result.

4 Related Work

General NER Systems. Traditional NER systems are mostly based on statistical models with hand-crafted features [25]. To alleviate the heavy feature-engineering work, later studies applied recurrent neural networks to automatically extract features. For example, the BiLSTM is first introduced in [14] to capture word-level information. Paper [15,21] further integrated character-level features into the BiLSTM model. However, these methods only focused on learning context-independent representations. To enhance the generalization of learned features, some works combined the pre-trained language model with the RNN-based NER systems and gained considerable improvement [24]. More recently, BERT [7] is designed to pre-train bidirectional transformers and achieve state-of-the-art results on NER. Specifically, we also enhance BiGCNN with the pre-trained BERT and obtain further performance boost.

Chinese NER Systems. Compared with general NER, recognizing Chinese entities is more challenging as there are no word boundaries in Chinese text. Previous works proposed to preform segmentation first and then conduct word-level NER [12,23]. These methods are vulnerable to segmentation errors and thus later works are mostly based on character-level features. In particular, a position-sensitive model [20] is presented to train character representations. The character-level BiLSTM is used in [30] to extract context features. However, these models lack necessary word-level features. Recent studies introduced LatticeLSTM [32] and WC-LSTM [19] to integrated potential words information into character-level features, and gained greatly improvements. In addition, considering the typical structure of Chinese characters, radical information and historical glyph scripts/styles are further leveraged in [8,29]. Particularly, Glyce [29] collected extensive glyphs to pre-train character representations. We serve the glyph as a simple feature to better encode characters without any pre-training and manual work, which is totally different from Glyce.

CNN-based Networks for NER. As RNNs are limited in parallel processing, some works utilized CNNs to improve the computational efficiency for NER. ID-CNN [26] stacked layers of dilated convolutions to capture long-term context features. GRN [3] introduced a gated relation network to model the local contexts and the global relations in a sentence. LR-CNN [11] presented lexicon rethinking CNN to integrate lexicons and tackle conflicts between potential words. All of these methods are based on traditional CNNs, which use one convolution to model both preceding and subsequent context information simultaneously. Differently, we propose a novel bidirectional CNN structure using two independent convolutions to capture these two different contexts separately, which can better distinguish the entity-related information.

5 Conclusion

In this paper, we propose a bidirectional gated convolutional neural network (BiGCNN) for Chinese NER, which employs two independent CNNs to better differentiate the entity-related information between preceding and subsequent contexts. We also present an effective gated convolutional unit to control context information, and introduce additional glyph feature to further improve the representation of Chinese characters. Experimental results on four datasets demonstrate that BiGCNN outperforms both RNN-based and CNN-based models while enjoying a remarkable efficiency acceleration both in training and testing.

Acknowledgments. This work was supported in part by the National Natural Science Foundation of China (Grant Nos. U1636211, 61672081,61370126), the Beijing Advanced Innovation Center for Imaging Technology (Grant No. BAICIT-2016001), and the Fund of the State Key Laboratory of Software Development Environment (Grant No. SKLSDE-2019ZX-17).

References

1. Bai, S., Kolter, J.Z., Koltun, V.: An empirical evaluation of generic convolutional and recurrent networks for sequence modeling. ArXiv (2018)
2. Cao, P., Chen, Y., Liu, K., Zhao, J., Liu, S.: Adversarial transfer learning for Chinese named entity recognition with self-attention mechanism. In: EMNLP, pp. 182–192 (2018)
3. Chen, H., Lin, Z., Ding, G., Lou, J., Zhang, Y., Karlsson, B.: Grn: gated relation network to enhance convolutional neural network for named entity recognition. In: AAAI, pp. 6236–6243 (2019)
4. Chen, Y., Xu, L., Liu, K., Zeng, D., Zhao, J.: Event extraction via dynamic multi-pooling convolutional neural networks. In: ACL, pp. 167–176 (2015)
5. Chung, J., Gulcehre, C., Cho, K., Bengio, Y.: Empirical evaluation of gated recurrent neural networks on sequence modeling. In: NIPS Workshop (2014)
6. Dauphin, Y.N., Fan, A., Auli, M., Grangier, D.: Language modeling with gated convolutional networks. In: ICML , pp. 933–941 (2017)
7. Devlin, J., Chang, M.W., Lee, K., Toutanova, K.: Bert: pre-training of deep bidirectional transformers for language understanding. In: NAACL-HLT (2019)
8. Dong, C., Zhang, J., Zong, C., Hattori, M., Di, H.: Character-based LSTM-CRF with radical-level features for Chinese named entity recognition. In: Lin, C.-Y., Xue, N., Zhao, D., Huang, X., Feng, Y. (eds.) ICCPOL/NLPCC -2016. LNCS (LNAI), vol. 10102, pp. 239–250. Springer, Cham (2016). https://doi.org/10.1007/978-3-319-50496-4_20
9. Fragkou, P.: Applying named entity recognition and co-reference resolution for segmenting English texts. Prog. Artif. Intell. **6**(4), 325–346 (2017)
10. Gehring, J., Auli, M., Grangier, D., Yarats, D., Dauphin, Y.N.: Convolutional sequence to sequence learning. In: ICML (2017)
11. Gui, T., Ma, R., Zhang, Q., Zhao, L., Jiang, Y.G., Huang, X.: CNN-based Chinese NER with lexicon rethinking. In: IJCAI (2019)
12. He, H., Sun, X.: F-score driven max margin neural network for named entity recognition in Chinese social media. In: EACL (2017)
13. He, K., Zhang, X., Ren, S., Sun, J.: Delving deep into rectifiers: Surpassing human-level performance on imagenet classification. In: ICCV (2015)
14. Huang, Z., Xu, W., Yu, K.: Bidirectional LSTM-CRF models for sequence tagging. ArXiv (2015)
15. Lample, G., Ballesteros, M., Subramanian, S., Kawakami, K., Dyer, C.: Neural architectures for named entity recognition. In: NAACL-HLT (2016)
16. Levow, G.A.: The third international Chinese language processing bakeoff: word segmentation and named entity recognition. In: SIGHAN Workshop, pp. 108–117 (2006)
17. Li, S., Zhao, Z., Hu, R., Li, W., Liu, T., Du, X.: Analogical reasoning on Chinese morphological and semantic relations. In: ACL (2018)
18. Liu, L., et al.: Empower sequence labeling with task-aware neural language model. In: AAAI (2018)
19. Liu, W., Xu, T., Xu, Q., Song, J., Zu, Y.: An encoding strategy based word-character LSTM for Chinese NER. In: NAACL-HLT, pp. 2379–2389 (2019)
20. Lu, Y., Zhang, Y., Ji, D.H.: Multi-prototype Chinese character embedding. In: LREC, pp. 855–859 (2016)
21. Ma, X., Hovy, E.: End-to-end sequence labeling via bi-directional LSTM-CNNS-CRF. In: ACL (2016)

22. Peng, N., Dredze, M.: Named entity recognition for Chinese social media with jointly trained embeddings. In: EMNLP, pp. 548–554 (2015)
23. Peng, N., Dredze, M.: Improving named entity recognition for Chinese social media with word segmentation representation learning. In: ACL (2016)
24. Peters, M.E., et al.: Deep contextualized word representations. In: NAACL-HLT (2018)
25. Sekine, S., Nobata, C.: Definition, dictionaries and tagger for extended named entity hierarchy. In: LREC, Lisbon, Portugal, pp. 1977–1980 (2004)
26. Strubell, E., Verga, P., Belanger, D., McCallum, A.: Fast and accurate entity recognition with iterated dilated convolutions. In: EMNLP (2017)
27. Vaswani, A., et al.: Attention is all you need. In: NIPS (2017)
28. Weischedel, R., et al.: Ontonotes release 4.0. LDC2011T03 (2011)
29. Wu, W., et al.: Glyce: glyph-vectors for Chinese character representations. ArXiv (2019)
30. Yang, Y., Zhang, M., Chen, W., Zhang, W., Wang, H., Zhang, M.: Adversarial learning for Chinese NER from crowd annotations. In: AAAI (2018)
31. Zhang, L., Wang, H., Sun, X., Mansur, M.: Exploring representations from unlabeled data with co-training for Chinese word segmentation. In: EMNLP (2013)
32. Zhang, Y., Yang, J.: Chinese NER using lattice LSTM. In: ACL (2018)
33. Zheng, H., Li, Z., Wang, S., Yan, Z., Zhou, J.: Aggregating inter-sentence information to enhance relation extraction. In: AAAI (2016)
34. Zhu, Y., Wang, G.: Can-NER: convolutional attention network for Chinese named entity recognition. In: NAACL-HLT (2019)

Dynamical User Intention Prediction via Multi-modal Learning

Xuanwu Liu[1,2], Zhao Li[1(✉)], Yuanhui Mao[1], Lixiang Lai[1], Ben Gao[1], Yao Deng[1], and Guoxian Yu[2]

[1] Alibaba Group, Hangzhou, China
{lizhao.lz,yuanhui.myh,lixiang.llx,gaoben.gb,dengyao.dy}@alibaba-inc.com
[2] College of Computer and Information Science, Southwest University, Chongqing, China
{alxw1007,gxyu}@swu.edu.cn

Abstract. Predicting the intention of users for different commodities has been receiving more and more attention in many applications, such as the decision of awarding bonus and the recommendation of commodity in E-commerce. Existing methods treat customer-to-commodity data as a flat data sequence while ignoring intrinsic multi-modality nature. Observing that different modalities (e.g., click ratios, collection, and purchase amount), as well as the elements within each modality, contribute differently to the prediction of purchasing intention. Besides, existing methods cannot handle the sparsity problem well. As a result, they cannot predict the user intention with sparse data. To address these issues, in this paper, we present a novel Dynamical User Intention Prediction via Multi-modal Learning (DUIPML) method to integrate different types of data to dynamically predict the user intention while reducing the impact of sparse data, which can be well applied on the practical bonus awarding scenario. Specifically, we firstly design a multi-modal fusion strategy to integrate different types of behavior information to obtain the initial user intention of each customer for each category. Next, we treat different clients with different preferences as two modalities and proposed a multi-modal alignment strategy to explore the latent correlations between different clients. After that, we communicate knowledge between the two clients based on the correlations to complete and enrich the user intention for each other, and thus to alleviate the issue of data sparsity. We apply the enriched user intention for the practical bonus awarding scenario on the Taobao platform in Alibaba group. Experiments on benchmark multi-modal datasets and the realistic E-commodity scenarios show that our method significantly outperforms related representative approaches both on effectiveness and adaptability.

Keywords: User intention prediction · Multi-modal learning · Bonus awarding.

Y. Nah et al. (Eds.): DASFAA 2020, LNCS 12112, pp. 519–535, 2020.
https://doi.org/10.1007/978-3-030-59410-7_35

1 Introduction

Predicting the intention of users is one of the most popular issues for recommended system and E-commerce platform [7,13,34,36]. The goal of user intention prediction is to recommend for each customer what they may be most interested in, so as to drive more turnover or attract more customers by the exactly bonus awarding [9,28]. In addition, predicting the user intention can also be used in many domains, such as precise bonus awarding for different customers and recognizing cheating behaviors. For these reasons, intention prediction has been heavily studied in E-commerce in recent years [1,5,12].

In many domains, the collected data can have different modalities. For example, a web page includes not only a textual description, but also images or videos to supplement its content. These different types of data are called *multi-modal* data [20,25]. Similarly, in e-commerce scenarios, there are also many different types of behaviours between customers and commodity. For example, the behaviour information includes clicking, buying, collecting and adding to shopping cart, these different behaviours can be treated as different multi-modal data. With the rapid growth of multi-modal data, efficient multi-modal learning solutions are in demand. Given the multi-modal behaviour information of a customer, we need to integrate all these information sources to obtain a more comprehensive and precise intention for each customer. Therefore, how to accomplish multi-modal learning based on these multi-type data in e-commerce is an interesting problem [11,26,33].

Due to the ability of dealing with different types of multi-modal data, Multi-modal Learning (MML) has been recently investigated and applied in many domains [21,31]. MML aims to integrate different multi-modal data to obtain a consistent expression for each instance and explore the latent correlations among modalities, by leveraging the techniques of multi-modal fusion and multi-modal alignment. Depending on using the category labels or not, existing multi-modal approaches can be roughly divided into unsupervised, supervised, and semi-supervised (see Table 1) [19]. The unsupervised methods only utilize co-occurrence(sharing similar semantic) multi-modal information to learn common representations across multi-modal data [11,35]. Supervised methods exploit label information, which provides a much better separation between classes in the multi-modal learning process. In many tasks, however, it can be difficult to attain strong supervision information due to the high cost of data labeling process. Thus, semi-supervised multi-modal learning are proposed to work with both labeled and unlabeled data [16,17,24], by integrating part or all of the available unlabeled data in its supervised learning.

Although MML methods [6,30] have achieved an impressive performance, they still have following limitations: (1) They typically assume that all the multi-modal data is sufficient, as a result that they cannot perform well as the collected data is sparse. (2) MML cannot recognize and predict new categories, which do not appear in the training data. (3) Existing multi-modal learning methods are designed only for static data, they cannot be applied in dynamical multi-modal data. However, in E-commerce scenarios, since the categories which customers have behaved on are far fewer than all the categories, and the different behaviours are insufficient (such as only clicking without buying, only buying without collecting), which may cause the collected multi-modal data too sparse to obtain a promising performance. Moreover, the practical scenarios typically demand that we can predict the intention of new categories which the users have never clicked or bought before. Given these observations, we need to design a more flexible and practical multi-modal strategy to handle the comprehensive request in the widely e-commerce scenarios.

To address the above limitations, we propose a general Dynamical User Intention Prediction via Multi-modal Learning (DUIPML) solution (as illustrated in Fig. 1) to achieve efficient user intention prediction. Specifically, DUIPML firstly designs a novel multi-modal fusion strategy to combine different types of behaviours information and to obtain the initial user-to-category intention. Next, to address the sparse and new category problem, it treats different clients as two modalities and designs a multi-modal alignment strategy to explore the correlations between clients. After that, DUIPML transfers the knowledge to complete and enrich the user intention for each other based on the correlations between clients. These enriched user intention can be utilized for bonus awarding by observing the change of intention as the bonus impact. The three components are separately explained in Sects. 3.2, 3.3 and 3.4, and the practical application on bonus awarding is elaborated in Sect. 3.5. Our main contributions are summarized as follows:

1. We propose a general Dynamical User Intention Prediction via Multi-modal Learning (DUIPML) framework, which includes multi-modal fusion, multi-modal alignment and knowledge transfer to efficiently leverage multi-modality data in the E-commerce scenarios.
2. Unlike existing methods, our proposed DUIPML can not only handle the sparse multi-modal data, but also predict the new categories which do not exist in the training data.
3. DUIPML can be utilized to decide the bonus awarding based on the enriched user intention with the dynamic multi-modal data, and DUIPML shows significantly better results than other related and representative approaches [8,10,14,16,21,23,33] both on benchmark multi-modal datasets and the realistic scenarios.

The remainder of this paper is organized as follows. We briefly review representative multi-modal learning and bonus awarding solutions in Sect. 2, and then elaborate on the proposed algorithm in Sect. 3. Section 4 provides experimental results and analysis on three real-world datasets and practical E-commerce scenarios. Conclusions and future work are given in Sect. 5.

2 Related Work

Our work is closely related to multi-modal learning (MML) and bonus awarding (BA) solutions (as illustrated in Table 1. Dynamical User Intention Prediction

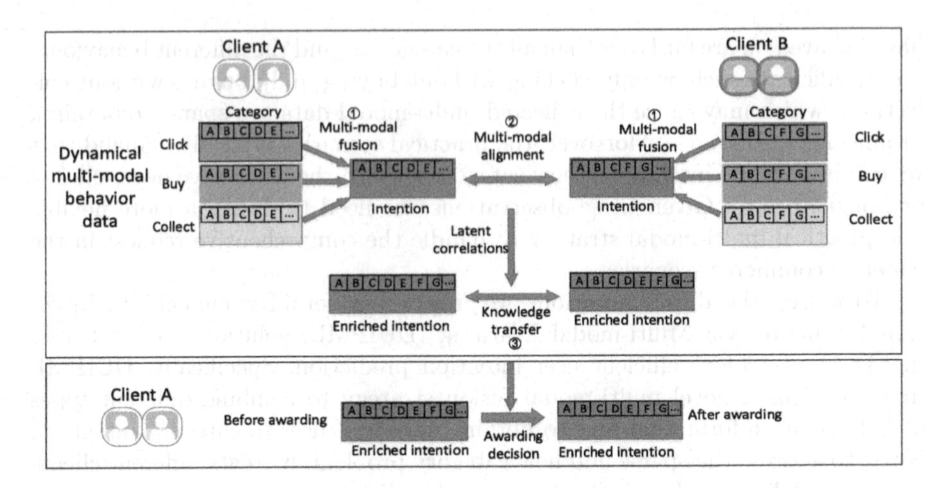

Fig. 1. The architecture of the proposed Dynamical User Intention Prediction via Multi-modal Learning (DUIPML). DUIPML mainly includes three steps: (1) Integrating different modalites to obtain user-to-category intention using a multi-modal fusion strategy; (2) Aligning multi-modal data to explore the correlations between different clients; (3) Obtaining the enriched user intention by transferring the knowledge from other clients based on the correlations to reduce the sparsity and predicting intention of customers for new categories. In addition, these dynamically enriched user intention can be efficiently applied on practical scenarios such as award bonus, by observing the change of enriched intention as the bonus reflect.

via Multi-modal Learning (DUIPML) can be viewed as an application of multi-modal learning on bonus awarding situation. Interested readers can refer to [2,25]. In the following, we give a brief review of related work, and emphasize how our approach differs.

Multi-modal Learning: Existing multi-modal learning methods can be divided into three categories: unsupervised, supervised, and semi-supervised. Unsupervised methods typically learn multi-modal functions by taking into account underlying data distributions or correlations. To name a few, Canonical Correlation Analysis (CCA) [8] achieves the multi-modal learning by maximizing the correlation between the projections of two data modalities. Collective Component Analysis (CoCA) [22] is proposed to handle dimensionality reduction on a heterogeneous feature space. On the other hand, supervised multi-modal learning approaches utilize semantic labels (or ranking order) of training data, to improve the performance. For example, Learning Coupled Feature Spaces(LCFS) [27] unifies coupled linear regressions and trace norm into a generic minimization formulation so that subspace learning and coupled feature selection can be performed simultaneously. Semantic Correlation Maximization (SCM) [33] maximizes the correlation between two modalities with respect to the semantic labels to optimize the learning functions. Semantics Preserving Hashing (SePH) [14] utilizes kernel logistic regression with a sampling strategy to learn the non-linear projections to generate one unified hash code for all observed views of any instance. Composite Correlation Quantization(CCQ) [18] finds correlation-maximal mappings that transform different modalities into isomorphic latent space. Recently, deep learning has also been incorporated with multi-modal learning. Deep cross-modal hashing (DCMH) [10] is an end-to-end learning framework with deep neural network, which combines hashing learning and deep feature learning by preserving the semantic similarity between modalities. In addition, semi-supervised multi-modal learning approaches leverage both labeled and unlabeled data to train the model. For example, Ranking-based deep cross-modal hashing (RDCMH) [16] is a semi-supervised method which can preserve multi-level semantic similarity between labeled and unlabeled multi-label objects by jointly optimizing the cross-modal semantic ranking loss and the quantization loss for cross-modal retrieval. Semi-supervised multi-view distance metric learning (SSM-DML) [32] constructs an accurate metric to precisely measure the dissimilarity between different examples associated with multiple modalities.

Bonus Awarding Algorithm: Existing bonus awarding algorithm aims to drive more customers and can be typically divided into two categories: statistical based and no-statistical based. The former typically exploit online data for training by the statistical methods, such as Thompson Sampling (TS) [23] and Upper Confidence Bounds(UCB) [21]. Thompson Sampling (TS) regards the profit distribution as a beta distribution and selects the item with respect to the largest random value for each time. Upper Confidence Bounds (UCB) calculates the standard deviation of the mean for all items and selects the item with respect to the largest one, which can explore the items with small profit. Both TS and UCB utilize the results of next day as the excitation for this day's decision. The no-statistical approaches treat this issue as a classical classification problem. For instance, DNN+classify [4] utilizes the deep neural network to train the model and solves it by classifying the users into two categories, which is now adopted in the Taobao platform.

Unlike existing MML and bonus awarding solutions, our DUIPML has to handle three open but general **problems**: (1) sparse problem; (2) new categories problem; (3) dynamical multi-modal data problem. The first problem implies that the collected multi-modal data of user-to-category is insufficient and sparse, which would reduce the prediction performance as the sparse data impacted. The second problem indicates that existing MML methods cannot recognize and predict new categories, which do not appear in the training data. The third one implies that existing multi-modal learning methods are designed only for static completely-paired data which cannot applied in dynamical unpaired multi-modal data. The distinctions between our DUIPML and existing multi-modal and bonus awarding solutions are listed in Table 1. Clearly, our DUIPML approach is more challenging and general than the existing solutions.

Table 1. Multiplicity data handled by different methods. 'un', 'su' and 'se' are short for 'unsupervised', 'supervised' and 'semi-supervised' MML, respectively.

Category		Methods	Multi-modal	Sparse data	New categories	Dynamical data
MML	un	CCA [8]	√	×	×	×
		CoCA [22]	√	×	×	×
	su	LCFS [27]	√	×	×	×
		SCM [15]	√	×	×	×
		SePH [15]	√	×	×	×
		CCQ [18]	√	×	×	×
		DCMH [10]	√	×	×	×
	se	RDCMH [16]	√	×	×	×
		SSM-DML [32]	√	×	×	×
BA		TS [23]	×	×	×	√
		UCB [21]	×	×	×	√
		DNN+classify [4]	×	×	×	√
Ours		**DUIPML**	√	√	√	√

3 Proposed Method

3.1 Notation and Problem Definition

The notation used in this paper is given in Table 2. Without loss of generality, we assume that the collected user-to-category behaviour information have l modalities data $\mathbf{X}^l \in \mathbb{R}^{n \times d}$, which represents the amount of different behaviours with respect to the d categories. As for the divided two clients with different preferences $\tilde{\mathbf{X}}^m \in \mathbb{R}^{n \times d_1}$, $\tilde{\mathbf{X}}^{m'} \in \mathbb{R}^{n \times d_2}$, the categories of the two clients are the subset of the whole categories, so that d_1 and d_2 are smaller or equal to d. DUIPML aims to complete and enrich the user-to-category intention so that outputs of DUIPML are $\tilde{\mathbf{X}}^m_{new}$, $\tilde{\mathbf{X}}^{m'}_{new} \in \mathbb{R}^{n \times d}$. As for the application on bonus awarding, DUIPML aims to obtain the decision of bonus awarding $\mathbf{A} \in \mathbb{R}^{n \times 1}$ by observing the change of the user intention as the bonus impact.

Table 2. Used Notation.

c	Number of selected candidates
l	Number of modalities
$\mathbf{X}^l \in \mathbb{R}^d$	User behaviors on different categories
$\tilde{\mathbf{X}}^m \in \mathbb{R}^{n \times d_1}$, $\tilde{\mathbf{X}}^{m'} \in \mathbb{R}^{n \times d_2}$	User intention of the two clients
$\mathbf{A} \in \mathbb{R}^{n \times 1}$	Decision of bonus awarding matrix
$\mathbf{Y} \in \mathbb{R}^{n \times d}$	Label matrix
$\tilde{\mathbf{X}}^m_{new}, \tilde{\mathbf{X}}^{m'}_{new} \in \mathbb{R}^{n \times d}$	Enriched user intention of the two clients
$\mathbf{W}^m \in \mathbb{R}^{d^1 \times d}$ and $\mathbf{W}^{m'} \in \mathbb{R}^{d^2 \times d}$	Projection matrix
$\mathbf{Z}^m \in \mathbb{R}^{n \times d}$ and $\mathbf{Z}^{m'} \in \mathbb{R}^{n \times d}$	Representation matrix in the common space

DUIPML mainly involves three parts: (1) Integrating different modalites to obtain user-to-category intention using a multi-modal fusion strategy; (2) Aligning multi-modal data to explore the correlations between different clients; (3) Transferring the knowledge for each other based on the latent correlations to reduce the impact of sparse data and predicting the intention of new categories. These three components are separately explained in Sects. 3.2, 3.3 and 3.4. In addition, we also apply the enriched user intention on the bonus awarding scenarios on the Taobao platform in Alibaba group for practical application. The overall workflow of DUIPML is shown in Fig. 1.

3.2 Modalities Fusion

In e-commerce, there are always different types of behaviour information between customer and commodity, such as purchasing, clicking, adding to purchasing car and collecting. To predict the user intention comprehensively, we need to consider combining these complex and different information. Unlike previous commonly adopted strategy which simply combines the different information using a linear combination, in which the weight parameters are hard to optimize and may cause the feature redundancy. We design a multi-modal fusion strategy, which aims to integrate all the different types of behaviour to obtain the customers' intention for commodities. Specifically, we first normalize these multi-modal behavior information to the range $[0, 1]$ using the amount of behaviours, which represents the user's latent intention probability on purchasing, clicking, adding to purchasing cart and collecting for different categories of products, the formulation is defined as follows:

$$\mathbf{X}^l = (1 + \exp^{-\mathbf{X}^l})^{-1} \tag{1}$$

where $\{\mathbf{X}^l\}_{l=1}^4$ represent the four different modalities (purchasing, clicking, adding to purchasing cart and collecting), respectively. The above equation indicates that the more user's behaviours on products, the more corresponding latent intention will be.

After obtaining the latent intention of different modalities, we then optimize a central modal by mapping all modalities to the central modal to ensure the consistency among different modalities, which is formulated as follows:

$$\tilde{\mathbf{x}}_i = \tilde{\mathbf{x}}_i . * \sum_{l=1}^{4} exp^{(\mathbf{x}_i^l - \tilde{\mathbf{x}}_i)} \tag{2}$$

where $\tilde{\mathbf{x}}_i$ represent the i-th samples of the central modality and can be initialized by the mean of $\mathbf{x}_i^l, \{l = 1, 2, 3, 4\}$. The above equation indicates that all the multi-modal data are useful to supply the latent preference. The larger $\mathbf{x}_i^l, \{l = 1, 2, 3, 4\}$ are, the larger $\tilde{\mathbf{x}}_i$ is. Using this equation, all the different behaviour information of the four modalities can be integrated to the central modality, which reflects the user's potential preference for different categories of goods. Generally, the obtained potential preference can be used for commodity recommendation or bonus awarding. However, in practical e-commerce cases, the multi-modal data are always sparse as a result that existing methods could not perform well, and they could not predict for the new categories which do not appear in the training data. To address these above limitations, we need to complete and enrich the sparse data and predict for the new categories by transferring different knowledge of preferences from different clients.

3.3 Modalities Alignment

To simultaneously achieve enriching the sparse data and obtaining the prediction of new categories, the strategy we adopted is exploring the correlations and transferring knowledge between different clients, so that the sparsity can be reduced and the intention of new categories can be predicted. Specifically, we treat the obtained sparse potential intention from previous step of two clients with different preferences as two modalities $\tilde{\mathbf{X}}^m \in \mathbb{R}^{n \times d_1}$ and $\tilde{\mathbf{X}}^{m'} \in \mathbb{R}^{n \times d_2}$, aiming at exploring the latent correlations between them and then transferring the intentions for each other based on the latent correlations. Now we introduce the proposed multi-modal alignment strategy for exploring the correlations between two modalities.

In classical multi-modal scenarios, they typically assume that the correspondence between samples of training data is completely known and can be used for helping to learn cross-modal functions. However, in realistic E-commerce scenarios, the dynamical multi-modal data of two clients are always weakly-paired even unpaired, whereas the correspondence is only partially known or totally unknown, it's a hard work to quantify the similarity between samples from different modalities. A basic idea is to map the two different modalities into the common subspace and then comparing the similarity between each sample pair, so that the similarity of similar samples will be high and will be low otherwise. In other words, all the samples will obtain the most similar samples in other modalities. Motivated by this idea, we propose a *subspace ranking* approach for multi-modal alignment. Subspace learning have been studied maturely and can be solved by lots of approaches such as PCA [29] and CCA [8]. Here we define the subspace learning as follows:

$$min||\tilde{\mathbf{X}}^m \mathbf{W}^{m^T} - \mathbf{Y}||_F^2 + ||\tilde{\mathbf{X}}^{m'} \mathbf{W}^{m'^T} - \mathbf{Y}||_F^2 \tag{3}$$

where $\mathbf{W}^{m^T}$ and $\mathbf{W}^{m'^T}$ are the projection matrices, $\mathbf{Y}$ is the common label subspace. We define $\mathbf{Z}^m = \tilde{\mathbf{X}}^m \mathbf{W}^{m^T}$ and $\mathbf{Z}^{m'} = \tilde{\mathbf{X}}^{m'} \mathbf{W}^{m'^T}$ as the representations in the common space with respective to the two modalities. The above equation aims to map the feature of different modalities to the common label space so that the structure similarity and category information could be well preserved in the common subspace. And the projection matrices can be solved easily by differentiating the objective function in Eq. 3 with respect to $\mathbf{W}^{m^T}$ and $\mathbf{W}^{m'^T}$ and setting them to zero.

Differently from previous methods exploring the correlations of two modalities by computing the similarity between each samples, which requests the square-level computation complexity, we proposed the subspace ranking strategy to achieve efficient and effective multi-modal alignment. The strategy we adopt is reordering the indexes of samples by ranking the intention of each category and directly comparing the order of samples in the common space. Specifically, for each category, we reorder the indexes of samples of $\mathbf{Z}^m$ and $\mathbf{Z}^{m'}$ by ranking the intention in an descending order, and obtain the reordered sequences $\{s_1^1, s_2^1, ..., s_d^1\}$ and $\{s_1^2, s_2^2, ..., s_d^2\}$. In this way, we can effectively achieve the multi-modal alignment by directly comparing the order of the two sequences with respect to each category.

Next, we aim to obtain c candidates which are the most similar in the other modality for each samples. Specifically, for each samples $\mathbf{x}_i^m$, we firstly select the top c pre-selections of each category in the other modality by comparing the order, therefore we can have $c \times d$ pre-selections $\{\{c_1\}, \{c_2\}, ..., \{c_d\}\}$, and the final candidates are generated as follows:

$$\mathbf{x}_j^{m'} \text{ is candidate if} \begin{cases} \mathbf{z}_j^{m'} \in \{c_1\} \bigcap \{c_2\} \bigcap ... \bigcap \{c_d\}, or \\ sim(\mathbf{z}_i^m, \mathbf{z}_j^{m'}) \text{ is the top } c \text{ largest} \end{cases} \tag{4}$$

where $sim(\mathbf{z}_i^m, \mathbf{z}_j^{m'})$ represents the cosine similarity of representations $\mathbf{z}_i^m$ and $\mathbf{z}_j^{m'}$ in the common space. The above equation indicates that a sample $\mathbf{x}_j^{m'}$ will be a candidate of $\mathbf{x}_i^m$ if $\mathbf{z}_j^{m'}$ exist in all the pre-selection sets or the similarity is the top c largest in all the pre-selections. The more similar two samples are, the more probably they become the candidates for each other. In this way, we can obtain the samples matching information and efficiently select the top c most similar candidates of each sample, whereas the computation complexity is much less than square-level. These obtained candidates can be used for the next knowledge transferring.

3.4 Knowledge Transfer

After obtaining the samples matching information, the next step is utilizing the sample-to-sample correlations to achieve knowledge transferring between different modalities, which aims to reduce the impact of sparse data and predict the intention of new categories for customs. Based on the multi-modal alignment strategy in the previous step, we then transfer the intention of the selected c candidates which represent the top c most relevant samples in the other modality with respect to each samples.

Specifically, first we extend $\tilde{\mathbf{X}}^m$ and $\tilde{\mathbf{X}}^{m'}$ to $\mathbb{R}^{n \times d}$ by setting the elements of missing part to 0, and then transferring knowledge for each sample as follows:

$$\tilde{\mathbf{x}}_{i_{new}}^m = \tilde{\mathbf{x}}_i^m * \sum_{j=1}^{c} exp^{(\tilde{\mathbf{x}}_j^{m'} - \tilde{\mathbf{x}}_i^m)} \tag{5}$$

where $\tilde{\mathbf{x}}_{i_{new}}^m$ represents the enriched user-to-category intention, $\tilde{\mathbf{X}}_j^{m'}$ is the selected top c candidates in the m'-th modality with respective to the sample $\tilde{\mathbf{x}}_i^m$ in the m-th modality. Similarly, we also can transfer the different intentions and preferences from the m to m' modality for each user to obtain $\tilde{\mathbf{X}}^{m'}_{new}$. Using the enriched user intention, not only the sparsity of intention can be reduced, but also the intention of new categories which a customer may never behaved can be predicted.

3.5 Application on Bonus Awarding

In this section, we introduce how to leverage the enriched user intention for realistic applications in E-commerce. These enriched preferences can be utilized to achieve accurate bonus awarding, while other applications such as commodity recommendation can also be achieved. Specifically, as shown in bottom of the Fig. 1, we estimate whether we should award bonus for customers based on the change of enriched preference as the bonus impacted, which can be formulated as:

$$\mathbf{A}_i = \begin{cases} 1, & \tilde{\mathbf{x}}^m_{a_i} - \tilde{\mathbf{x}}^m_{b_i} > \beta \\ 0, & otherwise \end{cases} \tag{6}$$

where $\mathbf{A}_i$ represents whether we should award bonus or not, and $\tilde{\mathbf{x}}^m_{b_i}$, $\tilde{\mathbf{x}}^m_{a_i}$ represent the dynamic enriched user intention *before* and *after* bonus awarding, respectively. β is the threshold and can be obtained by the proportion of the received and not received bonus users. Specifically, for a user, it should be awarded bonus if the change of corresponding intention is more than the threshold β, and not be awarded otherwise. To this end, we can make the decision of bonus awarding based on the learned enriched user intention.

4 Experiments

In this Section, we conduct a set of experiments to study the effectiveness of DUIPML under different scenarios, and compare its performance against related methods.

4.1 Experimental Setup

Datasets:
We conduct experiments on three benchmark datasets, Nus-wide, Wiki, and Mirflicker. Besides, we also applied our method on different realistic scenarios with collected data on the Taobao platform of Alibaba.

Nus-wide[1] is a large-scale dataset crawled from Flickr. It contains 260,648 labeled images associated with user tags, which can be considered as image-text pairs. Each image is annotated with one or more labels taken from 81 concept labels. Wiki[2] is generated from a group of 2,866 Wikipedia documents. Each document is an image-text pair. Mirflickr[3] originally includes 25,000 instances

[1] http://lms.comp.nus.edu.sg/research/NUS-WIDE.htm.
[2] https://www.wikidata.org/wiki/Wikidata.
[3] http://press.liacs.nl/mirflickr/mirdownload.html.

collected from Flicker. Each instance consists of an image and its associated textual tags, and is manually annotated with one or more labels, from a total of 24 semantic labels.

Realistic application mainly includes two scenarios:

(1) Taobao: using the 90-days behaviours of customer on Taobao platform for training and the next two-weeks clicking information for test, whereas the test data includes new categories which donot exist in the training data.
(2) YKS: utilizing 30-days behaviours of customer with awarding coupons on Taobao platform for training and whether they will accept coupons for test.

Comparing Methods:
Several related and representative methods are adopted for comparison, including CCA (Classical Correlation Analysis) [8], SCM (Semantic Correlation Maximization) [33], SePH (Semantics Preserving Hashing) [14], RDCMH (Ranking-based deep cross-modal hashing) [16], CCQ (Composite Correlation Quantization) [18], TS (Thompson Sampling) [23] and DNN+classify [4]. The first two methods are classical multi-modal learning methods for multi-modal representation or classification, and the next three methods are cross-modal hashing methods which are utilized to identify the performance of cross-modal retrieval, TS and DNN+classify are adopted in Taobao platform for bonus awarding. We adopted the available code for the comparing methods, and tuned the parameters based on the suggestions given in the respective code or papers. As for DUIPML, the parameter analysis of DUIPML is provided in Subsect. 4.3. All the experiments are conducted on the Alibaba distributed platform (ODPS) and the proposed DUIPML is able to efficiently generate taxonomy for 200 millions of item entities within 2 h.

Evaluation Metric:
We adopt the widely used Mean Average Precision (MAP) [3,14,33] to measure the retrieval performance of all multi-modal methods. As for the practical scenarios, we paint the ROC curves and chose the widely used AUC value to identify the effectiveness with respective to different methods. A larger MAP and AUC value correspond to a better retrieval or prediction performance.

4.2 Results

To verify the efficiency and adaptability of our proposed DUIPML, we design the following three parts of experiments:

(1) Comparing with existing multi-modal methods and DUIPML's variants on three benchmark datasets (Wiki, Mirflickr and Nus-wide) to verify the accuracy of proposed multi-modal alignment strategy of DUIPML on cross-modal retrieval tasks;

Table 3. Results (MAP) on three benchmark datasets with unpaired data.

		Mirflickr		Nus-wide		Wiki	
	Methods	$I \rightarrow T$	$T \rightarrow I$	$I \rightarrow T$	$T \rightarrow I$	$I \rightarrow T$	$T \rightarrow I$
	all the image-text pairs are unpaired, all methods use the totally unpaired data						
Image vs. Text	CCA	0.3216	0.3121	0.1715	0.1563	0.1011	0.0989
	SCM-seq	0.3398	0.3211	0.1953	0.1855	0.1107	0.1118
	SCM-orth	0.3404	0.3235	0.2343	0.2211	0.1126	0.1206
	SePH	0.3456	0.3456	0.2721	0.2612	0.1575	0.1437
	RDCMH	0.3778	0.3631	0.3117	0.2872	0.2342	0.2132
	CCQ	0.3784	0.3715	0.3312	0.3213	0.2342	0.2365
	DUIPML(nC)	0.3421	0.3224	0.2878	0.2832	0.2231	0.2235
	DUIPML(nR)	0.3259	0.3183	0.2618	0.2115	0.2015	0.2056
	DUIPML	**0.3925**	**0.3812**	**0.3421**	**0.3426**	**0.2612**	**0.2512**

(2) Comparing with multi-modal methods and the intention prediction method collaborative filtering (CF) to identify that our proposed DUIPML can simultaneously achieve enriching the sparse data and effective intention prediction for new categories on Taobao dataset;
(3) Comparing with bonus awarding methods (TS and DNN) on YKS dataset to verify that our DUIPML can achieve accurately bonus awarding in practical scenarios.

Results on Benchmark Datasets: Specifically, for the experiments (1), we set that the correlations between different modalities are totally unknown by randomly disturb the order of samples. In Table 3, '$I \rightarrow T$' and '$T \rightarrow I$' means Image retrieve Text and Text retrieve Image, respectively. And we denote two variants of DUIPML to further identify the efficiency of our proposed multi-modal alignment strategy: DUIPML(nC) denotes the variant learning the subspace using CCA; DUIPML(nR) is the variant without our proposed ranking strategy and directly compute the cosine similarity for each samples.

From Table 3, we can observe that CCQ and DUIPML give better results than other comparing methods. That is because they adopt different techniques to augment matched samples, which boost the performance of cross-modal retrieval. Although CCQ and DUIPML all augment paired samples, DUIPML still outperforms CCQ, which is due to its subspace ranking-based approach of multi-modal alignment for exploring the correlations among samples and selecting the most similar samples as candidates for cross-modal retrieval. In addition, DUIPML also outperforms its variants DUIPML(nC) and DUIPML(nR), which indicates that not only the superiority of the subspace learning strategy comparing to CCA, but also the effectiveness and necessity of proposed subspace ranking strategy for multi-modal alignment.

Results on Taobao Dataset: For the experiments (2), we use the 90-days behaviours of customer for training and the next two-weeks clicking information for test. We firstly compare the sparsity between before and after multi-modal learning in Table 4, whereas 'Average/Total categories' means that the proportion of average intention categories with all the categories for each customer and 'Average/Total customers' represents that the proportion of average customers who clicking this category with all the customers for each category. In addition, we further compare with representative user intention prediction method item-based Collaborative filtering(CF), which aims to simultaneously identify the effective accuracy and robustness of our proposed DUIPML on sparse multi-modal data.

Table 4. Results of sparsity on Taobao.

		Average/Total categories	Average/Total customers
Before	Original	3.72%	3.41%
After	CCA	5.31%	5.11%
	SCM-seq	6.23%	6.03%
	SCM-orth	5.46%	5.21%
	CCQ	7.11%	6.65%
	DUIPML	**7.82%**	**7.37%**

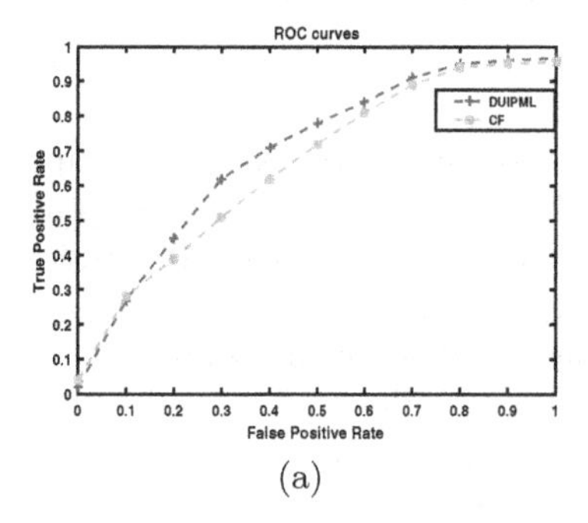

(a)

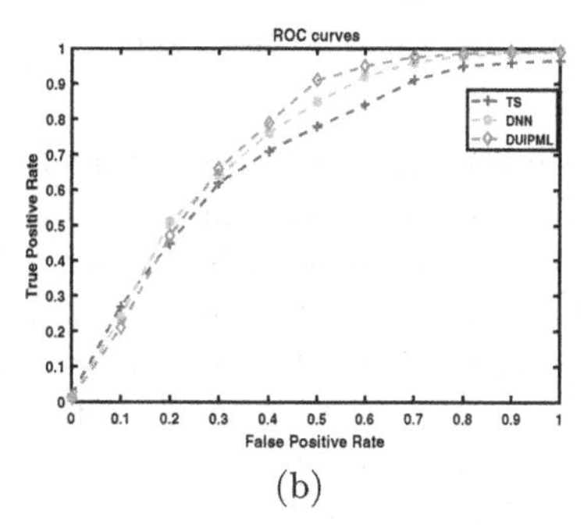

(b)

Fig. 2. (a): DUIPML vs. CF on Taobao datasets. (b): DUIPML vs. TS and DNN on YKS datasets.

From Table 4, we can see that all the MML methods have a more abundant result than original data, which identify that multi-modal learning could enrich the multi-modal data. While the results of our DUIPML outperforms those of other multi-modal methods, this is because our multi-modal alignment strategy can explore more accurate and various correlations than other methods, as a result that our DUIPML can enrich more abundant information for the sparse multi-modal data. This observation identifies again the efficiency of our proposed multi-modal alignment strategy than other MML approaches.

To further verify the efficiency and accuracy of the enriched information, we conducted experiments which the next two-weeks data compared with CF, and the results are shown in Fig. 2(a). We can indicate from Fig. 2(a) that our DUIPML have a more promising results than CF, which identifies that not only the enriched data is accurate enough to predict the latent intention of customers, but also our DUIPML can also achieve predicting the intention of new categories which do not appear in the training data. These observations in Fig. 2(a) prove that our method can be efficiently applied on practical scenarios in E-commerce.

Comprehensively analyzing the results in Table 4 and Fig. 2(a), our proposed DUIPML can achieve both accurately enriching the sparse multi-modal data and effectively predicting the latent intention of customers.

Results on YKS Dataset: For the experiments (3), we utilize 30-days behaviours of customer for training and whether these customers will receive coupons for test. As for our DUIPML, we set that the user should be awarded bonus if the change of enriched user intention is more than the threshold 0.15 and not be awarded bonus otherwise.

From Fig. 2(b), we can see that our DUIPML still outperforms other comparing methods and the AUC value of these three methods are 0.626, 0.682 and 0.695, respectively. This observation identify that our proposed DUIPML can be efficiently applied into realistic scenarios such as the bonus awarding, this is because that our DUIPML can dynamically observe the impact of the awarded bonus on latent preferences, so that we can more accurately estimate whether the customers will receive bonus.

Summarization: Comprehensively analyzing the above results, our method outperforms other comparing methods both on effectiveness and adaptability. Specifically, compared with existing MML methods, our DUIPML proposed a novel multi-modal alignment strategy to explore more accurate latent correlations among different modalities, which does not require the completely paired correlations for training. And DUIPML can dynamically predict the latent intention with the dynamical multi-modal data for the e-commerce applications such as the Bonus Awarding. Compared with existing Bonus Awarding methods, DUIPML has following advantages: (1) DUIPML can achieve more comprehensive prediction for user intention by integrating different multi-modal behaviour information; (2) DUIPML can reduce the impact of sparse data and predict the new categories not appeared in the training data by enriching the latent intention of users with different preferences; (3) DUIPML can dynamically observe the impact of the awarded bonus on latent preferences to achieve more accurately decision of bonus awarding.

In summary, our proposed DUIPML can achieve promising performance on benchmark datasets and practical scenarios.

4.3 Parameters and Rumtime Analysis

In this section, we firstly study the sensitivity of the key input parameters c(the number of selected candidates). We can see from Fig. 3 that DUIPML is slightly sensitive to c when $c \in [50, 200]$, and achieves the best performance when $c = 50$. Too less number of candidates has a negative impact on the performance, which indicates that the selected candidates can achieve both enriching the sparse data and efficient predicting the preference simultaneously. In summary, an effective c can be easily selected for our DUIPML.

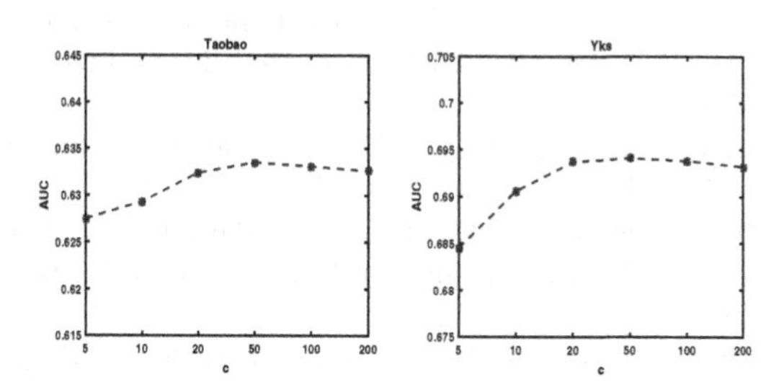

Fig. 3. Sensitivity of c on different datasets.

In addition, we also compare the runtime with TS and DNN, and the results are statistically shown in the Table 5. We can observe that the training time of our DUIPML is typically faster than TS and DNN. This is becauase our DUIPML adopts the subspace ranking strategy to reduce the time complexity to $O(n)$. DNN needs much more time because it requires a lot of time to train the deep neural networks. Both the training and predicting time can indicate the efficiency of our DUIPML.

Table 5. Running times (in seconds) on YKS datasets.

	TS	DNN	DUIPML
Training	7984.64	14356.55	7011.36
Predicting	94.31	178.69	75.72

5 Conclusion

In this paper, we proposed a general dynamical multi-modal learning for user intention prediction (DUIPML). Specifically, we firstly design a novel multi-modal fusion strategy to integrate different types of multi-modal data, to obtain a comprehensive preference of customers to different categories, next we proposed a multi-modal alignment strategy to explore the latent correlations between different clients and then transfer knowledge for each other to obtain the enriched intention. Finally, we utilize these enriched intention to achieve accurate bonus awarding by observing the change of intention as the bonus impact. Experiments on three benchmark datasets and practical e-commerce scenarios show that our DUIPML outperforms other state-of-art comparing methods. In the future work, we will focus more general multi-modal learning approaches for more open practical scenarios in E-commerce.

References

1. Akoglu, L., Tong, H., Koutra, D.: Graph based anomaly detection and description: a survey. Data Min. Knowl. Dis. **29**(3), 626–688 (2014). https://doi.org/10.1007/s10618-014-0365-y
2. Baltruaitis, T., Ahuja, C., Morency, L.P.: Multimodal machine learning: a survey and taxonomy. TPAMI **41**(2), 423–443 (2019)
3. Bronstein, M.M., Bronstein, A.M., Michel, F., Paragios, N.: Data fusion through cross-modality metric learning using similarity-sensitive hashing. In: CVPR, pp. 3594–3601 (2010)
4. Graves, A., Mohamed, A.r., Hinton, G.: Speech recognition with deep recurrent neural networks. In: ICASSP, pp. 6645–6649 (2013)
5. Guo, Q., et al.: Securing the deep fraud detector in large-scale e-commerce platform via adversarial machine learning approach. In: The World Wide Web Conference, pp. 616–626 (2019)
6. Guo, Y., Ding, G., Han, J., Gao, Y.: Sitnet: discrete similarity transfer network for zero-shot hashing. In: IJCAI, pp. 1767–1773 (2017)
7. Hooi, B., et al.: Fraudar: bounding graph fraud in the face of camouflage. In: KDD, pp. 895–904 (2016)
8. Hotelling, H.: Relations between two sets of variates. Biometrika **28**(3/4), 321–377 (1936)
9. Hu, Z., Zhang, J., Li, Z.: General robustness evaluation of incentive mechanism against bounded rationality using continuum-armed bandits. AAAI **33**, 6070–6078 (2019)
10. Jiang, Q.Y., Li, W.J.: Deep cross-modal hashing. In: CVPR, pp. 3270–3278 (2017)
11. Kumar, S., Udupa, R.: Learning hash functions for cross-view similarity search. In: IJCAI, pp. 1360–1365 (2011)
12. Lei, C., Ji, S., Li, Z.: Tissa: a time slice self-attention approach for modeling sequential user behaviors. In: The World Wide Web Conference, pp. 2964–2970 (2019)
13. Li, Z., et al.: Fair: fraud aware impression regulation system in large-scale real-time e-commerce search platform. In: ICDE, pp. 1898–1903 (2019)
14. Lin, Z., Ding, G., Han, J., Wang, J.: Cross-view retrieval via probability-based semantics-preserving hashing. IEEE Trans. on Cybern. **47**(12), 4342–4355 (2017)
15. Lin, Z., Ding, G., Hu, M., Wang, J.: Semantics-preserving hashing for cross-view retrieval. In: CVPR, pp. 3864–3872 (2015)
16. Liu, X., Yu, G., Domeniconi, C., Wang, J., Ren, Y., Guo, M.: Ranking-based deep cross-modal hashing. In: AAAI, pp. 4400–4407 (2019)
17. Liu, X., Yu, G., Domeniconi, C., Wang, J., Xiao, G., Guo, M.: Weakly-supervised cross-modal hashing. IEEE Trans. on Big Data **6**(99), 1–12 (2019)
18. Long, M., Yue, C., Wang, J., Yu, P.S.: Composite correlation quantization for efficient multimodal retrieval. In: SIGIR, pp. 579–588 (2016)

19. Ranjan, V., Rasiwasia, N., Jawahar, C.V.: Multi-label cross-modal retrieval. In: ICCV, pp. 4094–4102 (2015)
20. Shao, W., He, L., Lu, C.T., Wei, X., Philip, S.Y.: Online unsupervised multi-view feature selection. In: ICDM, pp. 1203–1208 (2016)
21. Shen, F., Zhou, X., Yu, J., Yang, Y., Liu, L., Shen, H.T.: Scalable zero-shot learning via binary visual-semantic embeddings. TIP **99**(1), 1–10 (2019)
22. Shi, X., Yu, P.: Dimensionality reduction on heterogeneous feature space. In: ICDM, pp. 635–644 (2012)
23. Thompson, D.J.: A theory of sampling finite universes with arbitrary probabilities (1952)
24. Wang, J., Kumar, S., Chang, S.F.: Semi-supervised hashing for large-scale search. TPAMI **34**(12), 2393–2406 (2012)
25. Wang, J., Zhang, T., Sebe, N., Shen, H.T.: A survey on learning to hash. TPAMI **40**(4), 769–790 (2018)
26. Wang, J., Liu, W., Kumar, S., Chang, S.F.: Learning to hash for indexing big data - a survey. Proc. IEEE **104**(1), 34–57 (2016)
27. Wang, K., He, R., Wang, W., Wang, L., Tan, T.: Learning coupled feature spaces for cross-modal matching. In: ICCV, pp. 2088–2095 (2013)
28. Weng, H., et al.: Cats: cross-platform e-commerce fraud detection. In: ICDE, pp. 1874–1885 (2019)
29. Wold, S., Esbensen, K., Geladi, P.: Principal component analysis. Chemom. Intell. Lab. Syst. **2**(1–3), 37–52 (1987)
30. Xu, Y., Yang, Y., Shen, F., Xu, X., Zhou, Y., Shen, H.T.: Attribute hashing for zero-shot image retrieval. In: ICME, pp. 133–138 (2017)
31. Yang, Y., Luo, Y., Chen, W., Shen, F., Jie, S., Shen, H.T.: Zero-shot hashing via transferring supervised knowledge. In: ACM MM, pp. 1286–1295 (2016)
32. Yu, J., Wang, M., Tao, D.: Semisupervised multiview distance metric learning for cartoon synthesis. TIP **21**(11), 4636–4648 (2012)
33. Zhang, D., Li, W.J.: Large-scale supervised multimodal hashing with semantic correlation maximization. In: AAAI, pp. 2177–2183 (2014)
34. Zhang, Y., Ahmed, A., Josifovski, V., Smola, A.: Taxonomy discovery for personalized recommendation. In: WSDM, pp. 243–252 (2014)
35. Zhou, J., Ding, G., Guo, Y.: Latent semantic sparse hashing for cross-modal similarity search. In: ACM SIGIR, pp. 415–424 (2014)
36. Ziegler, C.N., Lausen, G., Schmidt-Thieme, L.: Taxonomy-driven computation of product recommendations. In: CIKM, pp. 406–415 (2004)

Graph Convolutional Network Using a Reliability-Based Feature Aggregation Mechanism

Yanling Wang[1,2], Cuiping Li[1,2](✉), Jing Zhang[1,2], Peng Ni[1,2], and Hong Chen[1,2]

[1] Key Laboratory of Data Engineering and Knowledge Engineering, Renmin University of China, Beijing, China
[2] School of Information, Renmin University of China, Beijing, China
{wangyanling,licuiping,zhang-jing,nipeng,chong}@ruc.edu.cn

Abstract. Graph convolutional networks (GCNs) have been proven extremely effective in a variety of prediction tasks. The general idea is to update the embedding of a node by recursively aggregating features from the node's neighborhood. To improve the training efficiency, modern GCNs usually sample a fixed-size set of neighbors uniformly or sample according to nodes' importance, instead of using the full neighborhood. However, both the sampling strategies ignore the reliability of a link between the target node and its neighbor, which can be implied by the graph structure and may seriously impact the performance of GCNs. To deal with this problem, we present a Graph Convolutional Network using a Reliability-based Feature Aggregation Mechanism called GraphRFA, where we sample the neighbors for each node according to different kinds of link reliability and further aggregate feature information from different reliability-specific neighborhoods by a dual feature aggregation scheme. We also theoretically prove that our aggregation scheme is permutation invariant for the graph data, and provide two simple but effective instantiations satisfying such scheme. Experimental results demonstrate the effectiveness of GraphRFA on different datasets.

Keywords: Graph convolutional network · Link reliability · Neighborhood definition · Dual feature aggregation scheme

1 Introduction

Node embedding techniques learn low-dimensional representations for nodes in graphs and effectively preserve the structure information [6,12]. Among related algorithms [1,7,8,10,13,15,21,22,24], graph convolutional networks (GCNs) adopt deep learning on the graph data and achieve outstanding performance on various tasks, such as node classification, community detection, link prediction and so on [13,15,26,29]. The general idea behind GCNs is to update the representation of each node by aggregating the features from the node's neighborhood. The feature information of a node can be initialized by extracting

Y. Nah et al. (Eds.): DASFAA 2020, LNCS 12112, pp. 536–552, 2020.
https://doi.org/10.1007/978-3-030-59410-7_36

the side information of the node. In traditional models, all the n-hop neighbors form the neighborhood (i.e., the receptive field) of a given node, where n is less than or equal to the number of layers in the model. Since GCNs learn the node representations through recursive neighborhood aggregation, the large number of neighbors will lead to computational inefficiency. To make the memory and runtime of a single batch training controllable, modern GCNs usually sample a fixed-size set of neighbors for each node, instead of using the full neighborhood.

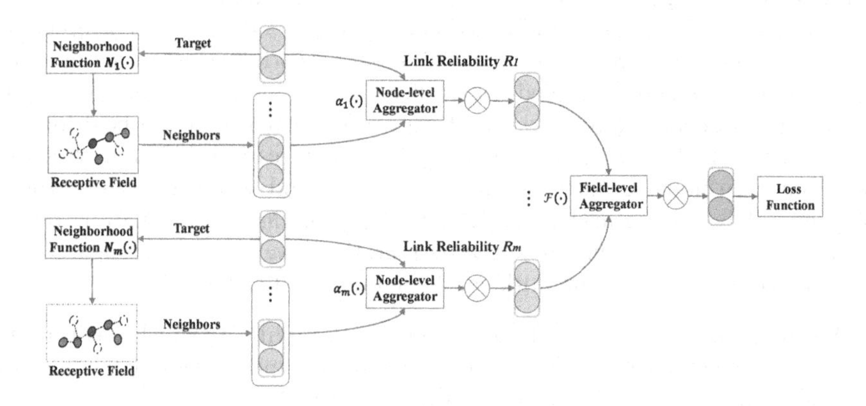

Fig. 1. The overview framework of GraphRFA.

The uniform sampling strategy treats every node equally [4,13]. Intuitively, not all the neighbors are useful for the target node, some non-uniform strategies are thereby designed to sample a set of neighbors according to the global node-importance such as pagerank centrality [3,5,25]. However, most sampling methods do not consider the correlations between the target node and its neighbors, making some important neighbors of the target node ignored. For example, in a social network, if one of your close friends is not an influential user in the whole social network, her/his influence on you will be weakened. Besides, a graph contained spurious connections can lead to ineffective graph convolution caused by the unreliable neighbors [27].

This paper aims to propose a link-reliability based GCN model, which pays attention to the likelihood that a link truly exists in a graph, i.e., the link reliability [11]. A higher score of link reliability indicates a more reliable relationship between two nodes. To achieve the goal, we try to answer the following questions: how to effectively measure the link reliability? And how to incorporate the receptive fields resulted by different measures together? To address these challenges, we present GraphRFA, a Graph Convolutional Network using a Reliability-based Feature Aggregation Mechanism. Figure 1 illustrates the overview framework of our model. Specifically, the model contains two main components:

Link-Reliability Based Neighborhood Definition: To measure the link reliability, we adopt random-walk methods to capture the characteristics of the graph structure from two views, i.e., the similarity between two nodes and the centrality of a link. Then we use a drop-bottom approach to sample neighborhood for each node, which drops the most unreliable neighbors and uniformly samples a fixed-size set of neighbors from the remaining ones. Compared with previous strategy, we allow the moderately reliable neighbors to be sampled. Moreover, we can adjust the ratio of dropped neighbors to trade off between uniform sampling and non-uniform sampling.

Dual Feature Aggregation: Since we measure the link reliability from different views such as similarity and centrality, we design a dual feature aggregation scheme to perform convolution across different receptive fields, which are composed by the neighbors sampled from different views. More specifically, a node-level aggregator aggregates the node features from each receptive field for each node, then a field-level aggregator merges the features from different receptive fields for each node. Unlike learning over images and sentences, the neighbors of a node have no natural order [13]. So we give theoretical analysis on our aggregation scheme to guarantee the permutation invariant requirement.

The contributions of our work can be summarized as follows:

- We propose a graph convolutional network GraphRFA, which incorporates the link reliability in neighborhood definition to capture the reliable information from the graph structure and the node features.
- In GraphRFA, different measures of link reliability can be supported simultaneously, and a dual feature aggregation scheme is proposed to aggregate feature information from the different kinds of reliability-specific receptive fields. We also give the theoretical analysis to guarantee our aggregation scheme is permutation invariant for the graph data.
- We evaluate our algorithm by downstream tasks. Experimental results show the effectiveness of GraphRFA on different graphs.

2 Related Work

Graph Convolutional Networks. Recently, a large number of graph convolutional networks are proposed, which generalize the convolution operation to the permutation invariant graph data and achieve outstanding performance on various tasks over graphs. As one of the pioneer works, [15] is designed based on the full graph Laplacian, thus it is hard to generate embeddings for previously unseen nodes efficiently. GraphSAGE [13] solves the efficiency problem by a sampling-based inductive learning method. Graph Attention Network (GAT) [22] pays different attentions to different neighbors by adding a self-attention layer. Many advanced variants are proposed [18,24]. For example, GIN [24] formally characterizes the properties of a powerful graph neural network and is as powerful as Weisfeiler-Lehman graph isomorphism test (WL test) [23].

Neighborhood Definition. The basic idea behind existing graph convolutional networks is to aggregate information from the neighborhood of the target node. Here, we summarize two kinds of approaches to define the neighborhood.

Adaptive Learning Approaches aim to learn the weights of different nodes in the receptive field. GAT [22] and GeniePath [18] are the representative approaches. GAT uses the self-attention mechanism to determine the importance of each one-hop neighbor. As an extension of GAT, GeniePath uses the LSTM-like function to filter signals from the n-hop neighbors. Nevertheless, both GAT and GeniePath do not emphasize the neighborhood sampling.

Sampling-style Approaches aim to reduce the size of the receptive field. Typically, existing approaches reduce the receptive field by sampling a set of neighbors for each node uniformly [13] or sampling according to the nodes' importance [3,5,25]. The authors of [4] aim to guarantee the convergence of the stochastic training algorithm with neighbor sampling. Thus, they propose a novel control variate based algorithm, denoted as CV-GCN in this paper, which achieves a similar convergence with the exact algorithm.

Discussion. In summary, neighborhood definition is a fundamental step in GCNs. To ensure the scalability, we aim to design a sampling-style algorithm. Distinct from the previous sampling-based algorithms, we consider the local correlations between target node and its neighbors from different views, hence we incorporate different kinds of link reliability in the sampling process.

3 Problem Formulation

3.1 Problem Definition

Let $\mathcal{G} = (\mathcal{V}, \mathcal{E})$ denote a graph, where each $v \in \mathcal{V}$ denotes a node and each $e \in \mathcal{E}$ denotes a link. Notation $X_v \in \mathbb{R}^{d_0}$ indicates the initial feature vector of node v. In real datasets, there are usually noisy links contained in $\mathcal{E}$.

The problem is to learn a representation for each node v in $\mathcal{G}$, such that the representation can not only capture the characteristics of the local graph structure of v, but also emphasize the reliability of the links between target node and its neighbors. In other words, we aim to learn the embeddings $h_i, h_j \in \mathbb{R}^d$ for nodes v_i and v_j such that h_i, h_j can be well distinguished if their local neighborhoods are structurally different or the link between v_i and v_j are unreliable.

3.2 Background

In the deep graph representation learning, hidden feature of node v at the l-th layer of the model is denoted by $h_v^{(l)} \in \mathbb{R}^{d_l}$, and $h_v^{(0)} = X_v$.

Graph convolution in [15] requires to calculate the propagation matrix $\hat{A} = \widetilde{D}^{-\frac{1}{2}}\widetilde{A}\widetilde{D}^{-\frac{1}{2}}$. $\widetilde{A}$ is the adjacency matrix of $\mathcal{G}$ with self-connections. $\widetilde{D}_{ii} = \sum_j \widetilde{A}_{ij}$. The matrix of hidden features at the l-th layer is updated by:

$$H^{(l)} = \sigma\left(\hat{A}H^{(l-1)}W^{(l)}\right) \tag{1}$$

where the weight matrix $W^{(l)}$ is learnable, and σ is an activation function.

GraphSAGE [13] trains a series of aggregators to aggregate information from the node's neighborhood, so that the trained aggregators can generate embeddings for previously unseen nodes effectively. [24] summarizes the strategy of neighborhood aggregation by AGGREGATE and COMBINE steps, given the one-hop neighborhood $\mathcal{N}(v)$ of node v, the l-th layer updates $h_v^{(l)}$ by:

$$h_{\mathcal{N}(v)}^{(l)} = AGGREGATE^{(l)}\left(\{h_u^{(l-1)}|u\in\mathcal{N}(v)\}\right), h_v^{(l)} = COMBINE^{(l)}\left(h_{\mathcal{N}(v)}^{(l)}, h_v^{(l-1)}\right) \tag{2}$$

Among different kinds of aggregators, GCN-aggregator [13] is an inductive variant of [15], and is essentially a modified mean-based function:

$$h_v^{(l)} = \sigma\left(W^{(l)} \cdot MEAN\left(h_v^{(l-1)} \cup \{h_u^{(l-1)}, \forall u \in \mathcal{N}(v)\}\right)\right) \tag{3}$$

4 The Proposed Approach

In this section, we describe the structure of our model. GraphRFA consists of two major components: *neighborhood definition* and *dual feature aggregation scheme.* Distinct from existing studies on the neighborhood definition, we sample neighbors for each node based on link reliability and devise a drop-bottom strategy to trade off between uniform sampling and non-uniform sampling. To further simultaneously support different measures of link reliability, a dual feature aggregation scheme is proposed to aggregate feature information from the receptive fields resulted by different reliability measures.

4.1 Neighborhood Definition Based on Link Reliability

We define the fixed-size neighborhood based on the measures of link reliability. The neighbors connected by more reliable links should propagate more useful information to the target nodes.

Random-Walk Based Measures for Link Reliability. Different measures of link reliability can be used in GraphRFA. To propose a general algorithm, we measure the link reliability based on graph structure without using any attribute of a node or a link. Specially, *similarity* and *centrality* can be regarded as the views of designing the link-reliability measures [20,28].

Under the similarity hypothesis, links connecting similar nodes are supposed to have high existence-likelihoods [20]. Following such an assumption, many well-known path-dependent indices are proposed, such as CN (Common Neighbors) [16], LP (Local Path) [30] and Katz index [14]. The number of paths between two nodes is used to quantify the link reliability. Intuitively, the highly interconnected nodes usually tend to be similar.

Under the centrality hypothesis, links with higher centrality is more reliable [28]. Specifically, these links directly or indirectly connect a large number of nodes in a graph. Both PA (Preferential Attachment) [2] and EB (Edge Betweenness) [9] are typical centrality-based indices.

Despite their success, most of the above measures are not very computational efficient. Our goal is to adopt the random-walk based methods, as random walk is computationally efficient in terms of both space and time requirements. In this paper, we focus on the undirected graphs, although the method can be extended to other types of graphs.

Similarity-Based Link Reliability: In traditional path-dependent indices such as LP and Katz, the reliability of the link (v_i, v_j) is proportional to the number of paths between v_i and v_j. From the perspective of random walk, the more paths between v_i and v_j exist, the more likely the random-walker walks between v_i and v_j. Therefore, we simulate local random walks with walk-length r starting from the initial node for t times, then compute the normalized visit-count to formulate the similarity-based link reliability:

$$R_{ij} = \frac{c_{i \to j} + c_{j \to i}}{\sum_{v_k \in \mathcal{V}} c_{i \to k} + \sum_{v_k \in \mathcal{V}} c_{j \to k}} \tag{4}$$

where $c_{i \to j}$ is the visit-count of v_j from the initial node v_i. Specially, in the neighborhood definition, we emphasize the neighbors that the target node tends to follow, so we sample neighbors according to the normalized visit-count from the target node to each of its neighbors through the local random walks.

Turn to traditional similarity-based measures, a unified framework $R_{ij} = \sum \beta^l |paths^l_{ij}|$ can formulate CN, LP and Katz, where for CN, $l = 2$, for LP, $l = 2, 3$, and for Katz, $l = 1, 2, 3, \cdots, +\infty$ [19]. $|paths^l_{ij}|$ represents the number of paths between v_i and v_j with path-length l. Free parameter β^l adjusts the weight of a path with path-length l. Particularly, the walk-length r in our method corresponds to the maximal path-length.

Centrality-Based Link Reliability: We put a walker on the initial node v_i and simulate random walks over the graph. Let P be the transition probability matrix, the probabilities that the walker locates at any other nodes at the t-th step is computed by:

$$\pi_i(t) = P^T \pi_i(t-1) \tag{5}$$

At the stationary state, the probability that the walker locates at v_j is $\pi_{ij} = \frac{d_j}{2|E|}$, where d_j is the degree of v_j, $|E|$ is the number of links in graph [17]. In other words, the stationary probability that the random-walker starting from any node locates at node v_i is $\pi_{*i} = \frac{d_i}{2|E|}$. A higher probability π_{*i} indicates that it is more likely to locate at v_i. Thus, if both π_{*i} and π_{*j} are high probabilities, the walker can easily walk to both v_i and v_j from any node, which means a large number of nodes can be connected by the link (v_i, v_j). According to the above analysis, the centrality-based link reliability can be formulated as:

$$R_{ij} = \pi_{*i} \cdot \pi_{*j} = \frac{d_i}{2|E|} \cdot \frac{d_j}{2|E|} \propto d_i \cdot d_j \tag{6}$$

Specially, Eq. 6 precisely corresponds to the Preferential Attachment (PA).

Neighborhood Function. Given a node in a graph, the neighborhood function $\mathcal{N}(\cdot)$ draws S_l one-hop neighbors for the node at the l-th layer. We provide two approaches to define $\mathcal{N}(\cdot)$. The core ideas behind them are shown in Fig. 2. Both approaches rank all one-hop neighbors for each node according to the link reliability between the node and its neighbors, while the difference is that:

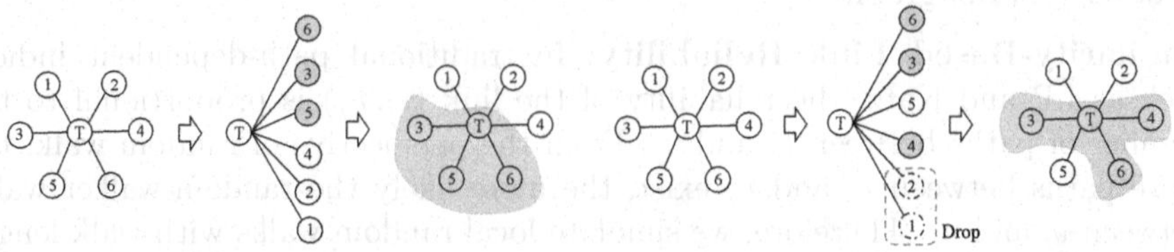

(a) Select-Top Approach (b) Drop-Bottom Approach

Fig. 2. Illustration of the link-reliability based neighborhood definition. In the example, the green nodes denote the sampled neighbors, the dashed nodes denote the dropped unreliable neighbors, and the grey areas indicate the resulted receptive fields.

Select-Top selects the top neighbors to form a fixed-size neighborhood. This straightforward approach has been utilized in previous attempts [3], which rank the one-hop neighbors according to nodes' importance rather than link reliability.

Drop-Bottom firstly drops the most unreliable $k\%$ neighbors and then uniformly samples from the remaining ones. $k\%$, named as dropping rate, is a hyper-parameter to be tuned to trade off between uniform sampling and non-uniform sampling.

4.2 Dual Feature Aggregation

Based on different measures of link reliability, we assign different receptive fields to each node. So we devise a dual feature aggregation scheme to perform convolution across different receptive fields from the node-level and the field-level.

Node-Level Aggregator. The node-level aggregator acts on each receptive field. Theoretically, any candidate function satisfying the permutation invariant requirement can be applied. The default node-level aggregator in our model follows state-of-the-art model GIN [24], in which the graph convolution for node v_i at layer l is defined as:

$$h_i^{(l)} = MLP^{(l)}\left(\left(1+\epsilon^{(l)}\right)\cdot h_i^{(l-1)} + \sum\nolimits_{v_j\in\mathcal{N}(v_i)} h_j^{(l-1)}\right) \quad (7)$$

where $MLP^{(l)}$ is the multi-layer perceptron at the l-th layer, and ϵ is a learnable parameter or a fixed scalar.

Since different measures of link reliability are adopted in Sect. 4.1, we derive a similarity-based neighborhood function $\mathcal{N}^s(\cdot)$ and a centrality-based neighborhood function $\mathcal{N}^c(\cdot)$ to produce the receptive field respectively. That means we have two node-level aggregators $\alpha^s(\cdot)$ and $\alpha^c(\cdot)$ to compute the reliability-specific embeddings h^s, h^c for each node in a graph.

Field-Level Aggregator. To further aggregate reliability-specific embeddings from different receptive fields, we develop the field-level aggregator. A basic requirement for the field-level aggregator $\mathcal{F}$ is the permutation invariant property, which is formally defined as:

Definition 1. *Let H be the feature set of nodes in graph G, and let σ be a permutation of $\{1, \cdots, n\}$. $\sigma(H)$ is a new feature set of nodes given by $[\sigma(H)]_i = H_{\sigma(i)}$. Mapping function $\mathcal{F}$ is said to be permutation invariant, if all the permutations σ of input H satisfy: $\mathcal{F}(\sigma(H)) = \sigma(\mathcal{F}(H))$.*

Associative property of permutation invariant manner has been discussed in [18]:

Remark 1. If the output of function α is invariant to the permutations of the input H, and function ρ is independent of the orders of H, $\rho \circ \alpha$ is still permutation invariant with respect to H.

Here, the function α corresponds to the node-level aggregator. Suppose we have m permutation invariant node-level aggregators, and each of them corresponds to a certain measure of link reliability, we need to prove that if we integrate different node-level aggregators, the result still satisfies the property of permutation invariance.

Theorem 1. *If the field-level aggregator $\mathcal{F}$ can be decomposed by Eq. 8, $\mathcal{F}$ is permutation invariant.*

$$\mathcal{F}(H) = \sum\nolimits_{i=1}^{m} \rho_i(\alpha_i(H)) \tag{8}$$

where the node-level aggregation function α_i is permutation invariant, and the function ρ_i is independent of the orders of input H.

Proof. We need to show that $\mathcal{F}$ in Theorem 1 satisfies the condition of Definition 1 (i.e., the permutation invariant property). In other words, for any permutation σ of input H, $\mathcal{F}(\sigma(H)) = \sigma(\mathcal{F}(H))$. Hence, we write $\mathcal{F}(\sigma(H))$ using Eq. 8 and derive the following equality:

$$\mathcal{F}(\sigma(H)) = \sum\nolimits_{i=1}^{m} \rho_i \circ \alpha_i(\sigma(H)) \tag{9}$$

The argument of α_i is invariant under the permutation σ, and ρ_i is independent of the orders of H. According to Remark 1, we can derive $\rho_i \circ \alpha_i(\sigma(H)) = \sigma(\rho_i \circ \alpha_i(H))$. Then, Eq. 9 can be rewritten as:

$$\mathcal{F}\left(\sigma\left(H\right)\right)=\sum\nolimits_{i=1}^{m}\sigma\left(\rho_{i}\circ\alpha_{i}\left(H\right)\right)=\sigma\left(\sum\nolimits_{i=1}^{m}\rho_{i}\circ\alpha_{i}\left(H\right)\right)=\sigma\left(\mathcal{F}\left(H\right)\right) \quad (10)$$

where the equality follows Definition 1. Thus, we prove that Eq. 8 is one kind of field-level aggregator that implies the permutation invariance.

In this paper, we mainly explore two simple but effective instantiations of Theorem 1.

Mean-Pooling Aggregation: We take the weighted-mean of node's reliability-specific embeddings as the candidate field-level aggregator:

$$h_i = \lambda \cdot \left[\alpha_s\left(H\right)\right]_i + \left(1-\lambda\right) \cdot \left[\alpha_c\left(H\right)\right]_i \quad (11)$$

where λ is learnable or fixed. Equation 11 is a simplest formalization of Theorem 1[1].

Concatenation Aggregation: We concatenate the reliability-specific embeddings of node v_i, then perform a linear transformation to reduce the dimension. The weight and bias of transformation are shared across all nodes, which helps reduce over-fitting. Formally, the concatenation aggregator is defined as:

$$h_i = \left[\left[\alpha_s\left(H\right)\right]_i, \left[\alpha_c\left(H\right)\right]_i\right] \cdot W + b \quad (12)$$

Next, we show that Eq. 12 satisfies Theorem 1. Equation 12 can be composed into the form of $\left(\left[\alpha_s\left(H\right)\right]_i \cdot W_s + b_s\right) + \left(\left[\alpha_c\left(H\right)\right]_i \cdot W_c + b_c\right)$, i.e., $\left[\rho_s \circ \alpha_s\left(H\right) + \rho_c \circ \alpha_c\left(H\right)\right]_i$, where α is the reliability-specific node-level aggregator and ρ is the linear transformation.

4.3 Optimization

We try two kinds of optimizations. The first one leverages the labels of a downstream supervised task such as multi-label classification to optimize the cross-entropy loss function. The second one directly optimizes the popular graph-based loss function [13], so that the nearby nodes tend to have similar embeddings, while disparate nodes tend to have highly distinct embeddings. We optimize the loss function using Adam optimizer.

4.4 Complexity Analysis

When quantifying the link reliability, the centrality-based measure is proportional to the node degree, while we need to simulate local random walks to compute the similarity-based measure. We set r to be the walk length, and we repeat the random walks on $\mathcal{G}=(\mathcal{V},\mathcal{E})$ from each initial node for t times. The computation complexity of local random walk is $\mathcal{O}\left(|\mathcal{E}|+|\mathcal{V}|rt\right)$ on a weighted

[1] We set $\lambda = 0.5$ in our experiments.

graph while $\mathcal{O}(|\mathcal{V}|rt)$ on an unweighted graph. We rank all one-hop neighbors according to the link reliability. The computation complexity is $\mathcal{O}(|\mathcal{V}|d \cdot logd)$, where d is the average degree in $\mathcal{G}$. Due to the sparsity of most real graphs (i.e., $d \ll |\mathcal{V}|$), we simplify $\mathcal{O}(|\mathcal{V}|d \cdot logd)$ to $\mathcal{O}(|\mathcal{V}|)$. As for the graph convolution, the computation complexity is $\mathcal{O}\left(|\mathcal{V}| \cdot \prod_{l=1}^{L} S_l\right)$, where $S_l, l \in \{1, 2, 3, \cdots, L\}$ is the neighbor sampling size at the l-th layer, and $S_l \ll |\mathcal{V}|$. The convolution complexity of GraphRFA is the same as that of other sampling-style GCNs.

Table 1. Summary of the datasets used in our experiments

Dataset	#Nodes	#Edges	#Labels	#Features	#Aver Degree	#Max Degree
PPI-S	14,755	228,431	121	50	31	722
PPI	56,944	818,716	121	50	29	722
Retweet	18,470	48,053	2	79	5	786
Reddit	232,965	57,307,946	41	602	492	21657

5 Experiments

5.1 Experimental Settings

Datasets. The experiments are conducted on the following undirected graphs, detailed summary is shown in Table 1:

- **PPI**[2]: This is a protein-protein interaction network in human tissue. In this dataset, 4/5 nodes are used for training, and 1/5 nodes are used for testing (with 54% for validation). Moreover, we examine on a smaller PPI network containing 14,755 nodes, denoted as **PPI-S**[3].
- **Retweet**[4]: This is a political retweet network. We treat this dataset as an undirected graph. We randomly select 2/3 nodes for training and 1/3 nodes for testing (with 30% for validation).
- **Reddit**[5]: This is a post network from Reddit made in the month of September, 2014. The data in the first 20 days is used for training, and the remaining is used for testing (with 30% for validation).

For PPI and Reddit, the train-validation-test splits and initial node features are provided by the original datasets. For the Retweet network without node attributes, we use one-hot encoding of int(degree/10) as the initial node feature.

[2] http://snap.stanford.edu/graphsage/.
[3] https://github.com/williamleif/GraphSAGE.
[4] http://networkrepository.com/soc-political-retweet.php.
[5] http://snap.stanford.edu/graphsage/.

Baselines. Among the following baselines, GraphCSC-M adopts importance based sampling to define the neighborhood for each node. The others adopt the uniform sampling. GeniePath-lazy learns the importance of different neighbors.

- **GraphSAGE:** This is a classic model for the inductive graph embedding, and it uses uniform sampling to control the size of receptive field.
- **GIN:** This is a state-of-the-art method designed for both node classification and graph classification. Emphatically, it performs as powerful as WL test.
- **GeniePath-lazy:** This algorithm is an extension of GAT and outperforms GAT shown in [18]. A path layer is designed for both breadth and depth exploration. Compared with GeniePath, GeniePath-lazy postpones the evaluations of the adaptive depth function. Through the experimental evaluation, GeniePath-lazy converges faster than GeniePath in most cases [18].
- **GraphCSC-M:** This is a typical GCN algorithm that defines the neighborhood according to nodes' importance, which is quantified by the node centrality in the whole graph, and this algorithm applies the attention mechanism to the graph convolution. The authors provide five possible centrality measures. Considering solvability and computation cost, we adopt degree-centrality and pagerank-centrality in the following experiments.
- **CV-GCN:** This sampling-style algorithm presents a control variate based method to achieve a similar convergence with the exact GCN [15], even using a small neighbor sampling size. CV and CVD are two estimators proposed in CV-GCN, so we choose the better result to present the model's performance.

Parameter Settings. We implement our algorithm in TensorFlow with Adam optimizer. All models are implemented under the inductive framework proposed by [13] except CV-GCN. All models adopt neighbor sampling to guarantee the scalability. We set the number of layers, neighbor sampling size, output dimension and batch size to be the same for different models. In detail, we set the number of layers to be 2, the neighbor sampling size to be $S_1 = 25, S_2 = 10$, the output dimension to be 256, the training batch size to be 512 and the validation batch size to be 256. Following [13], we subsample edges so that the maximum degree is 125. Specially, we adjust the neighbor sampling size and maximum degree when comparing with CV-GCN. Referring to the experimental setups in [24], we fix ϵ in Eq. 7 to be 0. In GIN and GraphRFA, we concatenate the node's previous layer representation with the aggregated feature vector computed by Eq. 7, and we find the concatenation operation can lead to a better prediction result. We simulate random walks with walk-length 3 starting from each initial node for 100 times to measure the similarity-based link reliability. Then we use the drop-bottom strategy to sample neighbors and perform a parameter sweep on dropping rates $\{0.2, 0.4, 0.6, 0.8\}$.

Evaluation Settings. We optimize GraphRFA under the supervised setting and unsupervised setting. Under the unsupervised setting, we put the learned embeddings to a logistic regression classifier for node classification. Specially for the Reddit, we randomly sample 10000 nodes from the training data and 6000 nodes from the testing data after finishing the unsupervised representation learning, and conduct the logistic regression based on the sampled nodes. We treat the binary classification on the Retweet as a special case of multi-class classification, and use Micro F1 score and Macro F1 score for evaluation on all datasets. Analogous trends hold for F1 score on the Retweet. Please note the theoretical guarantee of CV-GCN is proved under the supervised setting [4], so we do not evaluate CV-GCN under the unsupervised setting.

5.2 Experimental Results

Comparison with the Traditional Sampling-Style Baselines. In this section, we ran all models for 300 times. Note that GraphCSC-M proposed by [3] incorporates the node-centrality in loss function, however we focus on evaluating the methods of neighborhood definition and graph convolution. Therefore, we need to keep loss function the same for all models. The candidate loss functions are introduced in Sect. 4.3. In light of this, the examined model is essentially a variant of GraphCSC-M, denoted as GraphCSC-M*.

Table 2. Comparison under the supervised setting. Detail testing results of the node classification are shown (epoch=300).

	Micro F1				Macro F1			
	PPI-S	PPI	Retweet	Reddit	PPI-S	PPI	Retweet	Reddit
GraphSAGE-GCN	0.5455	0.5525	0.8889	0.9255	0.3624	0.3738	0.8812	0.8865
GIN	0.7058	0.8099	0.8995	0.9460	0.6177	0.7656	0.8924	0.9195
GeniePath-lazy	0.7908	0.8808	0.8983	0.9491	0.7364	0.8559	0.8903	0.9216
GraphCSC-M*	0.5918	0.7264	0.9374	0.8823	0.4967	0.6635	0.9328	0.8229
GraphRFA (mean)	0.7357	0.8491	0.9331	0.9510	0.6633	0.8166	0.9286	0.9274
GraphRFA (concat)	0.7588	0.8624	0.9353	0.9515	0.6949	0.8338	0.9307	0.9292
GraphRFA* (concat)	0.7671	0.8883	0.9353	0.9515	0.7090	0.8662	0.9307	0.9292

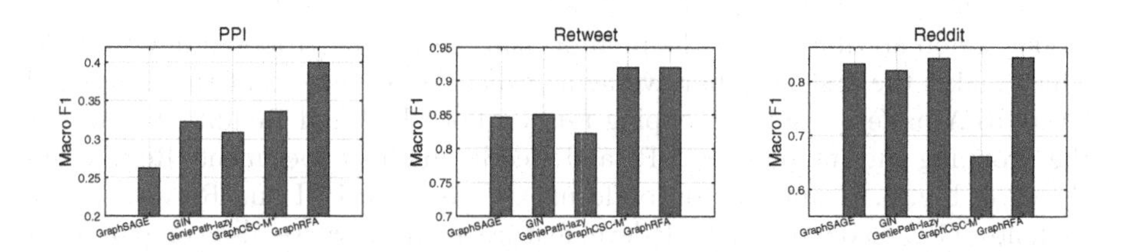

Fig. 3. Comparison under the unsupervised setting. Detail testing results of the node classification are shown (epoch=300). Here GraphSAGE denotes GraphSAGE-GCN. Similar trends hold for the Micro F1 score.

Supervised Learning: We perform a parameter sweep on initial learning rates {0.01, 0.001, 0.0001}. Experimental results are shown in Table 2. As an extension of GIN, our model consistently outperforms GIN. Among these baselines, GeniePath-lazy learns the importance of different neighbors and aggregates node features weighted by the learned importance, however such scheme will perform better on the algorithm using node's full neighborhood [18], i.e., the neighbor sampling may affect the effectiveness of GeniePath-lazy, while using the full neighborhood will lead to a large receptive field. Besides, we can see that GeniePath-lazy obtains better performance under the supervised setting. That is to say the task labels are necessary for learning the neighbors' importance. GraphCSC-M* uses centrality-based sampling to form the receptive field. Compared with this algorithm, our model performs more stable, especially on PPI and Reddit. In our model, the concatenation aggregator works better, which can be explained by the more sufficient feature interaction, so we use GraphRFA (concat) for evaluation in the following experiments. Moreover, we permit the dropping rates about different link reliability to be distinct, denoted as GraphRFA*. Compared with GraphRFA, GraphRFA* obtains better performance on the PPI.

Unsupervised Learning: We perform a parameter sweep on initial learning rates {0.01, 0.001, 0.0001}. From Fig. 3, we can see that GraphCSC-M* outperforms other baselines on PPI and Retweet, while this algorithm is not very effective on the Reddit. In other words, the information of node-centrality is insufficient for the neighborhood definition. Thus, we choose to measure the correlations between target node and its neighbors. GeniePath-lazy aims to learn such correlations, while the learned correlations can not support the neighbor sampling. In addition, we observe that the uniform-sampling based algorithms (i.e., GraphSAGE, GIN and GeniePath-lazy) perform well on the Reddit, which means more diverse neighbors can provide more benefits, and that is why the drop-bottom strategy prefers a smaller dropping rate on the Reddit [6]. Conversely, a larger dropping rate leads to a better prediction result on both PPI and Retweet. More details about tuning the dropping rate will be discussed later.

Comparison with CV-GCN. CV-GCN enjoys a similar convergence with the exact algorithm even using two neighbors per node, so we set its neighbor sampling size to be 2. The initial learning rate of this model is provided in [4]. The neighbor sampling size S_l of GraphRFA is constrained within 50. In our model, we do not implement the preprocessing technique proposed in CV-GCN, which makes the first neighbor averaging exact. We observe that the supervised GraphRFA prefers a small dropping rate on both PPI and Reddit, thus we set the dropping rate to be 0 on PPI and Reddit in this experiment. Results are shown in Fig. 4. GraphRFA outperforms CV-GCN on PPI and Retweet. Since we follow GIN to define the node-level feature aggregator, we also compare with GIN, and find that our model consistently outperforms GIN on every dataset.

[6] We set the dropping rate to be 0.2 on the Reddit.

Perhaps benefiting from the powerful feature aggregator, GIN performs better than CV-GCN on the Retweet.

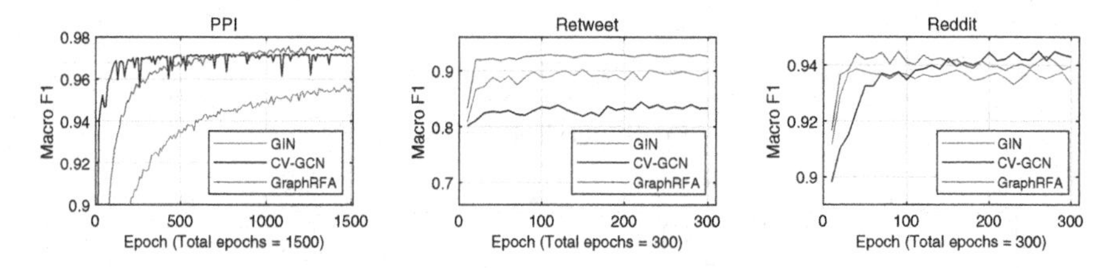

Fig. 4. Comparison with CV-GCN. Validation results of the supervised node classification are shown. (best seen in color).

Parameter Sensitivity: Insight into How the Drop-Bottom Works. The drop-bottom strategy trades off between uniform sampling and non-uniform sampling by adjusting the dropping rate, where the uniform sampling provides more diverse node features, while the non-uniform sampling aims to capture features from the more reliable neighbors. Here we explore how different dropping rates influence the performance of GraphRFA. We run all models for 300 times.

Figure 5 shows the results of node classification. Generally, we can see that supervised model prefers a smaller dropping rate than the unsupervised model. Since the supervised model is capable of learning a suitable aggregation function to weight different neighbors by the task labels, the more diverse neighbors can provide more benefits. On the contrary, the unsupervised model largely depends on the graph structure, thus highly reliable neighbors deserve more attentions due to the missing of the downstream labels. We can also observe the opposite choosing of dropping rate on Retweet and Reddit. Because in the political retweet network, users usually have clear attitude, which results in the more important role taken by the close neighbors than random neighbors on node representations. However, users in Reddit discuss various topics, and their interests are diverse. Only keeping the close neighbors may harm the performance of user representation learning.

Necessity of the Field-Level Feature Aggregation. In this section, we only consider one kind of measures (i.e., the similarity/centrality-based link reliability) then use the node-level aggregator to generate node representations, denoted as GraphRFA-sim and GraphRFA-cen. Specially, the adjusted dropping rates of GraphRFA-sim and GraphRFA-cen are directly applied to GraphRFA*. We also compare with GIN which adopts the uniform sampling. Taking the supervised setting as an example, Fig. 6 shows that the centrality-based link reliability plays a key role on both PPI and Retweet, and the similarity-based link reliability is also necessary for the Retweet. GIN performs better than GraphRFA-sim and GraphRFA-cen on the Reddit, that is because uniform sampling provides more benefits on this dataset discussed in the previous section. The field-level aggregator improves the prediction result on the Reddit.

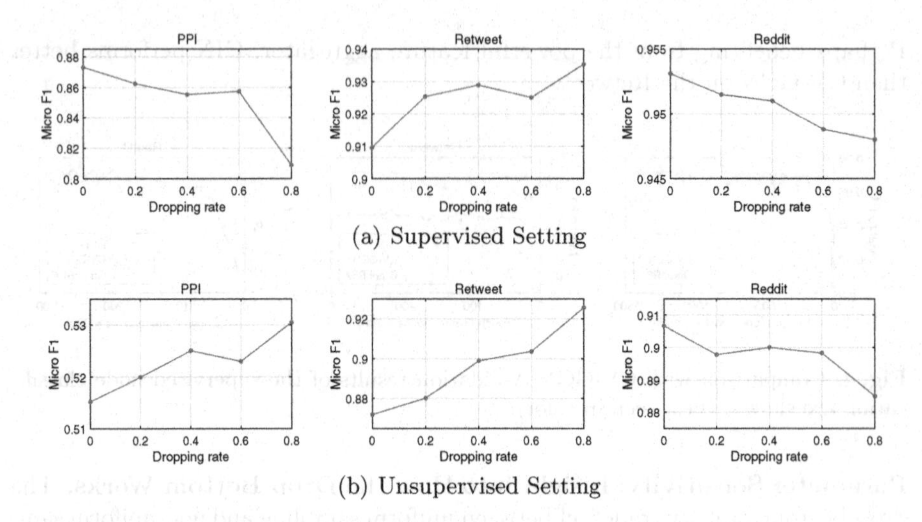

Fig. 5. Sensitivity analysis of dropping rate in the drop-bottom strategy. Testing results under the supervised and unsupervised settings are shown.

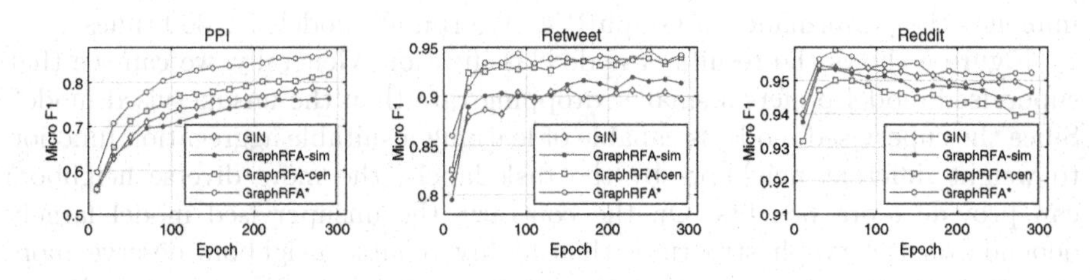

Fig. 6. Necessity of the field-level feature aggregation. Validation results of the supervised node classification are shown.

6 Conclusion

In this paper, we introduce an graph convolutional network, called GraphRFA, that captures reliable features from the node's neighborhood by a reliability-based feature aggregation mechanism. Theoretical analysis provides insight into how our model guarantees the permutation invariant property for graph data. In the experiments, we evaluate the effectiveness of GraphRFA on different datasets through supervised learning and unsupervised learning. Nonetheless, some extensions are possible too, such as learning an optimal dropping rate and extending GraphRFA to the heterogeneous graphs.

Acknowledgments. This work is supported by National Key R & D Program of China (No.2018YFB1004401) and NSFC under the grant No. 61532021, 61772537, 61772536, 61702522.

References

1. Ahmed, A., Shervashidze, N., Narayanamurthy, S.M., Josifovski, V., Smola, A.J.: Distributed large-scale natural graph factorization. In: WWW (2013)
2. Barabási, A., Albert, R.: Emergence of scaling in random networks. Science **286**(5439), 509–512 (1999)
3. Chen, H., Yin, H., Chen, T., Nguyen, Q.V.H., Peng, W., Li, X.: Exploiting centrality information with graph convolutions for network representation learning. In: ICDE (2019)
4. Chen, J., Zhu, J., Song, L.: Stochastic training of graph convolutional networks with variance reduction. In: ICML (2018)
5. Chen, J., Ma, T., Xiao, C.: FastGCN: fast learning with graph convolutional networks via importance sampling. In: ICLR (2018)
6. Cui, P., Wang, X., Pei, J., Zhu, W.: A survey on network embedding. IEEE Trans. Knowl. Data Eng. **31**(5), 833–852 (2019)
7. Dave, V.S., Zhang, B., Chen, P., Hasan, M.A.: Neural-brane: neural Bayesian personalized ranking for attributed network embedding. Data Sci. Eng. **4**(2), 119–131 (2019). https://doi.org/10.1007/s41019-019-0092-x
8. Defferrard, M., Bresson, X., Vandergheynst, P.: Convolutional neural networks on graphs with fast localized spectral filtering. In: NeurIPS (2016)
9. Girvan, M., Newman, M.E.J.: Community structure in social and biological networks. Proc. Nat. Acad. Sci. U.S.A. **99**(12), 7821–7826 (2002)
10. Grover, A., Leskovec, J.: node2vec: scalable feature learning for networks. In: KDD (2016)
11. Guimerà, R., Sales-Pardo, M.: Missing and spurious interactions and the reconstruction of complex networks. Proc. Nat. Acad. Sci. U.S.A. **106**(52), 22073–22078 (2009)
12. Hamilton, W.L., Ying, R., Leskovec, J.: Representation learning on graphs: methods and applications. IEEE Data Eng. Bull. **40**(3), 52–74 (2017)
13. Hamilton, W.L., Ying, Z., Leskovec, J.: Inductive representation learning on large graphs. In: NeurIPS (2017)
14. Katz, L.: A new status index derived from sociometric analysis. Psychometrika **18**(1), 39–43 (1953). https://doi.org/10.1007/BF0228902610.1007/BF02289010.1007/BF02289026
15. Kipf, T.N., Welling, M.: Semi-supervised classification with graph convolutional networks. In: ICLR (2017)
16. Liben-Nowell, D., Kleinberg, J.: The link-prediction problem for social networks. J. Am. Soc. Inform. Sci. Technol. **58**(7), 1019–1031 (2007)
17. Liu, W., Lü, L.: Link prediction based on local random walk. EPL (Europhys. Lett.) **89**(5), 58007 (2010)
18. Liu, Z., et al.: GeniePath: graph neural networks with adaptive receptive paths. In: AAAI (2019)
19. Lü, L., Jin, C., Zhou, T.: Similarity index based on local paths for link prediction of complex networks. Phys. Rev. E **80**(4 Pt 2), 046122 (2009)
20. Lü, L., Zhou, T.: Link prediction in complex networks: a survey. Physica A **390**(6), 1150–1170 (2011)
21. Perozzi, B., Al-Rfou, R., Skiena, S.: DeepWalk: online learning of social representations. In: KDD (2014)
22. Velickovic, P., Cucurull, G., Casanova, A., Romero, A., Liò, P., Bengio, Y.: Graph attention networks. In: ICLR (2018)

23. Weisfeiler, B., Lehman, A.: A reduction of a graph to a canonical form and an algebra arising during this reduction. Nauchno-Technicheskaya Informatsia **2**(9), 12–16 (1968)
24. Xu, K., Hu, W., Leskovec, J., Jegelka, S.: How powerful are graph neural networks? In: ICLR (2019)
25. Ying, R., He, R., Chen, K., Eksombatchai, P., Hamilton, W.L., Leskovec, J.: Graph convolutional neural networks for web-scale recommender systems. In: KDD (2018)
26. Ying, Z., You, J., Morris, C., Ren, X., Hamilton, W.L., Leskovec, J.: Hierarchical graph representation learning with differentiable pooling. In: NeurIPS (2018)
27. Yun, S., Jeong, M., Kim, R., Kang, J., Kim, H.J.: Graph transformer networks. In: NeurIPS (2019)
28. Zeng, A., Cimini, G.: Removing spurious interactions in complex networks. Phys. Rev. E **85**(3 Pt 2), 036101 (2012)
29. Zhang, M., Chen, Y.: Link prediction based on graph neural networks. In: NeurIPS (2018)
30. Zhou, T., Lü, L., Zhang, Y.: Predicting missing links via local information. Eur. Phys. J. B **71**(4), 623–630 (2009)

A Deep-Learning-Based Blocking Technique for Entity Linkage

Fabio Azzalini[1,2](✉), Marco Renzi[1], and Letizia Tanca[1]

[1] Politecnico di Milano, Milan, Italy
{fabio.azzalini,marco.renzi,letizia.tanca}@polimi.it
[2] Center for Analysis Decisions and Society, Human Technopole, Milan, Italy

Abstract. Nowadays, data integration must often manage noisy data, also containing attribute values written in natural language such as product descriptions or book reviews.

Entity Linkage has the role of identifying records that contain information referring to the same object. Modern Entity Linkage methods, in order to reduce the dimension of the problem, partition the initial search space into "Blocks" of records that can be considered similar according to some metrics, greatly reducing the overall complexity of the algorithm.

We propose a Blocking strategy that, differently from the traditional methods, aims at capturing the semantic properties of data by means of recent Deep Learning frameworks. This paper is mainly inspired by a recent work on Entity Linkage whose authors were among the first to investigate the application of tuple embeddings to data integration problems. We extend their method adopting an unsupervised approach: our blocking model is trained on an external corpus and then used on new datasets, exploiting a "transfer learning" paradigm. Our choice is motivated by the fact that, in most data integration scenarios, no training data is actually available. Using a semi-automatic approach to blocking, our model, after being trained on an external corpus, can be directly applied to any data integration problem.

We tested our system on six popular datasets and compared its performance against five traditional blocking algorithms. The test results demonstrated that our deep-learning-based blocking solution outperforms standard blocking algorithms, especially on textual and noisy data.

1 Introduction

Data integration is recognized as a crucial activity for companies, government agencies and in general for those who deal with data coming from multiple sources. However, with the increasing role of internet-based services like e-commerce, web sites for comparing products or online libraries, data integration is becoming even more challenging. These services deal with data that is typically noisy and that contains attribute values written in natural language, such as product descriptions or book reviews. Integrating such data is hard, because of the difficulties in managing dirty values and in extracting semantics out of long textual values written in natural language. In particular, the challenging part in such settings lies in identifying which records from the several source datasets

Y. Nah et al. (Eds.): DASFAA 2020, LNCS 12112, pp. 553–569, 2020.
https://doi.org/10.1007/978-3-030-59410-7_37

represent the same concept or the same entity, i.e., the activity known as *Entity Linkage*, or *Entity Resolution*. In the past, this task has been addressed applying pairwise matching algorithms over the Cartesian product of the records provided by two input sources. However, the current disruptive growth in dataset sizes makes the problem intractable. Modern Entity Resolution methods, in order to reduce the dimension of the problem, partition the initial search space into *blocks* within which the comparisons are performed, greatly reducing the number of matches and thus the overall complexity of the algorithm. Blocking methods apply functions and algorithms to filter out from the potential comparisons all those tuple pairs that are clearly not matching.

Traditional blocking schemes use hand-tuned functions to generate these buckets, and place the tuples inside them accordingly. In other words, all the records are passed through a function(s), also called *blocking function(s)* and each tuple is assigned to a bucket based on its *blocking key value* (BKV). One can clearly understand that the quality of the entity linkage (and thus ultimately of the entire data integration activity) is profoundly influenced by the blocking phase, both in efficiency, as the blocking phase should grant better time and memory consumptions with respect to a naïve Cartesian product approach, and effectiveness, since this phase should find as many true matching pairs as possible.

To better understand the traditional blocking process, let us consider the following sample datasets (Tables 1 and 2):

Table 1. Database A

ID	Name	Surname	Gender
a1	John	Smith	Man
a2	Robert	Sandy	Man
a3	Christine	Faulkner	Woman

Table 2. Database B

ID	Name	Surname	Gender
b1	Robertt	Sandy	Male
b2	Kristine	Fawkner	Female
b3	Johnny	Smith	Male

When using as blocking function $BKV = f(Gender) = Gender$, traditional blocking methods would group together only the tuples belonging to the same source database, since the two datasets A and B use different terms for values of the attribute *Gender*. This example shows a big problem that affects traditional blocking schemes: they ignore the semantics of the attribute values and leverage only the lexicon. Other shortcomings of traditional blocking methods include: *(i)* the sensitivity to morphological variations and data quality issues, *(ii)* the time-consuming and cumbersome activity of designing appropriate blocking functions, *(iii)* the necessity to design dedicated blocking strategies for each new dataset. With the ever-increasing need to process, analyze and integrate large datasets that contain noisy and long textual values (especially if in natural language) new solutions that exploit also semantic information can be of great help.

The system that we propose is based on the fact that, in case the datasets to be merged contain noisy values or texts written in natural language, a method that leverages only the morphological aspects of the attributes is not enough and not effective; therefore we aim to capture word and sentence meanings. Our algorithm consists of two key phases: first the records of the datasets to be integrated are transformed into real-valued vectors exploiting techniques such as word and sentence embeddings (Natural Language Processing (NLP) well-know techniques that encode semantic information in the numeric representation they produce), and then, to generate the blocks we apply to these vectors an innovative blocking technique, based on Locality Sensitive Hashing (LSH) [1].

In this work we want to keep the entire blocking phase in an unsupervised setting, thus without training the neural networks on the actual datasets to be merged. This choice stems from the evidence that a labeled training set generated on the actual data to be merged is typically difficult to find, and thus an unsupervised approach makes the task more feasible. To achieve this scope we exploit pre-trained word embeddings such as *GloVe* [2] and *fastText* [3], and train the recurrent neural network (RNN) responsible for composing the sentence embeddings on a separate available big dataset, namely the *SNLI corpus* [4], in a "transfer learning" fashion.

Our work is mainly inspired by DeepER [5] where the authors were among the first to investigate the application of tuple embeddings for the entire entity linkage pipeline. Our system, however, is focused on the blocking phase, and adopts a different paradigm for this specific task. More specifically, our contributions are:

- Differently from DeepER [5], we apply to blocking an unsupervised approach: the models are trained on an external corpus and then used on new datasets thanks to a "transfer learning" paradigm.
- Differently from traditional blocking algorithms, our system implements a semi-automated approach to blocking: our embedding models, after being trained on the external corpus, simply require to scan the input tuples to produce the blocks. Our solution, consequently, eases the difficult and cumbersome definition of the blocking functions that are used in traditional blocking schemes and makes the system more portable across the datasets.
- We compare the performances of our system with some of the most consolidated blocking paradigms on six popular real datasets, commonly used to evaluate blocking schemes. Additionally, we provide a wide range of experimental results to study empirically the differences of the architectural choices of our system. As a matter of fact, we believe that our experimental study is a major contribution of our work.

2 State of the Art

Standard Blocking is the easiest of the traditional algorithms. Once the blocking function is defined, this method positions inside the same bucket al.l the records with equal BKV. The peculiarity of this method is that each tuple is inserted into one bucket only, while the other traditional techniques can potentially put a record into several blocks.

This method presents two main problems: *(i)* it is not robust w.r.t. noisy values *(ii)* it is not suitable when the distribution of BKVs is very skewed as the buckets sizes would be too diverse.

Sorted Neighbourhood Blocking. [6] uses the BKVs generated by the blocking function(s) to sort the tuples in the databases. Once the datasets are sorted, a sliding window with a fixed size w ($w > 1$) is passed over the source records and the blocks are generated by the tuples that fall in the window step by step. This method presents two main problems: *(i)* dataset sorting is sensitive to the prefixes of the BKVs *(ii)* choosing a proper value for w is not trivial.

Q-gram Blocking is specifically designed to face the errors and variations in attribute values [7]. To account for them the method generates variations of the BKVs. Each record is then inserted into both its original bucket (the original BKV) and in all the derived buckets (variations on the original BKV). To build the new BKVs, the algorithm generates all the q-grams (substring of length q) of the original BKVs. This algorithm, even if it can capture variations in attribute values, is very expensive in both time and memory consumption, especially when the original BKVs are long.

Suffix Blocking [8], is based on ideas similar to the q-gram blocking. Here, however, instead of considering all the q-grams, only some suffixes of the original values are used. To build the new blocks two parameters are needed: l_{min} which corresponds to the minimum length of the suffix substrings that are generated, and b_{max}, used to discard blocks whose number of records exceed this threshold.

Canopy Cluster Blocking [9] adopts a clustering approach to blocking: it groups together the records based on how similar their BKVs are. A similarity function commonly used is the *Jaccard* measure, which basically indicates the percentage of q-grams two BKVs have in common.

All algorithms presented so far can be considered traditional blocking methods, and they all suffer, to different extent, from the problems presented in the introduction.

Recently, other blocking techniques have been presented in the literature, such as BLS [10], FALCON [11] and DEEPER [5], however we will not compare our system with these methods because they all are based on supervised learning approaches and thus somehow not comparable to ours.

3 Methodology

Our blocking scheme is logically composed of two phases: first, an embedding module is responsible for transforming the records into real-valued vectors, and second, a blocking module actually produces the buckets from these numerical vectors (see Fig. 1). The architecture we present is very similar to the one presented in [5], however there are important variations in the implementation of the embedding layer, the main difference being that our RNNs are trained on an external corpus, the SNLI corpus, and not on the actual datasets to be merged. This unsupervised approach is motivated by the following considerations. First, as noticed, in many applications no training data is available, and this choice makes our technique more realistic. Secondly, this paradigm ensures fairer evaluation tests because the traditional blocking methods to which we compare ours are not tuned on the datasets.

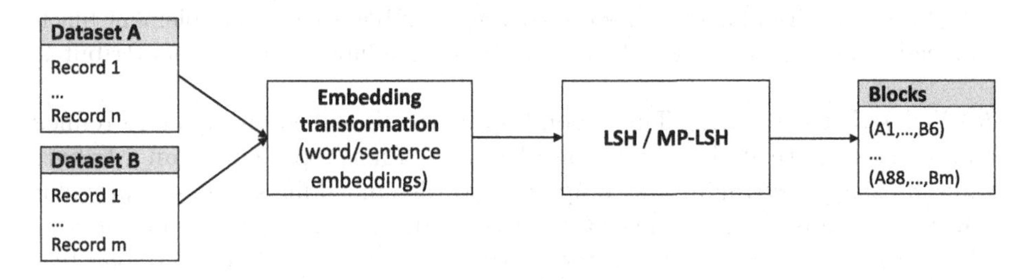

Fig. 1. Structure of our blocking method

3.1 Embedding Architectures

We now illustrate the two solutions that we adopt to implement the embedding layer. This layer receives as input the records of the datasets to be merged and outputs their vectorial representation.

In the rest of the paper we use the following notation, inherited from [5]: given a tuple t with n attributes $\{A_i\}_{i=1,\dots,n}$, the symbol $\mathbf{v}(t[A_k])$ represents the numerical representation of the attribute value $t[A_k]$, while the numerical value of the entire tuple is $\mathbf{v}(t)$; given a generic word w, its embedding representation will be indicated with $\mathbf{v}(w)$.

At their core, both the embedding architectures we propose make use of pre-trained word embeddings, i.e. fastText or GloVe: each word w of each attribute value $t[A_k]$ is transformed into a real-valued vector $\mathbf{v}(w)$. The fastText model we use is *crawl-300d-2M-subword* [3] where each word is represented as a 300-dimensional vector and the corpus on which it was trained is *Common Crawl*[1]. Regarding GloVe we use the model *glove.840B.300d* [2] and again each word is represented as a 300-dimensional vector and the training corpus was still *Common Crawl.* Once the single words are transformed, what differentiates the two architectures concerns how the single word embeddings are composed first to represent the attribute value (i.e. how to build $\mathbf{v}(t[A_k])$) and eventually the entire record (how to produce $\mathbf{v}(t)$).

Average-Based Architecture: With this approach, adopted in [5], when a tuple t with n attributes arrives at the embedding layer, each of its attributes $\{A_k\}_{k=1,\dots,n}$ is treated independently from the others. First we use a tokenizer

[1] http://commoncrawl.org/the-data/.

to split the attribute value $t[A_k]$ into its component words $\{w_i\}_{i=1,\dots,p}$. In our implementation we use the standard *NLTK* tokenizer available in the *NLTK* module[2]. In case one attribute is empty (i.e. there is a missing value) we insert the placeholder word "*unk*". Then each of these words is mapped to a 300-dimensional vector by applying the pre-trained word embedding model. The attribute vector representation $\mathbf{v}(t[A_k])$ is given by simply averaging the vectors of the component words. The tuple vector $\mathbf{v}(t)$ is finally obtained by the concatenation of all the n attribute vectors, $(\mathbf{v}(t[A_1]), \dots, \mathbf{v}(t[A_n]))$.

This composition method has the advantage of being easy to implement but it can lead to huge vector sizes when the data sources have a long list of attributes.

RNN-Based Architecture: This second approach to embedding is more refined and its use is motivated by some considerations. First, representing an attribute value by simply averaging its component words embeddings – as the previous solution – means ignoring the order of the words. This is generally not an issue when there are atomic values or few words in an attribute, but when long textual elements are present it becomes more important. Indeed, the order of words helps when there are attributes that encapsulate multi-word content such as descriptions or specifications of products. Second, the previous embedding approach cannot capture semantic relationships among attributes: to build the final tuple vector $\mathbf{v}(t)$, it employs a simple concatenation of the component attribute vectors. Third, another shortcoming of average-based solution that motivates the RNN based approach concerns the embedding sizes. As said, when using an average approach the tuple embeddings are proportional to the number of attributes n of the sources and this can easily lead to very big embedding vectors. With the RNN-based architecture the size of the resulting tuple vector is fixed a priori and is independent of the number of attributes.

As suggested in [5], we use both uni- and bi-directional recurrent neural networks (RNNs) with long short-term memory (LSTM) cells. In the following discussion we refer to this family of models with the term RNN-LSTM architectures. However, to actually implement these nets we do not follow the approach described in the paper [5] because, as anticipated, the authors train the models directly on the datasets to be integrated in a supervised fashion.

To devise our models we are instead inspired by the recent studies made by the Facebook AI research team in their paper [12] about sentence embedding models. Among the several solutions they investigate, they propose a model, named *Infersent*, a bi-LSTM net which transforms sentences written in natural language into their corresponding numerical representation. The authors show the strong results of this model on several tasks and in particular for semantic textual similarity applied in unsupervised settings. Our uni- and bi-LSTM nets are consequently implemented with the same architectures described in that paper for analogous models, and partially readapting the code the authors kindly release at [13]. In our work, however, we extend the analysis of these models with both fastText and GloVe and with a greater set of network sizes.

[2] https://www.nltk.org/_modules/nltk/tokenize.html.

This is interesting because *fastText* and *GloVe* handle words not present in their vocabulary in different ways. *GloVe* replaces the missing words with the default value "*unk*", while *fastText*, exploiting sub-word embeddings, is able to assign a numeric representation also to words not present in its vocabulary; for this reason *fastText* is usually regarded as a *character level embedding* method.

Given these premises, it is now possible to describe how these models actually perform our embedding task. When a tuple t with n attributes arrives at the embedding layer, each attribute value $t[A_k]$ undergoes the tokenizer which splits it into its component words $\{w_i\}_{i=1,\ldots,p}$. Each of these words is mapped to a 300-dimensional vector by applying the pretrained word embedding model (fastText or GloVe) which yields the words embeddings $\{\mathbf{v}(w_i)\}_{i=1,\ldots,p}$. After each word vector $\mathbf{v}(w)$ is generated it is given as input to the RNN-LSTM, which processes it and produces an internal hidden state. At the end of the current attribute value $t[A_k]$, the process is repeated from the first word of the adjacent attribute A_{k+1} seamlessly. This continuous feeding of the net ensures the first two appealing "properties" discussed above: since the tuple is processed as a single long sentence the process guarantees to encode both the words order and the possible semantic relationships among adjacent attributes. The final tuple embedding vector $\mathbf{v}(t)$ is obtained after the scan and process of the last word of the last attribute value. For a deeper understanding of how uni- and bi-directional LSTM nets work we refer the reader to [5,12].

Another important advantage with RNN-LSTM architecture is that it is possible to control and fix the size of the embedding tuples a priori. This holds because the dimensionality of the vectors is determined by the number of LSTM cells we put in the RNN hidden layer and this is a design decision one takes before actually applying the model. In our work we implement RNNs with 300, 1024 and 2048 LSTM cells and so in the tests we will evaluate the results with tuple embeddings whose sizes are 300, 1024, 2048 when applying the uni-LSTM nets and 600, 2048 and 4096 when using bi-LSTMs.

In order to use these RNN-LSTM nets, a prior training phase is necessary to estimate their parameters. The RNN-LSTM nets of our embedding layer are trained on the labeled external *SNLI* corpus [4] before being applied to the test datasets. This dataset is one of the largest labeled sources specifically designed to force semantics understanding, and thus a good candidate for our transfer-learning solution.

3.2 Blocking Algorithms

Now that we have a numerical representation we need to detect similar tuples and put them in the same block. This is the goal of the second step of our blocking system, where, exploiting two algorithms, we actually group records into the buckets. Since the first step transforms the records into vectors with the appealing property that tuples related in the meaning are projected onto vectors that are close to each other, the algorithms we use exploit this geometrical proximity to block them. These algorithms are Locality Sensitive Hashing (LSH) [1] and Multiprobe LSH [14], two popular techniques that belong to the family of algorithms used in approximate nearest neighbour (ANN) search problems.

Locality Sensitive Hashing: This algorithm aims at finding a hash function h that satisfies two requirements: *(i)* items that are (semantically) close should have the same hash value $h(x)$ with "high" probability P_1, *(ii)* points that are distant should have the same value $h(x)$ with "low" probability P_2. The ultimate goal of the h function is indeed to group together (with the same hash value) those items that are close, or equivalently, similar.

We generate K distinct hash functions $h_1, h_2, ..., h_K$ with $h_i \in \mathcal{H}$ and apply these functions in order, to each tuple x to be analyzed. Each tuple x is consequently assigned to the sequence of the outputs $g(x) = \{h_1(x), h_2(x), ..., h_K(x)\}$. This ordered sequence $g(x)$ is named the *hash code* of tuple x and it is used to identify that record.

Since for large values of K the likelihood that similar record get the same K-dimensional hash code is reduced, typically several hash codes are computed for each tuple. To do so, one builds L hashes $g_1(x), g_2(x), ..., g_L(x)$, where each $g_i(x)$ is a K-sized vector obtained applying the ith sequence of hash functions on tuple x. The set of hash codes $\{g_i(y)\}|_{y=1,...,N}$ applied to entire set of N vectors is generally named *hash table*.

Given the hash tables, then all the record sharing the same K-dimensional code in any of the L tables are considered "similar" and put into the same group.

Multiprobe LSH: In the standard LSH implementation described above, one considers as similar all the records that fall in the same group (that is those having the same hash code in any of the L tables). Using several hash tables increases the likelihood that similar records fall into the same block but it also comes with the side effect that more tuples, possibly distant or unrelated, are put in the same bucket.

Multiprobe LSH [14] is a variation of standard LSH that aims at reducing the number L of hash tables but without the loss of nearest neighbours.

4 Experimental Results

We tested our blocking system on six popular datasets with two objectives: first to compare its performances against five traditional blocking algorithms and secondly to understand how the several architectural variants we implemented differ among each other. Since every blocking algorithm has its own set of parameters we also define how these are set for the tests.

4.1 Datasets

The blocking algorithms are tested on six publicly available datasets that are commonly adopted to evaluate entity linkage and blocking schemes [5,15–17]. These are: *Restaurant* dataset [18], *Census* dataset [18], *Cora* dataset [18], *DBLP-ACM* datasets [19], *Amazon-Google Products* datasets [19] and *Abt-Buy* datasets [19].

The first two datasets represent the "easy" task: they are structured and mostly clean with very few typos and missing values. On these datasets traditional blocking techniques show strong results as reported in [15].

Cora is still a well structured dataset but it has some quality issues as will be discussed later in the results.

Restaurant is a set of tuples with restaurant names, addresses, cities, phones and food styles taken from Fodor and Zagat restaurant guides.

Census is a dataset generated by the US Census Bureau including information such as first and last names, middle initials, zip codes and street addresses.

Cora contains records about machine learning articles with many attributes as publication name, publication year, authors name, venue, etc.

DBLP-ACM are well-structured bibliographic data sources regarding computer science conferences. Among the shared fields there is a relatively long textual attribute *title* but its values are fixed and the same between the two sources so on these sets traditional blocking models are expected to perform well anyways.

The last two pairs of data sources are "challenging": they have long textual attributes (such as product description) written in natural language with plenty of variations. Both tasks deal with e-commerce products and the single sources have also duplicates inside each of them.

In Table 3 we summarize some of the key characteristics of the datasets.

Table 3. Specifications of the target datasets

Data set	Task	#Tuples	#True matches	#Attributes	Cartesian size
Restaurant	Deduplication	864	112	5	372816
Cora	Deduplication	1295	17184	12	837865
Census	Deduplication	841	327	5	353220
DBLP-ACM	Linkage	2616–2294	2224	5	6001104
AMZN-GP	Linkage	1363–3226	1300	5	4397038
ABT-BUY	Linkage	1081–1092	1097	4	1180452

The "Task" column specifies the type of activity to be performed: *(i) Deduplication* for finding duplicate tuples that all belong to a single source *(ii) Linkage* for finding related tuples across two or more data sources.

The "Cartesian size" column contains the number of comparisons one would perform without blocking ($n * m$ for the (entity) linkage task, $\frac{n*(n-1)}{2}$ for the deduplication case, where n and m are the number of tuples in the (two) source(s)).

4.2 Metrics

Two metrics are considered to assess the performances of the blocking algorithms: *Reduction ratio* (*RR*) and *Pair completeness* (*PC*), which are the measures typically used in the literature [7,15,20].

Following the notation of [20] RR and PC are defined as:

$$RR = 1.0 - \frac{s_M + s_N}{n_M + n_N} \qquad PC = \frac{s_M}{n_M} \tag{1}$$

where

- n_M is the number of matching pairs generated without blocking.
- n_N is the number of non-matching pairs generated without blocking.
- s_M is the number of matching pairs generated with blocking.
- s_N is the number of non-matching pairs generated with blocking.

RR measures how much the blocking technique can reduce the number of pair comparisons with respect to a naïve full comparison approach.

The second metric, PC, measures instead the effectiveness of the blocking method at not removing true matches from the set of comparisons.

These metrics range in $[0, 1]$ and they are typically in a trade-off [20]: an high reduction ratio may come at the cost of some missed true matches or viceversa a large pair completeness may require to compare also many unnecessary pairs.

In some works RR and PC are combined to assess the overall blocking result. In [17], for example, the metrics are multiplied $\alpha = RR \cdot PC$. In our tests to select the best result we use the *harmonic mean* of the two, $\alpha = 2 \cdot \frac{RR \cdot PC}{RR + PC}$.

4.3 Design of the Evaluation

For what concerns traditional blocking algorithms, parameters setting is done by hand, the best results are obtained experimentally and then used in the comparison with our blocking algorithm.

Traditional algorithms require a *blocking function* to operate, and in our tests, for each dataset, we run the experiments by applying iteratively one of the following functions: *attribute value*, *first n chars*, *last n chars* (with $n \in \{2, 3, 4\}$) and *Soundex* phonetic encoding. The blocking function is applied to every single attribute of the current dataset. In *standard* blocking, the blocking function is the only variable to be tested as no further parameters are defined for this algorithm. In *sorted neighbourhood* blocking, for each dataset the parameter window size w is varied in the range $w \in \{2, 3, 4, 5\}$. *Q-gram* blocking is tested with q-gram size $q \in \{3, 4, 5\}$ and threshold $t \in \{0.6, 0.8\}$. In *suffix* blocking the minimum suffix length $l_{min} \in \{3, 4, 5\}$ and maximum block size $b_{max} \in \{20, 50\}$. Finally, the *canopy cluster* blocking algorithm is applied with the q-gram size $q = 3$, the high threshold $t_h = 0.9$ and low threshold $t_l \in \{0.6, 0.7\}$.

Regarding our blocking system, we evaluate both the average-based and the RNN-LSTM-based models described in the methodology above, with the following parameters: the number of LSTM cells $n_{cells} \in \{300, 1024, 2048\}$ and as word embeddings both GloVe and fastText. To set the hash code size K and the number of hash tables L of the LSH algorithm, for each dataset we apply the theoretical formulae presented in [1]. While for choosing the hash functions we proceed similar to [5].

All the algorithms are implemented in Python and the tests are run on a Google Compute Engine[3] instance with the following specifications: Ubuntu 16.04.1 LTS (Xenial Xerus), 4 hyper-threaded cores Intel Xeon Processor @2 GHz CPU, 32 GB of RAM and 60 GB of SSD storage.

4.4 Test Results

We now first provide the results of the tests between our blocking system and the traditional methods and then we investigate the impact of different architectural choices of our model.

Our System vs. Traditional Blocking Algorithms

Reduction Ratio and Pair Completeness: Tables 4 and 5 illustrate the best results in terms of *RR* and *PC* obtained by each blocking algorithm on each of the six target datasets. Our method is named *Embedding*.

Table 4. Blocking algorithms comparison on restaurant, Cora and census datasets

	Restaurant			Cora			Census		
	RR	PC	α	RR	PC	α	RR	PC	α
Standard	0,9848	0,9464	**0,9652**	**0,9659**	0,7881	0,8679	**0,9829**	0,7703	0,8637
Suffix	**0,99**	0,875	0,9289	0,989	0,4441	0,6129	**0,9829**	0,7703	0,8637
Sorted N.	0,9848	0,9464	**0,9652**	**0,9659**	0,7881	0,8679	0,8976	0,8721	0,8847
Q-gram	0,9848	0,9464	**0,9652**	0,9464	**0,8322**	**0,8856**	**0,9829**	0,7703	0,8637
Canopy	0,9848	0,9464	**0,9652**	**0,9659**	0,7881	0,8679	**0,9829**	0,7703	0,8637
Embedding	0,9432	**0,9792**	0,9608	0,9233	0,7626	0,8352	0,8945	**0,9542**	**0,9234**

Table 5. Blocking algorithms comparison on DBLP-ACM, AMZN-GP and ABT-BUY datasets

	DBLP-ACM			AMZN-GP			ABT-BUY		
	RR	PC	α	RR	PC	α	RR	PC	α
Standard	**0,9996**	0,8826	0,9374	0,9865	0,4661	0,6331	0,9793	0,6272	0,7646
Suffix	0,9964	0,9528	0,9741	**0,9979**	0,1792	0,3038	**0,9935**	0,4011	0,5714
Sorted N.	0,9974	0,9834	0,9903	0,9604	0,4308	0,5947	0,9371	0,8386	0,8851
Q-gram	0,0	0,0	0,0	0,9865	0,4661	0,6331	0,9424	0,8049	0,8682
Canopy	0,9959	0,9645	0,9799	0,9865	0,4661	0,6331	0,9793	0,6272	0,7646
Embedding	0,9873	**0,9946**	**0,9909**	0,9436	**0,7885**	**0,8591**	0,9343	**0,9088**	**0,9213**

The experiments on *Restaurant* and *Census* datasets show competitive results between traditional methods and our blocking system, proving that embedding-based models keep up on clean and simple datasets. On these datasets traditional blocking algorithms seem to have an edge on *RR* but on the other

[3] https://cloud.google.com/compute/.

hand embedding based models provide better *PC*. The only exception is on *Cora* dataset where the q-gram blocking algorithm achieves a 6.9% better score for *PC* with respect to the embedding model. We explain this result by inspecting the dataset. This data source has serious quality issues on the majority of the attributes: the column *volume* has 76% of missing values, the attribute *address* 77%, field *note* reaches 90% of missing values. The only clean attribute is *title* and its values are mainly fixed with very little variations. All the traditional blocking methods select this field to filter the tuples, and without a prior imputation strategy it becomes really hard to leverage semantic information out of so many blank values.

Overall the performances of traditional blocking algorithms on these sets are aligned with those reported in [15] and in [17].

Even though on *DBLP-ACM* the results are still balanced, *Q-gram* shows null values because it was not possible to conclude the test: time and memory consumptions were prohibitive. This is a known limitation of this type of blocking algorithm, as computing all the q-grams of long textual values is expensive.

The most significant differences between our blocking system and the traditional approaches are on the tests on the last two "challenging" datasets: *AMZN-GP* and *ABT-BUY*. Our embedding-based model is slightly less efficient at reducing the number of pair comparisons but *PC* is much higher than traditional methods. Even though a *PC* of 0.78 (obtained by our blocking method on the *AMZN-GP* dataset) is still a relatively poor result the analysis suggests that our model is good at capturing the semantic information out of data.

Supporting this insight are also the results on *ABT-BUY* datasets on which we record the best performances in two scenarios:

- Traditional models are free to choose the attribute and blocking function giving the best result (see Table 5). In this case they all go for the *name* attribute. This attribute is relatively clean and easy for them to block on.
- Traditional models are forced to use as attribute the *description* of products (see Fig. 2). The values of this attribute are long textual descriptions written in natural language plenty of variations.

As can be seen in Fig. 2 the *PC* performances are completely different, with an absolute win for the embedding-based models.

Overall we consistently obtain the best *PC* except for one case, while maintaining a *RR* close to the best result obtained by traditional approaches. This is confirmed by the α score, where our system ranks first on four of the six datasets. We also think that often a high *PC* should be preferred w.r.t. a high *RR*, since performing a few more comparison could be less expensive than actually missing data that were positioned in the wrong block.

The results match our intuition that traditional blocking solutions work poorly on noisy and textual datasets because of their inability to leverage semantic information. By contrast, embedding-based models exploit the meaning of words and sentences and overall provide a more appropriate solution.

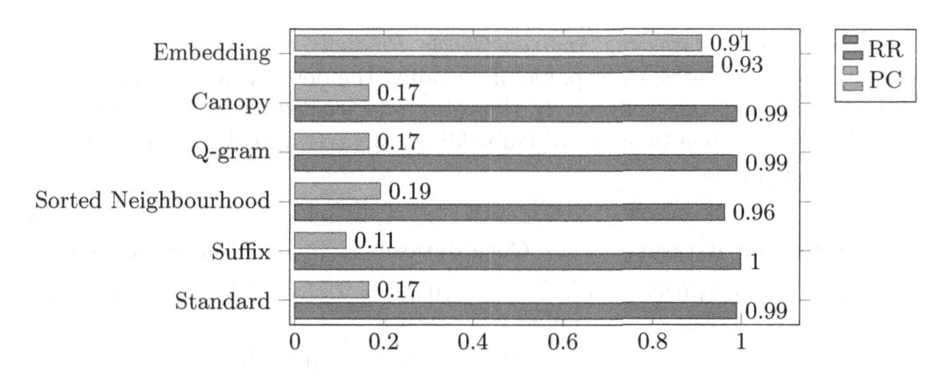

Fig. 2. RR and PC on ABT-BUY dataset forcing on description attribute

Execution Time: Despite the good results of our blocking scheme, tests confirmed that deep-learning-based models are very expensive in terms of time (and memory consumption); this is a well known pitfall of these models [21].

We show in Table 6 the average time needed (in seconds) to complete the blocking phase on each dataset.

Since our tests are run on a single computing instance, we believe that by adopting a parallel and distributed computation paradigm this blocking scheme can increase its efficiency by a wide margin. Additionally, the integration of the data sources is typically done at the beginning of a data analysis project, as a result we believe that it is worth spending more time (few minutes) on the integration phase, with the scope of having more precise and complete data, and hence significantly better performance during the data analysis task.

Table 6. Execution times of the blocking methods

Method	Restaurant	Cora	Census	DBLP-ACM	AMZN-GP	ABT-BUY
Standard	0.68880	2.73599	0.66381	3.15014	3.72337	1.33364
Suffix	0.70334	2.74421	0.67965	4.00995	3.10155	1.14573
Sorted N	0.70558	2.6667	0.73028	3.36834	6.14874	2.10252
Q-gram	7.59385	2.93346	7.66572	Unbounded	10.59905	1.87669
Canopy	0.29250	2.99539	0.69930	1449.90151	4.19712	2.73809
Embedding	19.12214	163.64995	142.29576	1080.66823	815.40971	402.29103

Results for Different Architectural Choices

RNN-LSTM vs. Average: One of the clearest results is the difference in performances when comparing the RNN-LSTM architectures with the simpler average scheme (see Fig. 3). The average based architecture provides sufficient results only on *Restaurant* and *Census* datasets where the attribute values are clean and atomic or at most composed of few words. When the textual values are longer, however taking into account words order guarantees more refined embeddings. This observation is particularly evident by considering the outcomes on the "challenging" datasets, *Amazon-Google Products* and *Abt-Buy.* On those sources the neural nets are capable of encoding the dependencies among words and adjacent fields more effectively, thus obtaining a substantial improvement on *PC.* Conversely when many word embeddings are averaged the resulting vector is less discriminative.

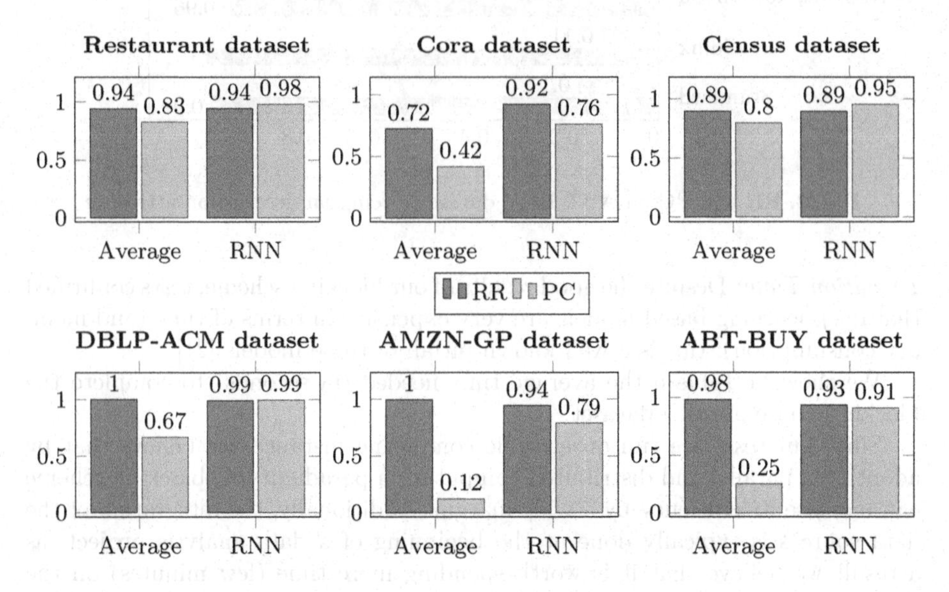

Fig. 3. RNN-LSTM vs. Average architecture performances

LSTM vs. Bi-LSTM: Another key evidence resulting from the tests concerns the superiority of the bi-LSTM architecture over the uni-LSTM model for the current task (see Figure 4). Bi-LSTM nets outperform the single LSTMs on every dataset, especially in terms of *PC* suggesting that they generate better tuple representations. Similarly to the previous set of tests, the gain in *PC* is more significant when dealing with the complex datasets.

fastText vs. GloVe: The two word embedding approaches show similar results on *Restaurant, DBLP-ACM* and *Census.* On the remaining datasets, however, GloVe is ahead regarding the *PC* (see Fig. 5). We explain this trend by recognizing greater generalization capabilities of GloVe on these data sources.

The datasets about e-commerce products in particular contain codes, commercial names and brands which are handled differently by the two embedding paradigms: GloVe ignores the majority of them because they are not in the dictionary of known words whereas fastText constructs new words embeddings by considering their n-grams. However, brand names rarely convey semantics about real-world entities and this should explain why fastText is not able to enrich the expressiveness of the embeddings.

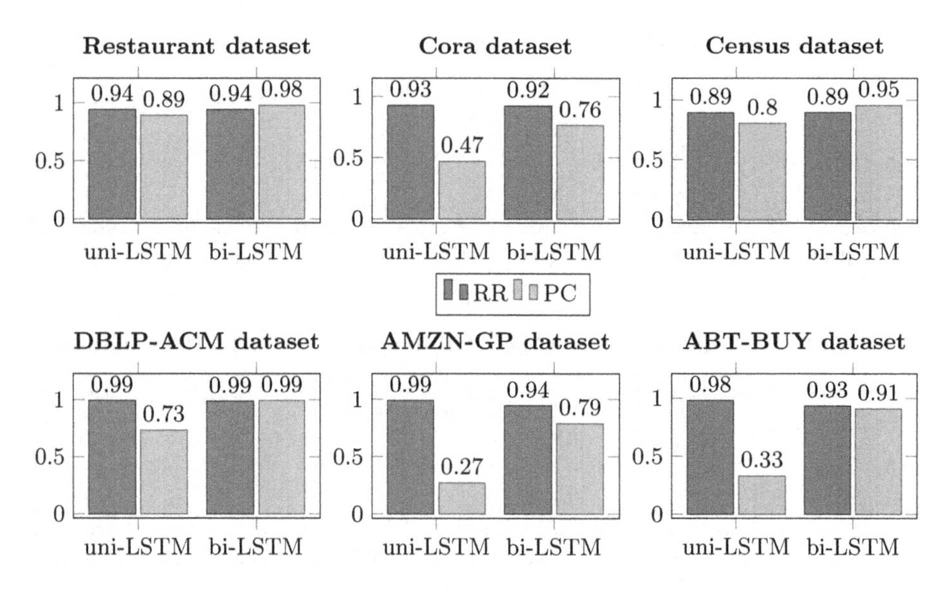

Fig. 4. Uni-LSTM vs. Bi-LSTM performances

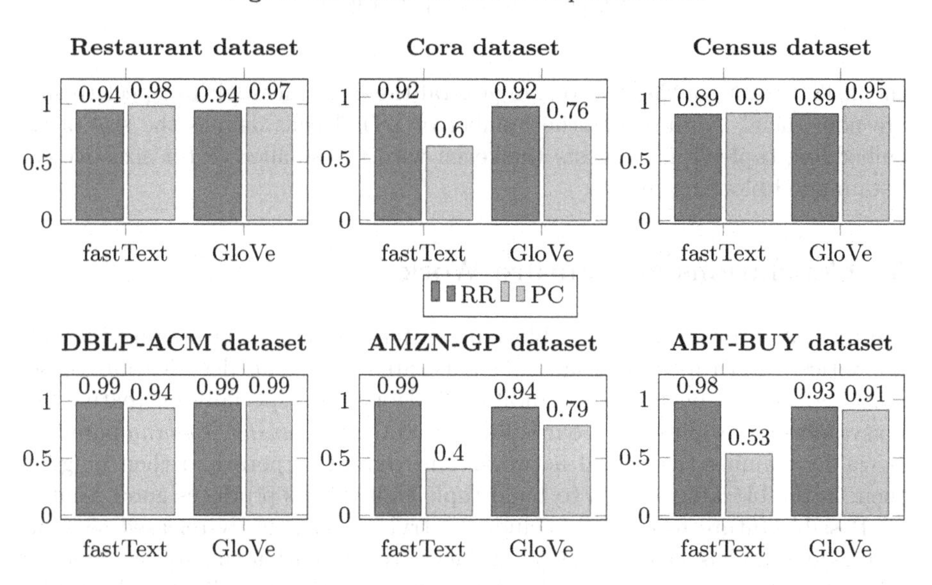

Fig. 5. fastText vs. GloVe performances

In other words, on these typologies of datasets a more relaxed model that promotes generalization seems to be more appropriate.

RNN-LSTM Sizes: Test results presented in Fig. 6 show the best performances obtained by using RNN-LSTM nets with the following number of LSTM cells: 300, 1024 and 2048. As can be seen, most of the top scores are associated with

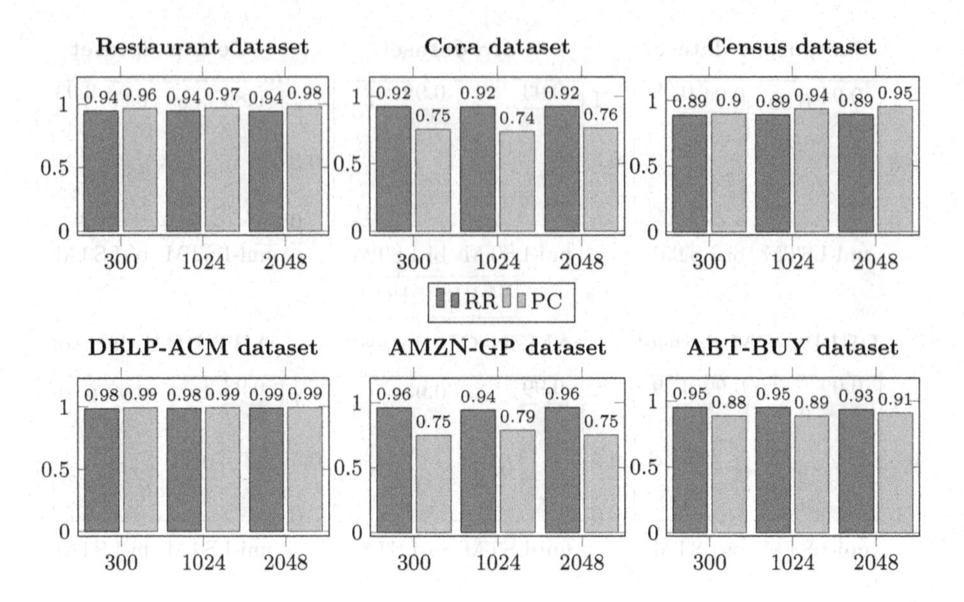

Fig. 6. RNN sizes performances

the 2048 variants of the nets but in general the differences with respect to sizes are not crucial. Noting that the number of LSTM cells defines the size of the embedding tuples this suggests that even with the smaller vector sizes we can obtain good blocking results.

5 Conclusions and Future Work

We presented an unsupervised blocking system capable of leveraging the data semantics. Experimental results demonstrated that our deep-learning-based blocking solution outperforms traditional algorithms especially on textual and noisy datasets. Additionally, our tests showed that a *transfer learning* paradigm is viable: training the neural networks on external corpora and then plugging them in the blocking system to build tuple embeddings produces good results.

Possible future work may include *(i)* trying other newly-released sentence embedding models such as [22] *(ii)* reducing the execution time of our blocking scheme adopting a parallel and distributed computation paradigm *(iii)* studying the applicability of *transfer learning* to a broader range of scenarios [23].

References

1. Wang, J., Shen, H.T., Song, J., Ji, J.: Hashing for similarity search: a survey. arXiv:1408.2927 (2014)
2. Pennington, J., Socher, R., Manning, C.D.: https://github.com/stanfordnlp/GloVe

3. Mikolov, T., Grave, E., Bojanowski, P., Puhrsch, C., Joulin, A.: https://fasttext.cc/docs/en/english-vectors.html
4. Bowman, S.R., Potts, C., Manning, C.D., Angeli, G.: A large annotated corpus for learning natural language inference. https://nlp.stanford.edu/projects/snli/
5. Ebraheem, M., Thirumuruganathan, S., Joty, S., Ouzzani, M., Tang, N.: Distributed representations of tuples for entity resolution. Proc. VLDB Endowment (2018)
6. Hernandez, M.A., Stolfo, S.J.: The merge-purge problem for large databases. In: ACM SIGMOD 95 (1995)
7. Baxter, R., Christen, P., Churches, T.: A comparison of fast blocking methods for record linkage. In: KDD (2003)
8. Aizawa, A., Oyama, K.: A fast linkage detection scheme for multi-source information integration. In: WIRI 2005 (2005)
9. McCallum, A., Nigam, K., Ungar, L.H.: Efficient clustering of high-dimensional data sets with application to reference matching. In: ACM SIGKDD 2000 (2000)
10. Michelson, M., Knoblock, C.A.: Learning blocking schemes for record linkage. In: AAAI, Boston (2006)
11. Das, S., et al.: Falcon: scaling up hands-off crowdsourced entity matching to build cloud services. In: ACM SIGMOD 2017 (2017)
12. Conneau, A., Kiela, D., Schwenk, H., Barrault, L., Bordes, A.: Supervised learning of universal sentence representations from natural language inference data. arXiv:1705.02364 (2018)
13. Conneau, A., Kiela, D., Schwenk, H., Barrault, L., Bordes, A.: https://github.com/facebookresearch/InferSent
14. Lv, Q., Josephson, W., Wang, Z., Charikar, M., Li, K.: Multi-probe LSH: efficient indexing for high-dimensional similarity search. In: Proceedings of the VLDB Endowment (2007)
15. Christen, P.: A survey of indexing techniques for scalable record linkage and deduplication (2011)
16. Koepcke, H., Thor, A., Rahm, E.: Evaluation of entity resolution approaches on real-world match problems. In: Proceedings of the VLDB Endowment (2010)
17. Papadakis, G., Svirsky, J., Gal, A., Palpanas, T.: Comparative analysis of approximate blocking techniques for entity linkage. In: Proceedings of the VLDB Endowment (2016)
18. RIDDLE repository. www.cs.utexas.edu/users/ml/riddle/data.html
19. Creative Commons license. https://dbs.uni-leipzig.de/en/research/projects/object_matching/benchmark_datasets_for_entity_resolution
20. Christen, P.: Data Matching. Concepts and Techniques for Record Linkage, Entity Resolution, and Duplicate Detection. Springer, Heidelberg (2012). https://doi.org/10.1007/978-3-642-31164-2
21. Mudgal, S., et al.: Deep learning for entity matching: a design space exploration. In: ACM SIGMOD 2018 (2018)
22. Perone, C.P., Silveira, R., Paula, T. S.: Evaluation of sentence embeddings in downstream and linguistic probing tasks. arXiv:1806.06259 (2018)
23. Thirumuruganathan, S., Parambath, S.A.P., Ouzzani, M., Tang, N., Joty, S.: Reuse and adaptation for entity resolution through transfer learning. arXiv:1809.11084 (2018)

PersonaGAN: Personalized Response Generation via Generative Adversarial Networks

Pengcheng Lv, Shi Feng(✉), Daling Wang, Yifei Zhang, and Ge Yu

School of Computer Science and Engineering, Northeastern University, Shenyang, China
ricardolvpc@foxmail.com, {fengshi,wangdaling,zhangyifei,yuge}@cse.neu.edu.cn

Abstract. Current personalized dialogue systems do not thoroughly model the context to capture richer information, and still tend to generate short, incoherent and boring responses. To tackle these problems, in this paper we propose a generative adversarial network model PersonaGAN for personalized dialogue generation. In addition to hierarchical modeling of context, we introduce a speaker-aware encoder in the generator to capture richer context information. Besides, we apply adversarial training to personalized dialogue generation task via using a transformer-based matching model as discriminator. The discriminator could give higher rewards for the responses which look like human written and lower rewards for machine generated responses. Such training strategy encourages the generator to generate responses which are grammatically fluent, informative and logically coherent with context. We evaluate the proposed model on a public available dataset and yield promising results on both automatic and human evaluation, which show that our model can generate more coherent and personalized responses while ensuring fluency.

Keywords: Response generation · Personalization · Generative adversarial network

1 Introduction

With the successful applications in intelligent assistant, online customer service, intelligent speaker, entertainment chat and other fields, the dialogue system has gradually become a research hotspot both in the academic and business communities. Recently, the availability of large-scale dialogue data has promoted the development of data-driven methods. Among them, sequence-to-sequence (Seq2Seq) model [15] has been widely used for dialogue generation. Seq2Seq model utilizes a multilayered Recurrent Neural Network (RNN) as encoder to map the input sequence to a vector of a fixed dimensionality, and then another deep RNN (decoder) to decode the target sequence from the vector.

Y. Nah et al. (Eds.): DASFAA 2020, LNCS 12112, pp. 570–586, 2020.
https://doi.org/10.1007/978-3-030-59410-7_38

In spite of the simplicity and effectiveness of Seq2Seq model, the lack of a consistent personality makes it impossible for current dialogue systems to behave coherently and engagingly. How to endow data-driven systems with the consistent "persona" is one of the major challenges. Previous persona based models tended to employ user distributed embeddings to capture individual implicit persona information [5]. Further studies attempted to introduce explicit persona information (such as a group of key-value profile pairs [10] or a series of natural and descriptive sentences [23]) as input into the Seq2Seq framework. Compared with the implicit persona information, this type of persona data alleviates the issues of data sparsity and poor interpretability.

However, the persona based approaches mentioned above did not pay enough attention to context modeling. Multi-turn context has a hierarchical structure, where words constitute utterance and utterances constitute context [19]. Most of previous studies did not distinguish different speakers in the context, who played different roles in the conversation. Accordingly, it is usually not enough to simply concatenating the utterances or only modeling context hierarchically. An example is given in Table 1. The context shows that two speakers A and B are discussing the occupation and hobby, where B is a singer. When the current statement (A) asks if B's most respected person is Bob Dylan, R2 expresses that "father" is his most admired man, which does take persona set into account, but it incorporates wrong persona. This is because when choosing personality, the whole context contains too much redundant information (including context information of both A and B). Therefore, the speakers in context should be given different weights when encoding, so as to retain more important information in subsequent steps. In addition, current dialogue systems tend to generate short, incoherent and boring responses, such as "I don't know", as shown in R1. The major reason is that the Seq2Seq model is trained with the maximum likelihood estimation (MLE) objective function, which encourages the model to produce high-frequency words [20] as much as possible and prevents the model from generating diverse and informative responses.

To tackle these problems, we propose a generative adversarial network model PersonaGAN for personalized dialogue generation. In our encoder, in addition to hierarchical modeling of context, we also deal with two speakers separately to capture richer context information. Moreover, we use a transformer based matching model as discriminator to replace the MLE objective function in the adversarial training stage. As a general rule, the responses written by human tend to be grammatically fluent, personalized and logically coherent with the context, while the responses generated by machine are irrelevant and dull. We label human written responses as positive examples and the generated responses as negative examples to train the discriminator, in order to encourage it to give higher rewards for the responses that look like human written, thus promote the generator's ability of generating coherent and personalized response.

We evaluate the proposed model on ConvAI2 persona-chat dataset, which is an extended version of PERSONA-CHAT [23], and compare our model with state-of-the-art personalized models using both automatic evaluation and human

Table 1. An example of context, persona set, a groundtruth response (TARGRT), and two responses (R1 and R2) generated by different models. Words in response are in the same color with the corresponding words in persona set, where the blue color means that these words are related to context, while the red one is not.

CONTEXT	A: What do you do for a living? B: I sing in the bar. you? A: Oh, that sounds really cool B: It is, I like Bob Dylan. lol someday i guess A: Is that the person you look up to?
Persona set (of B)	i am a musician and hope to make it big some day i play the **piano** and **guitar** my favorite type of music to sing is **folk** music my father is a hero.
TARGET	B: One of them, I **play instruments** like him and **folk** songs too
R1	B: I don't know. I am not sure
R2	B: Absolutely, my father is the one I admire most

evaluation. The results on both metrics show that our model can generate more coherent and personalized responses while ensuring fluency.

Our contributions are summarized as follows:

- We devise a novel adversarial training framework for personalized response generation. Our method employs adversarial mechanism for guiding model to generate fluent and coherent responses, and applies a persona enhanced discriminator to produce personalized dialogues.
- We propose a speaker-aware context encoder to capture richer context information and utilize a persona manager to select relevant persona dynamically based on attention mechanism.
- The experiment results on a benchmark dataset show that PersonaGAN outperforms previous methods in producing more fluent and personalized responses.

2 Related Work

Conventional dialog systems are usually built using rule based methods [17] or template based methods [18], which are difficult to scale up [25] and fail to produce novel response. Recently, the availability of large-scale conversational data [14] has promoted the development of data-driven methods. Seq2Seq model [15] is a popular and fundamental framework used for the dialogue generation task. It consists of a encoder and a decoder, where a encoder is responsible for mapping the input sequence to a vector with fixed dimension, and a decoder to generate a variable-length sequence conditioned on this dense vector.

In recent years, the researchers realize that it is vital to endow dialogue system with a consistent personality. Li et al. [5] employed user embeddings to represent implicit persona information of each speaker. Qian et al. [10] defined a group of key-value profile pairs as persona. Zhang et al. [23] constructed a dataset named PERSONA-CHAT by asking randomly paired crowd workers to chat based on some given personae, which consist of a series of natural and descriptive sentences. Yavuz et al. [21] explored the effectiveness of copy mechanism on PERSONA-CHAT. Lian et al. [7] leveraged Posterior Knowledge Selection to guide persona selection.

Although previous studies have made great progress, the current personalized dialogue systems are usually single-turn, or simply concatenate the multi-turn context into a sentence and treat it as single-turn, which do not thoroughly model the context, and may lose important information thus generate incoherent responses. Serban et al. [11] proposed Hierarchical Recurrent Encoder Decoder (HRED) to model the hierarchical structure of context. Furthermore, Xing et al. [19] proposed Hierarchical Recurrent Attention Network (HRAN) which attended to important parts within and among utterances. However, these two models did not take into account the differences of the speakers, and dealt with the utterances of different speakers in the same process. Inspired by Majumder et al. [8], we apply speaker-aware mechanism to deal with different speakers separately for choosing the right persona more accurately.

In addition, Seq2Seq model that trained via MLE object function tends to generate short, repetitive and boring texts. Li et al. [4] used Maximum Mutual Information (MMI) to replace MLE, in order to reduce the generation probability of boring responses. The other line of research focused on improve neural architecture by using conditional variational auto-encoder [13] or generative adversarial network [20] to enhance the informativeness and diversity.

In this paper, we apply generative adversarial network (GAN) [1] to personalized dialogue generation task. GAN has not achieved comparable improvement in text generation comparing in computer vision, for the text generation task is a process of sampling in discrete space [20], which makes the loss outputted from the discriminator non-differentiable and hard to backpropagate to the generator. To tackle this issue, Kusner et al. [3] applied gumbel-softmax trick to make the sampling process differentiable. Lamb et al. [2] provided the discriminator with the continuous intermediate hidden vectors of the generator. Yu et al. [22] viewed the text generation as a reinforcement learning problem, and used policy gradient algorithm to compute the loss.

Similarly, we use a binary classifier as our discriminator to compute the probability that the response is human generated. In addition, persona information is provided to enhance the discriminator for better distinguishing a response.

3 Model

In this section, we describe the components of the proposed PersonaGAN model in detail. As Fig. 1 shows, PersonaGAN consists of two main components: a generator to generate the response, and a discriminator to distinguish the generated response and human written response.

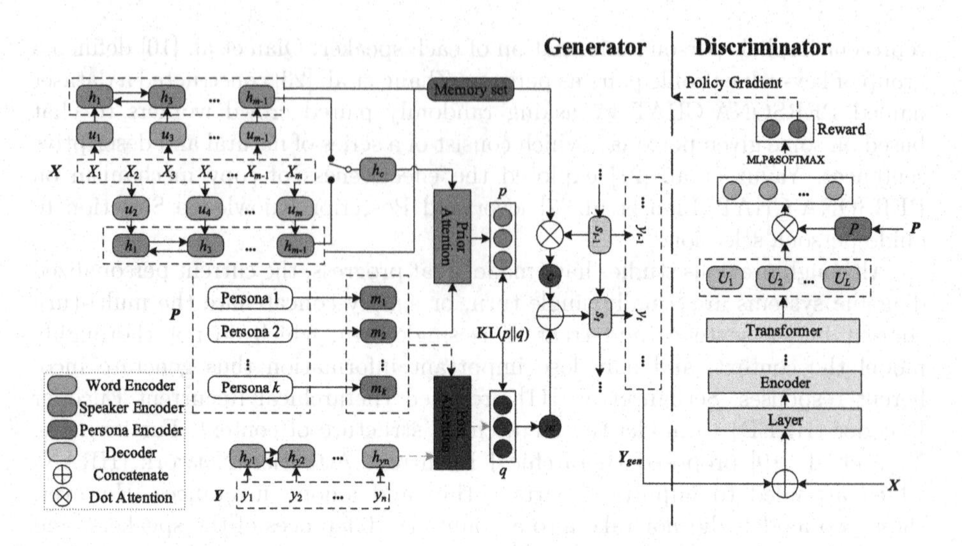

Fig. 1. Framework of PersonaGAN. The generator includes speaker-aware context encoder, persona encoder, persona manager and decoder. The speaker-aware context encoder is composed of word encoder, speaker encoder and the following gate. The persona manager consists of prior attention and post attention.

3.1 Overview

The generator G_θ is based on a Seq2Seq structure. Given context and a set of persona texts as input, the generator is capable of selecting appropriate persona from the persona set and then, to generate a response by incorporating the selected persona properly. The task can be formally defined as: given context $X = \{X_1, X_2, \ldots, X_m\}$ and persona set $P = \{P_1, P_2, \ldots, P_k\}$, the goal is to generate response $Y = y_1 y_2 \ldots y_n$, based on both context and persona set.

The discriminator D_ϕ is a matching model based on transformer [16] encoder layer. It receives context, generated response and persona set as input, then outputs a score (which is a comprehensive metric, including the probability that the response is human-written, matching degree with context and the amount of persona information contained) as reward to train the generator.

3.2 PersonaGAN

The goal of the discriminator is to distinguish whether a response Y is human written or machine generated. We expect the machine generated response, which is dull and incoherent can be identified by the discriminator. In addition, we want to encourage the discriminator to give high reward for the response that looks like the human generated data, because human written response not only has a higher matching degree with context, but contains more persona information. Therefore, we maximize the reward of human written response and minimize that of machine generated response to train the discriminator. The loss function of the discriminator is formulated as follows:

$$J(\phi) = -E_{Y \sim p_{data}} \log D_\phi(Y|X, P) - E_{Y \sim G_\theta} \log(1 - D_\phi(Y|X, P)) \quad (1)$$

where ϕ is the parameters of the discriminator, $D_\phi(Y|X, P)$ is the probability that the response is human generated given context X and persona set P.

In order to fool the discriminator, the generator tries to produce response which has a high similarity with human written response. That is, the goal of the generator is to maximize the expected reward. The loss function (in adversarial training stage) of the generator is:

$$J(\theta) = -E_{Y \sim G_\theta}(D_\phi(Y|X, P)) \quad (2)$$

To tackle the non-differentiable problem of generator, we use gradient policy algorithm to compute loss instead of negative log likelihood (NLL). The gradient of Eq. 2 can be derived as follows:

$$\nabla J(\theta) = -E_{Y \sim G_\theta}(D_\phi(Y|X, P) \nabla \log G_\theta(Y|X, P)) \quad (3)$$

where θ is the parameters of the generator, the $G_\theta(Y|X, P)$ is the probability of generating Y given X and P.

Both the generator and the discriminator are pre-trained before adversarial training. The generator is pre-trained on the training set with MLE loss. The discriminator is pre-trained using human written responses (real responses in dataset) as positive samples and responses produced by the pre-trained generator as negative samples. The overall algorithm of the PersonaGAN is summarized as Algorithm 1.

Algorithm 1. PersonaGAN

Require: generator G_θ; discriminator D_ϕ; dialog dataset $\mathcal{D} = (X, P, Y)$
1: Initialize G_θ, D_ϕ with random weights θ, ϕ
2: Pre-train G_θ using MLE on a training set $\mathcal{D}$
3: Generate responses using the pre-trained G_θ
4: Pre-train D_ϕ using generated responses as negative samples and real responses as positive samples via Equation 1
5: **for** epoch in number of epochs **do**
6:     **for** g in g-steps **do**
7:         Update G_θ via Equation 3
8:     **end for**
9:     **for** d in d-steps **do**
10:         Generate samples using G_θ
11:         Train discriminator D_ϕ via Equation 1
12:     **end for**
13: **end for**

3.3 Generator

We use random variables to represent each dyadic two-person conversation: the context X, the target response Y, the persona set P. X can be denoted as $X = \{X_1, X_2, \ldots, X_m\}$, where X_i is the ith utterance spoken by person P_λ alternately, $\lambda \in \{A, B\}$. Further, $X_i = x_{i,1}x_{i,2}\ldots x_{i,l}(x_{i,j}$ is a single word). Similarly, P consists of several unstructured persona texts: $P = \{P_1, P_2, \ldots, P_k\}$, where $P_i = p_{i,1}p_{i,2}\ldots p_{i,l}$. Note that we only use the persona set of the respondent as input, not that of both the poster and the respondent.

The generator consists of a context encoder, a persona encoder, a persona manager and a decoder. The following is the thorough details of these parts.

Encoder. The function of the encoder is to encode a sequence of tokens into a vector. Our encoders include speaker-aware context encoder and persona encoder to deal with multi-turn context and persona texts respectively.

Speaker-Aware Context Encoder. In order to fully capture the information in the multi-turn context, we apply a hierarchical encoding strategy in our context encoder, which consists of two parts: word level encoder and speaker encoder.

For each utterance X_i in context, we firstly employ word level encoder on it for feature extraction. The word level encoder is composed of convolutional neural networks (CNN) and Gated Recurrent Unit (GRU). Firstly, pre-trained GloVe embeddings [9] are used to represent X_i to a sentence matrix W_i. Then we use three different sized (2, 3, 4 respectively) filters to perform convolution on W_i. The outputs are then fed into a max-pooling layer followed by a concatenation of them. In addition, we also employ bidirectional GRU, which consists of a forward GRU and a backward GRU, to extract sequential characteristics of X_i. Forward GRU cell computes a left-to-right hidden state $\overrightarrow{h_t} = \text{GRU}\left(\overrightarrow{h}_{t-1}, x_t\right)$, while backward GRU reads X_i in a reverse order and obtains a right-to-left hidden state $\overleftarrow{h_t}$. We concatenate the last hidden states of both forward and backward GRU as the features of X_i, formulized as:

$$h = [\overrightarrow{h_l}; \overleftarrow{h_1}] = \left[\text{GRU}\left(\overrightarrow{h}_{l-1}, x_t\right); \text{GRU}\left(\overleftarrow{h}_{2}, x_t\right)\right] \tag{4}$$

where $[\cdot;\cdot]$ represents a vector concatenation, x_t is the current input. Eventually, we get a representation set $U = \{u_1, u_2, \ldots, u_m\}$, where u_i is a concatenation of the features from CNN and GRU for each X_i in context X.

As mentioned earlier, context is multi-turn and produced by two speakers alternatively, utterances from different speakers may interfere with each other. To capture the self-influence [8] of each speaker separately, we first divide U into two groups according to different speaker: $U_A = \{u_{A,1}, u_{A,3}, \ldots, u_{A,m-1}\}$ and $U_B = \{u_{B,2}, u_{B,4}, \ldots, u_{B,m}\}$, then we use two independent speaker GRU_λ (where $\lambda \in \{P_A, P_B\}$, GRU_A and GRU_B don't share any parameters) to encode U_A and U_B into $H_A = \{h_{A,1}, h_{A,3}, \ldots, h_{A,m-1}\}$ and $H_B = \{h_{B,2}, h_{B,4}, \ldots, h_{B,m}\}$ respectively. Eventually, we combine H_A and H_B into a memory set

$H = \{h_1, h_2, \ldots, h_m\}$ that will be used in the decoding step. In addition, we need a single vector representing the context as query for persona selection (described in next subsection). Following traditional practices, we use the last time step state from GRU_λ ($h_{A,m-1}$ and $h_{B,m}$) as representation of each speaker's context. Taking the dialogue data in Table 1 as an example, in the process of persona selection, using the dialogue history of speaker B alone may make it easier to select the right persona. Of course, in some cases, it is more appropriate to use that of A. Therefore, we devise a gate to balance the context information of A and B, as follows:

$$h_c = \alpha_c \cdot \mathrm{W}_a h_{A,m-1} + (1 - \alpha_c) \cdot \mathrm{W}_b h_{B,m} \tag{5}$$

$$\alpha_c = \sigma\left(\mathrm{V} \cdot h_{A,m-1}\right) \tag{6}$$

where W_a, W_b and V are learnable parameters. h_c will be used as context vector to select persona.

Persona Encoder. In this paper, persona set P is composed of a series of relatively simple texts compared with dialog context, and we do not need to considerate the order information among them. Therefore, our persona encoder follows the same architecture as the word level encoder, but they do not share any parameters. Specifically, it uses a bidirectional GRU to encode persona text set $P = \{P_1, P_2, \ldots, P_k\}$ into persona vector representation set $\{m_1, m_2, \ldots, m_k\}$, which will be use in the persona manager later.

Persona Manager To select appropriate persona information from persona set, we employ the posterior persona distribution, which is proposed by Lian et al. [7], to guide persona selection. During the training phase, real response is available. The real response can not only be used as a label to calculate NLL loss with generated response, but also assist in persona selection, because the correct persona can be selected with the help of real response. Thus the real response can be regarded as a kind of "label" to directly supervise the persona selection process.

Firstly, we use attention mechanism to compute the weight of every representation in persona set by conditioning them on Y as posterior distribution and on X as prior distribution respectively, specifically:

$$p_{prior} = p\left(m = m_i | h_c\right) = \frac{\exp\left(m_i \cdot h_c\right)}{\sum_{j=1}^{k} \exp\left(m_j \cdot h_c\right)} \tag{7}$$

$$p_{posterior} = p\left(m = m_i | h_y\right) = \frac{\exp\left(m_i \cdot h_y\right)}{\sum_{j=1}^{k} \exp\left(m_j \cdot h_y\right)} \tag{8}$$

where context hidden h_c represents context information, and h_y is representation of real response Y encoded by persona encoder. Here we use the dot attention, and apply softmax for normalization.

Then we use KullbackLeibler divergence as an auxiliary loss in addition to NLL, and force p_{prior} to approximate $p_{posterior}$. In test phase, even without $p_{posterior}$, p_{prior} can effectively calculate a reasonable attention weight so as to select correct persona. Eventually the linear combination m of elements in persona set will be use in decoder:

$$m = \sum_{i=1}^{k} \alpha_i m_i \tag{9}$$

where α_i is the ith value in $p_{posterior}$ (training phase) or p_{prior} (test phase).

Decoder. The decoder is a language model which generates response word by word sequentially, conditioned with the context and selected persona information.

Let s_t be the hidden state of decoder in t^{th} time step, y_{t-1} be the embedding of word generated in the last time step, m_t be the persona vector computed by persona manager and c_t be the attention-based context vector on the memory set $H = \{h_1, h_2, \ldots, h_k\}$ of the encoder. The current hidden state is:

$$s_t = \text{GRU}\left([y_{t-1}; m_t; c_t], s_{t-1}\right) \tag{10}$$

The word generated in current time step y_t is obtained after linear projection and softmax operation on hidden state s_t.

3.4 Discriminator

The discriminator is a binary classifier, which regards the judgment of response quality as a matching problem of generated response and given context. In addition, we introduce persona set into the discriminator to enhance it for evaluating personalized informativeness contained in the response. We use transformer encoder based matching model as our discriminator for its advantage in processing long sequence.

Firstly, we concatenate context and response as a single long sequence, then convert it into sequence of embeddings. In addition, we add positional encoding and type embedding (word from context or response).

The discriminator is composed of a stack of N (here we set N to 4) identical transformer encoder layers. Each layer has two sub-layers: a multi-head self-attention layer and a fully connected feed-forward network. The input and output of each sub-layer are added and layer normalized. That is, the output of each sub-layer is:

$$\text{LayerNorm}(\mathit{input} + \text{self\_attn}(\mathit{input})) \tag{11}$$

And then we get a set of representations for each word in input sequence, because the self attention mechanism can make the elements of each location get the information from all the other location elements, we use the first of them as the representation of the whole sequence.

To calculate interaction information between response and persona, we simply concatenate all the persona texts in persona set into a single sequence and encode it by bidirectional GRU, we concatenate the hidden states of both direction as the representation of each word. And then we use attention mechanism to generate a linear combination c of representations in persona set by conditioning them on response representation. Finally, we put c into multi-layer perceptron (MLP) to get the probability distribution on two categories(generated response or human written response):

$$r = \text{softmax}\left(\mathrm{W}_p\left[c\right]\right) = \text{MLP}\left(c\right) \tag{12}$$

where W_r is a weight matrix. r is a 2-dimension vector, which indicates the probability that the sentence is human written and that the sentence is machine generated, respectively. In adversarial training stage, we use the value of the last dimension of r as reward to train the generator.

4 Experiments

4.1 Dataset

We conduct experiments on ConvAI2 persona-chat dataset, which is an extended version of PERSONA-CHAT [23]. Each dialogue in this dataset was constructed from a randomly pair of crowd-workers, who were instructed to chat to know each other. To produce meaningful and interesting conversations, each worker was assigned a persona profile, describing their characteristic, such as age, gender, occupation and hobby. There are 18,878 dialogues in the Persona-chat dataset, which we divide into 17,878 for training, 1,000 for testing[1].

4.2 Baselines

We compared the proposed PersonaGAN model with several strong baselines.

Seq2Seq [12]: a Seq2Seq model with attention mechanism which does not have persona information as input.
Per-Seq2Seq: it has a similar architecture as Seq2Seq model. We additionally add a persona encoder used for incorporating persona information.
PAA&PAB [24]: a persona based generative network with persona-aware mechanism, which includes two approaches: Persona-aware Attention (PAA) and Persona-aware Bias (PAB). PAA uses persona representation to compute the attention weights to obtain the context vector at each decoding position, and PAB applies a gate mechanism to estimate the word generation distribution.

[1] Here we convert a session data to several dialogue data. For example, for a session $S = \{X_1, X_2, \ldots, X_m\}$, we convert it to $\{X_1, X_2\}$, $\{X_1, X_2, X_3\}, \ldots$, $\{X_1, X_2, \ldots, X_m\}$. Eventually, we have 131,428 for training and 7,799 for testing.

DP-GAN [20]: a Diversity-Promoting Generative Adversarial Network, which utilizes a language model based discriminator to better distinguish novel text from repeated and dull text. Based on its published code[2], we also implement a persona encoder for incorporating persona information.
Per-CVAE: a variant of Per.-CVAE [13]. On the basis of CVAE framework, it proposes a memory-augmented architecture, which is dedicated to generating diversified and personalized responses.
PostKS [7]: a persona based generative network with Posterior Knowledge Selection to guide knowledge selection. It regards persona as knowledge.

For the sake of fairness, we implement HRAN [19] to process context for Seq2Seq, Persona Seq2Seq, PAA&PAB and PostKS. Among them, Seq2Seq is compared for demonstrating the effect of introducing persona in dialogue generation while the PAA&PAB and PostKS are compared to verify that our model is more capable of incorporating correct persona than the existing persona based generative models. DP-GAN and per-CVAE are compared to prove that our model can deliver more coherent and engaging responses.

4.3 Implementation Details

In the generator, RNN is single layer bidirectional GRU with 800 hidden unit. The number of convolutional filters is set 32 and the filter sizes are set as 2, 3 and 4. The dimension of word embedding is set to 300 initialized using GloVe pre-trained word vector. The vocabulary size is 20,401. We pre-train the generator with a mini batch size of 16 and use Adam optimizer with an initial learning rate of 0.0005. In the discriminator, the number of transformer encoder layer is 4, head number of multi-head self-attention is set to 8. We use Adam optimizer whose learning rate is initialized to 0.001. Both our generator and discriminator are implemented[3] in PyTorch.

In adversarial training phase, the discriminator and the generator are trained alternately. To train the generator stably, we borrow trick form Suragnair[4] and Hungyi Lee, namely, training discriminator a lot more than generator (generator is trained only for one batch of examples); don't train the discriminator "too well", because a strong discriminator can make most generated responses receive reward around zero for the discriminator can identify them with high confidence [20]. Therefore we need our discriminator to be as smooth as possible. After repeated experiments, we find that the discriminator with accuracy of 0.6–0.7 performs best. Besides, following previous work [6], we also use teacher forcing technology. That is to say, in adversarial training phase, we provide the discriminator human written response and generated response alternately. Such a strategy can give the generator more direct access to the golden targets response, and prevents generator from breakdown. Here, we do not employ the REGS (reward for every generation step) strategy [6], for Monte-Carlo roll-out

[2] https://github.com/lancopku/DPGAN.
[3] Code available at: https://github.com/pancraslv/Persona-GAN.
[4] https://github.com/suragnair/seqGAN.

is quite time-consuming and the accuracy of a discriminator trained on partially decoded sequences is not as good as that trained on complete sequences [25].

4.4 Automatic Evaluation

Automatically evaluating an open-domain generative dialogue model is still an unsolved research challenge. Here we use automatic metrics such as Perplexity (PPL), BLEU, embedding-based metrics (Vector Extrema, Embedding Average, Greedy Matching), Distinct and Persona R/P/F1 to evaluate our model from multiple aspects. Smaller Perplexity score indicates that the model can generate real response with a higher probability. Distinct is used for evaluating the n-gram diversity of generated responses. BLEU analyzes the co-occurrences of n-grams in the ground truth and the generated response, while embedding-based metrics calculate the semantic similarity between them. Persona R/P/F1 [7] measure the uni-gram recall/precision/F1 scores between the generated response and the persona set. Specifically, non-stopword set in generated response is denoted by W_Y, and in predefined persona texts is denoted by W_P, the calculation formulas of Persona R and Persona P are as follows respectively:

$$\frac{|W_Y \cap W_P|}{|W_P|} \text{ and } \frac{|W_Y \cap W_P|}{|W_Y|} \tag{13}$$

and Persona F1 $= 2 \cdot$ (Recall $\cdot$ Precision) / (Recall + Precision).

Table 2. Automatic evaluation results in Perplexity (PPL), BLEU-1/2, Vector Extrema/Embedding Average/Greedy Matching (Embed E/A/G), Distinct-1/2 (Dist-1/2), Persona Recall/Precision/F1 (Per R/P/F1), the best results are in bold.

Model	PPL	BLEU-1/2 %	Embed E/A/G %	Dist-1/2 %	Per R/P/F1 %
Seq2Seq	**43.81**	19.32/8.14	53.12/92.00/73.86	0.49/1.83	0.84/3.16/1.29
Per-Seq2Seq	45.11	19.17/8.12	53.59/91.43/73.56	0.71/2.51	0.98/4.03/1.54
PAA& PAB	71.88	14.91/5.71	53.19/90.78/72.98	0.47/1.49	1.08/**6.48**/1.81
DP-GAN	58.72	13.52/5.16	51.53/90.08/71.54	**1.80**/9.12	0.38/1.88/0.62
Per-CVAE	47.28	8.37/2.80	48.75/90.72/73.34	1.79/**13.9**	1.30/2.49/1.65
PostKS	49.01	20.00/8.38	52.91/92.27/73.81	0.61/2.61	1.11/3.98/1.68
PersonaGAN	45.52	**20.10/8.55**	**53.65/92.38/74.29**	0.86/4.44	**1.71**/5.91/**2.58**

Table 2 shows the results of automatic evaluation. Our model outperforms baselines by achieving the highest or competitive performance in most of the automatic metrics. Compare to Seq2Seq model, Distinct metric of most persona based models have been dramatically improved, indicating that introducing persona is quite helpful in generating diverse responses. Besides, PAA&PAB gets a highest Persona Precision. We observe that some repeated and meaningless persona words account for a relatively large part of the generated responses, resulting in a higher Persona Precision, but a smaller Persona Recall, indicating that the PAA&PAB model does not make full use of the persona set. DP-GAN and

Per-CVAE have an extremely high Distinct score but their BLEU, embedding metrics and Persona R/P/F1 are quite low, indicating that it can generate substantially more diverse responses, yet it ignores fluency. Considering the results of human evaluation, we find that DP-GAN generates more grammatical errors in the responses, while Per-CVAE tends to produce long sentences, which usually have no grammatical errors, but express a vague meaning. Besides, comparing to PostKS, we achieve higher Persona, BLEU and embedding metrics. It indicates that GAN mechanism can more effectively encourage model to incorporate persona while ensuring fluency. The embedding metrics can better show that the responses generated by our method are more semantically relevant to the real responses, which can indirectly indicates that PerosnaGAN can generate more logically coherent responses.

Note that the PPL metric of Seq2Seq is the lowest of all models, because the data set is relatively persona-sparse. In the construction process of PERSONA-CHAT dataset, to produce high quality dialogue, trivially copying the predefined persona text into the messages is not encouraged [23]. So the persona word expressed in dialog is relatively sparse. Therefore, when generating a target word sequence, the probability of generating a persona word is very small, which will cause a large PPL. But the Seq2Seq is persona free, it will generate other non personal words with a large probability, so PPL will be diluted to a smaller value. From the metrics of Per-Seq2Seq we can see, when a simple mechanism is used to introduce persona information into the Seq2Seq framework, its Persona metrics and Distinct increase and its PPL gets worse. Furthermore, we conduct additionally experiment on a persona-dense dataset, which is constructed via a simple rule: If the word overlapping rate between response and persona set reaches a threshold (here we set it to 0.4), we consider this data as persona-dense data, then we get persona-dense testset containing 425 data. We only compute the PPL scores of four main models on this testset, and Table 3 shows the results, which verify our conjecture.

Table 3. Perplexity result on persona-dense testset.

Model	Seq2Seq	Per-Seq2Seq	PostKS	PersonaGAN
PPL	59.796	59.283	52.739	48.422

4.5 Human Evaluation

We randomly sample 100 messages from the testset to conduct the human evaluation. Three annotators[5] are recruited to judge a response from three aspects:

Fluency: generated response is fluent and free of grammatical errors.

[5] All annotators are fluent English speakers and are familiar with annotating rules.

Personality: generated response conforms to the given speaker's personalized characteristics.
Coherency: generated response is logically coherent with its context.

Table 4. Human evaluation results.

Model	Fluency	Personality	Appropriateness
Seq2Seq	1.36	0.71	0.80
Per-Seq2Seq	1.37	0.87	0.88
PAA+PAB	0.93	0.42	0.58
DP-GAN	0.72	0.28	0.26
Per-CVAE	1.35	0.78	0.60
PostKS	1.55	0.91	1.13
PersonaGAN	**1.71**	**1.15**	**1.26**

For each aspect, the rating level ranges from 0 to 2, where 0 means bad and 2 means excellent. According to the result in Table 4, our model performs remarkably better than all baselines in terms of all subjective metrics, demonstrating that PersonaGAN is capable of incorporating appropriate persona information and delivering fluency and coherent responses.

4.6 Ablation Experiments

In order to investigate the influence of context encoder and adversarial training, we implement several model variants for ablation experiments. The results are shown in Table 5.

w/o PED: We replace our persona enhanced discriminator (PED) with a persona free discriminator.
w/o GAN: After pre-training the generator, we don't conduct adversarial training.
w/o SAE: For simplicity, we call our context encoder SAE (Speaker-aware Encoder). Here we replace SAE with HRAN.

Table 5. Ablation experiments.

Model	PPL	BLEU-1/2 %	Embed E/A/G %	Dist-1/2 %	Per R/P/F1 %
PersonaGAN	**45.52**	20.10/**8.55**	53.65/92.38/74.29	0.86/4.44	**1.71/5.91/2.58**
w/o PED	48.43	**20.22**/8.39	**53.78**/92.49/**74.38**	**1.09/5.38**	1.63/5.57/2.46
w/o GAN	46.233	20.21/8.45	53.42/**92.59**/74.29	0.80/4.06	1.62/5.82/2.46
w/o SAE	49.020	20.15/8.44	52.87/92.23/73.83	0.52/2.16	1.31/4.37 /1.97

As shown in Table 5, we can observe that after removing the SAE, PPL increases a lot and Distinct, Persona R/P/F1 decrease a lot, indicating SAE is very helpful to choose the relevant persona information. In addition, compared with w/o GAN, w/o PED's Distinct is improved, which indicates that GAN mechanism is effective, but without the enhancement of persona, w/o PED's Persona metrics are not improved. However, adversarial training and SAE also demonstrate a negative effect on BLEU. This may be because the average length of responses generated by PersonaGAN is slightly shorter, which will result in brevity penalty and decrease BLEU slightly.

4.7 Case Study

Table 6 shows an example of responses generated by different models along with the input message and persona set. From the context we can see, current utterance is asking which kind of wine (white wine or red wine) the chat-bot likes. Because other baselines do not thoroughly model the context to capture long range history information, they fail to answer the type of wine, just reply the color they liked. In addition, the answer of our model reflects more personalized information, such as "i am a dancer", which corresponds to "i love to drink wine and dance in the moonlight" in persona set.

Table 6. Examples of the generated responses on PERSONA-CHAT.

Model	Utterance
CONTEXT	A: that's true. my favorite way to unwind is reading a book.
	B: you sound very like minded, just add some wine
	A: which do you prefer white or red?
Persona set(of B)	i love to drink wine and dance in the moonlight.
	i am very strong for my age.
	i'm 100 years old.
	i feel like i might live forever.
TARGET	B: fermented. i prefer well aged and fermented. i could bathe in it!
Seq2Seq	B: i love the color blue! i love it!
Per-Seq2Seq	B: i love the color of the stars!
PAA&PAB	B: i have a lot of the favorite color. i love to drink coffee
DP-GAN	B: that is your favorite? mine is red
Per-CVAE	B: i love it! i like sleeping on the moon! but it is my favorite color . . . lol.
PostKS	B: i love the color of moonlight. what is your favorite color?
PersonaGAN	B: red, i like to drink wine and i am a dancer, i am not really good at it.

5 Conclusion and Future Work

In this paper we propose a perosna based generative adversarial network named PersonaGAN for personalized dialogue generation. We evaluate the proposed model on a public available dataset and yield promising results on both automatic and human evaluation, which show that our model can generate more coherent and personalized responses while ensuring fluency. For the future work, we will explore modeling the persona information of speakers in multi-participant conversation.

Acknowledgement. The work was supported by the National Key R&D Program of China under grant 2018YFB1004700, National Natural Science Foundation of China (61872074, 61772122), and the CETC Joint fund.

References

1. Goodfellow, I.J., et al.: Generative adversarial nets. In: NIPS, pp. 2672–2680 (2014)
2. Goyal, A., Lamb, A., Zhang, Y., Zhang, S., Courville, A.C., Bengio, Y.: Professor forcing: A new algorithm for training recurrent networks. In: NIPS, pp. 4601–4609 (2016)
3. Kusner, M.J., Hernández-Lobato, J.M.: GANS for sequences of discrete elements with the Gumbel-softmax distribution. CoRR abs/1611.04051 (2016)
4. Li, J., Galley, M., Brockett, C., Gao, J., Dolan, B.: A diversity-promoting objective function for neural conversation models. In: NAACL, pp. 110–119 (2016)
5. Li, J., Galley, M., Brockett, C., Spithourakis, G.P., Gao, J., Dolan, W.B.: A persona-based neural conversation model. In: ACL, pp. 994–1003 (2016)
6. Li, J., Monroe, W., Shi, T., Jean, S., Ritter, A., Jurafsky, D.: Adversarial learning for neural dialogue generation. In: EMNLP, pp. 2157–2169 (2017)
7. Lian, R., Xie, M., Wang, F., Peng, J., Wu, H.: Learning to select knowledge for response generation in dialog systems. In: IJCAI, pp. 5081–5087 (2019)
8. Majumder, N., Poria, S., Hazarika, D., Mihalcea, R., Gelbukh, A.F., Cambria, E.: DialogueRNN: an attentive RNN for emotion detection in conversations. In: AAAI, pp. 6818–6825 (2019)
9. Pennington, J., Socher, R., Manning, C.D.: Glove: global vectors for word representation. In: EMNLP, pp. 1532–1543 (2014)
10. Qian, Q., Huang, M., Zhao, H., Xu, J., Zhu, X.: Assigning personality/profile to a chatting machine for coherent conversation generation. In: IJCAI, pp. 4279–4285 (2018)
11. Serban, I.V., Sordoni, A., Bengio, Y., Courville, A.C., Pineau, J.: Building end-to-end dialogue systems using generative hierarchical neural network models. In: AAAI, pp. 3776–3784 (2016)
12. Shang, L., Lu, Z., Li, H.: Neural responding machine for short-text conversation. In: ACL, pp. 1577–1586 (2015)
13. Song, H., Zhang, W., Cui, Y., Wang, D., Liu, T.: Exploiting persona information for diverse generation of conversational responses. In: IJCAI, pp. 5190–5196 (2019)
14. Soria-Comas, J., Domingo-Ferrer, J.: Big data privacy: challenges to privacy principles and models. Data Sci. Eng. **1**(1), 21–28 (2016). https://doi.org/10.1007/s41019-015-0001-x
15. Sutskever, I., Vinyals, O., Le, Q.V.: Sequence to sequence learning with neural networks. In: NIPS, pp. 3104–3112 (2014)
16. Vaswani, A., et al.: Attention is all you need. In: NIPS, pp. 5998–6008 (2017)
17. Weizenbaum, J.: ELIZA - a computer program for the study of natural language communication between man and machine. Commun. ACM **9**(1), 36–45 (1966)
18. Williams, J.D., Young, S.J.: Partially observable Markov decision processes for spoken dialog systems. Comput. Speech Lang. **21**(2), 393–422 (2007)

19. Xing, C., Wu, Y., Wu, W., Huang, Y., Zhou, M.: Hierarchical recurrent attention network for response generation. In: AAAI, pp. 5610–5617 (2018)
20. Xu, J., Ren, X., Lin, J., Sun, X.: Diversity-promoting GAN: a cross-entropy based generative adversarial network for diversified text generation. In: EMNLP, pp. 3940–3949 (2018)
21. Yavuz, S., Rastogi, A., Chao, G., Hakkani-Tür, D.: Deepcopy: grounded response generation with hierarchical pointer networks. CoRR abs/1908.10731 (2019)
22. Yu, L., Zhang, W., Wang, J., Yu, Y.: SeqGan: sequence generative adversarial nets with policy gradient. In: AAAI, pp. 2852–2858 (2017)
23. Zhang, S., Dinan, E., Urbanek, J., Szlam, A., Kiela, D., Weston, J.: Personalizing dialogue agents: i have a dog, do you have pets too? In: ACL, pp. 2204–2213 (2018)
24. Zheng, Y., Chen, G., Huang, M., Liu, S., Zhu, X.: Personalized dialogue generation with diversified traits. CoRR abs/1901.09672 (2019)
25. Zhu, Q., Cui, L., Zhang, W., Wei, F., Liu, T.: Retrieval-enhanced adversarial training for neural response generation. In: ACL, pp. 3763–3773 (2019)

Motif Discovery Using Similarity-Constraints Deep Neural Networks

Chuitian Rong[1,2], Ziliang Chen[2], Chunbin Lin[3(✉)], and Jianming Wang[1,2]

[1] School of Computer Science and Technology, Tiangong University, Tianjin, China
{chuitian,ziliang}@tjpu.edu.cn
[2] Tianjin Key Laboratory of Autonomous Intelligence Technology and Systems, Tianjin, China
[3] Amazon AWS, Atlanta, USA
lichunbi@amazon.com

Abstract. Discovering frequently occurring patterns (or motifs) in time series has many real-life applications in financial data, streaming media data, meteorological data, and sensor data. It is challenging to provide efficient motif discovery algorithms when the time series is big. Existing motif discovery algorithms trying to improve the performance can be classified into two categories: (i) reducing the computation cost but keeping the original time series dimensions; and (ii) applying feature representation models to reduce the dimensions. However, both of them have limitations when scaling to big time series. The performance of the first category algorithms heavily rely on the size of the dimension of the original time series, which performs bad when the time series is big. The second category algorithms cannot guarantee the original similarity properties, which means originally similar patterns may be identified as dissimilar. To address the limitations, we provide an efficient motif discovery algorithm, called FastM, which can reduce dimensions and maintain the similarity properties. FastM extends the deep neural network stacked AutoEncoder by introducing new central loss functions based on labels assigned by clustering algorithms. Comprehensive experimental results on three real-life datasets demonstrate both the high efficiency and accuracy of FastM.

Keywords: Time series · Motif discovery · Deep neural networks

1 Introduction

Time series is a sequence of data points in successive temporal order, which can be easily obtained from various applications, including financial data analysis, medical and health monitoring, scientific measurement, meteorological observation, music and motion capture. For example, [(2019-12-17 10:00, 1784.86), (2019-12-17 10:30, 1780.94), (2019-12-17 11:00, 1783.97)] is a time series representing the Amazon stock prices in three time points. Motif discovery is an

Y. Nah et al. (Eds.): DASFAA 2020, LNCS 12112, pp. 587–603, 2020.
https://doi.org/10.1007/978-3-030-59410-7_39

important subroutines that aims to find frequent occurring patterns (motifs) in time series. Since the discovered motifs can reveal useful latent information, so motif discovery is widely applied in time series data mining tasks and applications, such as association rule mining, classification, clustering, anomaly detection and activity recognition. The problem of motif discovery has been received extensive attentions in recent years [2,4,4–6,17,24]. Figure 1 shows the motifs discovered in a stock time series (with 12-year data ranging from 2007 to 2019). The discovered motif contains two subsequences whose similarity is 0.9. The motifs found in the stock time series can help analysts to analyze the trend of the stock, find the potential risks, and make the decisions for investment at the most favorable time.

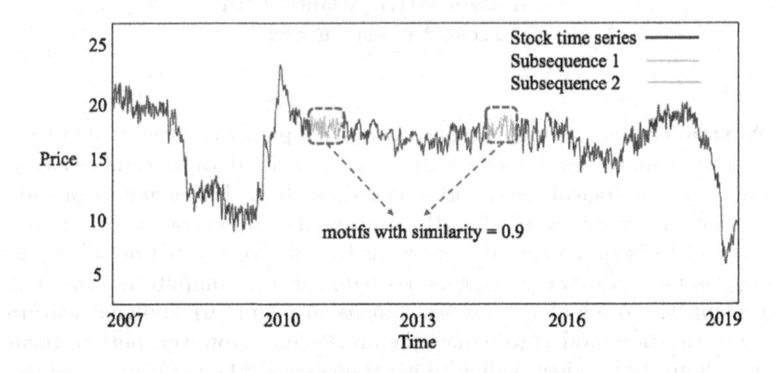

Fig. 1. Motifs discovered from stock time series.

The increasing of large scale time series raises the need for efficient motif discovery algorithms. Many motif discovery algorithms have been proposed [1,2,13,14,18,24,25], which tried to improve the performance in two directions: (i) reducing computation cost, which focuses on reducing the time cost of similarity computation among subsequence pairs from the raw time series data; and (ii) reducing dimensions, which focuses on exploiting new feature representation techniques to transform the time series into lower dimensions.

Existing Work on Reducing Computation Cost. The methods focus on reducing the similarity computation cost without feature representations had been proposed in early years. For example, MK algorithm [14], STAMP algorithm [24], and STOMP algorithm [25]. These algorithms are performed on the raw data set without dimension reduction. Although they have made important progress in improving the motif discovery task more efficient, the performance is limited by the long length feature of the time series. For big time series, the existing algorithms in this category do not scale well.

Existing Work on Reducing Dimensions. In Recent years, considering the high dimensions of the time series, researchers have proposed a widely used dimension reduction method, called SAX (Symbolic Aggregate Approximation) [1]. Due to its strong ability in feature representation, many representative methods have been proposed based on SAX , such as Enumeration MOEN [13], grammar-based method [11], Mr. Motif [2], the latest algorithm HIME [4] and so on. Although the existing algorithms in this category can achieve high performance, they may fail on discovering motifs as the similarity properties in the raw data may have been lost.

In order to provide a high performance motif discovery algorithm that can guarantee the similarity property on the original data, we proposed a framework called FastM, which extends the deep neural network AutoEncoder to maintain similarity property after dimension reduction. The traditional stacked AutoEncoder only considers the one-to-one reconstruction error, which ignores the intrinsic structure of the data and causes the loss of latent geometric information. This limitation prevents the possibility of applying AutoEncoder in motif discovery. To the best of our knowledge, we are the first one trying to extend AutoEncoder to solve time series motif discovery problem. In order to utilize the stacked AutoEncoders as the feature learning technique for motif discovery, the major challenge is to guarantee the new learned representations maintain the same similarity constraints as they are in the raw data. To address the challenge, we first assign a pseudo label for each subsequence in the time series by utilizing clustering method. Then, we propose new loss functions to improve the stacked AutoEncoder by taking the pseudo labels into consideration. Finally, we employ well-trained model on top of the features output by AutoEncoder to find motifs.

In summary, the main contributions of this work are as follows:

1. We proposed a new feature representation method, called FastM, for motif discovery using deep neural networks.
2. We applied clustering methods to assign labels to time series subsequences, and extended the stacked AutoEncoder to keep the similarity constraints of representation codes by minimizing the errors of the learned features with their labels.
3. We optimized the network parameters and applied the well-trained model to perform feature representations for time series.
4. We conducted extensive experiments on real-life datasets to evaluate our feature representation models.

2 Related Works

Time series motif discovery approach was first proposed in [10], which combined with a hash function to discover time series motifs. After that, time series motif discovery research task has received extensive attentions and many algorithms have been proposed in [15,18,21,22]. Motif discovery has become the native operations in time series analysis. The existing time series motif discovery algorithms are mainly divided into two categories: fixed-length oriented algorithms and variable-length oriented algorithms.

Fixed-Length Oriented Algorithms. The fixed-length algorithms proposed in [21] are based on SAX (Symbolic Aggregate approXimation) [9] to represent the time series with symbols. After that, it found the motifs using the string

similarity measures as in [12,19]. MK algorithm was proposed in [14], which aims to find the most similar subsequences pair as motifs by exploiting a pruning method to speed up the brute force algorithm. In [8], authors introduced a Quick-Motif algorithm, whose calculation speed is improved three orders of magnitude compared with the traditional fixed-length algorithm in [14]. In recent years, several new works was proposed. The algorithm STAMP [24] introduced an algorithm combined with MASS algorithm to find exact motifs for a given length. Then, the authors proposed an algorithm STOMP [26], which can reduce the time complexity of STAMP from $O(n^2 \log n)$ to $O(n^2)$.
Variable-Length Oriented Algorithms. As the length of motif cannot be predicted in advance and the variable-length motifs can reveal important latent information, many research works have focused on variable-length motif discovery. In 2011, the VLMD algorithm [17] was proposed based on calling fixed-length motif discovery algorithms to find K pair-motifs with variable-length. In [13], the authors proposed an algorithm by using a lower-bound to reduce the computing time of variable-length motifs discovery, this method can only use he Euclidean distance to measure the similarities between Z-Normalized segments. In [7,20], the authors proposed an alternative framework, which proposed grammar-based algorithms based on SAX discretization and the Sequitur grammar inference algorithm [16]. Its running time is faster than other algorithms. However, the idea of this algorithm is based on grammar induction, this method may be limited to some applications. [2] proposes a method based on discretization and the subsequences are not overlapping, which may lead to loss some real motifs.

In summary, existing works either (i) perform the motif discovery task on the original time series without dimension reduction, which scales bad in big time series or (ii) use dimension reduction techniques to improve the efficiency but miss the original similarity property. To address the limitations, we propose a feature representation method, called FastM, based on stacked AutoEncoder (a deep neural network technique) for motif discovery task. FastM can preserve similarity constraints among learned features and has high performance in big time series.

3 Background and Problem Definitions

3.1 Background: Auto-Encoder

AutoEncoder is an unsupervised machine learning algorithm, which has only one hidden layer and the network layer is a fully connected structure among neurons. The Stacked AutoEncoder (see Fig. 2) adopts a multi-layer structure, which is hierarchically stacked by a series of AutoEncoders. The network determines the weights of parameters by greedy learning manner layer-by-layer.

As shown in Fig. 2, we obtain the network parameters W_1, b_1 and the first order feature representation h_1 after we train the original data X_i. Then, using the first order feature representation h_1 as the training data, we can obtain the network parameters W_2, b_2 and the second order feature representation h_2. After all the layers are trained, we get the final feature representation h_n. The classic

AutoEncoder was widely used to reduce dimension. It is a neural network that considers the reconstruction error between the vector input x and vector input $\widetilde{x}$. It contains two main parts, *encoder* and *decoder*. The vector input x is took and mapped into a hidden representation z by the encoder. The form of the encoder is

$$z = f(W_e x + b_e) \tag{2}$$

where $f(x)$ is a non-linear activation function, W_e is the encoding matrix and b_e is a vector of bias parameters. z is the result of the encoder which is the hidden codings used to reconstruct the vector input x. The reconstruction process used a decoder is

$$\widetilde{x} = g(W_d z + b_d) \tag{3}$$

where $g(x)$ is also a non-linear activation function, W_d is the decoding matrix and b_e is a vector of bias parameters. AutoEncoder is to minimize the reconstruction error, which is produced by the a distance between the vector input x and vector input $\widetilde{x}$. The mean squared error is the most used distance and has the form

$$l(x, \widetilde{x}) = \|x - \bar{x}\|_2^2 = \|x - g(W_d z + b_d)\|_2^2 \tag{4}$$

where the dimensions of z is often less than the vector input x. Therefore, z may be forced to learn the most important information of the vector input x. When we set z less than the vector input x, the AutoEncoder is suitable for dimension reduction.

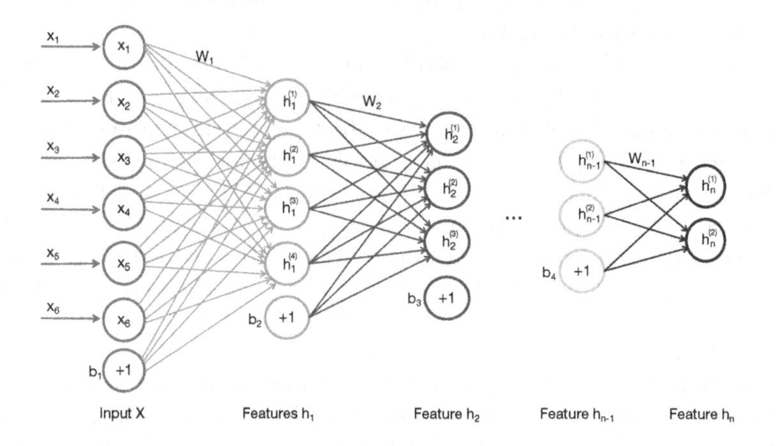

Fig. 2. A Stacked AutoEncoder with several hidden layers

3.2 Definitions

Definition 1 *(Time Series). Time series T is a sequence of real number data points ordered in temporal order and usually formalized as $T = [(t_1, d_1), (t_2, d_2), ..., (t_n, d_n)]$, where d is the data item at time t, and n is the length of the time series.*

Definition 2 *(Subsequence). A subsequence of time series T is a number of continuous data points can be defined as* $T[j : j+m] = [(t_j, d_j), (t_{j+1}, d_{j+1}), ..., (t_{j+m-1}, d_{j+m-1})]$. $T[j : j+m]$ *is a subsequence with length m starting from the* j^{th} *point.*

Definition 3 *(Time Series Motif). The time series motif is a frequent pattern in a time series. It also often described as repeated patterns, frequent trends, approximately repeated sequences, or frequent subsequences.*

Definition 4 *(Top-K Motifs). Given a time series T and the minimum length of motif* L_{min}, *Top-K motifs of T is defined as a set of subsequences clusters* $M = \{C_1, C_2, ..., C_n\}$, *in which each cluster* C_i *is the K most similar subsequences to its seed subsequence* T_i^{seed}, $T_i^{seed} \in C_i$ *and* $T_i^{seed} \in T$.

In this paper, our goal is to provide efficient algorithms to get top-k motifs.

4 Similarity-Constraints Representation Learning

In the tasks of motif discovery, the feature representation technique plays significant role to improve the efficiency. Without feature representation, the time complexity of motif discovery is proportional to the time series length n. However, if one feature representation technique was adopted, the time complexity can be reduced vastly. In this section, we exploit the stacked autoencoder as the feature representation technique.

As the classic stacked auto-encoder only considered the one-to-one reconstruction error between the input vectors and ignored the geometric structure of the data, the learned features may lose latent geometric information. Take the three subsequences T_1, T_2 and T_3 as example. The Euclidean distances between each pair of subsequences are $E(T_1, T_2) = d_1$, $E(T_1, T_3) = d_2$ and $E(T_2, T_3) = d_3$. The learned feature representations for these three subsequences are $T_1^{'}$, $T_2^{'}$ and $T_3^{'}$, respectively. The Euclidean distances between the learned features are represented as $d_1^{'}$, $d_2^{'}$ and $d_3^{'}$ correspondingly. The ideal autoencoder should preserve the relationships among $d_1^{'}$, $d_2^{'}$ and $d_3^{'}$ as the same among d_1, d_2 and d_3 correspondingly.

In order to preserve the similarity constraints among the learned features, we proposed our feature representation method based on stacked autoencoder. The feature representation method will learn a mapping from the inputs to a compact Euclidean space, where the similarity relationships of the corresponding input vectors are preserved or even reinforced. Next, we will introduce our proposed method that can preserve the similarity constraints after the feature representation.

4.1 Data Preprocessing

Data preprocessing is often a necessary step for Neural network training. In this work, we are interested on the data point values and focus on finding motifs from time series. Thus, we normalized the time series and converted it to a subsequence set. The subsequence started from the position p in time series T with length m can be represented as $T_p = (d_p, d_{p+1}, ..., d_{p+m-1}) \in R^m$ and regarded as a vector. So, the time series T can be segmented into multiple subsequences and represented as $S = (T_1, T_2, ..., T_q)^T \in R^{m*q}$, in which q is the number of subsequences.

4.2 Subsequence Labeling Method

The time series used for motif discovery dose not contain any similarity relationships or supervised signals in advance. As the stacked autoencoder is an unsupervised deep learning algorithm, we should incorporate the supervised signals to guide its unsupervised learning process to keep similarity constraints among the learned feature representations. So, we exploited the clustering method to assign pseudo labels for subsequences using their class labels(class centers).

In Algorithm 1, we listed the details of labeling method for subsequences based on clustering techniques. For the input subsequence set S, we will apply the clustering method to get a number of clusters (line 1). For each cluster C_i, its centroid c_{y_i} is computed as the cluster's label (lines 3–6). Then, we assign the centroid c_{y_i} as its class label for each subsequence $T_i \in C_i$. The labeled subsequence (T_i, c_{y_i}) will be grouped together as S^L.

Algorithm 1: *PseudoLabel*$(\mathcal{S}, k)$

Input : $\mathcal{S}$: the subsequence set $S = \{T_1, T_2, ..., T_q\}$, k: the cluster number
Output: M: labeled subsequence set, $S^L = \{(T_1, c_{y_i}), (T_2, c_{y_i}), ..., (T_q, c_{y_i})\}^T$

```
1 C ← cluster(S, k);
2 foreach cluster C_i ∈ C do
3     c_{y_i}=0;
4     foreach subsequence T_i ∈ C_i do
5         c_{y_i} = c_{y_i} + T_i;
6     c_{y_i} = c_{y_i}/|C_i|;
7     foreach subsequence T_i ∈ C_i do
8         S^L.add((T_i, c_{y_i}));
```

4.3 Basic Feature Representation Model

In this paper, we choose the Stacked Auto-Encoder as the feature representation tools for the time series.

In the classic autoencoder, the reconstruction error is computed as in Formula (4). In this section, we use the x_i to represent the i-th input vector, z_i

as the hidden vector and $\widetilde{x}_i$ as the output vector, respectively. Therefore, the reconstruction error in Formula (4) can be formalized as:

$$L_r = \sum_{i=1}^{n} \|x_i - \widetilde{x}_i\|_2^2 \tag{5}$$

The classic autoencoder is no suitable for us because it only consider the reconstruction error L_r and focus on reconstructing the x_i. As we all know, data set has latent distribution, which can be learned by the autoencoder. However, the classic autoencoder cannot preserve the geometric structure of the data. Therefore, we proposed our feature representation tools based on Stacked Auto-Encoder to solve the problems existing in classic autoencoders. We improved the Stacked Auto-Encoder by exploiting the clustering techniques to guide its unsupervised learning process.

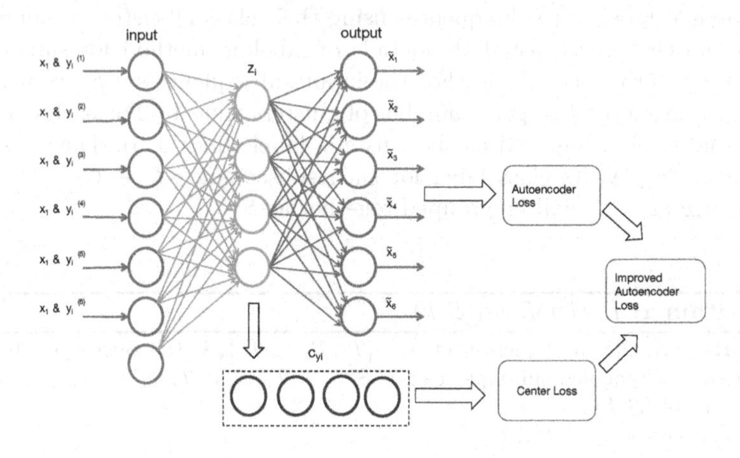

Fig. 3. The computation framework of center loss

Inspired by [23], we utilize the center loss to improve the ability of the neural network at preserving the similarity constraints among learned representations. So, we segmented the time series into subsequences set S and assign each subsequences a pseudo label by using clustering technique as shown in Algorithm 1. After that, we get a labeled subsequences set S^L. The input vector x_i of the stacked autoencoder is the subsequence $T_i \in S^L$, and c_{y_i} is its label. The computation framework of center loss is shown in Fig. 3. So,the formula of center loss can be formalize as follow:

$$L_C = \sum_{i=1}^{n} \|z_i - c_{y_i}\|_2^2 \tag{6}$$

Based on Formula (6), the gradients of L_C with respect to z_i can be reduced as:

$$\frac{\partial Lc}{\partial z_i} = z_i - c_{y_i} \tag{7}$$

Further more, the objective optimization function can be formalized as below:

$$L = L_r + \lambda L_c \tag{8}$$

In Formula (8), L_r is the restruction error and it can be computed by Formula (5). λ is a hyper parameter which was use to keep the balance between L_c and L_r. L_c in the Formula (6) can guarantee the input vectors and their labels to preserve the similar relative distances among their representations. For convenience, we incorporated the similarity constraints on the embedding layer and embedded it to the top of the hidden codes. Thus, the feature learned by the autoencoder are close to each other for the similar subsequences pairs. The basic feature learning algorithm is shown in Algorithm 2.

Algorithm 2: `Basic Feature Learning`

Input : $\{x_n\}_{n=1}^N$: training examples, k: cluster number for agglomerative clustering, I: iterative number, N: stacked autoencoder number
Output: $\{W_t\}_{n=1}^N$: parameters of stacked autoencoder
1 Initialize training example labels $\{y_n\}_{n=1}^N = 0$
2 $\{y_n\}_{n=1}^N \leftarrow PseudoLabel(\{x_n\}_{n=1}^N, k)$
3 **for** $n = 1 \rightarrow N$ **do**
4   **for** $i = 1 \rightarrow I$ **do**
5   Compute the stochastic gradient according to Eq(8);
6   update$\{W_t\}$;
7   Compute the hidden vector $\{z_n\}_{n=1}^N$ with current learned autoencoder;
8   Update $\{x_n\}_{n=1}^N = \{z_n\}_{n=1}^N$;

In Algorithm 2, we listed the details of the training steps of the AutoEncoders. Before we trained the AutoEncoders, the pseudo labels for input vectors are assigned (line 1–2). The main training steps of the AutoEncoders are shown in lines 3–8. As we can see, in each layer, the AutoEncoders learns the minimum loss of Formula (8) and update the W_t (lines 4–6). Then, we can obtain the hidden code z_i from this well-trained layer and use z_i as the new input vector for the next layer(lines 7–8). After all the layers had been trained, we can obtain the optimized model's parameters.

4.4 Improved Feature Representation Model

In this part, we proposed our second method. Review Algorithm 2, although this method may solve the problem and can obtain a good learning model for time series feature representation. The clustering algorithm may not be perfect and it may cause the bad codings produced by the autoencoder. Therefor, inspired by [3], in each layer of the autoencoder, we cluster the hidden codes z_i and update the c_{y_i} when the hidden codes changed. The update equation of c_{y_i} in this method is computed as:

$$\Delta c_j = \frac{\sum_{i=1}^{n} 1\left[y_i = j\right](c_j - z_i)}{1 + \sum_{i=1}^{n} 1\left[y_i = j\right]} \tag{9}$$

No matter which method we use, the objective function we use can be described as Formula (8). In order to obtain a good model for feature representation, in each hidden layer we cluster the hidden coding z_i and set it as the next layer's input vector. Then, after we train every layer of the autoencoder, we can obtain the parameter of the autoencoder. The algorithm may shown as Algorithm 3.

Algorithm 3: `Improved Feature Learning (FastM)`

Input : $\{x_n\}_{n=1}^{N}$: training examples, k: cluster number for clustering, I: the number of batch, L: the layer of the stacked autoencoders

Output: $\{W_t\}_{n=1}^{N}$: parameters of stacked autoencoder

1 Initialize training example labels $\{y_n\}_{n=1}^{N} = 0$;
2 $\{y_n\}_{n=1}^{N} \leftarrow PseudoLabel(\{x_n\}_{n=1}^{N}, k)$;
3 **for** $n = 1 \rightarrow L$ **do**
4   **for** $i = 1 \rightarrow I$ **do**
5   Compute the minimum loss in Eq(8);
6   updateW_t;
7   Compute the hidden vector $\{z_n\}_{n=1}^{N}$ with current learned autoencoder;
8   Update $\{x_n\}_{n=1}^{N} = \{z_n\}_{n=1}^{N}$;
9     $\{y_n\}_{n=1}^{N} \leftarrow PseudoLabel(\{x_n\}_{n=1}^{N}, k)$;

Different from the Algorithm 2, we run clustering algorithm for hidden each time. This method may force the hidden codings keep the similarity constraints of the forward codings. Each time before the hidden codes was set to the next layer, we cluster the hidden codes z_i and update the class center c_{y_i} just as the line 10. Although the clustering algorithm may perform not perfect, each time the data may cluster again and be forced to close to its true cluster center.

4.5 Model Validation

In this part, we introduced our method to testify the model whether or not it is suitable for feature representations for the time series. The model validation details are given in Algorithm 4. S is the subsequence set used for model validation. $S^{'}$ is the feature set generated by the stacked autoencoders model for S (line 1). In order to validate the accuracy of the model, we used these two subsequence sets S and S' to perform motif discovery task. We used the subsequence set S to generate the K-motif matrix M (lines 2–5). K-motif matrix stored the top-k motifs of the subsequences in S. For example, the first line of the matrix M stored the first top-k motifs of the subsequence T_1. Also, we can use the feature set $S^{'}$ to generate another K-motif matrix $M^{'}$. Then, we can compute the accuracy of the model by counting the number of the same motifs in the two matrices. The accuracy can be calculated as

$$p = \frac{m}{n} * 100\% \tag{10}$$

where m is the number of top K-frequent motifs, which is match to the original method, n is the number of top K-frequent motifs found by the original method, p is the accuracy.

Algorithm 4: `Model Validation Algorithm`

Input : S: the subsequence set used for test, w: parameters of stacked autoencoder model, k: the number of the motifs
Output: p: the accuracy of the model

1 $S^{'} \leftarrow$ FeatureLearning(S, w);
2 **foreach** *subsequence* $T_i \in \mathcal{S}$ **do**
3   **foreach** *subsequence* $T_j \in \mathcal{S}$ **do**
4         $d_{ij} \leftarrow$ EuclideanDistance (T_i, T_j);
5     $M_i \leftarrow$ findKMotifs (d_i, k);
6 **foreach** *subsequence* $T_i^{'} \in \mathcal{S}^{'}$ **do**
7   **foreach** *subsequence* $T_j^{'} \in \mathcal{S}^{'}$ **do**
8         $d_{ij}^{'} \leftarrow$ EuclideanDistance $(T_i^{'}, T_j^{'})$;
9     $M_i^{'} \leftarrow$ findKMotifs $(d_i^{'}, k)$;
10 $n = k * |S|$;
11 $m \leftarrow compared(M, M^{'})$;
12 $p = m/n$;

---

5 Experimental Evaluation

We evaluated our algorithms on three real-life time series datasets[1]. The distributions of the time series are shown in Fig. 4. We employed the well-known deep learning framework Tensorflow[2] to run on a high performance server equipped with NVIDIAGTX1080Ti. To fully evaluate the algorithms, we analyzed the following impacts:

- The impact of parameters (Sect. 5.1). We vary the following parameters: (i) the balance parameter λ, (ii) the number of clusters, and (iii) the value of K (the K from the top-K). The default values of the depth of AutoEncoder network.

[1] https://www.cs.ucr.edu/~eamonn/time_series_data.
[2] https://www.tensorflow.org.

- The impact of different clustering methods (Sect. 5.2).
- The impact of the depth of the AutoEncoder (Sect. 5.3).

Finally, we also analyze the different behaviors in different datasets due to different data distributions (Sect. 5.4), and the modify discovery results and performance (Sect. 5.5).

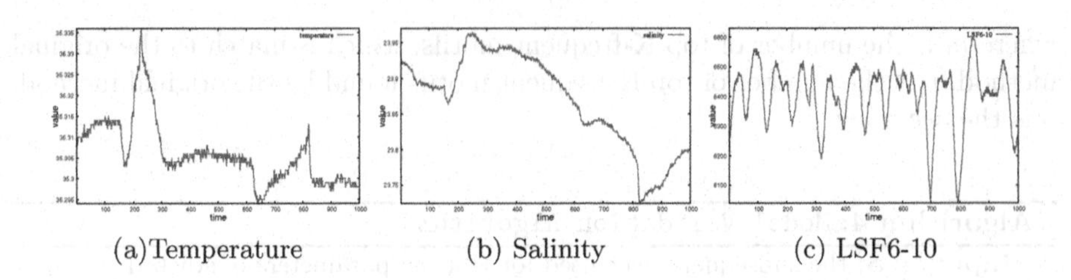

(a)Temperature (b) Salinity (c) LSF6-10

Fig. 4. The distribution of time series

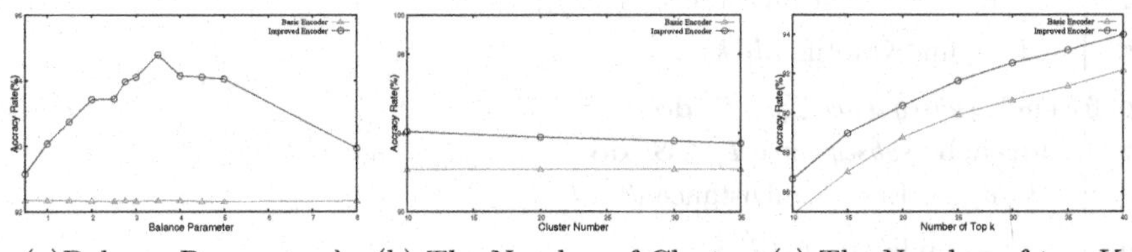

(a)Balance Parameter λ (b) The Number of Clusters (c) The Number of top-K

Fig. 5. The Variation of accuracy effected by autoEncoder parameters

5.1 The Effects of AutoEncoder Parameters

We compared the two different approaches of feature learning, as shown in the Algorithm 2 and Algorithm 3, with three different parameters, including the balance parameter, the number of clusters and the number of k.

(1) Effect of Balance Parameter λ. We changed the balance parameter λ from 0 to 8 with the value of K fixed to 40 and the number of clusters fixed to 15. The results are given in Fig. 5(a). Seen from the results, we can observe that the accuracy of the motif discovery based on the improved feature learning method is always higher than that based on the basic one. We can also observe that the basic model is not sensitive to the variations of λ. While, the improved model is sensitive to the variations of λ. That is because we adjusted the pseudo labels in each layer of the stacked AutoEncoder in the improved model. From the variation trend of accuracy of the improved model, we can find there exists a balance between the reconstruction error and the center loss. And, the improved model get its highest accuracy 95% when λ is 3.5.

(2) Effect of Cluster Number. We changed the number of cluster from 10 to 35 while keeping the balance parameter λ to be 2.75 and the value of k to be 40. The experimental results are given in Fig. 5(b). As shown, the improved model gets higher accuracy compared to the basic one and its accuracy is near the 94%. However, neither of the two models is sensitive to the number of cluster. That is due to that the loss function computation is determined by the distances among the hidden vectors and their respective centroid vector. Based on this experimental results, the number of cluster is not the key factor to effect the accuracy.

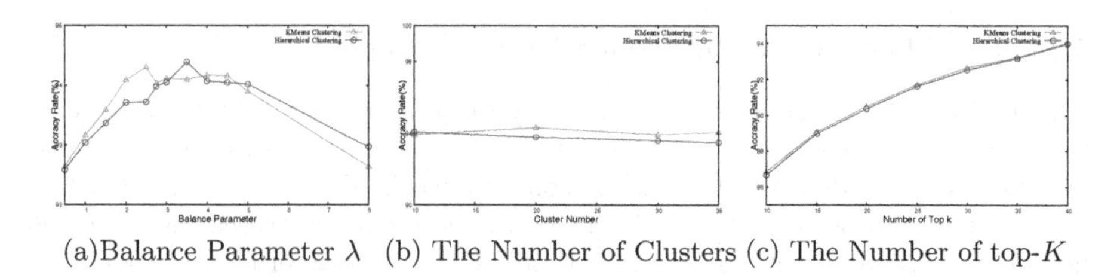

(a)Balance Parameter λ (b) The Number of Clusters (c) The Number of top-K

Fig. 6. The comparisons on different clustering methods

(3) Effect of the K Value. We vary the K value from 10 to 40 with the number of clusters fixed to 15 and the balance parameter fixed to 2.75. The results are shown in Figure 5(c). As shown, the improved model always get higher accuracy than the basic model under different values of Top-K. For the two models, the accuracy is increasing as the increasing of Top-K values. This is because for one seed subsequence its possible similar patterns is limited. If selected more similar subsequences according to their similarity, the accuracy will be increased.

5.2 The Comparisons on Different Clustering Methods

Here we evaluate the impact of applying different clustering methods to assign pseudo labels for subsequences. We compared the K-Means clustering and the Hierarchical clustering under different conditions. The results are presented in Fig. 6. When we changed the balance parameter λ, the accuracy of the improved models based on two different clustering methods have the similar variation trends. Although there are differences in the two variation trends, each of them can be used to realize the pseudo labeling task and can guarantee the high accuracy. When the number of clusters is changed from 10 to 35, we observe that the effects of clustering on the motif discovery accuracy is not quite large (See Fig. 6(b)). In addition, we observe the similar trends in 6(c) when changing the value of K in top-K. Thus, we can drop a conclusion that different clustering methods have little effects on improving the motif discovery accuracy.

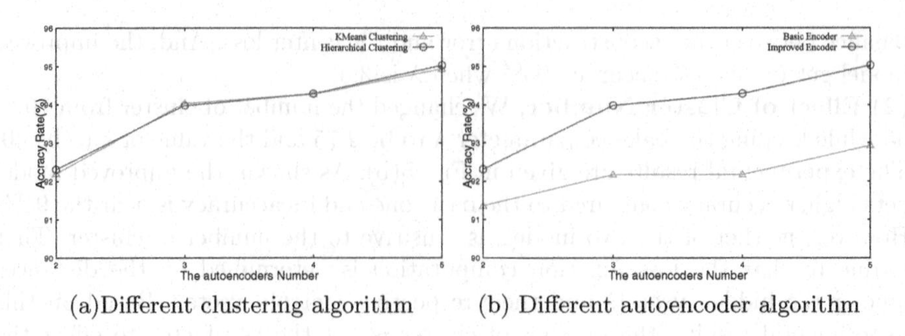

(a)Different clustering algorithm (b) Different autoencoder algorithm

Fig. 7. The Effects of AutoEncoder Network Depth

5.3 The Effects of AutoEncoder Network Depth

The depth of the AutoEncoder network plays an important role in model's accuracy. We conducted two experiments to verify the relationships between the accuracy and the depth of the AutoEncoder network. We changed the depth of autoencoder network from 2 to 5. In the first experiment, we tested the improved model with two different clustering methods, K-Means clustering and Hierarchical clustering. In the second experiment, we compared the improved model with the basic model with different depths. Seen from Fig. 7(a), we observe that the improved models that applying two different clustering methods for pseudo labeling almost have the same accuracy when the depths are the same. When the depth increases, the accuracy of motif discovery based on two models can all reach up to 95%. Comparing with the two models in Fig. 7, we can find that the clustering method applying to pseudo labels dose not have manifest effects on accuracy. This result is consisted with the result shown in Fig. 5(a). We also compared the accuracy of motif discovery using the two different models with the variations of autoencoder network depth. The results are shown in Fig. 7(b). From the results, we can observe that the accuracy of the motif discovery using two models are all increased when the depth is increased. The improved model has the higher accuracy at the same layer compared with the basic model. When the depth changed from 2 to 5, the accuracy of the motif discovery using the improved model is changed from 92.3% to 95%.

Table 1. Comparisons on different time series

Dataset	TotalLength	Size	TrainLength	TestLength	Mean	Std	Accuracy
Temperature	2,324,134	41.8M	1,394,000	116,206	35.126	0.1947	93.98%
Salinity	2,324,134	41.8M	1,394,000	116,206	24.960	1.190	89.23%
LSF6-10	180,191	3.2M	126,133	52,255	6412.78	108.23	72.90%

5.4 The Impact of Using Different Datasets

In order to verify the usability of our approach, we conducted experiments on three different data sets, whose distributions are given in Fig. 4. The statistic information of the three data sets and the experimental results are listed in Table 1. In this experiment, the balance parameter λ is set to 2.75 and the depth of AutoEncoder network is set to 3. Based on Fig. 4 and Table 1, we can observe that these three different data sets are different in many aspects, including the length, the mean and the standard deviation. After feature representation using our proposed method, the motif discovery task is performed on the learned features. The accuracy on three data sets is different. As the parameters λ and the depth of AutoEncoder network are optimized on the Temperature data set. So, the accuracy on the Temperature data set is the highest 93.98%. As the distribution of the Salinity data set is similar to the Temperature, the accuracy is 89.23%. While, the LSF6-10 data set is very different to other two data sets, the accuracy is only 72.90%. If we want to get higher accuracy on LSF6-10 data set, we should trained the model on this data set and optimize the related parameters.

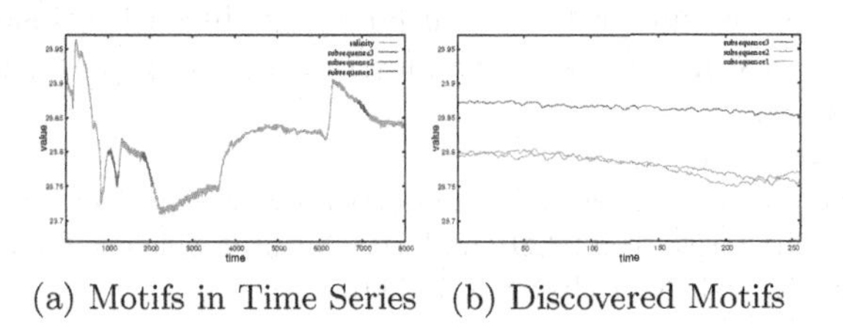

(a) Motifs in Time Series (b) Discovered Motifs

Fig. 8. Motif discovered

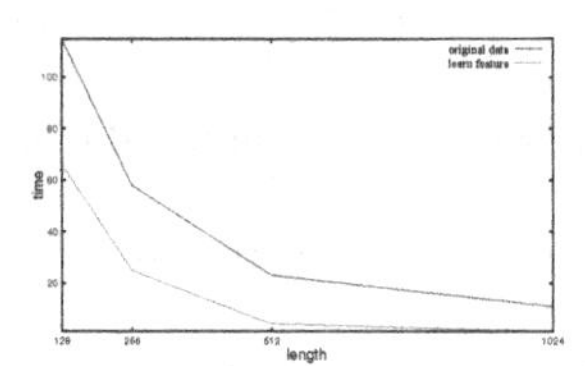

Fig. 9. Time costs

5.5 Motif Discovery

We present the motifs that discovered by our proposed approach. The results are given in Fig. 8 and Fig. 9. When perform the motif discovery task, we utilized the MASS Algorithm incorporated in Matrix Profile[3], which is a popular framework for time series processing. Figure 8(a) shows one of the discovered motifs, in which there are three similar patterns(subsequences) discovered in the Salinity

[3] https://www.cs.ucr.edu/~eamonn/MatrixProfile.html.

time series. The length of the motif is 256 and the Euclidean distances between each other are 0.57 0.58 and 0.79. In order to make comparisons, we also showed the three subsequences in Fig. 8(b). Form the Fig. 8(b), we can find that three subsequences are similar to each other. In order to prove the functions of our approach in improving the efficiency of motif discovery, we also conducted the experiments using the MASS Algorithm. In the Fig. 9, there are two time costs variation trends. One is running on the time series without dimension reduction, another is running on the learned representation features, which is 64 dimensions. When the subsequences are represented using the 64 dimension representations, the accuracy is more than 92%.

6 Conclusion

In this paper, we proposed a new time series motif discovery algorithm, called FastM, which significantly reduces the time series dimension and still maintains the similarity property. FastM is based on an extended stacked AutoEncoder deep neural network. We extended stacked AutoEncoder by (i) proposing new center loss function, and (ii) integrating with labels produced by clustering methods. Experiments on real-life datasets demonstrate the high performance and high accuracy of FastM.

Acknowledgment. This work was supported by the project of Natural Science Foundation of China (No.61402329) and the Natural Science Foundation of Tianjin(No.19JCYBJC15400, No.18JCYBJC15300).

References

1. Buhler, J., Tompa, M.: Finding motifs using random projections. J. Comput. Biol. **9**(2), 225–242 (2000)
2. Castro, N., Azevedo, P.J.: Multiresolution motif discovery in time series. In: SDM, pp. 665–676 (2010)
3. Chu, W., Cai, D.: Stacked similarity-aware autoencoders. In: IJCAI, pp. 1561–1567 (2017)
4. Gao, Y., Lin, J.: Efficient discovery of variable-length time series motifs with large length range in million scale time series. CoRR abs/1802.04883 (2018)
5. Gao, Y., Lin, J., Rangwala, H.: Iterative grammar-based framework for discovering variable-length time series motifs. In: ICMLA, pp. 7–12 (2016)
6. Lam, H.T., Calders, T., Pham, N.: Online discovery of top-k similar motifs in time series data. In: ICDM, pp. 1004–1015 (2010)
7. Li, Y., Lin, J., Oates, T.: Visualizing variable-length time series motifs. In: ICDM, pp. 895–906 (2012)
8. Li, Y., U, L.H., Yiu, M.L., Gong, Z.: Quick-motif: an efficient and scalable framework for exact motif discovery. In: ICDE, pp. 579–590 (2015)
9. Lin, J., Keogh, E., Li, W., Lonardi, S.: Experiencing sax: a novel symbolic representation of time series. Data Min. Knowl. Discov. **15**, 107–144 (2007). https://doi.org/10.1007/s10618-007-0064-z

10. Lin, J., Keogh, E., Lonardi, S., Patel, P.: Finding motifs in time series. In: Proceedings of 2nd Workshop on Temporal Data Mining at KDD, pp. 53–68 (2002)
11. Lin, J., Li, Y.: Finding approximate frequent patterns in streaming medical data. In: CBMS, pp. 13–18 (2010)
12. Lu, J., Lin, C., Wang, W., Li, C., Wang, H.: String similarity measures and joins with synonyms. In: SIGMOD, pp. 373–384 (2013)
13. Mueen, A.: Enumeration of time series motifs of all lengths. In: ICDM, pp. 547–556 (2013)
14. Mueen, A., Keogh, E.J., Zhu, Q., Cash, S., Westover, M.B.: Exact discovery of time series motifs. In: SDM, pp. 473–484 (2009)
15. Narang, A., Bhattacherjee, S.: Real-time approximate range motif discovery & data redundancy removal algorithm. In: EDBT, pp. 485–496 (2011)
16. Nevill-Manning, C.G., Witten, I.H.: Identifying hierarchical structure in sequences: a linear-time algorithm. J. Artif. Intell. Res. **7**, 67–82 (1997)
17. Nunthanid, P., Niennattrakul, V., Ratanamahatana, C.A.: Discovery of variable length time series motif. In: EEE, pp. 472–475 (2011)
18. Patel, P., Keogh, E.J., Lin, J., Lonardi, S.: Mining motifs in massive time series databases. In: ICDM, pp. 370–377 (2002)
19. Rong, C., Lin, C., Silva, Y.N., Wang, J., Lu, W., Du, X.: Fast and scalable distributed set similarity joins for big data analytics. In: ICDE, pp. 1059–1070 (2017)
20. Senin, P., et al.: GrammarViz 2.0: a tool for grammar-based pattern discovery in time series. In: Calders, T., Esposito, F., Hüllermeier, E., Meo, R. (eds.) ECML PKDD 2014. LNCS (LNAI), vol. 8726, pp. 468–472. Springer, Heidelberg (2014). https://doi.org/10.1007/978-3-662-44845-8_37
21. Tanaka, Y., Iwamoto, K., Uehara, K.: Discovery of time-series motif from multi-dimensional data based on MDL principle. Mach. Learn. **58**(2–3), 269–300 (2005). https://doi.org/10.1007/s10994-005-5829-2
22. Tang, H., Liao, S.S.: Discovering original motifs with different lengths from time series. Knowl. Based Syst. **21**, 666–671 (2008)
23. Wen, Y., Zhang, K., Li, Z., Qiao, Yu.: A discriminative feature learning approach for deep face recognition. In: Leibe, B., Matas, J., Sebe, N., Welling, M. (eds.) ECCV 2016. LNCS, vol. 9911, pp. 499–515. Springer, Cham (2016). https://doi.org/10.1007/978-3-319-46478-7_31
24. Yeh, C.C.M., Yan, Z., Ulanova, L., Begum, N., Keogh, E.: Matrix profile i: All pairs similarity joins for time series: a unifying view that includes motifs, discords and shapelets. In: ICDM, pp. 1317–1322 (2016)
25. Yeh, C.-C.M., et al.: Time series joins, motifs, discords and shapelets: a unifying view that exploits the matrix profile. Data Min. Knowl. Disc. **32**(1), 83–123 (2017). https://doi.org/10.1007/s10618-017-0519-9
26. Zhu, Y., Zimmerman, Z., Senobari, N.S., Yeh, C.M.: Matrix profile II: exploiting a novel algorithm and gpus to break the one hundred million barrier for time series motifs and joins. In: ICDM, pp. 739–748 (2016)

An Empirical Study on Bugs Inside TensorFlow

Li Jia[1], Hao Zhong[1](✉), Xiaoyin Wang[2], Linpeng Huang[1], and Xuansheng Lu[1]

[1] Department of Computer Science and Engineering, Shanghai Jiao Tong University, Shanghai, China
{insanelung,zhonghao,huang-lp,luxuansheng}@sjtu.edu.cn

[2] Department of Computer Science, University of Texas at San Antonio, San Antonio, USA
xiaoyin.wang@utsa.edu

Abstract. In recent years, deep learning has become a hot research topic. Although it achieves incredible positive results in some scenarios, bugs inside deep learning software can introduce disastrous consequences, especially when the software is used in safety-critical applications. To understand the bug characteristic of deep learning software, researchers have conducted several empirical studies on deep learning bugs. Although these studies present useful findings, we notice that none of them analyze the bug characteristic inside a deep learning library like TensorFlow. We argue that some fundamental questions of bugs in deep learning libraries are still open. For example, what are the symptoms and the root causes of bugs inside TensorFlow, and where are they? As the underlying library of many deep learning projects, the answers to these questions are useful and important, since its bugs can have impacts on many deep learning projects. In this paper, we conduct the first empirical study to analyze the bugs inside a typical deep learning library, *i.e.*, TensorFlow. Based on our results, we summarize 5 findings, and present our answers to 2 research questions. For example, we find that the symptoms and root causes of TensorFlow bugs are more like ordinary projects (*e.g.*, Mozilla) than other machine learning libraries (*e.g.*, Lucene). As another example, we find that most TensorFlow bugs reside in its interfaces (26.24%), learning algorithms (11.79%), and how to compile (8.02%), deploy (7.55%), and install (4.72%) TensorFlow across platforms.

1 Introduction

In recent years, deep learning has been a hot research topic, and researchers have used deep learning techniques to solve the problems in various research fields such as computer vision [36], speech recognition [31], natural language processing [23,45], software analysis [50] and graph classification [35]. Specifically, in the research community of databases, deep learning techniques and database techniques have promoted each other in various perspectives. On one hand, deep learning techniques are employed to handle various database problems (*e.g.*, tuning database configurations with a deep reinforcement learning

Y. Nah et al. (Eds.): DASFAA 2020, LNCS 12112, pp. 604–620, 2020.
https://doi.org/10.1007/978-3-030-59410-7_40

model [37]). On the other hand, database techniques are be used to improve deep learning systems (*e.g.*, optimizing neural networks [49]).

When implementing deep learning applications, instead of reinventing wheels, programmers often build their applications on mature libraries. Among these libraries, TensorFlow [20] is the most popular, and a recent study [53] shows that more than 36,000 applications of GitHub are built upon TensorFlow. As they are popular, one bug inside deep learning libraries can lead to bugs in many applications, and such bugs can lead to disastrous consequences. For example, Pei *et al.* [41] report that a Google self-driving car and a Tesla sedan crash, due to bugs in their deep learning software.

To better understand bugs of deep learning programs, researchers have conducted empirical studies on such bugs. In particular, Zhang *et al.* [53] conduct an empirical study to understand the bugs of TensorFlow clients. Here, a client of TensorFlow is a program that calls the APIs of TensorFlow. While Zhang *et al.* [53] analyze only TensorFlow clients, Islam *et al.* [33] analyze the clients of more deep learning libraries such as Caffe [34], Keras [19], Theano [24], and Torch [25]. Although their results are useful to improve the quality of a specific client, to the best of our knowledge, no prior studies have ever explored the bugs inside popular deep learning libraries. Although the bugs inside TensorFlow influence thousands of its clients, many questions on such bugs are still open. For example, what are the symptoms and the root causes of such bugs, and where are they? A better understanding on such bugs will improve the quality of many clients, but it is challenging to conduct the desirable empirical study, since TensorFlow implements many complicated algorithms and is written in multiple programming languages. In this paper, we conduct the first empirical study to analyze the bugs and their fixes inside TensorFlow, and we present our answers to the following research questions:

- **RQ1. What are the symptoms and causes of bugs in TensorFlow?**
  Motivation. The symptom and the root cause of a bug are important to understand and to fix the bug. For deep learning bugs, the results of the prior studies [33,53] are incomplete, because they analyze only deep learning clients. As the prior studies do not analyze bugs inside a deep learning library like TensorFlow, the answers to the above research question are still unknown.
  Major results. In total, we identify six symptoms and eleven root causes for the bugs inside TensorFlow. We find that root causes are more determinative than symptoms, since several root causes have dominated symptoms (Finding 1). In addition, we find that the symptoms and the root causes of TensorFlow bugs are more like those of ordinary projects (*e.g.*, Mozilla) than other machine learning libraries (Finding 2). For the symptoms, build failures have correlations with inconsistencies, configurations and referenced type errors, and warning-style bugs have correlation with inconsistencies, processing, and type confusions. For the root causes, dimension mismatches lead to functional errors, and type confusions have correlation with functional errors, crashes, and warning-style errors (Finding 3).

- **RQ2. Where are the bugs inside TensorFlow?**
  Motivation. From the perspective of TensorFlow developers, the locations of its bugs are important to improve the quality of TensorFlow. From the perspectives of the programmers of TensorFlow clients, they can be more careful to call TensorFlow, if they know such locations. From the perspective of researchers, they can design better detection techniques for our identified bugs, after the locations of target bugs are known. The prior studies [33,53] do not explore this research question.
  Major results. To explore the bug characteristics in different library components, we analyze TensorFlow bugs by location. We find that major reported bugs reside in deep learning algorithms (*kernel*, 11.79%) and their interfaces (*API*, 26.42%). The two categories of bugs are followed by bugs in the deployment such as compiling (*lib*, 8.02%), deploying (*platform*, 7.55%), and installing (*tools*, 4.72%). The other components such as *runtime* (3.77%), *framework* (0.94%) and *computation graph* (0.94%) have fewer bugs.

This paper presents an empirical research. The purpose of an empirical study is to answer open questions on important issues, which can enrich the knowledge and motivate the follow-up research. Empirical studies have been widely conducted in various research fields such as software error analysis [27,29], database management [21,39] and information security [26,28]. As open questions are too complicated to be automated, like ours, some empirical studies are conducted manually, especially for those on the bug characteristics [46,53].

2 Methodology

2.1 Dataset

We select TensorFlow as the subject of our study, since Zhang *et al.* [53] report that more than 36,000 GitHub projects call the APIs of TensorFlow. As a result, the bugs inside TensorFlow influence thousands of its clients. In total, we analyze 202 TensorFlow bug fixes repaired between December 2017 and March 2019, and 84 of them have corresponding bugs reports. The number is comparable to other empirical studies. For example, Thung *et al.* [47] analyze 500 bugs from machine learning projects such as Mahout, Lucene, and OpenNLP. For each project, they analyze no more than 200 bugs. As another example, Zhang *et al.* [53] analyze 175 bugs from TensorFlow clients. Indeed, for deep learning programs, libraries are typically much larger than clients. As a result, our analyzed bugs are much more complicated than the bugs in the prior studies [47,53].

We apply the following steps to extract bugs:

Step 1. Filtering pull requests by labels. To avoid problems which have not been handled correctly and to collect accurate information about fixed bugs, we start with *closed* pull requests with label "*ready to pull*". We notice that finished pull requests before a specific date are not tagged, so we also collect cases from earlier closed pull requests by searching keywords as described in Step 2. We manually check each collected pull request to ensure that its commit is already approved by reviewers and is merged into the master branch.

Step 2. Searching pull requests by keywords. From closed pull requests, we use the keywords such as "bug", "fix" and "error" to identify the ones that fix bugs. From bug fixes, we use the keywords such as "typo" and "doc" to remove the ones that fix superficial bugs. From the remaining bug fixes, we manually inspect them to select real ones, by reading their pull requests carefully. In total, we selected 202 bug fixes for latter analysis.

Step 3. Extracting bug reports and code changes. For each one of the 202 bug fixes, we extract its bug report and code changes. The extracted results and corresponding pull requests are used to determine their symptoms, root causes (RQ1), and locations (RQ2). We introduce the detailed analysis in Sect. 2.2.

2.2 Manual Analysis

In our study, we employ two graduate students to manually inspect all bugs. The two students major in computer science, and both are familiar with deep learning algorithms. They have experience in developing deep learning applications (*e.g.*, mining on business data) based on TensorFlow. Following our protocols, the two students inspect the bugs independently, and compare the results. If we cannot reach a consensus on a TensorFlow bug, they discuss it on our group meetings.

Protocol of RQ1. When they build their own taxonomy of bug symptoms and their root causes, they refer to the taxonomies of the prior studies [22,46]. In particular, they add an existing category into their taxonomy, if they find a TensorFlow bug falls into this category. If a TensorFlow bug does not belong to an existing category, they try to modify a similar category of the prior studies [22,46]. If they fail to find such a similar category, they add a new one.

For bug classifying, if a pull request has a corresponding bug report, they first read its report to identify its symptoms and root causes. If a pull request does not provide a report, they manually identify its symptom and root cause from the description, bug-related discussion, code changes and comments of the pull request. For example, a bug fix without report [18] is titled "*Fix for stringpiece build failure*". Based on the title, they determine that the symptom of the bug is build failure. They notice that the only code modification of this bug fix is:

```
1 void Append(StringPiece s) {
2 -       key_.append(s.ToString());
3 +       key_.append(string(s));
4         key_.append(1, delimiter); }
```

The `ToString()` method that is called to build the key in the buggy version is removed. In the fixed version, the `string(StringPiece)` method should be called to build the correct key, but in the old location, the method call is not updated. Considering this, they determine that the root cause of the bug is the inconsistency introduced by API change.

After the symptoms and root causes of all the bugs are extracted, the two students further classify them into categories, and use the *lift* function [30] to measure the correlations between symptoms and root causes. According to the definition, the *lift* between different categories (A and B) is computed as:

$$lift(A,B) = \frac{P(A \cap B)}{P(A) \cdot P(B)} \tag{1}$$

where $P(A)$, $P(B)$, $P(A \cap B)$ are the probabilities that a bug belongs to category A, category B, and both A and B. If a *lift* value is greater than one, category A and B are correlated, which means a symptom is correlated to a root cause. If it is equal to or less than one, a symptom is not correlated to a root cause.

Protocol of RQ2. In this research question, the two students analyze the locations of bugs. As an open source project, TensorFlow does not officially list its components, but like other projects, TensorFlow puts its source files into different directories, by their functionalities. When determining their functionalities, they refer to various sources such as official documents, TensorFlow tutorials, and forum discussions. Their identified components are as following:

1. **Kernel.** The kernel implements the core deep learning algorithms (*e.g.*, the `conv2d` algorithm), and its source files are located in the `core/kernels` directory.
2. **Computation graph.** TensorFlow uses computation graphs to define and to manage its computation tasks. The graph implements the definition, construction, partition, optimization, operation, and execution of computations. Most source files of this component are located in the `core/graph` directory; its data operations are located in the `core/ops` directory; and its optimization-related source files are located in the `core/grappler` directory.
3. **API.** TensorFlow provides APIs in various programming languages (*e.g.*, Python, C++, Java), which are located in the `python`, `c`, `cc` and `java` directories.
4. **Runtime.** The runtime implements the management of sessions, thread pools, and executors. TensorFlow has a common runtime (`core/common_runtime`) and a distribution runtime (`core/distributed_runtime`). Common runtime supports the executions on a local machine, and distribution runtime allows to deploy TensorFlow on distributed ones. For simplicity, we merge them into one component.
5. **Framework.** The framework implements basic functionalities (*e.g.*, logging, memory, and files). Most source files of this component are located in `core/framework` directory, and the serialization is located in `core/protobuf` directory.
6. **Tool.** The tool implements utilities. For example, `tools/git` and `tools/pip_package` directories implement the utilities to install TensorFlow;

the `core/debug` directory provides a tool to debug TensorFlow clients; and the `core/profile` directory provides a tool to profile the execution of TensorFlow and its clients.
7. Platform. The platform allows to deploy TensorFlow on various platforms. The `core/platform` directory contains the source files to handle hardware issues (*e.g.*, CPU and GPU); the `core/tpu` directory allows executing on TPU; the `lite` directory allows executing TensorFlow on mobile devices; and the `compiler` directory allows compiling to native code for various architectures.
8. Contribution. The `contrib` directory contains extensions that are often implemented by outside contributors. For example, the `contrib/seq2seq` directory contains a sequence-to-sequence model that is widely used in neural translation. After they become mature, they can be merged into other directories. In our study, we define a component for this directory.
9. Library. The library includes the API libraries. Most libraries are located in the `third-party` directory, and some libraries are located in other directories (*e.g.*, `core/lib`, `core/util` and some files under the root directory of `tensorflow`).
10. Documentation. The documentation includes samples, which are located in the `examples` and `core/example` directories. It also includes other types of documents. For example, the `security` directory stores security guidelines.

We use the *lift* metric as defined in Eq. 1 to measure the correlation between a bug location and a symptom or a root cause. Here, if a bug involves more than one directory, we count them once for each directory to ensure that each location does not lose a symptom and a root cause.

3 Empirical Result

This section presents the results of our study. More details are listed on our anonymized project website: https://github.com/fordataupload/tfbugdata/.

3.1 RQ1. Symptoms and Root Causes

The Categories of Symptoms. 1. Functional error (35.64%). If a program does not function as designed, we call it a functional error. For example, we find that a bug report [2] complains the functionality of the `tf.Print` method:

If you print a tensor of shape [n, 4] with tf.Print, by default (summarize=3 is the default value), you get: [[9 21 55]...], which wrongly looks like your tensor is of shape [n, 3]. The correct output should be: [[9 21 55...]...].

The method is designed to print the details of tensors. The bug report complains that it prints incorrect output, when the shape is `[n, 4]`. As the result is not as expected, it is a functional error.
2. Crash (26.73%). A crash occurs, when a program stops and exits irregularly. When it happens, the program often throws an error message. For example, a bug report [4] describes a crash caused by an unsupported operand type:

Using a TimeFreqLSTMCell in a dynamic_rnn without providing optional parameter frequency_skip results in an exception: TypeError: unsupported operand type(s) for /: 'int' and 'NoneType'.

3. Hang (1.49%). A hang occurs, when a program keeps running without stopping or responding. A bug report [17] provides description as below:

When running the above commands (Inception V3 synchronized data parallelism training with 2 workers and 1 external ps), the tf_cnn_benchmarks application hangs forever after some iterations (usually in warm up).

4. Performance degradation (1.49%). A performance degradation occurs, when a program does not return results in expected time. For example, we find a performance degradation in a bug report [16]:

There is a performance regression for TF 1.6 comparing to TF 1.5 for cifar 10.

5. Build failure (23.76%). A build failure occurs in the compiling process. For example, we find that a bug report [3] describes a build failure, which is caused by a missing header file:

Build failing due to missing header files "tensorflow/contrib/tpu/proto/tpu_embedding_config.pb.h".

6. Warning-style error (10.89%). Warning-style error means the running of a program is not disturbed, but modifications are still needed to get rid of risk or improve code quality, including interfaces to be deprecated, redundant code and bad code style. Most bugs in this category are shown by warning messages, while a few others do not provide visible messages which are found by code review or other events. For example, we find a bug in such category [11], since it calls a method with a deprecated argument:

According to tf.argmax, dimension argument was deprecated, it will be removed in a future version.

The Categories of Root Causes. 1. Dimension mismatch (3.96%). We put a bug into this category if it is caused by dimension mismatch in tensor computations and transformations. A bug fix [14] describes the cause of a bug in this category as:

Wrongly "+1" for output shape, that will cause CopyFrom failure in MklToTf op because of tensor size and shape mismatch.

The buggy code sets the dimension of an output tensor:

```
1 output_tf_shape.AddDim((output_pd->get_size() / sizeof(T)) + 1);
```

The fixed code sets the correct dimension:

```
1 output_tf_shape.AddDim((output_pd->get_size() / sizeof(T)));
```

2. Type confusion (12.38%). Type confusions are caused by the mismatches of types. A sample report [12] is as below:

CRF decode can fail when default type of "0" (as viewed by math_ops.maximum) does not match the type of sequence_length.

After the bug was fixed, programmers modified a test case to ensure that the method accepts more types of input values:

```
1  np.array(3, dtype=np.int32),
2  -          np.array(1, dtype=np.int32)
3  +          np.array(1, dtype=np.int64)
```

3. Processing (22.28%). We put a bug into this category, if it is caused by wrong assignment or initialization of variables, wrong formats of variables, or other wrong usages that are related to data processing. For example, we find a bug report [8] in such category as follow:

ConvNDLSTMCell class in tensorflow.contrib.rnn cannot pass the name attribute correctly when created, because of the missing parameter in constructor.

The constructor of `ConvNDLSTMCell` has no parameters to define their names:

```
1 super(Conv1DLSTMCell, self).__init__(conv_ndims=1, **kwargs)
```

The bug is fixed in a latter version:

```
1 super(Conv1DLSTMCell, self).__init__(conv_ndims=1, name=name, **
      kwargs)
```

4. Inconsistency (16.83%). We put a bug into this category, if it is caused by incompatibility due to API change or version update. For example, a bug report [7] complains that a removed `ops` is called:

NotFoundError: Op type not registered 'KafkaDataset' in binary. is returned from kafka ops. The issue was that the inclusion of kafka ops was removed due to the conflict merge from the other PR.

The above compilation error was caused by a conflict merge of two commits. One removed `kafka ops`, but the other added a call to the operator.

5. Algorithm (2.97%). We put a bug into the algorithm category, if it is caused by wrong logic in algorithms. For example, a bug report [5] complains that a method returns wrong values:

Input labels = tf.constant([[0., 0.5, 1.]]), predictions = tf.constant([[1., 1., 1.]]), the result of tf.losses.mean_pairwise_squared_error(labels, predictions) should be $[(0-0.5)^2+(0-1)^2+(0.5-1)^2]/3=0.5$, but TensorFlow returns different value 0.333333.

According to the code document, the mean pairwise squared error is incorrectly calculated. In the process of deduction, the denominators of two intermediate variables are wrong. A developer replaces an assignment and changes a method with corresponding parameters to fix denominators as below:

```
1 -   num_present_per_batch)
2 +   num_present_per_batch -1)
3  ...
4 +   math_ops.square(num_present_per_batch))
5 -   math_ops.multiply(num_present_per_batch, num_present_per_batch -1))
```

6. Corner case (15.35%). We put a bug into this category, if it is caused by erroneous handling of corner cases. A bug of this kind is reported [15] as:

When batch_size is 0, max pooling operation seems to produce an unhandled cudaError_t status. It may cause subsequent operations fail with odd error message.

As the reporter says, a crash happens when `batch_size` of the input is 0, which belongs to corner cases.

7. Logic error (9.90%). We put a bug into this category, if mistakes happen in the logic of a program. A logic error can be an incorrect program flow or a wrong order of actions. A bug report [10] provides the description as:

When a kernel Variable is shared by two Conv2Ds, ... there will be only one Conv2D getting the quantized kernel.

TensorFlow implements a mechanism called quantization to shrink tensors. The reporter complains that when a tensor shares two Conv2D, the second one cannot obtain the right quantized kernel. The logic of the code is flawed, in that the program in complex flow does not behave as expected.

8. Configuration error (7.43%). We put a bug into this category, if it is caused by wrong configuration. A sample bug [6] is as follow:

Linking of rule '...toco' fails because LD_LIBRARY_PATH is not configured.

To repair the bug, in a configuration file, programmers add the following statement to initiate `LD_LIBRARY_PATH`:

```
1 if 'LD_LIBRARY_PATH' in environ_cp and environ_cp.get('
    LD_LIBRARY_PATH')!='1':
2   write_action_env_to_bazelrc('LD_LIBRARY_PATH', ...)
```

9. Referenced types error (4.95%). We put a bug into this category, if it is caused by missing or adding unnecessary `include` or `import` statements. A bug [13] triggers the following error message:

The compiler couldn't find std::function, because header file #include <functional> is missing.

Programmers forget to add the said `include` statement, which causes the bug.

10. Memory (2.97%). We put a bug into the memory category, if it is caused by incorrect memory usages. For example, a bug report [9] describes a possible memory leak, which can be triggered by an exception, because of missing deconstruction operation.

11. Concurrency (0.99%). We put a bug into this category, if it is caused by synchronization problems. A bug report [1] describes a deadlock as follow:

notify_one was used to notify inserters and removers waiting to insert and remove elements into Staging Areas. This could result in deadlock when many removers were waiting for different keys.

As the reporter says, when multiple removers wait for keys but `notify_one` only notifies one of them, a deadlock may occur.

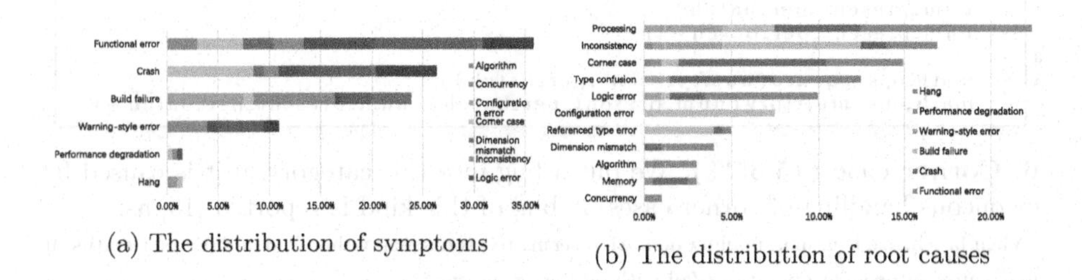

(a) The distribution of symptoms

(b) The distribution of root causes

Fig. 1. Distribution of bug symptoms and root causes

Distribution. Figure 1a shows the distribution of symptoms. Its horizontal axis shows symptom categories, and its vertical axis shows the percentage of corresponding symptom. For each symptom, we refine its bugs by their root causes. Tan *et al.* [46] report the distributions of Mozilla, Apache, and the Linux kernel. We find that the distribution of TensorFlow is close to their distributions. Figure 1a shows that functional errors account for 39%, which are the most common bugs of TensorFlow. Tan *et al.* [46] show that in Mozilla, Apache, and the Linux kernel, function errors vary from 50% to 70%. We find that crashes account for 26.5% TensorFlow bugs, which are close to Linux (27.2%), and hangs account for 1% bugs, which are close to Mozilla (2.1%).

Figure 1b shows the distribution of root causes. Its horizontal axis shows cause categories, and its vertical axis shows the percentage of corresponding causes. For each root cause, we refine its bugs by symptoms. We find that all the symptoms have multiple and evenly distributed root causes, but the distribution of root causes are not so evenly. The distributions lead to our first finding:

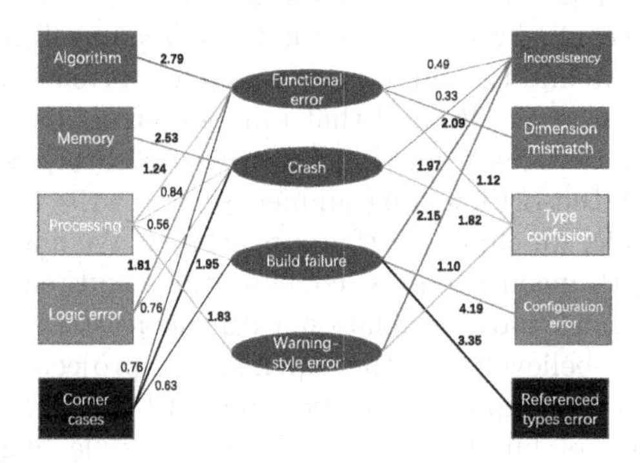

Fig. 2. Correlation between symptoms and root causes

Finding 1. Compared to symptoms, root causes are more determinative, since several root causes have dominated symptoms.

Tan *et al.* [46] show that in Mozilla, Apache, and the Linux kernel, the dominant root cause is semantic (80%). In our taxonomy, memory, configuration and referenced types errors belong to semantic bugs (85%), which are close to Tan *et al.* Thung *et al.* [47] show that in machine learning systems, algorithm errors are the most common bugs (22.6%). The above observations lead to another finding:

Finding 2. The symptoms and causes of TensorFlow are more like an ordinary software system (*e.g.*, Mozilla) than a machine learning system (*e.g.*, Lucene).

A machine learning system typically provide many algorithms for users to invoke. For example, although Lucene is also large (554,036 lines of code), the symptoms and root causes of its bugs are more different from TensorFlow than ordinary software systems like Mozilla. We find that Lucene provides numerous APIs to handle natural language texts in different ways (*e.g.*, tokenization).

In the contrast, TensorFlow provides much fewer interfaces to invoke, which is more like a traditional software system.

Correlation of Bug Categories. Figure 2 shows the correlation of bug categories. The rectangles on the left side denote symptoms, the ovals on the right side denote root causes. We choose different colors to distinguish the correlations, and the root causes of the same color are not related. We ignore categories whose bugs are fewer than three, since they are statistically insignificant. For example, we ignore hangs, since only two bugs are hangs. The lines denote correlations, and we highlight correlations whose values are greater than one.

Both Tan *et al.* [46] and we find that crashes have correlations with memory bugs and corner cases. Tan *et al.* [46] find that crashes also have correlations with concurrency, but we do not consider it, since only two of our analyzed bugs are concurrency. Instead, our study shows that crashes of TensorFlow have correlations with type confusions, which are not identified by Tan *et al.* In addition, Tan *et al.* [46] and we find that function errors have correlations with processing and logic errors. Tan *et al.* [46] find that function errors have correlations with missing features by defining a missing feature as a feature is not implemented yet. As we find that TensorFlow programmers seldom write their unimplemented features in their code, we eliminate this subcategory. We find that build failures have correlation with inconsistencies, configurations and referenced type errors, and warning-style bugs have correlation with inconsistencies, processing, and type confusions. We believe that other open source projects (*e.g.*, Mozilla) also have the two types of symptoms, but are ignored by Tan *et al.* [46]. We identify the correlations of build failures and warning-style bugs, complementing the study of Tan *et al.* [46]. For our identified root causes and symptoms of TensorFlow, our observations lead to the following finding:

Finding 3. For symptom of TensorFlow bugs, build failures have correlation with inconsistencies, configurations and referenced type errors, and warning-style bugs have Correlation with inconsistencies, processing, and type confusions. For the root causes of TensorFlow bugs, dimension mismatches lead to functional errors, and type confusions have correlation with functional errors, crashes, and warning-style errors.

3.2 RQ2. Bug Locations

Distribution. As it contains immature implementations, it is not surprising that *contribution* is the buggies component (34.91%). The next two buggy components are *kernel* (11.79%), and *API* (26.42%). The two components implement the main functionalities of TensorFlow. In particular, kernel implements deep learning algorithms, and API implements their interfaces. Following them, the next three buggy components are *library* (8.02%), *platform* (7.55%), and *tool*

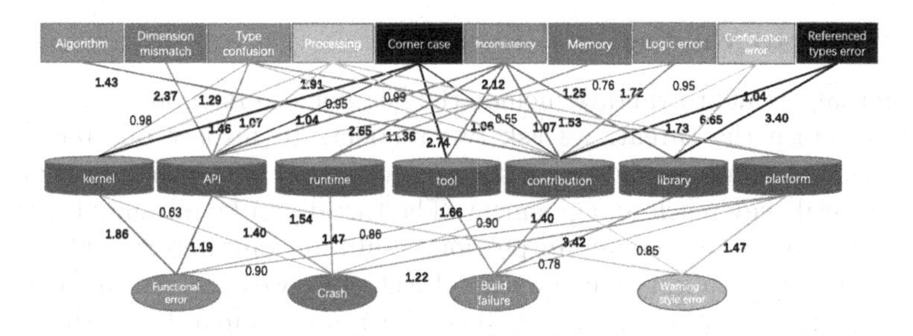

Fig. 3. Correlation between locations

(4.72%). The three components implement features to support the compilation, the execution, and the installation on different platforms. The other components are less buggy. Even though computation graphs define the process of TensorFlow computing, we find that only 0.94% bugs locate in this component. Our observations lead to a finding:

Finding 4. In TensorFlow, the major reported bugs are in deep learning algorithms and their interfaces, and the bugs in compiling, deploying, and installing TensorFlow on different platforms occupy a smaller proportion.

Correlation of Bug Categories. Figure 3 shows the correlations among symptoms, root causes, and bug locations. In this figure, the rectangles denote root causes; the ovals denote symptoms; and the cylinders denote bug locations. We ignore bug locations, if their bugs are fewer than three. The lines denote correlations, and we highlight correlations whose values are greater than one.

For root causes, we find that inconsistencies are popular, and for symptoms, crashes and build failures are popular among the components. From the perspective of components, we find that *kernel* has strong correlation with functional errors and corner cases, which indicates semantic bugs are dominant in this component. Meanwhile, we find that *API* has strong correlation with root causes related to tensor computations such as dimension mismatches and type confusions. For *library* and *tool*, their symptoms have strong correlations with build failures, and their root causes have strong correlations with inconsistences. The above observations lead to a finding:

Finding 5. Crashes and build failures are popular symptoms, and inconsistencies are a popular root cause among components. For those buggiest components, we find that *kernel* contains many sematic bugs, and *API* contains root causes related to tensor computations such as dimension mismatches and type confusions. In the related components such as *library* and *tool*, build failures are popular, and most bugs are caused by inconsistencies.

3.3 Threat to Validity

The internal threats to validity include the possible errors of our manual inspection. To reduce the threat, we ask two students to inspect our bugs. When they encounter controversial cases, they discuss them with others on our group meeting, until they reach an agreement. The threat can be mitigated with more researchers, so we release our inspection results on our website. The internal threats to external validity include our subject, since we analyzed the bugs inside only TensorFlow. Although our analyzed bugs are comparable with the prior studies and other studies (*e.g.* [53]) also analyzed only TensorFlow bugs, they are limited. The threat can be reduced by analyzing more libraries in future.

4 The Significance of Our Findings

Improving the Quality of Deep Learning Libraries. For every root cause of TensorFlow bug, we find several major symptoms occupy a large proportion (Finding 1), and the correlations between root cause and symptom can also suggest possible links (Finding 3), which can help developers to diagnose the cause of a bug according to its symptom. Since TensorFlow bug characteristics show strong similarity to traditional software (Finding 2), the experience and tools of bug repairing in other software can also be transferred to TensorFlow. Since the proportion of bugs in different component varies obviously (Finding 4), developers should pay more attention to safety check and test case design, when adding new features or making modifications to bug-prone components. Moreover, as the integration of libraries is common in deep learning software, the connection of different libraries should obtain higher priority in development. To overcome this problem, developing unified APIs can be helpful.

Combining the Results of the Prior Studies. From two different perspectives of deep learning software, the prior studies [33,53] analyze the bugs of deep learning clients, but our study analyzes the bugs of deep learning libraries. The bugs inside deep learning libraries can have impacts on the bugs of their clients. For example, Islam *et al.* [33] find that 11% of TensorFlow client bugs are caused by incorrect usages of deep learning APIs. From the perspective of deep learning libraries, such bugs can be caused by the inconsistency bugs in our study. As another example, the prior studies [33,53] show that unaligned tensors and the absences of type checking are common causes of deep learning client bugs. We suspect that such bugs are related to dimension mismatches and type confusions, which are found in our study. In future work, we plan to combine the results of the prior studies and ours and explore more advanced techniques to detect deep learning bugs.

The Inspiration to Databases and Their Applications. We notice that some bugs in deep learning libraries are common in database systems (*e.g.*, memory bugs 2.97% and concurrency bugs 0.99%), and detecting such bugs has been a hot research topic in the research community of databases [38,43,44]. As advocated by Wang *et al.* [51], the existing techniques in the database community can be tailored to handle similar bugs in deep learning libraries [51]. Additionally, as more and more deep learning techniques and frameworks are applied to solve database problems [37,52], our revealed bugs in side such libraries are also important for database researchers and programmers.

5 Related Work

Empirical Studies on Bug Characteristics. There has been a number of recent studies studying bugs from open source repositories. Tan *et al.* [46] analyze the bug characteristics of open source projects such as the Linux kernel and Mozilla. Thung *et al.* [47] analyze the bugs of machine learning systems such as Mahout, Lucene, and OpenNLP. Zhang *et al.* [53] analyze the client code that calls TensorFlow. Islam *et al.* [33] analyze the clients of more deep learning bugs. Humbatova *et al.* [32] introduce a taxonomy of faults in deep learning systems. Compared with all existing works, we analyze bugs inside a representative deep-learning library *i.e.*, TensorFlow, which is a different angle from theirs.

Detecting Deep Learning Bugs. Pei *et al.* [41] propose a whitebox framework to test real-world deep learning systems. Ma *et al.* [40] propose a set of multi-granularity criteria to measure the quality of test cases prepared for deep learning systems. Tian *et al.* [48] and Pham *et al.* [42] introduce differential testing to discover bugs in deep learning software. Our empirical study reals new types of bugs, which cannot be effectively detected by the above approaches. Our findings are useful for researchers, when they design detection approaches for such bugs.

6 Conclusion and Future Work

Although researchers have conducted empirical studies to understand deep learning bugs, these studies focus on bugs of its clients, and the nature of bugs inside a deep library is still largely unknown. To deepen the understanding of such bugs, we analyze 202 bugs inside TensorFlow. Our results show that (1) its root causes are more determinative than its symptoms; (2) bugs in traditional software and TensorFlow share various common characteristics; and (3) inappropriate data formatting (dimension and type) is bug prone and popular in API implements while inconsistent bugs are common in other supporting components. In future work, we will analyze bugs from more deep-learning libraries to obtain a more comprehensive understanding of bugs in deep learning frameworks, and we plan to design automatic tools to detect bugs in deep-learning libraries.

Acknowledgement. We appreciate the anonymous reviewers for their insightful comments. This work is sponsored by the National Key R&D Program of China No. 2018YFC083050.

References

1. Fix deadlocks in staging areas (2017). https://github.com/tensorflow/tensorflow/pull/13684
2. Bug in tf.print summarized formatting (2018). https://github.com/tensorflow/tensorflow/issues/20751
3. Cannot opened include file "tensorflow/contrib/tpu/proto/tpu_embedding_config.pb.h": no such file or directory (2018). https://github.com/tensorflow/tensorflow/issues/16262
4. Exception when not providing optional parameter frequency_skip in timefreqlstmcell (2018). https://github.com/tensorflow/tensorflow/issues/16100
5. Fix an imperfect implementation of tf.losses.mean_pairwise_squared_error (2018). https://github.com/tensorflow/tensorflow/pull/16433
6. Fix broken python3 build (2018). https://github.com/tensorflow/tensorflow/pull/16130
7. Fix build issue with KafkaDataset (2018). https://github.com/tensorflow/tensorflow/pull/17418
8. Fix error: ConvNDLSTMCell does not pass name parameter (2018). https://github.com/tensorflow/tensorflow/pull/17345
9. Fix possible memory leak (2018). https://github.com/tensorflow/tensorflow/pull/21950
10. Fix routing of quantized tensors (2018). https://github.com/tensorflow/tensorflow/pull/19894
11. Fix tf.argmax warnings on dimension argument by using axis instead (2018). https://github.com/tensorflow/tensorflow/pull/18558
12. Fix var type issue which breaks crf_decode (2018). https://github.com/tensorflow/tensorflow/pull/21371
13. Fixed build error on gcc-7 (2018). https://github.com/tensorflow/tensorflow/pull/21017
14. [INTEL MKL] fix bug in MklSlice op when allocating output tensor (2018). https://github.com/tensorflow/tensorflow/pull/22822
15. Max pooling cause error on empty batch (2018). https://github.com/tensorflow/tensorflow/issues/21338
16. MKL DNN: fix the TF1.6 speed issue by fixing MKL DNN LRN taking the optimum path (2018). https://github.com/tensorflow/tensorflow/pull/17605
17. tf_cnn_benchmarks.py stuck when running with multiple GPUs and ImageNet data with protocol grpc+verbs (2018). https://github.com/tensorflow/tensorflow/issues/11725
18. Fix for stringPiece build failure (2019). https://github.com/tensorflow/tensorflow/pull/21956
19. Keras (2019). https://keras.io
20. Abadi, M., et al.: TensorFlow: a system for large-scale machine learning. In: Proceedings of OSDI, pp. 265–283 (2016)
21. Anand, S.S., Bell, D.A., Hughes, J.G.: An empirical performance study of the Ingres search accelerator for a large property management database system. In: Proceedings of VLDB, pp. 676–685 (1994)
22. Avizienis, A., Laprie, J., Randell, B., Landwehr, C.E.: Basic concepts and taxonomy of dependable and secure computing. IEEE Trans. Dependable Sec. Comput. **1**(1), 11–33 (2004)

23. Bengio, Y., Ducharme, R., Vincent, P., Janvin, C.: A neural probabilistic language model. J. Mach. Learn. Res. **3**, 1137–1155 (2003)
24. Bergstra, J., et al.: Theano: deep learning on GPUs with Python. In: Proceedings of the NIPS, BigLearning Workshop (2011)
25. Collobert, R., Bengio, S., Marithoz, J.: Torch: a modular machine learning software library (2002)
26. Derr, E., Bugiel, S., Fahl, S., Acar, Y., Backes, M.: Keep me updated: an empirical study of third-party library updatability on android. In: Proceedings of the CCS, pp. 2187–2200 (2017)
27. Endres, A.: An analysis of errors and their causes in system programs. IEEE Trans. Software Eng. **1**(2), 140–149 (1975)
28. Florêncio, D.A.F., Herley, C.: A large-scale study of web password habits. In: Proceedings of the WWW, pp. 657–666 (2007)
29. Glass, R.L.: Persistent software errors. IEEE Trans. Software Eng. **7**(2), 162–168 (1981)
30. Han, J., Kamber, M., Pei, J.: Data Mining: Concepts and Techniques. Morgan Kaufmann Publishers, Burlington (2011)
31. Hinton, G., et al.: Deep neural networks for acoustic modeling in speech recognition: the shared views of four research groups. IEEE Signal Process. Mag. **29**(6), 82–97 (2012)
32. Humbatova, N., Jahangirova, G., Bavota, G., Riccio, V., Stocco, A., Tonella, P.: Taxonomy of real faults in deep learning systems. In: Proceedings of the ICSE (2020, to appear)
33. Islam, M.J., Nguyen, G., Pan, R., Rajan, H.: A comprehensive study on deep learning bug characteristics. In: Proceedings of the ESEC/FSE, pp. 510–520 (2019)
34. Jia, Y., et al.: Caffe: convolutional architecture for fast feature embedding. In: Proceedings of the MM, pp. 675–678 (2014)
35. Kipf, T.N., Welling, M.: Semi-supervised classification with graph convolutional networks. In: Proceedings of the ICLR (2017)
36. Krizhevsky, A., Sutskever, I., Hinton, G.E.: ImageNet classification with deep convolutional neural networks. In: Proceedings of the NIPS, pp. 1106–1114 (2012)
37. Li, G., Zhou, X., Li, S., Gao, B.: Qtune: a query-aware database tuning system with deep reinforcement learning. PVLDB **12**(12), 2118–2130 (2019)
38. Lin, Q., Chen, G., Zhang, M.: On the design of adaptive and speculative concurrency control in distributed databases. In: Proceedings of the ICDE, pp. 1376–1379 (2018)
39. Lockemann, P.C., Nagel, H., Walter, I.M.: Databases for knowledge bases: empirical study of a knowledge base management system for a semantic network. Data Knowl. Eng. **7**, 115–154 (1991)
40. Ma, L., et al.: DeepGauge: multi-granularity testing criteria for deep learning systems. In: Proceedings of the ASE, pp. 120–131 (2018)
41. Pei, K., Cao, Y., Yang, J., Jana, S.: DeepXplore: automated whitebox testing of deep learning systems. In: Proceedings of the SOSP, pp. 1–18 (2017)
42. Pham, H.V., Lutellier, T., Qi, W., Tan, L.: CRADLE: cross-backend validation to detect and localize bugs in deep learning libraries. In: Proceedings of the ICSE, pp. 1027–1038 (2019)
43. Ren, K., Thomson, A., Abadi, D.J.: VLL: a lock manager redesign for main memory database systems. VLDB J. **24**(5), 681–705 (2015)
44. van Renen, A., et al.: Managing non-volatile memory in database systems. In: Proceedings of the SIGMOD, pp. 1541–1555 (2018)

45. Sutskever, I., Vinyals, O., Le, Q.V.: Sequence to sequence learning with neural networks. In: Proceedings of the NIPS, pp. 3104–3112 (2014)
46. Tan, L., Liu, C., Li, Z., Wang, X., Zhou, Y., Zhai, C.: Bug characteristics in open source software. Empirical Softw. Eng. **19**(6), 1665–1705 (2014)
47. Thung, F., Wang, S., Lo, D., Jiang, L.: An empirical study of bugs in machine learning systems. In: Proceedings of the ISSRE, pp. 271–280 (2012)
48. Tian, Y., Pei, K., Jana, S., Ray, B.: DeepTest: automated testing of deep-neural-network-driven autonomous cars. In: Proceedings of the ICSE, pp. 303–314 (2018)
49. Wang, L., et al.: Superneurons: dynamic GPU memory management for training deep neural networks. In: Proceedings of the PPoPP, pp. 41–53 (2018)
50. Wang, S., Liu, T., Nam, J., Tan, L.: Deep semantic feature learning for software defect prediction. IEEE Trans. Softw. Eng., 1 (2018, early access). https://doi.org/10.1109/TSE.2018.2877612
51. Wang, W., Zhang, M., Chen, G., Jagadish, H.V., Ooi, B.C., Tan, K.: Database meets deep learning: challenges and opportunities. SIGMOD Record **45**(2), 17–22 (2016)
52. Xu, B., et al.: NADAQ: natural language database querying based on deep learning. IEEE Access **7**, 35012–35017 (2019)
53. Zhang, Y., Chen, Y., Cheung, S., Xiong, Y., Zhang, L.: An empirical study on TensorFlow program bugs. In: Proceedings of the ISSTA, pp. 129–140 (2018)

Partial Multi-label Learning with Label and Feature Collaboration

Tingting Yu[1,2], Guoxian Yu[1,2(✉)], Jun Wang[1], and Maozu Guo[3]

[1] School of Software, Shandong University, Jinan, China
[2] College of Computer and Information Science, Southwest University, Chongqing, China
ttyu@swu.edu.cn, {gxyu,kingjun}@sdu.edu.cn
[3] College of Electrical and Information Engineering, Beijing University of Civil Engineering and Architecture, Beijing, China
guomaozu@bucea.edu.cn

Abstract. Partial multi-label learning (PML) models the scenario where each training instance is annotated with a set of candidate labels, and only some of the labels are relevant. The PML problem is practical in real-world scenarios, as it is difficult and even impossible to obtain precisely labeled samples. Several PML solutions have been proposed to combat with the prone misled by the irrelevant labels concealed in the candidate labels, but they generally focus on the smoothness assumption in feature space or low-rank assumption in label space, while ignore the negative information between features and labels. Specifically, if two instances have largely overlapped candidate labels, irrespective of their feature similarity, their ground-truth labels should be similar; while if they are dissimilar in the feature and candidate label space, their ground-truth labels should be dissimilar with each other. To achieve a credible predictor on PML data, we propose a novel approach called PML-LFC (Partial Multi-label Learning with Label and Feature Collaboration). PML-LFC estimates the confidence values of relevant labels for each instance using the similarity from both the label and feature spaces, and trains the desired predictor with the estimated confidence values. PML-LFC achieves the predictor and the latent label matrix in a reciprocal reinforce manner by a unified model, and develops an alternative optimization procedure to optimize them. Extensive empirical study on both synthetic and real-world datasets demonstrates the superiority of PML-LFC.

Keywords: Partial Multi-label Learning · Feature and label collaboration · Confidence estimation · Smoothness assumption · Low-rank

1 Introduction

Multi-label learning (MLL) deals with the scenario where each instance is annotated with a set of discrete non-exclusive labels [6,29]. Recent years have

This work is supported by NSFC (61872300 and 61871010).

Y. Nah et al. (Eds.): DASFAA 2020, LNCS 12112, pp. 621–637, 2020.
https://doi.org/10.1007/978-3-030-59410-7_41

witnessed an increasing research and application of MLL in various domains, such as image annotation [20], cybersecurity [7], gene functional annotation [24], and so on. Most MLL methods have an implicit assumption that each training example is precisely annotated with all of its relevant labels. However, it is difficult and costly to obtain fully annotated training examples in most real-word MLL tasks [17]. Therefore, recent MLL methods not only focus on how to assign a set of appropriate labels to unlabeled examples using the label correlations [27,31], but also on replenishing missing labels for incompletely labeled samples [14,16,20].

Existing MLL solutions still overlook another fact that naturally arises in real-world scenarios. For example, in Fig. 1, the image was crowdly-annotated by workers with 'Seaside', 'Sky', 'Sandbeach', 'Cloud', 'Tree', 'People', 'Sunset' and 'Ship'. Among these labels, the first five are relevant, and the last three are irrelevant of this image. Obviously, the training procedure is prone to be misled by irrelevant labels concealed in the candidate labels of training samples. To combat with such major difficulty, some pioneers term learning on such training data with irrelevant labels as *Partial Multi-label Learning* (PML) [21,23], and proposed several PML approaches [5,13,19] to identify the irrelevant labels concealed in the candidate labels of annotated samples, and to achieve a predictor robust (or less prone) to irrelevant labels of training data.

Fig. 1. An exemplary partial multi-label learning scenario. The image is annotated with eight candidate labels, only the first five (in red) are relevant, and the last three (in black) are irrelevant. (Color figure online)

However, contemporary approaches either mainly focus on the smoothness assumption that the (dis-)similar instances should have (dis-)similar ground-truth labels [5,19,21], or the low-rank assumption that the ground-truth label matrix should be low-rank [13,23]. While these two assumptions can not well handle the case that two instances without any common candidate label but with high feature similarity, and two instances with some overlapped candidate labels but with a low feature similarity. As a result, the smoothness-based methods [5,19,21] ignore the *negative* information that two instances with high (low) feature similarity but with a low (high) semantic similarity from the candidate labels. In other words, these methods do not make a good collaborative use of the information from the label and feature space for PML. For example, in Fig. 2(a) and Fig. 2(d), two instances ($\mathbf{x}_1$ and $\mathbf{x}_2$) not only have a high (low) feature similarity, but also a high (low) semantic similarity due to largely overlapped (non-overlapped) candidate labels. From the setting of PML, these two instances are likely to have overlapped (non-overlapped) ground-truth labels. On the other hand, if two instances are without any overlapped candidate label (say zero semantic similarity), their ground-truth labels should be non-overlapped (as Fig. 2(b) show). Besides, if these two instances have a low feature similarity but with a high semantic similarity (as Fig. 2(c) show), their ground-truth labels can be overlapped to some extent.

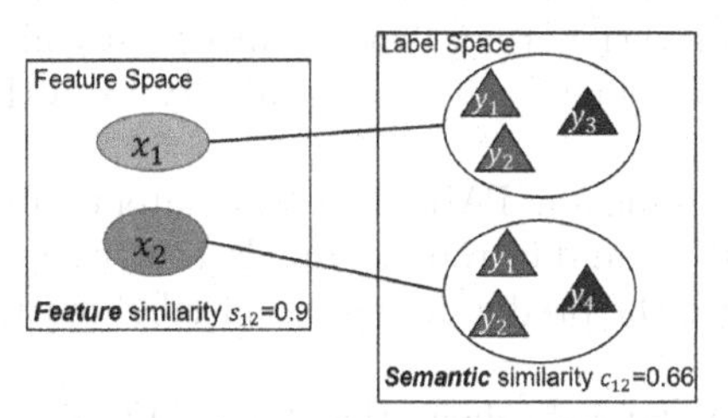

(a) Two instances with both high feature and semantic similarity.

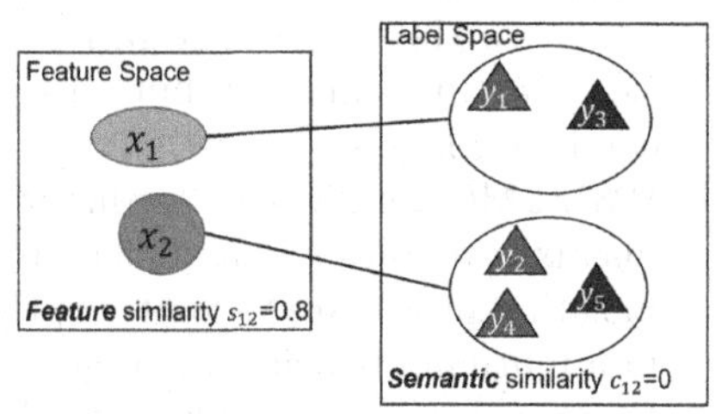

(b) Two instances with high feature similarity but with low (zero) semantic similarity.

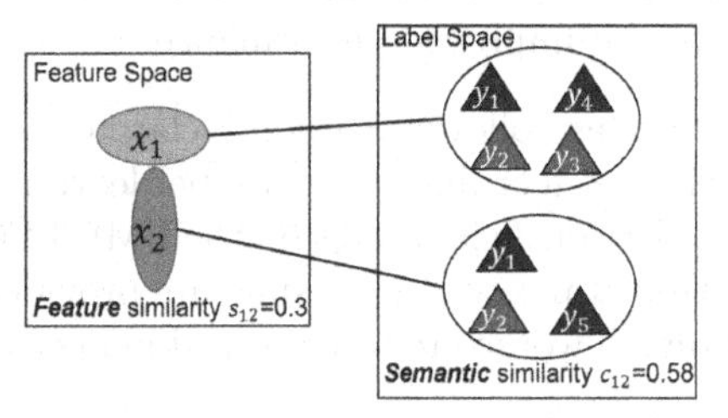

(c) Two instances with low (moderate) feature similarity but with high semantic similarity.

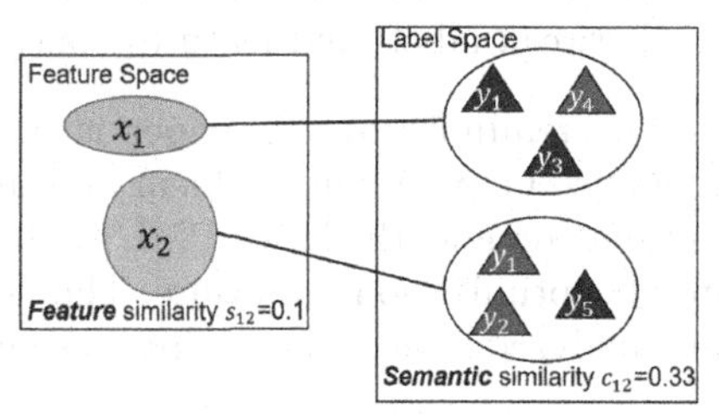

(d) Two instances with low feature similarity and low semantic similarity.

Fig. 2. The feature and label information of PML data in four scenarios. The semantic similarity is computed on the candidate labels of instances, the feature similarity is computed on the feature vectors of instances. Ground-truth labels are highlighted in red, while other candidate labels are in grey. (Color figure online)

Given these observations, we introduce the Partial Multi-label Learning with Label and Feature Collaboration (PML-LFC). PML-LFC firstly learns a linear predictor with respect to a *latent* ground-truth label matrix, and induces a low-rank constraint on the coefficient matrix of the predictor to account for the label correlations of multi-label data. Then, it computes the *feature* similarity between instances and the *semantic* similarity between them using the candidate label vectors, and forces the inner product similarity of latent label vectors consistent with the feature similarity and semantic similarity. In this way, both the label and feature information are collaboratively used to induce the latent label vectors, and the four scenarios illustrated in Fig. 2 are jointly modelled. PML-LFC finally achieves the predictor and the latent label matrix in a reciprocal reinforce manner by a unified model, and develops an alternative optimization procedure to optimize them.

The main contributions of this paper are summarized as follows:

(i) We introduce PML-LFC to jointly leverage the label and feature information of partial multi-label data to induce a credible multi-label classifier, where existing PML solutions isolate the usage of label and feature information, or ignore the usage of negative information that two instances with high (low) feature similarity but with a low (high) semantic similarity from the candidate labels.

(ii) PML-LMC unifies the predictor training on PML data and latent label matrix exploration in a unified objective, and introduces an alternative optimization procedure to jointly optimize the predictor and latent label matrix in a mutually beneficial manner.

(iii) Empirical study on public multi-label datasets shows that PML-LFC significantly outperforms the related and competitive methods: fPML [23], PML-LRS [13], DRAMA [19], PRACTICLE [5], and two classical multi-label classifiers (RankSVM [4], and ML-KNN [28]). In addition, the collaboration between labels and features contributes an improved performance.

The reminder of this paper is organized as follows. Section 2 clarifies the difference between our problem and multi-label learning, partial-label learning, and then reviews the latest PML methods. Section 3 elaborates on the PML-LFC and its optimization procedure. The experimental setup and results are provided and analyzed in Sect. 4. Conclusions and future works are summarized in Sect. 5.

2 Related Work

PML is different from multi-label crowd consensus learning [9,17,25], which wants to obtain high-quality consensus annotations of repeatedly annotated instances, while PML does not have such repeated annotations of the same instances. PML is unlike multi-label weak-label learning [14,16], which focus on learning from annotated training data with incomplete (missing) labels. PML is also different from the popular partial-label learning (PLL) [3,26], which assumes only one label from the candidate labels of the sample is the ground truth and aims to induce a multi-class predictor to assign one label for unseen sample. PLL can be viewed as a degenerated version of PML. We observe that PML is more difficult than the typical MLL and PLL problems, since the ground truth labels of samples are not directly accessible to train the predictor and a set of discrete non-exclusive labels should be carefully assigned. To be self-inclusive and help reader being informed, we give a brief review of popular PML solutions.

[21] introduced two PML approaches (PML-fp and PML-lc) to elicit the ground-truth labels by minimizing the confidence weighted ranking loss between candidate and non-candidate labels. PML-fp focuses on the utilization of feature information of training data, while PML-lc focuses on the label correlations. To mitigate the negative impact of irrelevant labels in the training phase, [5] proposed a two-stage approach (PARTICLE), which firstly elicits credible labels via iterative label propagation, and then takes the elicited labels to induce a multi-label classifier with virtual label splitting (PARTICLE-VLS) or maximum a posteriori reasoning (PARTICLE-MAP). [19] introduced another two-stage PML approach (DRAMA) that firstly estimates the confidence value for each label by utilizing the feature manifold that indicates how likely a label is correct, and then induces a gradient boosting model to fit the label confidences by exploring the label correlations with the previously elicited labels in each boosting round. Due to the isolation between label elicitation and the classifier training, the elicited labels maybe not compatible with the classifier. [23] introduced a feature-induced PML solution (fPML), which coordinately factorizes the observed sample-label association matrix and the sample-feature matrix into low-rank matrices and then reconstructs the sample-label matrix to identify irrelevant labels. At the same time, fPML optimizes a compatible predictor based on the reconstructed sample-label matrix. Similarly, [13] assumed the observed label matrix is the linear combination of a ground-truth label matrix with low-rank and an irrelevant label matrix with sparse constraints, and introduced a solution called PML-LRS.

These aforementioned state-of-the-art PML solutions either mainly focus on the usage of feature manifold that similar instances will have similar labels, or on the usage of ground-truth label matrix being low-rank. They still isolate the joint effect of features and labels for effective partial multi-label learning to some extent. The (latent) labels of instances are dependent on the features of these instances [10,30], and the semantic similarity derived from the label sets of multi-label instances are positively correlated with the feature similarity between them [15,18,24]. Both the label and feature information of multi-label data should be well accounted for effective learning on PML data. Given that, we introduce PML-LFC to collaboratively use the feature and label information, which will be detailed in the next Section.

3 Proposed Method

Let $\mathcal{X} \in \mathbb{R}^d$ denote the d-dimensional feature space, and $\mathcal{L} = \{0|1\}_{c=1}^{q}$ denote the label space with respect to q distinct labels. Given a PML dataset $\mathcal{D} = \{(\mathbf{x}_i, \mathcal{L}_i)|1 \leq i \leq n\}$, where $\mathbf{x}_i \in \mathcal{X}$ is the feature vector of the i-th sample, and $\mathcal{L}_i \subset \mathcal{L}$ is the set of candidate labels currently annotated to $\mathbf{x}_i$. The key characteristic of PML is that only a subset labels $\tilde{\mathcal{L}}_i \subset \mathcal{L}_i$ are the ground-truth labels of $\mathbf{x}_i$, while the others $(\mathcal{L}_i - \tilde{\mathcal{L}}_i)$ are irrelevant for $\mathbf{x}_i$. However, $\tilde{\mathcal{L}}_i$ is not directly accessible to the predictor. The target of PML is to induce a multi-label classifier $F : \mathcal{X} \to 2^{\mathcal{L}}$ from $\mathcal{D}$. A naive PML solution is to divide

PML problem into q binary sub-problems, and then adopt a noisy label resistant learning algorithm [11,12]. But this naive solution generally suffers from the label sparsity issue of multi-label data, where each instance typically is only annotated with several labels of whole label space and each label is only annotated to a small portion of instances. Moreover, it disregards the correlations between labels. Another straightforward solution is to take candidate labels as the ground-truth labels and then apply off-the-shelf MLL algorithms [6] to train the predictor. However, the predictor will be seriously misled by the false positive labels in the candidate labels.

To bypass the difficulty of the lack of known ground-truth labels of training instances, we take $\mathbf{P} = [\mathbf{p}_1, \cdots, \mathbf{p}_n]^\top \in [0,1]^{n \times q}$ as the latent label confidence matrix, where p_{ic} reflects the confidence of the c-th label being the ground-truth for the i-th instance. Unlike existing two-stage approaches [5,19] that firstly estimate the credible labels and then train predictor using the estimated labels. We integrate the estimation of label confidence matrix $\mathbf{P}$ and predictor learning into a unified framework as follows:

$$\min \sum_{i=1}^{n} L(\mathbf{x}_i, \mathbf{p}_i, \mathbf{f}) + \alpha \Omega(\mathbf{f}) + \beta \Phi(\mathbf{P}) \tag{1}$$

where L denotes the loss function, Ω controls the complexity of the prediction model $\mathbf{f}$ and Φ is the regularization term for label confidence matrix $\mathbf{P}$, α and β are the trade-off parameters for the last two terms. In this unified formulation, the model is learned from the confidence label matrix $\mathbf{P}$ rather than the original noisy label matrix $\mathbf{Y}$. Therefore, the key is how to obtain reliable confidence matrix $\mathbf{P}$.

In this paper, we propose to train the predictor based on the widely-used the least square loss to fit the confidence label matrix $\mathbf{P}$ as follows:

$$L(\mathbf{x}_i, \mathbf{p}_i, \mathbf{f}) = \sum_{i=1}^{n} \|\mathbf{x}_i \mathbf{W} - \mathbf{p}_i\|^2 \tag{2}$$

where $\mathbf{W} = [\mathbf{w}_1, \cdots, \mathbf{w}_q]^\top \in \mathbb{R}^{d \times q}$ is the coefficient matrix for the predictor. It is recognized the labels of multi-label instances are correlated and the label data matrix of instances should be a low rank one [22,23]. Given that, we instantiate the regularization on the predictor with low-rank constraint on $\mathbf{W}$ as follows:

$$\Omega(\mathbf{f}) = \mathbf{rank}(\mathbf{W}) \tag{3}$$

The main bottleneck of PML problem is the lack of the ground-truth labels of training instances. To overcome this bottleneck, most efforts operate in the feature space based on the assumption that similar (dissimilar) instances have similar (dissimilar) label assignments. They adopt manifold regularization [1] derived from the feature similarity to refine the labels of PML data, and then induce a predictor on the refined labels [5,19]. Some efforts work in the label space using the knowledge that the latent ground-truth label matrix should be low-rank [13,23], or that the relevant labels of an instance are hidden in the candidate label set and should be ranked ahead of irrelevant ones outside of the candidate set [21]. Although these efforts leverage the feature information to identify the relevant/irrelevant labels of training instances to some extent, they are still inclined to assign similar labels to two instances with high feature similarity but without any common candidate label. From the definition of PML, it is easy to observe that the ground-truth labels of training instances are hidden in the collected candidate label set. In other words, if two annotated instances do not share any candidate label, then there is no overlap between the individual ground-truth label sets of the two instances (as Fig. 2(b) show). Besides, they also prefer to assign different label sets to two instances without sufficient large feature similarity but with largely overlapped candidate labels (as shown in Fig. 2(c). In summary, contemporary PML approaches do not sufficiently use the negative information that two instances with high (low) feature similarity but with a low (high) semantic similarity from the candidate labels, since they do not collaboratively use the feature and label information in a coherent way.

To remedy this issue, we specify the last term in Eq. (1) as follows:

$$\begin{aligned} \varPhi(\mathbf{P}) &= \sum_{i,j,i\neq j} \left(\mathbf{s}_{ij}\mathbf{c}_{ij} - \mathbf{p}_i^\top \mathbf{p}_j\right)^2 \\ &\text{s.t. } \mathbf{P} \geq \mathbf{0},\ \sum_{c=1}^{q} p_{ic} = 1, \forall i = 1,2,3,\cdots,n \end{aligned} \tag{4}$$

where $\mathbf{s}_{ij}$ represents the feature similarity between $\mathbf{x}_i$ and $\mathbf{x}_j$, $\mathbf{c}_{ij}$ reflects the semantic similarity derived from candidate labels of these two instances, respectively. The first constraint guarantees that each candidate label has a non-negative confidence value, and the second constraint restricts the confidence value is within [0, 1], and the sum of them equal to 1. We can find that: (i) if two instances have both high values of $\mathbf{s}_{ij}$ and $\mathbf{c}_{ij}$, then $\mathbf{p}_i$ and $\mathbf{p}_j$ should be close to each other; (ii) if two instances have a large (or moderate) value of $\mathbf{c}_{ij}$, then $\mathbf{p}_i$ and $\mathbf{p}_j$ can still have some overlaps; (iii) if two instances have a zero (or low) value of $\mathbf{c}_{ij}$ and a low value of $\mathbf{s}_{ij}$, then $\mathbf{p}_i$ and $\mathbf{p}_j$ should be not overlapped. Our minimization of Eq. (4) jointly considers the above cases. In contrast, contemporary PML methods ignore the semantic similarity between instances. They do not make effective use of the negative information in the last two cases stated above. We want to remark that given the existence of irrelevant labels of training instances, it is not an easy job to quantify and leverage the important label correlation for partial multi-label learning. Thanks to the semantic similarity, which quantifies the similarity between instances based on the pattern that two (or more) labels co-annotate to the same instances, this pattern is also transferred to the latent confident label matrix $\mathbf{P}$. In addition, this pattern transfer is also coordinated by the low-rank constraint on the coefficient matrix and by the feature similarity, which alleviates the negative impact of irrelevant labels on quantifying the semantic similarity. In this way, the information sources from the feature and label spaces are jointly used to guide the latent label matrix learning, which rewards a credible multi-label predictor.

Here, we initialize the label confidence matrix $\mathbf{P}$ as:

$$\forall 1 \leq i \leq n: \quad p_{i,c} = \begin{cases} \frac{1}{|\mathcal{L}_i|}, & \text{if } c \in \mathcal{L}_i \\ 0, & \text{otherwise} \end{cases} \tag{5}$$

To quantify the feature similarity between instances, we adopt the widely-used Gaussian heat kernel similarity as follows:

$$\mathbf{s}_{ij} = exp^{-\frac{\|\mathbf{x}_i - \mathbf{x}_j\|_2^2}{t^2}} \tag{6}$$

where t denotes the kernel width and is empirically set to $t = \sum_{i,j,i\neq j}^{n} \frac{\|\mathbf{x}_i - \mathbf{x}_j\|}{n-1}$. Clearly, $\mathbf{s}_{ij} \in (0,1)$ when there are no two identical instances. We want to remark that other similarity metrics can also be adopted here. Our choice of Gaussian heat kernel is for its simplicity and wide application.

Diverse similarity metrics can also be adopted to quantify the semantic similarity between multi-label instances [18,24], here we use the cosine similarity as follows:

$$\mathbf{c}_{ij} = \frac{\mathbf{y}_i^\top \mathbf{y}_j}{\|\mathbf{y}_i\| \, \|\mathbf{y}_j\|} \tag{7}$$

where $\mathbf{y}_i$ is the one-hot coding label vector for $\mathbf{x}_i$, $\mathbf{y}_{ic} = 1$ if the c-th label is annotated to $\mathbf{x}_i$, $\mathbf{y}_{ic} = 0$ otherwise. Obviously, $\mathbf{c}_{ij} \in [0,1]$, it has a large value when two instances have a large portion of overlapped candidate labels, moderate value when they share some overlapped candidate labels, and zero value when they do not have any overlapped candidate label.

Based on the above analysis, we can instantiate the PML-LFC as follows:

$$\begin{aligned} &\min_{\mathbf{W},\mathbf{P}\geq\mathbf{0}} \|\mathbf{XW} - \mathbf{P}\|_2^2 + \alpha \mathbf{rank}(\mathbf{W}) + \beta \sum_{i,j,i\neq j} \left(\mathbf{s}_{ij}\mathbf{c}_{ij} - \mathbf{p}_i^\top \mathbf{p}_j\right)^2 \\ &\text{s.t.} \sum_{c=1}^{q} p_{ic} = 1, \forall i = 1,2,3,\cdots,n \end{aligned} \tag{8}$$

The problem above can be further rewritten as follows:

$$\begin{aligned} &\min_{\mathbf{W},\mathbf{P}\geq\mathbf{0}} \|\mathbf{XW} - \mathbf{P}\|_2^2 + \alpha \mathbf{rank}(\mathbf{W}) + \beta \left\|\mathbf{H} \odot (\mathbf{S} \odot \mathbf{C} - \mathbf{PP}^\top)\right\|_2^2 \\ &\text{s.t.} \sum_{c=1}^{q} p_{ic} = 1, \forall i = 1,2,3,\cdots,n \end{aligned} \tag{9}$$

where $\mathbf{H} \in \mathbb{R}^{n\times n}$, $\mathbf{H}_{ij} = 0$ if $i = j$; and $\mathbf{H}_{ij} = 1$ otherwise, $\odot$ is the Hadamard product. However, the rank function in Eq. (9) is hard to optimize, the nuclear norm $\|\bullet\|_*$ is suggested to surrogate the rank function. Therefore, Eq. (9) is reformulated as follows:

$$\begin{aligned} &\min_{\mathbf{W},\mathbf{P}\geq\mathbf{0}} \|\mathbf{XW} - \mathbf{P}\|_2^2 + \alpha \|\mathbf{W}\|_* + \beta \left\|\mathbf{H} \odot (\mathbf{S} \odot \mathbf{C} - \mathbf{PP}^\top)\right\|_2^2 \\ &\text{s.t.} \sum_{c=1}^{q} p_{ic} = 1, \forall i = 1,2,3,\cdots,n \end{aligned} \tag{10}$$

3.1 Optimization

Since the optimization problem in Eq. (10) is non-convex with respect to $\mathbf{W}$ and $\mathbf{P}$ at the same time. We apply the alternative optimization procedure to approximate them. Specifically, we alternatively optimize one variable while fixing the other one as a constant. The detailed procedure is presented below.

Update W: With $\mathbf{P}$ fixed, Eq. (10) with respect to $\mathbf{W}$ is equivalent to the following problem:

$$\min_{\mathbf{W}} \|\mathbf{XW}-\mathbf{P}\|_2^2 + \alpha \|\mathbf{W}\|_* \tag{11}$$

The minimization of Eq. (11) is a trace norm minimization problem, which is time-consuming. To reduce the computation time of Eq. (11), we use the Accelerated Gradient Descent (AGD) algorithm [8] to optimize $\mathbf{W}$ as summarized in Algorithm 1.

In particular, $F(\mathbf{W})$, $p_{l_t}(\mathbf{Z}_t)$ and $Q_{l_t}(\mathbf{W}_t, \mathbf{Z}_t)$ in Algorithm 1 are defined as follows:

$$F(\mathbf{W}) = \|\mathbf{XW}-\mathbf{P}\|_2^2 + \alpha \|\mathbf{W}\|_* \tag{12}$$

$$p_{l_t}(\mathbf{Z}_t) = \frac{l_t}{2}\left\|\mathbf{W} - \left(\mathbf{Z}_t - \frac{1}{l_t}\nabla f(\mathbf{Z}_t)\right)\right\|_2^2 + \alpha \|\mathbf{W}\|_* \tag{13}$$

$$f(\mathbf{Z}_t) = \|\mathbf{Z}_t\mathbf{X}-\mathbf{P}\|_2^2 \tag{14}$$

$$Q_{l_t}(\mathbf{W}_t, \mathbf{Z}_t) = f(\mathbf{Z}_t) + \langle \mathbf{W}_t - \mathbf{Z}_t, \nabla f(\mathbf{Z}_t)\rangle + \frac{l_t}{2}\|\mathbf{W}_t - \mathbf{Z}_t\|_2^2 + \alpha\|\mathbf{W}\|_* \tag{15}$$

Update P: With $\mathbf{W}$ fixed, Eq. (10) with respect to $\mathbf{P}$ reduces to:

$$\begin{aligned} &\min_{\mathbf{P}\geq \mathbf{0}} \|\mathbf{XW}-\mathbf{P}\|_2^2 + \beta \left\|\mathbf{H}\odot\left(\mathbf{S}\odot\mathbf{C}-\mathbf{PP}^\top\right)\right\|_2^2 \\ &\text{s.t.} \sum_{c=1}^{q} p_{ic} = 1, \forall i = 1,2,3,...,n \end{aligned} \tag{16}$$

By introducing the Lagrange multiplier λ, Eq. (16) is equivalent to:

$$\min_{\mathbf{P}\geq \mathbf{0}} \|\mathbf{XW}-\mathbf{P}\|_2^2 + \beta \left\|\mathbf{H}\odot\left(\mathbf{A}-\mathbf{PP}^\top\right)\right\|_2^2 + \lambda\|\mathbf{P1}_q - \mathbf{1}_n\|_2^2 \tag{17}$$

where $\mathbf{A} = \mathbf{S}\odot\mathbf{C}$, $\mathbf{1}_q$ $(\mathbf{1}_n)$ are the $q(n)$-dimensional column vector with all ones. The gradient with respect to $\mathbf{P}$ is:

$$\begin{aligned} \nabla\mathbf{P} =& \mathbf{P} - \mathbf{XW} - \beta\mathbf{H}\odot\left(\mathbf{A}-\mathbf{PP}^\top\right)\mathbf{P} \\ &- \beta\mathbf{H}^\top\odot\left(\mathbf{A}^\top - \mathbf{PP}^\top\right)\mathbf{P} + \lambda\left(\mathbf{P1}_q - \mathbf{1}_n\right)\mathbf{1}_q^\top \end{aligned} \tag{18}$$

We can use the Karush-Kuhn-Tucker (KKT) conditions [2] for the nonnegativity of $\mathbf{P}$ as:

$$\begin{aligned} &(\mathbf{P} - \mathbf{XW} - \beta\mathbf{H}\odot\mathbf{AP} + \beta\mathbf{H}\odot\mathbf{B} - \beta\mathbf{H}^\top\odot\mathbf{A}^\top\mathbf{P} \\ &+ \beta\mathbf{H}^\top\odot\mathbf{B} + \lambda\mathbf{PQ} - \lambda\mathrm{N})_{ij}\mathbf{P}_{ij} = 0 \end{aligned} \tag{19}$$

where $\mathbf{B} = \mathbf{P}\mathbf{P}^\top\mathbf{P}$, $\mathbf{Q}$ and $\mathbf{N}$ are all-one matrices with $q \times q$ dimensions and $n \times q$ dimensions, respectively. Let $\mathbf{XW} = \mathbf{XW}^+ - \mathbf{XW}^-$, where $\mathbf{M}_{ij}^+ = \frac{|\mathbf{M}|_{ij}+\mathbf{M}_{ij}}{2}$ and $\mathbf{M}_{ij}^- = \frac{|\mathbf{M}|_{ij}-\mathbf{M}_{ij}}{2}$ for any matrix $\mathbf{M}$, Eq. (19) can be rewritten as:

$$
\begin{aligned}
&(\mathbf{P} - \mathbf{XW}^+ + \mathbf{XW}^- - \beta\mathbf{H} \odot \mathbf{AP} + \beta\mathbf{H} \odot \mathbf{B} - \beta\mathbf{H}^\top \odot \mathbf{A}^\top\mathbf{P} \\
&+ \beta\mathbf{H}^\top \odot \mathbf{B} + \lambda\mathbf{PQ} - \lambda\mathbf{N})_{ij}\mathbf{P}_{ij} = 0
\end{aligned} \tag{20}
$$

Equation (20) leads to the following update formula:

$$
\mathbf{P}_{ij} = \mathbf{P}_{ij}\sqrt{\frac{(\mathbf{XW}^+ + \beta\mathbf{H} \odot \mathbf{AP} + \beta\mathbf{H}^\top \odot \mathbf{A}^\top\mathbf{P} + \lambda\mathbf{N})_{ij}}{(\mathbf{P} + \mathbf{XW}^- + \beta\mathbf{H} \odot \mathbf{B} + \beta\mathbf{H}^\top \odot \mathbf{B} + \lambda\mathbf{PQ})_{ij}}} \tag{21}
$$

Algorithm 2 summarizes the pseudo-code of PML-LFC. We observe that PML-LFC only needs at most ten iterations to converge on our used datasets.

Algorithm 1. Optimization of $\mathbf{W}$

Input: $\mathbf{X}$, $\mathbf{P}$, α.
Output: $\mathbf{W}$.

1: Initialize $\mathbf{W}_0 = \mathbf{Z} \in \mathbb{R}^{d\times q}$, $l_0 > 0$, $\gamma > 0$, $\delta_1 = 1$, $t = 1$
2: **Iterate:**
3: Set $\bar{l} = l_{t-1}$
4: **While** $F(p_{\bar{l}}(\mathbf{Z}_{t-1})) > Q_{\bar{l}}(p_{\bar{l}}(\mathbf{Z}_{t-1}), \mathbf{Z}_{t-1})$
5: Set $\bar{l} = \gamma\bar{l}$
6: Set $l_t = \bar{l}$ and update
7: $\mathbf{W}_t = p_{l_t}(\mathbf{Z}_t)$
8: $\delta_{t+1} = \frac{1+\sqrt{1+4\delta_t^2}}{2}$
9: $\mathbf{Z}_{t+1} = \mathbf{W}_t + \left(\frac{\delta_t - 1}{\delta_t + 1}\right)(\mathbf{W}_t - \mathbf{W}_{t+1})$

---

4 Experiments

4.1 Experimental Setup

Dataset: For a quantitative performance evaluation, five synthetic and three real-world PML datasets are collected for experiments. Table 1 summarizes characteristics of these datasets. Specifically, to create a synthetic PML dataset, we take the current labels of instances as ground-truth ones. For each instance $\mathbf{x}_i$, we randomly insert the irrelevant labels of $\mathbf{x}_i$ with $a\%$ number of the ground-truth labels, and we vary $a\%$ in the range of $\{10\%, 50\%, 100\%, 200\%\}$. All the datasets are randomly partitioned into 80% for training and the rest 20% for testing. We repeat all the experiments for 10 times independently, report the average results with standard deviations.

Algorithm 2. PML-LFC: Partial Multi-label Learning with Label and Feature Collaboration

Input:

$\mathbf{X}$: $n \times d$ instance-feature matrix;

$\mathbf{Y}$: $n \times q$ instance-label association matrix;

α, β: scalar input parameters.

Output:

Prediction coefficients $\mathbf{W}$.

1: Initialize $\mathbf{P}$ by Eq.(5);
2: **Do:**
3: Seek the optimal $\mathbf{W}$ by optimizing Eq. (11) and Algorithm 1;
4: Fix $\mathbf{W}$, update $\mathbf{P}$ by optimizing Eq. (16);
5: **While** not convergence or within the allowed number of iterations

Comparing Methods: Four representative PML algorithms, including fPML [23], PML-LRS [13], DRAMA [19] and PARTICLE-VLS [5] are used as the comparing methods. DRAMA and PARTICLE-VLS mainly utilize the feature similarity between instances, while fPML and PML-LRS build on low-rank assumption of the label matrix, and fPML additionally explores and uses the coherence between the label and feature data matrix. In addition, two representative MLL solutions (ML-KNN [28] and Rank-SVM[4]) are also included as baselines for comparative analysis. The last two comparing methods directly take the candidate labels as ground-truths to train the respective predictor. For these comparing methods, parameter configurations are fixed or optimized by the suggestions in the original codes or papers. For our PML-LMC, we fix $\alpha = 10$ and $\beta = 10$. The parameter sensitivity of α and β will be analyzed later.

Evaluation Metrics: For a comprehensive performance evaluation and comparison, we adopt five widely-used multi-label evaluation metrics: *hamming loss* (HammLoss), *one-error* (OneError), *coverage* (Coverage), *ranking loss* (RankLoss) and *average precision*(AvgPrec). The formal definition of these metrics can be founded in [6,29]. Note here *coverage* is normalized by the number of distinct labels, thus it ranges in [0, 1]. Furthermore, the larger the value of *average precision*, the better the performance is, while the opposite holds for the other four evaluation metrics.

4.2 Results and Analysis

Table 2 reports the detailed experimental results of six comparing algorithms with the noisy label ratio of 50%, while similar observation can be found in terms of other noisy label ratios. The first stage of DARAM and PARTLCE-VLS utilizes the feature similarity to optimize the ground-truth label confidence in different ways. However, due to the features of three real-world PML datasets are protein-protein interaction networks, we directly use the network structure to optimize the ground-truth label confidence matrix in the first stage

Table 1. Characteristics of the experimental dataset. '#Instance' is the number of Examples, '#Features' is the number of features, '#Labels' is the number of distinct labels, 'avgGLs' is the average number of ground-truth labels of each instance, and '#' is the number of noise labels of the dataset.

Data set	#Instances	#Features	#Labels	avgGLs	#Noise
slashdot	3782	1079	22	1.181	-
scene	2407	294	6	1.074	-
enron	1702	1001	53	3.378	-
medical	978	1449	45	1.245	-
Corel5k	5000	499	374	0.245	-
YeastBP	6139	6139	217	5.537	2385
YeastCC	6139	6139	50	1.348	260
YeastMF	6139	6139	39	1.005	234

by respective algorithms. In the second stage, PARTICLE-VLS introduces a virtual label technique to transform the problem into multiple binary training sets, and results in the class-imbalanced problem and causes computation exception due to the sparse biological network data. For this reason, its results on the last three datasets can not be reported. Due to page limit, we summarize the win/tie/loss counts of our method versus the other comparing method in 23 cases (five datasets × four ratios of noisy labels and three PML datasets) across five evaluation metrics in Table 3.

Based on the results in Table 2 and 3, we can observe the following: (i) On the real-word PML datasets *YeastBP*, *YeastCC* and *YeastMF*, PML-LFC achieve the best performance in most cases except ML-KNN on *ranking loss* evaluation. (ii) On the synthic datasets, PML-LFC frequently outperforms other methods and slightly loses to RankSVM and DARAM on medical dataset. (iii) Out of 115 statistical tests PML-LFC achieves much better results than the popular PML methods PML-LRS, fPML, DARAM and PARTICLE-VLS in 91.30%, 85.22%, 80.87% and 85.22% cases, respectively. PML-LFC also significantly outperforms two classical MLL approaches RankSVM and ML-KNN in 92.17% and 88.70% cases, respectively. Which proves the necessity of accounting for irrelevant labels of PML training data. PML-LFC outperforms PML-LRS in most cases because PML-LRS mainly operates in the label space. fPML is similar to PML-LRS, while it uses feature information to guide the low-rank label matrix approximation. As a result, it sometimes obtains similar results as PML-LFC. PML-LFC also performs better than DARAM and PARTICLE-VLS, which mainly use information from the feature space. Another cause for the superiority of PML-LFC is that other comparing methods do not make a concrete use of the negative information between the label and feature space. From these results, we can draw a conclusion that PML-LFC well accounts the negative information between features and labels for effective partial multi-label learning.

Table 2. Experiment results on different datasets with noisy labels (50%). •/◦ indicates whether PML-LFC is statistically (according to pairwise t-test at 95% significance level) superior/inferior to the other method.

Metric	RankSVM	ML-KNN	PML-LRS	fPML	DARAM	PARTICLE-VLS	PML-LFC
	slashdot						
HammLoss	0.078±0.005•	0.184±0.006•	0.044±0.000◦	0.043±0.000◦	0.052±0.000◦	0.053±0.001◦	0.073±0.000
RankLoss	0.161±0.002•	0.053±0.000◦	0.153±0.006•	0.127±0.006•	0.118±0.000•	0.305±0.032•	0.110±0.007
OneError	0.534±0.005•	0.680±0.015•	0.446±0.012•	0.480±0.013•	0.404±0.001•	0.769±0.074•	0.393±0.028
Coverage	0.182±0.022•	0.120±0.006◦	0.165±0.007•	0.139±0.003•	0.133±0.001•	0.305±0.031•	0.128±0.007
AvgPrec	0.582±0.017•	0.472±0.011•	0.639±0.007•	0.627±0.007•	0.686±0.001•	0.375±0.036•	0.696±0.010
	scene						
HammLoss	0.272±0.012•	0.110±0.013◦	0.148±0.005•	0.167±0.001•	0.121±0.000◦	0.123±0.017◦	0.146±0.003
RankLoss	0.259±0.015•	0.097±0.008•	0.124±0.011•	0.145±0.005•	0.118±0.002•	0.110±0.020•	0.094±0.004
OneError	0.553±0.009•	0.260±0.009•	0.314±0.027•	0.362±0.009•	0.265±0.003•	0.251±0.044◦	0.258±0.007
Coverage	0.232±0.017•	0.109±0.011•	0.118±0.009•	0.136±0.006•	0.114±0.001•	0.097±0.018•	0.093±0.002
AvgPrec	0.635±0.038•	0.838±0.016•	0.804±0.016•	0.774±0.005•	0.830±0.001•	0.828±0.033•	0.843±0.005
	enorn						
HammLoss	0.109±0.006•	0.108±0.006•	0.060±0.001•	0.104±0.002•	0.068±0.001•	0.064±0.008•	0.051±0.001
RankLoss	0.189±0.037•	0.054±0.000◦	0.145±0.009•	0.197±0.009•	0.143±0.002•	0.238±0.037•	0.099±0.008
OneError	0.476±0.047•	0.323±0.032•	0.326±0.036•	0.416±0.030•	0.260±0.004•	0.453±0.102•	0.254±0.013
Coverage	0.481±0.038•	0.285±0.005•	0.369±0.014•	0.331±0.016•	0.354±0.002•	0.451±0.071•	0.284±0.010
AvgPrec	0.504±0.053•	0.611±0.019•	0.613±0.015•	0.659±0.008•	0.613±0.002•	0.466±0.088•	0.683±0.009
	medical						
HammLoss	0.482±0.008•	0.070±0.009•	0.343±0.034•	0.022±0.002◦	0.015±0.000◦	0.021±0.001◦	0.024±0.000
RankLoss	0.018±0.003◦	0.042±0.006•	0.075±0.027•	0.046±0.005•	0.036±0.003	0.113±0.021•	0.036±0.005
OneError	0.169±0.004◦	0.270±0.020•	0.420±0.013•	0.216±0.008•	0.193±0.008◦	0.220±0.082•	0.199±0.013
Coverage	0.276±0.025•	0.095±0.011•	0.114±0.027•	0.065±0.010•	0.058±0.001•	0.116±0.020•	0.052±0.009
AvgPrec	0.854±0.024◦	0.766±0.015•	0.665±0.018•	0.831±0.007•	0.839±0.007◦	0.730±0.022•	0.834±0.012
	Corel5k						
HammLoss	0.081±0.007•	0.161±0.005•	0.051±0.009•	0.010±0.000	0.554±0.000•	0.019±0.000•	0.010±0.000
RankLoss	0.281±0.006•	0.134±0.000•	0.063±0.005◦	0.210±0.008•	0.277±0.001•	0.326±0.056•	0.120±0.006
OneError	0.802±0.007•	0.740±0.010•	0.639±0.017•	0.649±0.008•	0.801±0.002•	0.855±0.073•	0.631±0.010
Coverage	0.391±0.007•	0.372±0.010•	0.403±0.007•	0.470±0.017•	0.539±0.003•	0.547±0.041•	0.281±0.013
AvgPrec	0.292±0.008•	0.230±0.003•	0.393±0.006◦	0.286±0.005•	0.199±0.008•	0.144±0.052•	0.312±0.002
	YeastBP						
HammLoss	0.052±0.004•	0.316±0.005•	0.329±0.012•	0.071±0.004•	0.062±0.000•	–	0.024±0.000
RankLoss	0.193±0.007•	0.025±0.000◦	0.331±0.007•	0.208±0.009•	0.161±0.000•	–	0.143±0.002
OneError	0.643±0.007•	0.757±0.008•	0.743±0.013•	0.682±0.004•	0.796±0.002•	–	0.523±0.013
Coverage	0.308±0.005•	0.407±0.010•	0.374±0.008•	0.312±0.005•	0.295±0.002•	–	0.281±0.012
AvgPrec	0.389±0.007•	0.232±0.007•	0.242±0.011•	0.394±0.012•	0.214±0.001•	–	0.411±0.012
	YeastCC						
HammLoss	0.046±0.008•	0.318±0.016•	0.351±0.012•	0.093±0.005•	0.071±0.000•	–	0.027±0.000
RankLoss	0.188±0.004•	0.026±0.000◦	0.308±0.009•	0.179±0.007•	0.178±0.000•	–	0.173±0.008
OneError	0.555±0.004•	0.639±0.018•	0.658±0.014•	0.524±0.007•	0.832±0.003•	–	0.448±0.014
Coverage	0.117±0.009•	0.173±0.010•	0.150±0.007•	0.112±0.004•	0.111±0.002•	–	0.103±0.003
AvgPrec	0.516±0.010•	0.398±0.018•	0.386±0.012•	0.535±0.009•	0.193±0.002•	–	0.590±0.014
	YeastMF						
HammLoss	0.055±0.005•	0.338±0.004•	0.348±0.004•	0.044±.008•	0.077±0.001•	–	0.026±0.000
RankLoss	0.253±0.009•	0.025±0.000◦	0.386±0.008•	0.269±0.006•	0.251±0.000•	–	0.243±0.012
OneError	0.681±0.010•	0.785±0.005•	0.761±0.012•	0.693±0.009•	0.878±0.001•	–	0.661±0.017
Coverage	0.123±0.008•	0.172±0.006•	0.168±0.007•	0.124±0.003•	0.137±0.001•	–	0.121±0.005
AvgPrec	0.421±0.008•	0.330±0.006•	0.302±0.010•	0.442±0.009•	0.160±0.000•	–	0.457±0.011

4.3 Further Analysis

We perform ablation study to further study the effectiveness of PML-LFC. For this purpose, we introduce three variants of PML-LFC, namely, PML-LMF(oF),

Table 3. Win/Tie/Lose counts (pairwise t-test at 95% signification level) of PML-LFC against each other comparing algorithm with different ratios of noisy labels {10%, 50%, 100%, 200%} on different datasets across five evaluation criteria.

Metric	PML-LFC against					
	RankSVM	ML-KNN	PML-LRS	fPML	DARAM	PARTICLE-VLS
HammLoss	21/0/2	17/2/4	18/2/3	16/3/4	16/2/5	15/3/5
RankLoss	20/1/2	16/2/5	22/1/0	19/1/3	18/2/3	22/0/1
OneError	22/0/1	23/0/0	23/0/0	21/0/2	19/0/4	20/0/3
Coverage	21/1/1	23/0/0	22/0/1	23/0/0	21/1/1	20/0/3
AvgPrec	22/0/1	23/0/0	20/1/2	19/1/3	19/0/4	21/0/2
Total (Win/Tie/Lose)	106/2/7	102/4/9	105/4/6	98/5/12	93/5/17	98/3/14

PML-LFC(oL) and PML-LFC(nJ). PML-LFC(oF) only uses feature similarity, PML-LFC(oL) only utilizes the semantic similarity. PML-LFC(nJ) does not jointly optimize the latent label matrix and the predictor in a unified objective function, it firstly optimizes the latent label matrix and then the multi-label predictor. Figure 3 shows the results of these variants and PML-LFC on the *slashdot* dataset. All the experimental settings are the same as previous section.

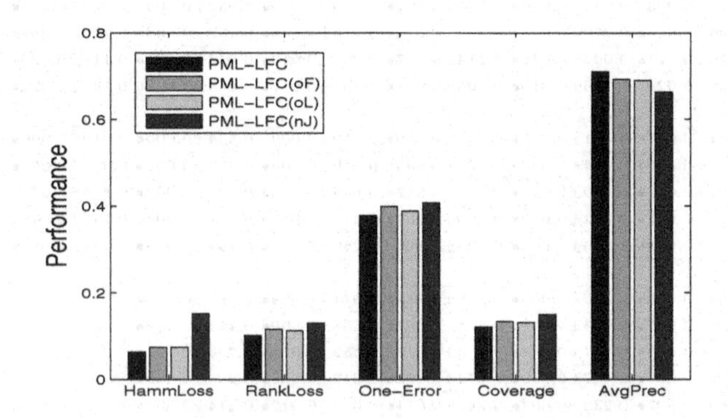

Fig. 3. The performance of PML-LFC and its degenerated variants on the *slashdot* dataset. For the first four evaluation metrics, the lower the value, the better the performance is. For AvgPrec, the higher the value, the better the performance is.

From Fig. 3, we can find that PML-LFC has the lowest HammLoss, RankLoss, One-Error, Coverage, and the highest AvgPrec among the four comparing methods. Neither the feature similarity nor the semantic similarity alone induces a comparable multi-label predictor with PML-LFC. In addition, PML-LFC(oF) and PML-LFC(oL) have similar performance with each other, which indicate that both the feature and label information can be used to induce a multi-label predictor. PML-LFC leverages both the label and feature information, it induces a less error-prone multi-label classifier and achieves a better classification performance than these two variants. PML-LFC(nJ) has the lowest performance across the five evaluation metrics, which corroborates the disadvantage of isolating the confident matrix learning and multi-label predictor training. This study further confirms that both the feature and label information of multi-label data should be appropriately used for effective partial multi-label learning, and our alternative optimization procedure has a reciprocal reinforce effect for the predictor and the latent label matrix.

To investigate the sensitivity of α and β, we vary α and β in the range of $\{0.001, 0.01, 0.1, 1, 10, 100\}$ for PML-LFC on the medical dataset. The experimental results (measured by the five evaluation metrics) are shown in Fig. 4. The results on other datasets give similar observations. From Fig. 4(a), we can observe that, when $\alpha = 10$, PML-LFC achieves the best performance. This observation suggests that it's necessary to consider the low-rank label correlation for partial multi-label learning. When α is too large or too small, the label correlation is underweighted or overweighted, thus the performance manifests a reduce. From Fig. 4(b), we can see that PML-LFC achieves the best performance when $\beta = 10$. When β is too small, the feature similarity and semantic similarity of multi-label instances are not well accounted, which leads to a poor performance. When β is too large (i.e., 100), PML-LFC also achieves a poor performance, as it excessively overweights the feature similarity and semantic similarity, but underweights the prediction model. From this analysis, we adopt $\alpha = 10$ and $\beta = 10$ for experiments.

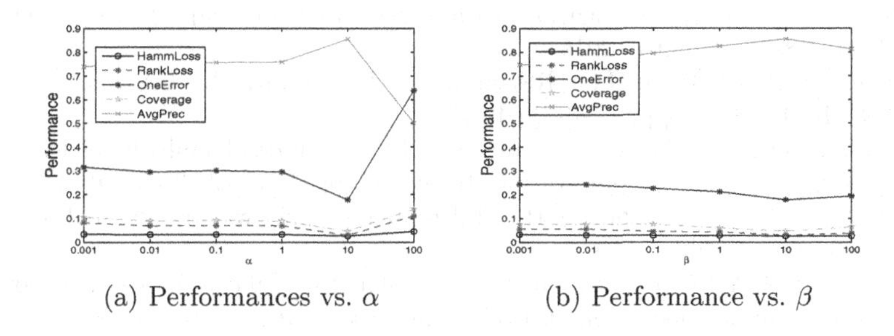

(a) Performances vs. α (b) Performance vs. β

Fig. 4. Results of PML-LFC under different input values of α and β.

5 Conclusions

We investigated the partial multi-label learning problem and proposed an approach called PML-LFC, which leverages the feature and label information for effective multi-label classification. PML-LFC takes into account the negative information between labels and features of partial multi-label data. Extensive experiments on PML datasets from different domains demonstrate the effectiveness of PML-LFC. We are planning to incorporate the abundant unlabeled data for effective extreme partial multi-label learning with a large label space.

References

1. Belkin, M., Niyogi, P., Sindhwani, V.: Manifold regularization: a geometric framework for learning from labeled and unlabeled examples. JMLR **7**(11), 2399–2434 (2006)
2. Boyd, S., Vandenberghe, L.: Convex Optimization. Cambridge University Press, Cambridge (2004)
3. Cour, T., Sapp, B., Taskar, B.: Learning from partial labels. JMLR **12**(5), 1501–1536 (2011)
4. Elisseeff, A., Weston, J.: A kernel method for multi-labelled classification. In: NeurIPS, pp. 681–687 (2002)
5. Fang, J.P., Zhang, M.L.: Partial multi-label learning via credible label elicitation. In: AAAI, pp. 3518–3525 (2019)
6. Gibaja, E., Ventura, S.: A tutorial on multilabel learning. ACM Comput. Surv. **47**(3), 52 (2015)
7. Han, Y., Sun, G., Shen, Y., Zhang, X.: Multi-label learning with highly incomplete data via collaborative embedding. In: KDD, pp. 1494–1503 (2018)
8. Ji, S., Ye, J.: An accelerated gradient method for trace norm minimization. In: ICML, pp. 457–464 (2009)
9. Li, S.Y., Jiang, Y., Chawla, N.V., Zhou, Z.H.: Multi-label learning from crowds. TKDE **31**(7), 1369–1382 (2019)
10. Li, Y.F., Hu, J.H., Jiang, Y., Zhou, Z.H.: Towards discovering what patterns trigger what labels. In: AAAI, pp. 1012–1018 (2012)
11. Liu, T., Tao, D.: Classification with noisy labels by importance reweighting. TPAMI **38**(3), 447–461 (2016)
12. Natarajan, N., Dhillon, I.S., Ravikumar, P.K., Tewari, A.: Learning with noisy labels. In: NeurIPS, pp. 1196–1204 (2013)
13. Sun, L., Feng, S., Wang, T., Lang, C., Jin, Y.: Partial multi-label learning by low-rank and sparse decomposition. In: AAAI, pp. 5016–5023 (2019)
14. Sun, Y.Y., Zhang, Y., Zhou, Z.H.: Multi-label learning with weak label. In: AAAI, pp. 593–598 (2010)
15. Tan, Q., Liu, Y., Chen, X., Yu, G.: Multi-label classification based on low rank representation for image annotation. Remote Sens. **9**(2), 109 (2017)
16. Tan, Q., Yu, G., Domeniconi, C., Wang, J., Zhang, Z.: Incomplete multi-view weak-label learning. In: IJCAI, pp. 2703–2709 (2018)
17. Tu, J., Yu, G., Domeniconi, C., Wang, J., Xiao, G., Guo, M.: Multi-label answer aggregation based on joint matrix factorization. In: ICDM, pp. 517–526 (2018)
18. Wang, C., Yan, S., Zhang, L., Zhang, H.J.: Multi-label sparse coding for automatic image annotation. In: CVPR, pp. 1643–1650 (2009)
19. Wang, H., Liu, W., Zhao, Y., Zhang, C., Hu, T., Chen, G.: Discriminative and correlative partial multi-label learning. In: IJCAI, pp. 2703–2709 (2019)
20. Wu, B., Jia, F., Liu, W., Ghanem, B., Lyu, S.: Multi-label learning with missing labels using mixed dependency graphs. IJCV **126**(8), 875–896 (2018)
21. Xie, M.K., Huang, S.J.: Partial multi-label learning. In: AAAI, pp. 4302–4309 (2018)
22. Xu, L., Wang, Z., Shen, Z., Wang, Y., Chen, E.: Learning low-rank label correlations for multi-label classification with missing labels. In: ICDM, pp. 1067–1072 (2014)
23. Yu, G., et al.: Feature-induced partial multi-label learning. In: ICDM, pp. 1398–1403 (2018)

24. Yu, G., Fu, G., Wang, J., Zhu, H.: Predicting protein function via semantic integration of multiple networks. TCBB **13**(2), 220–232 (2016)
25. Zhang, J., Wu, X.: Multi-label inference for crowdsourcing. In: KDD, pp. 2738–2747 (2018)
26. Zhang, M.L., Yu, F., Tang, C.Z.: Disambiguation-free partial label learning. TKDE **29**(10), 2155–2167 (2017)
27. Zhang, M.L., Zhang, K.: Multi-label learning by exploiting label dependency. In: KDD, pp. 999–1008 (2010)
28. Zhang, M.L., Zhou, Z.H.: ML-KNN: a lazy learning approach to multi-label learning. Pattern Recogn. **40**(7), 2038–2048 (2007)
29. Zhang, M.L., Zhou, Z.H.: A review on multi-label learning algorithms. TKDE **26**(8), 1819–1837 (2014)
30. Zhang, Y., Zhou, Z.H.: Multilabel dimensionality reduction via dependence maximization. TKDD **4**(3), 14 (2010)
31. Zhu, Y., Kwok, J.T., Zhou, Z.H.: Multi-label learning with global and local label correlation. TKDE **30**(6), 1081–1094 (2017)

Optimal Trade Execution Based on Deep Deterministic Policy Gradient

Zekun Ye[1], Weijie Deng[1], Shuigeng Zhou[1(✉)], Yi Xu[1], and Jihong Guan[2]

[1] Shanghai Key Lab of Intelligent Information Processing, and School of Computer Science, Fudan University, Shanghai 200433, China
{zkye16,14307110195,sgzhou,yxu17}@fudan.edu.cn

[2] Department of Computer Science and Technology, Tongji University, 4800 Caoan Road, Shanghai 201804, China
jhguan@tongji.edu.cn

Abstract. In this paper, we address the *Optimal Trade Execution* (OTE) problem over the limit order book mechanism, which is about how best to trade a given block of shares at minimal cost or for maximal return. To this end, we propose a *deep reinforcement learning* based solution. Though reinforcement learning has been applied to the OTE problem, this paper is the first work that explores deep reinforcement learning and achieves state of the art performance. Concretely, we develop a *deep deterministic policy gradient* framework that can effectively exploit comprehensive features of multiple periods of the real and volatile market. Experiments on three real market datasets show that the proposed approach significantly outperforms the existing methods, including the *Submit & Leave* (SL) policy (as baseline), the Q-learning algorithm, and the latest hybrid method that combines the Almgren-Chriss model and reinforcement learning.

1 Introduction

In quantitative finance area, *optimal trade execution* (OTE), also called *optimal liquidation*, is a critical issue in many investment activities [7,16,19,24]. It is defined as how to sell a certain number of shares of a given stock within a fixed time while maximizing the revenue, or how to buy a certain number of shares within a fixed time so that the trade cost is minimized. Here, the cost is usually measured by the adverse deviation of actual transaction prices from an arrival price baseline, which can be interpreted as Perold's *implementation shortfall* [18].

In addition to its significance in financial industry, OTE has also attracted much research interest of the academia. Bertsimas and Lo are the pioneers of this problem, they treated it as a stochastic dynamic programming problem, and proposed a dynamic optimization method [4]. Later, an influential work is the Almgren-Chriss model [2], which gives a closed-form solution for a trading trajectory. Following that, some works employed *reinforcement learning* (RL) to tackle the OTE problem [8,15,20]; Some other works tried to model the limit order book dynamics as a Markov process [5,6,9], In addition, there are also

Y. Nah et al. (Eds.): DASFAA 2020, LNCS 12112, pp. 638–654, 2020.
https://doi.org/10.1007/978-3-030-59410-7_42

research works that predict future parameters of the limit order book [16,17]. A recent work [1] proposed an online learning method based on the Almgren-Chriss model.

This paper focuses on RL based solutions for the OTE problem as RL has achieved great success in many areas, especially when it is combined with deep learning. RL is mainly based on Markov Decision Process (MDP) [3], it tries to learn a policy that maps states to optimal actions. [12] presents an overview of reinforcement learning applied to market microstructure and high-frequency trading. Comparing to dynamic programming, RL has obvious advantages: the agents can learn online and adapt constantly while performing the given task. Furthermore, function approximation can be employed to represent knowledge.

Up to now, there are two works that applied RL to the OTE problem. In [15], the authors proposed a modified Q-learning method where elapsed time and remaining inventory were used as states. And in [8], a hybrid method was introduced, which first uses the Almgren-Chriss model to solve the optimization problem for predicting the volume to be traded in each time slot, then adopts Q-learning to find the best actions. These two works have the following common drawbacks: 1) The remaining inventory—an important state component, is discretized into a few fixed intervals, which simplifies computation but also degrades model accuracy. 2) They all consider only the information of a single time point, which cannot comprehensively exploit the temporal correlation and dynamics of the real market.

To overcome the drawbacks of the above RL based approaches and further boost the performance of OTE, here we develop a deep reinforcement learning (DRL) solution for the OTE problem. Concretely, we propose a new framework based on deep deterministic policy gradient (DDPG) [14].

DDPG is a combination of policy gradient and Q-learning. The core idea of Q-learning is to maximum the value function as follows:

$$a^* = argmax_a Q(s,a), a^* = \pi^*(s) \tag{1}$$

where a, s, $Q(s,a)$ and π indicate *action*, *state*, the *value function*, and *policy*, respectively. While policy gradient (PG) [22] adopts another way of thinking, directly maximizes the expectation of long-term return:

$$\pi^* = argmax_\pi E_{\tau \sim \pi(\tau)}[r(\tau)] \tag{2}$$

where τ is a trajectory obtained when using strategy π, and $r(\tau)$ represents the overall return of this trajectory. DDPG uses a neural network as its function approximation based on the Deterministic Policy Gradient Algorithm [21], which maximizes the expectation of long-term return and the Q-value at the same time.

In our framework, the feature extraction network consists of two branches: one takes the current market information as input, the other uses the order book data of several consecutive time slots preceding the current time point as input. Features from the two network branches are combined as input of the training process. Therefore, our framework exploits both current and historical market information and consequently boost performance substantially. Furthermore, our framework can be efficiently trained in end-to-end style.

Note that in recent years, there are some works [10,13,23] that apply reinforcement learning for portfolio selection in the financial field. Especially, in [23], the authors used DDPG in the portfolio selection problem. However, we should point out that the problem of portfolio selection is completely different from the OTE problem. The aim of OTE is about how to sell/buy a certain number of shares of a given stock within a fixed time while maximizing the revenue. Here, "within a fixed time" often means a very short period of time (2 min in our paper) and the operation is preformed over the limit order book mechanism. While the goal of portfolio selection is to pursue a long-term (say a month, several months or several years) total income over a number of rounds of selling/buying. Furthermore, in portfolio selection the low-frequency data collected daily are processed, while the OTE problem handles high-frequency data collected in seconds, which faces more challenges.

Contributions of this paper are as follows:

1. We propose a deep reinforcement learning based solution for the OTE problem, which is the first of such work.
2. We design a deep deterministic policy gradient based framework where both current and historical market information are exploited for model training.
3. We conduct extensive experiments on three real market datasets, which show that our method substantially outperforms the existing methods.

2 Preliminaries

Here, we briefly introduce some basic concepts, including limit order trading and market simulation, which are closely related to the OTE problem.

In the context of OTE, trades are preformed over the limit order book mechanism. The traders can specify the volume and a limit on the price of shares that they are desired to sell/buy. The selling side is called the *ask side* and the buying side is called the *bid side*. A *limit order* means that both volume and price are prespecified. An order on one side is executed only when the order matches a previously-submitted or newly-arrived order on the other side. Another type of orders is the *market order*, where the traders specify only the volume, and the orders are executed against the best offers available on the other side.

Considering a trading episode, we want to sell (we can also say to buy. If not explicitly stated, selling is the default case in this paper) V shares of a stock within a time period H. We divide H into L time slots of equal size. The time duration of each slot can be several seconds or minutes. At the beginning of each time slot, the trader submits a limit order to specify the price and volume for selling. The trader gets r_i turnover for selling h_i shares at the end of each time slot. After L time slots (at the end of the episode), the remaining inventory must be submitted as a market order. Suppose the market order obtains r_f turnover, then the total turnover is $(\sum_{i=1}^{i=L} r_i) + r_f$. Note that the trader cannot compute

Table 1. A toy example of limit order book.

Buy orders		Sell orders	
Shares	Price	Shares	Price
700	26.19	800	26.25
300	26.17	1100	26.31
1000	26.15	600	26.32
1500	26.14	500	26.32
100	26.06	900	26.45
300	25.98	2000	26.58
400	25.89	400	26.61
300	25.84	100	26.65
500	25.55	300	26.73
600	25.09	900	27.01

the optimal trading strategy beforehand, because s/he has not the information of market dynamics, and the distribution of future bid prices.

In what follows, we introduce the detailed transaction process in a trading time slot. A toy example of limit order book is shown in Table 1. Suppose it is at the beginning of time slot t. The left two columns show the buy orders, which are ordered by price, with the highest price at the top, denoted as $bidp_1^t$, the corresponding volume is denoted as $bidv_1^t$. Similarly, we denote the k-th order in the buy orders as $bidp_k^t$ with volume $bidv_k^t$. The right two columns are the sell orders, which are also ordered by price, but with the lowest price at the top. We denote the k-th order in the ask orders as $askp_k^t$ with volume $askv_k^t$. We call $(askp_1^t - bidp_1^t)$ the *spread*, and $(askp_1^t + bidp_1^t)/2$ the *mid-price* of time-slot t.

If the price of a trader's order is higher than the highest price in the buy orders, then the trader is a *passive trader*, who has to wait for other buy orders. For example, say we submit a limit sell order with price \$26.45 and volume 1000. This order will be placed in the sell order, immediately after the existing order of 900 shares at \$26.45. Though their prices are the same, that order was submitted earlier than ours.

Otherwise, if the price of a trader's order is not higher than the highest price in the buy orders, s/he is an *active trader*, and the order will get executed immediately. For example, say we submit a limit sell order with price \$26.16 and volume 1500, it will be executed with the limit buy orders of 700 shares at \$26.19, and 300 shares at \$26.17. Thus a total of 1000 shares are executed. The remaining (unexecuted) 500 shares of our selling order turns to a new $askp_1$ at \$26.16.

An assumption in this work is that the trading activity does not affect the market attributes. In our settings, a trading episode is two minutes, and a trading time slot is only several seconds, so the impact of trading activity is trivial for the resilient market. Such assumption is also widely used in existing works [8,15,20]. The validity of this assumption will be further examined in our future research.

3 Methodology

In this section, we first briefly introduce *policy gradient* (PG), then present the framework of our method. Following that, we describe the technical details of our method, including the setting of *Action*, *State* and *Reward*, the *feature extraction network* (FEN), and the model training process.

3.1 Policy Gradient

The goal of *policy gradient* is to maximize the expectation of long-term return. Let us use $J(\theta)$ to represent the objective function as follows:

$$J(\theta) = E_{\tau\sim\pi(\tau))}[r(\tau)] = \int_{\tau\sim\pi(\tau)} \pi_\theta(\tau)r(\tau)d\tau \tag{3}$$

Then, we take the derivative of the objective function and have

$$\begin{aligned} \nabla_\theta J(\theta) &= \int_{\tau\sim\pi(\tau)} \pi_\theta(\tau)\nabla_\theta log\pi_\theta(\tau)r(\tau)d\tau \\ &= E_{\tau\sim\pi(\tau)}[\nabla_\theta log\pi_\theta(\tau)r(\tau)] \end{aligned} \tag{4}$$

Assume the time horizon is finite, and the total length of the trajectory is T, we can write the policy in product form as follows:

$$\begin{aligned} \pi(\tau) &= \pi(s_0, a_0, s_1, a_1, ..., s_T, a_T) \\ &= p(s_0)\prod_{t=0}^{T} \pi_\theta(a_t|s_t)p(s_{t+1}|s_t, a_t) \end{aligned} \tag{5}$$

We take its derivative and obtain:

$$\begin{aligned} \nabla_\theta log(\pi(\tau)) &= \nabla_\theta log[p(s_0)\prod_{t=0}^{T} \pi_\theta(a_t|s_t)p(s_{t+1}|s_t, a_t)] \\ &= \nabla_\theta[logp(s_0)) + \sum_{t=0}^{T} log(\pi_\theta(a_t|s_t)) \\ &+ \sum_{t=0}^{T} log(p(s_{t+1}|s_t, a_t))] \\ &= \sum_{t=0}^{T} \nabla_\theta log(\pi_\theta(a_t|s_t)) \end{aligned} \tag{6}$$

We substitute Eq. (6) into Eq. (4), and get

$$\nabla_\theta J(\theta) = E_{\tau\sim\pi(\tau)}[(\sum_{t=0}^{T} \nabla_\theta log\pi_\theta(a_t|s_t))(\sum_{t=0}^{T} r(s_t, a_t))] \tag{7}$$

When the solution of the gradient is obtained, the rest is to update the parameters. So PG can summarized into the following two steps:

- calculate $\nabla_\theta J(\theta)$,
- update $\theta = \theta + \alpha\nabla_\theta J(\theta)$.

3.2 The Framework

In this paper, we try to combine *reinforcement learning* (RL) and state-of-the-art *deep learning* (DL) to solve the OTE problem. In RL, the agent interacts with the environment with little prior information and learns from the environment by trial-and-error while improving its strategy. Its little requirement for modeling and feature engineering is suitable for dealing with complex financial markets. On the other hand, DL has witnessed rapid progress in speech recognition and image identification. Its outperforming over conventional machine learning methods has proven its outstanding capability to capture complex, non-linear patterns. To the best of our knowledge, this work is the first to employ deep reinforcement learning method to solve the OTE problem.

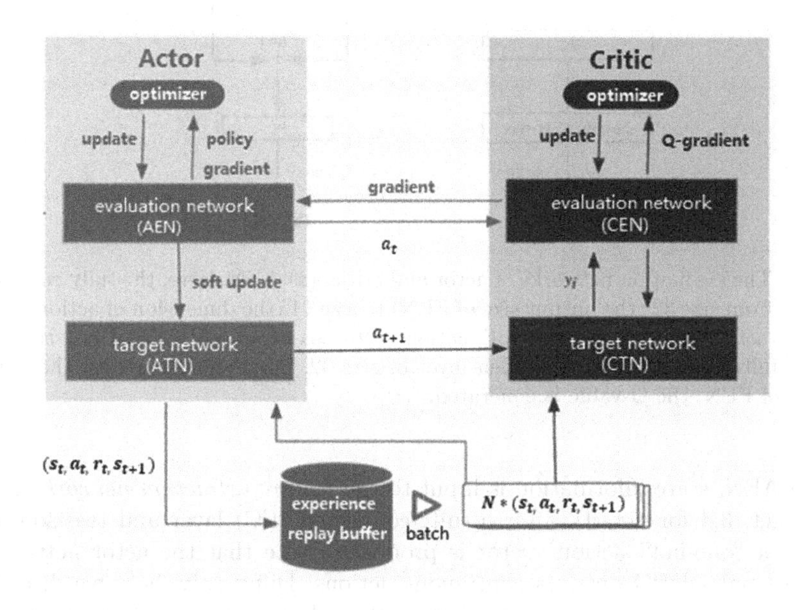

Fig. 1. The DDPG based OTE framework.

Figure 1 shows the framework of Deep Deterministic Policy Gradient (DDPG) based OTE. DDPG is a deep RL algorithm that concurrently learns a Q-function and a policy. It uses off-policy data and the Bellman equation to learn the Q-function, and uses the Q-function to learn the policy. The framework has two roles: the *actor* and the *critic*, and consists of four corresponding networks: the *actor evaluation network* (AEN), the *critic evaluation network* (CEN), the *actor target network* (ATN) and the *critic target network* (CTN). The actor takes the state as input, and outputs the action. The critic takes both action and state as input, and outputs the Q-value. The actor produces a tuple (s_t, a_t, r_t, s_{t+1}) when interacting with the environment at time t, where s_t and s_{t+1} are the current state and the next state, a_t is the action, and r_t is the corresponding reward. This tuple is put into an experience replay buffer. A mini-batch of the replay buffer is used to train the actor and critic networks when the replay buffer is full. Figure 2 shows the evaluation networks of the actor and the critic. Their target networks have similar structure to that of the evaluation networks.

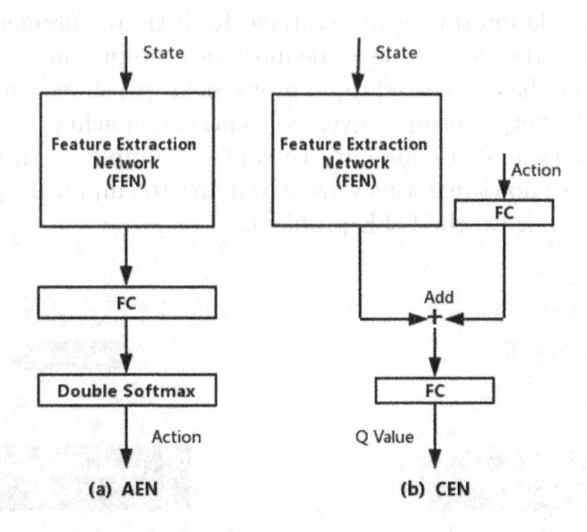

Fig. 2. The evaluation networks of actor and critic. (a) AEN. Here, the fully connected layer is from size 32 (the output size of FEN) to size 11 (the dimension of action). After the two soft-max layers, a "one-hot" action vector is output. (b) CEN. Here, the input action fully connects with a hidden layer of size 32, in combination with the output vector of FEN, the Q-value is generated.

In AEN, state information is input to the *feature extraction network* (FEN) (see Sect. 3.4 for detail), after a fully-connected (FC) layer and two soft-max layers, a "one-hot" action vector is produced. Note that the actor network in the original DDPG outputs continuous actions. Differently, here we discretize actions for fast convergence (see Sect. 3.3), and employ two soft-max layers to approximate the output to a one-hot vector. If arg-max, instead of soft-max is used, the objective function will be non-differentiable.

In CEN, the input consists of state and action, which pass a FEN module and a FC layer respectively, their outputs are combined and then injected into a FC layer, finally the Q value is gotten. Note that the two FENs for actor and critic are totally independent, i.e., they do not share any parameter.

The target networks are not trainable, "soft" updated by the evaluation networks using exponential weighted average. This means that the target values are constrained to change slowly, which can greatly improve the stability of learning and the robustness of the networks.

3.3 State, Action and Reward

State. Unlike existing works [8,15] using only the current time information as state, we combine both historical information and the current time information as state.

Concretely, the state consists of two parts. The first part is a three-dimensional tensor with shape (c, k, w), where $c=4$ is the number of channels, each of which corresponds to a moving window over the data of ask price ($askp$), ask volume ($askv$), bid price ($bidp$) and bid volume ($bidv$), respectively. k and w stand for the height and width of the moving windows. Here, we set $k=5$ and $w=10$, that is, we use the top-5 price values and corresponding volume values in the limit order book across 10 time points: the current time point t, and $t-1$, ..., $t-9$. For convenience, we denote the tensor as *PV* tensor. The second part is a vector consisting of the following information: current time t, remaining volume (RV), $askp_1^t$, $askv_1^t$, current *mid-price* (MP), the average price (AP), the ask price submitted last time (LAP) and its rank in the order book (LR). Figure 3 shows the information composition of state.

We use the *PV* tensor to provide temporal correlation and dynamics information of the market, while the vector provides the latest market information. Since the temporal correlation and dynamics in the *PV* tensor are most informative for OTE, we normalize the data of the four channels by the top-1 prices and corresponding volumes, i.e., $askp_1^t$ and $askv_1^t$, $bidp_1^t$ and $bidv_1^t$, respectively. This also contributes to the convergence of our algorithm.

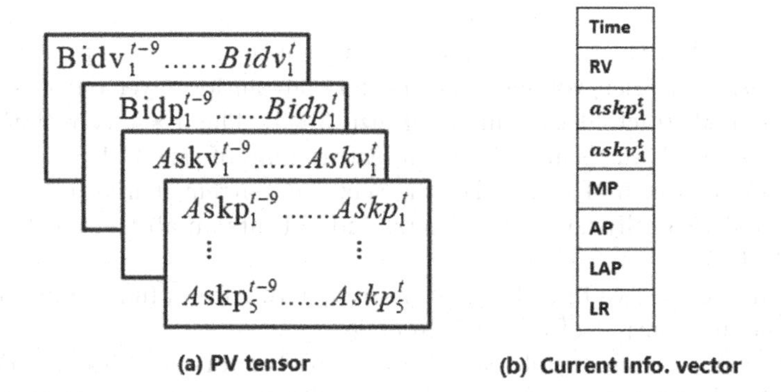

Fig. 3. Illustration of state information composition. (a) The *PV* tensor containing historical data; (b) A vector of combined market features of current time.

Action. As in [15], our action is to determine the price of a new limit order based on $askp_1^t$, and reposition all remaining inventory. We can effectively withdraw any previous outstanding limit order and replace it with the new limit order (or leave it alone if their prices are equal), which is indeed supported by the actual exchange institutes. Because the unit of price is 0.01, we discretize Action a into $-0.05, -0.04... - 0.01, 0, 0.01, ..., 0.05$. So $Action_t = a$ corresponds to the price $askp_1^t + a$. A negative a corresponds to crossing the *mid-price* towards the buyers, and a positive a corresponds to placing the new order in sell book.

Reward. As the market is dynamic, simply setting reward as the turnover cannot reflect the real situation of the market, so we define the reward at time t as follows:

$$reward_t = r_t - h_t * MP_0 \tag{8}$$

where MP_0 is the *mid-price* at the beginning of the whole episode, also as the baseline price. When we get r_t turnover for selling h_t shares by the end of time slot t, the reward $reward_t$ is the difference between r_t and the turnover at the baseline price. Since we submit a market order in the end, we have $\sum_i^L h_t + h_f = V$. Finally, we compute the total reward by the end of the episode, which is denoted as *implementation reward* (*IR*). We also evaluate the *averaged IR* (*AIR*) over the amount of executed volume. Formally, we have

$$IR = \sum_t^L r_t + r_f - V * MP_0 = \sum_t^L reward_t + reward_f, \tag{9}$$

$$AIR = IR/V. \tag{10}$$

Actually, these two indexes are natural measures of OTE performance.

3.4 Feature Extraction Network (FEN)

In what follows, we introduce the feature extraction network (FEN). FEN consists of two independent input branches, as shown in Fig. 4.

The upper branch of FEN consists of three convolution layers, and takes the *PV* tensor as input. The first convolution layer outputs 2 feature maps of size $3*8$, which reflect the temporal characteristics of the market and the correlation between ask/bid price/volume. The second convolution layer outputs 32 more diverse and abstract feature maps of size $3*1$. The $1*1$ convolution layer compresses the channels and obtains a feature map of size $3*1$.

The lower branch of FEN takes the vector of current market information as input, it has two fully-connected layers, and outputs an abstract feature vector of size $3*1$.

Outputs of the two branches are concatenated. After a fully connected layer, we get the final output of the FEN module.

FEN plays an important role in our DDPG actor-critic framework, which transforms the raw state into the abstract features, combining both historical and current information of the market. Both the actor network and the critic network take the output of their FENs as input.

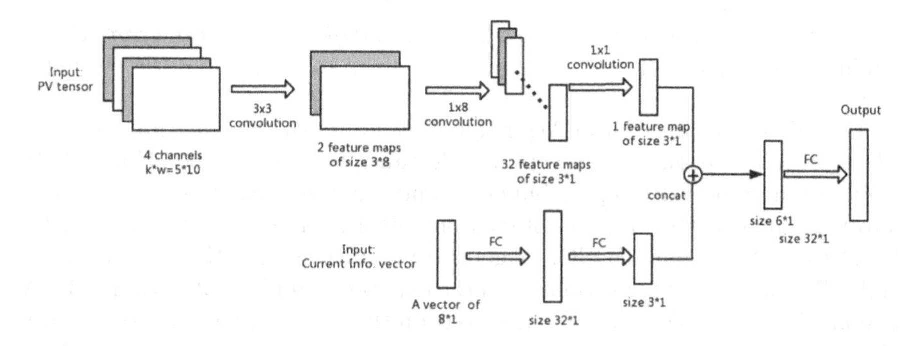

Fig. 4. Architecture of the feature extraction network (FEN). The PV tensor input consists of four channels of size $5 * 10$, after the first convolution with two $3 * 3$ filters, we get $3 * 8 * 2$ feature maps. Then, via 32 filters of $1 * 8$, we get $3 * 1 * 32$ feature maps, which go through a $1 * 1$ filter, we obtain a $3 * 1$ feature map. On the other hand, the vector input goes through two fully connected layers, we also get a $3 * 1$ feature map. By concatenating the two $3 * 1$ feature maps, and after a fully connected layer, we finally get the combined feature as output.

3.5 Training

DDPG is a combination of policy gradient [22] and Q-learning, it uses deep neural network for its function approximation based on the Deterministic Policy Gradient Algorithms [21].

We denote the deterministic policy (the actor network) as $\mu(s|\theta^{\mu})$, the value function (the critic network) as $Q(s,a|\theta^{Q})$, the corresponding target networks as $\mu'(s|\theta^{\mu'})$ and $Q'(s,a|\theta^{Q'})$, respectively. Here, θ^{μ}, θ^{Q}, $\theta^{\mu'}$ and $\theta^{Q'}$ are the parameters of the corresponding networks. The objective function for the actor is

$$J(\theta^{\mu}) = maxE_{\pi}[Q(s_t, \mu(s_t|\theta^{\mu})|\theta^{Q})]. \tag{11}$$

This objective function takes state as input and gets the action by the actor evaluation network. By inputting the state and the action into the critic evaluation network, we get the output Q-value. We maximize the Q-value to train the actor evaluation network with fixed parameters of the critic evaluation network.

The objective function for the critic is

$$J(\theta^{Q}) = minE_{\pi}[\frac{1}{2}(y_t - Q(s_t, a_t|\theta^{Q}))^2], \tag{12}$$

where

$$y_t = reward_t + \gamma Q'(s_{t+1}, \mu'(s_{t+1}|\theta^{\mu'})|\theta^{Q'}). \tag{13}$$

This objective function takes the state and action at time t as input, and gets the Q-value of time t by the critic evaluation network. Meanwhile, with the next state s_{t+1} and the actor target network, we get the next action a_{t+1}. By inputting s_{t+1} and a_{t+1} into the critic target network, we get the Q-value of time $t+1$. γ in Eq. (12) is the reward discount, and $reward_t$ is the reward at time t. The difference in Eq. (12) is also called *temporal-difference error* [11]. We minimize the square of *temporal-difference error* to train the critic evaluation network.

Our algorithm is given in Algorithm 1. Suppose the training process covers M episodes (Line 4), and we replicate K times for each episode (Line 5). We maintain an experience replay buffer R, and put the transition (s_t, a_t, r_t, s_{t+1}) into the replay buffer for each interaction with the environment (Line 11). Different from the original DDPG, we generate a series of transitions in an episode, and collect a batch of transitions from the experience replay buffer (Line 13). We minimize the loss (based on Eq. (12)) to train the critic evaluation network (Line 15), and train the actor evaluation network based on Eq. (11) (Line 16). Then, the target networks of actor and critic are soft updated by their evaluation networks (Line 17). Unlike [8,15] that use backward simulation, which means that their simulations are from the last time-slot to the first time-slot during training, while our method uses forward simulation (simulating from the first time-slot to the last time-slot). This is more consistent with the real situation and is easy to be adapted to online style.

Algorithm 1 DDPG-OTE training

1: Randomly initialize critic network $Q(s,a|\theta^Q)$ and actor network $\mu(s|\theta^\mu)$ with weights θ^Q and θ^μ
2: Initialize target network Q' and μ' with weights $\theta^{Q'} \leftarrow \theta^Q$, $\theta^{\mu'} \leftarrow \theta^\mu$
3: Initialize replay buffer R
4: **for** episode = 1, M **do**
5:   **for** replicate = 1, K **do**
6:     Initialize a random process ψ for action exploration
7:     Receive initial observation state S_0
8:     **for** t = 1, L **do**
9:       Select action $a_t = \mu(s_t|\theta^\mu) + \psi_t$ according to the current policy and exploration noise
10:       Execute action a_t and observe reward r_t and observe new state s_{t+1}.
11:       Store transition (s_t, a_t, r_t, s_{t+1}) in R
12:     **end for**
13:     Sample a random minibatch of N transitions (s_t, a_t, r_t, s_{t+1}) from R
14:     Set $y_i = r_i + \gamma Q'(s_{i+1}, \mu'(s_{i+1}|\theta^{\mu'})|\theta^{Q'})$
15:     Update critic by minimizing the loss:
    $Loss = \frac{1}{N}\sum_i (y_i - Q(s_i, a_i|\theta^Q))^2$
16:     Update the actor policy using the sampled gradient:
    $\nabla_{\theta^\mu} J \simeq \frac{1}{N}\sum_i \nabla_a Q(s,a|\theta^Q)|_{s=s_i,a=\mu(s_i)} \nabla_{\theta^\mu}\mu(s|\theta^\mu)|_{s_i}$
17:     Update the target networks:
    $\theta^{Q'} \leftarrow \tau\theta^Q + (1-\tau)\theta^{Q'}$
18:     $\theta^{\mu'} \leftarrow \tau\theta^\mu + (1-\tau)\theta^{\mu'}$
19:   **end for**
20: **end for**

---

4 Performance Evaluation

4.1 Datasets

We have evaluated the proposed method with real data of several hundreds of stocks from *Shenzhen Security Exchange* (SZSE) and *Shanghai Security Exchange* (SSE). Here we present only the results of three randomly selected stocks to show the effectiveness of our method and its advantage over the existing ones. The three stocks have different liquidities and belong to different fields (including finance, industry and technology), their codes are 600030, 300124 and 000049 respectively, as shown in Table 2. The stock 600030 is included in the *MSCI* indexes, from *Shanghai Security Exchange* (SSE), 300124 and 000049 are from *Shenzhen Security Exchange* (SZSE). Their *daily trading volumes* (DTVs) are about 100M, 10M and 1M. In the experiments, we use 20 months (2017/1–2017/11, 2018/3–2018/11) of historical snapshots of these stocks from the database of *QTG Capital Management Co., Ltd.* Every snapshot includes 10 levels of depth in the order book, and the frequency is 3 s. The data are split to three sets in chronological order for training, validation and testing:

- Training set: 2017/1–2017/11, 2018/3–2018/6;
- Validation set: 2018/7–2018/8;
- Testing set: 2018/9–2018/11.

Table 2. The three stocks used in this paper.

	Stock1	Stock2	Stock3
Code	600030	300124	000049
Name	CITIC Securities	INOVANCE	DESAY
Exchange	SSE	SZSE	SZSE
Field	Finance	Industry	Technology
DTV	100M	10M	1M

4.2 Experimental Settings

In our experiments, the goal is to sell V shares in H minutes. We consider two order sizes: $V = 5000$ and 10000 shares, and set $L = 8$ (the number of time slots in an episode), corresponding to $H = 2$ min. The traders take actions at the beginning of each time slot. So there are 5 snapshots in each time slot (lasting 15 s). The batch size (B) for training is 64, and the *size of replay buff* (SRB) is 512. In our algorithm, the softupdate rate $\tau = 0.01$, the reward discount $\gamma = 0.99$, the number of replicates $K = 8$, and the probability of exploration for taking action is decreased linearly from 0.2 to 0.01. We set the learning rates for the actor and the critic to 10^{-4}.

We use the common policy, *Submit&Leave (SL)* policy as the baseline. The SL policy takes MP_0 as the price of sell limit order with volume of V in the beginning. After L time slots, SL policy submits the market order for any unexecuted shares. In addition, we compare our method with two existing methods. One is the approach proposed in [15], which is based on Q-learning. For short, we denote it by QL; Another represents the state of the art, which combines the AC model and Q-learning [8]. Here, we denote it by AC_QL. For these two methods, we use the optimized parameters suggested in the original papers. For simplicity, we denote our method by DDPG.

4.3 Experimental Results

To compare our method DDPG with AC_QL, QL and SL, we evaluate the mean and standard deviation of AIR over the testing datasets of three stocks. Furthermore, we also compute the *win rate* (*WR*) of DDPG over the other three policies, which indicates the winning probability of DDPG over the other policies. Note that our simulations consider other active traders in the market (which is more close to the real trading market). But, AC_QL submits only market order.

Table 3. The results of μ_{AIR} and σ_{AIR} on 3 stocks. For the two cases $V = 5{,}000$ and 10,000, the best values are in **bold**.

Code	Policy	$V = 5{,}000$		$V = 10{,}000$	
		μ_{AIR}	σ_{AIR}	μ_{AIR}	σ_{AIR}
600030	DDPG	**0.0183**	**0.3516**	**0.0178**	**0.3517**
	AC_QL	0.0085	0.3530	0.0082	0.3530
	QL	0.0167	0.3521	0.0164	0.3521
	SL	0.0137	0.3519	0.0133	0.3519
300124	DDPG	**−0.0017**	**0.0279**	**−0.0055**	**0.0325**
	AC_QL	−0.0203	0.0599	−0.0256	0.0598
	QL	−0.0064	0.0450	−0.0082	0.0421
	SL	−0.0084	0.0305	−0.0121	0.0361
000049	DDPG	**−0.0044**	**0.2178**	**−0.0120**	**0.2196**
	AC_QL	−0.0209	0.2257	−0.0410	0.2275
	QL	−0.0139	0.2244	−0.0230	0.2249
	SL	−0.0103	0.2184	−0.0169	0.2202

Results of AIR. Table 3 shows the results of the mean of *AIR* and the standard deviation of *AIR* of the four methods SL, QL, AC_QL and DDPG on three stocks.

For the mean of AIR, the larger the better; For the standard deviation of AIR, the smaller the better. We can see that our method DDPG outperforms the other methods in terms of the mean of AIR. Notice that in financial market, the unit of order price is 0.01 and consider that the fluctuation of price is usually very limited within several minutes, so the advantage of even 1 unit in averaged price actually means a large performance improvement. It is interesting to notice that as the stock liquidity decreases, QL gradually loses its advantage over SL. Meanwhile, our DDPG keeps the edge over SL with little fluctuation. It indicates that our DDPG is insensitive to stock liquidity and has better adaptability to the dynamic market.

A mature agent is supposed to learn to stop-profit and stop-loss, thus has steady performance and is more resistant to risk. Here, we use the standard deviation of AIR to measure the robustness of different policies. We can see that our method has the smallest standard deviation of AIR among all the four policies, so it is the most robust method, while AC_QL has the largest standard deviation of AIR in all cases. SL has relatively low standard deviation of AIR because it usually has $IR = 0$ when all orders are executed at MP_0. The comparison of the standard deviation of AIR with the existing methods indicates that comprehensively exploiting historical market information is helpful to avoid market risk.

Table 4. The wr of DDPG over the other policies on 3 stocks.

Code	Policy	$V = 5000$	$V = 10000$
600030	DDPG vs AC_QL	77.02%	77.00%
	DDPG vs QL	54.81%	55.32%
	DDPG vs SL	93.10%	90.50%
300124	DDPG vs AC_QL	71.47%	74.80%
	DDPG vs QL	55.30%	63.59%
	DDPG vs SL	77.55%	74.64%
000049	DDPG vs AC_QL	73.18%	85.05%
	DDPG vs QL	66.04%	72.68%
	DDPG vs SL	75.11%	71.36%

Results of win rate (wr). To further validate the advantages of our policy, we also evaluate the win rate (wr) of DDPG over the other three policies on three stocks. Formally, let us consider the SL policy for example, and suppose that the experiment lasts NE_t episodes, and in NE_w episodes DDPG outperforms SL, then the wr of DDPG over SL is NE_t/NE_w. Table 4 presents the results, which clearly show that our method overwhelmingly surpasses the other policies (with $wr > 50\%$), regardless of stock liquidity and trading volume.

Convergence. To examine the convergence of our DDPG model, we run the model and record the implementation reward (IR) after each episode of training, and then evaluate the average IR of each of the most recent 1000 episodes. Figure 5 shows the difference between the average IR results of DDPG and that of the baseline policy SL by setting $V = 5000$. From Fig. 5, we can see that all the three curves (corresponding to the three stocks) tend to converge as the number of episodes increases. Note that in Fig. 5, we show the results from the 1000th episode, as we compute the average IR of each of the most recent 1000 episodes.

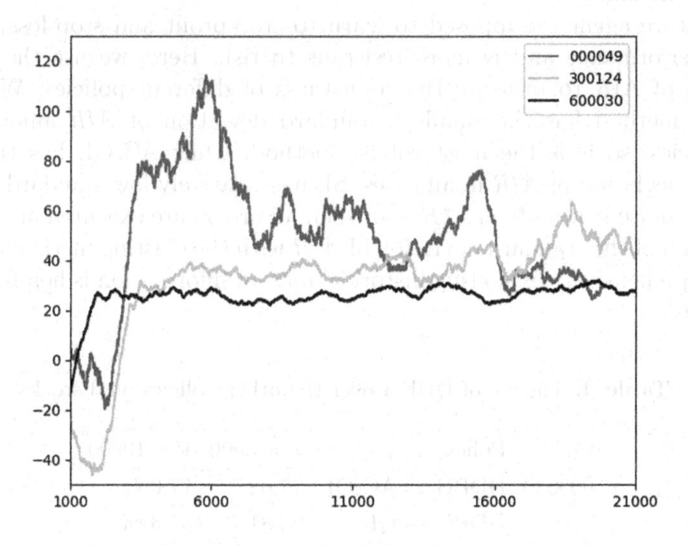

Fig. 5. Convergence of DDPG training on 3 stocks.

5 Conclusion

In the paper, we propose a deep reinforcement learning based method to solve the optimal trade execution problem, where we use the DDPG algorithm and design a feature extraction network to extract deep features from both historical and current market data. We evaluate the proposed method on three real stocks from two China security exchanges. The experimental results show that our method significantly outperforms the existing methods. In the future, we plan to explore more advanced feature selection methods and try to give up the discretization of action so that the agent will have real-time response to the market. We will also consider to deploy the method on the real market for more comprehensive evaluation.

Acknowledgement. This work was supported in part by Science and Technology Commission of Shanghai Municipality Project (#19511120700). Jihong Guan was partially supported by the Program of Science and Technology Innovation Action of Science and Technology Commission of Shanghai Municipality under Grant No. 17511105204 and the Special Fund for Shanghai Industrial Transformation and Upgrading under grant No. 18XI-05, Shanghai Municipal Commission of Economy and Informatization.

References

1. Akbarzadeh, N., Tekin, C., van der Schaar, M.: Online learning in limit order book trade execution. IEEE Trans. Signal Process. **66**(17), 4626–4641 (2018). https://doi.org/10.1109/TSP.2018.2858188
2. Almgren, R., Chriss, N.: Optimal execution of portfolio transactions. J. Risk **3**, 5–40 (2001)
3. Barto, A.G., Mahadevan, S.: Recent advances in hierarchical reinforcement learning. Discrete Event Dyn. Syst. **13**(4), 341–379 (2003)
4. Bertsimas, D., Lo, A.: Optimal control of execution costs - a study of government bonds with the same maturity date. J. Financ. Mark. **1**, 1–50 (1998)
5. Cont, R., Larrard, A.D.: Price dynamics in a Markovian limit order market. SIAM J. Financ. Math. **4**(1), 1–25 (2013)
6. Cont, R., Stoikov, S., Talreja, R.: A stochastic model for order book dynamics. Oper. Res. **58**(3), 549–563 (2010)
7. Feng, Y., Palomar, D.P., Rubio, F.: Robust optimization of order execution. IEEE Trans. Signal Process. **63**(4), 907–920 (2015)
8. Hendricks, D., Wilcox, D.: A reinforcement learning extension to the Almgren-Chriss framework for optimal trade execution. In: 2104 IEEE Conference on Computational Intelligence for Financial Engineering & Economics (CIFEr), pp. 457–464. IEEE (2014)
9. Huang, W., Lehalle, C.A., Rosenbaum, M.: Simulating and analyzing order book data: the queue-reactive model. J. Am. Stat. Assoc. **110**(509), 107–122 (2015)
10. Jiang, Z., Xu, D., Liang, J.: A deep reinforcement learning framework for the financial portfolio management problem. arXiv preprint arXiv:1706.10059 (2017)
11. Johnson, J.D., Li, J., Chen, Z.: Reinforcement learning: an introduction-RS Sutton, AG Barto, MIT Press, Cambridge, MA 1998, 322 pp. ISBN 0-262-19398-1. Neurocomputing **35**(1), 205–206 (2000)
12. Kearns, M., Nevmyvaka, Y.: Machine learning for market microstructure and high frequency trading. High Frequency Trading: New Realities for Traders, Markets, and Regulators (2013)
13. Liang, Z., Jiang, K., Chen, H., Zhu, J., Li, Y.: Deep reinforcement learning in portfolio management. arXiv preprint arXiv:1808.09940 (2018)
14. Lillicrap, T.P., et al.: Continuous control with deep reinforcement learning. arXiv preprint arXiv:1509.02971 (2015)
15. Nevmyvaka, Y., Feng, Y., Kearns, M.: Reinforcement learning for optimized trade execution. In: Proceedings of the 23rd International Conference on Machine Learning, pp. 673–680. ACM (2006)
16. Palguna, D., Pollak, I.: Non-parametric prediction in a limit order book. In: 2013 IEEE Global Conference on Signal and Information Processing (GlobalSIP), p. 1139. IEEE (2013)

17. Palguna, D., Pollak, I.: Mid-price prediction in a limit order book. IEEE J. Sel. Topics Signal Process. **10**(6), 1 (2016)
18. Perold, A.F.: The implementation shortfall. J. Portfolio Manag. **33**(1), 25–30 (1988)
19. Rosenberg, G., Haghnegahdar, P., Goddard, P., Carr, P., Wu, K., De Prado, M.L.: Solving the optimal trading trajectory problem using a quantum annealer. IEEE J. Sel. Top. Signal Process. **10**(6), 1053–1060 (2016)
20. Sherstov, A.A., Stone, P.: Three automated stock-trading agents: a comparative study. In: Faratin, P., Rodríguez-Aguilar, J.A. (eds.) AMEC 2004. LNCS (LNAI), vol. 3435, pp. 173–187. Springer, Heidelberg (2006). https://doi.org/10.1007/11575726_13
21. Silver, D., Lever, G., Heess, N., Degris, T., Wierstra, D., Riedmiller, M.: Deterministic policy gradient algorithms. In: Proceedings of the 31st International Conference on International Conference on Machine Learning, ICML 2014, vol. 32, pp. I-387–I-395. JMLR.org (2014)
22. Sutton, R.S., McAllester, D.A., Singh, S.P., Mansour, Y.: Policy gradient methods for reinforcement learning with function approximation. In: Advances in Neural Information Processing Systems, pp. 1057–1063 (2000)
23. Xiong, Z., Liu, X.Y., Zhong, S., Walid, A., et al.: Practical deep reinforcement learning approach for stock trading. In: NeurIPS Workshop on Challenges and Opportunities for AI in Financial Services: The Impact of Fairness, Explainability, Accuracy, and Privacy (2018)
24. Ye, Z., Huang, K., Zhou, S., Guan, J.: Gaussian weighting reversion strategy for accurate on-line portfolio selection. In: 2017 IEEE 29th International Conference on Tools with Artificial Intelligence (ICTAI), pp. 929–936 (2017)

A Fast Automated Model Selection Approach Based on Collaborative Knowledge

Zhenyuan Sun[1,3], Zixuan Chen[1,3], Zhenying He[2,3](✉), Yinan Jing[2,3](✉), and X. Sean Wang[1,2,3,4](✉)

[1] School of Software, Fudan University, Shanghai, China
{zysun17,chenzx17,xywangCS}@fudan.edu.cn
[2] School of Computer Science, Fudan University, Shanghai, China
{zhenying,jingyn}@fudan.edu.cn
[3] Shanghai Key Laboratory of Data Science, Shanghai, China
[4] Shanghai Institute of Intelligent Electronics and Systems, Shanghai, China

Abstract. Great attention has been paid to data science in recent years. Besides data science experts, plenty of researchers from other domains are conducting data analysis as well because big data is becoming more easily accessible. However, for those non-expert researchers, it can be quite difficult to find suitable models to conduct their analysis tasks because of their lack of expertise and the existence of excessive models. In the meantime, existing model selection approaches rely too much on the content of data sets and take quite long time to make the selection, which makes these approaches inadequate to recommend models to non-experts online. In this paper, we present an efficient approach to conducting automated model selection efficiently based on analysis history and knowledge graph embeddings. Moreover, we introduce exterior features of data sets to enhance our approach as well as address the cold start issue. We conduct several experiments on competition data from Kaggle, a well-known online community of data researchers. Experimental results show that our approach can improve model selection efficiency dramatically and retain high accuracy as well.

Keywords: Model selection · Data analysis history · Knowledge graph

1 Introduction

Nowadays data are easily accessible, and plenty of researchers who are not data science experts also have the need to use data analysis as a tool in their work. However, these non-experts are not equipped with enough expertise, and cannot pick a suitable model from lots of machine learning and deep learning models.

Z. Sun and Z. Chen—Joint first authors, who contributed equally to this research.

Y. Nah et al. (Eds.): DASFAA 2020, LNCS 12112, pp. 655–662, 2020.
https://doi.org/10.1007/978-3-030-59410-7_43

Several approaches such as Auto-sklearn [2], Azure Auto-ML [3] and Oboe [11] have been proposed to address model selection problems. However, recommending models for non-experts raises three new challenges for the model selection problem. First, the high efficiency of data analysis is crucial for those non-experts, because they depend on the results of data analysis to run their whole business workflows efficiently during the process of agile business analysis while existing approaches need quite long time to conduct model selection because they depend on internal data of data sets. Second, these non-experts lack adequate expertise while existing approaches are too difficult for them to use. Third, to satisfy business needs, various models for different tasks need to be selected while existing model selection approaches focus on a few limited tasks.

Therefore, we consider it significant to design a fast and unconstrained automated model selection approach without the need to analyze the content of data sets. In this paper, we propose a novel Fast Automatic Model Selection approach, *FAMS*, which combines data analysis history and data science knowledge collaboratively to recommend suitable models for different data sets to non-experts. We utilize the analysis history on each data set to convert the model selection problem into a recommendation problem based on the fact that the models perform well on one data set have high possibility to have good performance on data sets which are similar to this data set and introduce exterior features to address the cold start issue. A great advantage of FAMS is that it can provide fine-grained recommendation of all kinds of models over different problems as long as these models have been used in history. Moreover, to offset noise in analysis records and present relationships between models better, we manually build a model knowledge graph, namely *MKG*, and leverage this knowledge base to learn latent representations of models for enhancing the quality of recommender systems.

Figure 1 illustrates the workflow of FAMS. When a user chooses a data set, FAMS utilizes analysis records and the knowledge support of MKG to recommend models. This process which is based on collaborative filtering strategy is implemented by matrix factorization and enhanced by exterior features, history records and knowledge support.

To summarize, in this paper, we make the following contributions:

- We propose a highly efficient and unconstrained automated model selection framework, namely *FAMS* based on enhanced collaborative filtering given the setting of the model selection problem.
- We build a model knowledge graph, namely *MKG*, which consists of the relationship and attributes of models to enhance recommendation performance.
- We conduct several experiments on authentic data sets, proving that FAMS is much more efficient than other existing approaches, while retaining its effectiveness, and can support much more models.

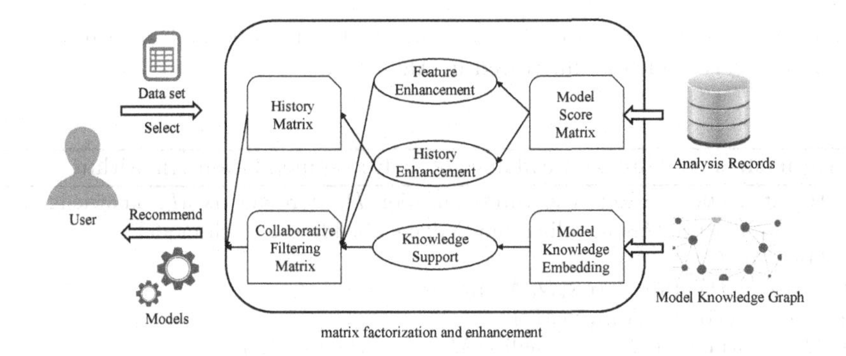

Fig. 1. The Workflow of FAMS

2 Model Recommendation

In this section, we introduce our FAMS approach in detail. Different from most traditional model selection approaches, to eliminate the dependence on the data of data sets, we regard the model selection problem as a recommendation problem, and employ collaborative filtering (CF for short) based on analysis records. In this way, we can exploit the latent features extracted by CF to help select suitable models instead of merely exploring data sets themselves. Therefore, we propose a threefold enhanced collaborative filtering, introducing features of data sets and knowledge support into recommendation.

In traditional recommendation tasks, such as user-item recommendation systems, a score matrix is used to illustrate to what extent a user is satisfied with an item, which is often obtained from scores, stars, click rates, etc. In this model selection setting, the score of one model on a data set is denoted by one element in the matrix, representing to what extent the model is suitable for this data set. The score is assigned by the sum of record weight of all analysis history records where this model is used. For each record of the analysis history, a *record weight* is the value to represent the quality of this analysis record. In this way, a model score for each model on each data set can be calculated and all these scores form a matrix, namely *score matrix*.

It is no denying that by employing collaborative filtering, many traditional recommendation problems can be solved excellently. However, in the setting of model selection, merely employing CF strategy cannot guarantee satisfactory performance because of two main causes. First, recommendation for model selection encounters a difficulty of data sparseness because a large number of records are on popular data sets while many data sets have few analysis records. Second, the quality of these data is varied because we cannot ensure that all analysis records have satisfactory performance. Some models although performing quite poorly on one data set are still misused, hence introducing noise in the score matrix.

The noise in the score matrix and the sparseness can negatively influence the performance of CF. Therefore, to resolve these issues, we employ a threefold enhanced collaborative filtering algorithm to use knowledge from various ways collaboratively, which is illustrated in Algorithm 1.

Algorithm 1: Threefold Enhanced CF Recommendation Algorithm

Input: Data set exterior features collection D, score matrix M_S, knowledge matrix M_K including knowledge embedding of each model
Output: CF Matrix M_{CF}
1 $M_F \leftarrow$ getFeatureMatrix(D, M_S);
2 $M_H \leftarrow$ getHistroyMatrix(M_S);
3 $M \leftarrow$ normalize(M) + normalize(M_F) + normalize(M_H);
4 Decompose M into data set latent factor matrix M_{data} and model latent factor matrix M_{model} ;
5 $M_{model} \leftarrow M_{model} + M_K$;
6 $M_{CF} \leftarrow M_{data} \bullet M_{model}$;
7 $M_{ij} = \alpha \cdot M_{CFij} + (1-\alpha) \cdot Conf(i) \cdot M_{Hij}$;
8 Rank M_{ij} for each data set i and recommend top models

To get better recommendation performance, we need to increase the quality of the input matrix. All data sets can be divided into two categories. The first category is data sets with ample analysis records. We call these data sets *hot data sets*. And the other category is called *cold data sets*, which do not have enough analysis records. We conduct feature enhancement to solve the issue of recommendation on cold data sets. For cold data sets, suffering from lack of analysis records, CF performs quite poorly. It is the so-called cold start issue. To recommend models for a cold data set, we use its neighbors to offer some help collaboratively and build a feature matrix (Line 1). To fulfill this idea, we employ the KNN algorithm to utilize exterior features of data sets to calculate the feature matrix of a cold data set by the embeddings of its neighbors. For hot data sets, we build a history matrix from the score matrix by only retaining the score of k best models and set other scores to be 0 for each data set in the matrix in order to eliminate noise in data (Line 2). By using this matrix for history enhancement we can make top models on records easier to be recommended. This matrix can be used both to enhance CF and recommend models directly as a single recommendation strategy.

These two matrices are added into the score matrix to increase the quality of data before matrix decomposition (Line 3). Then we decompose the matrix into a data set latent factor matrix and a model latent factor matrix by using Truncated SVD approach [10] (Line 4).

And knowledge embedding from a manually built model knowledge graph, MKG is used to enhance the model latent factor matrix to quantify the similarity between each model more accurately and offset the negative influence of noise. MKG consists of 171 vertexes including 126 model vertexes that represent a specific model and 712 edges with a corresponding label. To use the model knowledge graph to help enhance CF, we employ a graph embedding model named Complex [7] to generate graph embedding of each model. And this graph embedding of models is added to the model latent factor matrix to represent models better (Line 5).

Finally, we get the inner product of these two matrices as the CF matrix (Line 6). After we get the CF matrix, a two-way mixed recommendation calculates the recommendation matrix M, combining the CF matrix and the history matrix (Line 7). In the formula, α is a parameter that controls how much we rely on CF to recommend models, and $Conf(i)$ means the confidence of data set i, which represents the extent of difference between good models and incompetent models and is used to decide to what extent we rely on history records to recommend. For data sets with high confidence, suitable models can be selected from history directly so we tend to rely more on the top models of history records to recommend. On the contrary, for data sets with low confidence, simply using the record cannot lead to suitable models so we rely more on CF. Eventually, we recommend models by ranking models and selecting top models (Line 8).

3 Experiments

3.1 Experimental Settings

Kaggle is an online data science community, where plentiful competitions are held, and competitors tend to share their models. Therefore, we can collect adequate analysis records on the data sets of these competitions (One analysis record shared by a user is called a notebook in the community, and we will use this term in this paper. And we may use the term competition and data set interchangeably). We collected 95,975 notebooks from 337 competitions. Among them, 25,333 notebooks mentioned specific models. 95 models have been used, which are all included in MKG.

For each notebook, there exist three measures to evaluate the performance of this notebook, which are all used in our experiments. The first one is its evaluation standard in this competition (e.g. accuracy for a classification competition). The second one is a medal mark showing whether this notebook is an excellent notebook. And the last one is votes from other users. In our experiments, we calculate the record weight for each notebook based on its evaluation standard on corresponding competition and its votes.

3.2 Efficiency Evaluation

We use an experiment of varying data set sizes to show FAMS is useful, and existing approaches (mentioned in the introduction) are not adequate when users need model recommendation immediately online. Experimental results in Fig. 2 shows that as the size of data sets increases, the time cost of all the other three approaches increases rapidly. (Note that all these three approaches are workflow frameworks containing not only model selection but the time cost of model selection cannot be separated so the response time in this experiment contains the time to run the whole workflow.) On the contrary, FAMS retains its efficiency because it does not rely on the data in data sets.

In this experiment, the largest data set is only 100 MB. However, when we are doing data analysis, much larger data sets with GBs or even TBs are used. It may take several days to run these existing approaches. Therefore, for plenty of researchers who want to know suitable models for the given data set immediately to begin their data analysis, these existing work is not adequate.

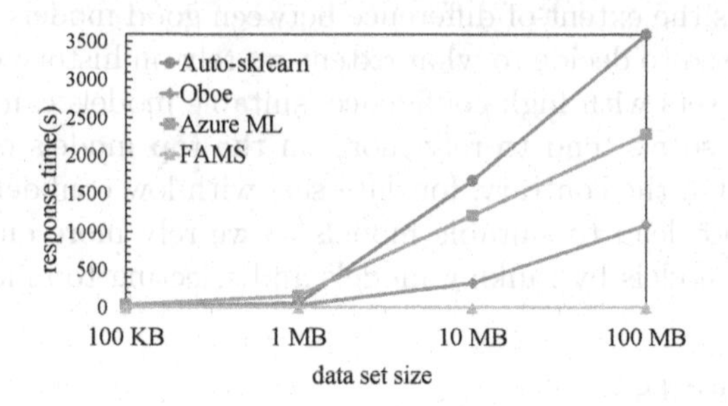

Fig. 2. Efficiency experiment

3.3 Effectiveness Evaluation

Support Evaluation. One important advantage of FAMS is that it can support model selection on all kinds of data sets with analysis records and data set descriptions. Table 1 presents the count of supported problem types and models. Because of the high flexibility of FAMS, it can be applied to all kinds of models and problem types. On the contrary, other approaches can only resolve limited kinds of problems. All statistics of these three approaches are obtained from their corresponding documents and open source projects. Auto-sklearn and Azure focus on classification and regression problems while Oboe can only deal with classification problems. Among 337 Kaggle data sets we use in the experiments, FAMS can be used for all of them while Auto-sklearn, as well as Azure, can be executed on no more than 200 data sets, and Oboe can only deal with 53 data sets due to the limitation of these three model selection approaches.

Experiment on Medal Match. This experiment shows that FAMS can make accurate recommendations to match excellent models of medal notebooks. We devise a baseline approach which can be the intuitive idea of non-experts for whom we are recommending models. Note that it is impossible to compare FAMS with other existing model selection approach in this experiment because FAMS is much more fine-grained (i.e. can support a lot more models) and is in a different setting for online fast model selection.

In this experiment, for a data set, we use all medal notebooks as the test set because they are acknowledged to be accurate and inspirational, and other notebooks as the training set. If a model is used by a number of medal notebooks, it is considered to be an excellent model for this competition. We observed whether selected models by FAMS can match models used by these medal notebooks. In detail, we calculate the model score for a model on a data set by the count of its use in medal notebooks multiplied by the average normalized evaluation value in which this model is used. And then all models are ranked and top models are selected to be test results.

To compare, we devise a baseline approach that intuitively calculates model score in the top n (which is set to be 12 in the experiment) notebooks with the highest evaluation values.

There are two sets of experiments (i.e. 5v8 and 8v5). For example, 5v8 means 5 models are recommended and 8 models are the answers to match. In this setting, accuracy represents the ratio of matched models to 5 recommended models, and recall represents the ratio of matched models to 8 medal models. Experimental results in Table 2 show that FAMS outperforms the baseline approach a lot in both two conditions. Thus this experiment can prove the effectiveness and usefulness of FAMS.

Table 1. Coverage statistics

	FAMS	Auto-sklearn	Azure	Oboe
Ttypes	**20**	5	6	2
Models	**95**	27	15	12

Table 2. Medal match

	5 v 8		8 v 5	
	Baseline	FAMS	Baseline	FAMS
Accuracy	0.701	**0.820**	0.412	**0.499**
Recall	0.438	**0.513**	0.659	**0.799**
F-value	0.540	**0.631**	0.507	**0.614**

4 Related Work

Various approaches have been used in the domain of model selection. Several papers [1,4,5] regard model selection problem as a classification problem. Nonetheless, when models are in great numbers, classification approaches are not good enough anymore. Recent researches [6,9] in model selection begin to employ the collaborative filtering strategy. Nevertheless, most of them are limited to a very small domain. Different from former researches, Chengrun Yang et al. [11] have a more extensive view and propose Oboe, aimed at machine learning model selection. Oboe forms a matrix of the cross-validated errors of models on data sets and fits a rank model to learn the low-dimensional feature vectors for models and data sets to make recommendation.

Besides the above model selection work, there exist several researches of automated machine learning [2,3,8]. Nicoló Fusi et al. [3] propose Azure, which uses probabilistic matrix factorization based on history machine learning workflows to predict the performance of workflow on new data sets. This work also tries to solve the cold start issue by extracting numerous meta-features from data sets. Matthias Feurer et al. [2] propose Auto-sklearn, an automated machine learning approach. They use Bayesian optimization to find optimal models and parameters.

5 Conclusion and Future Work

We propose a novel fast and unconstrained automated model selection approach, FAMS, which utilizes analysis history and knowledge support collaboratively to conduct highly efficient model recommendation without expertise as input. Experimental results show the high efficiency and effectiveness of FAMS. There still exist several improvements that can be done in the future. We can take temporal information of analysis records into account so as to increase the possibility of new models to be recommended, and extend this work to an automated data mining pipeline.

Acknowledgement. This work was supported by the National Key R&D Program of China (NO. 2018YFB 1004404 and 2018YFB1402600) and the Shanghai Sailing Program (NO. 18YF1401300).

References

1. Cifuentes, C.G., Sturzel, M., Jurie, F., Brostow, G.J.: Motion models that only work sometimes. In: BMVC, pp. 1–12 (2012)
2. Feurer, M., Klein, A., Eggensperger, K., Springenberg, J.T., Blum, M., Hutter, F.: Auto-sklearn: efficient and robust automated machine learning. In: Automated Machine Learning 2019, pp. 113–134 (2019)
3. Fusi, N., Sheth, R., Elibol, M.: Probabilistic matrix factorization for automated machine learning. In: NIPS 2018, pp. 3352–3361 (2018)
4. Guo, G., Wang, C., Ying, X.: Which algorithm performs best: algorithm selection for community detection. In: WWW 2018, pp. 27–28 (2018)
5. Mac Aodha, O., Brostow, G.J., Pollefeys, M.: Segmenting video into classes of algorithm-suitability. In: CVPR 2010, pp. 1054–1061 (2010)
6. Matikainen, P., Sukthankar, R., Hebert, M.: Model recommendation for action recognition. In: CVPR 2012, pp. 2256–2263 (2012)
7. Trouillon, T., Welbl, J., Riedel, S., Gaussier, É., Bouchard, G.: Complex embeddings for simple link prediction. In: ICML 2016, pp. 2071–2080 (2016)
8. Wang, C., Wang, H., Mu, T., Li, J., Gao, H.: Auto-model: utilizing research papers and HPO techniques to deal with the cash problem. In: ICDE 2020
9. Wang, Y., Hebert, M.: Model recommendation: generating object detectors from few samples. In: CVPR 2015, pp. 1619–1628 (2015)
10. Xu, P.: Truncated SVD methods for discrete linear ill-posed problems. Geophys. J. R. Astronom. Soc. **135**(2), 505–514 (1998)
11. Yang, C., Akimoto, Y., Kim, D.W., Udell, M.: OBOE: collaborative filtering for AutoML model selection. In: KDD 2019, pp. 1173–1183 (2019)

Attention with Long-Term Interval-Based Gated Recurrent Units for Modeling Sequential User Behaviors

Zhao Li[1(✉)], Chenyi Lei[1], Pengcheng Zou[1], Donghui Ding[1], Shichang Hu[1], Zehong Hu[1], Shouling Ji[2], and Jianliang Gao[3]

[1] Alibaba-Group, Hangzhou, China
{lizhao.lz,chenyi.lcy,xuanwei.zpc,donghui.ddh,shichang.hsc, zehong.hzh}@alibaba-inc.com

[2] Zhejiang University, Hangzhou, Zhejiang, China
sji@zju.edu.cn

[3] Central South University, Changsha, China
gaojianliang@csu.edu.cn

Abstract. Recommendations based on sequential User behaviors have become more and more common. Traditional methods depend on the premise of Markov processes and consider user behavior sequences as interests. However, they usually ignore the mining and representation of implicit features. Recently, recurrent neural networks (RNNs) have been adopted to leverage their power in modeling sequences and consider the dynamics of user behaviors. In order to better locate user preference, we design a network featuring **A**ttention with **L**ong-term **I**nterval-based **G**ated **R**ecurrent **U**nits (ALI-GRU) to model temporal sequences of user actions. In the network, we propose a time interval-based GRU architecture to better capture long-term preferences and short-term intents when encoding user actions rather than the original GRU. And a specially matrix-form attention function is designed to learn weights of both long-term preferences and short-term user intents automatically. Experimental results on two well-known public datasets show that the proposed ALI-GRU achieves significant improvement than state-of-the-art RNN-based methods.

Keywords: Attention mechanism · Recurrent neural networks · User modeling

1 Introduction

Traditional personalized recommendation methods, such as item to item collaborative filtering did not consider the dynamics of user behaviors. For example, to predict user's next action such as the next product to purchase, the profiling of both long-term preferences and short-term intents of user are required. Therefore, modeling the user's behaviors as sequences provides key advantages.

Y. Nah et al. (Eds.): DASFAA 2020, LNCS 12112, pp. 663–670, 2020.
https://doi.org/10.1007/978-3-030-59410-7_44

Nonetheless, modeling sequential user behaviors raises even more challenges than modeling them without the temporal dimension. How to identify the correlation and dependence among actions is one of the difficult issues. Recently, many different kinds of RNN algorithms have been proposed for modeling user behaviors to leverage their powerful descriptive ability for sequential data [5,8]. However, there are several limitations that make it difficult to apply these methods into the wide variety of applications in the real-world. One inherent assumption of these methods is that the importance of historical behaviors decreases over time, which is also the intrinsic property of RNN cells such as gated recurrent units (GRU) and long- and short-term memory (LSTM). This assumption does not always apply in practice, where the sequences may have complex cross-dependence. In this paper, we propose a network featuring **A**ttention with **L**ong-term **I**nterval-based **G**ated **R**ecurrent **U**nits (ALI-GRU) for modeling sequential user behaviors to predict user's next action. We adopt a series of bi-directional GRU to process the sequence of items that user had accessed. The GRU cells in our network consist of time interval-based GRU, where the latter is to reflect the short-term information of time intervals. In addition, the features extracted by bi-directional GRU are used to drive an attention model, where the attention distribution is calculated at each timestamp. We have performed a series of experiments using well-known public datasets. Experimental results show that ALI-GRU outperforms the state-of-the-art methods by a significant margin.

2 Related Work

The related work is given at two aspects, modeling of sequential user behaviors and attention mechanism.

Modeling Sequential User Behaviors. Due to the significance to user-centric tasks such as personalized search and recommendation, modeling sequential user behaviors has attracted great attention from both industry and academia. Most of pioneering work relies on model-based Collaborative Filtering (CF) to analyze user-item interaction matrix. For the task of sequential recommendation, Rendle *et al.* [10] propose Factorizing Personalized Markov Chain to combine matrix factorization of user-item matrix with Markov chains. He *et al.* [3] further integrate similarity-based methods [7] into FPMC to tackle the problem of sequential dynamics. But the major problems are that these methods independently combine several components, rely on low-level hand-crafted features of user or item, and have difficulty to handle long-term behaviors. To the contrary, with the success of recurrent neural networks (RNNs) in the past few years, a paucity of work has made attempts to utilize RNNs [4]. The insight that RNN-based solutions achieve success in modeling sequential user behaviors is that the well demonstrated ability of RNN in capturing patterns in the sequential data. Recent studies [9,11,14] also indicate that time intervals within sequential signal are a very important clue to update and forget information in RNN architecture. But in practice, there is complex dependence and correlation between sequential user behaviors, which requires deeper analysis of relation among behaviors rather than simply modeling the presence, order and time intervals. To summarize, how to design an effective RNN architecture to model sequential user behaviors effectively is still a challenging open problem.

Attention Mechanism. The success of attention mechanism is mainly due to the reasonable assumption that human beings do not tend to process the entire signal at once, but only focus on selected portions of the entire perception space when and where needed [6]. Recent researches start to leverage different attention architectures to improve performance of related tasks. For example, Yang *et al.* [13] propose a hierarchical attention network at word and sentence level, respectively, to capture contributions of different parts of a document. Vaswani *et al.* [12] utilize multi-head attention mechanism to improve performance. Nevertheless, most of previous work calculates attention distribution according to the interaction of every source vector with a single embedding vector of contextual or historical information, which may lead to information loss caused by early summarization, and noise caused by incorrect previous attention.

Indeed, the attention mechanism is very sound for the task of modeling sequential user behaviors. However, to the best of our knowledge, there is few work concentrating on this paradigm, except a recent study [1], which considers the attention mechanism into a multimedia recommendation task with multilayer perceptron. An effective solution with attention mechanism for better modeling sequential user behaviors is to be investigated in this paper.

3 ALI-GRU

We start our discussion with some definition of notations. Let $\mathcal{U}$ be a set of users and $\mathcal{I}$ be a set of items in a specific service such as products in online shopping websites. For each user $u \in \mathcal{U}$, his/her historical behaviors are given by $\mathcal{H}^u = \{(i_k^u, t_k^u) | i_k^u \in \mathcal{I}, t_k^u \in \mathcal{R}^+, k = 1, 2, \ldots, N_u\}$, where (i_k^u, t_k^u) denotes the interaction between user u and item i_k^u at time t_k^u, interaction has different forms in different services, such as clicking, browsing, adding to favorites, etc. The objective of modeling sequential user behaviors is to predict the conditional probability of the user's next action $p(i_{N_u+1}^u | \mathcal{H}^u, t_{N_u+1}^u)$ for a certain given user u.

As illustrated in the left part of Fig. 1, our designed network features an attention mechanism with long-term interval-based gated recurrent units for modeling sequential user behaviors. This network architecture takes the sequence of items as raw signal. There are four stages in our network. The embedding layer maps items to a vector space to extract their basic features. The bi-directional GRU layer is designed to capture the information of both long-term preferences and short-term intents of user, it consists of normal GRUs and time interval-based GRUs. The attention function layer reflects our carefully designed attention mechanism, which is illustrated in the right part of Fig. 1. Finally, there is an output layer to integrate the attention distribution and the extracted sequential features, and utilize normal GRUs to predict the conditional probability of next item.

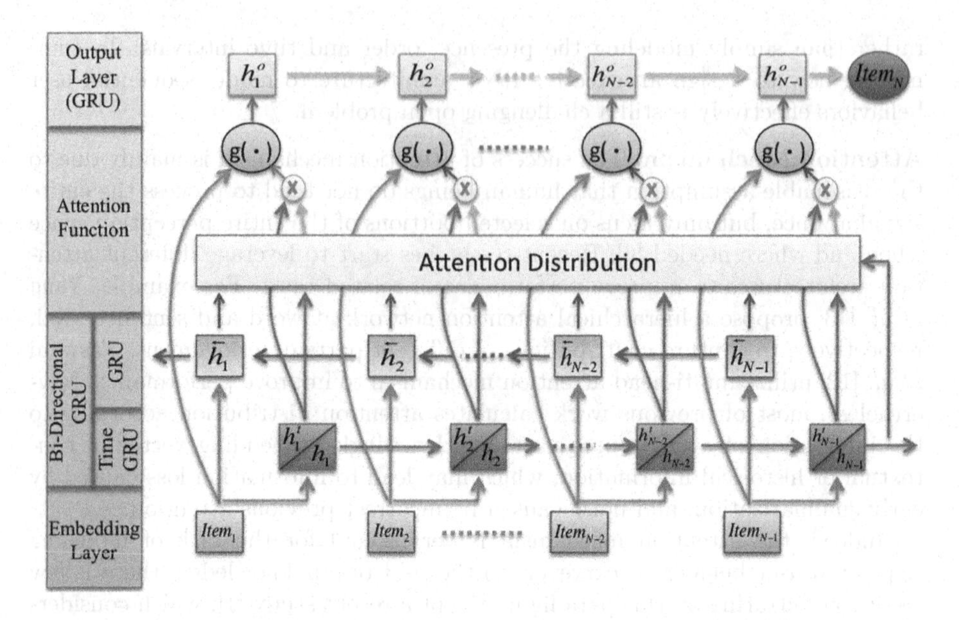

Fig. 1. (Best view in color). The proposed framework for modeling sequential user behaviors. (Color figure online)

Bi-directional GRU Layer with Time-GRU. This layer is designed to extract driven signals from input sequence and to refine the long-term memory by contextual cues. We propose a network structure with time-GRU to extract short-term dynamics of user intents as driven signal of the attention function.

The structure of time-GRU is different from the normal GRU. For the input I_N, the normal GRU computes linear interpolation between the last state h_{N-1} and the candidate activation $\tilde{h}_N$,

$$h_N = (1 - z_N) \odot h_{N-1} + z_N \odot \tilde{h}_N \tag{1}$$

where $\odot$ is an element-wise multiplication.

Since GRU is originally designed for NLP tasks, there is no consideration of time intervals within inputs. To include the short-term information, we augment the normal GRU with a time gate T_N

$$\begin{aligned} T_N = &\sigma(W_t \Delta t_N + U_t I_N + b_t) \\ &\text{s.t. } W_t < 0 \end{aligned} \tag{2}$$

where Δt_N is the time interval between adjacent actions. The constraint $W_t < 0$ is to utilize the simple assumption that smaller time interval indicates larger correlation. Moreover, we generate a time-dependent hidden state h^t_N in addition to the normal hidden state h_N, i.e.

$$h^t_N = (1 - z_N \odot T_N) \odot h^t_{N-1} + z_N \odot T_N \odot \tilde{h}^t_N \tag{3}$$

where we utilize the time gate as a filter to modify the update gate z_N so as to capture short-term information more effectively.

In addition, to utilize contextual cues to extract long-term information, we propose to combine the output of forward normal GRU (h_N in Eq. (1)) with all the outputs of backward GRU at different steps (the output of backward GRU at step k is denoted by $\overleftarrow{h}_k$ in Fig. 1). Specifically, we produce concatenated vectors $[h_{N-1}, \overleftarrow{h}_{N-1}], [h_{N-1}, \overleftarrow{h}_{N-2}], \ldots, [h_{N-1}, \overleftarrow{h}_1]$, as shown in the right part of Fig. 1, where $[,]$ stands for concatenation of vectors. This design effectively captures the contextual cues as much as possible.

Attention Function Layer. Unlike previous attention mechanisms, we do not simply summarize the contextual long-term information into individual feature vectors. We design to attend the driven signals at each time step along with the embedding of contextual cues.

Specifically, as shown in the right part of Fig. 1, we use $\mathbf{H}_k = [h_{N-1}, \overleftarrow{h}_k] \in \mathcal{R}^{2d}, k = 1, 2, \ldots, N-1$, where d is the dimension of GRU states, to represent the contextual long-term information. $h_k^t \in \mathcal{R}^d$ denotes the short-term intent reflected by item i_k. We then construct an attention matrix $A \in \mathcal{R}^{(N-1)*(N-1)}$, whose elements are calculated by

$$A_{ij} = \alpha(\mathbf{H}_i, h_j^t) \in \mathcal{R} \tag{4}$$

where the attention weight

$$\alpha(\mathbf{H}_i, h_j^t) = v^T \tanh(W_a \mathbf{H}_i + U_a h_j^t) \tag{5}$$

is adopted to encode the two input vectors. There is a pooling layer along the direction of long-term information, and then a softmax layer to normalize the attention weights of each driven signal. Let a_k be the normalized weight on h_k^t, then the attended short-term intent vector is $\hat{h}_k^t = a_k h_k^t \in \mathcal{R}^d$. At last, we use $g(i_k, \hat{h}_k^t) = [i_k, \hat{h}_k^t, |i_k - \hat{h}_k^t|, i_k \odot \hat{h}_k^t] \in \mathcal{R}^{4d}$ as the output to the next layer, where i_k is the embedded vector of the item at the k-th step.

Our carefully designed attention mechanism described above is to reduce the loss of contextual information caused by early summarization. Furthermore, since driven signals are attended to the long-term information at different steps, the attentions can obtain the trending change of user's preferences, being more robust and less affected by the noise in the historical actions.

4 Experiments

To verify the performance of ALI-GRU, we conduct a series of experiments on two well-known public datasets (LastFM[1] and CiteULike[2]). We compare ALI-GRU with the following state-of-the-art approaches for performance evaluation:

[1] http://www.dtic.upf.edu/~ocelma/MusicRecommendationDataset/lastfm-1K.html.
[2] http://www.citeulike.org/faq/data.adp.

Basic GRU/Basic LSTM [2], **Session RNN** [4], **Time-LSTM** [14]. All RNN-based models are implemented with TensorFlow. Training was done on a single GeForce Tesla P40 GPU with 8 GB graphical memory.

In this experiment, we use the datasets as adopted in [14], i.e. LastFM (987 users and 5000 items with 818767 interactions) and CiteULike (1625 users and 5000 items with 35834 interactions). Both datasets can be formulated as a series of tuples <user_id, item_id, timestamp>. Our target is to recommend songs in LastFM and papers in CiteULike for users according to their historical behaviors.

For the fair of comparison, we follow the segmentation of training set and test set in [14]. Specifically, 80% users are randomly selected for training. The remaining users are for testing. For each test user u with k historical behaviors, there are $k-1$ test cases, where the k-th test case is to perform recommendations at time t^u_{k+1} given the user's previous k actions, and the ground-truth is i^u_{k+1}. The recommendation can also be regarded as a multi-class classification problem.

We use one-hot representations of items as inputs to the network, and one fully-connected layer with 8 nodes for embedding. The length of hidden states of GRU-related layers including both normal GRU and Time-GRU is 16. A softmax function is used to generate the probability prediction of next items. For training, we use the AdaGrad optimizer, which is a variant of Stochastic Gradient Descent (SGD). Parameters for training are minibatch size of 16 and initial learning rate of 0.001 for all layers. The training process takes about 8 h.

Table 1. Recall@10 comparison results on LastFM & CiteULike

	LastFM	CiteULike
Basic-LSTM	0.2451	0.6824
Session-RNN	0.3405	0.7129
Time-LSTM	0.3990	0.7586
ALI-GRU	**0.4752**	**0.7764**

The results of the sequential recommendation tasks on LastFM and CiteULike are shown in Table 1, where Recall@10 is used to measure whether the ground-truth item is in the recommendation list. It obviously shows that our approach performs the best on both LastFM and CiteULike. Meanwhile, ALI-GRU obtains significant improvement over Time-LSTM, which is the best baseline, averagely by 10.7% for Recall@10. It owes to the superiority of introducing attention mechanism into RNN-based methods especially in capturing the contribution of each historical action. Therefore, the experimental results demonstrate the effectiveness of our proposed ALI-GRU.

Due to the importance of cold-start performance for recommendation system, We also analyze the influence of cold-start on the LastFM dataset. Actually, cold-start refers to the lacking of enough historical data for a specific user, and it often decreases the efficiency of making recommendations. So, in this experiment test cases are separately counted for different numbers of historical actions, and small number refers to cold-start. The results given in Fig. 2 show that ALI-GRU performs just slightly worse than the state-of-the-art methods for cold users with only 5 actions. This is due to the fact that it considers short-term information as driven signals, which averages source signal to some extent and leads to less accurate modeling for cold users. And with the increasing of historical actions, the ALI-GRU performs the best.

Therefore, the experimental results demonstrate the effectiveness of our proposed ALI-GRU. Both bi-directional GRU and attention mechanism can better model the user preferences for making recommendations.

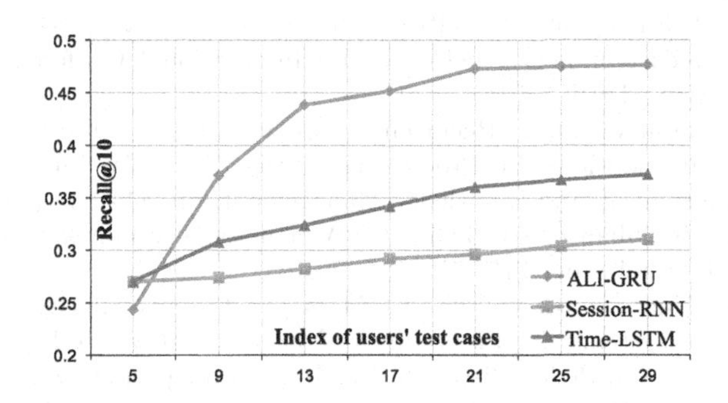

Fig. 2. Recall@10 evaluated on different indexes of users' test cases in LastFM.

5 Conclusions

In this paper, we propose to integrate a matrix-form attention mechanism into RNNs for better modeling sequential user behaviors. Specifically, we design a network featuring Attention with Long-term Interval-based Gated Recurrent Units to model temporal sequences of user actions, and using a Time-GRU structure to capture both long-term preferences and short-term intents of users as driven signals for better robustness. The empirical evaluations on two public datasets for sequential recommendation task show that our proposed approach achieves better performance than several state-of-the-art RNN-based solutions. One limitation of this work is the lack of user profiling in providing personalized content, which will be addressed in our future work.

Acknowledgment. This work was partly supported by the National Key Research and Development Program of China under No. 2018YFB0804102, NSFC under No. 61772466, U1936215, and U1836202, the Zhejiang Provincial Natural Science Foundation for Distinguished Young Scholars under No. LR19F020003, the Provincial Key Research and Development Program of Zhejiang, China under No. 2019C01055, the Ant Financial Research Funding, the Alibaba-ZJU Joint Research Institute of Frontier Technologies and Zhejiang Lab under No. 2019KE0AB01.

References

1. Chen, J., Zhang, H., He, X., Nie, L., Liu, W., Chua, T.: Attentive collaborative filtering: multimedia recommendation with reature- and item-level attention. In: Proceedings of the 40th International ACM SIGIR Conference on Research and Development in Information Retrieval, pp. 335–344. ACM (2017)
2. Chung, J., Gulcehre, C., Cho, K., Bengio, Y.: Empirical evaluation of gated recurrent neural networks on sequence modeling. In: NIPS Workshop on Deep Learning. MIT Press (2014)
3. He, R., McAuley, J.: Fusing similarity models with Markov chains for sparse sequential recommendation. In: International Conference on Data Mining (ICDM), pp. 191–200. IEEE (2016)
4. Hidasi, B., Karatzoglou, A., Baltrunas, L., Tikk, D.: Session-based recommendations with recurrent neural networks. In: International Conference on Learning Representations. IEEE (2016)
5. Hidasi, B., Karatzoglou, A.: Recurrent neural networks with top-k gains for session-based recommendations. In: Proceedings of the 27th ACM International Conference on Information and Knowledge Management, pp. 843–852 (2018)
6. Hubner, R., Steinhauser, M., Lehle, C.: A dual-stage two-phase model of selective attention. Psychol. Rev. **117**, 759–784 (2010)
7. Kabbur, S., Ning, X., Karypis, G.: FISM: factored item similarity models for top-n recommender systems. In: Proceedings of the 19th ACM SIGKDD International Conference on Knowledge Discovery and Data Mining, pp. 659–667. ACM (2013)
8. Liu, Q., Wu, S., Wang, L.: Multi-behavioral sequential prediction with recurrent log-bilinear model. Trans. Knowl. Data Eng. **291**, 1254–1267 (2017)
9. Neil, D., Preiffer, M., Liu, S.: Phased LSTM: accelerating recurrent network training for long or event-based sequences. In: Advances in Neural Information Processing Systems (NIPS), pp. 3882–3890. MIT Press (2016)
10. Rendle, S., Freudenthaler, C., Schmidt-Thieme, L.: Factorizing personalized Markov chains for next-basket recommendation. In: Proceedings of the 19th International Conference on World Wide Web, pp. 811–820. ACM (2010)
11. Vassøy, B., Ruocco, M., de Souza da Silva, E., Aune, E.: Time is of the essence: a joint hierarchical RNN and point process model for time and item predictions. In: Proceedings of the Twelfth ACM International Conference on Web Search and Data Mining, pp. 591–599 (2019)
12. Vaswani, A., et al.: Attention is all you need. In: Advances in Neural Information Processing Systems, pp. 5998–6008 (2017)
13. Yang, Z., Yang, D., Dyer, C., He, X., Smola, A., Hovy, E.: Hierarchical attention networks for document classification. In: The 2016 Conference of the North American Chapter of the Association for Computational Linguistics: Human Language Technologies (NAACL), pp. 1480–1489. NAACL (2016)
14. Zhu, Y., et al.: What to do next: modeling user behaviors by time-LSTM. In: Proceedings of the 26th International Joint Conference on Artificial Intelligence, pp. 3602–3608. AAAI Press (2017)

Latent Space Clustering via Dual Discriminator GAN

Heng-Ping He[1,2], Pei-Zhen Li[1,2], Ling Huang[1,2], Yu-Xuan Ji[1,2], and Chang-Dong Wang[1,2(✉)]

[1] School of Data and Computer Science, Sun Yat-sen University, Guangzhou, China
hehp6@mail2.sysu.edu.cn, sysuLiPeizhen@163.com, huanglinghl@hotmail.com, jiyx6@mail2.sysu.edu.cn, changdongwang@hotmail.com

[2] Guangdong Province Key Laboratory of Computational Science, Guangzhou, China

Abstract. In recent years, deep generative models have achieved remarkable success in unsupervised learning tasks. Generative Adversarial Network (GAN) is one of the most popular generative models, which learns powerful latent representations, and hence is potential to improve clustering performance. We propose a new method termed CD2GAN for latent space **C**lustering via dual discriminator GAN (**D2GAN**) with an inverse network. In the proposed method, the continuous vector sampled from a Gaussian distribution is cascaded with the one-hot vector and then fed into the generator to better capture the categorical information. An inverse network is also introduced to map data into the separable latent space and a semi-supervised strategy is adopted to accelerate and stabilize the training process. What's more, the final clustering labels can be obtained by the cross-entropy minimization operation rather than by applying the traditional clustering methods like K-means. Extensive experiments are conducted on several real-world datasets. And the results demonstrate that our method outperforms both the GAN-based clustering methods and the traditional clustering methods.

Keywords: Latent space clustering · Unsupervised learning · D2GAN

1 Introduction

Data clustering aims at deciphering the inherent relationship of data points and obtaining reasonable clusters [1,2]. Also, as an unsupervised learning problem, data clustering has been explored through deep generative models [3–5] and the two most prominent models are Variational Autoencoder (VAE) [6] and Generative Adversarial Network (GAN) [7]. Owing to the belief that the ability to synthesize, or "create" the observed data entails some form of understanding, generative models are popular for the potential to automatically learn disentangled representations [8]. A WGAN-GP framework with an encoder for clustering is proposed in [9]. It makes the latent space of GAN feasible for clustering by carefully designing input vectors of the generator. Specifically, Mukherjee et al.

Y. Nah et al. (Eds.): DASFAA 2020, LNCS 12112, pp. 671–679, 2020.
https://doi.org/10.1007/978-3-030-59410-7_45

proposed to sample from a prior that consists of normal random variables cascaded with one-hot encoded vectors with the constraint that each mode only generates samples from a corresponding class in the original data (i.e., clusters in the latent space are separated). This insight is key to clustering in GAN's latent space. During the training process, the noise input vector of the generator is recovered by optimizing the following loss function:

$$J = ||E(G(z_n)) - z_n||_2^2 + L(E(G(z_c)), z_c) \tag{1}$$

where z_n and z_c are Gaussian and one-hot categorical components of the input vector of the generator respectively and $L(\cdot)$ denotes the cross-entropy loss function.

However, as pointed in [10], one can not take it for granted that pretty small recovery loss means pretty good features. So we propose a CD2GAN method for latent space clustering via dual discriminator GAN. The problem of mode collapse can be largely eliminated by introducing the dual discriminators. Then by focusing on recovering the one-hot vector we can obtain the final clustering result by the cross-entropy minimization operation rather than by applying the traditional clustering techniques like K-means.

2 The Proposed Method

First of all, we will give key definitions and notational conventions. In what follows, K represents the number of clusters. D_1 and D_2 denote two discriminators respectively, where D_1 favors real samples while D_2 prefers fake samples. G denotes the generator that generates fake samples and E stands for the encoder or the inverse network that serves to reduce dimensionality and get the latent code of the input sample. Let $z \in \mathbb{R}^{1\times\tilde{d}}$ denote the input of the generator, which consists of two components, i.e., $z = (z_n, z_c)$. z_n is sampled from a Gaussian distribution, i.e., $z_n \sim \mathcal{N}(\mu, \sigma^2 I_{d_n})$. In particular, we set $\mu = 0$, $\sigma = 0.1$ in our experiments and it is typical to set σ to be a small value to ensure that clusters in the latent space are separated. And $z_c \in \mathbb{R}^{1\times K}$ is a one-hot categorical vector. Specifically, $z_c = e_k, k \sim \mathcal{U}\{1, 2, \cdots, K\}$, e_k is the k-th elementary vector.

2.1 Latent Space Clustering via Dual Discriminator GAN

From Eq. (1) we know that in [9], in order to get the latent code corresponding to the real sample, it recovers the vector z_n which is sampled from a Gaussian distribution and offers no useful information for clustering. In addition, as pointed in [10], small recovery loss does not necessarily mean good features. So here comes the question, what are good features? Intuitively, they should contain the unique information of the latent vector. The above two aspects show that the latent vector corresponding to the real sample should not contain the impurity (i.e., z_n). In other words, the degree of dimension reduction is not enough. Motivated by this observation, our method only needs to perfectly recover the

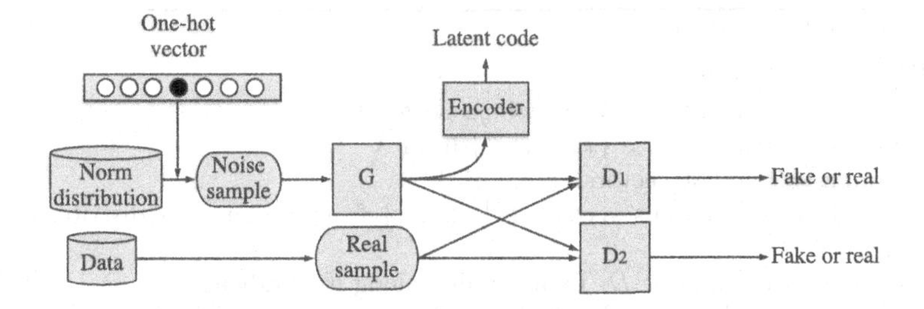

Fig. 1. The CD2GAN framework.

most unique information (i.e., z_c) of the latent vector corresponding to the real sample, and utilizes only this part for clustering.

Motivated by all above facts, we propose a dual discriminator GAN framework [11] with an inverse network architecture for latent space clustering, which is shown in Fig. 1. From [11], the Kullback-Leibler (KL) and reverse KL divergences are integrated into a unified objective function for the D2GAN framework as follows:

$$\begin{aligned}\min_{G}\max_{D1,D2} L_1(G,D_1,D_2) = & \alpha E_{x\sim P_{data}}[log(D_1(x))] + E_{z\sim P_{noise}}[-D_1(G(z))] \\ & + E_{x\sim P_{data}}[-D_2(x)] + \beta E_{z\sim P_{noise}}[log(D_2(G(z))]\end{aligned} \quad (2)$$

where P_{data} and P_{noise} denote the unknown real data distribution and the prior noise distribution respectively. α and β are two hyperparameters, which control the effect of KL and reverse KL divergences on the optimization problem. It is worth mentioning that the problem of mode collapse encountered in GAN can be largely eliminated since the complementary statistical properties from KL and reverse KL divergences can be exploited to effectively diversify the estimated density in capturing multi-modes.

In addition, the input noise vector (i.e., z) of the generator is z_n (sampled from a Gaussian distribution without any cluster information) concatenated with a one-hot vector $z_c \in \mathbb{R}^{1\times K}$ (K is the number of clusters). To be more precise, $z = (z_n, z_c)$. It is worth noting that each one-hot vector z_c corresponds to one type of mode provided z_n is well recovered [9]. The inverse network, i.e., the encoder, has exactly opposite structure to the generator and serves to map data into the separable latent space. The overall loss function is:

$$\min_{E,G}\max_{D1,D2} L_1(G,D_1,D_2) + L_2(E,G) + L_3(E,G) \quad (3)$$

where

$$L_2(E,G) = \lambda||E(G(z_c)) - z_c||_2^2 \quad (4)$$

$$L_3(E,G) = \epsilon\mathcal{L}_{CE}(z_c, E(G(z_c))) \quad (5)$$

and $E(G(z_c))$ denotes the last K dimensions of $E(G(z))$, $\mathcal{L}_{CE}(\cdot)$ is the cross-entropy loss. Similar to GAN [7], the D2GAN with an inverse network model

Algorithm 1. CD2GAN

Input: Dataset X, Parameters α, β, ϵ, λ, K

1: **for** $t = 1, 2, \cdots, number\_of\_training\_epochs$ **do**
2: Sample m real samples $\{x^1, ..., x^m\}$ from X
3: Sample m one-hot vectors $\{z_c^1, ..., z_c^m\}$
4: Sample m continuous vectors $\{z_n^1, ..., z_n^m\}$ from a Gaussian distribution ($\mu = 0, \sigma = 0.1$)
5: Update discriminator D_1 by ascending along its gradient: $\bigtriangledown_{parameter D_1} \frac{1}{m} \sum_{i=1}^{m} [\alpha \log D_1(x^i) - D_1(G(z^i))]$
6: Update discriminator D_2 by ascending along its gradient: $\bigtriangledown_{parameter D_2} \frac{1}{m} \sum_{i=1}^{m} [\beta \log D_2(G(z^i)) - D_2(x^i)]$
7: Sample m one-hot vectors $\{z_c^1, ..., z_c^m\}$
8: Sample m continuous vectors $\{z_n^1, ..., z_n^m\}$ from a Gaussian distribution ($\mu = 0, \sigma = 0.1$)
9: Update generator G and the inverse network E by descending along their gradient: $\bigtriangledown_{parameter G,E} \frac{1}{m} \sum_{i=1}^{m} [\beta \log D_2(G(z^i)) - D_1(G(z^i)) + L_2(E, G) + L_3(E, G)]$
10: **end for**
11: **for** $i = 1, 2, \cdots, number\_of\_real\_samples$ **do**
12: Get the label set Y of all x^i by **equation** (6)
13: **end for**

Output: Label set Y

can be trained by alternatively updating D_1, D_2, G and E, where Adam [12] is adopted for optimization. It is worth noting that the one-hot vector, i.e., $z_c \in \mathbb{R}^{1\times K}$ can be regarded as the cluster label indicator of the fake sample generated from G since each one-hot vector corresponds to one mode. In addition, through adequate training, data distribution of samples generated from G will become much more identical to the real data distribution P_{data}. From this perspective, the inverse network (i.e., the encoder E) can be regarded as a muti-class classifier, which is the key reason why we introduce the cross-entropy loss $L_3(E, G)$ to the overall loss function.

When it comes to clustering, we can utilize the well trained encoder E. Formally, given a real sample x, the corresponding cluster label $\tilde{y}$ can be calculated as follows:

$$\tilde{y} = \arg\min_t \epsilon \mathcal{L}_{CE}(one\_hot(t), E(x)[\tilde{d} - K :]) + \lambda || one\_hot(t) - E(x)[\tilde{d} - K :]||_2^2 \quad (6)$$

where $t = 0, 1, \cdots, K-1$, $one\_hot(t) \in \mathbb{R}^{1\times K}$ is the t-th elementary vector and $E(x)[\tilde{d} - K :]$ returns a slice of $E(x)$ consisting of the last K elements. λ and ϵ remain the same as in the model training phase. For clarity, the proposed CD2GAN method is summarized in Algorithm 1.

2.2 Semi-supervised Strategy

In this section, we propose a semi-supervised strategy to accelerate and stabilize the training process of our model.

Table 1. Summary of the five datasets in the experiments.

Datasets	#Data point	Dimension	#Cluster
MNIST	70000	28×28	10
Fashion	70000	28×28	10
Pendigit	10992	1×16	10
10_x73k	73233	1×720	8
Pubmed	19717	1×500	3

From [9] we know that the one-hot component can be viewed as the label indicator of the fake samples. With the training going on, we get lots of real samples with their labels, and the encoder can be seen as a multi-class classifier. Like semi-supervised classification problem. Specifically, given a dataset with ground-truth cluster labels, a small portion (1%–2% or so) of it will be sampled along with the corresponding ground-truth cluster labels. Let $\tilde{x} = (x, y)$ denote one of the real samples for semi-supervised training, where $x \in \mathbb{R}^{1\times d}$ represents the observation while y is the corresponding cluster label. During the training process, y is encoded to a one-hot vector just like z_c, which will then be concatenated with one noise vector z_n to serve as the input of the generator G. Afterwards, the latent code of $G(z)$, i.e. $E(G(z))$, can be obtained through the encoder. Subsequently, the corresponding x will be fed into the encoder E to get another latent code $E(x)$. As mentioned above, the last K dimensions of the latent code offer the clustering information, and intuitively, one could expect better clustering performance if the the last K dimensions of $E(G(z))$ and $E(x)$ are more consistent. In light of this, a loss function can be designed accordingly:

$$L_4(E, G) = \gamma||(E(x_r) - E(G(z_n, one\_hot(x_l))))[\tilde{d} - K:]||_2^2) \tag{7}$$

where γ is a hyperparameter. In essence, we compute the Euclidean distance between the last K dimensions of $E(G(z))$ and $E(x)$, which is easy to be optimized. In this way, the overall objective function for the semi-supervised training can be defined as follows:

$$\min_{E,G} \max_{D_1,D_2} L_1(G, D_1, D_2) + L_2(E, G) + L_3(E, G) + L_4(E, G) \tag{8}$$

From the perspective of semi-supervised learning, we should sample a fixed small number of real samples when updating the parameters of E and G. In practice, faster convergence and less bad-training can be achieved when the semi-supervised training strategy is adopted.

Table 2. Comparison results for unsupervised clustering. The best result in each measure is highlighted in bold.

		MNIST	Fashion	Pendigit	10_x73k	Pubmed
NMI	ClusterGAN	0.885	0.611	0.729	0.731	0.125
	GAN with bp	0.873	0.488	0.683	0.544	0.061
	AAE	0.895	0.554	0.654	0.617	0.132
	GAN-EM	0.905	0.577	0.722	0.734	0.157
	InfoGAN	0.844	0.541	0.709	0.563	0.127
	SC	-	-	0.701	-	0.104
	AGGLO	0.677	0.565	0.681	0.599	0.112
	NMF	0.411	0.491	0.554	0.695	0.061
	CD2GAN	**0.911**	**0.638**	**0.773**	**0.783**	**0.210**
ARI	ClusterGAN	0.893	0.487	0.651	0.677	0.117
	GAN with bp	0.884	0.332	0.602	0.398	0.072
	AAE	0.906	0.425	0.590	0.546	0.112
	GAN-EM	0.902	0.399	0.642	0.659	0.132
	InfoGAN	0.825	0.398	0.632	0.401	0.102
	SC	-	-	0.598	-	0.097
	AGGLO	0.502	0.460	0.563	0.452	0.096
	NMF	0.344	0.322	0.421	0.557	0.076
	CD2GAN	**0.924**	**0.501**	**0.709**	**0.701**	**0.187**

3 Experiments

3.1 Experimental Setting

Datasets and Evaluation Measures. We adopt five well-known datasets (i.e. MNIST [13], Fashion [14], Pendigit [15], 10_x73k [9] and Pubmed [16]) in the experiments and the basic information are listed in Table 1. Two commonly used evaluation measures, i.e., normalized mutual information (NMI) and adjusted Rand index (ARI) are utilized to evaluate the clustering performance [17].

Baseline Methods and Parameter Setting. We adopt 8 clustering methods as our baseline methods including both GAN-based methods and traditional clustering methods. They are ClusterGAN [9], GAN with bp [9], adversarial autoencoder (AAE) [18], GAN-EM [19], InfoGAN [8], spectral clustering (SC) [20], Agglomerative Clustering (AGGLO) [21] and Non-negative Matrix Factorization (NMF) [22].

We set $\alpha = 0.1$, $\beta = 0.1$, $\epsilon = 1$, $\gamma = 1$ for all the datasets. As for λ, we set $\lambda = 0$ for **10_x73k** and **MNIST** and set $\lambda = 0.1$ for the other datasets. When it comes to the network architecture, we adopt some techniques as recommended in [23] for image datasets, and for the other datasets we use the full connected networks. The learning rate of **Pubmed** is set to be 0.0001 while 0.0002 for the other datasets.

Table 3. Comparison results with adopting the semi-supervised strategy for model training. The best result in each measure is highlighted in bold.

		MNIST	Fashion	Pendigit	10_x73k	Pubmed
NMI	AAE-Semi	0.943	0.721	0.810	0.823	0.299
	GAN-EM-Semi	**0.951**	0.726	0.804	0.794	0.297
	CD2GAN-Semi	0.945	**0.741**	**0.860**	**0.895**	**0.371**
ARI	AAE-Semi	0.944	0.661	0.789	0.804	0.304
	GAN-EM-Semi	**0.955**	0.653	0.754	0.788	0.311
	CD2GAN-Semi	**0.955**	**0.690**	**0.831**	**0.880**	**0.411**

3.2 Comparison Results

Comparison results are reported in Table 2 for unsupervised clustering, where the values are averaged over 5 normal training (GAN-based methods typically suffer from bad-training). As can be seen, the proposed CD2GAN method beats all the baseline methods on all the datasets in terms of the two measures. Particularly, CD2GAN is endowed with the powerful ability of representation learning, which accounts for the superiority over traditional clustering methods, i.e., SC, AGGLO and NMF. What's more, we also conduct experiments to validate the effectiveness of the semi-supervised training strategy and the results are reported in Table 3. As shown in the table, the proposed CD2GAN-Semi (CD2GAN with semi-supervised training strategy) beats AAE-Semi (AAE with semi-supervised training strategy) and GAN-EM-Semi (GAN-EM with semi-supervised training strategy) on almost all the datasets except **MNIST**.

4 Conclusion

In this paper, we propose a method termed CD2GAN for latent space clustering via D2GAN with an inverse network. Specifically, to make sure that the continuity in latent space can be preserved while different clusters in latent space can be separated, the input of the generator is carefully designed by sampling from a prior that consists of normal random variables cascaded with one-hot encoded vectors. In addition, the mode collapse problem is largely eliminated by introducing the dual discriminators and the final cluster labels can be obtained by the cross-entropy minimization operation rather than by applying traditional clustering method like K-means. What's more, a novel semi-supervised strategy is proposed to accelerate and stabilize the training process. Extensive experiments are conducted to confirm the effectiveness of the proposed methods.

Acknowledgments. This work was supported by NSFC (61876193) and Guangdong Natural Science Funds for Distinguished Young Scholar (2016A030306014).

References

1. Mai, S.T., et al.: Evolutionary active constrained clustering for obstructive sleep apnea analysis. Data Sci. Eng. **3**(4), 359–378 (2018)
2. Chen, M., Huang, L., Wang, C., Huang, D.: Multi-view spectral clustering via multi-view weighted consensus and matrix-decomposition based discretization. In: DASFAA, pp. 175–190 (2019)
3. Zhou, P., Hou, Y., Feng, J.: Deep adversarial subspace clustering. In: CVPR, pp. 1596–1604 (2018)
4. Yu, Y., Zhou, W.J.: Mixture of GANs for clustering. In: IJCAI, pp. 3047–3053 (2018)
5. Min, E., Guo, X., Liu, Q., Zhang, G., Cui, J., Long, J.: A survey of clustering with deep learning: from the perspective of network architecture. IEEE Access **6**, 39501–39514 (2018)
6. Kingma, D.P., Welling, M.: Auto-encoding variational bayes. arXiv preprint arXiv:1312.6114 (2013)
7. Goodfellow, I., et al.: Generative adversarial nets. In: NIPS, pp. 2672–2680 (2014)
8. Chen, X., Duan, Y., Houthooft, R., Schulman, J., Sutskever, I., Abbeel, P.: Infogan: interpretable representation learning by information maximizing generative adversarial nets. In: NIPS, pp. 2172–2180 (2016)
9. Mukherjee, S., Asnani, H., Lin, E., Kannan, S.: ClusterGAN: latent space clustering in generative adversarial networks. In: AAAI, pp. 4610–4617 (2019)
10. Hjelm, R.D., et al.: Learning deep representations by mutual information estimation and maximization. arXiv preprint arXiv:1808.06670 (2018)
11. Nguyen, T., Le, T., Vu, H., Phung, D.: Dual discriminator generative adversarial nets. In: NIPS, pp. 2670–2680 (2017)
12. Kingma, D.P., Ba, J.: Adam: a method for stochastic optimization. arXiv preprint arXiv:1412.6980 (2014)
13. LeCun, Y., Bottou, L., Bengio, Y., Haffner, P., et al.: Gradient-based learning applied to document recognition. Proc. IEEE **86**(11), 2278–2324 (1998)
14. Xiao, H., Rasul, K., Vollgraf, R.: Fashion-MNIST: a novel image dataset for benchmarking machine learning algorithms. arXiv preprint arXiv:1708.07747 (2017)
15. Alpaydin, E., Alimoglu, F.: Pen-based recognition of handwritten digits data set. University of California, Irvine. Mach. Learn. Repository. Irvine: University California **4**(2) (1998)
16. Sen, P., Namata, G., Bilgic, M., Getoor, L., Gallagher, B., Eliassi-Rad, T.: Collective classification in network data. AI Mag. **29**(3), 93–106 (2008)
17. Wang, C.D., Lai, J.H., Suen, C.Y., Zhu, J.Y.: Multi-exemplar affinity propagation. IEEE Trans. Pattern Anal. Mach. Intell. **35**(9), 2223–2237 (2013)
18. Makhzani, A., Shlens, J., Jaitly, N., Goodfellow, I., Frey, B.: Adversarial autoencoders. arXiv preprint arXiv:1511.05644 (2015)
19. Zhao, W., Wang, S., Xie, Z., Shi, J., Xu, C.: GAN-EM: GAN based EM learning framework. arXiv preprint arXiv:1812.00335 (2018)
20. Shi, J., Malik, J.: Normalized cuts and image segmentation. IEEE Trans. Pattern Anal. Mach. Intell. **22**(8), 888–905 (2000)
21. Zhang, W., Wang, X., Zhao, D., Tang, X.: Graph degree linkage: agglomerative clustering on a directed graph. In: Fitzgibbon, A., Lazebnik, S., Perona, P., Sato, Y., Schmid, C. (eds.) ECCV 2012. LNCS, vol. 7572, pp. 428–441. Springer, Heidelberg (2012). https://doi.org/10.1007/978-3-642-33718-5_31

22. Lee, D.D., Seung, H.S.: Learning the parts of objects by non-negative matrix factorization. Nature **401**(6755), 788 (1999)
23. Radford, A., Metz, L., Chintala, S.: Unsupervised representation learning with deep convolutional generative adversarial networks. arXiv preprint arXiv:1511.06434 (2015)

Neural Pairwise Ranking Factorization Machine for Item Recommendation

Lihong Jiao[1], Yonghong Yu[2(✉)], Ningning Zhou[1(✉)], Li Zhang[3], and Hongzhi Yin[4]

[1] School of Computer Science, Nanjing University of Posts and Telecommunications, Nanjing, China
zhounn@njupt.edu.cn

[2] Tongda College, Nanjing University of Posts and Telecommunications, Nanjing, China
yuyh@njupt.edu.cn

[3] Department of Computer and Information Sciences, Northumbria University, Newcastle, UK
li.zhang@northumbria.ac.uk

[4] School of Information Technology and Electrical Engineering, The University of Queensland, Brisbane, Australia
db.hongzhi@gmail.com

Abstract. The factorization machine models attract significant attention from academia and industry because they can model the context information and improve the performance of recommendation. However, traditional factorization machine models generally adopt the point-wise learning method to learn the model parameters as well as only model the linear interactions between features. They fail to capture the complex interactions among features, which degrades the performance of factorization machine models. In this paper, we propose a neural pairwise ranking factorization machine for item recommendation, which integrates the multi-layer perceptual neural networks into the pairwise ranking factorization machine model. Specifically, to capture the high-order and nonlinear interactions among features, we stack a multi-layer perceptual neural network over the bi-interaction layer, which encodes the second-order interactions between features. Moreover, the pair-wise ranking model is adopted to learn the relative preferences of users rather than predict the absolute scores. Experimental results on real world datasets show that our proposed neural pairwise ranking factorization machine outperforms the traditional factorization machine models.

Keywords: Recommendation algorithm · Factorization machine · Neural networks

1 Introduction

With the development of information technology, a variety of network applications have accumulated a huge amount of data. Although the massive data

Y. Nah et al. (Eds.): DASFAA 2020, LNCS 12112, pp. 680–688, 2020.
https://doi.org/10.1007/978-3-030-59410-7_46

provides users with rich information, it leads to the problem of "information overload". The recommendation systems [1] can greatly alleviate the problem of information overload. They infer users latent preferences by analyzing their past activities and provide users with personalized recommendation services. Factorization machine (FM) [2] model is very popular in the field of recommendation systems, which is a general predictor that can be adopted for the prediction tasks working with any real valued feature vector.

Recently, deep learning techniques have shown great potential in the field of recommendation systems, some researchers also have utilized deep learning techniques to improve the classic factorization machine models. Typical deep learning based factorization machine models include NFM [3], AFM [4], and CFM [5]. Neural Factorization Machine (NFM) [3] seamlessly unifies the advantages of neural networks and factorization machine. It not only captures the linear interactions between feature representations of variables, but also models nonlinear high-order interactions. However, both FM and NFM adopt a pointwise method to learn their model parameters. They fit the user's scores rather than learn the user's relative preferences for item pairs. In fact, common users usually care about the ranking of item pairs rather than the absolute rating on each item. Pairwise ranking factorization machine (PRFM) [6,7] makes use of the bayesian personalized ranking (BPR) [8] and FM to learn the relative preferences of users over item pairs. Similar to FM, PRFM can only model the linear interactions among features.

In this paper, we propose the Neural Pairwise Ranking Factorization Machine (NPRFM) model, which integrates the multi-layer perceptual neural networks into the PRFM model to boost the recommendation performance. Specifically, to capture the high-order and nonlinear interactions among features, we stack a multi-layer perceptual neural network over the bi-interaction layer, which is a pooling layer that encodes the seconde-order interactions between features. Moreover, the bayesian personalized ranking criterion is adopted to learn the relative preferences of users, which makes non-observed feedback contribute to the inference of model parameters. Experimental results on real world datasets show that our proposed neural pairwise ranking factorization machine model outperforms the traditional recommendation algorithms.

2 Preliminaries

Factorization Machine is able to model the interactions among different features by using a factorization model. Usually, the model equation of FM is defined as follows:

$$\hat{y}(\mathbf{x}) = w_0 + \sum_{i=1}^{n} w_i x_i + \sum_{i=1}^{n} \sum_{j=i+1}^{n} \langle \mathbf{v}_i, \mathbf{v}_j \rangle x_i x_j \tag{1}$$

where $\hat{y}(\mathbf{x})$ is the predicted value, and $\mathbf{x} \in R^n$ denotes the input vector of the model equation. x_i represents the i-th element of $\mathbf{x}$. $w_0 \in R$ is the global bias, $\mathbf{w} \in R^n$ indicates the weight vector of the input vector $\mathbf{x}$. $\mathbf{V} \in R^{n \times k}$ is the latent feature matrix, whose $\mathbf{v}_i$ represents the feature vector of x_i. $\langle \mathbf{v}_i, \mathbf{v}_j \rangle$ is the dot product of two feature vectors, which is used to model the interaction between x_i and x_j.

3 Neural Pairwise Ranking Factorization Machine

To model the high-order interaction behaviors among features as well as learn the relatively preferences of user over item pairs, we propose the neural pairwise ranking factorization machine (NPRFM) model, whose underlying components are NFM and PRFM. Figure 1 presents the framework of our proposed neural pairwise ranking factorization machine, which consists of four layers, i.e. embedding layer, Bi-interaction layer, hidden layer and prediction layer.

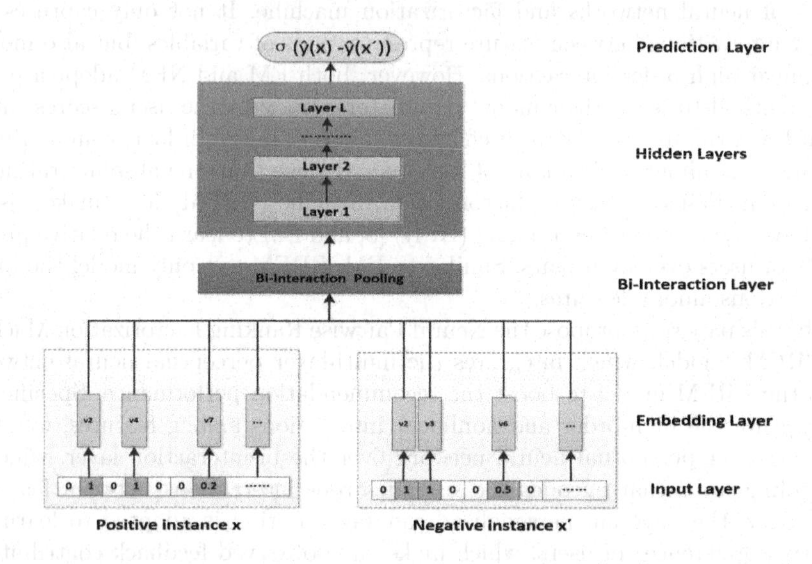

Fig. 1. The framework of neural pairwise ranking factorization machine

The input of NPRFM includes positive and negative instances. Both positive or negative instance contain user, item and context information. By using one-hot encoding, the positive and negative instances are converted into sparse feature vectors $\mathbf{x} \in R^n$ or $\mathbf{x}' \in R^n$, respectively.

3.1 Embedding Layer

After one-hot encoding, we use the embedding table lookup operation to obtain the embedded representations of features included in the input instance. Formally, the embedded representation of $\mathbf{x}$ is,

$$\mathbf{V}_x = \mathbf{V}.onehot(\mathbf{x}) \tag{2}$$

where $\mathbf{V}_x$ is a set of embedding vectors, i.e., $\mathbf{V}_x = \{x_1\mathbf{v}_1, ..., x_n\mathbf{v}_n\}$, and $\mathbf{v}_i \in R^k$ is the embedded representation of the i-th feature.

3.2 Bi-interaction Layer

The Bi-Interaction layer is a pooling operation, which converts the set of embedding vectors $\mathbf{V}_x$ to one vector $f_{BI}(\mathbf{V}_x)$:

$$f_{BI}(\mathbf{V}_x) = \sum_{i=1}^{n} \sum_{j=i+1}^{n} x_i \mathbf{v}_i \odot x_j \mathbf{v}_j \tag{3}$$

where $\odot$ represents the element-wise product of two vectors. As shown in Eq. (3), the Bi-interaction layer captures the pair-wise interactions among the low dimensional representations of features. In other words, the Bi-Interaction pooling only encodes the second-order interactions among features.

3.3 Hidden Layers and Prediction Layer

Since the Bi-interaction layer only captures the second-order interactions among features, and can not model the complexity interactive patterns among features, we utilize the multi-layer perceptron (MLP) to learn the interaction relationships among features, which endows our proposed model with the ability of capturing the high-order interactions. Specifically, in the hidden layers, we stack multiple fully connected hidden layers over the Bi-interaction layer, where the output of a hidden layer is fed into the following hidden layer that makes use of the weighted matrix and non-linear activation function, such as sigmoid, tanh and ReLU, to nonlinearly transform this output. Formally, the MLP is defined as,

$$\begin{aligned} \mathbf{z}_1 &= \sigma_1\left(\mathbf{W}_1 f_{BI}\left(\mathbf{V}_x\right) + \mathbf{b}_1\right), \\ \mathbf{z}_2 &= \sigma_2\left(\mathbf{W}_2 \mathbf{z}_1 + \mathbf{b}_2\right), \\ &\ldots \\ \mathbf{z}_L &= \sigma_L\left(\mathbf{W}_L \mathbf{z}_{L-1} + \mathbf{b}_L\right) \end{aligned} \tag{4}$$

where L denotes the number of hidden layers, $\mathbf{W}_l$, $\mathbf{b}_l$ and σ_l represent the weight matrix, bias vector and activation function for the l-th layer, respectively.

The prediction layer is connected to the last hidden layer, and is used to predict the score $\hat{y}(\mathbf{x})$ for the instance $\mathbf{x}$, where $\mathbf{x}$ can be positive or negative instances. Formally,

$$\hat{y}(\mathbf{x}) = \mathbf{h}^T z_L \tag{5}$$

where $\mathbf{h}$ is the weight vector of the prediction layer. Combining the Eq. (4) and (5), the model equation of NPFFM is reformulated as:

$$\hat{y}(\mathbf{x}) = \sum_{i=1}^{n} w_i x_i + \mathbf{h}^T \sigma_L(\mathbf{W}_L(\ldots\sigma_1(\mathbf{W}_1 f_{BI}(\mathbf{V}_x) + \mathbf{b}_1)\ldots) + \mathbf{b}_L) \tag{6}$$

3.4 Model Learning

We adopt a ranking criterion, i.e. the Bayesian personalized ranking, to optimize the model parameters of NPRFM. Formally, the objective function of NPRFM is defined as:

$$\mathcal{L}^{NPRFM} = \sum_{(\mathbf{x},\mathbf{x}')\in\chi} -ln\sigma(\hat{y}(\mathbf{x}) - \hat{y}(\mathbf{x}')) + \frac{\lambda}{2}(\|\Theta\|_F^2) \tag{7}$$

where $\sigma(.)$ is the logistic sigmoid function. And $\Theta = \{w_i, \mathbf{W}_l, \mathbf{b}_l, \mathbf{v}_i, \mathbf{h}\}$, $i \in (1...n)$, $l \in (1...L)$ denotes the model parameters. χ is the set of positive and negative instances. We adopt the Adagrad [9] optimizer to update model parameter because the Adagrad optimizer utilizes the information of the sparse gradient and gains an adaptive learning rate, which is suitable for the scenarios of data sparse.

4 Experiments

4.1 DataSets and Evaluation Metrics

In our experiments, we choose two real-world implicit feedback datasets: Frappe[1] and Last.fm[2], to evaluate the effectiveness of our proposed model.

The **Frappe** contains 96,203 application usage logs with different contexts. Besides the user ID and application ID, each log contains eight contexts, such as weather, city and country and so on. We use one-hot encoding to convert each log into one feature vector, resulting in 5382 features.

The **Last.fm** was collected by Xin et al. [5]. The contexts of user consist of the user ID and the last music ID listened by the specific user within 90 min. The contexts of item include the music ID and artist ID. This dataset contains 214,574 music listening logs. After transforming each log by using one-hot encoding, we get 37,358 features.

We adopt the leave-one-out validation to evaluate the performance of all compared methods, which has been widely used in the literature [4,5,10]. In addition, we utilize two widely used ranking based metrics, i.e., the Hit Ratio (HR) and Normalized Discounted Cumulative Gain ($NDCG$), to evaluate the performance of all comparisons.

4.2 Experimental Settings

In order to evaluate the effectiveness of our proposed algorithm, we choose FM, NFM, PRFM as baselines.

- **FM**: FM [2,11] is a strong competitor in the field of context-aware recommendation, and captures the interactions between different features by using a factorization model.

[1] http://baltrunas.info/research-menu/frappe.
[2] http://www.dtic.upf.edu/ocelma/MusicRecommendationDataset.

- **NFM**: NFM [10] seamlessly integrates neural networks into factorization machine model. Based on the neural networks, NFM can model nonlinear and high-order interactions between latent representations of features.
- **PRFM**: PRFM [7] applies the BPR standard to optimize its model parameters. Different from FM and NFM, PRFM focuses on the ranking task that learns the relative preferences of users for item pairs rather than predicts the absolute ratings.

For all compared methods, we set dimension of the hidden feature vector $k = 64$. In addition, for FM, we set the regularization term $\lambda = 0.01$ and the learning rate $\eta = 0.001$. For NFM, we set the number of hidden layers is 1, the regularization term $\lambda = 0.01$ and the learning rate $\eta = 0.001$. For PRFM, we set the regularization term $\lambda = 0.001$ and the learning rate $\eta = 0.1$. For NPRFM, we set the regularization term $\lambda = 0.001$, the learning rate $\eta = 0.1$, and the number of hidden layers $L = 1$. In addition, we initialize the latent feature matrix V of NPRFM with the embedded representations learned by PRFM.

4.3 Performance Comparison

We set the length of recommendation list $n = 3, 5, 7$ to evaluate the performance of all compared methods. The experimental results on the two datasets are shown in Tables 1 and 2.

Table 1. Performance comparison on the Frappe dataset ($k = 64$)

Recommendation algorithm	$n = 3$		$n = 5$		$n = 7$	
	HR	NDCG	HR	NDCG	HR	NDCG
FM	0.2445	0.1795	0.3050	0.2107	0.3422	0.2216
NFM	0.2510	0.1797	0.3702	0.2199	0.4686	0.2504
PRFM	0.4650	0.3868	0.5654	0.4280	0.6383	0.4533
NPRFM	0.4786	0.3962	0.5751	0.4358	0.6469	0.4607

Table 2. Performance comparison on the Last.fm dataset ($k = 64$)

Recommendation algorithm	$n = 3$		$n = 5$		$n = 7$	
	HR	NDCG	HR	NDCG	HR	NDCG
FM	0.0770	0.0584	0.1064	0.0706	0.1344	0.0803
NFM	0.0972	0.0723	0.1372	0.0886	0.1702	0.1000
PRFM	0.1828	0.1374	0.2545	0.1667	0.3094	0.1857
NPRFM	0.1855	0.1402	0.2624	0.1715	0.3219	0.1921

From Tables 1 and 2, we have the following observations: (1) On both datasets, FM performs the worst among all the compared methods. The reason is that FM learns its model parameters by adopting a point-wise learning scheme, which usually suffers from data sparsity. (2) NFM is superior to FM with regards to all evaluation metrics. One reason is that the non-linear and high-order interactions among representations of features are captured by utilizing the neural networks, resulting in the improvement of recommendation performance. (3) On both datasets, PRFM achieves better performance than those of FM and NFM. This is because PRFM learns its model parameters by applying the BPR criterion, in which the pair-wise learning method is used to infer the latent representations of users and items. To some extent, the pair-wise learning scheme is able to alleviate the problem of data sparsity by making non-observed feedback contribute to the learning of model parameters. (4) Our proposed NPRFM model consistently outperforms other compared methods, which demonstrates the effectiveness of the proposed strategies. When $n = 3$, NPRFM improves the HR of PRFM by 2.9% and 1.5% on Frappe and Last.fm, respectively. In terms of $NDCG$, the improvements of NPRFM over PRFM are 2.4% and 2.0% on Frappe and Last.fm, respectively.

4.4 Sensitivity Analysis

Impact of the Depth of Neural Networks. In this section, we conduct a group of experiments to investigate the impact of the depth of neural networks on the recommendation quality. We set $n = 5$, $k = 64$, and vary the depth of neural networks from 1 to 3.

In Table 3, NPRFM-i denotes that NPRFM model with i hidden layers, especially, NPRFM-0 is equal to PRFM. We only present the experimental results on $HR@5$ in Table 3 and the experimental results on $NDCG@5$ show similar trends.

Table 3. Impact of L

Methods	Frappe	Lastfm
NPRFM-0	0.5654	0.2545
NPRFM-1	0.5751	0.2624
NPRFM-2	0.5592	0.2572
NPRFM-3	0.5654	0.2077

Table 4. Impact of k

k	Frappe	Lastfm
16	0.4650	0.1641
32	0.5515	0.2027
64	0.5751	0.2624
128	0.5692	0.2514

From Table 3, we observe that NPRFM has the best performance when the number of the hidden layer is equal to one, and the performance of NPRFM degrades when the number of the hidden layer increases. This is because the available training data is not sufficient for NPRFM to accurately learn its model parameters when the number of hidden layers is relatively large. By contrast, if the number of layers is small, NPRFM has limited ability of modeling the complex interactions among embedded representations of features, resulting in the sub-optimal recommendation performance.

Impact of *k*. In this section, we conduct another group of experiments to investigate the impact of the dimension of embedded representations of features k on the recommendation quality.

As shown in Table 4, our proposed NPRFM model is sensitive to the dimension of embedded representation of feature. We find that the performance of NPFFM is optimal when the dimension of embedded representation of feature is equal to 64. A possible reason is that the proposed model already has enough expressiveness to describe the latent preferences of user and characteristics of items when $k = 64$.

5 Conclusion

In this paper, we propose the neural pairwsie ranking factorization machine model, which integrates the multi-layer perceptual neural networks into the PRFM model to boost the recommendation performance of factorization model. Experimental results on real world datasets show that our proposed neural pairwise ranking factorization machine model outperforms the traditional recommendation algorithms.

Acknowledgments. This work is supported in part by the Natural Science Foundation of the Higher Education Institutions of Jiangsu Province (Grant No. 17KJB520028), NUPTSF (Grant No. NY217114), Tongda College of Nanjing University of Posts and Telecommunications (Grant No. XK203XZ18002) and Qing Lan Project of Jiangsu Province.

References

1. Adomavicius, G., Tuzhilin, A.: Toward the next generation of recommender systems: a survey of the state-of-the-art and possible extensions. TKDE **17**(6), 734–749 (2005)
2. Rendle, S.: Factorization machines. In: ICDM, pp. 995–1000. IEEE (2010)
3. He, X., Chua, T.-S.: Neural factorization machines for sparse predictive analytics. In: SIGIR, pp. 355–364. ACM (2017)
4. Xiao, J., Ye, H., He, X., Zhang, H., Wu, F., Chua, T.-S.: Attentional factorization machines: learning the weight of feature interactions via attention networks. arXiv preprint arXiv:1708.04617 (2017)
5. Xin, X., Chen, B., He, X., Wang, D., Ding, Y., Jose, J.: CFM: convolutional factorization machines for context-aware recommendation. In: IJCAI, pp. 3926–3932. AAAI Press (2019)
6. Yuan, F., Guo, G., Jose, J.M., Chen, L., Yu, H., Zhang, W.: LambdaFM: learning optimal ranking with factorization machines using lambda surrogates. In: CIKM, pp. 227–236. ACM (2016)

7. Guo, W., Shu, W., Wang, L., Tan, T.: Personalized ranking with pairwise factorization machines. Neurocomputing **214**, 191–200 (2016)
8. Rendle, S., Freudenthaler, C., Gantner, Z., Schmidt-Thieme, L.: BPR: Bayesian personalized ranking from implicit feedback. In: UAI, pp. 452–461. AUAI Press (2009)
9. Lu, Y., Lund, J., Boyd-Graber, J.: Why adagrad fails for online topic modeling. In: EMNLP, pp. 446–451 (2017)
10. He, X., Liao, L., Zhang, H., Nie, L., Hu, X., Chua, T.-S.: Neural collaborative filtering. In: WWW, pp. 173–182 (2017)
11. Rendle, S.: Factorization machines with libFM. TIST **3**(3), 57 (2012)

Reward-Modulated Adversarial Topic Modeling

Yuhao Feng, Jiachun Feng, and Yanghui Rao(✉)

School of Data and Computer Science, Sun Yat-sen University, Guangzhou, China
{fengyh3,fengjch5}@mail2.sysu.edu.cn, raoyangh@mail.sysu.edu.cn

Abstract. Neural topic models have attracted much attention for their high efficiencies in training, in which, the methods based on variational auto-encoder capture approximative distributions of data, and those based on Generative Adversarial Net (GAN) are able to capture an accurate posterior distribution. However, the existing GAN-based neural topic model fails to model the document-topic distribution of input samples, making it difficult to get the representations of data in the latent topic space for downstream tasks. Moreover, to utilize the topics discovered by these topic models, it is time-consuming to manually interpret the meaning of topics, label the generated topics, and filter out interested topics. To address these limitations, we propose a Reward-Modulated Adversarial Topic Model (RMATM). By integrating a topic predictor and a reward function in GAN, our RMATM can capture the document-topic distribution and discover interested topics according to topic-related seed words. Furthermore, benefit from the reward function using topic-related seed words as weak supervision, RMATM is able to classify unlabeled documents. Extensive experiments on four benchmark corpora have well validated the effectiveness of RMATM.

Keywords: Neural topic model · Generative Adversarial Net · Reward function

1 Introduction

Topic modeling aims to discover the hidden topic information from documents. As an unsupervised learning method, Latent Dirichlet Allocation (LDA) [1] is widely used for topic modeling. However, it relies on mean-field variational inference or Gibbs sampling for parameter estimation. Variational inference often involves complex derivation, while Gibbs sampling costs substantial computing resources. To address these limitations, neural topic models have been developed and attracted increasing attentions due to their advantage of high efficiency in training. Neural variational document model (NVDM) [9] is proposed based on variational auto-encoder (VAE) [6]. It constructs a deep neural network to build up a variational inference approximating the continuous latent distributions of a corpus. The model shows significant performance in generalization

Y. Nah et al. (Eds.): DASFAA 2020, LNCS 12112, pp. 689–697, 2020.
https://doi.org/10.1007/978-3-030-59410-7_47

through learning the posterior probability of data. To discover discrete latent topics, GSM [8] is proposed based on NVDM by introducing Gaussian softmax. Meanwhile, NVLDA and ProdLDA [12] are proposed by constructing a Laplace approximation to approximate the Dirichlet prior for topic discovery. Different from VAE, Generative Adversarial Net (GAN) [3] captures the accurate distributions of data, instead of approximative ones. Adversarial-neural Topic Model (ATM) [15] is proposed as the first GAN-based neural topic model. Though showing great effectiveness, ATM is unable to generate the document-topic distributions for input documents. How to address the problem of lacking document-topic distributions in ATM still stays open. Furthermore, after obtaining the distribution of a topic over words, we need to manually interpret the meaning of each topic, label the generated topics, and filter out interested ones. It can be quite time-consuming. Therefore, discovering interested topics according to predefined information automatically can receive much labor-saving. However, the topic models mentioned above are not capable of detecting predefined interested topics, leaving the meaning of topics on human's judgement.

In light of these considerations, we propose a Reward-Modulated Adversarial Topic Model (RMATM) based on the GAN framework in conjunction with a topic predictor and a reward function, so as to address the issue of the existing ATM and discover interested topics. The generator in RMATM learns the continuous latent distribution of a corpus to generate corresponding fake samples. The discriminator distinguishes the real samples from the generated ones. The topic predictor aims to capture the document-topic distribution. For the purpose of guiding the generator to discover interested topics, we build up a reward function, in which word embedding methods are used to measure the associations between topics and words. Benefit from the reward function that uses seed words of each topic, RMATM can classify documents without human efforts in labeling training samples. To summarize, the main contributions of this study are as follows: First, RMATM can discover interested topics by integrating the reward function in GAN. Second, our model can infer the document-topic distributions of input documents explicitly by introducing a topic predictor, which could address the problem of lacking it in the existing GAN-based topic model. Third, RMATM can classify documents without any labeled samples in training by introducing topic-related seed words as weak supervision.

2 Methodology

In this section, we describe the RMATM whose structure is shown in Fig. 1.

Reward Function. By constructing an appropriate reward function, our model can guide the direction of generator. Firstly, given K topic-related seed words as interested topics $T = \{t_1, t_2, ..., t_K\}$, we can use the word embedding methods to get the word embedding expression where each word corresponds to a low-dimensional vector. Secondly, by calculating the cosine similarity between embeddings of a word in the vocabulary and each topic-related seed word in T, we

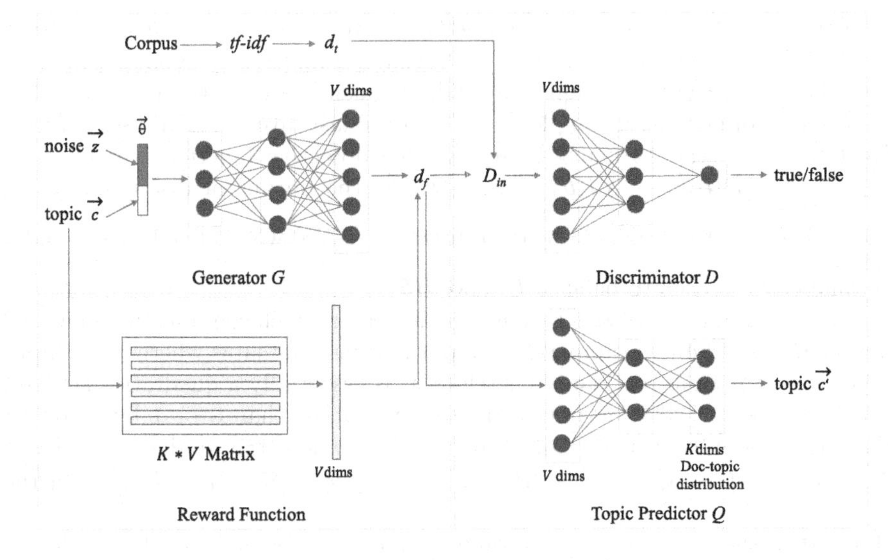

Fig. 1. The structure of RMATM.

can measure the association between each word and a certain topic. The process is detailed as $W_{m,n} = \frac{\sum_{i=1}^{L}(T_{mi} \times V_{ni})}{\sqrt{\sum_{i=1}^{L} T_{mi}^2} \times \sqrt{\sum_{i=1}^{L} V_{ni}^2}}$ and $R_{m,n} = W_{m,n} - \frac{1}{V}\sum_{j=1}^{V} W_{m,j}$, where $W_{m,n}$ means the association between the m-th topic and the n-th word in the vocabulary, T_{mi} and V_{ni} are the i-th items of $\boldsymbol{T}_m$ and $\boldsymbol{V}_n$, in which $\boldsymbol{T}_m$ is the embedding of the m-th topic-related seed word with a dimension of L, and $\boldsymbol{V}_n$ is the embedding of the n-th word in the vocabulary with a dimension of L. Furthermore, $W \in \mathbb{R}^{K \times V}$ is the reward matrix. Using the averaged reward value of each row as the baseline, and subtracting the baseline from the reward values of each row, we can get $R \in \mathbb{R}^{K \times V}$ which contains the positive and negative reward values. The reward function is $J(\boldsymbol{\theta}) = \frac{1}{V}\sum_{n=1}^{V} R_{m,n} \cdot \boldsymbol{d}_f^n$, where $\boldsymbol{\theta}$ is the input of generator G, $\boldsymbol{d}_f$ is the generated sample, $\boldsymbol{d}_f^n$ is the n-th value in $\boldsymbol{d}_f$, and $R_{m,n}$ represents the m-th row and the n-the column in matrix R. To maximize $J(\boldsymbol{\theta})$, we use the policy gradient by following [13].

The reward function can be built up in other ways which represent the relationship between two words. In our study, we focus on exploring the similarity of two words determined by word embeddings. Besides, we can use large-scale pre-trained embedding models and embeddings trained on the local corpus simultaneously to calculate rewards.

Generative Adversarial Net Architecture. After converting each document to vectors, we apply GAN to learn the true distribution of documents $\mathbb{P}_t$. In the structure of GAN, there is a generator G and a discriminator D, which both utilize the Multi-Layer Perception (MLP) structure.

The generator takes a discrete one-hot vector $\boldsymbol{c}$ and a continuous noise $\boldsymbol{z}$ as input $\boldsymbol{\theta}$, and tries to generate V-dimensional samples corresponding to topic $\boldsymbol{c}$ so that the generated samples are indistinguishable from the real ones. The distribution of generated samples is $\mathbb{P}_g$, and the last layer of MLP uses softmax function to get the normalized vectors. The objective function is defined by $L_G = \underset{\boldsymbol{d}_f \sim \mathbb{P}_g}{\mathbb{E}} [D(\boldsymbol{d}_f)] + \lambda_1 J(\boldsymbol{\theta})$, where $D(\cdot)$ is the output of discriminator D, $\underset{\boldsymbol{d}_f \sim \mathbb{P}_g}{\mathbb{E}} [D(\boldsymbol{d}_f)]$ is the original objective function of WGAN, defined as L_g, and a parameter λ_1 is used to balance L_g and $J(\boldsymbol{\theta})$.

The discriminator takes a V-dimensional vector which contains the generated sample $\boldsymbol{d}_f$ and the real sample $\boldsymbol{d}_t$ as input and discriminates whether the input is generated or from real data. Therefore, it is essentially a quadratic classifier. The last layer of MLP outputs a probability rather than 0 or 1 by following Wasserstein GAN with a gradient penalty [4], where the gradient penalty is described as $\hat{\boldsymbol{d}} = \varepsilon \boldsymbol{d}_f + (1 - \varepsilon)\boldsymbol{d}_t$ and $L_{gp} = \underset{\hat{d} \sim \mathbb{P}_{\hat{d}}}{\mathbb{E}} [(\|\nabla_{\hat{d}} D(\hat{d})\|_2 - 1)^2]$. In the above, $\boldsymbol{d}_t$ is the real sample, ε is a random number between 0 and 1, therefore $\hat{\boldsymbol{d}}$ can be sampled uniformly with a straight line between $\boldsymbol{d}_f$ and $\boldsymbol{d}_t$. Consequently, we describe the objective function of discriminator D as $L_D = \underset{\boldsymbol{d}_f \sim \mathbb{P}_g}{\mathbb{E}} [D(\boldsymbol{d}_f)] - \underset{\boldsymbol{d}_t \sim \mathbb{P}_t}{\mathbb{E}} [D(\boldsymbol{d}_t)] + \lambda_2 L_{gp}$, where λ_2 is the gradient penalty coefficient.

Topic Predictor $\boldsymbol{Q}$. Our topic predictor Q is also based on the MLP. It takes the generated samples $\boldsymbol{d}_f$ as input and outputs a topic category which corresponds to topic $\boldsymbol{c}$ in the training process. In the testing process, it takes the real samples as input and outputs its topics. The last layer of MLP will be considered as the document-topic distribution. Q uses cross-entropy as the loss function. The objective function is $L_Q = -\boldsymbol{c} \cdot \log Q(\boldsymbol{d}_f)$, where $Q(\cdot)$ is the output of module Q. We train the generator G, discriminator D, and topic predictor Q simultaneously, which can make the topic predictor Q learn more different samples.

3 Experiments

3.1 Datasets

In our experiments, four benchmark corpora are used after tokenization and eliminating non UTF-8 characters and stop words. Firstly, two publicly available datasets Grolier[1] and NYtimes[2] are used for topic coherence evaluation and a case study. Secondly, to evaluate the model performance on the tasks

[1] https://cs.nyu.edu/~roweis/data/.
[2] http://archive.ics.uci.edu/ml/datasets/Bag+of+Words.

of text clustering and weakly-supervised text classification, we employ 20newsgroups[3] and DBpedia[4] which contain labels. The statistics of datasets are shown in Table 1.

Table 1. The statistics of corpora.

Datasets	#Documents	#Words	#Classes
Grolier	30,991	15,276	-
NYtimes	300,000	44,442	-
20newsgroups	18,854	2,000	20
DBpedia	560,000	10,000	14

In our study, the following models are used as baselines: **LDA** [1], **GSM**[5] [8], **NVLDA**[6] [12], **ProdLDA**[6] [12], **ATM** [15], and **WeSHClass**[7] [7].

3.2 Experimental Setting and Results

Topic Evaluation. In this task, six topic coherence metrics, i.e., C_P, C_V, C_A, C_{NPMI}, C_{UCI}, and C_{UMass} [11] are used. Coherence score over top 10 words of each topic is evaluated by using the Palmetto library[8]. The numbers of topics are set to 20, 40, 60, 80, 100 and the averaged coherence scores are calculated as the final results, as shown in Table 2. A higher value indicates a better performance. In the case of RMATM, we randomly choose 20, 40, 60, 80, and 100 words from the lists of topics in news[9] and wikipedia[10] as interested topic-related seed words. It shows that the performance of LDA is stable, and the similar results between ATM and GSM indicate that the methods based on VAE or GAN have similar performance in topic mining. In addition, the results in the topic coherence task indicate that RMATM performs better than others. Based on RMATM's ability of discovering interested topics, we choose 8 predefined topics on NYtimes dataset, and top 10 words in each topic are shown in Table 3.

Text Clustering Performance Comparison. Here, we evaluate the model performance on text clustering using the actual labels as ground truth. The classical evaluation criteria are employed: Adjusted Rand index (ARI) [5], Adjusted

[3] http://qwone.com/~jason/20Newsgroups/.
[4] https://wiki.dbpedia.org/.
[5] https://github.com/linkstrife/NVDM-GSM.
[6] https://github.com/akashgit/autoencoding_vi_for_topic_models.
[7] https://github.com/yumeng5/WeSHClass.
[8] http://aksw.org/Projects/Palmetto.html.
[9] https://www.usnews.com/topics/subjects.
[10] https://en.wikipedia.org/wiki/Wikipedia:WikiProject_Lists_of_topics.

Table 2. Averaged topic coherence on Grolier and NYtimes datasets.

Dataset	Model	C_P	C_V	C_A	C_{NPMI}	C_{UCI}	C_{UMass}
Grolier	LDA	0.2001	0.4017	0.2264	0.0698	0.0842	−2.5350
	GSM	0.2178	0.4212	0.2158	0.0724	0.1752	−2.4335
	NVLDA	0.0024	0.4027	0.2012	0.0152	−0.7224	−3.3924
	ProdLDA	0.1026	0.4134	0.2255	0.0332	−0.3526	−3.1274
	ATM	0.2197	0.4226	0.2048	0.0679	0.1228	−2.7265
	RMATM	**0.2384**	**0.4352**	**0.2475**	**0.0874**	**0.2125**	**−2.3281**
NYtimes	LDA	0.3135	0.4058	0.2317	0.0815	0.4898	−2.3465
	GSM	0.3482	0.4450	0.2417	0.0831	0.6240	−2.0568
	NVLDA	0.0353	0.4082	0.2129	0.0264	−0.6286	−3.2932
	ProdLDA	0.0724	0.4858	0.2031	0.0325	−0.5851	−3.0283
	ATM	0.3245	0.4965	0.2395	0.0824	0.5065	−2.3475
	RMATM	**0.3612**	**0.5136**	**0.2672**	**0.1023**	**0.7233**	**−1.8563**

Table 3. Top 10 words which represent each topic generated by RMATM under the topic-related seed words of "computer", "political", "space", "energy", "disease", "movie", "art", and "music" on NYtimes dataset.

Topic	Words
Computer	Signal processing data research digital system device application information electrical
Political	Election coalition democratic premier parliament minister communist government leader party
Space	Plane planet ocean aircraft surface satellite telescope orbit solar earth
Energy	Electrical earth oil pollution fuel coal temperature pressure heat gas water
Disease	Treatment system humans brain bacteria skin symptom infection blood cell
Movie	Writer love screen character fiction actor novel story Hollywood film
Art	Painting abstract poetry gallery color artist photography sculpture collection museum
Music	Conductor performed theater symphony jazz director orchestra composer piano opera

Mutual information (AMI) [14], and Fowlkes-Mallows index (FMI) [2]. The values of ARI, AMI, and FMI are between 0 and 1, and a higher value indicates a larger similarity between the clusters and the benchmark labels. Topic numbers of the compared models are set to 20 on 20newsgroups and 14 on DBpedia. In RMATM, the numbers of topics are also respectively set to 20 and 14 and interested topic-related seed words are chosen as aforementioned. The results are shown in Table 4, which indicate that RMATM achieves the best performance in text clustering by introducing the external information about topics of documents. Furthermore, the existing GAN-based model ATM is unable to capture the document-topic distributions, which has been addressed by our GAN-based method for topic modeling.

Table 4. Performance of text clustering in 20newsgroups and DBpedia datasets.

Method	20newsgroups			DBpedia		
	ARI	AMI	FMI	ARI	AMI	FMI
LDA	0.207	0.3589	0.257	0.383	0.5012	0.423
GSM	0.177	0.3182	0.235	0.532	0.6236	0.563
NVLDA	0.153	0.2730	0.197	0.497	0.6044	0.524
ProdLDA	0.128	0.2454	0.174	0.455	0.5541	0.494
RMATM	**0.248**	**0.3719**	**0.287**	**0.569**	**0.6513**	**0.602**

Weakly-Supervised Text Classification. By exploiting class-related keywords as weak supervision and generating pseudo documents, WeSHClass is able to classify unlabeled documents. In this part, we set up the same keywords and use macro-F1 and micro-F1 to evaluate the model performance on weakly-supervised text classification. Due to space constraint, the total keywords are not detailed here. Following WeSHClass, the same keywords are used as topic-related seed words for the interested topics in RMATM. The result in Table 5 indicates that RMATM achieves better performance on 20newsgroups and DBpedia datasets when compared to WeSHClass. The reason may be that based on the strong generalisation ability of GAN, our RMATM is able to generate higher quality pseudo documents which correspond to real ones.

Table 5. Weakly-supervised text classification results on 20newsgroups and DBpedia.

Method	20newsgroups		DBpedia	
	Macro-F1	Micro-F1	Macro-F1	Micro-F1
WeSHClass	0.3924	0.4182	0.6288	0.7014
RMATM	**0.4137**	**0.4317**	**0.7378**	**0.7441**

Impact of Word Embeddings on the Reward Function. In RMATM, we use word embedding models[11] pre-trained on different corpora to evaluate that how different word embeddings in the reward function will influence the effectiveness of RMATM. Meanwhile, we train word embeddings [10] on 20newsgroups. The result is shown in Table 6, which indicates that word embedding

[11] https://github.com/3Top/word2vec-api.

trained with the similar type of corpus achieves better performance. In addition, by applying different word embeddings to calculate the averaged reward values, we can further improve the performance of weakly-supervised text classification.

Table 6. Micro-F1 in weakly-supervised text classification using different word embeddings in the reward function.

Dataset	Embedding corpus	Dimensions	Micro-F1
20newsgroups	Google News	300	0.3875
	DBpedia	1000	0.2686
	20newsgroups	100	0.4056
	20newsgroups+Google News	100(300)	**0.4317**
DBpedia	Google News	300	0.6988
	DBpedia	1000	0.7256
	Google News+DBpedia	300(1000)	**0.7441**

4 Conclusions

In order to discover meaningful information in documents without labels, we propose a method named RMATM based on GAN in conjunction with a topic predictor and a reward function. Our method can guide the generator to discover interested topics for weakly-supervised text classification. In the result of topic coherence and text clustering, our model achieves better performance than various baselines, and the reward function helps for discovering more meaningful topics. In the task of weakly-supervised text classification, our model also achieves a better performance than WeSHClass.

Acknowledgment. The research described in this paper was supported by the National Natural Science Foundation of China (61972426).

References

1. Blei, D.M., Ng, A.Y., Jordan, M.I.: Latent Dirichlet allocation. J. Mach. Learn. Res. **3**, 993–1022 (2003)
2. Fowlkes, E.B., Mallows, C.L.: A method for comparing two hierarchical clusterings. J. Am. Stat. Assoc. **78**(383), 553–569 (1983)
3. Goodfellow, I.J., et al.: Generative adversarial nets. In: NIPS, pp. 2672–2680 (2014)
4. Gulrajani, I., Ahmed, F., Arjovsky, M., Dumoulin, V., Courville, A.C.: Improved training of Wasserstein GANs. In: NIPS, pp. 5767–5777 (2017)
5. Hubert, L., Arabie, P.: Comparing partitions. J. Classif. **2**(1), 193–218 (1985)
6. Kingma, D.P., Welling, M.: Auto-encoding variational Bayes. In: ICLR (2014)

7. Meng, Y., Shen, J., Zhang, C., Han, J.: Weakly-supervised hierarchical text classification. In: AAAI, pp. 6826–6833 (2019)
8. Miao, Y., Grefenstette, E., Blunsom, P.: Discovering discrete latent topics with neural variational inference. In: ICML, pp. 2410–2419 (2017)
9. Miao, Y., Yu, L., Blunsom, P.: Neural variational inference for text processing. In: ICML, pp. 1727–1736 (2016)
10. Mikolov, T., Chen, K., Corrado, G., Dean, J.: Efficient estimation of word representations in vector space. In: ICLR (2013)
11. Röder, M., Both, A., Hinneburg, A.: Exploring the space of topic coherence measures. In: WSDM, pp. 399–408 (2015)
12. Srivastava, A., Sutton, C.A.: Autoencoding variational inference for topic models. In: ICLR (2017)
13. Sutton, R.S., McAllester, D.A., Singh, S.P., Mansour, Y.: Policy gradient methods for reinforcement learning with function approximation. In: NIPS, pp. 1057–1063 (1999)
14. Vinh, N.X., Epps, J., Bailey, J.: Information theoretic measures for clusterings comparison: variants, properties, normalization and correction for chance. J. Mach. Learn. Res. **11**, 2837–2854 (2010)
15. Wang, R., Zhou, D., He, Y.: ATM: adversarial-neural topic model. Inf. Process. Manage. **56**(6) (2019)

Link Inference via Heterogeneous Multi-view Graph Neural Networks

Yuying Xing[1,2], Zhao Li[1(✉)], Pengrui Hui[1], Jiaming Huang[1], Xia Chen[1], Long Zhang[1], and Guoxian Yu[2]

[1] Alibaba Group, Hangzhou 311121, China
{lizhao.lz,pengrui.hpr,jimmy.hjm,xia.cx,james.zl}@alibaba-inc.com
[2] College of Computer and Information Science, Southwest University, Chongqing 400715, China
{yyxing4148,gxyu}@swu.edu.cn

Abstract. Graph Neural Network (GNN) has been attaching great attention along with its successful industry applications, such as social network, recommendation system and so on. Most existing GNN algorithms for link inference tasks mainly concentrate on homogeneous network where single typed nodes and edges are considered. Besides, they are transductive, incapable of handling unseen data, and are difficult to generalize to big graph data. In this paper, we introduce a new idea, i.e. Heterogeneous Multi-view Graph Neural Network (HMGNN), to remedy these problems. A more complex and unstudied heterogeneous network structure where multiple node and edge types co-exist, and each of them also contains specific attributes, is learned in this framework. The proposed HMGNN is end-to-end and two stages are designed: i) The first stage extends the widely-used GraphSAGE model to the studied heterogeneous scenario to generate the vector embedded representations for each type of nodes. ii) The second stage develops a novel and inductive subspace-based strategy for link inference by aggregating multi-typed node and edge feature views. Comprehensive experiments on large-scale spam detection and link prediction applications clearly verify the effectiveness of our model.

Keywords: Heterogeneous · Graph Neural Network · Multi-view

1 Introduction

Graph Neural Network (GNN) [15,21] is a popular deep learning framework for graph data modeling and analysis. Recently, it has attached great attention due to its superior performance, and was widely used in various fields, such as recommendation systems [14] and anomaly detection [12]. Traditional GNN-based methods are generally used for node classification, link inference (e.g. edge classification and link prediction), community detection and many other downstream network learning tasks. Despite that current GNN models (GNNs) show

Y. Nah et al. (Eds.): DASFAA 2020, LNCS 12112, pp. 698–706, 2020.
https://doi.org/10.1007/978-3-030-59410-7_48

their great success on link inference applications [15], however, there remain several challenging issues they face: (i) Most existing GNNs [7,18] mainly focus on homogeneous graphs (used interchangeably with *networks*), where single typed nodes and edges are considered. Although several heterogeneous works [1,13] are presented, but they do not generalize to more complex multiplex heterogeneous networks. Figure 1 gives a real-world example of the spam detection scenario, where multi-typed nodes and edges recording different operations co-exist in the user-item bipartite graph. Moreover, each type of nodes and edges both have specific contents (given as attribute vectors). Unfortunately, none of the state-of-the-art approaches can handle such a intricate heterogeneous network as far as we know; (ii) It remains a challenge to scale up most of them since the limitation of core assumptions underlying their designs are violated when working in a big data environment; (iii) Most of these approaches [5,11] are transductive, which are incapable of dealing with previously unseen edges. Factually, one application platform can generate billions of data every day, which causes the scalability of homogeneous/heterogeneous GNNs become more difficult, and it is unrealistic to conduct repetitive training to capture the evolving network.

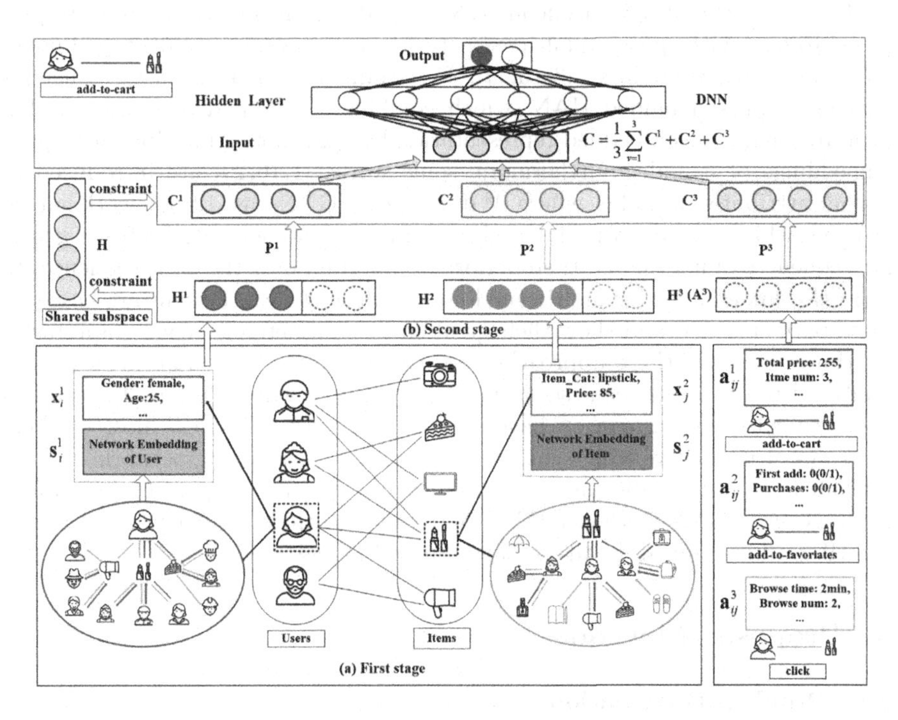

Fig. 1. An example of heterogeneous multi-view graph neural network. Two kinds of node objects (i.e. users and items), and three edge types (i.e. click, add-to-favourites and add-to-cart) co-exist in the user-item bipartite graph.

To remedy these challenges, this paper introduces a novel idea named HMGNN for addressing multi-typed nodes and edges with attributes in heterogeneous multi-view graphs based on graph neural network. The presented HMGNN is end-to-end and two stages are designed in this framework. The first stage extends the widely-used GraphSAGE algorithm to the above mentioned heterogeneous scenario to learn embeddings for each type of nodes. The second stage develops an inductive multi-view generator based on subspace learning strategy to fuse multiple node and edge feature views (i.e. feature representations) for link inference. It also addresses the difficulty that most existing multi-view learning approaches challenge to scale to big heterogeneous graph data. Comprehensive experiments on large-scale spam detection and link prediction applications clearly verify the effectiveness of our model.

2 Related Works

Here we only focus our discussion of related work on heterogeneous GNNs, and a recent comprehensive review of GNN can be found at [15].

Recently, several heterogeneous GNN-based algorithms [3,4,13,17] are proposed to deal with specific multiplex network structures, e.g. multi-typed nodes [10] and multi-typed links [3,13,17]. For example, [13] introduced a heterogeneous approach named HAN, which combines GNN and attention mechanism to learn node embedded representation by aggregating multiple meta-path embeddings of a node. Furthermore, some works [1,16,20] are introduced to exploit attributes of nodes, where [1] developed a GNN-based framework named GATNE which uses meta-path based random walk and node attributes to generate sequences of nodes, and then performs skip-gram over the node sequences to learn the embeddings for multi-typed nodes. Most of these designs are based on meta-path strategy, and show their great success on heterogeneous graph tasks.

However, all of these approaches do not generalize to a more intricate but really existed heterogeneous scenario where not only various types of nodes and edges co-exist, but also they have specific attributes. Besides, most of them [5,11] are transductive which challenge to capture the evolving network, and are difficult to scale to graph tasks with billions of data. This paper develops a new heterogeneous GNN method (HMGNN) to address these challenges, and Sect. 3 describes the technical details of our proposed model.

3 Proposed Method

3.1 Problem Formulation

Denote $\mathcal{G} = (\mathcal{V}, \mathcal{E}, \mathcal{X}, \mathcal{A})$ is a heterogeneous network, which contains R node types and T types of edges between nodes. The following contents show the definitions of multi-typed nodes and edges.

- **Node Definition:** $\mathcal{V} = \{\mathcal{V}^r\}_{r=1}^R$ and $\mathcal{X} = \{\mathcal{X}^r\}_{r=1}^R$ collect R types of nodes and attributes of these nodes, respectively. $\mathcal{V}^r = \{v_1^r, v_2^r, ..., v_{n_r}^r\}(1 \leq r \leq R)$ represents the r-th type node set including n_r nodes, and $\mathcal{X}^r = \{\mathbf{x}_1^r, \mathbf{x}_2^r, ..., \mathbf{x}_{n_r}^r\}$ is the corresponding attribute set. v_i^r is the i-th node of $\mathcal{V}^r$, and $\mathbf{x}_i^r \in \mathbb{R}^{d_v^r}(1 \leq i \leq n_r)$ records the node features of v_i^r.
- **Edge Definition:** Similar to the node definition, we have $\mathcal{E} = \{\mathcal{E}^t\}_{t=1}^T$ and $\mathcal{A} = \{\mathcal{A}^t\}_{t=1}^T$, where $\mathcal{E}^t = \{e_{ij}^t\}(1 \leq t \leq T, 1 \leq i \leq n_{r_1}, 1 \leq j \leq n_{r_2})$ and $\mathcal{A}^t = \{\mathbf{a}_{ij}^t\}$ show the t-th type edge set including m_t links and corresponding edge contents, respectively. If two nodes from different node types, e.g., $v_i^{r_1}$ and $v_j^{r_2}$, are associated with the t-th type of edge, we have $e_{ij}^t \in \mathcal{E}^t$, and $\mathbf{a}_{ij}^t \in \mathbb{R}^{d_e^t}$ is attribute vector of e_{ij}^t. Vector $\mathbf{Y}_{ij}^t \in \mathbb{R}^q$ encodes the labels w.r.t e_{ij}^t where $\mathbf{Y}_{ij}^t(l) = 1(1 \leq l \leq q)$ represents e_{ij}^t is annotated with the l-th label, $\mathbf{Y}_{ij}^t(l) = 0$ otherwise.

Given $\mathcal{G}$, the objective of HMGNN is to make predictions for multiple types of edges, i.e $\mathcal{F} = \{\mathbf{F}^t\}_{t=1}^T$ by capturing complex heterogeneous structure of $\mathcal{G}$.

3.2 The Proposed HMGNN

The following contents show two stages of HMGNN for node embedding generation and data fusion of multi-typed nodes and edges, respectively.

Learning Node Embeddings from Heterogeneous Network. The first stage of HMGNN targets to learn the node embeddings for each type of nodes. For a certain typed node $v_i^r \in \mathcal{V}^r(1 \leq i \leq n_r, 1 \leq r \leq R)$ of $\mathcal{G}$, we split its overall representation $\mathbf{h}_i^r$ into two parts: the node embedding with structural information $\mathbf{s}_i^r$, and the known node attributes $\mathbf{x}_i^r$. To obtain $\mathbf{s}_i^r$, HMGNN first reconstructs the heterogeneous network $\mathcal{G}$ as shown in Fig. 2. Let v_i^r and $v_j^{r_1}(1 \leq r, r_1 \leq R, r_1 \neq r)$ be two nodes from different types of node sets, we construct a link if there exists at least one type of edges between these two nodes, i.e. if $\exists e_{ij}^t(1 \leq t \leq T)$, we have $\tilde{e}_{ij} \in \tilde{\mathcal{E}}$. Thus a new heterogeneous network $\tilde{\mathcal{G}}(\mathcal{V}, \tilde{\mathcal{E}})$ can be constructed based on the original $\mathcal{G}$.

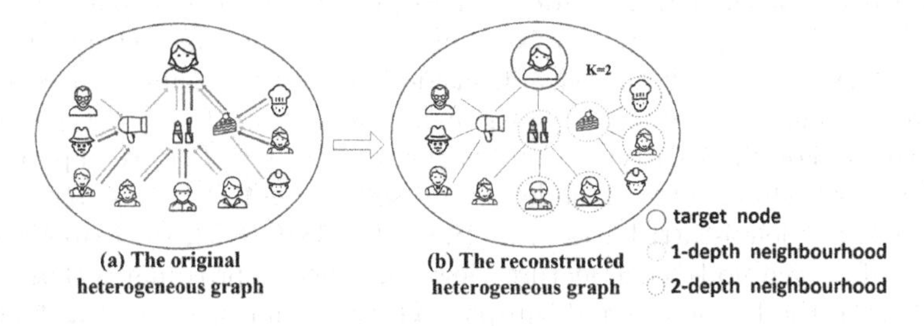

Fig. 2. Heterogeneous network reconstruction (The left figure shows the original heterogenous network $\mathcal{G}$, and the right one is the reconstructed graph $\tilde{\mathcal{G}}$).

Next, we extend the popular GraphSAGE algorithm [7] to the reconstructed heterogeneous graph $\tilde{\mathcal{G}}$ for the node embedding (i.e. $\mathbf{s}_i^r$) generation of v_i^r. We first sample fixed-size K depth heterogeneous neighbourhoods rather than homogeneous nodes in GraphSAGE for v_i^r as shown in Fig. 2. For example, for a user node in the bipartite graph, we sample specific number of items as its 1-depth neighbours, and users with different node type from 1-depth neighbourhood as 2-depth neighbours, and so on the K-th depth neighbours. To avoid an explosive increase of training complexity, not all heterogeneous neighbors are used for sampling. We define $\tilde{N}(v_i^r)$ as the sampled neighbourhood number of v_i^r drawn from the set $\{v_j^{r_1} \in \mathcal{V}^{r_1} : (v_i^r, v_j^{r_1}) \in \tilde{\mathcal{E}}\}$. After that, we adopt forward propagation algorithm for the generation of $\mathbf{s}_i^r$ by aggregating the neighbourhood features of the local sampled node sets. Specially, we utilize the mean aggregator [7] to calculate $\mathbf{s}_i^r$ in the k-th ($k \in 1, 2, ..., K$) iteration as follows:

$$\mathbf{s}_i^r(k) \longleftarrow \delta(\widetilde{\mathbf{W}} \cdot MEAN(\{\mathbf{s}_i^r(k-1)\} \cup \{\mathbf{s}_j^{r_1}(k-1), \forall v_j^{r_1} \in \tilde{N}(v_i^r))\}) \quad (1)$$

where $\mathbf{s}_i^r(k)(1 \leq k \leq K)$ represents node embedding of v_i^r in the k-th iteration, which incorporates features from local heterogeneous neighbourhoods of v_i^r, and the elementwise mean of the neighbourhood vectors simply are taken in $\{\mathbf{s}_j^{r_1}(k-1), \forall v_j^{r_1} \in \tilde{N}(v_i^r)\}$. $\delta(.)$ is the sigmoid function. $AGGREGATE_k = \widetilde{\mathbf{W}} \cdot MEAN(\cdot)$ is the mean aggregator function where $\widetilde{\mathbf{W}} = \{\widetilde{\mathbf{W}}^k\}_{k=1}^K$ includes the weight parameter matrices of K aggregators functions, which are used to propagate information between different layers. Considering supervised information of edges, this paper adopts cross-entropy loss function [19] for the parameter learning of K aggregator functions.

To this end, we can obtain the final embedded representation of v_i^r by exploiting the structural information and attributes of its K depth sampled neighbourhoods, and the overall representation through $\mathbf{h}_i^r = [\mathbf{s}_i^r(K), \mathbf{x}_i^r]$. The first stage of HMGNN not only considers complex relations of multi-typed nodes of $\mathcal{G}$ in an inductive manner, but also greatly saves the computing overhead of node embeddings generation. Experimental studies also show its effectiveness on capturing heterogeneous structure of $\mathcal{G}$.

Data Fusion of Multiple Node and Edge Views. The second stage introduces a multi-view generator to fuse multi-typed nodes and edges representations (i.e. views) for link inference. Considering the case of imbalanced number of different types of edges, this paper makes predictions for each type of edges respectively. For the t-th type of edges (i.e. $\mathcal{E}^t(1 \leq t \leq T)$), we suppose there are V views including $V-1$ node feature sets (i.e. $\mathbf{H}^r(1 \leq r \leq R)$) w.r.t $\mathcal{E}^t$, and the V-th view is attributes of $\mathcal{E}^t$ (i.e. $\mathbf{A}^t$). HMGNN first creates a virtual subspace for each node/edge feature view, and then approximates those virtual subspaces to the learned shared subspace $\mathbf{H}$ across multiple views as follows:

$$Loss_{MV} = \frac{1}{V}\sum_{v=1}^{V} \| \mathbf{C}^v - \mathbf{H} \|_2^2, \ where\ \mathbf{C}^v = \mathbf{H}^v\mathbf{P}^v \quad (2)$$

where $\mathbf{H}^v \in \mathbb{R}^{m_t \times d_v}$ is the v-th node $(1 \leq v \leq V-1)$/edge $(v = V)$ feature views, and m_t is the total edge number of $\mathcal{E}^t$. $\mathbf{P}^v \in \mathbb{R}^{d_v \times d}$ is the projection matrix of the v-th view, which can used during both the model training and testing process since the specific characteristics of each views. $\mathbf{C}^v = \mathbf{H}^v\mathbf{P}^v$ represents the target virtual subspace of the v-th view. $\mathbf{H} \in \mathbb{R}^{m_t \times d}$ is the shared subspace across multiple node feature views and the target edge feature view. Equation (2) aims to generate virtual spaces of multiple node and the target edge views by mapping V original node and edge feature spaces with high dimension to the learned low-dimensional shared subspace $\mathbf{H}$. By this means, HMGNN can predict $\mathcal{E}^t$ in an inductive manner, and reduce the time costs of making prediction for unseen edges by directly using the aggregated virtual subspace $\frac{1}{V}\sum_{v=1}^{V}\mathbf{C}^v$.

To predict the labels of edges of $\mathcal{E}^t$, this paper adopts the widely-used Deep Neural Network method (DNN [9]) as base classifier, and other classification approaches can also be applied here. Thus for a seen or unseen edge e_{ij}^t, we can make prediction w.r.t the l-th label by using $\mathbf{F}_{ij}^t(l) = f(\mathbf{w}^T\mathbf{c}_{ij}^t)(1 \leq l \leq q)$, where $f(.)$ is softmax function, $\mathbf{c}_{ij}^t$ represents the aggregated virtual subspace vector w.r.t e_{ij}^t, and $\mathbf{w}$ is the corresponding weighted vector of e_{ij}^t. By this means, HMGNN can infer the labels of specific type of links.

In summary, the unified objective function of the proposed HMGNN is defined as follows:

$$Loss = \lambda Loss_{MV} + Loss_{CE} \tag{3}$$

where $Loss_{MV}$ represents the data fusion loss of multiple node and target edge views. λ is a balance parameter. $Loss_{CE}$ is the cross entropy loss which contains two parts: one is for the parameter learning of multiple aggregator functions in the first stage, and the other is classification loss of used DNN model in the second stage. Equation (3) makes node embedding generation, data fusion of multi-typed nodes and edges, and link inference in a coordinate fashion, and we adopt the widely-used gradient descent method [8] for parameter optimization of HMGNN.

4 Experiments

4.1 Experimental Setup

Three datasets including two public heterogeneous network datasets [1] (i.e. Amazon: 10,166 single typed nodes, 148,865 edges form 2 edge types; Twitter: 100,000 single typed nodes, 331,899 edges form 4 edge types), and one real Taobao dataset (acquired from online Taobao e-commerce platform, which contains 1500,000,000 nodes from 2 node types and 4104,603,557 edges from 3 edge types) are used in this paper. We apply the presented HMGNN model to the large-scale e-commerce spam detection problem on Taobao dataset, and heterogeneous graphs on Amazon and Twitter datasets to explore the performance of HMGNN on edge classification and link prediction tasks, respectively. The widely-used binary classification metrics (i.e. the area under the Precision-recall curve [2] (PR-AUC) and F1-score) are adopted for evaluation.

4.2 Experimental Results and Analysis

Performances on Large-Scale Spam Detection Task. The bipartite graph of e-commerce spam detection scenario shown in Fig. 1 is a typical heterogeneous network. This paper models such complex data for large-scale fraudulent transactions (i.e. spam) detection by predicting the 'add-to-cart' edges. To investigate the performance of HMGNN on this task, two related deep learning frameworks [6,9] and two variants of HMGNN are used for comparison. 2 depth with 10 fixed-size neighbourhoods are sampled for each user/item node, the shared space dimension d and the learning rate of used DNN classifier are set as 128 and 0.01, respectively. Table 1 reports the experimental results. From this table, we can see that: (i) Both DNN, IFCM and HMGNN use DNN as base classifier, HMGNN outperforms them, which shows the effectiveness of HMGNN on capturing complex relations of heterogeneous user-item bipartite graph. (ii) HMGNN outperforms its two variants (i.e. HMGNN(Nh), HMGNN(Nm)), where HMGNN(Nh) only considers single typed 'add-to-cart' edges for node embedding generation, and HMGNN(Nm) simply splices multiple node and the target edge feature views into a single one. This fact reflects the effectiveness of HMGNN on integrating multi-typed edges and multi-view data fusion strategy, respectively.

Table 1. Results of HMGnn on the real EcomPlat dataset.

Datasets	Metrics	Methods				
		DNN	IFCM	**HMGNN(Nh)**	**HMGNN(Nm)**	**HMGNN**
Taobao	*PR*-AUC	77.23	79.04	79.15	79.13	**79.76**
	F1-score	72.08	73.79	75.28	75.81	**76.09**

Performances on Link Prediction Task. Link prediction is a key problem which targets to predict the possible links by learning the known network-structured data. To evaluate the performance of HMGNN on this task, this paper conducts it on two public heterogeneous networks. Four recently proposed approaches including one homogeneous network embedding method ANRL [18] and three heterogeneous network embedding algorithms (i.e. MNE [17], GATNE_T [1](transductive) and GATNE_I [1](inductive)), and one HMGNN variant HMGNN(Nm) are used for comparison. The parameter settings are as follows: We sample 2 depth 5 fixed-size neighborhoods for the single typed nodes. $d = 128$, and the learning rate of DNN classifier on Amazon and Twitter are set as 0.015 and 0.03, respectively. Table 2 reports the averaged results of multi-typed edges predictions. From this table, we can observe that: (i) HMGNN achieves the best performance compared to ANRL, MNE, GATNE_T and GATNE_I, this fact indicates the effectiveness of HMGNN on integrating the multi-typed nodes and edges. (ii) HMGNN outperforms its variants HMGNN(Nm), and this observation again proves the effectiveness of multi-view data fusion strategy of HMGNN.

Table 2. Results of HMGNN on different heterogeneous network datasets.

Datasets	Metrics	Methods					
		ANRL	MNE	GATNE_T	GATNE_I	**HMGNN(Nm)**	**HMGNN**
Amazon	*PR*-AUC	71.68	90.28	**97.44**	96.25	78.50	**97.44**
	F1-score	67.72	83.25	92.87	91.36	73.48	**95.35**
Twitter	*PR*-AUC	70.04	91.37	92.30	92.04	79.86	**93.98**
	F1-score	64.69	84.32	84.69	84.38	81.73	**92.83**

Besides, we also tested the sensitivity of the important balance parameter λ by varying it in the range of $\{0.001, 0.01, 0.1, 1, 10, 100, 1000\}$. The results on Amazon dataset show that an overall good performance can be achieved when $\lambda = 10$, and too small or too large λ will cause a low performance of HMGNN, which shows the importance of a reasonable λ.

5 Conclusion

This paper introduces a novel idea based on multi-view subspace learning into the heterogeneous GNN for link inference tasks. Two stages are included in this framework where the first one exploits both the structural information and attributes of multi-typed nodes for node embedding generation. The second stage develops an inductive multi-view generator for link inference by fusing multiple node and edge features views. Experimental studies on edge classification and link prediction applications both show the effectiveness of our model.

References

1. Cen, Y., Zou, X., Zhang, J., Yang, H., Zhou, J., Tang, J.: Representation learning for attributed multiplex heterogeneous network. In: KDD (2019)
2. Davis, J., Goadrich, M.: The relationship between precision-recall and ROC curves. In: ICML, pp. 233–240 (2006)
3. Dong, Y., Chawla, N.V., Swami, A.: metapath2vec: scalable representation learning for heterogeneous networks. In: KDD, pp. 135–144 (2017)
4. Fan, S., et al.: Metapath-Guided Heterogeneous Graph Neural Network for Intent Recommendation (2019)
5. Grover, A., Leskovec, J.: node2vec: scalable feature learning for networks. In: KDD, pp. 855–864 (2016)
6. Guo, Q., et al.: Securing the deep fraud detector in large-scale e-commerce platform via adversarial machine learning approach. In: WWW, pp. 616–626 (2019)
7. Hamilton, W., Ying, Z., Leskovec, J.: Inductive representation learning on large graphs. In: NIPS, pp. 1024–1034 (2017)
8. Hinton, G.E., Salakhutdinov, R.R.: Reducing the dimensionality of data with neural networks. Science **313**(5786), 504–507 (2006)
9. Kim, B., Kim, J., Chae, H., Yoon, D., Choi, J.W.: Deep neural network-based automatic modulation classification technique. In: ICTC, pp. 579–582 (2016)

10. Liu, Z., Chen, C., Yang, X., Zhou, J., Li, X., Song, L.: Heterogeneous graph neural networks for malicious account detection. In: CIKM, pp. 2077–2085 (2018)
11. Wang, D., Cui, P., Zhu, W.: Structural deep network embedding. In: KDD, pp. 1225–1234 (2016)
12. Wang, J., Wen, R., Wu, C., Huang, Y., Xion, J.: FdGars: fraudster detection via graph convolutional networks in online app review system. In: WWW, pp. 310–316 (2019)
13. Wang, X., et al.: Heterogeneous graph attention network. In: KDD, pp. 2022–2032 (2019)
14. Wu, S., Tang, Y., Zhu, Y., Wang, L., Xie, X., Tan, T.: Session-based recommendation with graph neural networks. In: AAAI, vol. 33, pp. 346–353 (2019)
15. Wu, Z., Pan, S., Chen, F., Long, G., Zhang, C., Yu, P.S.: A comprehensive survey on graph neural networks. arXiv preprint arXiv:1901.00596 (2019)
16. Zhang, C., Song, D., Huang, C., Swami, A., Chawla, N.V.: Heterogeneous graph neural network. In: KDD, pp. 793–803 (2019)
17. Zhang, H., Qiu, L., Yi, L., Song, Y.: Scalable multiplex network embedding. In: IJCAI, vol. 18, pp. 3082–3088 (2018)
18. Zhang, Z., et al.: ANRL: attributed network representation learning via deep neural networks. In: IJCAI, vol. 18, pp. 3155–3161 (2018)
19. Zhang, Z., Sabuncu, M.: Generalized cross entropy loss for training deep neural networks with noisy labels. In: NIPS, pp. 8778–8788 (2018)
20. Zheng, L., Li, Z., Li, J., Li, Z., Gao, J.: AddGraph: anomaly detection in dynamic graph using attention-based temporal GCN. In: IJCAI, pp. 4419–4425 (2019)
21. Zhou, J., Cui, G., Zhang, Z., Yang, C., Liu, Z., Sun, M.: Graph neural networks: a review of methods and applications. arXiv preprint arXiv:1812.08434 (2018)

SAST-GNN: A Self-Attention Based Spatio-Temporal Graph Neural Network for Traffic Prediction

Yi Xie[1,2], Yun Xiong[1,2,3(✉)], and Yangyong Zhu[1,2,3]

[1] School of Computer Science, Fudan University, Shanghai, China
{18110240043,yunx,yyzhu}@fudan.edu.cn
[2] Shanghai Key Laboratory of Data Science, School of Computer Science, Fudan University, Shanghai, China
[3] Shanghai Institute for Advanced Communication and Data Science, Fudan University, Shanghai, China

Abstract. Traffic prediction, which aims at predicting future traffic conditions based on historical observations, is of considerable significance in urban management. However, such tasks are challenging due to the high complexity of traffic flow. Traditional statistical learning methods perform badly in handling such regularities. Furthermore, these methods often split spatial information and temporal information as individual features, which cause extracted features hard to be fused. In this paper, we target on efficiently capturing such highly nonlinear dependencies and fusing features from both spatial and temporal dimensions to predict future traffic flow conditions accurately. To tackle this problem, we proposed the **S**elf-**A**ttention based **S**patio-**T**emporal **G**raph **N**eural **N**etwork (SAST-GNN). In SAST-GNN, we innovatively proposed to add a self-attention mechanism to more accurately extract features from the temporal dimension and the spatial dimension simultaneously. At the same time, for better fusing information from both dimensions, we improved the spatio-temporal architecture by adding channels(or heads) and residual blocks. Thus features can be fused from different aspects. Experiments on two real-world datasets illustrated that our model outperforms six baselines in traffic flow prediction tasks. Especially in short-term and mid-term prediction based on long-term dependencies, our model performs much better than other baselines.

Keywords: Traffic flow prediction · Spatio-temporal network · Self-attention

1 Introduction

Accurate prediction of future traffic conditions in urban roads can be widely used in urban management for alleviating congestion, planning routes, and controlling flow, *etc.* However, with the trend of global urbanization, traffic condition is

Y. Nah et al. (Eds.): DASFAA 2020, LNCS 12112, pp. 707–714, 2020.
https://doi.org/10.1007/978-3-030-59410-7_49

increasingly complicated. Earlier works mainly focus on statistical learning methods. For example, Support Vector Regression (SVR) [2] and Auto-Regressive and Moving Average model (ARMA) [1] for temporal features extracting, while k-Nearest Neighbor algorithm (KNN), Support Vector Machine (SVM) for spatial features. However, such statistical algorithms perform poorly due to the extracted features from both dimensions are split, without comprehensive consideration for fusing.

The neural networks enabled people the ability to capture information from both dimensions and fuse them. In subsequent studies, researchers formulate traffic flow prediction tasks as a graphic modeling problem in the spatial dimension, while using a sequential model [4] to capture temporal features [3]. Recent studies model such topology on a graph as non-Euclidean structure and apply Graph Convolutional Networks (GCN) [5,6] to extract spatial features [8–10,13,15]. Since GCN is designed to handle irregular structures, such improvements are more suitable for urban road networks. The illustration of traffic flow prediction tasks are shown in Fig. 1.

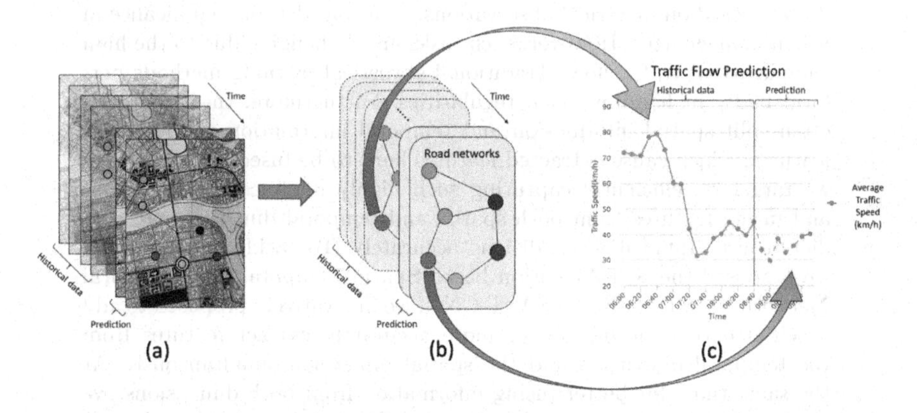

Fig. 1. (**a**) The sequence of urban maps, in which colored points represent sensors. (**b**) The sequence of urban maps is abstracted as a sequence of urban road networks. The topology of urban road networks is regarded as spatial features and formulated as a non-Euclidean structure, where nodes represent sensors while edges represent corresponding roads. The features of the urban road networks changing all the time are formulated as a time sequence. Our purpose is to accurately predict future traffic conditions based on historical data for each node. (**c**) The illustration of the traffic flow prediction tasks in a specific node.

However, problems in GCN and sequential models are also serious:

- Aggregating information by GCN might introduce noise since the learned weight matrix assigns each item of features a particular weight individually and ignores correlations between nodes;
- Classical recursive sequential models to extract temporal features might be limited by weak long-term dependencies.

For overcoming problems mentioned above, we propose SAST-GNN, an algorithm that uses the completely self-attention mechanism to capture information from both dimensions.

The main contributions of this paper are following:

- We incorporate self-attention blocks in spatial and temporal extractors. Such an attention-based model can efficiently filter noise from nodes with lower relationships and aggregate higher quality information. Besides, the self-attention mechanism can also enable our model to deal with inductive tasks.
- We propose an improved structure for traffic flow prediction tasks, in which spatial and temporal features are extracted by multiple heads or channels, thus making the captured features more diversified so that the learned representations more expressive.
- We conduct several analysis experiments with two real-world traffic datasets. The experimental results demonstrate our proposed model and its variants outperform six comparison methods.

2 Problem Definition

Definition 1 (Traffic flow graph sequence). In timestamp t, the traffic flow graph is represented as a weighted graph $G_t = (V, X_t, E, W)$, whose nodes are sensors in urban road networks to record traffic speed, denoted as $V \in R^{|V|}$; $X_t \in R^{|V| \times d_V}$ is the recorded traffic speed in timestamp t for each sensor; roads that connect sensors are regarded as edges in the graph, denoted as $E \in R^{|E|}$; length of the corresponding road is formulated as the weight of edges, denoted as $W \in R^{|E| \times d_E}$. V, E, W is seen as a part of the topological information of a traffic flow graph, which is static all the time. The only variable in the traffic flow graph is X_t. Therefore, graphs in each timestamp compose to a graph sequence $G = \{G_0, G_1, ..., G_T\}$, where T is the total recorded timesteps.

Definition 2 (Traffic flow prediction problems). Suppose that current time is $t-1$, and we have the history data $G = \{G_0, G_1, ..., G_{t-1}\}$, where $G_i = (V, X_i, E, W)$, in which $i \in R$. The problem is to predict future traffic flow sequence $G = \{G_t, G_{t+1}, ..., G_T\}$ from t to T, aiming at minimizing objective function F, where

$$F = min|\tilde{X_i} - X_i|, \quad (1)$$

in which $\tilde{X_i}$ is the ground truth of nodes features in time t. In this task, $\tilde{X_i}$ denotes average traffic speed in time t for all sensors.

3 Proposed Model

3.1 Model Architecture

The basic unit of our model is spatio-temporal block (ST-block). The overall architecture of our proposed ST-blocks is shown as Fig. 2. Inspired by several

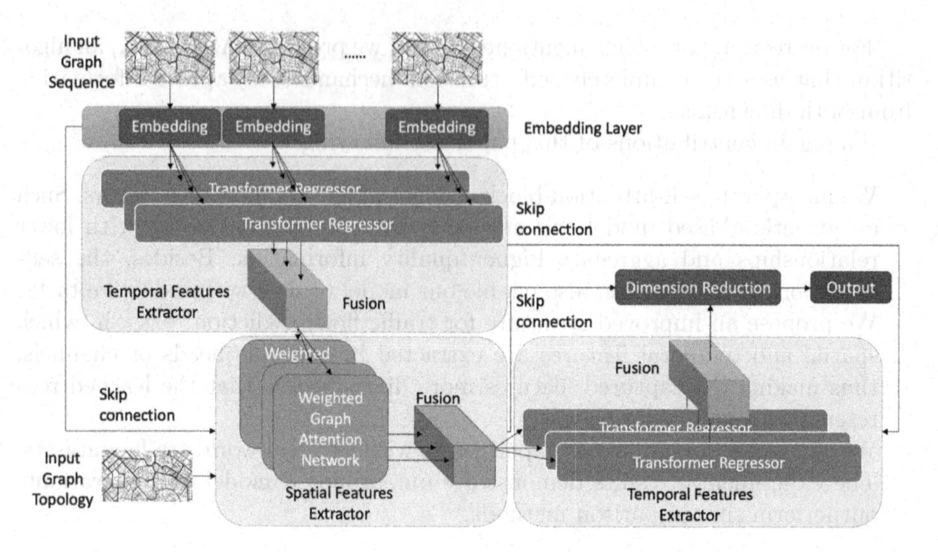

Fig. 2. The framework of an ST-block, which aims to obtain spatial and temporal information simultaneously, and then a linear transformation projects a high dimension feature vectors to a suitable dimension. An ST-block is composed of three parts: two temporal features extractors enclose one spatial features extractor for better capturing temporal information, which is harder to be caught.

recent works [14,15], in order to efficiently fuse extracted temporal features and spatial features, we applied the structure that two temporal extractors enclose one spatial extractor to strengthen long-term temporal dependencies.

For better capturing diverse features, we apply multiple channels (or heads) to each extractor block and fuse extracted features by concatenation and linear transformation. Furthermore, in order to prevent over-fitting and improve training efficiency, we add several dropout layers and residual blocks for each extractor. In our paper, for better fitting hidden representations, we applied a 1×1 convolutional kernel, a linear projection, and an identical transformation as the corresponding residual block [7], respectively.

3.2 Spatial Features Extractor

Spatial features extractor aims to handle spatial features in our model. In this paper, inspired by GAT [11], we handle spatial information by an improved graph attention networks, Weighted Graph Attention Networks (WGAT), instead of using GCN. In WGAT, we consider edges weights before normalizing attention coefficients. The node-wise attention coefficients in WGAT is calculated as:

$$\alpha_{ij} = \frac{\exp\left(w_{ij} LeakyReLU(\boldsymbol{a}^T[W\boldsymbol{h}_i||W\boldsymbol{h}_j])\right)}{\sum_{k \in N_i} \exp\left(w_{ik} LeakyReLU(\boldsymbol{a}^T[W\boldsymbol{h}_i||W\boldsymbol{h}_k])\right)}, \tag{2}$$

where w_{ij} is the weight of edge between node i and node j. In our task, weights of edges will be defined in Sect. 4.2.

For extracting diversified and stable features in the spatial dimension, we applied multi-head attention to learn diversified attention coefficient matrices simultaneously:

$$\boldsymbol{h}_i' = ||_{n=1}^{N}\sigma(\sum_{j\in N_i}\alpha_{ij}^n W_n \boldsymbol{h}_j), \tag{3}$$

where N represents the number of heads, $||$ is a concatenation operation.

3.3 Temporal Features Extractor

We use two temporal features extractors, located at the head and the tail of a ST-block respectively, to extract temporal features. The unit in temporal extractors is a variant of Transformer [12].

Original Transformer model could not directly applied in our temporal features extractor because it is a classification model essentially. For better matching our task, we proposed a regression model: Transformer Regressor (TR). In the Transformer Regressor, structures of the encoder and the decoder are same as Transformer. Details are shown as Fig. 3. Besides, similar to WGAT, we also applied the multi-head attention mechanism in TR.

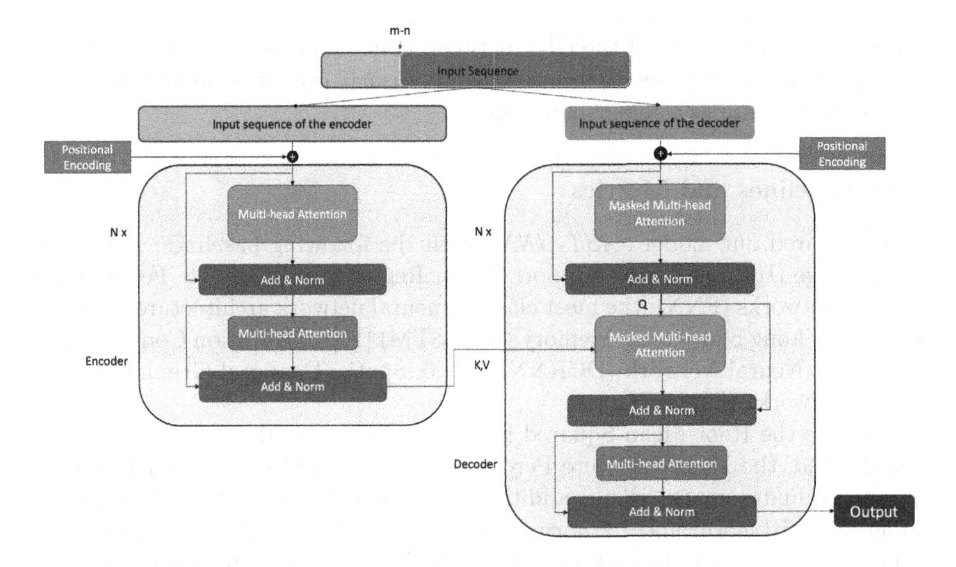

Fig. 3. The illustration of TR. Each part of TR is stacked by encoders or decoders, respectively. The length of the input sequence of the encoder equals the length of the historical observations m, and the length of the input sequence of the decoder equals the length of predicted sequence n. The input sequence of the decoder is part of the input sequence of the encoder.

4 Experiments

4.1 Datasets

We evaluate the performance of our proposed model using two real-world large-scale datasets PeMSD7-228 (228 sensors, 12672 timestamps) and PeMS-Bay (325 sensors, 51840 timestamps), both of which are collected from the Caltrans Performance Measurement System (PeMS) [16].

4.2 Data Pre-processing

The pre-processing of two datasets we referenced [14] and [13]. The same strategy pre-processes both datasets.

The adjacent matrix of the road network is computed based on the distances among sensors. In this paper, we follow the settings in [13] and [14], where edge-wise weights are defined as:

$$w_{ij} = \begin{cases} \exp(-\frac{d_{ij}^2}{\sigma^2}), (i \neq j) \cap (\exp(-\frac{d_{ij}^2}{\sigma^2}) \geq \epsilon) \\ 0, otherwise \end{cases}, \quad (4)$$

where w_{ij} is the weight of the edge between node i and node j, which is determined by the distance between sensor i and sensor j. σ is the standard deviation of distances, and ϵ is a hyper-parameter.

4.3 Baselines and Metrics

We compared our model *SAST-GNN*[1] with the following baselines: 1. Historical Average (HA); 2. Linear Support Vector Regression (LSVR); 3. Feedforward Neural Networks (FNN), the most classical neural network architecture; 4. Fully-connected Long Short-term Memory (FC-LSTM) [17]; 5. Diffusion Convolutional Recurrent Neural Network (DCRNN) [13]; 6. Spatio-Temporal Graph Convolutional Networks (ST-GCN) [14].

We use the Root Mean Squared Errors (RMSE), the Mean Absolute Errors (MAE) and, the Mean Absolute Percentage Errors (MAPE) as evaluation metrics to evaluate our model. In addition, the loss function is defined as a combination of L1 norm and L2 norm: $Loss = ||\tilde{X} - X||_1 + ||\tilde{X} - X||_2$, where $\tilde{X}$ represents the ground-truth vector, while X denotes the prediction vector.

[1] https://github.com/FANTASTPATR/SAST-GNN.

4.4 Performance Comparison

See Tables 1 and 2.

Table 1. Performance comparison on PeMSD7-228

Model	MAE			MAPE			RMSE		
	15 min	30 min	60 min	15 min	30 min	60 min	15 min	30 min	60 min
HA	4.01	4.01	4.01	10.61	10.61	10.61	7.20	7.20	7.20
LSVR	2.53	3.70	4.77	6.07	8.60	12.31	4.80	6.28	9.78
FNN	2.77	4.19	5.46	6.47	9.48	14.23	4.90	6.77	8.43
FC-LSTM	3.57	3.92	4.16	8.60	9.55	10.10	6.20	7.03	7.51
ST-GCN	2.25	3.03	4.02	5.26	7.33	9.85	4.04	5.70	7.64
DCRNN	2.25	2.98	3.83	5.30	7.39	9.85	4.04	5.58	7.19
SAST-GNN	**2.06**	**2.64**	**3.68**	**5.06**	**6.68**	**9.84**	**3.58**	**4.71**	**6.42**

Table 2. Performance comparison on PeMS-Bay

Model	MAE			MAPE			RMSE		
	15 min	30 min	60 min	15 min	30 min	60 min	15 min	30 min	60 min
HA	2.88	2.88	2.88	6.80	6.80	6.80	5.59	5.59	5.59
LSVR	1.97	2.76	3.56	4.01	5.58	8.23	3.87	5.48	6.92
FNN	1.60	1.98	3.03	3.12	4.50	7.20	3.23	4.90	5.87
FC-LSTM	2.20	2.34	2.55	4.85	5.30	5.84	4.28	4.74	5.31
ST-GCN	1.41	1.84	2.37	3.02	4.19	5.39	3.02	4.19	5.27
DCRNN	1.38	1.74	**2.07**	2.90	**3.90**	**4.90**	2.95	3.97	4.74
SAST-GNN	**1.33**	**1.71**	2.35	**2.88**	4.01	5.40	**2.42**	**3.41**	**4.55**

5 Conclusions

In this paper, we proposed a new model called SAST-GNN to predict future traffic conditions based on historical data. In SAST-GNN, we improved existing architectures by adding channels and replacing features extractors as self-attention models to better capture information. In the future, we aim to focus on the improvement of structure thus our model performs better in long-term dependencies.

Acknowledgement. This work is supported in part by the National Natural Science Foundation of China Projects No. U1936213, U1636207, the Shanghai Science and Technology Development Fund No. 19DZ1200802, 19511121204.

References

1. Williams, B.M., Hoel, L.A.: Modeling and forecasting vehicular traffic flow as a seasonal ARIMA process: theoretical basis and empirical results. J. Transp. Eng. **129**(6), 664–672 (2003)
2. Chen, R., Liang, C., Hong, W., Gu, D.: Forecasting holiday daily tourist flow based on seasonal support vector regression with adaptive genetic algorithm. Appl. Soft Comput. **26**(C), 435–443 (2015)
3. Shi, X., Chen, Z., Wang, H., Yeung, D., Wong, W., Woo, W.: Convolutional LSTM network: a machine learning approach for precipitation nowcasting. In: NIPS (2015)
4. Hochreiter, S., Schmidhuber, J.: Long short-term memory. Neural Comput. **9**(8), 1735–1780 (1997)
5. Defferrard, M., Bresson, X., Vandergheynst, P.: Convolutional neural networks on graphs with fast localized spectral filtering. In: NIPS (2016)
6. Kipf, T., Welling, M.: Semi-supervised classification with graph convolutional networks. In: ICLR (2017)
7. He, K., Zhang, X., Ren, S., Sun, J.: Deep residual learning for image recognition. In: CVPR (2015)
8. Rahimi, A., Cohn, T., Baldwin, T.: Semi-supervised user geolocation via graph convolutional networks. In: ACL (2018)
9. Chai, D., Wang, L., Yang, Q.: Bike flow prediction with multi-graph convolutional network. In: Proceedings of the 26th ACM SIGSPATIAL International Conference on Advances in Geographic Information Systems, pp. 397–400 (2018)
10. Li, R., Wang, S., Zhu, F., Huang, J.: Adaptive graph convolutional neural networks. In: AAAI (2018)
11. Veličković, P., Cucurull, G., Casanova, A., Romero, A., Liò, P., Bengio, Y.: Graph attention networks. In: ICLR (2018)
12. Vaswani, A., et al.: Attention is all you need. In: NIPS (2017)
13. Li, Y., Yu, R., Shahabi, C., Liu, Y.: Diffusion convolutional recurrent neural network: data-driven traffic forecasting. In: ICLR (2018)
14. Yu, B., Yin, H., Zhu, Z.: Spatio-temporal graph convolutional networks: a deep learning framework for traffic forecasting. In: IJCAI (2018)
15. Diao, Z., Wang, X., Zhang, D., Liu, Y., Xie, K., He, S.: Dynamic spatial-temporal graph convolutional neural networks for traffic forecasting. In: AAAI (2019)
16. Chen, C., Petty, K., Skabardonis, A., Varaiya, P., Jia, Z.: Freeway performance measurement system: mining loop detector data. Transp. Res. Rec.: J. Transp. Res. Board **1748**, 96–102 (2001)
17. Sutskever, I., Vinyals, O., Le, Q.V.: Sequence to sequence learning with neural networks. In: NIPS (2014)

Clustering and Classification

Enhancing Linear Time Complexity Time Series Classification with Hybrid Bag-Of-Patterns

Shen Liang[1,2], Yanchun Zhang[2,3(✉)], and Jiangang Ma[4]

[1] School of Computer Science, Fudan University, Shanghai, China
sliang11@fudan.edu.cn

[2] Cyberspace Institute of Advanced Technology (CIAT), Guangzhou University, Guangzhou, China

[3] Institute for Sustainable Industries and Liveable Cities, Victoria University, Melbourne, Australia
Yanchun.Zhang@vu.edu.au

[4] School of Science, Engineering and Information Technology, Federation University Australia, Ballarat, Australia
j.ma@federation.edu.au

Abstract. In time series classification, one of the most popular models is Bag-Of-Patterns (BOP). Most BOP methods run in super-linear time. A recent work proposed a linear time BOP model, yet it has limited accuracy. In this work, we present *Hybrid Bag-Of-Patterns (HBOP)*, which can greatly enhance accuracy while maintaining linear complexity. Concretely, we first propose a novel time series discretization method called *SLA*, which can retain more information than the classic SAX. We use a *hybrid* of SLA and SAX to expressively and compactly represent subsequences, which is our *most important* design feature. Moreover, we develop an efficient time series transformation method that is key to achieving linear complexity. We also propose a novel X-means clustering subroutine to handle subclasses. Extensive experiments on over 100 datasets demonstrate the effectiveness and efficiency of our method.

Keywords: Time series · Classification · Bag-Of-Patterns

1 Introduction

Time series classification (TSC) is an important research topic that has been applied to medicine [18], biology [12], etc. Many TSC methods [3,8,10,18,19, 21,22] are based on Bag-Of-Patterns (BOP) [10], which represents a time series with the histogram (word counts) of "words" (symbolic strings) converted from

---

This work is funded by NSFC Grant 61672161 and Dongguan Innovative Research Team Program 2018607201008. We sincerely thank Dr Hoang Anh Dau from University of California, Riverside for responding to our inquiries on the UCR Archive [5].

Y. Nah et al. (Eds.): DASFAA 2020, LNCS 12112, pp. 717–735, 2020.
https://doi.org/10.1007/978-3-030-59410-7_50

its subsequences by some discretization method [3,8,13,20]. BOP can capture local semantics [10], and is robust against noise [12,18] and phase shifts [10].

Despite their success, most BOP-based methods have super-linear training time (w.r.t. training set size, i.e. the number of training time series × time series length). Recently, a linear time model called BOPF has been proposed [12], yet its accuracy is relatively limited. This is (partly) due to two reasons. First, the discretization method used by BOPF is SAX [13], which depicts subsequence segments with only mean values. This can cause much information loss. Second, BOPF condenses each class with its centroid, failing to detect subclasses.

Aimed to enhance accuracy while maintaining linear complexity, we propose a novel model called *Hybrid Bag-Of-Patterns (HBOP)*, which involves a novel discretization method called *Symbolic Linear Approximation (SLA)* to retain more information. SLA depicts subsequences with Piecewise Linear Approximation (PLA) coefficients and is more expressive than SAX, yet its compactness is weaker (Sect. 4.1). Thus, as our *most important* design feature, we use a *hybrid* of SAX and SLA to make them complement each other. Plus, we develop a novel and efficient PLA transformation method to achieve linear time, and a novel X-means [16] clustering subroutine to handle subclasses.

Our main contributions in this paper are as follows.

- We introduce Symbolic Linear Approximation (SLA), a novel time series discretization method that is more expressive than the classic SAX [13].
- We present Hybrid Bag-Of-Patterns (HBOP), a novel TSC model featuring a hybrid of SLA and SAX. *SLA and its hybrid with SAX are our most important contributions in this paper.*
- We design an efficient PLA transformation method that is key to achieving linear time complexity.
- We develop a novel X-means [16] clustering subroutine to handle subclasses.
- We conduct extensive experiments on over 100 datasets from the well-known UCR archive [5] to demonstrate the effectiveness and efficiency of HBOP.

For the rest of the paper, Sect. 2 describes the preliminaries. Section 3 provides our SLA method, Sect. 4 presents our HBOP model. Section 5 reports the experiments. Section 6 reviews related work. Section 7 concludes the paper.

2 Preliminaries

Our paper deals with *time series classification (TSC)*, which uses a set of labeled training time series to build a model and predict the labels of future ones. Throughout this paper, we use n to denote the number of training time series, and m to denote the time series length. Here we assume all time series have the same length, yet our method can fit datasets with varied time series lengths.

In this paper, we utilize the Bag-Of-Patterns (BOP) model. Figure 1 shows the process of obtaining the BOP representation of a time series. First, we extract its subsequences by sliding window. Next, we transform each subsequence into a "word" (symbolic string) with a discretization method [3,8,13,20]. Finally, we

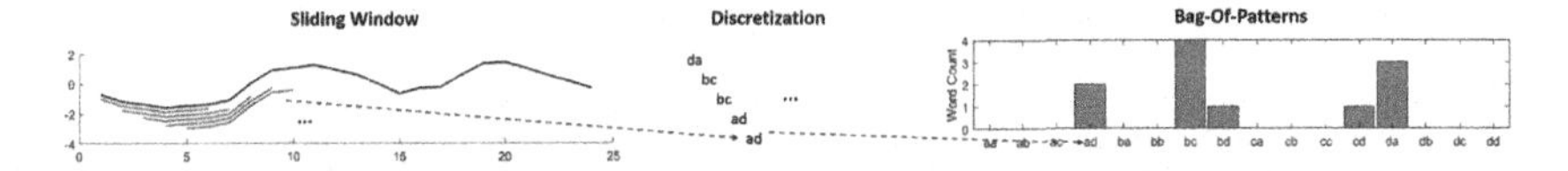

Fig. 1. The process of obtaining BOP representation. Left to right: extracting subsequences, discretizing them into symbolic words, obtaining word counts.

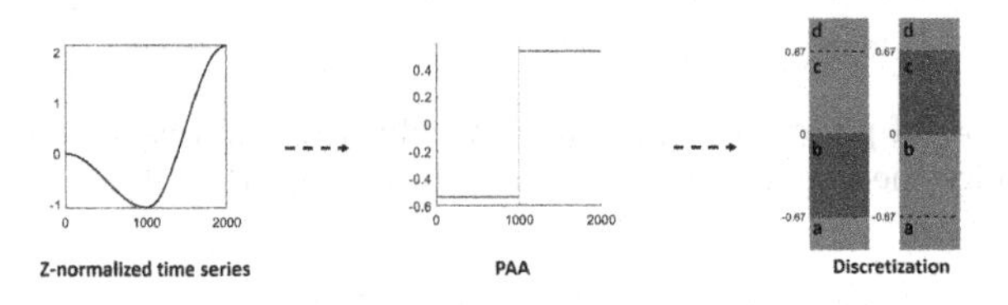

Fig. 2. SAX workflow. Left to right: a z-normalized time series, obtaining the PAA representation, obtaining the SAX word (which is *bc*).

obtain the histogram (word counts) of the words as the BOP representation. Note that if the same word corresponds to several consecutive subsequences, only the first is counted [10]. For example, if a time series is discretized to *ab, ab, ac, ac, ab, ab*, then the word counts of *ab* and *ac* are 2 and 1.

A key element of BOP is the discretization method. It transforms a time series into a low dimensional real-valued space $\mathbb{Q}$, and replaces the real values with discrete symbols. In this paper, we use δ to denote *dimensionality* of $\mathbb{Q}$, and γ to denote *cardinality* (number of possible symbols). A classic discretization method is Symbolic Aggregate approXimation (SAX) [13]. For a z-normalized time series, SAX first uses Piecewise Aggregate Approximation (PAA) [11] to represent it with the mean values of its δ equal-length segments. Next, SAX divides the real number space into γ "bins" with standard normal quantiles. Each bin is assigned a symbol. The segment mean values are mapped to the bins, leading to a SAX word. This is shown in Fig. 2, with $\delta = 2$ and $\gamma = 4$.

3 Symbolic Linear Approximation (SLA)

We now present our novel SLA discretization method. Following [20], we divide SLA into *preprocessing* and *discretization* phases. For *preprocessing* (Fig. 3), we use a static set of time series to decide the bin boundaries. We first transform each time series with Piecewise Linear Approximation (PLA), which divides it into $\delta/2$ equal-length segments and conducts linear regression on each segment. For a length-r segment $Y = y_0, y_1, \ldots, y_{r-1}$, we set up a Cartesian coordinate system such that the coordinates of Y are $(0, y_0), (1, y_1), \ldots, (r-1, y_{r-1})$. Let

$$\bar{x} = \frac{\sum_{i=0}^{r-1} i}{r} = \frac{r-1}{2}, \ \overline{x^2} = \frac{\sum_{i=0}^{r-1} i^2}{r} = \frac{(r-1)(2r-1)}{6}, \bar{y} = \frac{\sum_{i=0}^{r-1} y_i}{r}, \ \overline{xy} = \frac{\sum_{i=0}^{r-1} i y_i}{r} \tag{1}$$

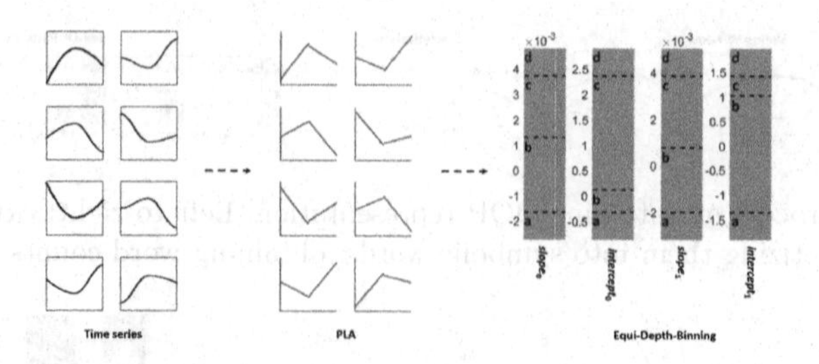

Fig. 3. Workflow of preprocessing with SLA, which transforms a time series set with PLA and decides the boundaries by equi-depth-binning. Here $\delta = \gamma = 4$.

the *slope* and *intercept* of the line obtained from linear regression are

$$a = \frac{\overline{xy} - \bar{x}\bar{y}}{\overline{x^2} - \bar{x}^2}, \ b = \bar{y} - a\bar{x} \tag{2}$$

We represent each of the $\delta/2$ segments with its PLA slope and intercept, characterizing each time series with δ PLA coefficients. For each of the δ dimensions, we decide the bin boundaries using equi-depth-binning [3,20]: For a static set of n' time series, there are n' PLA coefficients for each dimension. We set the bin boundaries such that each of the γ bins has n'/γ coefficients in it.

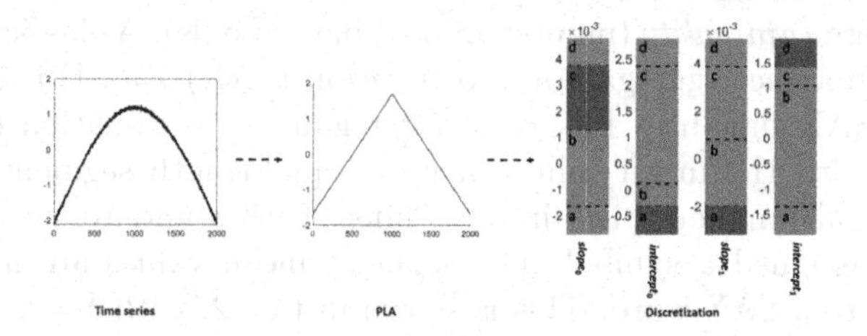

Fig. 4. Workflow of discretization with SLA, which transforms a time series with PLA and obtains its SLA word (*caad*)

In the *discretization* phase (Fig. 4), we transform a given time series with PLA, and then map each PLA coefficient to its corresponding bin, thus obtaining the SLA word. We will further compare SLA and SAX in Sect. 4.1.

4 Hybrid Bag-Of-Patterns (HBOP)

We now introduce our HBOP model. It has a similar workflow to BOPF [12], yet boasts its unique highlights (Fig. 5). For training, we pre-set some parameters called *discretization configurations*, which highlight our *most important* design: a *hybrid* of SAX and SLA. We also set a range of possible subsequence lengths

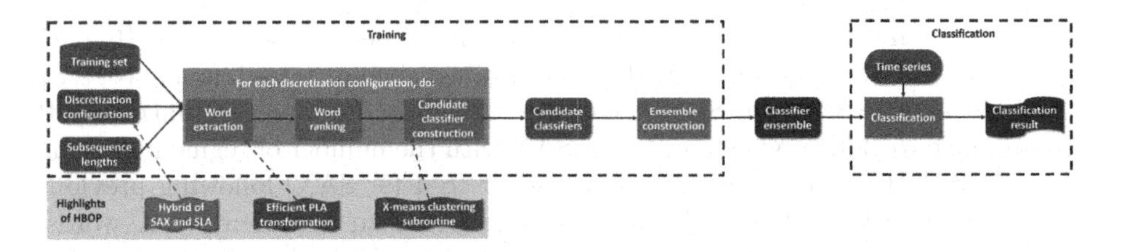

Fig. 5. The workflow and highlights of HBOP. Note that among the three highlights, the most important one is a *hybrid* of SLA and SAX.

LR. For *each* configuration, we extract words from all subsequences within LR. Here, we propose a novel linear-time PLA transformation method. Next, we rank the words, and build candidate classifiers to form an ensemble later. Here we develop a novel X-means [16] clustering subroutine to handle subclasses. We then use top-ranking classifiers to construct the final ensemble, with which we classify future time series. We now introduce each step in this workflow.

4.1 Discretization Configurations and Subsequence Lengths

As for discretization configurations, eac of them is a tuple $\langle \Phi, \delta, \gamma \rangle$, where Φ is a discretization method, δ is its dimensionality, and γ is its cardinality. Compared to existing works [3,8,10,18,19,21,22] which consider *only one* choice for Φ, HBOP's *most important* feature is using *both* SLA and SAX. Our intuition is as follows: Recall that SAX is based on PAA [11] and SLA is based on PLA. An advantage of PLA over PAA is that when a time series is divided into the same number of segments, PLA can retain more trend information than PAA. For example, for the *z-normalized* time series shown in Fig. 6, PAA cannot retain the original trends with two segments (*left*), while PLA can (*middle*).

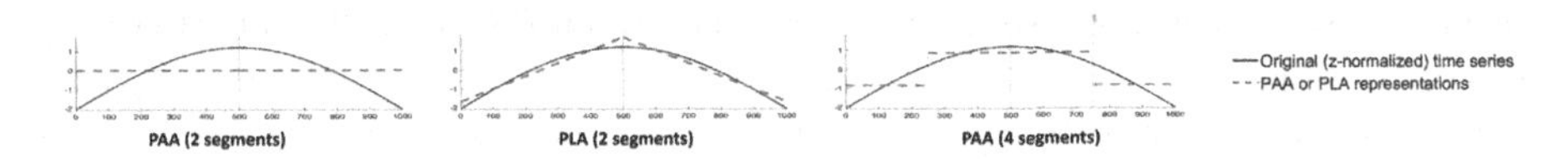

Fig. 6. A comparison of PAA and PLA. PAA with 2 segments fails to capture the trend of the original data (*left*). However, by replacing PAA with PLA (*middle*) or doubling the number of PAA segments (*right*), we can capture this trend. In the latter two cases, PLA and PAA have the same dimensionality 4.

However, the advantage of PLA comes at a price: its dimensionality is twice that of PAA. While PAA cannot retain the trend with the same *number of segments* as PLA, it can do so with the same *dimensionality* (Fig. 6 *right*). This inspires us to design HBOP such that *the PLA-based SLA is used when the number of segments is small, and the PAA-based SAX is used when the number of segments is large.* Thus, we can keep critical trends while having relatively low dimensionalities. Concretely, the number of segments is set among $\{6, 7, 8\}$ for SAX, and $\{3, 4, 5\}$ for SLA. Thus, the dimensionality δ is among $\{6, 7, 8\}$ for SAX, and $\{6, 8, 10\}$ for SLA. We choose these *specific* values to make HBOP comparable to BOPF, which only uses SAX with the number of segments among $\{3, 4, 5, 6, 7\}$ [12]. For cardinality γ, we set it to 4 for SAX, following previous works [12,18,19,21]. For SLA, an γ of 4 can lead to too many SLA words, undermining efficiency. Thus, we reduce γ to 3 for SLA. As will be shown in Sect. 5, even with a lower γ for SLA, HBOP can yield satisfactory accuracy.

In summary, we have a total of 6 discretization configurations, namely $\{\langle \text{SAX}, \delta, 4\rangle | \delta \in \{6, 7, 8\}\} \cup \{\langle \text{SLA}, \delta, 3\rangle | \delta \in \{6, 8, 10\}\}$. We also need a range of possible subsequence lengths. Here we use the same settings as BOPF [12], i.e. $\{0.025, 0.05, 0.075, \ldots, 1\}m$ (m is the time series length) which consists of 40 choices. Values less than 10 and duplicate values are removed [12].

4.2 Word Extraction

The workflow of word extraction for *each* discretization configuration is shown in Fig. 7. For each subsequence length l, we divide each length-m time series into $\lfloor m/l \rfloor$ non-overlapping length-l subsequences called *seeds*, and transform them with PAA or PLA depending on the discretization method Φ. We then decide the bin boundaries for Φ. For SAX, these are only linked to cardinality γ [13]. For SLA, these are decided with a static set of subsequences. Rather than use all subsequences, we use only the seeds to form this set to avoid overfitting [21]. Next, we discretize all seeds and then the non-seed subsequences. For the latter, we do not discretize all of them explicitly. Rather, if any two adjacent seeds share the same word ξ, then for all non-seeds between them, we directly set their words to ξ. Otherwise, we discretize these non-seeds in a *binary-search-like* manner. For example, suppose $l = 4$, let $S_0, \ldots, S_4$ be the $0, \ldots, 4$-th subsequences of a time series, thus S_0 and S_4 are two adjacent seeds. Suppose their words ξ and ϵ are different, we then explicitly discretize S_2. If its word is ϵ, then we need to explicitly discretize S_1 as the words of S_0 and S_2 are different. However, since the words of S_2 and S_4 are both ϵ, we directly set the word of S_3 as ϵ without explicitly discretizing it. The intuition is that overlapping subsequences tend to have similar shape [9] and thus share the same word. Note that we z-normalize each subsequence (seed or non-seed) that we explicitly discretize. This is mandatory for SAX and can mitigate shift and scaling [13] for SLA. After discretization, we obtain the word counts.

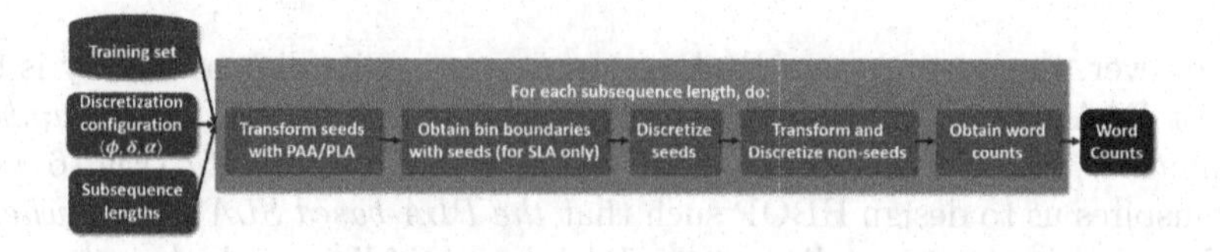

Fig. 7. The workflow of word extraction for a single discretization configuration.

Efficiency-wise, for a single subsequence length l, word extraction for SAX takes $O(nm)$ time with the method in [12]. Note that we follow [12] and treat δ and γ as constants, for they are manually set and not directly linked to dataset size. For SLA, we need to take the following steps. First, we transform $O(nm/l)$ seeds with PLA. Second, we decide the bin boundaries. Third, we discretize the seeds. Fourth, we transform a maximum of $O(nm)$ non-seeds with PLA. Fifth, we discretize the non-seeds. The second step takes $O(nm/l)$ time with a fast selection algorithm. The third and fifth steps take a total of $O(1) \times O(nm) = O(nm)$ time for $O(nm)$ subsequences. The first and the fourth steps take $O(nml)$ time with brute force. However, we now present an efficient PLA transformation method that reduces this to $O(nm)$, starting with the following theorem.

Theorem 1. *Given a time series $T = t_0, t_1, \ldots, t_{m-1}$ and its subsequence $S = t_p, t_{p+1}, \ldots, t_{p+l-1}$, let $Z = z_p, z_{p+1}, \ldots, z_{p+l-1}$ be the subsequence obtained by z-normalizing S. PLA further divides Z into several length-r segments. For a segment $Y = z_q, z_{q+1}, \ldots, z_{q+r-1}$ in Z, let*

$$\overline{x'} = \frac{\sum_{i=q}^{q+r-1} i}{r} = \frac{2q+r-1}{2},\ \overline{y} = \frac{\sum_{i=q}^{q+r-1} z_i}{r},\ \overline{x'y} = \frac{\sum_{i=q}^{q+r-1} iz_i}{r}$$

$$\overline{x'^2} = \frac{\sum_{i=q}^{q+r-1} i^2}{r} = \frac{(q+r-1)(q+r)[2(q+r)-1]}{6r} - \frac{(q-1)q(2q-1)}{6r}$$

then the SLA slope and intercept of segment Y are

$$a = \frac{\overline{x'y} - \overline{x'}\bar{y}}{\overline{x'^2} - \overline{x'}^2},\ b = \bar{y} - a(\overline{x'} - q) \tag{3}$$

Proof. Inheriting the notations $\overline{x}$, $\overline{x}^2$ from Eq. 1, and letting

$$\overline{xy} = \frac{\sum_{i=0}^{r-1} iz_{i+q}}{r} \tag{4}$$

by Eq. 2, we have

$$a = \frac{\overline{xy} - \bar{x}\bar{y}}{\overline{x^2} - \bar{x}^2},\ b = \bar{y} - a\bar{x} \tag{5}$$

Note that

$$\begin{aligned} r^2(\overline{x'^2} - \overline{x'}^2) = r\frac{(q+r-1)(q+r)(2q+2r-1)}{6} \\ -r\frac{(q-1)q(2q-1)}{6} - \left[\frac{r(2q+r-1)}{2}\right]^2 = \frac{r^2(r^2-1)}{12} \end{aligned} \tag{6}$$

Similarly,

$$r^2(\overline{x^2} - \overline{x}^2) = r\sum_{i=q}^{q+r-1}(i-q)^2 - \left[\sum_{i=q}^{q+r-1}(i-q)\right]^2 = \frac{r^2(r^2-1)}{12} = r^2(\overline{x'^2} - \overline{x'}^2) \quad (7)$$

Also,

$$\begin{aligned} r^2(\overline{x'y} - \overline{x'}\overline{y}) &= \sum_{i=q}^{q+r-1}\left(riz_i - z_i\sum_{j=q}^{q+r-1} j\right) = \sum_{i=q}^{q+r-1}\left[r(i-q)z_i - \frac{r(r-1)}{2}z_i\right] \\ &= r\sum_{i=0}^{r-1} iz_{i+q} - \sum_{j=0}^{r-1} j\sum_{i=q}^{q+r-1} z_i = r^2(\overline{xy} - \overline{x}\overline{y}) \end{aligned} \quad (8)$$

From Eq. 7, 8 and a in Eq. 5, we have a in Eq. 3. In addition,

$$r\overline{x} = \sum_{i=0}^{r-1} i = \sum_{i=q}^{q+r-1}(i-q) = r(\overline{x}' - q) \quad (9)$$

From Eq. 9 and b in Eq. 5, we have b in Eq. 3. □

Using Eq. 3, we develop a cumulative sum method [15] similar to that used in [12] for efficient PLA transformation. For a time series $T = t_0, t_1, \ldots, t_{m-1}$, we define $U = u_0, u_1, \ldots, u_m$, $V = v_0, v_1, \ldots, v_m$, and $W = w_0, w_1, \ldots, w_m$ where

$$u_j = \sum_{i=0}^{j-1} t_i,\ v_j = \sum_{i=0}^{j-1} t_i^2,\ w_j = \sum_{i=0}^{j-1} it_i \quad (10)$$

with $u_0 = v_0 = w_0 = 0$. For a given subsequence $S = t_p, t_{p+1}, \ldots, t_{p+l-1}$, its mean $\overline{s}$ and standard deviation σ_s are calculated as [12,15]

$$\overline{s} = \frac{u_{p+l} - u_p}{l},\ \sigma_s = \sqrt{\frac{v_{p+l} - v_p}{l} - \overline{s}^2} \quad (11)$$

For an SLA segment (after z-normalizing S) $Y = z_q, z_{q+1}, \ldots, z_{q+r-1}$, its mean value $\overline{y}$ [12,15] and $\overline{x'y}$ are

$$\overline{y} = \frac{\frac{u_{q+r} - u_q}{r} - \overline{s}}{\sigma_s},\ \overline{x'y} = \frac{\frac{w_{q+r} - w_q}{r} - \overline{s}\overline{x'}}{\sigma_s} \quad (12)$$

By pre-calculating U, V, W in $O(m)$ time, we can obtain $\overline{y}$ and $\overline{x'y}$ in constant time. Also, $\overline{x}'$ and $\overline{x}'^2$ can be calculated in constant time. Thus we can obtain the slope and intercept of any segment in constant time with Eq. 3. Hence, it takes $O(nm)$ time to conduct PLA for all $O(nm)$ subsequences. Combining this finding with the discussions above, we conclude that word extraction for both SAX [12] and SLA takes $O(nm)$ time for a fixed l. Since the maximum number of options for l is constant, the total time is $O(nm)$.

4.3 Word Ranking

We now rank the extracted words. For *each* discretization configuration, we score each word with χ^2 test [21], where a larger χ^2-value means the word is more discriminative among classes. For example, consider four time series with labels (L_0, L_0, L_1, L_1). Suppose the counts for word ξ in them are (1, 1, 0, 0), and those for word ϵ are (1, 0, 0, 0). Obviously ξ is more discriminative than ϵ. This is reflected by their χ^2-values, which are 2 and 1. We sort the words in χ^2-value descending order, thus obtaining the word ranks.

Efficiency-wise, it takes $O(n)$ time to score each word. With the number of words is below γ^δ which is constant, scoring all words takes $O(n)$ time. Plus, sorting the words takes $O(1)$ time. Therefore, the total time is $O(n)$.

4.4 Candidate Classifier Construction

We now build candidate classifiers for the final ensemble, whose workflow is in Fig. 8. For *each* discretization configuration, we build four nearest centroid classifiers (NCCs). Their differences are two-fold, on which we now elaborate.

The first difference is with the choice of labels. Existing BOP-based NCCs [12,19,22] use the original class labels and summarize each class with only one centroid, ignoring potential subclasses. By contrast, we exploit a novel X-means [16] clustering subroutine to detect subclasses. X-means extends K-means so as to automatically decide the number of clusters from a given range. For HBOP, we use the word counts of the top-h words as the feature vector of each example, and individually apply X-means to each class with more than 10 examples. Here h is fixed to 5000. and the range of possible number of clusters is set to 2–5. After clustering for all classes, we replace the original class labels with new cluster labels. For instance, for original labels L_1 and L_2, if L_1 has three clusters and L_2 has two, we relabel each example to one of $C_{11}, C_{12}, C_{13}, C_{21}, C_{22}$. Since subclasses do not always exist, we keep the option of inheriting the original class labels (namely directly using them as cluster labels) open, leading to two options: applying or not applying X-means.

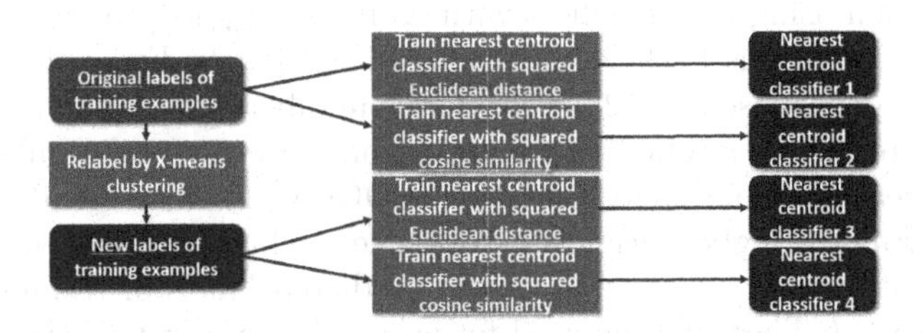

Fig. 8. Workflow of candidate classifier construction for a *single* discretization configuration.

The second difference involves the choice of similarity measure of the NCCs. Here we follow [12] and choose between two options. The first is the squared Euclidean distance (SED) [12]. For a word ξ, let $Ct_T(\xi)$ be its word count in a time series T. For a cluster C, let St_C be the set of time series in C, and $Ct_C(\xi)$ be the sum of $Ct_T(\xi)$ for all $T \in St_C$. Then the Euclidean centroid of ξ in C is $\overline{Ct}_C(\xi) = Ct_C(\xi)/|St_C|$ [12]. For top-d ranking words $\xi_0, \xi_1, \ldots, \xi_{d-1}$, the word count vector of T is $Ct_T = [Ct_T(\xi_0), Ct_T(\xi_1), \ldots, Ct_T(\xi_{d-1})]$ and the centroid vector is $\overline{Ct}_C = [\overline{Ct}_C(\xi_0), \overline{Ct}_C(\xi_1), \ldots, \overline{Ct}_C(\xi_{d-1})]$ [12]. Their SED is [12]

$$SED(T,C) = \sum_{i=0}^{d-1} (Ct_T(\xi_i) - \overline{Ct}_C(\xi_i))^2 \tag{13}$$

The smaller $SED(T,C)$ is, the more likely T falls into C.

The second option is the squared cosine similarity (SCS) [12,19,22] in vector space, where the term frequency of a word ξ in a time series T is [12,19]

$$tf(\xi,T) = \begin{cases} 1 + log(Ct_T(\xi)), if\ Ct_T(\xi) > 0 \\ 0, \qquad\qquad\qquad\quad otherwise \end{cases} \tag{14}$$

The term frequency $tf(\xi,C)$ of ξ in a cluster C can be obtained by replacing $Ct_T(\xi)$ with $Ct_C(\xi)$ in Eq. 14 [12,19]. Let c be the total number of clusters and c_ξ be the number of clusters containing ξ. The inverse document frequency of ξ, and the tf-idf weight of ξ and C are [12,19]

$$idf(\xi) = log\left(1 + \frac{c}{c_\xi}\right),\ tfidf(\xi,C) = tf(\xi,C) \cdot idf(\xi) \tag{15}$$

For top-d ranking words $\xi_0, \xi_1, \ldots, \xi_{d-1}$, the term frequency vector of time series T is $tf(T) = [tf(\xi_0,T), tf(\xi_1,T), \ldots, tf(\xi_{d-1},T)]$, and the tf-idf vector of cluster C is $tfidf(C) = [tfidf(\xi_0,C), tfidf(\xi_1,C), \ldots, tfidf(\xi_{d-1},C)]$. Their SCS is [12]

$$SCS(T,C) = \frac{(\sum_{i=0}^{d-1} tf(\xi_i,T) \cdot tfidf(\xi_i,C))^2}{\sum_{i=0}^{d-1} tf(\xi_i,T)^2 \cdot \sum_{i=0}^{d-1} tfidf(\xi_i,C)^2} \tag{16}$$

The larger $SCS(T,C)$ is, the more likely T falls into C.

In summary, with 2 choices for labels and 2 choices for similarity measure, we have $2 \times 2 = 4$ candidate classifiers for each discretization configuration. Note that when using the classifiers with X-means applied, we need to map the predicted cluster labels back to the original class labels. For each classifier, we use the method proposed in [12] to decide the number d of the top ranking words to use. Concretely, we gradually increase d from 1 to the total number of words, conducting leave-one-out cross validation (LOOCV) for each d and selecting the d value yielding the highest accuracy. We cache the highest CV accuracy for each classifier as its weight [2], as well as the predicted labels of training examples under the best d. These will be used for ensemble construction (Sect. 4.5). The CVs can be done incrementally (see [12] for more).

For efficiency, classifier construction takes $O(cn)$ [12] time with the number of K-means iterations in X-means considered constant. Following [12], we treat the number of original classes c' as constant, thus the maximum possible c value is $5c'$, and the total time complexity is $O(5c'n) = O(n)$.

4.5 Ensemble Construction and Online Classification

We now construct the final ensemble. With 6 discretization configurations and 4 classifiers for each of them, we have $6 \times 4 = 24$ candidate classifiers. We first rank them by their weights [2] (Sect. 4.4). We then decide the ensemble size by gradually adding classifiers into it. Each time a classifier is added, we conduct LOOCV where we classify each training example by weighted voting [2], using the cached weights and predicted labels (Sect. 4.4). The number of classifiers with the highest accuracy is set as the final ensemble size. Ties are broken by favoring the largest. Similar to CVs in Sect. 4.4, the CVs here are incremental. Each time a classifier is added, we only need to add its weighted votes to the cumulative votes by previous classifiers. The ensemble construction time is $O(n)$.

To classify a future time series, we individually obtain its word counts and make a prediction using each classifier in the ensemble. The final result is obtained by weighted voting [2]. With our efficient PLA transformation method (Sect. 4.2), the classification time is $O(m)$.

4.6 Summary

The entire HBOP method is shown in Algorithm 1. For training (lines 1–8), we first calculate the cumulative sums in Eq. 10 (line 1), and then pre-set discretization configurations and subsequence lengths (line 2). For each configuration (line 3), we conduct word extraction and ranking (lines 4–5). We then build candidate classifiers (lines 6–7), out of which we build the final ensemble (lines 8). For classification (lines 9–14), we first get the cumulative sums for the future time series (line 9). For each classifier in the ensemble (line 10), we extract the word counts (line 11) and predict with the current classifier (line 12). We use weighted voting [2] to obtain the final label (lines 13).

For efficiency, in training, word extraction, word ranking and candidate classifier construction take $O(nm)$, $O(n)$ and $O(n)$ time for each of the $O(1)$ discretization configurations. Plus, it takes $O(n)$ time to build the ensemble, thus the total training time is $O(nm)$, i.e. linear. We will validate this in Sect. 5.3. The online classification time is $O(m)$ per time series (Sect. 4.5).

With the same linear time complexity as BOPF, HBOP is unique in the following aspects. First, the main difference is that BOPF only uses SAX [13], while we apply both SLA and SAX. Second, BOPF does not tackle subclasses, while we do so with X-means [16]. Third, BOPF uses a single subsequence length for each discretization configuration, while we use multiple lengths. A recent study [21] suggests that this can improve accuracy. Fourth, BOPF uses simple majority vote for classification, while we use weighted vote [2]. Fifth, BOPF uses a fixed ensemble size of 30 [12], while our ensemble size is decided by CV.

5 Experimental Results

We now discuss the experiments. We begin by noting that all our source code and raw results are available at [1].

Algorithm 1: HBOP($TS, labels$)

Input : a set of training time series TS, their labels $labels$, a future time series T_f
Output: the predicted label $label_f$

```
   // Training
1  [U, V, W] = cumulativeSum(TS); allClassifiers = [];
2  [allConfig, allSubLengths] = setDiscConfigAndSubLengths();
3  foreach config in allConfig do
4      BOP = extractWords(TS, U, V, W, config, allSubLengths);
5      wordRank = rankWords(BOP, labels);
6      cls = buildClassifiers(BOP, wordRank, labels);
7      allClassifiers = [allClassifiers, cls];
8  ensemble = buildEnsemble(allClassifiers, labels);
   // Classification
9  [U_f, V_f, W_f] = cumulativeSum(T_f); allPredictions = [];
10 foreach classifier in ensemble do
11     BOP_f = getWords(T_f, U_f, V_f, W_f, classifier.config);
12     pre = predict(BOP_f, classifier); allPredictions = [allPredictions, pre];
13 return label_f = weightedVote(ensemble, allPredictions);
```

For experiments, we use 110 datasets from the well-known UCR Archive [5]. *See* [1] *for the reasons why we do not use the remaining datasets in this archive.* For brevity, we omit details of the datasets here. Their names can be found on [1], while their details are at [5]. Our parameter settings are in Sect. 4.1. We implemented HBOP in Python. The experiments were run on a PC with Intel Core i5-4590 CPU @3.3 GHz, 28 GB memory and Windows 10 OS.

5.1 Classification Accuracy

We compare the accuracy of HBOP with six baselines from three categories. *First*, we use three one-nearest-neighbor (1NN) classifiers with different distance measures: Euclidean distance (**ED**), constrained DTW with learned warping window width (**DTWCV**) and unconstrained DTW (**DTW**) [4]. *Second*, we use two BOP models: **BOPF** [12] which is our main rival method, and **BOSS-VS** [19] which is a relatively efficient BOP model. *Third*, we use a shapelet-based classifier named Fast Shapelets (**FS**) [17], which is a relatively efficient method exploiting short patterns called *shapelets* [7,9,15,17,24] (Sect. 6). The reason why we use these baselines is that the 1NN classifiers are the standard baselines for TSC [4], while the rest, like our HBOP, are designed with efficiency in mind.

We now compare the accuracies of all methods. As FS [17] and HBOP are randomized methods (in HBOP, the randomness is introduced by X-means [16]). We use their average accuracies over five runs. For brevity, we do not show the raw accuracies here (which are available at [1]). Rather, we use the critical difference diagram (CDD) [6] ($\alpha = 0.05$) in Fig. 9 to compare the accuracies across all datasets, following the guidelines in [4]. As is shown, HBOP significantly outperforms all baseline methods. Among the efficient baselines, BOPF and BOSS-VS have similar performances to DTWCV while FS performs poorly.

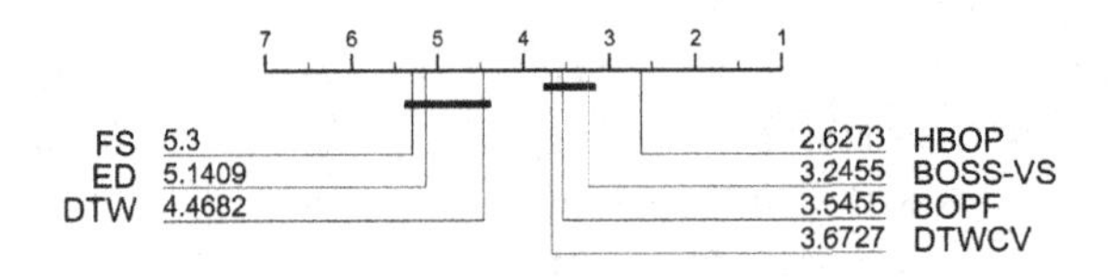

Fig. 9. Critical difference diagram (CDD) of all methods on all datasets.

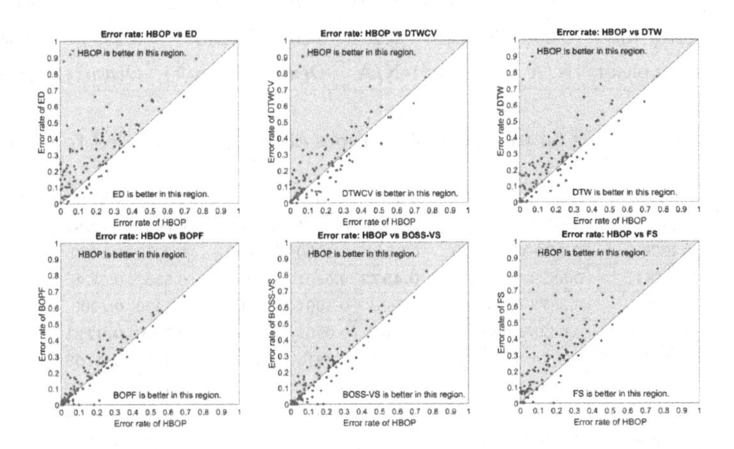

Fig. 10. Scatter plot of error rates of all methods on all datasets. Each red dot corresponds to a dataset.

We now conduct one-on-one comparison between HBOP and the baselines. The raw error rates of the former against the latter are visualized in Fig. 10, which are in turn summarized in Table 1. As for Wilcoxon signed-rank test p-values, with a threshold of $p = 0.05$, HBOP significantly outperforms all baselines, which is in line with Fig. 9. As with wins, ties and losses, HBOP is also the best. Next, we summarize the average error rates (over all datasets) and training time complexities (except ED and DTW as they do not require training) of all methods in Table 2, which shows that HBOP is the most accurate while having low complexity. Among the other efficient baselines, BOPF and BOSS-VS are both fast and relatively accurate, while FS ranks low in accuracy. Finally, we consider HBOP's accuracies in multiple applications. By the taxonomy on the UCR webpage [5], the datasets used here cover 14 types of data. We select one dataset from each type and show the error rates of all methods on them in Table 3 (the results on other datasets are at [1]). As is shown, HBOP has top or competitive accuracies in most cases, which showcases its applicability in multiple domains.

Table 1. One-on-one comparison of HBOP and the baselines

Methods	Wilcoxon p-value	#HBOP wins	#HBOP ties	#HBOP losses
HBOP vs ED	3.20×10^{-13}	88	2	20
HBOP vs DTWCV	1.72×10^{-4}	72	2	36
HBOP vs DTW	9.41×10^{-9}	81	2	27
HBOP vs BOPF	1.02×10^{-4}	71	6	33
HBOP vs BOSS-VS	8.32×10^{-3}	66	5	39
HBOP vs FS	5.70×10^{-15}	94	1	15

Table 2. A summarized comparison of all methods

	ED	DTWCV	DTW	BOPF	BOSS-VS	FS	HBOP
Avg. error rate	0.3007	0.2429	0.2640	0.2085	0.2067	0.2853	**0.1881**
Time complexity	N/A	$O(n^2m^2)$	N/A	$\boldsymbol{O(nm)}$	$O(nm^{\frac{3}{2}})$	$O(nm^2)$	$\boldsymbol{O(nm)}$

Table 3. Error rates on examples of 14 types of datasets (rounded to 4 decimals)

Dataset	Type	ED	DTW-CV	DTW	BOPF	BOSS-VS	FS	HBOP	HBOP Rank
RefDevices	Device	0.6053	0.5600	0.5360	0.4747	0.4933	0.6667	**0.4587**	1
ECG200	ECG	**0.1200**	**0.1200**	0.2300	0.1600	0.1600	0.2360	0.1600	3
EOGHor	EOG	0.5829	0.5249	**0.4972**	0.5801	0.5138	0.5552	0.5530	4
InsSmall	EPG	0.3373	0.3052	0.2651	0.1004	**0.0442**	0.1430	0.1205	3
PigAirway	HemoDy	0.9423	0.9038	0.8942	0.0962	0.1538	0.7010	**0.0721**	1
Fungi	HRM	0.1774	0.1774	0.1613	0.0215	0.4409	0.1731	**0.0108**	1
Adiac	Image	0.3887	0.3913	0.3964	0.3453	0.2967	0.4225	**0.2343**	1
Worms2Cls	Motion	0.3896	0.4156	0.3766	0.3247	0.2208	0.3403	**0.2104**	1
PowerCons	Power	**0.0667**	0.0778	0.1222	0.1000	0.0833	0.1189	0.1222	6
Car	Sensor	0.2667	0.2333	0.2667	0.2667	0.2500	0.2867	**0.1767**	1
CBF	Simulated	0.1478	0.0044	0.0033	**0.0000**	0.0022	0.0638	**0.0000**	1
OliveOil	Spectro	0.1333	0.1333	0.1667	0.1333	0.1333	0.2733	**0.1133**	1
Rock	Spectrum	0.1600	0.1600	0.4000	**0.1200**	0.1800	0.3880	0.1600	2
MelPed	Traffic	**0.1518**	**0.1518**	0.2094	0.2902	0.2265	0.2758	0.1561	3

5.2 Impact of Design Choices

We now explore the impact of our design choices on the accuracy of HBOP. We conclude the main design features of HBOP as follows.

(1) Our key feature is using both SAX and SLA, rather than only SAX.
(2) We keep both the options of using and not using X-mean [16] open to handle potential existence (or non-existence) of subclasses.
(3) We opt to adopt weighted vote [2] for classification, rather than simple majority vote which is used in BOPF [12].

To evaluate their effects, we test variants of HBOP in which one or more of these design features are removed. Concretely:

(1) For the first feature, while keeping the same numbers of segments in subsequences, we replace SLA-based discretization configurations with SAX-based ones. Recall that in the *original* HBOP, the number of SLA segments (half the dimensionality) is among $\{3, 4, 5\}$, while the number of SAX segments (same as the dimensionality) is among $\{6, 7, 8\}$. Replacing SLA with SAX, we have a set of *variants* which uses only SAX with the number of segments among $\{3, 4, 5, 6, 7, 8\}$. Even after this replacement, SAX still has an advantage over SLA, for SAX has a cardinality of 4 while SLA only has a lower cardinality of 3.
(2) For the second one, we consider two cases: the case where we only consider applying X-means, and the case where we only consider not applying it.
(3) For the third one, we replace weighted vote with majority vote [12].

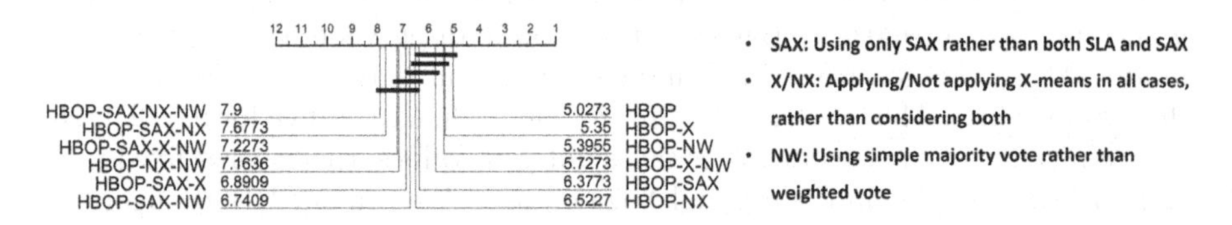

Fig. 11. Critical difference diagram of HBOP and its variants.

Figure 11 shows the resulting CDD [6] ($\alpha = 0.05$), which shows that (1) our key design logic, i.e. the hybrid of SAX and SLA, yields better accuracy than only using SAX; (2) using X-means is better than not doing so, while considering both options yields even better results; (3) weighted vote is better than majority vote; (4) with full design features, HBOP has the best results.

5.3 Efficiency

For efficiency, we compare HBOP with BOPF. We choose to do so (rather than use other methods) because BOPF is the only linear-time baseline. Both methods are implemented in Python. All running times are averaged over five runs. Figure 12 shows the running times of the two methods. For training, the ratio of training time of HBOP to BOPF is 5.184 ± 1.861 (*mean* $\pm$ *standard deviation*) on each dataset. The extra time of HBOP may be due to extra discretization configurations, extra subroutines, etc. For online classification[1], the ratio of average classification time of HBOP to BOPF is 7.905 ± 6.382 (*mean* $\pm$ *standard deviation*). The large standard deviation is likely due to the fact that BOPF uses a *fixed* ensemble size [12], while HBOP uses a *data-adaptive* ensemble size decided by CV (Sect. 4.5). In conclusion, HBOP is slower than BOPF, yet the longer running time comes with better accuracies (Sect. 5.1).

[1] Here we have ignored the *Fungi* dataset on which we have obtained abnormally short classification time for BOPF. See [1] for more on this.

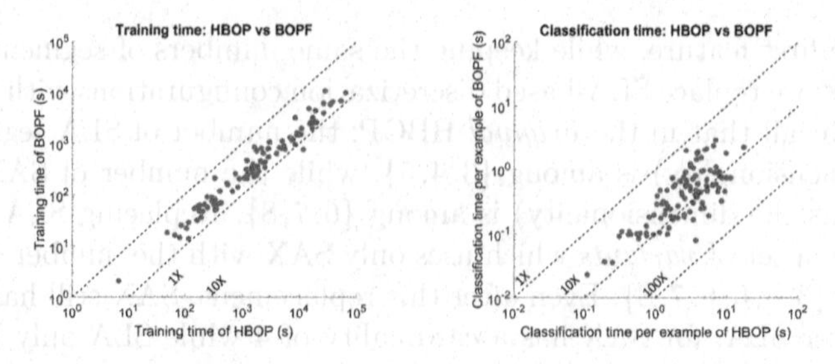

Fig. 12. Running time of HBOP and BOPF. *left)* Training time. *right)* Online classification time (per example).

We now verify that HBOP has linear training complexity by examining the relationships between HBOP training time with the number of training examples n and time series length m. Based on the Yoga dataset where the original n and m are 300 and 426 [5], we create a number of variant datasets by altering one of the two variables. The range for n is $25:25:300$, while the range for m is $25:25:425$. When one of the two variables takes different values, the other is fixed to its aforementioned value in the original Yoga dataset. We run our HBOP algorithm on these variant datasets and record the training time (averaged over five runs). The results are shown in Fig. 13. The adjusted coefficient of determination (R^2) value for different n settings is 0.992, indicating strong linear relationship. The adjusted R^2 value for different m settings is 0.912. Here the linear relationship is weaker, yet can still be considered strong. The linear relationship may have been weakened by the flexibility within HBOP. For example, our binary-search-like approach of discretization (Sect. 4.2) means that the proportion of subsequences *explicitly* discretized is data-specific. Also, the number of X-means clusters may vary when changing n and m.

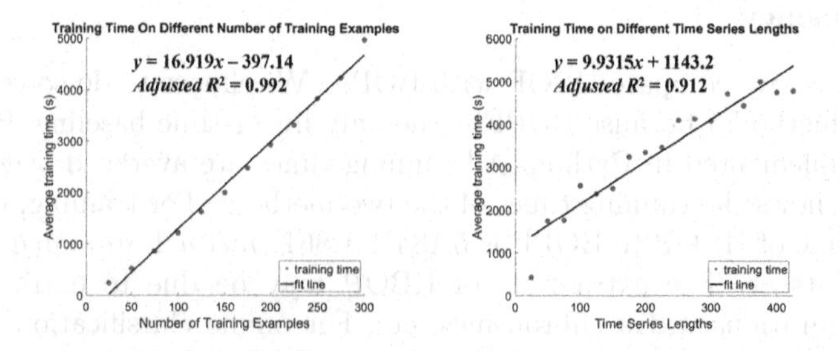

Fig. 13. HBOP training time on variants of the Yoga Dataset with different number of training examples n (*left*) and different time series lengths m (*right*).

6 Related Work

Time series classification (TSC) has attracted great attention from researchers. A classic TSC paradigm is the nearest neighbor (NN) classifier [23]. For the NN distance measure, a simple choice is Euclidean distance (ED), which is comparatively efficient yet fails to handle phase shifts and scaling [12]. Elastic distance measures [23] have been proposed to address these problems, among which the most classic one is DTW. Utilizing dynamic programming to align two time series, DTW has been shown to be a very competitive distance measure [23], inspiring numerous variants [25,26].

While NN classifiers utilize entire time series streams, another option is to exploit short patterns in time series [12]. Some pattern-based methods [7,9,15,17,24] exploit subsequences called *shapelets* that can distinguish among classes based on their distances to the time series. Another type of pattern-based methods [3,8,10,18,19,21,22] are based on the Bag-Of-Patterns (BOP) model. BOP represents a time series by the histogram of a set of symbolic words, which are discretized from subsequences in the time series. As with discretization method, popular choices include SAX [13] and SFA [20]. Also, [3] proposes a method based on linear approximation. However, it only uses slopes and fails to consider intercepts. [14] extends SAX with linear regression, yet the original authors failed to apply this method to TSC. [8] proposes a discretization method using polynomial coefficients, which is similar to our SLA. However, this method can only achieve linear time complexity when its BOP sliding windows are of constant sizes. [12] proposes a BOP-based method tailored to have linear time complexity. However, its classification accuracy can be relatively limited, which we attempt to improve upon in this paper.

7 Conclusions and Future Work

In this paper, we have proposed HBOP, an efficient time series classification algorithm with linear time complexity. The *most important* feature of HBOP is a *hybrid* of SAX and a novel discretization method we call SLA. Our other highlights include an efficient PLA transformation method that is key to achieving linear complexity, and a novel X-means [16] clustering subroutine to address potential subclasses. Extensive experiments have demonstrated the effectiveness and efficiency of our method. For future work, we plan to further incorporate the SFA [20] discretization method into our model. We also plan to explore the possibility of using SLA for time series indexing [20,23].

References

1. HBOP code. https://github.com/sliang11/Hybrid-Bag-Of-Patterns
2. Bagnall, A., Lines, J., Hills, J., Bostrom, A.: Time-series classification with COTE: the collective of transformation-based ensembles. IEEE Trans. Knowl. Data Eng. **27**(9), 2522–2535 (2015)

3. Calbimonte, J., Yan, Z., Jeung, H., Corcho, Ó., Aberer, K.: Deriving semantic sensor metadata from raw measurements. In: SSN 2012, pp. 33–48 (2012)
4. Dau, H.A., et al.: The UCR Time Series Archive. CoRR abs/1810.07758 (2018)
5. Dau, H.A., et al.: Hexagon-ML: The UCR Time Series Classification Archive, October 2018. https://www.cs.ucr.edu/~eamonn/time_series_data_2018/
6. Demsar, J.: Statistical comparisons of classifiers over multiple data sets. J. Mach. Learn. Res. **7**, 1–30 (2006)
7. Grabocka, J., Schilling, N., Wistuba, M., Schmidt-Thieme, L.: Learning time-series shapelets. In: KDD 2014, pp. 392–401 (2014)
8. Grabocka, J., Wistuba, M., Schmidt-Thieme, L.: Scalable classification of repetitive time series through frequencies of local polynomials. IEEE Trans. Knowl. Data Eng. **27**(6), 1683–1695 (2015)
9. Hills, J., Lines, J., Baranauskas, E., Mapp, J., Bagnall, A.: Classification of time series by shapelet transformation. Data Min. Knowl. Disc. **28**(4), 851–881 (2014). https://doi.org/10.1007/s10618-013-0322-1
10. Jessica, L., Rohan, K., Yuan, L.: Rotation-invariant similarity in time series using bag-of-patterns representation. J. Intell. Inf. Syst. **39**(2), 287–315 (2012)
11. Keogh, E., Chakrabarti, K., Pazzani, M., Mehrotra, S.: Dimensionality reduction for fast similarity search in large time series databases. Knowl. Inf. Syst. **3**(3), 263–286 (2001)
12. Li, X., Lin, J.: Linear time complexity time series classification with bag-of-pattern-features. In: ICDM 2017, pp. 277–286 (2017)
13. Lin, J., Keogh, E., Li, W., Lonardi, S.: Experiencing SAX: a novel symbolic representation of time series. Data Min. Knowl. Disc. **15**(2), 107 (2007). https://doi.org/10.1007/s10618-007-0064-z
14. Malinowski, S., Guyet, T., Quiniou, R., Tavenard, R.: 1d-SAX: a novel symbolic representation for time series. In: Tucker, A., Höppner, F., Siebes, A., Swift, S. (eds.) IDA 2013. LNCS, vol. 8207, pp. 273–284. Springer, Heidelberg (2013). https://doi.org/10.1007/978-3-642-41398-8_24
15. Mueen, A., Keogh, E., Young, N.: Logical-shapelets: an expressive primitive for time series classification. In: KDD 2011, pp. 1154–1162 (2011)
16. Pelleg, D., Moore, A.: X-means: extending K-means with efficient estimation of the number of clusters. In: ICML 2000, pp. 727–734 (2000)
17. Rakthanmanon, T., Keogh, E.: Fast shapelets: a scalable algorithm for discovering time series shapelets. In: SDM 2013, pp. 668–676 (2013)
18. Schäfer, P.: The BOSS is concerned with time series classification in the presence of noise. Data Min. Knowl. Disc. **29**(6), 1505–1530 (2015). https://doi.org/10.1007/s10618-014-0377-7
19. Schäfer, P.: Scalable time series classification. Data Min. Knowl. Disc. **30**(5), 1273–1298 (2016). https://doi.org/10.1007/s10618-015-0441-y
20. Schäfer, P., Högqvist, M.: SFA: a symbolic fourier approximation and index for similarity search in high dimensional datasets. In: EDBT 2012, pp. 516–527 (2012)
21. Schäfer, P., Leser, U.: Fast and accurate time series classification with WEASEL. In: CIKM 2017, pp. 637–646 (2017)
22. Senin, P., Malinchik, S.: SAX-VSM: interpretable time series classification using SAX and vector space model. In: ICDM 2013, pp. 1175–1180 (2013)
23. Wang, X., Mueen, A., Ding, H., Trajcevski, G., Scheuermann, P., Keogh, E.: Experimental comparison of representation methods and distance measures for time series data. Data Min. Knowl. Disc. **26**(2), 275–309 (2013). https://doi.org/10.1007/s10618-012-0250-5

24. Ye, L., Keogh, E.: Time series shapelets: a novel technique that allows accurate, interpretable and fast classification. Data Min. Knowl. Disc. **22**(1–2), 149–182 (2011). https://doi.org/10.1007/s10618-010-0179-5
25. Yuan, J., Lin, Q., Zhang, W., Wang, Z.: Locally slope-based dynamic time warping for time series classification. In: Proceedings of the 28th ACM International Conference on Information and Knowledge Management, CIKM 2019, pp. 1713–1722 (2019)
26. Zhao, J., Itti, L.: shapeDTW: shape dynamic time warping. Pattern Recogn. **74**, 171–184 (2018)

SentiMem: Attentive Memory Networks for Sentiment Classification in User Review

Xiaosong Jia, Qitian Wu, Xiaofeng Gao(✉), and Guihai Chen

Shanhai Key Laboratory of Scalable Computing and Systems,
Department of Computer Science and Engineering,
Shanghai Jiao Tong University, Shanghai 200240, People's Republic of China
{jiaxiaosong,ech0740}@sjtu.edu.cn
{gao-xf,gchen}@cs.sjtu.edu.cn

Abstract. Sentiment analysis for textual contents has attracted lots of attentions. However, most existing models only utilize the target text to mine the deep relations from text representation features to sentiment values, ignoring users' historicalally published texts, which also contain much valuable information. Correspondingly, in this paper we propose SentiMem, a new sentiment analysis framework that incorporates user's historical texts to improve the accuracy of sentiment classification.

In SentiMem, to exploit users' interests and preferences hidden in the texts, we adopt SenticNet to capture the concept-level semantics; as for users' temperaments, we combine multiple sentiment lexicons with multi-head attention mechanism to extract users' diverse characters. Then, two memory networks: *Interests Memory Network* and *Temperaments Memory Network* are used to store information about users' interests and temperaments respectively. Interests memory is updated in a first-in-first-out way and read by an attention mechanism to match the users' most recent interests with the target text. Temperaments memory is updated in a forgetting-and-strengthening manner to match the gradual shift of human's characteristics. Additionally, we learn a global matrix to represent these common features among human's temperaments, which is queried when classifying a user's new posted text. Extensive experiments on two real-world datasets show that SentiMem can achieve significant improvement for accuracy over state-of-the-art methods.

Keywords: Sentiment analysis · Memory network · Multi-head attention · Social network

---

This work was supported by the National Key R&D Program of China [2018YFB1004700]; the National Natural Science Foundation of China [61872238, 61972254]; the Shanghai Science and Technology Fund [17510740200], and the CCF-Huawei Database System Innovation Research Plan [CCF-Huawei DBIR2019002A].

Y. Nah et al. (Eds.): DASFAA 2020, LNCS 12112, pp. 736–751, 2020.
https://doi.org/10.1007/978-3-030-59410-7_51

1 Introduction

People frequently use texts to express their feelings on online platforms, such as social networks and e-commerce websites. The sentiment analysis is to recognize the sentiment attitudes behind texts and classify them into different categories like positive texts and negative texts. The sentiment scores could play a significant role in many applications and thus there has been an increasing number of researches on sentiment analysis in recent years.

1.1 Literature Review and Motivation

Recently, deep learning based methods have become the mainstream for sentiment analysis and achieved decent performances. There are models based on CNN/RNN. Wang et al. [24] proposed an aspect sentiment classification model with both word-level and clause-level networks. Ma et al. [15] utilized commonsense knowledge with an attentive LSTM for targeted aspect-based sentiment analysis. Vaswani et al. [23] proposed an entirely attention-based methods to explore the deep relationship between words and it outperforms RNN/CNN based models a lot. Now, start-of-the-art performances of sentiment analysis tasks are dominated by attention-based pretrained models like BERT [5] and GPT [19].

However, most of the above-mentioned methods **focus on** leveraging textual features from the target text but **ignore** the information from text promulgators. Indeed, in the real world individuals possess their own **characteristics** like interests and temperaments, which could greatly influence the sentiment attitudes of users' texts. E.g., in terms of users' **interests**, consider an example where a dog man always post texts about his love of dog. One day, he writes a text *My dog is a very bad guy!*. Even this text includes the word *bad*, we can still recognize its positivity because we know he loves his dog and the word *bad* reflects his cosset. Nevertheless, for sentiment analysis models based on single text, they may tend to give a negative label according to the word *bad*. No matter how powerful the single-text based models are, this kind of mistakes will happen because these models only have that text and can only interpret it in a general way without considering user interests. Additionally, in terms of users' **temperaments**, some people like to use exaggerated words like *great*, *excellent*, and *fantastic* to express their happiness. Then, if one of them posts a twitter *Today is fine.*, we can conclude that it may be a bad day for him because he uses the ordinary word *fine* instead of those exaggerated words this time. If we merely feed this text into a model, it would be definitely be classified as a positive one. This overestimation comes from the users' unknown temperaments.

Therefore, an accurate sentiment analysis model should take user' historical texts into consideration. Few researches pioneered such idea in this field. Tang et al. [22] learns a static embedding for each user and feeds both the user's embedding vector and the target text into the neural network to classify sentiment polarity of the text. However, this model ignores the order of texts, whose importance has been emphasized in [3]. Actually, since a user's characteristics may change over time, his recent texts may be more representative while a static embedding wrongly treats all texts in the same way. Dou et al. [6] adopted end-to-end memory networks [21] to store historical texts' embeddings and read the memories by the target text. Although it utilizes sequential information, the model is still weak because it only extracts basic semantic information from words by stacking memory networks. It uses pretrained word embeddings such as Word2Vec [16] and Glove [18] containing only general meanings of words. Consider the examples of users' interests and temperaments again: the mistake originates from the abstract and high-level characteristics of human. It is hard to extract them from general representations of words without domain knowledge.

Thus, there are several challenges to conduct sentiment classification in historical text sequence. **Firstly**, same words/sentences may have quite different sentiment attitudes for different people because of the diverse and abstract personal characteristics. **Second**, in most situations, we have no user profile information due to privacy issues and cost of investigations, so users' characteristics can only be deduced from their published texts. **Thirdly**, a user's *interests* and *temperaments* may change over time. For example, a person may be pessimistic some days and be optimistic later. Hence the model should dynamically capture user's time-variant characteristics instead of assuming some static features.

1.2 Our Contribution

To these ends, in this paper, we propose SentiMem, a new framework that models user's characteristics of *interests* and *temperaments* to help with the sentiment analysis. Specifically, we use a pre-trained model to extract general **semantic** embedding of texts first. To obtain users' **interests** from texts, we combines SenticNet [2], a knowledge base includes concepts of words With domain knowledge and concepts, SentiMem could process words with specific semantic easier to extract users' interests. As for users' **temperaments**, we let scores from multiple sentiment lexicons of a word as its embedding. By using sentiment polarity of words explicitly and directly, SentiMem could better extract users' temperaments from texts in terms of writing style and intensities of common used words.

Additionally, we design two memory networks to store users' characteristics. To capture the temporal dynamics in the historical sequence of user's publish texts, the updating method occurs sequentially for user's historically published texts so that it could capture user's time-variant characteristics. The reading and updating of the two memory networks are also based on an attention mechanism to dynamically weigh the influence of those significant and relevant historical texts. Additionally, we apply mutli-head attention mechanism to capture the very diverse characteristics of human in different embedding space [23].

To verify the model, we conduct experiments on two real-world datasets from Amazon reviews and compare with several state-of-the-art methods. The results show that SentiMem outperforms other competitors. Also, we do some ablation tests and verify the effectiveness of two memory networks for modeling the temporal dynamics in user's characteristics.

Our main contributions are summarized as follows:

- **New aspect:** We take a fresh view and combine users' historical texts with SenticNet and sentiment lexicons to capture users' interests and temperaments hidden behind texts.
- **Methodology:** We propose a new sentiment analysis framework with two memory networks, which can collaboratively capture temporal dynamics in user's interests and temperaments.
- **Experiment:** We conduct experiments on two Amazon reviews datasets to exhibit its superiority and effectiveness of SentiMEM.

2 Problem Formulation and Overall Model

We denote a text $\boldsymbol{x}$ of length m as a sequence of m words, i.e. $\boldsymbol{x} = \{\boldsymbol{w_1}, \boldsymbol{w_2}, ..., \boldsymbol{w_m}\}$ where $\boldsymbol{w_j}$ is the jth word. Then we define the problem as: Given a sequence of n texts in time order posted by user u: $X^u = \{\boldsymbol{x_1}, \boldsymbol{x_2}, ..., \boldsymbol{x_n}\}$ where $\boldsymbol{x_j}$ is the jth text the user posted, we want to predict the nth text's label y_n^u where $y_n^u \in \{1, 2, ..., K\}$ indicates the sentiment attitude of text with 1 as lowest and K as highest. In this paper, we call $\boldsymbol{x_n}$ as the **target text** and other texts of user u as **historical texts**.

In this paper, we propose SentiMem framework to exploit users' interests and preferences hidden in texts. Figure 1 depicts its overall architecture.

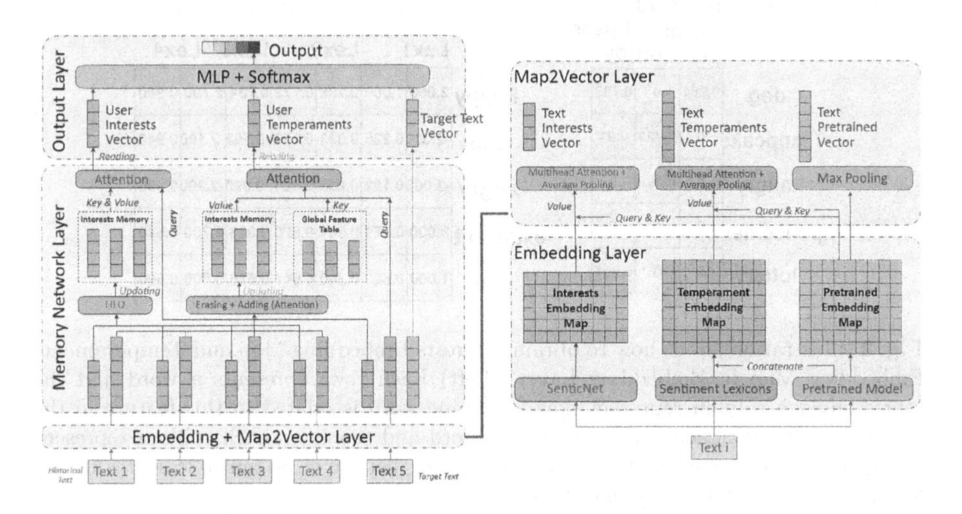

Fig. 1. Model framework of SentiMem with four layers: (1) Embedding Layer, (2) Map2Vector Layer, (3) Memory Network Layer, and (4) Output Layer. Note that Embedding Layer and Map2Vector Layer are used to produce three vectors for individual texts, while Memory Network Layer and Output Layer are for the sequential texts. (Color figure online)

3 SentiMem Framework

In this section, we introduce SentiMem model in a layer-by-layer way with four subsections sequentially.

3.1 Embedding Layer

In the Embedding Layer, we aims to obtain several embeddings of texts in a word-level manner, i.e. there would be a vector for each word. Thus there would be a matrix for each text.

We harness an encoder structure as a pretrained model to obtain basic semantic representations of texts. The pretrained model should keep the shape of sequences (like LSTM [9]/GRU [4]/BiGRU [11]) to maintain semantics of each word. Formally, for a input text $\boldsymbol{x} = \{\boldsymbol{w_1}, \boldsymbol{w_2}, ..., \boldsymbol{w_m}\}$, after the pretrained model, we will get a sequence of dense representation of the text $\boldsymbol{x^p} = \{\boldsymbol{w^p_1}, \boldsymbol{w^p_2}, ..., \boldsymbol{w^p_n}\}$ where $\boldsymbol{w^p_k}$ represents the denser and more abstract representation of word w_k. In this paper, we call the representations for a text after the pretrained model as **Pretrained Embedding Map** (shown as blue part of the Embedding Layer in Fig. 1).

To embed a single word to an interests vector, we use SenticNet [2], a knowledge base that contains 100,000 concepts with a huge set of affection properties.

	Is-a-pet	Kind-of-food	Arises-joy
dog	0.981	0	0.789
cupcake	0	0.922	0.91
angry	0	0.459	0
policemen	0	0	0
lottery	0	0	0.991

	Lex1		Lex2		Lex3	Lex4	
happy	2.000	1.000	0.735	0.772	0.734	2.700	0.900
sad	-2.000	0.225	0.333	0.149	-0.562	-2.100	0.943
angry	-3.000	0.122	0.830	0.604	-0.890	-2.300	0.900
excellent	3.000	0.970	0.587	0.870	0.823	2.700	0.640
fine	1.000	0.823	0.240	0.647	0.626	0.800	0.600

Fig. 2. Illustration about how to obtain Interests Embedding Map and Temperaments Embedding Map in Embedding Layer. (Left) Each row represents a word and the scores on each column represent how much one word is related to this feature in the SenticNet. (Right) Each row represents a word and scores on each column represent scores on different sentiment lexicons. Note that there may be multiple scores in one lexicon like valence, arousal, and dominance in [17].

As we can see in Fig. 2 (Left), features of a word including the interests and knowledge information are scored between 0 and 1. For example, *dog* is scored highly at *IsA-pet* and *Arises-joy* which means we can capture word-level interests information from SenticNet. Nevertheless, SenticNet is a high-dimensional and sparse embedding which makes it hard to use directly. Followed by [1,15], we adopt a dimension reduction technique to map the results of SenticNet to a relatively low-dimensional (100) continuous space with little loss of the semantic and relativeness. For a text $\boldsymbol{x} = \{\boldsymbol{w}_1, ..., \boldsymbol{w}_m\}$, we can get its **Interests Embedding Map** (shown as green part of Embedding Layer in Fig. 1) $\boldsymbol{x}^i = \{\boldsymbol{w}_1^i, ..., \boldsymbol{w}_n^i\}$ where $\boldsymbol{w}_k^i$ is word w_k's SenticNet embedding.

To obtain a single word's temperaments vector, we utilize sentiment lexicons, a traditional sentiment analysis tool, which has been shown powerful when coupled with word embeddings [20]. Specifically, as shown in Fig. 2 (right), we concatenate scores of a word obtained from multiple open-source sentiment lexicons with the word's pretrained embedding to get the *temperaments vector* of this word. The reason why using concatenation is that sentimental scores are just sentiment polarity and using them singly would lose too much semantic information. With the concatenation of scores and pretrained embeddings, it can better capture the temperaments characteristic of a word. Finally, we can get **Temperaments Embedding Map** (shown as orange part of Embedding Layer in Fig. 1) of each text as $\boldsymbol{x}^t = \{\boldsymbol{w}_1^t, \boldsymbol{w}_2^t, ..., \boldsymbol{w}_n^t\}$ where $\boldsymbol{w}_k^t$ is the concatenation mentioned before.

In conclusion, by embed words in different ways, they are represented from a different perspective. Specifically, pretrained embedding is based on Glove/Word2Vec and basic sequential models (LSTM/GRU/BiGRU) which are generally used in NLP area to extract semantics of texts. As for SenticNet [2], it can extract concepts of words (as shown in Fig. 2 Left) so that we can capture users' interests and knowledge from texts. Shin et al. [20] has shown that with multiple open-source sentiment lexicons, the words can be represented in a more diverse way. With the sentiment scores, valence, arousal, and dominance [17], the users' temperaments behind the words can be better represented.

3.2 Map2Vector Layer

In the Map2Vector layer, the matrices (Pretrained Embedding Map, Interests Embedding Map, Temperaments Embedding Map) obtained from Embedding Layer are represented in a dense form - document-level vectors. It has been shown to be effective [6,14] for downstream tasks for several reasons like reducing the model complexity, representing in a more abstract and global way, and lower computational demands.

Target Text Vector. For Pretrained Embedding Map to merge the pretrained embedding map into one vector for target text (shown as **Target Text Vector** in the Output Layer, Fig. 1), we adopt the max-pooling to its pretrained embedding map, which is a widely used technique. Specifically, we take the maximum among all words for each dimension.

Text Interests Vector. To capture the interests and knowledge in one text of user, Map2Vector layer extracts **Text Interests Vector** from Interests Embedding Map. For example, if a user posted *This dog is so cute.*, the corresponding

Text Interests Vector should indicate she is a dog-person. Then, when dealing with his/her new text about dogs, the model should tend to classify it as a positive text.

We use multi-head attention mechanism, a powerful method involving lots of the state-of-the-art NLP models like Bert [5] and GPT [19]. The intuition is: multi-head attention allows the model to jointly attend to information from different representation subspaces at different positions (like kernels in the CNN) [23]. In our scene, considering that human have complex and diverse characteristics, with multiple head, SentiMem could capture writers' interests and temperaments from different perspective of texts.

The basic equation of multi-head attention mechanism is shown in (1). Considering it is not our contributions, please see [23] for technical details.

$$\text{Attention}(Q, K, V) = \text{softmax}(QK^T)V \tag{1}$$

Specifically, for a text $\boldsymbol{x}$ with its pretrained embedding map $\boldsymbol{x^p}$ and interests embedding map $\boldsymbol{x^i}$, we let $\boldsymbol{x^p} = Q = K$ and $\boldsymbol{x^i} = V$. Setting pretrained embedding as key and query is because pretrained model's embedding includes general and position information of words and it can give relatively precise weights. Setting interests embedding as value is because SenticNet model's embedding includes only words' specific interests knowledge and this information is what we want to extract. Finally, after taking average pooling over all the words' interests embedding in MultiHead, we can get **Text Interests Vector** (shown as green vector of Map2Vector Layer in Fig. 1). Figure 3 illustrates how this layer works.

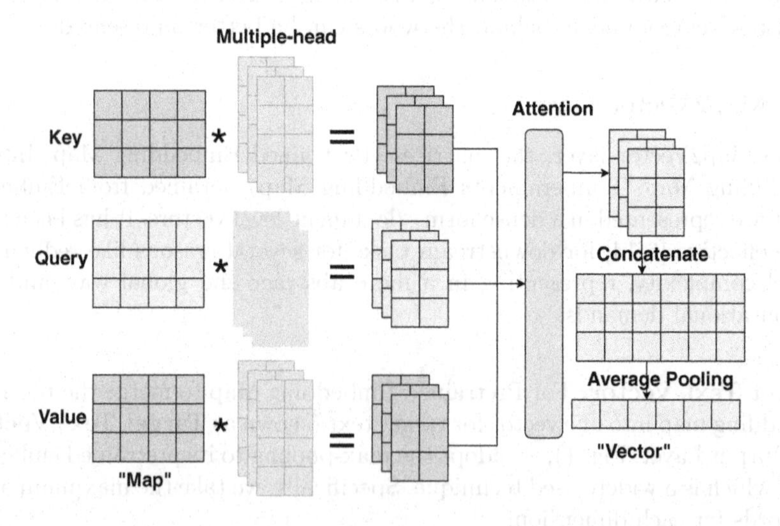

Fig. 3. Illustration for Multihead Attention+Average Pooling in Map2Vector layer. It takes Pretrained Embedding Map as query and key and Interests/Temperaments Embedding Map as value to obtain Text Interests/Temperaments Vector respectively.

Text Temperaments Vector. Similarly, we use multi-head attention to obtain Text Temperaments Vector from Temperaments Embedding Map, which can reflect user's temperaments hidden in the text. For example, if a user often posts texts to express his sadness, the temperaments vectors should contain the semantic information that he is an pessimistic person. Then, the model will be inclined to classify his/her new text as a positive one.

Formally, for a text $\boldsymbol{x}$ with its pretrained embedding map $\boldsymbol{x}^{\boldsymbol{p}}$ and temperaments embedding map $\boldsymbol{x}^{\boldsymbol{t}}$, we let $\boldsymbol{x}^{\boldsymbol{p}} = Q = K$ and $\boldsymbol{x}^{\boldsymbol{t}} = V$ as show in (1). The intuition behind it is also the same as Text Interests Vector. After taking average pooling over all the vector's in MutliHead, we can finally get **Text Temperaments Vector** (shown as orange vector of Map2Vector Layer in Fig. 1).

In conclusion, in the Map2Vector layer, we use mutli-head attention mechanism to extract three kinds of vectors (Pretrained, Temperaments, Interests) of each text from embedding maps. These vectors represent the basic semantic, temperaments, and interests features of each text respectively.

3.3 Memory Network Layer

In the Memory Network layer, we use memory network to store users' temporal interests and temperaments dynamically by writing operation and obtain how both of them influence the target text by reading operation. Additionally, by attention mechanism, SentiMem weighs the influences of those significant and relevant historical texts and characteristics.

Interests Memory Network (IMN). It aims to store users' interests and knowledge in sequential order and help to capture users' temporal dynamics of interests in the sequence.

Formally, we have *Text Interests Vectors* (from Map2Vector Layer) of n texts $X^{\text{inter}} = \{\boldsymbol{x}_1^{\text{inter}}, ..., \boldsymbol{x}_n^{\text{inter}}\}$ posted by one user u and need to get **User Interests Vector** for target text. Initially, there is a uniformly randomized memory matrix $M_u^i = \{\boldsymbol{m}_1^u, ..., \boldsymbol{m}_K^u\}$, where K is the number of memory slots and $\boldsymbol{m}_j^u$ is the jth memory vectors of user u, with the same dimension as text interests vector.

We **update** user's IMN so that it contains the information of historical texts. Intuitively, what a user recently has done have larger influence on their sentiments. So we adopt a first-in-first-out strategy for the updating of IMN, i.e. IMN only stores the most recent K interests vectors of a user. Formally, if IMN is $M_u^i = \{\boldsymbol{x}_{j-1}^{\text{inter}}, \boldsymbol{x}_{j-2}^{\text{inter}}, ...\boldsymbol{x}_{j-K}^{\text{inter}}\}$, after updating with j^{th} historical text's interest vector $\boldsymbol{x}_j^{\text{interests}}$, then $M_u^i = \{\boldsymbol{x}_j^{\text{inter}}, \boldsymbol{x}_{j-1}^{\text{inter}}, ...\boldsymbol{x}_{j-K+1}^{\text{inter}}\}$.

We **read** IMN with target text's interest vector to obtain User Interests Vector (shown as green vector of Output Layer in Fig. 1). Here we use attention mechanism and the intuition behind is that different aspects of interests are contained in different historical texts and they should have different levels of influences on the target text. For $M_u^i = \{\boldsymbol{m}_1^u, ..., \boldsymbol{m}_K^u\}$ and target text's interest vector $\boldsymbol{x}_n^{\text{interests}}$, weights z_u^j can be computed by (2). With weights z_u^j for user

u's j^{th} memory vectors in IMN, we can get **User Interests Vector** $\boldsymbol{u}^{i}$.

$$w_j^u = \boldsymbol{m}_j^{u\,T} \cdot \boldsymbol{x}_n^{\text{inter}}, \qquad z_j^u = \frac{\exp(w_j^u)}{\sum_{i=1}^{K} \exp(w_i^u)}, \qquad \boldsymbol{u}^i = \sum_{i=1}^{K} z_i^u \boldsymbol{m}_i^u \tag{2}$$

Temperaments Memory Network (TMN). It aims to model users' temperaments such as optimism, peevishness, and pessimism. Temperaments is more abstract than interests and knowledge since the former one is more about the internal characteristics of human and the latter one is a more external form. To be specific, in SentiMem, the temperaments embedding map is only derived from sentiment scores while the interests embedding map comes from the concept of SenticNet. Hence, we propose a more complicated memory network to extract users' temperaments. Though temperaments of human are abstract, there are also some common features among them. Inspired by latent factor models in recommender systems [13], we use **a global latent vector table** to represent the common features of human's temperaments and a memory matrix for each user to represent this user's temporal personal dispositions for the common features.

Formally, we have text temperaments vectors of n texts posted by one user $X^{\text{temper}} = \{\boldsymbol{x}_1^{\text{temper}}, ..., \boldsymbol{x}_n^{\text{temper}}\}$ and we aim to get *User Temperaments Vector* for the target text. We maintain a temperaments memory matrix for each user $M_u^t = \{\boldsymbol{m}_1 ..., \boldsymbol{m}_K\}$ where K is the number of memory slots. Also we have a global temperaments feature table $F = \{\boldsymbol{f}_1, ..., \boldsymbol{f}_K\}$ where $\boldsymbol{f}_j$ represents one common feature of human's temperaments and is a parameter vector with the same dimension as the memory vector.

Reading of TMN resorts to attention mechanism as well. For the target text's temperaments vector x_n^{temper}, weights z_j^u can be computed by (3) and **User Temperaments Vector** $\boldsymbol{u}^t$ can be given by weighed sum of memory matrix as in (3). The intuition behind this reading method is that using target text's temperaments vector as query and global temperaments feature vectors as key to capture which temperaments features this text is more about. Then using this user's own temperaments disposition vectors as value to obtain user's specific temperaments about target text.

$$w_j^u = \boldsymbol{f}_j{}^{\boldsymbol{T}} \cdot \boldsymbol{x}_n^{\text{temper}}, \qquad z_j^u = \frac{\exp(w_j^u)}{\sum_{i=1}^{K} \exp(w_i^u)}, \qquad \boldsymbol{u}^t = \sum_{i=1}^{K} z_i^u \boldsymbol{m}_i^u \tag{3}$$

For **updating** of TMN, as mentioned in Sect. 1, a user's temperaments may shift over time in a slight and slow manner. To model this nature, we need to update a user's personal dispositions after each posting. However, we should not use first-in-first-out strategy again because people's temperaments usually would not change a lot in a short time.

Here we adopt the *Erasing-Adding* updating method. Formally, for a historical text's temperaments vector $\boldsymbol{x}^{temper}$, we first obtain a erasing vector $\boldsymbol{e}$ with a linear layer and then we *erase* each of the memory vector $\boldsymbol{m}_j$ as (4).

$$\boldsymbol{e} = \delta(\boldsymbol{W}^{\text{erase}}\boldsymbol{x}^{temper} + \boldsymbol{b}^{\text{erase}}), \qquad \boldsymbol{m}_j \leftarrow \boldsymbol{m}_j \odot (\mathbf{1} - z_j^u \boldsymbol{e}), \tag{4}$$

where $j \in \{1, 2, ..., K\}$, δ is a non-linear function, $\mathbf{1}$ is a vector of all 1 and $\boldsymbol{z}_j^u$ is the weight. After erasing, we *add* new contents to each of the memory vector $\boldsymbol{m}_j$ with adding vector $\boldsymbol{a}$ obtained from the linear layer. (5) shows the details.

$$\boldsymbol{a} = \delta(\boldsymbol{W}^{\text{add}}\boldsymbol{x}^{temper} + \boldsymbol{b}^{\text{add}}), \qquad \boldsymbol{m}_j \leftarrow \boldsymbol{m}_j + z_j^u \boldsymbol{a} \tag{5}$$

The intuition behind the updating equations is that: the magnitude of updating for each disposition vector $\boldsymbol{m}_j$ is decided by how much this text has shown about it. With two additional linear layers for forgetting and strengthening, the model can update user's temperaments matrix more flexibly.

3.4 Output Layer

In the Output Layer, we concatenate **Target Text Vector**, **User Interests Vector** and **User Temperaments Vector** together as one vector and feed it into a fully-connected layer followed by a Softmax layer. We use the label with the highest probability as the prediction result. The loss function is the weighted cross-entropy loss defined as

$$L = -\sum_{i=1}^{M}\sum_{c=1}^{C} w_c y_i^c \log(p(\hat{y_i^c}|x)) + \frac{l}{2}\|\theta\|_2^2, \tag{6}$$

where M is the number of train data, C is the number of class, l is the weight of L2 regularization and w_c is the weight of class c. We set w_c as the reciprocal of class c's ratio to deal with the imbalanced-class problem.

4 Experiments

4.1 Experiment Setup

Datasets: We conduct experiments on Amazon review dataset [8] with product reviews and metadata from Amazon, including 142.8 million reviews spanning 1996–2014. We choose two product categories: *Cell Phones and Accessories* (5-core) and *Movies and TV* (5-core). Their statistics are shown in Table 1. For each user, we first sort their reviews in time order. For each review, we let it to be the target text, label it as the target and all reviews posted before it as historical texts. We randomly select 70% reviews (with its historical texts) as the training set and the rest as the test set.

Metric: We use the same metric and label processing method as [14]. Specifically, we use accuracy of prediction for labels as the metric. We evaluate the model on both 2-class task which divides labels into 2 kinds (positive and negative) and 5-class task which divides labels into 5 kinds (1–5).

Table 1. Statistics of datasets

Datasets	#Users	#Reviews	#Reviews/Users
Cell Phones and Accessories	27874	190821	6.85
Movies and TV	123952	1691396	13.65

Baseline Methods: As mentioned before, this model is a kind of generalized enhancement technique which needs a pretrained model. Thus, pretrained models could be viewed as baselines and our goal is to improve their performance. We conduct experiments with the following three pretrained models:

- **LSTM** [9]: Long short-term memory (LSTM) is an artificial recurrent neural network (RNN) architecture which is composed of a cell, an input gate, an output gate and a forget gate. The cell remembers values over arbitrary time intervals and the three gates regulate the flow of information into and out of the cell. It is widely used in NLP field.
- **GRU** [4]: Gated recurrent units (GRU) is like LSTM with forget gate but has fewer parameters than LSTM, as it lacks an output gate. GRUs have been shown to exhibit better performance on some tasks.
- **BiGRU** [11]: Bidirectional gated recurrent units(BiGRU) connect hidden layers of two GRU with opposite directions to the same output. It is especially useful when the context of the input is needed.

Models utilizing users' historical texts could also be baselines. As mentioned in Sect. 1, the proposed model aims to **only use texts** as inputs since in many scenarios other types of information is unavailable. Thus, to fairly compare the proposed model with other baseline models, the baselines should process texts relatively independently so that we can only keep the texts processing parts of them without modifying them a lot. After we surveyed papers about utilizing user' historical texts, we choose the following two representative and fair-to-compare-with models and modify them to satisfy our task:

- **UPNN** [22] uses CNN+average-pooling to obtain document-level vectors of texts and proposes to learn a static embedding vector for each user and each product from users' reviews and information about purchased products respectively. Here we only keep the part about processing users' reviews.
- **UPDMN** [6] uses LSTM to extract document-level vectors of texts and uses end-to-end memory networks [21] to store and read embeddings of users' reviews and purchased production. Similar to UPNN, we only keep the memory network for processing users' reviews.

Preprocessing and Word Embedding. We use public NLP tool torchtext to prepossess all texts, and GloVe (dimension 300) as word embedding for pretrained model. Four open-source sentiment lexicons are used to build temperaments embedding of words: AFINN-111 [7], MaxDiff Twitter Sentiment Lexicon [12], The NRC Valence, Arousal, and Dominance Lexicon [17], and VADER-Sentiment-Analysis [10]. All lexicons have scores for words and are scaled to

Table 2. 2-class task and 5-class task prediction accuracy

Cell phones and accessories	2-class	5-class
UPNN	81.7	70.5
UPDMN	82.1	71.0
LSTM	81.5	70.2
SentiMem+LSTM	82.6 (**+1.1**)	71.3 (**+1.1**)
GRU	81.5	70.4
SentiMem+GRU	82.5 (**+1.0**)	71.5 (**+1.1**)
BiGRU	82.0	70.8
SentiMem+BiGRU	83.0 (**+1.0**)	72.0 (**+1.2**)
Movies and TV	2-class	5-class
UPNN	76.8	67.3
UPDMN	77.2	67.5
LSTM	76.4	66.8
SentiMem+LSTM	77.6 (**+1.2**)	68.0 (**+1.2**)
GRU	76.6	66.9
SentiMem+GRU	77.7 (**+1.1**)	68.1 (**+1.2**)
BiGRU	77.1	67.2
SentiMem+BiGRU	78.5 (**+1.4**)	68.5 (**+1.3**)

* **+X** means SentiMem improves the performance of pretrained models by X.

range $[0, 1]$. Some lexicons also include detailed sentiment information such as valence, arousal, and dominance. For temperaments embedding (dimension 300+6) from lexicons and interests embedding (dimension 100) from SenticNet, there are missing features for some words, and we set the corresponding embeddings as 0 to make them neutral.

4.2 Experimental Results

We denote our model as *SentiMem+X* where X is the pretrained model and conduct experiments on two categories: *Cell Phones and Accessories*, Movies and TV. The results are shown in Table 2. We can find that:

- SentiMem model can always improve performances of pretrained models by about 1% with additional information from historical texts, sentiments lexicons, and SenticNet. There is no significant difference among the improvements for different pretrained models and different tasks. SentiMem outperforms UPNN and UPDMN as well.
- All models works worse on Movies and TV dataset than on Cell Phones and Accessories. It is because reviews about cell phones are more objective and easier to classify the sentiment attitudes, since they are usually representing whether or not be satisfied with products. While reviews about movies

usually contain more complicated feelings, which makes it harder to find the main sentiment attitudes. However, SentiMem can improve the performance of pretrained models slightly more on Movies and TV dataset.

- We do significance tests on the results of SentiMem and baseline methods, and get $p \leq 0.005$ on Cell Phones and Accessories-2 class and Movies and TV-5 class. We also have $p \leq 0.01$ on Cell Phones and Accessories-5 class and Movies and TV-2 class. They suggest that SentiMem has explicit improvements over all the baselines on all tasks and datasets specifically.

4.3 Ablation Test

To evaluate the effects of each component of SentiMem, we do ablation test by removing user interests vector and user temperaments vector respectively. The results are shown in Fig. 4. We can see that:

- Both user interests vector and user temperaments vector can improve the performance of pretrained models. It shows the effectiveness of capturing

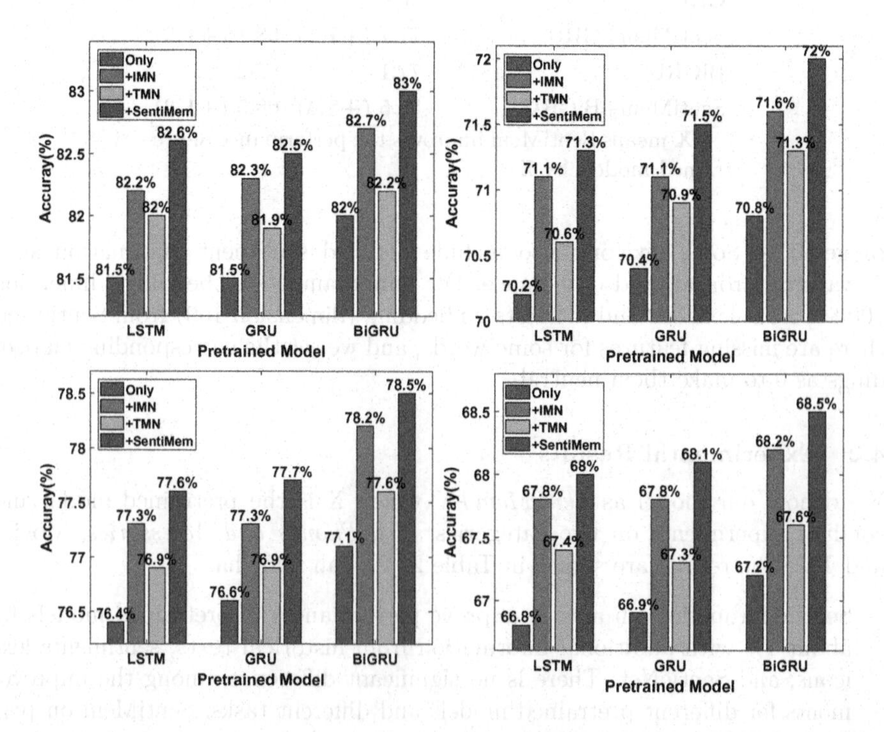

Fig. 4. Ablation Test of SentiMem Model on (a) Cell Phones and Accessories (2-class) (b) Cell Phones and Accessories (5-class) (c) Movies and TV (2-class) (d) Movies and TV (5-class) **Only** means only pretrained model. **+IMN** means adding User Interests Vector. **+TMN** means adding User Temperaments Vector. **+SentiMem** means adding both.

both users' interests and temperaments by multi-head attention, SenticNet/sentiment lexicons, and IMN/TMN.
- User interests vector can improve more than user temperaments vector does. It may be because SenticNet embedding is dimension 100 while sentiment lexicons embedding is dimension 6. Carrying more information, SenticNet can enhance the performances better.
- Adding both of them can achieve the best improvements, but they are smaller than the sum of two individual improvements. It may come from the fact that they both utilize historical texts and there is overlapped information.

4.4 Attention Visualization in Interests Memory Network

As shown in Fig. 5, we visualize the attention weights when reading from Interests Memory Network (IMN). What below target text is the 5 texts stored in IMN when reading. Deeper color on the left of the 5 texts means higher attention weight. Note that the texts shown is a part of the origin reviews, which we choose to show the main idea of reviews.

In the upper figure, the target texts is about ear phones. The 4th text has the highest weight since it is still about ear phones. The 2th and 5th are lighter because they are about headset which is similar to ear phone. We can know from historical texts that this user is a headset fan and is quite picky. IMN can capture this and help classify the sentiment attitudes of this user's review about headsets. In the lower figure, we can see that 2th text has the highest weight since

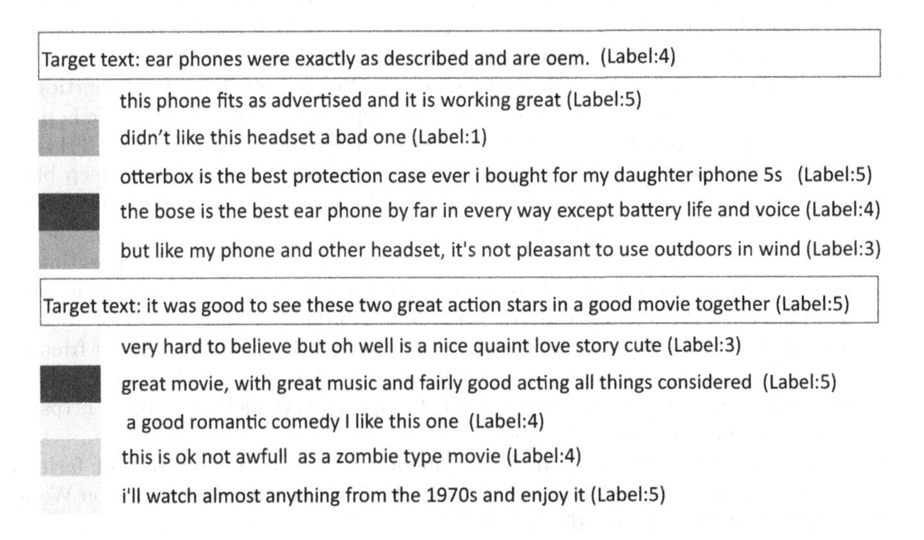

Fig. 5. Attention Visualization in Interests Memory Network (IMN). (Upper Figure) Cell Phones and Accessories; (Lower Figure) Movies and TV. What below target text is the 5 texts stored in IMN when reading. Deeper color on the right of texts means higher attention weight.

it is also about action movie just the same as the target text. Other texts are less concerned. IMN can capture this user's love for action movie and improve the prediction performance.

5 Conclusion

In this paper, we propose SentiMem to model user's characteristics including interests and temperaments, which can be helpful for sentiment classification of target text, by making use of user's historically published texts, sentiment lexicons, and SenticNet. We propose two kinds of memory networks (IMN and TMN) and utilize multi-head attention mechanism to capture the temporal dynamics of user's characteristics hidden in the historical texts. Experiments on two Amazon Review datasets show that SentiMem can explicitly improve the performances of sentiment classification over baseline methods.

References

1. Cambria, E., Fu, J., Bisio, F., Poria, S.: Affectivespace 2: enabling affective intuition for concept-level sentiment analysis. In: Association for the Advancement of Artificial Intelligence (AAAI), pp. 508–514 (2015)
2. Cambria, E., Poria, S., Hazarika, D., Kwok, K.: Senticnet 5: discovering conceptual primitives for sentiment analysis by means of context embeddings. In: Association for the Advancement of Artificial Intelligence (AAAI), pp. 1795–1802 (2018)
3. Chen, X., Xu, H., Zhang, Y., Tang, J., Cao, Y., Qin, Z., Zha, H.: Sequential recommendation with user memory networks. In: ACM International Conference on Web Search and Data Mining (WSDM), pp. 108–116 (2018)
4. Cho, K., van Merrienboer, B., Bahdanau, D., Bengio, Y.: On the properties of neural machine translation: encoder-decoder approaches. In: Eighth Workshop on Syntax, Semantics and Structure in Statistical Translation, pp. 103–111 (2014)
5. Devlin, J., Chang, M., Lee, K., Toutanova, K.: BERT: pre-training of deep bidirectional transformers for language understanding. CoRR abs/1810.04805 (2018). http://arxiv.org/abs/1810.04805
6. Dou, Z.Y.: Capturing user and product information for document level sentiment analysis with deep memory network. In: Conference on Empirical Methods in Natural Language Processing (EMNLP), pp. 521–526 (2017)
7. Hansen, L.K., Arvidsson, A., Nielsen, F.A., Colleoni, E., Etter, M.: Good friends, bad news - affect and virality in Twitter. In: Park, J.J., Yang, L.T., Lee, C. (eds.) FutureTech 2011. CCIS, vol. 185, pp. 34–43. Springer, Heidelberg (2011). https://doi.org/10.1007/978-3-642-22309-9_5
8. He, R., McAuley, J.: Ups and downs: modeling the visual evolution of fashion trends with one-class collaborative filtering. In: International Conference on World Wide Web (WWW), pp. 507–517 (2016)
9. Hochreiter, S., Schmidhuber, J.: Long short-term memory. Neural Comput. **9**(8), 1735–1780 (1997)
10. Hutto, C.J., Gilbert, E.: Vader: a parsimonious rule-based model for sentiment analysis of social media text. In: Association for the Advancement of Artificial Intelligence (AAAI) (2014)

11. Jagannatha, A.N., Yu, H.: Bidirectional RNN for medical event detection in electronic health records. In: Proceedings of NAACL-HLT, pp. 473–482 (2016)
12. Kiritchenko, S., Zhu, X., Mohammad, S.M.: Sentiment analysis of short informal texts. J. Artif. Intell. Res. (JAIR) **50**, 723–762 (2014)
13. Koren, Y., Bell, R.M., Volinsky, C.: Matrix factorization techniques for recommender systems. IEEE Comput. **42**(8), 30–37 (2009)
14. Ma, S., Sun, X., Lin, J., Ren, X.: A hierarchical end-to-end model for jointly improving text summarization and sentiment classification. In: International Joint Conference on Artificial Intelligence (IJCAI), pp. 4251–4257. AAAI Press (2018)
15. Ma, Y., Peng, H., Cambria, E.: Targeted aspect-based sentiment analysis via embedding commonsense knowledge into an attentive LSTM. In: Association for the Advancement of Artificial Intelligence (AAAI), pp. 5876–5883 (2018)
16. Mikolov, T., Chen, K., Corrado, G., Dean, J.: Efficient estimation of word representations in vector space. arXiv preprint arXiv:1301.3781 (2013)
17. Mohammad, S.: Obtaining reliable human ratings of valence, arousal, and dominance for 20,000 English words. In: Association for Computational Linguistics (ACL), pp. 174–184 (2018)
18. Pennington, J., Socher, R., Manning, C.D.: Glove: global vectors for word representation. In: Empirical Methods in Natural Language Processing (EMNLP), pp. 1532–1543 (2014)
19. Radford, A., Wu, J., Child, R., Luan, D., Amodei, D., Sutskever, I.: Language models are unsupervised multitask learners. OpenAI Blog (2019)
20. Shin, B., Lee, T., Choi, J.D.: Lexicon integrated CNN models with attention for sentiment analysis. In: Workshop on Computational Approaches to Subjectivity, Sentiment and Social Media Analysis (WASSA), pp. 149–158 (2017)
21. Sukhbaatar, S., Weston, J., Fergus, R., et al.: End-to-end memory networks. In: Advances in Neural Information Processing Systems (NeurIPS), pp. 2440–2448 (2015)
22. Tang, D., Qin, B., Liu, T.: Learning semantic representations of users and products for document level sentiment classification. In: Association for Computational Linguistics (ACL), pp. 1014–1023 (2015)
23. Vaswani, A., et al.: Attention is all you need. In: Advances in Neural Information Processing Systems (NIPS), pp. 5998–6008 (2017)
24. Wang, J., et al.: Aspect sentiment classification with both word-level and clause-level attention networks. In: International Joint Conferences on Artificial Intelligence (IJCAI), pp. 4439–4445 (2018)

AMTICS: Aligning Micro-clusters to Identify Cluster Structures

Florian Richter(✉), Yifeng Lu, Daniyal Kazempour, and Thomas Seidl

Ludwig-Maximilians-Universität München, Munich, Germany
{richter,lu,kazempour,seidl}@dbs.ifi.lmu.de.de

Abstract. OPTICS is a popular tool to analyze the clustering structure of a dataset visually. The created two-dimensional plots indicate very dense areas and cluster candidates in the data as troughs. Each horizontal slice represents an outcome of a density-based clustering specified by the height as the density threshold for clusters. However, in very dynamic and rapid changing applications a complex and finely detailed visualization slows down the knowledge discovery. Instead, a framework that provides fast but coarse insights is required to point out structures in the data quickly. The user can then control the direction he wants to put emphasize on for refinement. We develop AMTICS as a novel and efficient divide-and-conquer approach to pre-cluster data in distributed instances and align the results in a hierarchy afterward. An interactive online phase ensures a low complexity while giving the user full control over the partial cluster instances. The offline phase reveals the current data clustering structure with low complexity and at any time.

Keywords: Data streams · Hierarchical clustering · Density-based · Visual analysis

1 Introduction

Clustering is an essential task in the field of data mining and unsupervised machine learning. The initial data explorations are usually an important but difficult and exhausting step of the analysis. A huge performance bottleneck is the identification of parameter ranges to return interesting results.

In density-based clustering approaches, we are mainly interested in finding dense regions of neighboring objects. However, the definition of proximity has to be chosen mostly manually while investigating a new dataset. Probably the most prominent density-based clustering method is DBSCAN [3], which detects arbitrarily shaped clusters with similar densities while being robust against noise. Estimating proper parameters to distinguish between sparse and dense regions is a non-trivial task here, especially in dynamic applications.

A method called OPTICS [1] was designed to cope with this problem. As an extension of DBSCAN, it provides a hierarchical model of the cluster structure

Y. Nah et al. (Eds.): DASFAA 2020, LNCS 12112, pp. 752–768, 2020.
https://doi.org/10.1007/978-3-030-59410-7_52

in the dataset. The results are represented as a two-dimensional plot, such that analysts visually identify troughs as cluster candidates.

While being a suitable visualization tool for experts to estimate the number and size of clusters hidden in the dataset, it still contains the risk of poor parameter choices. In a worst-case scenario, no troughs can be identified due to a wide spectrum of densities in the dataset, leading to a flat OPTICS plot. Our novel approach AMTICS provides an interactive way to choose promising density levels for further investigation. Due to this explicit choice, attention on certain aspects is established and the coarser presentation of the dataset offers a broader comprehension of the data. The advantage is not only the efficient recomputations to allow a stream application of AMTICS. Instead of visualizing all minor density fluctuations, our novel method AMTICS provides a coarse estimation of the cluster structure for a fast explorative human-based visual analysis. An ensemble of density-based online clustering instances is the core of AMTICS. These efficient and distributed instances are interchangeable, so observation levels can be added or discarded. Analysts are free to choose which density levels are promising to increase the granularity there. This interactivity improves the analysis performance of the human-in-the-loop. For changing conditions or new analysis interests, the observation focus can be changed dynamically. In a final step, all instances are aligned to produce an approximated density plot of the recently observed objects which is a huge benefit for any further data analysis. AMTICS, as shown here, utilizes DenStream. However, the main contribution is the agglomeration of approximative online density-based clustering results, so DenStream can be replaced by other clustering techniques.

2 Preliminaries

Density-based clustering is a well-studied topic in data science. As most readers will already know, density is here defined by two parameters: The radius ε to define the neighborhood of each point $N_\varepsilon(x)$ and the minimal number of points $MinPts$ required for a dense neighborhood. Every point is

- a core point if it has a dense neighborhood. Neighbored core points establish clusters.
- a border point if its neighborhood contains a core point. It is also added to this core point's cluster.
- noise otherwise.

One of the most popular density-based methods is DBSCAN [3]. It selects points until a core point is found. All transitively neighboring core points are merged to a common cluster and neighboring border points are included. If no further reachable core points can be found, this strategy is repeated for the remaining yet untouched points until all points are classified as either core points, border points or noise. A major benefit of DBSCAN is its ability to detect arbitrarily shaped clusters while many other clustering approaches focus on elliptically shaped clusters. Second, it also includes robustness against noise due to the density property. The fixed ε-parameter, on the other hand, is a drawback as clusters with deviating densities are not detected in a single DBSCAN instance. Choosing a lower ε value will assign sparse clusters to noise while a higher ε value will more likely merge separate nearby clusters.

To overcome this issue and to assist in finding a suitable ε value in case of an initial data exploration task, Ankerst et al. developed OPTICS [1]. Given *MinPts*, this method determines for each point its core distance, the minimal distance needed such that the ε-neighborhood contains *MinPts* many points. For a point p let kNN be the k-th nearest neighbor and d a distance function. Then the core distance is defined by

$$\text{core}_{\varepsilon,MinPts}(p) = \begin{cases} d(p, \text{MinPtsNN}), |N_\varepsilon(p)| \geq \text{MinPts} \\ \text{undefined, otherwise} \end{cases}$$

The ε value is used as an upper bound for performance improvement. Using the core distance, the reachability distance can be defined as

$$\text{reach}_{\varepsilon,MinPts}(o,p) = \begin{cases} max(\text{core}_{\varepsilon,MinPts}(p), d(p,o)), & \text{if} |N_\varepsilon(p)| \geq \text{MinPts} \\ \text{undefined}, & \text{otherwise} \end{cases}$$

The set of data points gets ordered by its reachability distance. For each point, its successor is the point with the smallest reachability distance out of the unprocessed points. This ordering is not unique, due to start point ambiguity and potential choices between equidistant objects. Finally, a reachability plot is provided using the ordering on the x-axis and the reachability distance on the y-axis. Since dense object clusters in the data space have low pairwise reachability distances, they are accumulated in the plot and the cluster is identified as a trough in the reachability plot.

In an interactive online setting, we cannot apply OPTICS due to its high computational complexity. Hence, we propose a novel approach, which adapts the idea of DenStream [2] by utilizing micro-clusters. A micro-cluster is an aggregation of a group of data points, storing the number of aggregated points as the weight w, the center of the group c and the radius r. To enable incremental updates, instead of storing the center and radius, a linear sum LS and a squared sum SS are stored. For an update of a micro-cluster with point p, the procedure

$$w = w + 1, LS = LS + p, SS = SS + p^2$$

has to be performed. As a decay mechanism, the current values are determined after multiplying all three parameters with the factor $2^{-\lambda * \delta t}$ if δt is the time interval since the last update of the micro-cluster. $\lambda > 0$ has to be chosen to suit the desired rate of decay.

The center and the radius can be derived from the provided statistics as $c = LS/w$, $r = \sqrt{(SS/w^2 - LS^2/w)}$.

The radius is defined as the standard deviation of the aggregated points. DenStream uses two sets of micro-clusters, the outlier micro-clusters, and potential micro-clusters. Outlier micro-clusters contain few points such that their weight is below a certain threshold. If it decays without new points being merged into this cluster, it will disappear. Exceeding the weight threshold it will become a potential micro-cluster. For the final cluster result, only the p-micro-clusters are used by merging touching p-micro-clusters into macro-clusters.

3 AMTICS

Our clustering approach is a two-phase algorithm maintaining an online intermediate representation of clusters and providing an offline refinement step to construct the current cluster hierarchy. The online phase uses multiple instances of a density-based online clustering algorithm for various density levels $\varepsilon_1, \ldots, \varepsilon_k$. Traditionally OPTICS is applied first and interesting density levels are determined visually. In this work we reverse the application order by the application of cluster algorithms on different density levels and merging these information into an approximate OPTICS plot. The key points are the ensemble of single density clusterings, the alignment of micro-clusters and the final transformation into a reachability plot.

3.1 Online Stream Ensemble

We choose DenStream [2] as a starting point. Few required parameters make it suitable for user interaction. The maintained finite set of micro-clusters is necessary for the complexity constraints. Further it allows to compare micro-cluster structures of different density levels, such that clusters can be aligned in one model. In future work we will investigate which density-based online cluster methods, for example a grid-based approach, can be used instead of DenStream but this is not the focus in this work.

At all time a finite set of DenStream instances $\{DS_\varepsilon \mid 0 \leq \epsilon \leq \infty\}$ observe the stream and maintain their micro-clusters. Each instance can be deleted or initialized anytime during the stream except of two instances: DS_0 and DS_∞. DS_0 will classify every object as a different one-point cluster. DS_∞ builds exactly one large cluster containing all objects. Both instances define the boundaries of our final result.

Since DenStream guarantees to maintain a finite set of micro-clusters, keeping several but finitely many instances is within the complexity limitations of an online algorithm. In case that the user wants to introduce a new instance, we duplicate the denser neighboring instance. Due to the decay λ the new instance will quickly adapt to the recent points. Each ε-instance is a set of overlapping micro-clusters. Two micro-clusters are touching or overlapping if the distance between their centers is smaller than the sum of their radii, which is the standard deviation of its points. We show the intermediate result of four instances in Fig. 1 for the two moons dataset. From left to right and from top to bottom the ε level is decreasing. Note that the first instance contains always only one large cluster and the last one contains no cluster. Although the plot contains both types of micro-clusters, only the red potential micro-clusters are used for the following steps.

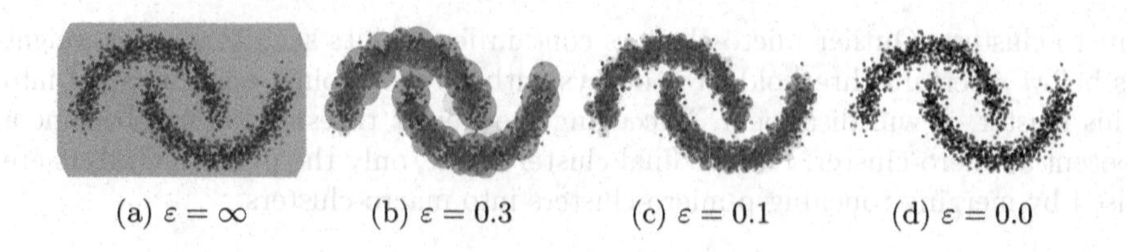

(a) $\varepsilon = \infty$ (b) $\varepsilon = 0.3$ (c) $\varepsilon = 0.1$ (d) $\varepsilon = 0.0$

Fig. 1. The two moons dataset clustered with different instances of DenStream. Potential micro-clusters are drawn red, outlier micro-clusters blue. (Color figure online)

Algorithm 1: AMTICS.getClusters

Data: Set of DenStream instances DS_ε
Result: Mapping on micro-cluster sets $m : (\varepsilon, i) \to MC$

```
1  initialization of empty mapping m;
2  foreach ε do
3      i = 0;
4      C = ∅;
5      while DS_ε contains potential micro-clusters do
6          i = i + 1;
7          get any potential micro-cluster mc ∈ DS_ε;
8          C = find all p-micro-clusters connected with mc;
9          remove all micro-clusters in C from DS_ε;
10         m(ε, i) = C;
11     end
12 end
```

3.2 Hierarchical Alignment

The previous online phase provides layers of clusterings for all ε-values and we need to align the clusters of each layer with the clusters of the layer below. A cluster is represented by a set of micro-clusters, which are points with a certain weight. We call two clusters C_1 and C_2 directly related if $C_1 \in DS_{\varepsilon_1}$ and $C_2 \in DS_{\varepsilon_2}$ with $\varepsilon_1 > \varepsilon_2$, there is no instance in between DS_{ε_1} and DS_{ε_2}, and $D(C_1, C_2) = \min(\{D(C_1, C_i) \mid C_i \in DS_{\varepsilon_2}\})$ for a suitable distance function D.

We model the alignment of two instances in the refinement as a transportation model, since we have to match micro-clusters of different weights and positions. Therefore, we apply the Earth Mover's Distance EMD [6] and extend it to compute the cluster distance.

The EMD is defined for two clusters $C_1 = \{(p_1, w_1), \ldots, (p_m, w_m)\}$ and $C_2 = \{(q_1, v_1), \ldots, (q_n, v_n)\}$. We use the Euclidean distance $d(x, y) = \|x - y\| = \sqrt{\sum_{i=1}^{n}(x_i - y_i)^2}$ as ground distance between two micro-cluster centers. The aim is to find the flow $F = (f_{i,j}) \in \mathbb{R}^{m \times n}$ of weight from C_1 to C_2 which minimizes the costs given by

$$c_F = \sum_{i=1}^{m} \sum_{j=1}^{n} f_{i,j} d(p_i, q_j)$$

in strict accordance with the following constraints:

- The flow has to be non-negative, so weights are only sent from C_1 to C_2 and not vice versa:
  $f_{i,j} \geq 0, \ \forall 1 \leq i \leq m, 1 \leq j \leq n$
- The sent flow is bounded by the weights in C_1:
  $\sum_{j=1}^{n} f_{i,j} \leq w_i, \ \forall 1 \leq i \leq m$
- The received flow is bounded by the weights in C_2:
  $\sum_{i=1}^{m} f_{i,j} \leq v_j, \ \forall 1 \leq j \leq n$
- All weights possible have to be sent:
  $\sum_{i=1}^{m} \sum_{j=1}^{n} f_{i,j} = \min\left(\sum_{i=1}^{m} w_i, \sum_{j=1}^{n} v_j\right)$

The distance is then defined as $EMD(C_1, C_2) = \frac{\sum_{i=1}^{m} \sum_{j=1}^{n} f_{i,j} d(p_i, q_j)}{\sum_{i=1}^{m} \sum_{j=1}^{n} f_{i,j}}$.

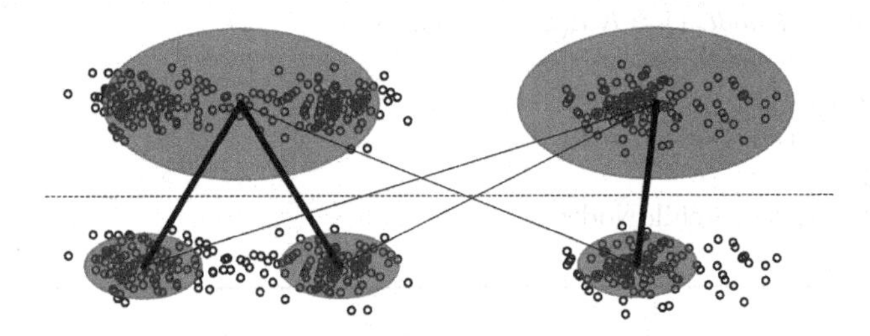

Fig. 2. The distances for all pairs of clusters of consecutive ε levels have to be computed to find the closest and most likely parent cluster.

We apply the EMD on all potential candidate pairs which are clusters of consecutive density levels. Eventually we aim for a tree of clusters such that the root is the cluster in DS_∞ containing all points. The height of each node represents the density level of this cluster. If two clusters are directly related then the corresponding tree nodes are connected. Our method determines for every cluster the parent cluster with the minimal EMD. In Fig. 2 a matching is performed for two ε layers. The strong lines indicate the best matching parent cluster. After application of the EMD we gain an alignment of all pairwise consecutive cluster layers connected by the smallest EMD distances. The desired outcome is a tree-like hierarchy, such that child nodes contain always less points than parent nodes, which is comparable to a max-heap. The reason is discussed in the following.

Algorithm 2: AMTICS.buildHierarchy

```
   Data: Mapping on micro-cluster sets m : (ε, i) → MC
   Result: Hierarchy of clusters H
 1 initialize empty graph H
 2 sort all ε descending
   /* create node N for m(∞,0)                                    */
 3 N = node(m(∞,0))
 4 parentNodes = list(N)
 5 H.root = N
   /* Generate a graph of related clusternodes                     */
 6 foreach (ε_prev, ε_next) do
 7 |   childNodes = list()
 8 |   foreach m(ε_next, i) do
 9 |   |   node_c = node(m(ε_next, i))
10 |   |   foreach node_p ∈ parentNodes do
11 |   |   |   N.d(node_p) = EMD(node_c, node_p)
12 |   |   end
   |   |   /* find closest parent nodes                            */
13 |   |   node_p1 = min(N.d)
14 |   |   node_p2 = min_second(N.d)
15 |   |   if N.d(node_p1) ≈ N.d(node_p2) then
16 |   |   |   merge(node_p1, node_p2)
17 |   |   end
18 |   |   Add (N, node_p1) to H
19 |   end
20 |   parentNodes = childNodes
21 end
```

3.3 Shared Micro-cluster Coverage

In an optimal case, large clusters split into smaller shards on the lower ε-levels and the constructed hierarchy represents a relation of subsets regarding the contained point set and the cluster nodes represent a max-heap structure regarding the cluster weights. For a maximal ε value all points are contained in a super-cluster which is also true for DS_∞. The lowest level classifies each point as noise and so does DS_0. This is not necessarily the case at this point, even for small synthetic datasets. The generation of micro-clusters is depending on the processing order of the points and we usually cannot guarantee a perfectly uniform distribution of stream objects, neither spatial nor temporal.

While OPTICS and DBSCAN consider ε neighborhoods for all points, Den-Stream trades this accuracy for its ability to aggregate points into micro-clusters. The more complex the dataset is and the higher the dimension of the points are the more likely is that the aligned micro-clusters do not form a hierarchy anymore. We identified two effects causing problems here by violating the heap condition, which is the case if a cluster in a smaller ε level is the children of a cluster with less weight. To ensure a valid hierarchy, clusters with larger neighborhoods have to cover clusters with smaller ε values.

As we do not store every singular point we assume the contained points to be mostly equally distributed within the defined circular area. Although this is a very useful assumption for the general case it has a drawback. Every point added to the micro-cluster shifts the center towards this point's coordinates. If all points are perfectly equally distributed regarding not only their position but also their sequential occurrence, the micro-cluster would not move. In reality and even for synthetic datasets, a small set of points can cause a cluster to shift apart from its potential connecting neighbor, causing a cluster to split. We call such events micro-trends and they are the more likely the larger a dataset is.

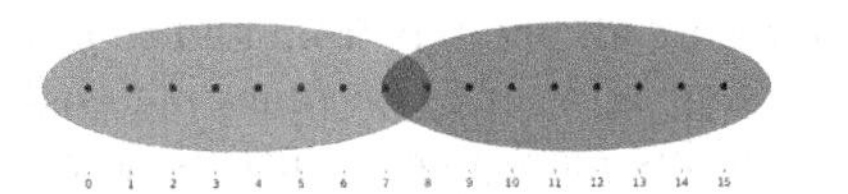

(a) Two touching larger micro-clusters, each containing half of the points.

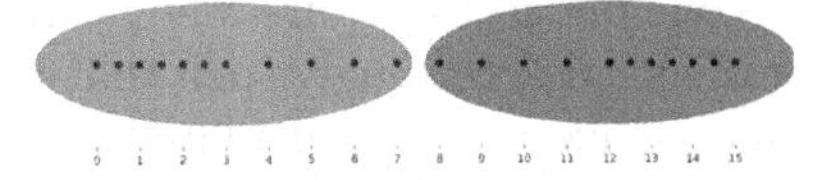

(b) After adding six points to the dataset, the centers drifted apart, splitting the cluster in two.

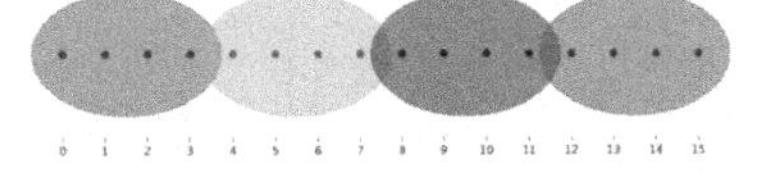

(c) Four touching smaller micro-clusters, each containing quarter of the points.

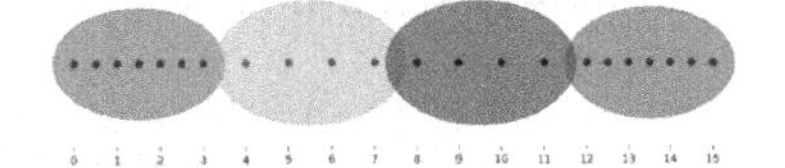

(d) After adding six points to the dataset, the centers stayed stable. Although the radii changed, the micro-clusters are still touching.

Fig. 3. A one-dimensional dataset of 16 points is clustered. Then 6 points are added. The larger micro-clusters are affected such that they disconnect the main cluster. The smaller micro-clusters do not disconnect.

The actual problem occurs by comparing micro-clusters of different radii which we do by aligning micro-clusters of different ε levels. Larger micro-clusters are much more affected by micro-trends. Let us assume two one-dimensional micro-clusters $mc_1 = \{0, \ldots, 7\}$ and $mc_2 = \{8, \ldots, 15\}$ as displayed in Fig. 3. The micro-clusters are touching so they form a cluster in the output result. If we add three additional points per micro-cluster the micro-clusters loose their connection as the points are not equally distributed anymore. In comparison to these large micro-clusters we clustered the same dataset with smaller micro-clusters. Each of the four micro-clusters contains four neighboring points. When adding the six additional points, the outer micro-clusters absorb three points each. However, the points within all clusters are equal distributed, so the centers are not shifted. The radius of each outer micro-cluster is reduced but all four micro-clusters are still connected.

In Fig. 5 we display two DenStream instances where the previously described effect occurs. Considering both gaps with smallest distance between the moons, the complete point set is connected in the lower instance $DS_{0.25}$ while disconnected in $DS_{0.3}$.

Starting from the initial assumption that smaller clusters with higher density on the lower hierarchy levels should always stay connected in higher ε levels, we suggest the following strategy. During the creation phase of the cluster nodes, we previously determined for each cluster the closest parent cluster considering the EMD. Instead we also compute the distance to the second nearest potential parent cluster. If the ratio of the nearest p_1 and the second nearest cluster p_2 as $ratio = EMD(p_1, n)/EMD(p_2, n)$ is close to 1.0, both parent clusters cover a dominant area of the child cluster.

We merge this pair of clusters by shifting all the weight and micro-clusters of the second one p_2 to the first one p_1. In addition, all pointers from and to p_2 have to be changed accordingly such that the parent node of p_2 becomes p_1. Possibly distances have to be recalculated as p_1 contains more micro-clusters now. It is also possible that this initializes a cascade of merging operations bottom-up until the root node is reached. This is repeated top-down until no merges are required anymore. This cascade is still no issue for the online applicability as the number of operations is limited by the number of ε layers above, which is a user-defined finite and mostly small number.

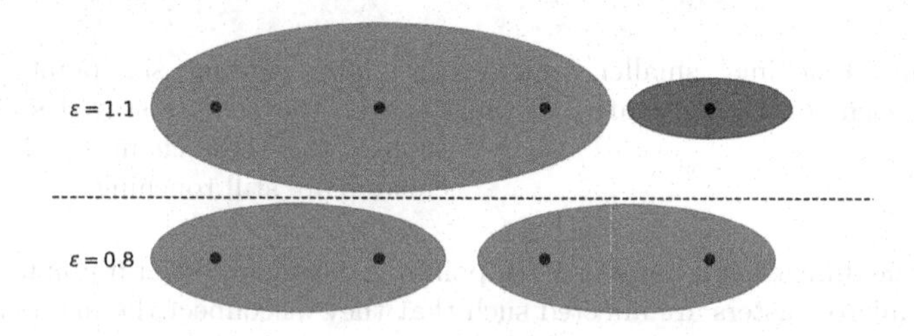

Fig. 4. Clusters of DenStream instances with higher ε values can split into clusters with a higher point coverage due to processing order and the micro-cluster architecture.

3.4 Local Outliers

Border points sometimes are not covered by potential micro-clusters of larger radius while being contained in smaller micro-clusters, see e.g. Fig. 4. Two DenStream instances are applied to this dataset with ε values 1.1 and 0.8. As described before we construct a hierarchy and the figure shows both levels and the established micro-cluster structure. As the larger ε value enables a micro-cluster to cover the first three points but not the fourth point, only three points are covered in the final clustering and the remaining point is treated as noise. In the level below the smaller neighborhood range allows only the first two points to be merged into one cluster. The third point is then processed and cannot be merged into the first micro-cluster. It establishes a second micro-cluster which starts as an outlier micro-cluster. Then it is merged with the remaining point, which allows this cluster to be raised into a potential micro-cluster. The alignment step will align the one cluster in the top level to all clusters in the bottom level, causing the larger micro-cluster to cover less points.

To repair the hierarchical structure and induce the monotonicity required for a reachability plot, we virtually add cluster points to the parent clusters such that their weight exceeds or equals the weight of the children clusters. Our assumption here relies on the better coverage quality of smaller micro-clusters for arbitrarily shaped clusters. Technically the hierarchy has to be processed bottom-up and for each parent cluster the weight sum of all directly related is calculated. If this sum exceeds the parent cluster weight, its weight is set to the sum. Otherwise nothing has to be done. Eventually the root DS_∞ is reached. It is very common that the weight sum of the second level will exceed the weight of the top cluster. However this cluster covers by definition all points of the dataset. The ratio of the number of points that it should have to the number of points it would have regarding the weights of its children clusters is then propagated down the hierarchy and used as a scaling factor for all weights.

After this operation the ε levels present a valid hierarchy ensuring monotonicity between weights of consecutive levels: For all $0 \leq \varepsilon < \varepsilon' \leq \infty$ if a pair of points p_1, p_2 is clustered in the same cluster in DS_ε, then p_1 and p_2 are also clustered in one cluster in $DS_{\varepsilon'}$.

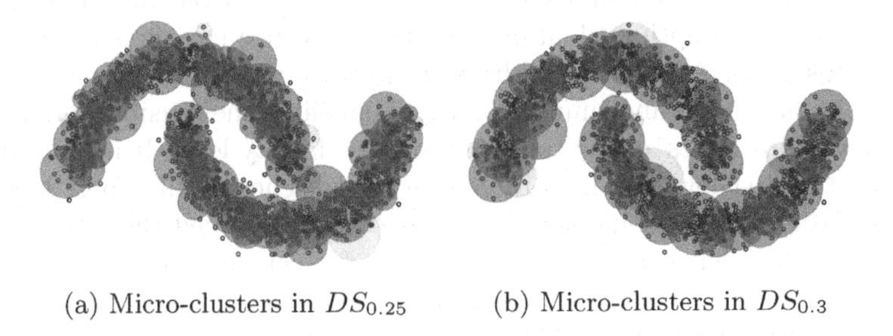

(a) Micro-clusters in $DS_{0.25}$ (b) Micro-clusters in $DS_{0.3}$

Fig. 5. Depending on the actual distribution of points a DenStream instance using smaller micro-clusters can sometimes connect point sets, which are segmented by larger micro-clusters.

3.5 Generating the Reachability Plot

To yield a reachability plot we construct a dendrogram of horizontal bars B in a first step. We approach top-down through the hierarchy and transform the hierarchy of clusters into a tree of line objects, where each line represents a cluster. For each cluster we plot a horizontal bar $b = (x, y, w) \in B$ with (x, y)

being the starting point and w being the bar width. The width is defined by the weight sum of all contained micro-clusters and represents the cluster size. For the height we define $y = \varepsilon$. The first bar for DS_∞ will be drawn with $y_0 = 1.25 \cdot \max_{\varepsilon \in \mathbb{R}}$ as $b_0 = (0, y_0, DS_\infty.weight) \in B$.

Recursively if a bar has been drawn as $b = (x, \varepsilon, w)$ and the according cluster has n children clusters $c_1, \ldots, c_n$, we first compute the remaining space by $rem = w - \sum_{i=1}^{n} c_i.weight$. All children bars are distributed equally, using rem/n as an intermediate space between them. As we ensured in the previous section that the sum of all children weights will not be larger than the parent weight, the intermediate space will be zero at least.

The final refinement is the definition of the reachability $r(z) = \min\{y > 0 \mid (x, y, w) \in B \wedge x \leq z \leq x + w\}$. Geometrically we sweep-line from left to right and choose always the lowest bar of all candidates at this point on the x-axis. As b_∞ spans the complete interval, a minimum can always be found.

3.6 Limitations and Complexity

The used Earth Mover‘s Distance is quite slow in performance. However the overall complexity is not changed as we are only looking on finitely many DenStream instances. The aim of the *EMD* is to distribute one set of objects onto a set of bins with the same capacity. In our case we explicitly use it as a distance between differently sized sets. Since the distance computation is a core step in our method it might be worthwhile to improve this step further, for example by introducing a suitable index structure to reduce the number of distance computations or lower distance bounds for candidate filtering. However, alternatives to EMD and more in-depth evaluations on other distance measures are not in the scope of this paper and will be addressed in future works.

A strong limitation and simultaneously a benefit is the possibility to initialize and remove ε instances at any time and for any density level. It is quite difficult to choose the first few instances in case of the absence of all expert knowledge over the dataset. After some key levels have been identified the interactiveness is quite useful to get a more accurate picture for certain density levels. Giving a reasonable start environment depending on the fed data automatically would improve the usability significantly.

Regarding the complexity, each DenStream instance keeps at most $W/MinPts$ many potential micro-clusters in memory as shown in [2]. As W is the overall weight of all clusters, it can be replaced by $W = v/(1 - 2^{-\lambda})$ with v being the number of objects observed in the stream per time unit. Since we keep k instances of DenStream in parallel, the complexity is given by $\mathcal{O}(kv/(1-2^{-\lambda}))$ or $\mathcal{O}(k/(1 - 2^{-\lambda}))$ per stream object. As we only introduce finitely many clustering instances, the complexity is constant for $\lambda > 0$. Solving optimizations like Earth Mover‘s distances is expensive, however as the number of micro-clusters is finite, the hierarchy is constructed in constant time. This also holds for the number of hierarchy nodes. Although it is not required for the final result, we can ensure a constant complexity for the whole chain of operations.

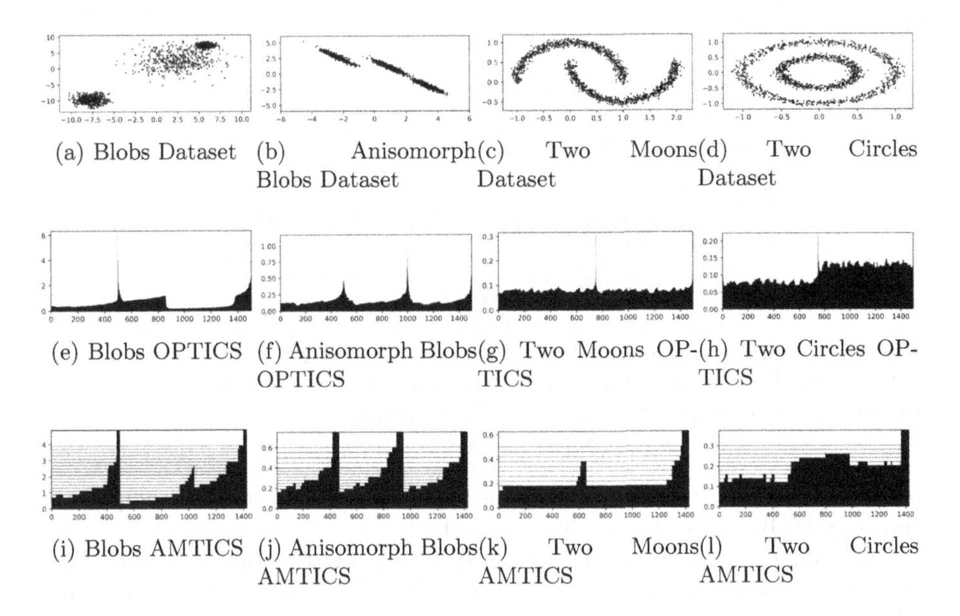

(a) Blobs Dataset (b) Anisomorph Blobs Dataset (c) Two Moons Dataset (d) Two Circles Dataset

(e) Blobs OPTICS (f) Anisomorph Blobs OPTICS (g) Two Moons OPTICS (h) Two Circles OPTICS

(i) Blobs AMTICS (j) Anisomorph Blobs AMTICS (k) Two Moons AMTICS (l) Two Circles AMTICS

Fig. 6. OPTICS and AMTICS in comparison on different synthetic datasets.

4 Evaluation

AMTICS is a data exploration tool to get first results of the clustering structure within a data stream. To be applicable to streams an online algorithm has to process each object in $\mathcal{O}(1)$. We empirically prove this claim to be correct by measuring the performance on a data stream. The two moons dataset was already mentioned in the previous section. We generated two moons datasets with various sample sizes between 500 and 5000 points. As a baseline we applied OPTICS to these datasets and measured the computation time. We repeated each computation three times and used the minimum to compare with our algorithm. For AMTICS we used four decaying factors $\lambda \in \{0.01, 0.02, 0.03, 0.04\}$ as we know that the decaying influences the number of micro-clusters and thus the performance. To get consistent results we applied AMTICS with the same settings on the streamed datasets repeatedly. The complete computation time over all stream data points is aggregated. For all AMTICS runs we used five ε instances.

In Fig. 7 we plotted the computation time in seconds for the different dataset sizes. All evaluations were performed on a workstation with an Intel Xeon CPU 3.10 GHz clock frequency on 16 GB memory. The results show a clear linear increase in computation time for AMTICS which is expected since the additional processing time for each arriving item has constant complexity. In comparison to OPTICS which shows a quadratic complexity in the number of data points, AMTICS has a reliable linear complexity over the number of data items. Obviously OPTICS can be faster for small datasets where point neighborhoods consist

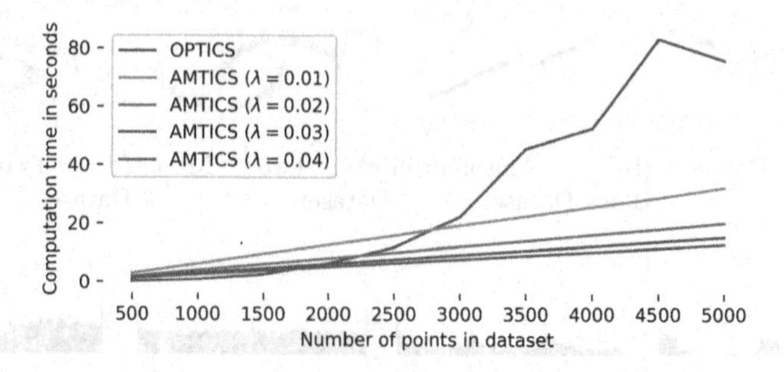

Fig. 7. Comparison of computation time for OPTICS and different AMTICS instances with varying decay factors. While OPTICS has a quadratic complexity, AMTICS shows its linear growth for increasing numbers of data points.

only of few points. However, already for medium sized datasets the computation time of OPTICS exceeds our efficient method.

To achieve the performance advantage, we trade time for accuracy. As we only produce an approximation of the OPTICS plot, the result can be coarse. This is especially true in the beginning of a stream and for a small set of ε instances. To get an impression of how AMTICS results look like compared to OPTICS and to compare the detected clusters we refer to some 2d example datasets in Fig. 6 and the popular chameleon dataset in Fig. 8. In the 4 test cases, AMTICS can compete with OPTICS by identifying the overall structure of the clusters, although the OPTICS results are slightly more accurate. The most difficult scenario for AMTICS seems to be the two circles dataset. Due to the curving, the micro-clusters in the inner circle are touching the outer circles rather early. This leads to a rather large transition phase between the detection of two clusters and one cluster only, which can be seen in the corresponding AMTICS plot.

To give AMTICS a little challenge we also clustered the chameleon dataset, which is tough due to the combination of solid clusters, sinoidal arranged points and much noise. Both OPTICS and AMTICS struggle with the detection of the solid clusters. However, AMTICS is still able to identify the cluster structure with only few ε instances. Some clusters are connected with dense noise, so they are aggregated in the result.

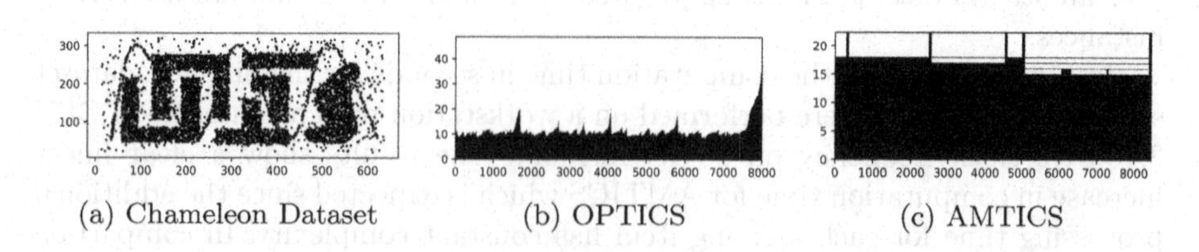

(a) Chameleon Dataset (b) OPTICS (c) AMTICS

Fig. 8. OPTICS and AMTICS in comparison on the Chameleon dataset.

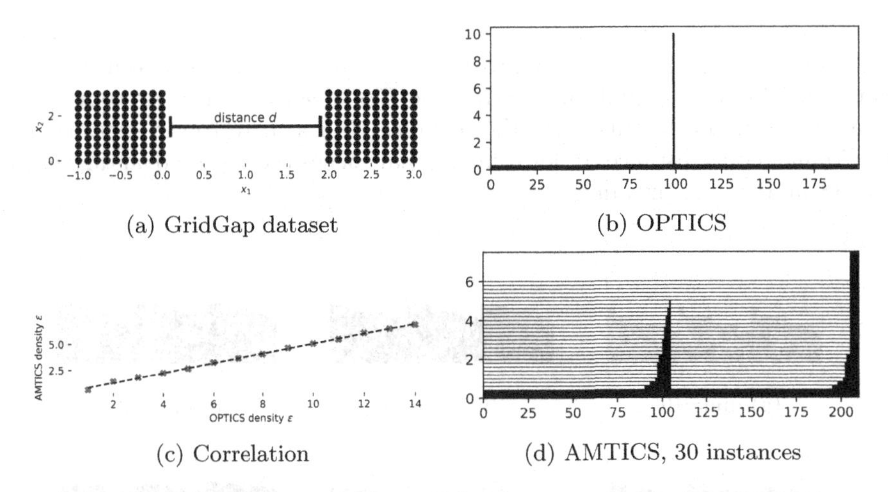

(a) GridGap dataset (b) OPTICS

(c) Correlation (d) AMTICS, 30 instances

Fig. 9. The evaluation setup to estimate the correlation between density in OPTICS and density in AMTICS. Two regular point grids are divided into two clusters. The gap distance in-between is increased incrementally. Comparison of two result plots of OPTICS and AMTICS on the GridGap dataset with $d = 10$.

OPTICS and AMTICS use a parameter ε with a synonymous meaning. We investigated the correlation between both values ε_{OPTICS} and ε_{AMTICS} and define a controllable test scenario 'GridGap' as sketched in Fig. 9. It consists of two regular grid-based point sets with a defined margin d in-between. If d is larger than the point distance in one of the grids, it corresponds directly to ε_{OPTICS}. The maximum ε for AMTICS to divide the dataset is ε_{AMTICS}. Since instances have to be defined in advance we choose a step width of 0.02.

In Fig. 9 we exemplarily gave the resulting plots for $d = 10$. The overall structure is represented. We compared the peak height of the OPTICS plot with the peak height of the needle-like pillar in the middle of the AMTICS plot. Evaluating several distances, we concluded that both approaches are correlated linearly by $f(x) = 0.47 \cdot x + 0.33$. The constant term is a result of the distance within both clusters. In case of OPTICS, the distance is constant 0.1 for a grid of 10×10 points. For AMTICS the distance between micro-clusters is used. Since micro-clusters represent a Gaussian distribution but we distributed the points equally, distances are stretched in comparison to OPTICS. Hence we assume that the actual distribution of the data points will influence this constant offset and it should not be expected that the same density level in both methods yield the same results.

In a last evaluation we show the applicability of AMTICS to datasets of higher dimensionality. The datasets contain 2000 datapoints partitioned into three circular clusters of equal size. The numbers of dimensions are 10, 20, 30, 40, 50, and 100. In Fig. 10 we give the AMTICS plots for the datasets. For higher dimensions the distances rise and obviously the density-based method will fail for really high dimensions due to the curse-of-dimensionality and the fact that differences between distances of object pairs will vanish. For lower dimensions and practical applications our approach is able to distinguish between the three clusters. This test is performed using the Euclidean distance. At some point it might be useful to substitute the distance with a more robust measure like the Mahalanobis distance. In low-dimensional spaces the speed advantage of the Euclidean distance prevails.

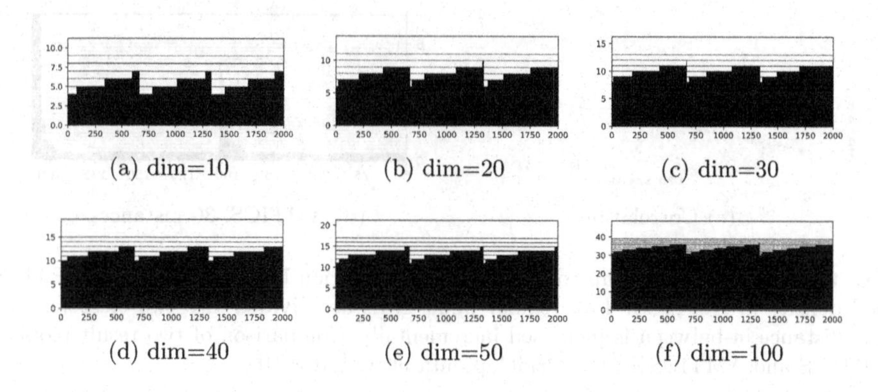

(a) dim=10 (b) dim=20 (c) dim=30

(d) dim=40 (e) dim=50 (f) dim=100

Fig. 10. AMTICS applied to three-blobs datasets with multiple dimensions.

5 Related Work

Since AMTICS is targeted at a density-based clustering model, we elaborate first on DBSCAN [3] as among the most prominent algorithms for density based clustering being followed by GDBSCAN [7], HDBSCAN [5] and DENCLUE [4]. In DBSCAN the users set an ε and minimum number of objects to be located within this range by which the expected density is characterized. Nevertheless DBSCAN also comes with certain weaknesses. One of them is that it is a challenging task to determine an adequate ε-range. A second weakness is its incapability to detect clusters of different density. A method which was constructed as a visualization and has been constructed with these two weak points in mind is OPTICS [1]. To overcome these weak points of DBSCAN, OPTICS orders the points of a given data set in a linear fashion, such that neighboring points in this ordering are following consecutively after each other. In the form of a reachability plot, users can spot valleys of different depths by which (1) the different densities of clusters become visible and (2) determining an adequate ε-range is facilitated.

Since data can occur in a stream setting, DBSCAN and OPTICS are not suitable for high velocity scenarios. As such a density based clustering algorithm tailored at data streams has been designed known as DenStream [2]. Here the authors introduce dense micro-clusters (core-micro-clusters) with the purpose to summarize dense clusters of arbitrary shapes. Having a density based method for the stream setting, works have emerged on providing an OPTICS-fashioned method for high-velocity scenarios. As one related work we have OpticsStream [8] which hybridizes the concept of a density based stream clustering with an extension of OPTICS. Their approach to a streaming version maintains, based on the reachability distance, an ordered list of core-micro-clusters. The maintenance over time is ensured through insertions and deletions in a separate micro-cluster list. As a result an OPTICS plot is generated based on the current micro-cluster structure. Further the authors introduce a 3-dimensional reachability plot where the third dimension is an axis representing the time. However the 3-dimensional construct renders it difficult to clearly identify the valleys and thus different density levels. One of the major differences between OpticsStream and AMTICS is that our approach does not start one single instance of an density based stream clustering but several. This gives the opportunity to detect clusters even though they are changing their density over time.

6 Conclusion

AMTICS is an efficient and interactive density-based online micro-clustering algorithm. It follows a divide-and-conquer strategy by clustering the same dataset on different density levels and merges the separate results into a hierarchy of clusters of various sizes and densities. The hierarchy is finally displayed visually as a reachability plot in which valleys refer to dense areas that are more likely to be clusters.

With AMTICS, it is possible to process data streams and explore the clustering structure visually. It provides the flexibility to shift the focus to certain density levels by incrementally adding or removing clustering instances. AMTICS produces an approximative reachability plot anytime in the stream on demand. Similar to OPTICS, the alignments of micro-clusters provide insights into the cluster structure while in contrast providing a coarse overview for a rapid visual analysis. The advantage is not only the agile construction of a reachability plot, which can be constructed with related methods like OpticsStream [8]. Using the layered hierarchy provides the desired level of cluster granularity which augments the analysis performance as well by reducing the complexity of the visual analysis for humans.

In future works, we are going to investigate heuristics for a useful set of starting ε levels. This will assist a human operator in the visual analysis process. As a further future topic, the distance computation can be improved. Although the Earth Mover's distance is very suitable for the application, we do not need to find the exact distance values. If we provide a sufficient ordering of super-clusters as potential parent nodes in the hierarchy, we can improve the performance in the offline-phase.

References

1. Ankerst, M., Breunig, M.M., Kriegel, H.P., Sander, J.: Optics: ordering points to identify the clustering structure. In: ACM SIGMOD Record, vol. 28, no. 2, pp. 49–60. ACM (1999)

2. Cao, F., Ester, M., Qian, W., Zhou, A.: Density-based clustering over an evolving data stream with noise. In: Proceedings of the 2006 SIAM International Conference on Data Mining, pp. 328–339. SIAM (2006)
3. Ester, M., Kriegel, H.P., Sander, J., Xu, X., et al.: A density-based algorithm for discovering clusters in large spatial databases with noise. In: KDD, vol. 96, no. 34, pp. 226–231 (1996)
4. Hinneburg, A., Keim, D.A., et al.: An efficient approach to clustering in large multimedia databases with noise. In: KDD, vol. 98, pp. 58–65 (1998)
5. McInnes, L., Healy, J., Astels, S.: HDBSCAN: hierarchical density based clustering. J. Open Source Softw. **2**(11), 205 (2017)
6. Rubner, Y., Tomasi, C., Guibas, L.J.: The earth mover's distance as a metric for image retrieval. Int. J. Comput. Vision **40**(2), 99–121 (2000)
7. Sander, J., Ester, M., Kriegel, H.P., Xu, X.: Density-based clustering in spatial databases: the algorithm GDBSCAN and its applications. Data Min. Knowl. Disc. **2**(2), 169–194 (1998)
8. Tasoulis, D.K., Ross, G., Adams, N.M.: Visualising the cluster structure of data streams. In: R. Berthold, M., Shawe-Taylor, J., Lavrač, N. (eds.) IDA 2007. LNCS, vol. 4723, pp. 81–92. Springer, Heidelberg (2007). https://doi.org/10.1007/978-3-540-74825-0_8

Improved Representations for Personalized Document-Level Sentiment Classification

Yihong Zhang and Wei Zhang(✉)

School of Computer Science and Technology,
East China Normal University, Shanghai, China
51184501185@stu.ecnu.edu.cn, zhangwei.thu2011@gmail.com

Abstract. Incorporating personalization into document-level sentiment classification has gained considerable attention due to its better performance on diverse domains. Current progress in this field is attributed to the developed mechanisms of effectively modeling the interaction among the three fundamental factors: users, items, and words. However, how to improve the representation learning of the three factors themselves is largely unexplored. To bridge this gap, we propose to enrich users, items, and words representations in the state-of-the-art personalized sentiment classification model with an end-to-end training fashion. Specifically, relations between users and items are respectively modeled by graph neural networks to enhance original user and item representations. We further promote word representation by utilizing powerful pre-trained language models. Comprehensive experiments on several public and widely-used datasets demonstrate the superiority of the proposed approach, validating the contribution of the improved representations.

Keywords: Sentiment classification · Representation learning · Graph neural network

1 Introduction

Document-level sentiment classification [15] becomes an increasingly popular research problem with the flourish of user-generated reviews occurring in various platforms, such as Yelp, Amazon, etc. The aim of this problem is to classify the sentiment polarities reflected by the review text. The research on this problem is beneficial since sentiment-labeled reviews are knowledgeable references for customers to decide whether to purchase some products, commercial platforms to manage business policies, and sellers to enhance their product quality. Most of the existing methods frame this problem as a special case of text classification,

This work was supported in part by the National Key Research and Development Program (2019YFB2102600), NSFC (61702190), Shanghai Sailing Program (17YF1404500), and Zhejiang Lab (2019KB0AB04).

Y. Nah et al. (Eds.): DASFAA 2020, LNCS 12112, pp. 769–785, 2020.
https://doi.org/10.1007/978-3-030-59410-7_53

whereby each discrete rating score is regarded as a sentiment label. They have witnessed the methodological development from traditional feature engineering based models (e.g., lexical terms, syntactic features) [10,22] to representation learning approaches, including recurrent neural networks [23] and convolutional neural networks [11].

Recent studies have started to consider the effects of users who write reviews and items corresponding to reviews. This is intuitive since the preference of users and characteristics of items affect the ratings to be given [4], which in turn influence the generation of review text. Some conventional approaches [13,18] build upon topic models to incorporate the user and item roles into the connection of text and sentiment labels from a probabilistic perspective. Recent studies are mainly motivated by the impressive success of deep representation learning for natural language processing. They [1,2,4,6,24,28] incline to leverage the representations of users and items to affect the document-level representations. In other words, the interactions among the representations of users, items, and words have been extensively investigated. They have been demonstrated to be beneficial for inferring sentiment.

However, the representations of users, items, and words themselves have not been treated seriously. As for the user and item representations, the most commonly adopted technique is to map each of their IDs to a real-valued vector in a low-dimensional space and trains them along with recommendation models. This simple strategy prohibits further utilizing the relations among users and items to enhance the basic representations, such as the co-occurrence relation between items (or users) and social relations between users. On the other hand, recent studies about pre-trained language models (e.g., BERT [5]) have exhibited better performance in different tasks (e.g., general sentiment classification) than pre-trained word embeddings (e.g., Word2vec [17]) due to its ability to produce contextualized word representation. Yet no efforts of applying pre-trained language models to personalized document-level sentiment classification have been observed to our knowledge. The above insights about the representations of the three factors pave the way for our study.

In this paper, we propose a novel neural network built on the state-of-the-art Hierarchical User Attention and Product Attention (HUAPA) [28] neural network for the studied task, named IR-HUAPA (short for Improved Representations for HUAPA). The main innovations of this approach lie in improving the users, items, and words representations. To be specific, we construct a user relation graph and an item relation graph, and leverage graph convolutional networks (GCN) [12] to encode the relations into user and item representations. We further utilize BERT to obtain contextualized word representations with the knowledge learned from a large-scale text corpus. These representations are seamlessly fed into HUAPA to replace the original representations. The whole model is trained in an end-to-end fashion, ensuring the optimization of GCN and fine-tuning of BERT. We summarize the main contributions as follows:

* Unlike existing studies which focus on modeling better interactions among users, items, and words for this problem, we aim to learn better representations of the three factors themselves, showing a different perspective.
* We present IR-HUAPA, a novel model that benefits from the powerful pre-trained language model and graph neural networks to obtain the ability of enriching users, items, and words representations.
* We conduct extensive experiments on popular and accessible datasets collected from Yelp and Amazon. The experiments show IR-HUAPA achieves the superior performance against the state-of-the-art HUAPA. The contribution of each type of improved representation is validated as well.

2 Related Work

In this section, we review the literature from three aspects, i.e., personalized document-level sentiment modeling, graph neural networks, and pre-trained language models.

2.1 Personalized Document-Level Sentiment Modeling

Personalized document-level sentiment modeling is a newly emerging interdisciplinary research direction that naturally connects the domains of natural language processing and recommender system. That is to say, the key to success is the combination of user modeling and text analysis. The research direction mainly contains two categories, 1) *personalized review-based rating prediction* [16,31] and 2) *personalized document-level sentiment classification* [24]. The former category argues the review text only exists in the model training stage and is unavailable when testing. The core part of the corresponding methods is to utilize review text information to enrich user and product representations which are later fed into recommender system methods to generate rating prediction. By contrast, personalized document-level sentiment classification assumes the review text could be seen in both training and testing stages. This fundamental difference makes the methods for personalized review-based rating prediction not so applicable for personalized document-level sentiment classification due to the ignorance of target review text.

As for the studied problem, Tang et al. [24] first addressed the effect of user preference and item characteristics in review text modeling. They developed a convolutional neural network (CNN) based representation learning approach. Its main idea is to modify word representation by multiplying them with user and item-specific matrices, and further concatenating user and item representations with document-level representation. Chen et al. [4] leveraged attention mechanism [3] which is shown effective for sentiment classification [14,29]. Specifically, user and item representations are used in the attention computation formulas to determine the importance of words and sentences for classifying sentiment labels. HUAPA [28] further decomposes its whole neural architecture into a hierarchical user attention network and hierarchical item attention network, hoping to differentiate the effect of users and items overlooked by [4]. More recently, [1] compares different manners to associate user and item representations with text representation learning.

Nevertheless, all the above studies concentrate on better characterizing the interactions among users, items, and words, while neglecting to pursue better representations themselves. Although Zhao et al. [32] learned user representations from social trust network for sentiment classification, the combination of document representations with users is simply realized by concatenation, with the limited capability of characterizing their interactions.

2.2 Graph Neural Networks

Graph neural networks GNNs [26] are powerful tools to learn from graph data and encode nodes into low-dimensional vectors. Many applications of GNNs have been observed. Wang et al. [27] applied GNNs to model the user-item bipartite graph and so as to recommend items to users. [30] constructs a text-corpus graph by regarding words and documents as nodes. Temporal relation is even incorporated into a graph for stock prediction [7]. In this paper, we exploit GCN [12], a concrete type of GNNs, to encode user relation and item relation into low-dimensional representations, which is beneficial for personalized document-level sentiment classification.

2.3 Pre-trained Language Models

Pretraining word representations has become an indispensable procedure for achieving state-of-the-art performance in diverse natural language processing tasks. Early standard pipelines in this regard utilize pre-trained word embeddings like Word2vec [17] and Glove [20]. However, they could only utilize static word representations without considering the specific context of different words. Thus they inevitably suffer from the issue of word ambiguity. To alleviate this, advanced pre-trained language models are developed. They include ELMO [21], BERT [5], XLNet [29], to name a few. Since most of the previous approaches for personalized document-level sentiment classification are only based on pre-trained word embeddings, we aim to apply pre-trained language models to the studied problem. Without loss of generality, we leverage BERT to obtain contextualized word representations.

3 Methods

In this section, we first formalize the studied problem of personalized document-level sentiment classification. Afterwards, we provide an overview of the proposed model IR-HUAPA. Then the concrete details of the base model HUAPA is elaborated. Finally, we illustrate the proposed IR-HUAPA model.

3.1 Problem Formulation

Assume we have a user set $\mathcal{U}$ and an item set $\mathcal{I}$. For a given user $u \in \mathcal{U}$ and an item $i \in \mathcal{I}$, we denote the document written by u for i as $d^{u,i}$. In the rest of this paper, we omit the superscript and directly use d for simplicity. We suppose $d = \{s_1, s_2, \cdots, s_{n_d}\}$ with n_d sentences. For the j-th sentence s_j, it consists of a sequence of words $\{w_1^j, w_2^j, \cdots, w_{l_j}^j\}$, where l_j is the length of the sentence. Each word w is a one-hot encoding on the vocabulary space, meaning $w \in \{0,1\}^{|\mathcal{V}|}$ and only one dimension of w corresponds to 1, where $\mathcal{V}$ denotes the predefined vocabulary. Similarly, we have one-hot representations for users and items, i.e. $u \in \{0,1\}^{|\mathcal{U}|}$ and $i \in \{0,1\}^{|\mathcal{I}|}$. Given this, personalized document-level sentiment classification aims to learn a function $f(d, u, i) \rightarrow r$ which can infer the sentiment label r.

3.2 Overview of IR-HUAPA

The overall architecture of our model IR-HUAPA is depicted in Fig. 1. The input to IR-HUAPA contains the user-to-user graph, item-to-item graph, and a target document. The improved user and item representations are obtained by propagating representations of neighbors in the graphs, shown in the top-left and top-right parts of the figure. The improved contextualized word representations are gained by BERT. These representations are further fed into the hierarchical user/item attention network. The whole model is learned in an end-to-end fashion by optimizing the hybrid loss function inherited from HUAPA.

3.3 The HUAPA Model

The proposed IR-HUAPA model is built upon the Hierarchical User Attention and Product Attention (HUAPA). In this section, we provide a concrete introduction to the three parts of HUAPA.

Input Representation. The original representations of users, items, and words in HUAPA are simply based on the lookup operation [24], a common way of utilizing the advantage of distributional representation. Taking word w as an example, the operation is defined as:

$$\mathbf{w} = \mathbf{E}_W w, \tag{1}$$

where $\mathbf{E}_w \in \mathbb{R}^{K \times |\mathcal{V}|}$ is a trainable word embedding matrix. Likewise, we could gain the user and item representations, i.e. $\mathbf{u}$ and $\mathbf{i}$.

Hierarchical User/Item Attention Network. HUAPA decomposes its text modeling into two parts, hierarchical user attention network and hierarchical item attention network. Basically, they have the same model architecture, except that the former considers user effect while the latter is for item. As such, we

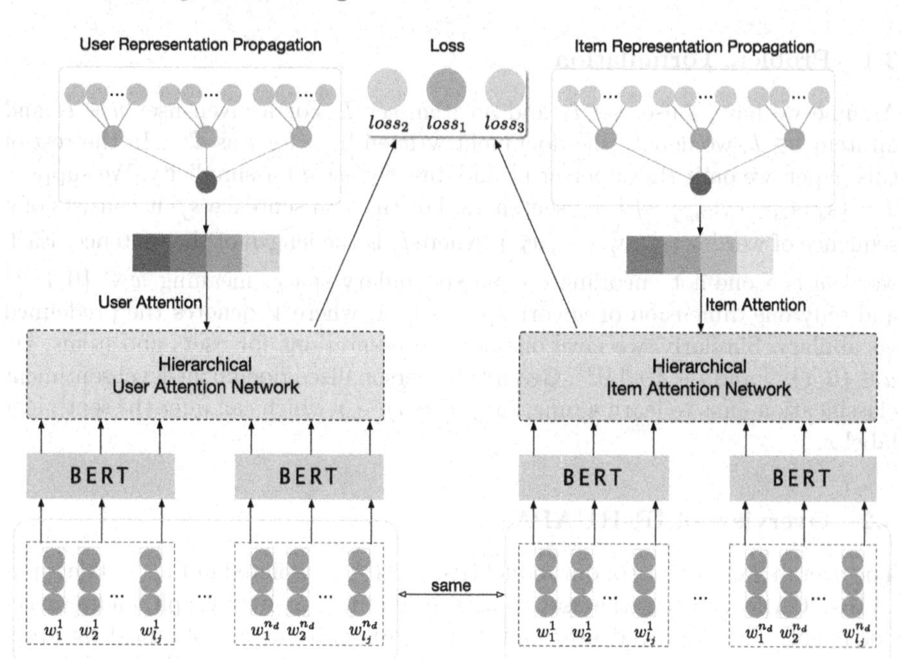

Fig. 1. The architecture of IR-HUAPA. User representation propagation and item representation propagation mean aggregating representations of neighbors to update target user and item representations by GCN, respectively.

elaborate the hierarchical user attention network and simplify the details of the hierarchical item attention network.

The network first adopts bi-directional LSTM [8] layers to connect word representations based on forward and backward recurrent modeling. Specifically, for the j-th sentence s_j, we define the following expression to calculate the hidden representation of each word,

$$\mathbf{h}_{j,1}, \mathbf{h}_{j,2}, ..., \mathbf{h}_{j,l_j} = \text{BiLSTM}(\mathbf{w}_1^j, \mathbf{w}_2^j, ..., \mathbf{w}_{l_j}^j; \Theta_{BiLSTM}). \tag{2}$$

HUAPA argues each word in a sentence should contribute differently for a specific user to form the sentence-level representation. It performs attention to words as follows:

$$a(\mathbf{h}_{j,z}, \mathbf{u}) = (\mathbf{v}_W^U)^{\text{T}} \tanh(\mathbf{W}_{W1}^U \mathbf{h}_{j,z} + \mathbf{W}_{W2}^U \mathbf{u} + \mathbf{b}_W^U), \quad z \in \{1, ..., l_j\}, \tag{3}$$

where $a(\mathbf{h}_{j,z}, \mathbf{u})$ denotes the importance score. $\mathbf{v}_W^U$, $\mathbf{W}_{W1}^U$, $\mathbf{W}_{W2}^U$, and $\mathbf{b}_W^U$ are trainable parameters. Based on this, the sentence-level representations $\mathbf{s}_j$ ($j \in \{1, ..., n_d\}$) are computed through the following ways:

$$\alpha_{j,z}^u = \frac{\exp(a(\mathbf{h}_{j,z}, \mathbf{u}))}{\sum_{z'=1}^{l_j} \exp(a(\mathbf{h}_{j,z'}, \mathbf{u}))}, \quad z \in \{1, ..., l_j\}, \tag{4}$$

$$\mathbf{s}_j = \sum_{z=1}^{l_j} \alpha_{j,z}^u \mathbf{h}_{j,z}. \tag{5}$$

For the gained sentence representations, bi-directional LSTM layers are utilized as well to first associate them to get the hidden sentence representations $\mathbf{h}_j$ ($j \in \{1, ..., n_d\}$). Based on the assumption that sentences contribute unequally for the user, attention based formulas similar as Eqs. 3, 4, and 5 are defined to have the final user-aware document-level representation $\mathbf{d}_u$:

$$a(\mathbf{h}_j, \mathbf{u}) = (\mathbf{v}_S^U)^{\mathrm{T}} \tanh(\mathbf{W}_{S1}^U \mathbf{h}_j + \mathbf{W}_{S2}^U \mathbf{u} + \mathbf{b}_S^U), \quad j \in \{1, ..., n_d\}, \tag{6}$$

$$\alpha_j^u = \frac{\exp(a(\mathbf{h}_j, \mathbf{u}))}{\sum_{j'=1}^{n_d} \exp(a(\mathbf{h}_{j'}, \mathbf{u}))}, \quad j \in \{1, ..., n_d\}, \tag{7}$$

$$\mathbf{d}_u^d = \sum_{j=1}^{n_d} \alpha_j^u \mathbf{h}_j. \tag{8}$$

From the view of items, the above procedures are repeated for the hierarchical item attention network to obtain item-aware document-level representation $\mathbf{d}_i^d$. The above attention computations achieve the interaction of users (items) and words, so as to capture the relations among them for representation learning.

Loss Functions. Based on these two document representations from different views, HUAPA derives a hybrid loss function consisting of three specific losses, i.e., $loss_1$, $loss_2$, and $loss_3$. Among them, $loss_1$ depends on the fusion of $\mathbf{d}_u$ and $\mathbf{d}_i$ and is defined as follows:

$$loss_1 = -\sum_d \mathbf{y}_d \log(\mathrm{softmax}(\mathbf{W}[\mathbf{d}_u^d; \mathbf{d}_i^d] + \mathbf{b})), \tag{9}$$

where $\mathbf{y}_d$ is the one-hot encoding of ground-truth label and [;] denotes the row-wise concatenation. $loss_2$ and $loss_3$ are defined in a similar manner except the input document representations are different. For $loss_2$, it has $\mathbf{d}_u^d$ as input while $loss_3$ takes $\mathbf{d}_i^d$ as input. Consequently, the optimization of HUAPA could be naturally conducted to minimize the above loss functions.

3.4 The IR-HUAPA Model

As aforementioned, IR-HUAPA aims to improve the input representations of the three factors (i.e., users, items, and words) for achieving better classification performance. The original input representations of HUAPA are obtained by mapping their IDs to low-dimensional vectors. We argue this simple treatment could not encode rich internal relations lying in each factor type.

To be more specific, semantic context information of words could not be captured by the currently adopted pre-trained word embeddings. Although the bi-directional LSTM layers in the upper level address the sequential modeling of word sequences, the long-range contextualized relations still could not be handled well. From the view of users or items, there might exist different relations between them that could be used. Taking users for illustration, co-occurrence is a common relation among different users. It means how often two users occur together in the same context window (e.g., a user sequence with all users commenting on a specific item). Another possible relation is user social relation that might exist in some review platforms such as Yelp. Later we will answer the question of how to boost input representations for pursuing better classification.

Improved User and Item Representations. Co-occurrence information among users or items is ubiquitous and could be easily derived from review corpus. Moreover, it does not involve privacy issues encountered by social relations. As a result, we regard co-occurrence as the main relation considered in this paper.

Graph Construction. We build a user-user graph $\mathcal{G}^U$ and an item-item graph $\mathcal{G}^I$. The number of the nodes is $|\mathcal{U}|$ in the user-user graph and $|\mathcal{I}|$ in the item-item graph. To determine the weights of user-user edges and item-item edges, we calculate their co-occurrence counts in a training dataset. Taking the computation of user-user edge weights for clarification, we first segment all the reviews into different sequences according to which item the reviews are written for. For example, we assume reviews d_1 and d_3 are written for item aa, and d_2 and d_4 are written for item ab. Based on our segmentation strategy, d_1 and d_3 should belong to the same sequence, and d_2 and d_4 belong to a different sequence. For each segmented review sequence, we extract their corresponding users to form a user sequence in chronological order. Based on the user sequence, we apply a fixed-size sliding window approach to count the co-occurrence of each user pair. In other words, if two users often appear in the same sequence and their position interval is small, their user-user edge weight would be large.

We denote $\mathbf{A}^U$ and $\mathbf{A}^I$ as the weight matrices of the user-user graph and item-item graph. Formally, the weights are calculated as follows:

$$\mathbf{A}^U_{u',u''} = \textit{Co-occurrence}(u', u''), (u', u'' \in \{1, ..., |\mathcal{U}|\}), \tag{10}$$

$$\mathbf{A}^I_{i',i''} = \textit{Co-occurrence}(i', i''), (i', i'' \in \{1, ..., |\mathcal{I}|\}), \tag{11}$$

where $\textit{Co-occurrence}(\cdot, \cdot)$ denotes the co-occurrence count of any pair. If a pair involves the same user or item, $\textit{Co-occurrence}(\cdot, \cdot)$ sets a pseudo count of co-occurrence to ensure self-connection. The complexity of building the graphs mainly depends on the length of the sequence and the sliding window size, which is acceptable in real situations.

Graph Convolutional Network. Given the two built graphs, we leverage GCN to perform user representation propagation and item representation propagation on their respective graphs. The basic intuition behind GCN is that the representation of a target node should be influenced by the representations of its neighbor nodes. Specifically, we denote the initialization of the user embedding matrix and item embedding matrix as $\mathbf{U}^{(0)}$ and $\mathbf{I}^{(0)}$, respectively, which are obtained by the mentioned lookup operation. Afterwards, we define the GCN based representation propagation as follows:

$$\mathbf{U}^{(t)} = \sigma(\mathbf{D}_U^{-\frac{1}{2}} \mathbf{A}^U \mathbf{D}_U^{-\frac{1}{2}} \mathbf{U}^{(t-1)} W_U^{(t)}), \quad (12)$$

$$\mathbf{I}^{(t)} = \sigma(\mathbf{D}_I^{-\frac{1}{2}} \mathbf{A}^I \mathbf{D}_I^{-\frac{1}{2}} \mathbf{I}^{(t-1)} W_I^{(t)}), \quad (13)$$

where $\mathbf{U}^{(t)}$ and $\mathbf{I}^{(t)}$ are the updated embedding matrices after performing the t-th propagation. $W_U^{(t)}$ and $W_I^{(t)}$ are the parameters of GCN which are optimized together with the whole model. σ is a non-linear activation function (e.g., ReLU) and $\mathbf{D}$ is a diagonal matrix with $\mathbf{D}_{j,j} = \sum_k \mathbf{A}_{j,k}$, playing a role of normalization.

Through the above manner, the internal relation between users or items could be encoded. After a fixed number of propagation, we could get the improved user and item representations. They are used in attention computation (e.g., Eq. 3 and 6) to learn document representations.

Improved Word Representation. As discussed previously, current approaches for personalized document-level sentiment classification like HUAPA directly employ pre-trained word embeddings as their input, which is not advantageous since they could not capture specific semantic context for each word. By contrast, we leverage BERT, a bidirectional pre-trained language model based on the transformer network (refer to the details in [25]). The core benefit of using BERT lie in two aspects: (1) we can acquire context-specific word representations; (2) fine-tuning of BERT for personalized document-level sentiment classification task is feasible.

In particular, for review d containing n_d sentences, we feed each sentence into BERT and get contextualized word representations encoding word-word relation. For simplicity, we define the procedure as follows:

$$\mathbf{w}_1^j, \mathbf{w}_2^j, ...\mathbf{w}_{l_j}^j = \mathrm{BERT}(w_1^j, w_2^j, ..., w_{l_j}^j; \Theta_{BERT}), \quad j \in \{1, ..., n_d\}, \quad (14)$$

where Θ_{BERT} represents the parameters of BERT. Afterwards, the obtained word representations are incorporated into Eq. 2 to replace the original word embeddings.

4 Experimental Setup

4.1 Datasets

To evaluate the adopted approaches, we conduct experiments on three datasets from two sources. The first dataset is gotten from the Yelp dataset challenge of the year 2019[1], which is named as Yelp19. The other two datasets are selected from the Amazon dataset collection [19]. Based on their categories, we name them as Toy and Kindle, respectively. Some basic text processing techniques are adopted, such as exploiting NLTK[2] for sentence segmentation.

[1] https://www.yelp.com/dataset/challenge.
[2] http://www.nltk.org/.

Table 1. Basic statistics of the adopted datasets.

Datasets	#Review	#User	#Item
Yelp19	45687	1342	1623
Toy	29135	1628	2417
Kindle	58324	1525	1183

The basic statistics of the three datasets are summarized in Table 1. For each dataset, we split it into the training set, validation set and test set with the ratios of 0.8, 0.1, and 0.1. Since Yelp19 provides social relations between users while Toy and Kindle do not have, we conduct additional tests on Yelp19 for utilizing social relation graph to learn improved user representations. It is worth noting we do not use the benchmarks [24] for testing since the co-occurrence graph construction needs the temporal information of reviews and requires reviews are arranged in a chronological order, while the benchmarks are built without considering the temporal order.

4.2 Evaluation Metrics

Following the previous studies, we adopt *Accuracy* and *RMSE* as the two evaluation metrics to evaluate all the adopted approaches. The computation of *Accuracy* is defined as follows:

$$Accuracy = \frac{\sum_{d \in \mathcal{D}_{test}} \mathbb{I}(y_d, \hat{y}_d)}{|\mathcal{D}_{test}|}, \tag{15}$$

where $\hat{y}_d$ is model-generated label. $\mathbb{I}(y_d, \hat{y}_d)$ is an indicator function, and if $\hat{y}_d$ equals to y_d, it takes value of 1. $|\mathcal{D}_{test}|$ is the size of testing data. Different from *Accuracy* which measures classification performance, *RMSE* characterizes the numerical divergence between true rating scores and predicted rating scores, which is formulated as follows:

$$RMSE = \sqrt{\frac{\sum_{d \in \mathcal{D}_{test}} (y_d - \hat{y}_d)^2}{|\mathcal{D}_{test}|}}. \tag{16}$$

4.3 Baselines

We adopt the following baselines, including the strong competitor HUAPA.

- **Majority**: It assigns the sentiment label that appears most in a training set to each review document tested.
- **TextFeature+UPF** [24]: The feature-based approach contains two categories of features, i.e., textual features which are extracted from n-grams and sentiment lexicons, and user and item features about leniency and popularity. SVM is exploited to take these features as input and classify them.

- **UPNN** [24]: UPNN firstly employs convolutional neural network to learn document representations from user- and item-aware word representations. Then it concatenates user and product representations with the document representation for sentiment classification.
- **NSC+LA** [4]: NSC+LA is a hierarchical attention network which does not consider the effect of user and item on sentiment classification.
- **NSC+UPA** [4][3]: NSA+UPA is a pioneering approach using user and item based attention mechanism in neural networks for personalized document-level sentiment classification. It exhibits better performance than CNN based representation learning approach [24].
- **HUAPA** [28][4]: HUAPA is currently the state-of-the-art model for our task. It decomposes attention computation in NSC+UPA into user attention computation and item attention computation. In this way HUAPA obtains multi-view document representations and refines the commonly used loss function.

4.4 Hyper-Parameter Setting

The version of BERT used is "bert-base-uncased" with 110M parameters. We set the embedding size of users and items to be 200 and initialize them by a uniform distribution U(−0.01, 0.01) before propagating representations in graphs. The dimensions of hidden states in an LSTM cell are set to 100, so as to get a 200-dimensional output because of considering bi-direction. We train IR-HUAPA in a mini-batch setting where we constrain each review to have at most 30 sentences, and each sentence has no more than 40 words. This ensures high-speed matrix computation could be run on a GTX 1080 GPU. We use Adam to optimize IR-HUAPA, with the initial learning rate to be 0.005 and other hyper-parameters by default. All the hyper-parameters of the baselines and IR-HUAPA are tuned on the validation sets for generalization.

5 Experimental Results

We perform extensive experiments to answer the following two questions:

Q1: Can IR-HUAPA achieve the state-of-the-art results for the task?
Q2: Does each type of improved representation indeed contribute to boost the classification performance?

5.1 Model Comparison Q1

Table 2 shows the results on the three datasets. Based on the performance comparison, we have the following key observations. (1) It conforms to the expectation that Majority performs consistently much worse since it does not consider

[3] https://github.com/thunlp/NSC.
[4] https://github.com/wuzhen247/HUAPA.

Table 2. Performance comparison of different approaches.

Model		Yelp19		Toy		Kindle	
		Accuracy	RMSE	Accuracy	RMSE	Accuracy	RMSE
Baselines	Majority	0.215	1.299	0.309	1.254	0.267	1.125
	TextFeature+UPF	0.563	1.085	0.514	1.097	0.471	0.998
	UPNN	0.591	0.942	0.519	1.043	0.496	0.924
	NSC+LA	0.634	0.678	0.530	0.956	0.532	0.860
	NSC+UPA	0.663	0.653	0.559	0.810	0.557	0.823
	HUAPA	0.674	0.634	0.564	0.804	0.578	0.763
Our model	IR-HUAPA	**0.693**	**0.612**	**0.579**	**0.780**	**0.592**	**0.738**

Table 3. Ablation study of IR-HUAPA.

Model		Yelp19		Toy		Kindle	
		Accuracy	RMSE	Accuracy	RMSE	Accuracy	RMSE
HUAPA		0.674	0.634	0.564	0.804	0.578	0.763
IR-HUAPA	-Bert	0.685	0.624	0.571	0.792	0.582	0.753
	-User Graph	0.678	0.626	0.569	0.793	0.579	0.749
	-Item Graph	0.684	0.628	0.572	0.789	0.584	0.745
	Full	**0.693**	**0.612**	**0.579**	**0.780**	**0.592**	**0.738**

any information about text, user, and item, which shows the necessity of modeling them for boosting performance. (2) All representation learning models outperform the feature-engineering approach TextFeature+UPF with large margins, indicating manually designed features do not easily lead to comparable performance than automatic learning feature representations. (3) Both HUAPA and NSC+UPA gain better performance than NSC+LA. This demonstrates incorporating user and item factors is beneficial for sentiment classification. (4) HUAPA performs better than NSC+UPA, revealing regarding user and item as different views to construct view-specific representation is suitable. Besides, the hybrid loss function might bring more informative signals for helping model training. (5) The proposed IR-HUAPA model achieves superior results, significantly better than other baselines in terms of both Accuracy and RMSE. In particular, IR-HUAPA outperforms the strong competitor HUAPA from 0.674 to 0.693 on Yelp19, from 0.564 to 0.579 on Toy, and from 0.578 to 0.592 on Kindle in terms of Accuracy. This validates the effectiveness of improving user, item, and word representations in an integrated model architecture.

5.2 Ablation Study Q2

In this part, we design several variants of IR-HUAPA to investigate whether each type of improved representation could promote the final performance. -BERT denotes removing the pre-trained language model from the full model.

Table 4. Results of user relation modeling with different methods.

Model	Yelp19	
	Accuracy	RMSE
HUA(BERT)	0.668	0.642
eSMF+HUA(BERT)	0.673	0.632
SG+HUA(BERT)	0.682	0.621
CG+HUA(BERT)	0.683	0.618

Similarly, -User Graph and -Item Graph represent removing the corresponding graph, respectively. The results are shown in Table 3. First and most important, by comparing -BERT, -User Graph, and -Item Graph with the full version of IR-HUAPA, we can see consistent performance reduction for the three variants. These ablation tests show each representation type indeed contributes. We further compare -BERT with HUAPA, the difference of which lies in the improved user and item representations. The performance gap demonstrates the advantage of using GCN for encoding co-occurrence relations into user and item representations. By comparing the performance of -BERT, -User Graph, and -Item Graph themselves, we can observe -User Graph suffers from marginally worse performance compared with the other two variants. This might be attributed to the fact that user factor plays a crucial role in determining sentiment labels.

5.3 Analysis of User Relation Modeling

We aim to study how different methods of user relation modeling affect sentiment classification. To achieve this, we present a new variant of IR-HUAPA, named HUA(BERT), which removes the hierarchical item attention network from the full model. As aforementioned, the dataset Yelp19 has user social relations. Thus we adopt GCN to model the social graph (SG) and combine it with HUA(BERT) to build the model SG+HUA(BERT). To evaluate the ability of GCN based modeling for user-user graphs, we consider an alternative strategy, i.e., eSMF [9]. It factorizes a social relation matrix to learn user representation. We incorporate this strategy into HUA(BERT) and obtain the baseline eSMF+HUA(BERT). At last, we use CG+HUA(BERT) to represent the idea of modeling user-user co-occurrence graph by GCN, just as that adopted by IR-HUAPA.

Table 4 presents the performance of the above methods. We observe that eSMF+HUA(BERT) improves the performance of HUA(BERT) by modeling user social relations. We also compare eSMF+HUA(BERT) with SG+HUA(BERT), and find SG+HUA(BERT) brings additional improvements in terms of Accuracy and RMSE. This shows that the improvements by our model are not only from considering user-user relations, but are also caused by the effective GCN based relation modeling. Moreover, CG+HUA(BERT) and SG+HUA(BERT) have comparable performance and are both effective, showing the proposed approach could be generalized to different user-user relations.

Table 5. Effect of embedding propagation layer number.

Layer number	Yelp19		Toy		Kindle	
	Accuracy	RMSE	Accuracy	RMSE	Accuracy	RMSE
1	0.690	**0.610**	0.576	0.782	0.589	**0.736**
2	**0.693**	0.612	**0.579**	**0.780**	**0.592**	0.738
3	0.685	0.618	0.566	0.784	0.587	0.742

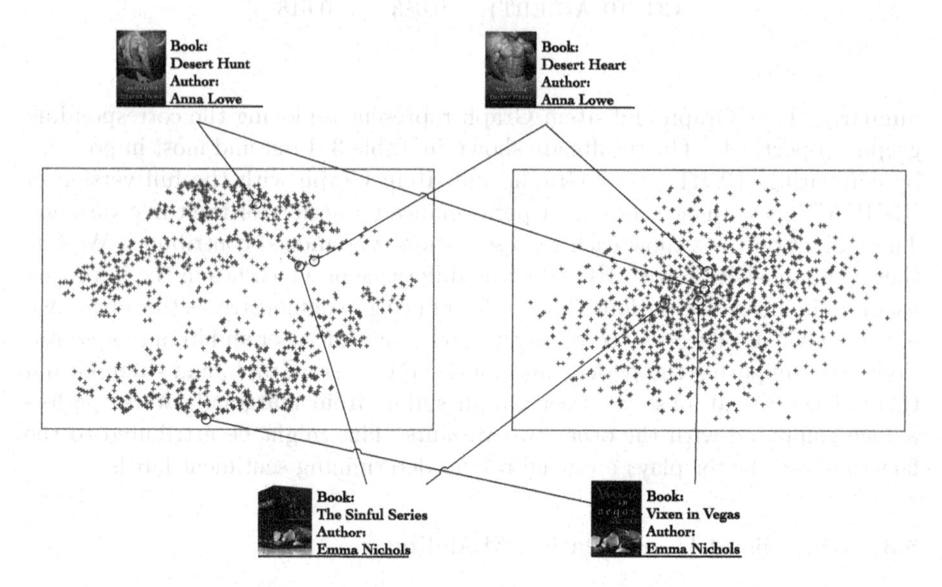

Fig. 2. Visualization of item representations learned by HUAPA and the proposed IR-HUAPA. The left figure corresponds to HUAPA while the right figure is for IR-HUAPA. Each point denotes one item in the Kindle dataset.

5.4 Effect of Layer Number for Propagation

We further analyze how the layer number used for embedding propagation affects the performance on the test sets. As shown in Table 5, the performance gap between one-layered propagation and two-layered propagation is not very significant. When the number of layers continues to increase, it seems that the performance would drop. In practice, we choose the layer number according to the results on the validation datasets, and it is set to 2 as default.

5.5 Case Study

In this part, we conduct a real case study to show the empirical advantage of IR-HUAPA by modeling the co-occurrence relation between items through GCN. In Fig. 2, we visualize the item representations learned by HUAPA and IR-HUAPA.

The tool t-SNE is utilized to visualize the high-dimensional vectors into the 2-dimensional plane. For illustration, four specific items (books) are selected. The first two books are "The Sinful Series" and "Vixen in Vegas", both of which are written by Emma Nichols[5]. We mark the two books in the figure as the red points. Another two books are "Desert Hunt" and "Desert Heart", and the author of them is Anna Lowe[6]. We utilize the yellow points to denote them.

As shown in the left part of Fig. 2, the distance between the book "The Sinful Series" and "Vixen in Vegas" is relatively large. This case does not conform to the real situation since these two books are written by the same author. They might have some correlations with each other and their semantic representations should be close. It shows that the simple way of mapping items based on their IDs could not lead to semantically meaningful representations, although they are learned along with sentiment classification neural models. On the contrary, the two representations learned by IR-HUAPA are much closer to each other. A similar phenomenon could be observed for the other two books i.e. "Desert Hunt" and "Desert Heart". In summary, although IR-HUAPA does not utilize the authors' information of the four books, it still can capture some correlations between the books belonging to the same author. This is welcomed for its ability to model the item-item co-occurrence relations by GCN.

6 Conclusion

In this paper, we aim to boost the performance of personalized document-level sentiment classification by improving the representations of users, items, and words. To achieve this, we have proposed to encode the item-item and user-user co-occurrence relations through graph convolutional network into user and item representations. Moreover, we have obtained contextualized word representations by the pre-trained language model BERT. We have conducted extensive experiments on three public datasets and shown IR-HUAPA achieves superior performance against the state-of-the-art models.

References

1. Amplayo, R.K.: Rethinking attribute representation and injection for sentiment classification. In: EMNLP, pp. 5601–5612 (2019)
2. Amplayo, R.K., Kim, J., Sung, S., Hwang, S.: Cold-start aware user and product attention for sentiment classification. In: ACL, pp. 2535–2544 (2018)
3. Bahdanau, D., Cho, K., Bengio, Y.: Neural machine translation by jointly learning to align and translate. In: ICLR (2015)
4. Chen, H., Sun, M., Tu, C., Lin, Y., Liu, Z.: Neural sentiment classification with user and product attention. In: EMNLP, pp. 1650–1659 (2016)

[5] http://emmanicholsromance.com/.
[6] http://www.annalowebooks.com/.

5. Devlin, J., Chang, M., Lee, K., Toutanova, K.: BERT: pre-training of deep bidirectional transformers for language understanding. In: NAACL, pp. 4171–4186 (2019)
6. Dou, Z.: Capturing user and product information for document level sentiment analysis with deep memory network. In: EMNLP, pp. 521–526 (2017)
7. Feng, F., He, X., Wang, X., Luo, C., Liu, Y., Chua, T.: Temporal relational ranking for stock prediction. ACM Trans. Inf. Syst. **37**(2), 27:1–27:30 (2019)
8. Hochreiter, S., Schmidhuber, J.: Long short-term memory. Neural Comput. **9**(8), 1735–1780 (1997)
9. Hu, G., Dai, X., Song, Y., Huang, S., Chen, J.: A synthetic approach for recommendation: Combining ratings, social relations, and reviews. In: IJCAI, pp. 1756–1762 (2015)
10. Jiang, L., Yu, M., Zhou, M., Liu, X., Zhao, T.: Target-dependent Twitter sentiment classification. In: ACL, pp. 151–160 (2011)
11. Kim, Y.: Convolutional neural networks for sentence classification. In: EMNLP, pp. 1746–1751 (2014)
12. Kipf, T.N., Welling, M.: Semi-supervised classification with graph convolutional networks. In: ICLR (2017)
13. Li, F., Wang, S., Liu, S., Zhang, M.: SUIT: a supervised user-item based topic model for sentiment analysis. In: AAAI, pp. 1636–1642 (2014)
14. Li, Y., Cai, Y., Leung, H., Li, Q.: Improving short text modeling by two-level attention networks for sentiment classification. In: DASFAA, pp. 878–890 (2018)
15. Liu, B.: Sentiment Analysis - Mining Opinions, Sentiments, and Emotions. Cambridge University Press, Cambridge (2015)
16. McAuley, J.J., Leskovec, J.: Hidden factors and hidden topics: understanding rating dimensions with review text. In: RecSys, pp. 165–172 (2013)
17. Mikolov, T., Sutskever, I., Chen, K., Corrado, G.S., Dean, J.: Distributed representations of words and phrases and their compositionality. In: NIPS, pp. 3111–3119 (2013)
18. Mukherjee, S., Basu, G., Joshi, S.: Joint author sentiment topic model. In: SDM, pp. 370–378 (2014)
19. Ni, J., Li, J., McAuley, J.J.: Justifying recommendations using distantly-labeled reviews and fine-grained aspects. In: EMNLP, pp. 188–197 (2019)
20. Pennington, J., Socher, R., Manning, C.D.: Glove: global vectors for word representation. In: EMNLP, pp. 1532–1543 (2014)
21. Peters, M.E., et al.: Deep contextualized word representations. In: NAACL, pp. 2227–2237 (2018)
22. Taboada, M., Brooke, J., Tofiloski, M., Voll, K., Stede, M.: Lexicon-based methods for sentiment analysis. Comput. Linguist. **37**(2), 267–307 (2011)
23. Tang, D., Qin, B., Liu, T.: Document modeling with gated recurrent neural network for sentiment classification. In: EMNLP, pp. 1422–1432 (2015)
24. Tang, D., Qin, B., Liu, T.: Learning semantic representations of users and products for document level sentiment classification. In: ACL, pp. 1014–1023 (2015)
25. Vaswani, A., et al.: Attention is all you need. In: NIPS, pp. 5998–6008 (2017)
26. Velickovic, P., Cucurull, G., Casanova, A., Romero, A., Liò, P., Bengio, Y.: Graph attention networks. In: ICLR (2018)
27. Wang, X., He, X., Wang, M., Feng, F., Chua, T.: Neural graph collaborative filtering. In: SIGIR, pp. 165–174 (2019)
28. Wu, Z., Dai, X., Yin, C., Huang, S., Chen, J.: Improving review representations with user attention and product attention for sentiment classification. In: AAAI, pp. 5989–5996 (2018)

29. Yang, Z., Yang, D., Dyer, C., He, X., Smola, A.J., Hovy, E.H.: Hierarchical attention networks for document classification. In: NAACL, pp. 1480–1489 (2016)
30. Yao, L., Mao, C., Luo, Y.: Graph convolutional networks for text classification. In: AAAI, pp. 7370–7377 (2019)
31. Zhang, W., Yuan, Q., Han, J., Wang, J.: Collaborative multi-level embedding learning from reviews for rating prediction. In: IJCAI, pp. 2986–2992 (2016)
32. Zhao, K., Zhang, Y., Zhang, Y., Xing, C., Li, C.: Learning from user social relation for document sentiment classification. In: DASFAA, pp. 86–103 (2019)

Modeling Multi-aspect Relationship with Joint Learning for Aspect-Level Sentiment Classification

Jie Zhou[1,2(✉)], Jimmy Xiangji Huang[3], Qinmin Vivian Hu[4], and Liang He[1,2]

[1] Shanghai Key Laboratory of Multidimensional Information Processing, East China Normal University, Shanghai 200241, China
jzhou@ica.stc.sh.cn

[2] School of Computer Science and Technology, East China Normal University, Shanghai 200241, China
lhe@cs.ecnu.edu.cn

[3] Information Retrieval and Knowledge Management Research Lab, York University, Toronto, ON M3J 1P3, Canada
jhuang@yorku.ca

[4] The School of Computer Science, Ryerson University, Toronto, ON M5B 2K3, Canada
vivian@ryerson.ca

Abstract. Aspect-level sentiment classification is a crucial branch for sentiment classification. Most of the existing work focuses on how to model the semantic relationship between the aspect and the sentence, while the relationships among the multiple aspects in the sentence is ignored. To address this problem, we propose a joint learning (Joint) model for aspect-level sentiment classification, which models the relationships among the aspects of the sentence and predicts the sentiment polarities of all aspects simultaneously. In particular, we first obtain the augmented aspect representation via an aspect modeling (AM) method. Then, we design a relationship modeling (RM) approach which transforms sentiment classification into a sequence labeling problem to model the potential relationships among each aspect in a sentence and predict the sentiment polarities of all aspects simultaneously. Extensive experiments on four benchmark datasets show that our approach can effectively improve the performance of aspect-level sentiment classification compared with the state-of-the-art approaches.

Keywords: Aspect-based sentiment classification · Joint · Neural networks

1 Introduction and Motivation

As a fundamental subtask of sentiment classification [21,28], aspect-level sentiment classification [3,4,30,35,54] is a central concern of academic communities and industries. The aim is to determine the sentiment polarity expressed for a given aspect in reviews as positive, neutral or negative. For example, as shown in

Y. Nah et al. (Eds.): DASFAA 2020, LNCS 12112, pp. 786–802, 2020.
https://doi.org/10.1007/978-3-030-59410-7_54

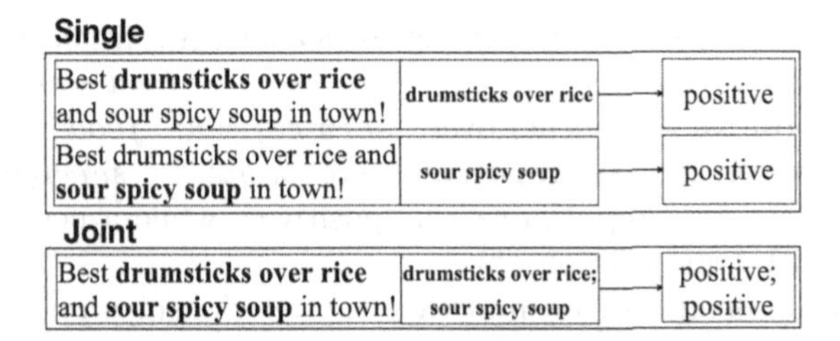

Fig. 1. An example of aspect-level sentiment classification.

Fig. 1, there are two aspects "***drumsticks over rice***" and "***sour spicy soup***" presented in the sentence *"Best **drumsticks over rice** and **sour spicy soup** in town!"*, where the user expresses positive sentiments.

In this paper, we focus on the relationships among the aspects in a sentence. The motivations are mainly from the preliminary experiments on four public benchmark datasets for aspect-level sentiment classification as follows.

First, the reviews usually contain multiple aspects and the sentiment polarities of the aspects in the sentence are usually the same. From Table 1, we find that the average number of aspects in each sentence over the four datasets is about 1.59 and the average percent of samples which contain more than one aspects over four datasets is about 37.88%. In addition, the average percent of instances which contain multiple aspects and express the same polarity is about 27.16%. For example, in sentence *"Nice **atmosphere**, the **service** was very pleasant and the **desert** was good."*, the sentiment polarities for the three aspects "***atmosphere***", "***service***" and "***desert***" are all positive.

Table 1. Statistics of aspect information over four benchmark datasets. *#AvgAspects* represents the average number of aspects in sentences. *#MultiAspects* denotes the number of samples which contain more than one aspects and the percent of them in the corresponding dataset. And *#SamePolarity* indicates the number of samples which contain more than one aspects that have the same sentiment polarity.

	Restaurant14	Laptop14	Restaurant15	Restaurant16	Average
#AvgAspects	1.83	1.58	1.47	1.47	1.59
#MultiAspects	1285(49.84%)	697(37.21%)	553(32.49%)	734(31.98%)	817(37.88%)
#SamePolarity	886(34.37%)	494(26.37%)	408(23.97%)	549(23.92%)	584(27.16%)

Second, the sentiment polarities of some aspects should be inferred by the descriptions of other aspects. For example, in sentence *"I also ordered for **delivery** and the restaurant forgot half the **order**."*, the sentiment polarity of "***delivery***" is influenced by "***order***", which has negative sentiment for the users express "forgot".

Third, prepositions are important to tell the aspects. Some prepositions (e.g., and, as well as, also and so on) express the same polarity and some (e.g. but, yet, however, nevertheless and so on) express a contrast sentiment. For example, in the sentence *"The* ***service*** *was excellent, and the* ***food*** *was delicious"*, the user expresses positive sentiments over "***service***" and "***food***" with a preposition "and". However, in *"the* ***appetizers*** *are ok but the* ***service*** *is slow"*, the expected sentiment on "***appetizers***" is "positive", while a negative sentiment on "***service***" with a preposition "but".

From these observations, we find that the relationships among the aspects are important for aspect-level sentiment classification. However, previous work mainly focused on how to enhance the representation of the sentence towards the given aspect for aspect-level sentiment classification [2,9,18,24,47]. Schmitt et al. [34] and Xu et al. [49] proposed to perform aspect extraction and aspect-level sentiment classification jointly. In this paper, we focus on modeling the relationships among the aspects in the task aspect-level sentiment classification. To the best of our knowledge, for aspect-level sentiment classification, most of the previous studies infer the sentiment polarity of the given aspect independently and the relationships among the aspects in a sentence is ignored.

To address this problem, we propose a joint learning (Joint) model for aspect-level sentiment classification. It models the relationships among the multiple aspects in a sentence and predicts the sentiments of these aspects jointly. In particular, an aspect modeling (AM) method is developed to obtain the augmented representation of the given aspect in a sentence. Then we build a relationship modeling (RM) approach to convert sentiment polarity judgement as a sequence labeling problem such that the relationships among the aspects of a sentence can be learned. The experimental results show the effectiveness of our proposed model.

In summary, the main contributions of this paper are as follows:

- We propose a joint learning (Joint) model for aspect-level sentiment classification to model the relationships among the multiple aspects in a sentence and predict the sentiment polarities of them simultaneously.
- An aspect modeling (AM) method is designed to obtain augmented aspect representation. Different from the existing work which obtains the sentence representation towards the given aspect, we obtain the aspect representation in a sentence.
- Furthermore, we propose a relationship modeling (RM) approach to convert aspect-level sentiment classification into a sequence labeling problem naturally, which can model the relationships among the aspects in a sentence.
- A series of experiments over four public benchmark datasets show the effectiveness of our proposed model. In particular, our model outperforms the state-of-the-art approaches in most cases.

The rest of this paper is structured as follows. In Sect. 2, we present an overview of the related work. Section 3 introduces the details of our proposed Joint model. The details of the experimental setup are described in Sect. 4, followed by the experimental results and analyses in Sect. 5. Finally, we conclude our work and show some ideas for future research in Sect. 6.

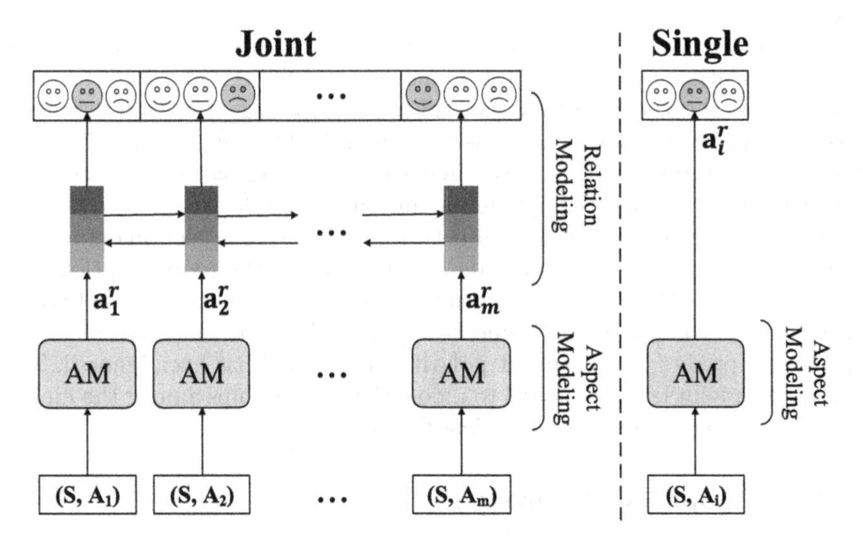

Fig. 2. The architecture of our joint learning (Joint) model for aspect-level sentiment classification. First, we obtain the augmented aspect representation via aspect modeling (AM). Then, we use a relationship modeling (RM) method to model the relationships among the aspects in a sentence. Finally, the sentiment polarities of the aspects in the sentence are predicted jointly like a sequence labeling problem.

2 Related Work

Aspect-level sentiment classification plays a significant role in sentiment classification [6,20,21,28]. Early work mainly used traditional machine learning methods, which highly depended on the quality of extensive handcraft features such as sentiment lexicon, dependency information and n-gram [14,16,27,42,43]. There are a large number of studies on the topic of aspect-level sentiment classification [10,19,32,33,37,44,46,48,50,53]. Here we mainly review the work which is most related to our research.

With the advances of deep learning models, various neural networks are of growing interest in aspect-level sentiment classification for their ability to learn sentence representation automatically for sentiment classification [1,2,7,11,18, 22,38,47]. Tang et al. [38] proposed a TD-LSTM model, which modeled the left and right contexts of the aspect to take the aspect information into account. To model the interaction between the aspect and its surrounding contexts, Zhang et al. [52] adopted a three-way gated neural network.

Attention mechanisms [1] have been proposed for their ability to capture the important parts related to the given aspect [2,9,13,23,24,45,47]. Wang et al. [47] designed an attention-based LSTM to explore the potential correlation of aspect and sentiment polarities. Chen et al. [2] adopted multiple layers of attention and fed the attentive representation into a RNN model. Ma et al. [24] utilized an interactive attention network to build the semantic representations.

In addition, deep memory networks [36] were proposed for aspect-level sentiment classification in [2,39,41,55]. Commonsense knowledge of sentiment-related concepts was incorporated into the end-to-end training of a deep neural network for aspect-level sentiment classification [25]. He et al. [11] transferred the knowledge from document-level sentiment classification dataset to aspect-level sentiment classification via pre-training and multi-task learning.

Despite these advances of neural networks in aspect-level sentiment analysis, many of existing methods focused on how to improve the representation of the sentence towards the aspect by capturing the intricate relatedness between an aspect and its context words, while how to model the relationships among the aspects and predict the sentiment polarities in parallel has been ignored. Thus, we propose a joint learning model to model the relationships among the multiple aspects for aspect-level sentiment classification.

3 Our Proposed Approach

In this section, we propose our Joint model for aspect-level sentiment classification in detail. Figure 2 shows the architecture of our Joint model. First, we obtain the augmented aspect representation via an aspect modeling (AM) method in Sect. 3.1. Then, in Sect. 3.2, we develop a relationship modeling (RM) strategy to model the relationships among the multiple aspects in a sentence. We predict the sentiment polarities of the aspects in sentence jointly, which can be transformed into a sequence labeling problem naturally.

Mathematically, we give a formal representation to define the problem as follows: given a sentence-aspects pair (S, A), where $A = \{A_1, A_2, ..., A_m\}$ consists of m aspects which orderly occur in the sentence S that contains n words $\{w_1, w_2, ..., w_n\}$. The ith aspect $A_i = \{w_{start_i}, w_{start_i+1}, ..., w_{end_i-1}, w_{end_i}\}$ is the sub-sequence of sentence S, where $start_i$ and end_i denote the starting index and ending index of the aspect A_i in sentence S respectively. The goal of aspect-level sentiment classification is to predict sentiment polarity $c \in \{N, O, P\}$ for the aspect A_i in the sentence S, where N, O, and P denote the "negative", "neutral", and "positive" sentiment polarities respectively.

3.1 Aspect Modeling (AM)

We adopt an aspect modeling (AM) method to obtain the augmented aspect representation a_i^r for the aspect $A_i, i \in [1, m]$. The details of the AM is shown in Fig. 3. It consists of four parts, namely aspect-specific embedding, aspect-specific encoder, aspect-specific attention and augmented aspect representation. Specifically, aspect-specific embedding and aspect-specific encoder are used to obtain the aspect-specific representation and aspect-oriented context-sensitive representation of each word respectively. Regarding aspect-specific attention, we use it to capture the important parts for a specific aspect. With these parts, an aspect can be projected into an augmented aspect representation, which can be further utilized for relationship modeling. Details of each component are introduced as follows.

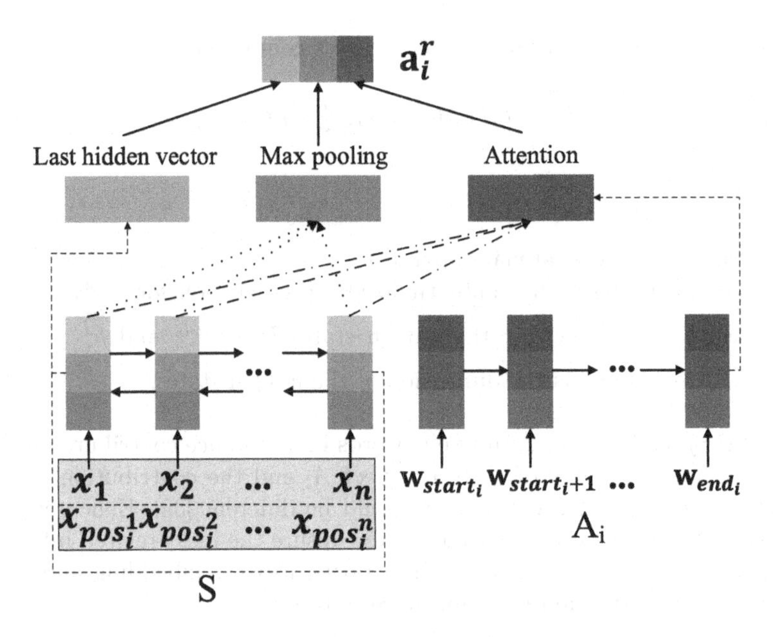

Fig. 3. The details of aspect modeling (AM). We use an aspect-specific attention-based Bi-LSTM model to obtain the augmented aspect representation a_i^r for the aspect $A_i, i \in [1, m]$.

Aspect-Specific Embedding. Each word w_j is mapped into a low-dimensional continuous vector space $x_j \in \mathbb{R}^{d_w}$, which is calculated by looking up the word embedding $E_w \in \mathbb{R}^{d_w \times |V|}$. Here d_w denotes the dimension of the word embedding and $|V|$ is the size of vocabulary. Inspired by [5,9,51], we append position representation into word embedding to obtain the aspect-specific embedding. We define a position index sequence to model the position information of the aspect A_i in its corresponding sentence.

$$pos_i^j = \begin{cases} |j - start_i|, & j < start_i \\ 0, & start_i \leq j \leq end_i \\ |j - end_i|, & j > end_i \end{cases} \tag{1}$$

where j represents the index of the word w_j in sentence, and pos_i^j can be viewed as the relative distance of the w_i in sentence to the i_{th} aspect A_i. And its corresponding position embedding $x_{pos_i^j}$ is calculated by looking up a position embedding matrix $E_p \in \mathbb{R}^{d_p \times L_m}$, where L_m denotes the max length of the sentences and d_p is the dimension of position embedding.

Aspect-Specific Encoder. Bi-directional long short-term memory (Bi-LSTM) [8] model is employed to accumulate the context information from aspect-specific embedding. The Bi-LSTM contains a forward $\overrightarrow{LSTM}$ which reads the sentence from w_1 to w_n and a backward $\overleftarrow{LSTM}$ from w_n to w_1. The aspect-specific contextualized representation for each word is computed as follows:

$$\overrightarrow{h_i^j} = \overrightarrow{LSTM}([x_j; x_{pos_i^j}]), j \in [1, n] \tag{2}$$

$$\overleftarrow{h_i^j} = \overleftarrow{LSTM}([x_j; x_{pos_i^j}]), j \in [n, 1] \tag{3}$$

where ; donates the concatenate operator.

The final aspect-specific contextualized representation for a given word w_j is obtained by concatenating the hidden states $\overrightarrow{h_i^j} \in \mathbb{R}^{d_h}$ and $\overleftarrow{h_i^j} \in \mathbb{R}^{d_h}$, i.e., $h_j = [\overrightarrow{h_i^j}, \overleftarrow{h_i^j}]$, where d_h is the dimension of the hidden states.

Aspect-Specific Attention. Intuitively, words in a sentence contribute differently to the sentiment polarity of a given aspect A_i and the contribution of a word for different aspects in the sentence should be different [39]. Hence, we design an aspect-specific attention mechanism to capture the most related information in response to the given aspect by integrating aspect embedding and position information into attention mechanism. Specifically,

$$u_i^j = tanh(W_A[h_i^j; e_i^a; x_{pos_i^j}] + b_A) \tag{4}$$

$$\alpha_i^j = \frac{exp(u_i^{j^T} u_A)}{\sum_k exp(u_k^T u_A)} \tag{5}$$

where $W_A \in \mathbb{R}^{(2d_h+d_a+d_p)\times(2d_h+d_a+d_p)}$, $b_A \in \mathbb{R}^{2d_h+d_a+d_p}$ and $u_A \in \mathbb{R}^{2d_h+d_a+d_p}$ are the learnable parameters. ; denotes the concatenate operator. By modeling the words contained in the aspect with LSTM, we obtain the last hidden state for aspect embedding which is denoted as $e_i^a \in \mathbb{R}^{d_a}$, where d_a is the dimension of the aspect embedding. Then, the attentive aspect representation $a_i^{r^A}$ is calculated as follows:

$$a_i^{r^A} = \sum_j \alpha_i^j h_i^j \tag{6}$$

Augmented Aspect Representation. We adopt the last hidden state and max pooling to improve the representation of the aspect A_i.

$$a_i^r = [a_i^{r^A}; a_i^{r^M}; a_i^{r^L}] \tag{7}$$

where ; represents the concatenate operator and $a_i^{r^L}$ indicates the last hidden vector. $a_i^{r^M}$ denotes the representation of max pooling, which can be calculated as follows:

$$[a_i^{r^M}]_k = \max_{1 \le j \le n} [h_{i,k}^j] \tag{8}$$

3.2 Relationship Modeling (RM)

To capture the relationships among the aspects in the sentence, we propose a relationship modeling (RM) approach. This approach judges the sentiment polarities of the multiple aspects in the sentence simultaneously, which is converted into a sequence labeling problem naturally. In particular, we employ a Bi-LSTM model to learn the potential relationships among the aspects. The Bi-LSTM model transforms the input a_i^r into the contextualized aspect representation h_i (i.e. hidden states of Bi-LSTM).

$$\overrightarrow{h_i} = \overrightarrow{LSTM}(a_i^r), i \in [1, m] \tag{9}$$

$$\overleftarrow{h_i} = \overleftarrow{LSTM}(a_i^r), i \in [m, 1] \tag{10}$$

Thus, the contextualized aspect representation h_i is obtained as $h_i = [\overrightarrow{h_i}; \overleftarrow{h_i}]$, which summarizes the information of the whole aspects centered around A_i. Hence we obtain the contextualized aspect representations $H = \{h_1, h_2, ..., h_m\} \in \mathbb{R}^{m \times 2d_h}$ from the augmented aspect representations $a^r = \{a_1^r, a_2^r, ..., a_m^r\}$.

Finally, we feed H into a softmax function and obtain the probability distribution of sentiment polarities over all the aspects by:

$$P_c(S, A_i) = Softmax(W_p h_i), i \in [1, m] \tag{11}$$

where $W_p \in \mathbb{R}^{2d_h \times |C|}$ is the trainable weight vector, and C is number of classes.

The loss function is defined by the cross-entropy of the predicted and true label distributions for training:

$$L = -\sum_{(S,A)\in D} \sum_{i}^{m} \sum_{c\in C} y_c(S, A_i) \cdot log P_c(S, A_i) \tag{12}$$

where D is the set of training set and C is the collection of sentiment labels. $y_c(S, A_i)$ is ground truth for sentence-aspect pair (S, A_i) and and $P_c(S, A_i)$ denotes the predicted probability of (S, A_i) labeled as the sentiment class c.

Table 2. Statistics of Restaurant14-16 and Laptop14 datasets.

	Positive		Negative		Neutral	
	Train	Test	Train	Test	Train	Test
Restaurant14	2,164	728	807	196	637	196
Laptop14	994	341	870	128	464	169
Restaurant15	1,178	439	382	328	50	35
Restaurant16	1,620	597	709	190	88	38

4 Experimental Setup

In this section, we first describe the datasets and the implementation details in Sect. 4.1. Then, we present the baseline methods in Sect. 4.2. Finally, the pre-training is introduced in Sect. 4.3.

4.1 Datasets and Implementation Details

Datasets. We conduct experiments to validate the effectiveness of our proposed approach on four standard datasets, taken from SemEval'14[1] [26], SemEval'15[2] [31] and SemEval'16[3] [30], namely Restaurant 2014, Laptop 2014, Restaurant 2015 and Restaurant 2016, which are the same as [11]. Full statistics of the datasets are summarized in Table 2. We also randomly sample 10% from the original training data as the development data which is used to tune algorithm parameters. Accuracy (Acc) and Macro-Average F1 (Macro-F1) are adopted to evaluate the model performance, which are the primary metrics used in aspect-level sentiment classification [11,17].

Implementation Details. In our experiments, word embedding vectors are initialized with 300-dimension GloVe [29] vectors and fine-tuned during the training, the same as [39]. The dimension of hidden state vectors and position embedding are 300 and 100 respectively. Words out of vocabulary GloVe, position embedding and weight matrices are initialized with the uniform distribution $U(-0.1, 0.1)$, and the biases are initialized to zero. Adam [15] is adopted as the optimizer. To avoid overfitting, dropout is used in our training model and we search the best dropout rate from 0.4 to 0.7 with an increment of 0.1. We obtain the best hyper-parameter learning rate and mini-batch size from {0.001, 0.0005} and {4, 8, 16, 32} respectively via grid search. We implement our neural networks with Pytorch[4]. We keep the optimal parameters based on the best performance on the development set and the optimal model is used for evaluation in the test set.

4.2 Baseline Methods

In order to comprehensively evaluate the performance of our model, we compare our models with the following baselines on the four benchmark datasets:

- **Majority** assigns the majority sentiment label in the training set to all examples in the test set, which was adopted as a basic baseline in [24,40].
- **Feature+SVM** trains a SVM classifier using manual features such as n-gram feature, parse feature and lexicon features. We adopt the results of this method reported in [16] and the results on Restaurant15-16 are not available.

[1] Available at: http://alt.qcri.org/semeval2014/task4/.
[2] Available at: http://alt.qcri.org/semeval2015/task12/.
[3] Available at: http://alt.qcri.org/semeval2016/task5/.
[4] https://pytorch.org/.

- **ContextAvg** feeds the average of word embeddings and the aspect vector to a softmax function, which was adopted as a baseline in [39].
- **LSTM** adopts one LSTM [12] network to model the sentence and the last hidden state is regarded as the final representation.
- **TD-LSTM** models the preceding content and following contexts of the aspect via a forward LSTM and backward LSTM respectively[38].
- **ATAE-LSTM** combines aspect embedding with word embedding and hidden states to strengthen the effect of aspect embedding and has an attention layer above the LSTM layer [47].
- **MemNet** proposes deep memory network which employs multi-hop attention [39].
- **IAN** generates the representations for aspects and contexts separately by interactive learning via attention mechanism [24].
- **RAM** utilizes LSTM and multiple attention mechanisms to capture sentiment features on position-weighted memory [2], which is the state-of-the-art approach before pre-training is applied.
- **PRET+MULT** incorporates knowledge from document-level corpus through pre-training and multi-task learning [11]. It is considered as a state-of-the-art aspect-level sentiment classification model with pre-training.

The existing state-of-the-art baselines are from He's [11] and we endorse all the baselines they have adopted. We rerun IAN to on our datasets and evaluate it with both accuracy and Macro-F1. We also run the release code of PRET+MULT[5] to generate results of each instance for significance tests.

4.3 Pre-training

He et al. [11] transferred the knowledge from document-level sentiment classification datasets to aspect-level sentiment classification by pre-training and multi-task learning. For fair comparison, we pre-train our model over subset of Yelp restaurant reviews[6] for restaurant datasets and Amazon Electronics reviews[7] for laptop dataset respectively, the same as [11]. Since the reviews are on a 5-point scale, reviews with rating < 3, > 3 and $= 3$ are labeled as negative, positive and neutral respectively. For each dataset, we sample 50k instances for each class label randomly. Since the aspect labels are not available in these datasets, we set the aspect as "null" and only the weights of aspect modeling model of the pre-trained are used to initialize our joint learning model.

5 Experimental Results and Analyses

We report our experimental results and conduct extensive analyses in this section. In particular, we first investigate the effectiveness of our joint learning

[5] Available at: https://github.com/ruidan/Aspect-level-sentiment.
[6] Available at: https://www.yelp.com/dataset/challenge.
[7] Available at: http://jmcauley.ucsd.edu/data/amazon/.

Table 3. The accuracy and Macro-F1 on Restaurant14-16 and Laptop14 datasets. The best results obtained on each collection is marked in bold. "Single" represents the model which predicts the sentiment polarity of the aspect independently via aspect modeling (AM). "Joint" denotes our joint learning model which infers the polarities jointly through relationship modeling. "PRET" represents the model with pre-training. The marker † and ‡ refer to p-values < 0.05 when comparing with the state-of-the-art approaches "RAM" and "PRET+MULT" respectively. The marker $_{\triangleleft}$ and $_{\triangleright}$ refer to p-values < 0.05 when comparing with "Single" and "Single+PRET", respectively.

	Restaurant14		Laptop14		Restaurant15		Restaurant16	
	Acc	Macro-F1	Acc	Macro-F1	Acc	Macro-F1	Acc	Macro-F1
Majority	65.00	26.26	53.45	23.22	54.74	23.58	72.36	27.99
Feature+SVM	80.89	-	72.10	-	-	-	-	-
ContextAvg	71.53	58.02	61.59	53.92	73.79	47.43	79.87	55.68
LSTM	74.49	59.32	66.51	59.44	75.40	53.30	80.67	54.53
TD-LSTM	78.00	68.43	71.83	68.43	76.39	58.70	82.16	54.21
ATAE-LSTM	78.60	67.02	68.88	63.93	78.48	62.84	83.77	61.71
MemNet	78.16	65.83	70.33	64.09	77.89	59.52	83.04	57.91
IAN	77.86	66.31	71.79	65.92	78.58	54.94	82.42	57.12
RAM	78.48	68.54	72.08	68.43	79.98	60.57	83.88	62.14
PRET+MULT	79.11	69.73	71.15	67.46	81.30	**68.74**	85.58	69.76
Single	78.13	66.17	68.34	61.63	78.91	60.04	83.12	60.90
Single+PRET	80.61	69.43	70.53	63.26	79.88	61.91	86.86	69.64
Joint	$80.27^{\dagger}_{\triangleleft}$	$69.90^{\dagger}_{\triangleleft}$	$72.57_{\triangleleft}$	$67.23_{\triangleleft}$	$80.11_{\triangleleft}$	$62.81_{\triangleleft}$	$85.45^{\dagger}_{\triangleleft}$	$67.88^{\dagger}_{\triangleleft}$
Joint+PRET	$\mathbf{81.96}^{\dagger\ddagger}_{\triangleright}$	$\mathbf{71.80}^{\dagger\ddagger}_{\triangleright}$	$\mathbf{73.04}^{\dagger\ddagger}_{\triangleright}$	$\mathbf{69.16}^{\ddagger}_{\triangleright}$	$\mathbf{82.72}^{\dagger\ddagger}_{\triangleright}$	$68.26^{\dagger}_{\triangleright}$	$\mathbf{88.36}^{\dagger\ddagger}_{\triangleright}$	$\mathbf{72.20}^{\dagger\ddagger}_{\triangleright}$

model for aspect-level sentiment classification in Sect. 5.1. Then, we investigate the effectiveness of relationship modeling in Sect. 5.2. After that, we provide an intuitive understanding of why our proposed joint learning model is more effective via case studies in Sect. 5.3.

5.1 Effectiveness of Our Model

As shown in Table 3, we report the performance of the classic baselines and our proposed joint learning model and its ablations with pre-training or not. We also conduct pairwise t-test on both Accuracy and Macro-F1 to verify if the improvements over the compared models are reliable.

According to this table, we obtain the following observations. **(1)** "Joint" outperforms the state-of-the-art methods that do not apply pre-training (e.g. RAM) in most cases. It shows that "Joint" can achieve superior performances even with a simple AM model for aspect-level sentiment classification. **(2)** When incorporating pre-training strategy into the training of our "Joint" model, the performance can be further boosted. In particular, "Joint+PRET" achieves significant improvements over the best baseline "PRET+MULT", which transfers the knowledge in document level through pre-training and multi-task learning [11], across all the four datasets in terms of accuracy. Regarding Macro-F1, our model achieves the best performance in Restaurant14, Laptop14 and Restaurant15. **(3)** Our full model "Joint" ("Joint+PRET") achieves significant improvements over the "Single" ("Single+PRET") across all the four datasets. It indicates the effectiveness of our proposed model by modeling the relationships among the aspects by incorporating relationship modeling. **(4)** "Joint+PRET" ("Single+PRET") outperforms "Joint" (Single+PRET) in terms of Accuracy and Macro-F1. All these observations show that pre-training can improve the performance significantly for our model.

Table 4. The accuracy and Macro-F1 on the subsets made up of test samples that contain more than one aspects. *Improvement* represents the relative improvement by comparing Joint (Joint+PRET) with Single (Single+PRET).

	Restaurant14		Laptop14		Restaurant15		Restaurant16	
	Acc	Macro-F1	Acc	Macro-F1	Acc	Macro-F1	Acc	Macro-F1
Without Pre-training								
Single	80.02	68.19	71.44	65.60	75.43	52.51	85.08	64.30
Joint	81.12	69.92	73.35	67.45	77.36	54.69	87.08	68.41
Improvement	1.37%	2.54%	2.67%	2.82%	2.56%	4.15%	2.35%	6.39%
With Pre-training								
Single+PRET	80.50	68.90	71.92	65.81	79.34	59.74	87.53	65.76
Joint+PRET	81.92	71.50	73.82	67.93	80.22	62.83	88.86	69.12
Improvement	1.76%	3.77%	2.64%	3.22%	1.11%	5.17%	1.52%	5.11%

5.2 Effectiveness of RM

In this section, we investigate the effectiveness of our relationship modeling strategy. As shown in Table 4, we report the results on the subsets made up of test samples that consist of more than one aspects. As shown in Table 1, the maximum percent of samples that contain the multiple aspects is 49.84% on the Restaurant14 dataset. It is observed that our proposed model "Joint" ("Joint+PRET") outperforms "Single" ("Single+PRET") over all the four datasets. Specifically, our "Joint" model which models the relationships among the multiple aspects, obtains the maximum improvement of 6.39% over the "Single" model in terms of Macro-F1 on the Restaurant16 dataset. These observations indicate that our relationship modeling method can capture the relationships among multiple aspects in the sentence more effectively.

5.3 Case Studies

To illustrate how our proposed model works, we sample some classic cases and visualize the attention weights based on our joint learning model in Fig. 4. We show the results of "Single" model, "Joint" model, and the one with pre-training (i.e "Joint+PRET"). The color depth denotes the importance degree of the weight. The deeper color means the higher weight.

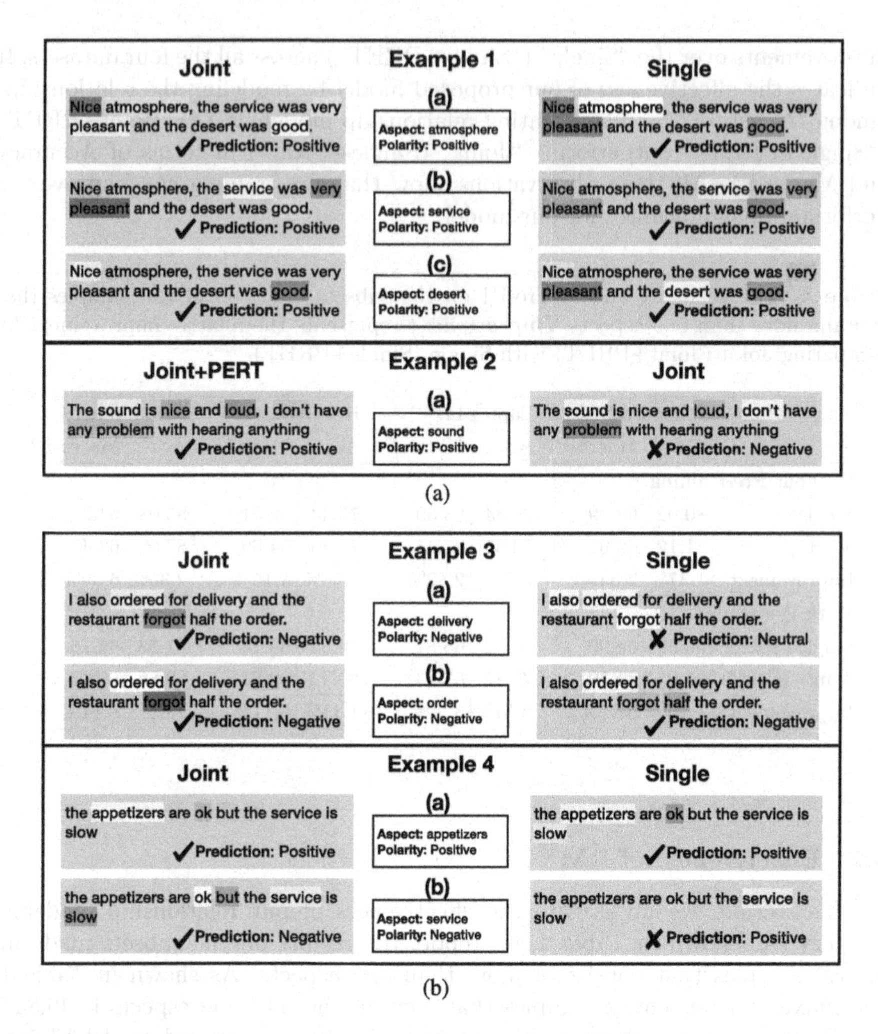

Fig. 4. Visualization of four typical examples. The color depth denotes the importance degree of the weight. The deeper color means the higher weight.

First, we observe that our joint learning model can capture the most related words towards a specific aspect more effectively. For example, in Fig. 4 (a) example 1, "Joint" provides higher weights for the related words "nice", "very pleasant" and "good" for aspect "atmosphere", "service" and "desert" respectively. However, "Single" focuses on all the opinion words for different aspects even though it determines the sentiment polarities rightly. All these observations indicate that our joint learning model can capture the aspect-specific opinions words more accurately.

Second, it is observed that our "Joint" model can model the relationships among the aspects in a sentence. For example 3 in Fig. 4 (b), "Single" makes a wrong prediction for aspect “delivery” because it ignores the influence of the aspect “order”. However, our “Joint” model judges the sentiment polarities accurately for it learns that the bad order causes the bad delivery. In addition, in example 4, our joint model can capture the important words towards “appetizers” and “service” accurately. In a word, our model can model the relationships among the multiple aspects by integrating relationship modeling.

Third, “Joint” does pay attention to the right aspect-related opinion words, but makes the wrong prediction. Taking the sentence in Fig. 4 (a) example 2 as an example, both “Joint” and “Joint+PRET” focus on the word “nice”, “loud” for aspect “sound”, while “Joint” predicts the sentiment polarity as negative, which shows the effectiveness of pre-training. To be specific, pre-training can learn the domain knowledge and sentiment information from large datasets to address this problem.

6 Conclusions and Future Work

In this paper, we propose a joint learning (Joint) model for aspect-level sentiment classification. The main idea of this model is to model the potential relationships among the multiple aspects in the sentence. A series of experimental results on the four benchmark datasets show the great advantages of our proposed model by integrating aspect modeling (AM) and relationship modeling (RM). In particular, our model outperforms all the state-of-the-art approaches across all the four datasets in terms of accuracy and can capture the relationships among the aspects effectively. It is also interesting to find that our RM method can improve the performance of samples which contain multi-aspects effectively. In addition, the case studies provide an insight of how our proposed approach works.

In the future, we would like to explore more effective aspect modeling frameworks to enhance the aspect representation. It is also worth of investigating how to model the relationships among the aspects in a sentence more effectively for aspect-level sentiment classification.

Acknowledgments. We greatly appreciate anonymous reviewers and the associate editor for their valuable and high quality comments that greatly helped to improve the quality of this article. This research is funded by the Science and Technology Commission of Shanghai Municipality (19511120200). This research is also supported by the Natural Sciences and Engineering Research Council (NSERC) of Canada, an NSERC CREATE award in ADERSIM (http://www.yorku.ca/adersim), the York Research Chairs (YRC) program and an ORF-RE (Ontario Research Fund-Research Excellence) award in BRAIN Alliance (http://brainalliance.ca).

References

1. Bahdanau, D., Cho, K., Bengio, Y.: Neural machine translation by jointly learning to align and translate. arXiv preprint arXiv:1409.0473 (2014)

2. Chen, P., Sun, Z., Bing, L., Yang, W.: Recurrent attention network on memory for aspect sentiment analysis. In: Proceedings of EMNLP, pp. 452–461 (2017)
3. Chen, T., Xu, R., He, Y., Wang, X.: Improving sentiment analysis via sentence type classification using BiLSTM-CRF and CNN. Expert Syst. Appl. **72**, 221–230 (2017)
4. Cheng, J., Zhao, S., Zhang, J., King, I., Zhang, X., Wang, H.: Aspect-level sentiment classification with heat (hierarchical attention) network. In: Proceedings of CIKM, pp. 97–106. ACM (2017)
5. Collobert, R., Weston, J., Bottou, L., Karlen, M., Kavukcuoglu, K., Kuksa, P.: Natural language processing (almost) from scratch. J. Mach. Learn. Res. **12**(Aug), 2493–2537 (2011)
6. Du, J., Gui, L., He, Y., Xu, R., Wang, X.: Convolution-based neural attention with applications to sentiment classification. IEEE Access **7**, 27983–27992 (2019)
7. Fan, C., Gao, Q., Du, J., Gui, L., Xu, R., Wong, K.F.: Convolution-based memory network for aspect-based sentiment analysis. In: Proceedings of SIGIR, pp. 1161–1164. ACM (2018)
8. Graves, A., Schmidhuber, J.: Framewise phoneme classification with bidirectional LSTM and other neural network architectures. Neural Netw. **18**(5–6), 602–610 (2005)
9. Gu, S., Zhang, L., Hou, Y., Song, Y.: A position-aware bidirectional attention network for aspect-level sentiment analysis. In: Proceedings of COLING, pp. 774–784 (2018)
10. He, R., Lee, W.S., Ng, H.T., Dahlmeier, D.: Effective attention modeling for aspect-level sentiment classification. In: Proceedings of COLING, pp. 1121–1131 (2018)
11. He, R., Lee, W.S., Ng, H.T., Dahlmeier, D.: Exploiting document knowledge for aspect-level sentiment classification. In: Proceedings of ACL, pp. 579–585 (2018)
12. Hochreiter, S., Schmidhuber, J.: Long short-term memory. Neural Comput. **9**(8), 1735–1780 (1997)
13. Huang, B., Ou, Y., Carley, K.M.: Aspect level sentiment classification with attention-over-attention neural networks. In: Thomson, R., Dancy, C., Hyder, A., Bisgin, H. (eds.) SBP-BRiMS 2018. LNCS, vol. 10899, pp. 197–206. Springer, Cham (2018). https://doi.org/10.1007/978-3-319-93372-6_22
14. Jiang, L., Yu, M., Zhou, M., Liu, X., Zhao, T.: Target-dependent Twitter sentiment classification. In: Proceedings of ACL, pp. 151–160 (2011)
15. Kingma, D.P., Ba, J.: Adam: a method for stochastic optimization. In: Proceedings of ICLR, vol. 5 (2015)
16. Kiritchenko, S., Zhu, X., Cherry, C., Mohammad, S.: NRC-Canada-2014: detecting aspects and sentiment in customer reviews. In: Proceedings of SemEval, pp. 437–442 (2014)
17. Li, X., Bing, L., Lam, W., Shi, B.: Transformation networks for target-oriented sentiment classification. In: Proceedings of ACL, pp. 946–956 (2018)
18. Li, X., Lam, W.: Deep multi-task learning for aspect term extraction with memory interaction. In: Proceedings of EMNLP, pp. 2886–2892 (2017)
19. Li, Z., Wei, Y., Zhang, Y., Zhang, X., Li, X., Yang, Q.: Exploiting coarse-to-fine task transfer for aspect-level sentiment classification. In: Proceedings of AAAI (2019)
20. Lin, C., He, Y.: Joint sentiment/topic model for sentiment analysis. In: Proceedings of the 18th ACM Conference on Information and Knowledge Management, pp. 375–384. ACM (2009)
21. Liu, B.: Sentiment analysis and opinion mining. Synth. Lect. Hum. Lang. Technol. **5**(1), 1–167 (2012)

22. Liu, J., Zhang, Y.: Attention modeling for targeted sentiment. In: Proceedings of ACL, vol. 2, pp. 572–577 (2017)
23. Liu, Q., Zhang, H., Zeng, Y., Huang, Z., Wu, Z.: Content attention model for aspect based sentiment analysis. In: Proceedings of WWW, pp. 1023–1032. International World Wide Web Conferences Steering Committee (2018)
24. Ma, D., Li, S., Zhang, X., Wang, H.: Interactive attention networks for aspect-level sentiment classification. In: Proceedings of IJCAI, pp. 4068–4074 (2017)
25. Ma, Y., Peng, H., Cambria, E.: Targeted aspect-based sentiment analysis via embedding commonsense knowledge into an attentive LSTM. In: Proceedings of AAAI, pp. 5876–5883 (2018)
26. Manandhar, S.: Semeval-2014 task 4: aspect based sentiment analysis. In: Proceedings of SemEval (2014)
27. Pang, B., Lee, L., Vaithyanathan, S.: Thumbs up?: sentiment classification using machine learning techniques. In: Proceedings of EMNLP, pp. 79–86 (2002)
28. Pang, B., Lee, L., et al.: Opinion mining and sentiment analysis. Found. Trends® Inf. Retrieval **2**(1–2), 1–135 (2008)
29. Pennington, J., Socher, R., Manning, C.: Glove: global vectors for word representation. In: Proceedings of EMNLP, pp. 1532–1543 (2014)
30. Pontiki, M., et al.: Semeval-2016 task 5: aspect based sentiment analysis. In: Proceedings of SemEval, pp. 19–30 (2016)
31. Pontiki, M., Galanis, D., Papageorgiou, H., Manandhar, S., Androutsopoulos, I.: Semeval-2015 task 12: aspect based sentiment analysis. In: Proceedings of SemEval, pp. 486–495 (2015)
32. Ruder, S., Ghaffari, P., Breslin, J.G.: A hierarchical model of reviews for aspect-based sentiment analysis. In: Proceedings of EMNLP, pp. 999–1005 (2016)
33. Saeidi, M., Bouchard, G., Liakata, M., Riedel, S.: Sentihood: targeted aspect based sentiment analysis dataset for urban neighbourhoods. In: Proceedings of COLING, pp. 1546–1556 (2016)
34. Schmitt, M., Steinheber, S., Schreiber, K., Roth, B.: Joint aspect and polarity classification for aspect-based sentiment analysis with end-to-end neural networks. In: Proceedings of EMNLP, pp. 1109–1114 (2018)
35. Schouten, K., Frasincar, F.: Survey on aspect-level sentiment analysis. Proc. IEEE TKDE **28**(3), 813–830 (2016)
36. Sukhbaatar, S., Weston, J., Fergus, R., et al.: End-to-end memory networks. In: Proceedings of NIPS, pp. 2440–2448 (2015)
37. Sun, C., Huang, L., Qiu, X.: Utilizing BERT for aspect-based sentiment analysis via constructing auxiliary sentence. In: Proceedings of NAACL (2019)
38. Tang, D., Qin, B., Feng, X., Liu, T.: Effective LSTMs for target-dependent sentiment classification. In: Proceedings of COLING, pp. 3298–3307 (2016)
39. Tang, D., Qin, B., Liu, T.: Aspect level sentiment classification with deep memory network. In: Proceedings of EMNLP, pp. 214–224 (2016)
40. Tay, Y., Luu, A.T., Hui, S.C.: Learning to attend via word-aspect associative fusion for aspect-based sentiment analysis. In: Proceedings of AAAI, pp. 5956–5963 (2018)
41. Tay, Y., Tuan, L.A., Hui, S.C.: Dyadic memory networks for aspect-based sentiment analysis. In: Proceedings of CIKM, pp. 107–116. ACM (2017)
42. Vo, D.T., Zhang, Y.: Target-dependent twitter sentiment classification with rich automatic features. In: Proceedings of IJCAI, pp. 1347–1353 (2015)
43. Wagner, J., et al.: DCU: aspect-based polarity classification for SemEval task 4. In: Proceedings of SemEval, pp. 223–229 (2014)

44. Wang, B., Liakata, M., Zubiaga, A., Procter, R.: TDParse: multi-target-specific sentiment recognition on Twitter. In: Proceedings of ACL, vol. 1, pp. 483–493 (2017)
45. Wang, J., et al.: Aspect sentiment classification with both word-level and clause-level attention networks. In: Proceedings of IJCAI, pp. 4439–4445 (2018)
46. Wang, S., Mazumder, S., Liu, B., Zhou, M., Chang, Y.: Target-sensitive memory networks for aspect sentiment classification. In: Proceedings of ACL, vol. 1, pp. 957–967 (2018)
47. Wang, Y., Huang, M., Zhao, L., et al.: Attention-based LSTM for aspect-level sentiment classification. In: Proceedings of EMNLP, pp. 606–615 (2016)
48. Xu, H., Liu, B., Shu, L., Yu, P.S.: Bert post-training for review reading comprehension and aspect-based sentiment analysis. In: Proceedings of NAACL (2019)
49. Xu, X., Tan, S., Liu, Y., Cheng, X., Lin, Z.: Towards jointly extracting aspects and aspect-specific sentiment knowledge. In: Proceedings of CIKM, pp. 1895–1899. ACM (2012)
50. Xue, W., Li, T.: Aspect based sentiment analysis with gated convolutional networks. In: Proceedings of ACL, pp. 2514–2523 (2018)
51. Zeng, D., Liu, K., Lai, S., Zhou, G., Zhao, J.: Relation classification via convolutional deep neural network. In: Proceedings of COLING, pp. 2335–2344 (2014)
52. Zhang, M., Zhang, Y., Vo, D.T.: Gated neural networks for targeted sentiment analysis. In: Proceedings of AAAI, pp. 3087–3093 (2016)
53. Zhou, J., Chen, Q., Huang, J.X., Hu, Q.V., He, L.: Position-aware hierarchical transfer model for aspect-level sentiment classification. Inf. Sci. **513**, 1–16 (2020)
54. Zhou, J., Huang, J.X., Chen, Q., Hu, Q.V., Wang, T., He, L.: Deep learning for aspect-level sentiment classification: survey, vision and challenges. IEEE Access **7**, 78454–78483 (2019)
55. Zhu, P., Qian, T.: Enhanced aspect level sentiment classification with auxiliary memory. In: Proceedings of COLING, pp. 1077–1087 (2018)

Author Index